BASIC ENGINEERING

PLASTICITY

BASIC ENGINEERING

PLASTICITY

An Introduction with Engineering and Manufacturing Applications

D. W. A. Rees

School of Engineering and Design,

Brunel University, UK

ELSEVIER

AMSTERDAM • BOSTON • HEIDELBERG • LONDON • NEW YORK • OXFORD • PARIS
SAN DIEGO • SAN FRANCISCO • SINGAPORE • SYDNEY • TOKYO
Butterworth-Heinemann is an imprint of Elsevier

Butterworth-Heinemann is an imprint of Elsevier
Linacre House, Jordan Hill, Oxford OX2 8DP
30 Corporate Drive, Suite 400, Burlington, MA 01803

First edition 2006

British Library Cataloguing in Publication Data
A catalogue record for this book is available from the British Library

Library of Congress Cataloging-in-Publication Data
A catalog record for this book is available from the Library of Congress

ISBN-13: 978-0-7506-8025-7
ISBN-10: 0-7506-8025-3

For information on all Butterworth-Heinemann publications
visit our web site at http://books.elsevier.com

CONTENTS

CHAPTER 3

YIELD CRITERIA

CHAPTER 4

NON-HARDENING PLASTICITY

CHAPTER 5

ELASTIC-PERFECT PLASTICITY

CHAPTER 6

SLIP LINE FIELDS

CHAPTER 7

LIMIT ANALYSIS

CHAPTER 8

CRYSTAL PLASTICITY

CHAPTER 9

THE FLOW CURVE

CHAPTER 10

PLASTICITY WITH HARDENING

CHAPTER 11

ORTHOTROPIC PLASTICITY

CHAPTER 12

PLASTIC INSTABILITY

CHAPTER 13

STRESS WAVES IN BARS

CHAPTER 14

PRODUCTION PROCESSES

CHAPTER 15

APPLICATIONS OF FINITE ELEMENTS

PREFACE

This book brings together the elements of the mechanics of plasticity most pertinent to engineers. The presentation of the introductory material, the theoretical developments and the use of appropriate experimental data appear within a text of 15 chapters. A textbook style has been adopted in which worked examples and exercises illustrate the application of the theoretical material. The latter is provided with appropriate references to journals and other published sources. The book thereby combines the reference material required of a researcher together with the detail in theory and application expected from a student. The topics chosen are primarily of interest to engineers as undergraduates, postgraduates and practitioners but they should also serve to capture a readership from among applied mathematicians, physicists and materials scientists. There is not a comparable text with a similar breath in the subject range. Within this, much new work has been drawn from the research literature. The package of topics presented is intended to complement, at a basic level, more advanced monographs on the theory of plasticity. The unique blend of topics given should serve to support syllabuses across a diversity of undergraduate courses including manufacturing, engineering and materials.

The first two chapters are concerned with the stress and strain analyses that would normally accompany a plasticity theory. Both the matrix and tensor notations are employed to emphasise their equivalence when describing constitutive relations, co-ordinate transformations, strain gradients and decompositions for both large and small deformations. Chapter 3 outlines the formulation of yield criteria and their experimental confirmation for different initial conditions of material, e.g. annealed, rolled, extruded etc. Here the identity between the yield function and a plastic potential is made to provide flow rules for the ideal plastic solids examined in Chapters 4 and 5. Chapter 4 compares the predictions from the total and incremental theories of classical plasticity with experimental data. Differences between them have been attributed to a strain history dependence lying within non-radial loading paths. Chapter 5 compiles solutions to a number of elastic-perfect plastic structures. Ultimate loads, collapse mechanisms and residual stress are among the issues considered from a loading beyond the yield point.

In Chapter 6 it is shown how large scale plasticity in a number of forming processes can be described with slip line fields. For this an ideal, rigid-plastic, material is assumed. The theory identifies the stress states and velocities within a critical deformation zone. The rolling Mohr's circle and hodograph constructions are particulary useful where a full field description of the deformation zone is required. Alternative upper and lower bound analyses of the forming loads for metal forming are given in Chapter 7. Bounding methods provide useful approximations and are more rapid in their application.

Chapters 8 - 10 allow for material hardening behaviour and its influence upon practical plasticity problems. Firstly, in Chapter 8, a description of hardening on a micro-scale is given. It is shown from the operating slip processes and their directions upon closely packed atomic planes, that there must exist a yield criterion and a flow rule. There follows from this the concept of an initial and a subsequent yield surface, these being developed further in later chapters. The measurement and description of the flow curve (Chapter 9) becomes an essential requirement when the modelling the observed, macro-plasticity behaviour. The simplest isotropic hardening model is outlined in chapter 10. Also discussed here is the model of kinematic hardening for when a description of the Bauschinger effect is required.

In Chapter 11 the theory of orthotropic plasticity for rolled sheet metals and extruded tubes is given. These two models of hardening behaviour are extended in Chapter 12 to provide predictions to plastic instability in structures and necking in sheet metal forming.

A graphical analysis of the plasticity induced by longitudinal impact of bars is given in Chapter 13. The plasticity arising from high impact stresses is shown to be carried by a stress wave which interacts with an elastic wave to distribute residual stress in the bar. Chapter 14 considers the control of plasticity arising in conventional production processes including: forging, extrusion, rolling and machining. Here, the detailed analyses of ram forces, roll torques and strain rates employ the principles of force equilibrium and strain compatibility. This approach recognises that there are alternatives to slip lines and bounding methods, all of which are complementary when describing plasticity in practice.

Thanks are due to the author's past teachers, students and conference organisers who have kept him active in this area. The subject of plasticity continues to develop with many solutions provided these days by various numerical techniques. In this regard, the material presented here will serve to provide the essential mechanics required for any numerical implementation of a plasticity theory. Examples of this are illustrated within the final Chapter 15, where my collaborations with the University of Liége (Belgium) and the Warwick Manufacturing Centre (UK) are gratefully acknowledged.

ACKNOWLEDGEMENTS

The figures listed below have been reproduced, courtesy of the publishers of this author's earlier articles, from the following journals:

Acta Mechanica, Springer-Verlag (Figs 3.11, 3.13, 11.6)
Experimental Mechanics, Society for Experimental Mechanics (Figs 10.5, 11.10, 11.11)
Journal of Materials Processing Technology, Elsevier (Fig. 12.19, 12.23)
Journal Physics IV, France, EDP Sciences (Fig. 11.15)
Research Meccanica, Elsevier (Figs 8.13, 10.9, 10.13)
Proceedings of the Institution of Mechanical Engineers, Council I. Mech. E. (Fig. 10.15)
Proceedings of the Royal Society, Royal Society (3.7, 3.14, 4.1, 4.5, 9.20, 10.7, 10.8, 10.12)
Zeitschrift für Angewandte Mathematik und Mechanik, Wiley VCH (Figs 5.15, 5.16)

and from the following conference proceedings:

Applied Solid Mechanics 2 (eds A. S. Tooth and J. Spence) Elsevier Applied Science, 1988, Chapter 17 (Figs 4.8, 4.9, 4.10).

LIST OF SYMBOLS

The intention within the various theoretical developments given in this book has been to define each new symbol where it first appears in the text. In this regard each chapter should be treated as self-contained in its symbol content. There are, however, certain symbols that re-appear consistently throughout the text, such as those representing force, stress and strain. These symbols are given in the following list along with others most commonly employed in plasticity theory.

α, β	curvilinear co-ordinates (slip lines)
α_{ij}	kinematic hardening translations
$\alpha_{ij}^{s}, \beta_{ij}^{s}$	Schmidt's orientation factors
β, ϕ	friction and shear angles
δ	rolling draft
ε, γ	normal and shear strains
$\dot{\varepsilon}, \dot{\gamma}$	normal and shear rates of strain
ε_{ij}^{P*}	micro-plastic strain tensor
$\bar{\varepsilon}^{P}$	equivalent plastic strain
σ, τ	direct and shear stress
$\sigma_1, \sigma_2, \sigma_3$	principal stresses
σ_m	mean or hydrostatic stress
σ_{ij}^{*}	micro-stress tensor
σ_{ij}'	transformed stress
σ_t, σ_c	tensile and compressive strengths
$\bar{\sigma}$	equivalent stress
μ	friction coefficient
μ, v	Lode's parameters
ϕ, λ	scalar multipliers
θ	angular twist
θ, α	die angles
ψ	hardening measure
v	Poisson's ratio
ρ	density
λ	extension (stretch) ratio
χ, H, F	hardening functions
a, b, l, z	lengths
A	section or surface area
b, t	breadth and thickness
c	propagation velocity
C, T	torque
e_1, e_2, e_3	principal engineering strains
e_{ij}	distortions
ep	subscripts denoting elastic-plastic
E	superscript denoting elastic

E, G, K	elastic constants
f	yield function (plastic potential)
$F, G, H, L..$	anisotropy parameters
F, P	force
H_{ij}, C_{ijkl}	orthotropic tensors
H_{ijkl}, H_{ijklmn}	orthotropic tensors continued
I, J	second moments of area
I_1, I_2, I_3	strain invariants
J_1, J_2, J_3	stress invariants
J_1', J_2', J_3'	stress deviator invariants
K	buckling coefficient
l, m, n	direction cosines
$L_1.., Q_1.., T_1..$	anisotropy parameters continued
m, n	half-waves in buckling
M	bending moment
n	hardening exponent
p	pressure
P	superscript denoting plastic
Q	stress ratio
Q, S	shape and safety factors
$r\ \theta, z$	polar co-ordinates
r_1, r_2	incremental strain ratios (r values)
R	extrusion ratio
R_1, R_2	radii of curvature
t_1, t_2	back tensions
u, v, w	displacements
U	strain energy
v, ω	linear and angular velocities
V	volume
W	work done
x, y, z	Cartesian co-ordinates
x_i	spacial co-ordinates
X_i	material co-ordinates
X, Z	equivalence coefficients
$Y (= \sigma_o), k$	tensile and shear yield stresses
z	Considére's subtangent

$\mathbf{\Omega} (= \omega_{ij})$	rotation tensor/matrix
$\mathbf{B}, \mathbf{C}, \mathbf{G}, \mathbf{L}$	deformation tensors
$\mathbf{E} (= \varepsilon_{ij})$	infinitesimal strain tensor/matrix
$\mathbf{F}, \mathbf{H}$	deformation gradients
$\mathbf{m}, \mathbf{n}, \mathbf{u}$	unit vectors
$\mathbf{M} (= l_{ij})$	rotation matrix
$\mathbf{S}$	nominal stress tensor
$\mathbf{T} (=\sigma_{ij})$	stress tensor/matrix
$\mathbf{T}' (=\sigma_{ij}')$	deviatoric stress tensor/matrix
$\mathbf{U}, \mathbf{V}$	stretch tensors

CHAPTER 1

STRESS ANALYSIS

1.1 Introduction

Before we can proceed to the study of flow in a deforming solid it is necessary to understand what is meant by the term stress. Various definitions of stress have been used so it is pertinent to begin with explanations as to how it arises and is quantified. Firstly, it is essential that the tensorial nature of stress is appreciated. It will be shown that stress is a symmetrical second order Cartesian tensor. Where deformation is small (infinitesimal) we can represent stress in both the tensor component and matrix notations. Stress is first introduced for simple uniaxial and shear loadings. A combination of these loadings gives both normal and shear stress, these comprising two of the six independent components that are possible within a stress tensor. The transformation properties of stress are to be examined following a rotation in the orthogonal co-ordinates chosen to define the stress state at a point. Alternative stress definitions are given when it becomes necessary to distinguish between the initial and current areas for large (finite) deformations. Finite deformation will affect the definition of stress because the initial and current areas can differ appreciably. The chosen definition of stress becomes important when connecting the stress and strain tensors within a constitutive relationship for elastic and plastic deforming solids.

The following analyses will alternate between the engineering and mathematical co-ordinate notations listed in Table 1.1. This will enable the reader to interchange between notations in recognition of the equivalence between them.

Table 1.1 Symbol Equivalence in Engineering and Mathematical Notations

Quantity	Engineering Notation	Mathematical Notation
Material co-ordinates	x, y, z	x_1, x_2, x_3
Spacial co-ordinates	X, Y, Z	X_1, X_2, X_3
Material displacements	u, v, w	u_1, u_2, u_3
Spacial displacements	U, V, W	U_1, U_2, U_3
Unit co-ordinate vectors	$\mathbf{u}_x, \mathbf{u}_y, \mathbf{u}_z$	$\mathbf{u}_1, \mathbf{u}_2, \mathbf{u}_3$
Direction cosines	l, m, n	l_1, l_2, l_3
Unit normal equation	$\mathbf{u}_n = l\mathbf{u}_x + m\mathbf{u}_y + n\mathbf{u}_z$	$\mathbf{n} = l_1\mathbf{u}_1 + l_2\mathbf{u}_2 + l_3\mathbf{u}_3$
Unit normal column matrix	$\mathbf{u}_n = \{l \ \ m \ \ n\}^T$	$\mathbf{n} = \{l_1 \ \ l_2 \ \ l_3\}^T$
Normal stress	$\sigma_x, \sigma_y, \sigma_z$	$\sigma_{11}, \sigma_{22}, \sigma_{33}$
Shear stress	$\tau_{xy}, \tau_{xz}, \tau_{yz}$	$\sigma_{12}, \sigma_{13}, \sigma_{23}$
Normal strain (see Ch. 2)	$\varepsilon_x, \varepsilon_y, \varepsilon_z$	$\varepsilon_{11}, \varepsilon_{22}, \varepsilon_{33}$
Shear strain (see Ch. 2)	$\frac{1}{2}\gamma_{xy}, \frac{1}{2}\gamma_{xz}, \frac{1}{2}\gamma_{yz}$	$\varepsilon_{12}, \varepsilon_{13}, \varepsilon_{23}$
Stresses on oblique plane	σ, τ	$\sigma_{11}', \sigma_{21}', \sigma_{31}'$

Note that a rotation matrix **M** employs the direction cosines in the above table for a co-ordinate transformation between Cartesian axes 1, 2 and 3, in each notation as follows:

$$\mathbf{M} = \begin{bmatrix} l_{11} & l_{12} & l_{13} \\ l_{21} & l_{22} & l_{23} \\ l_{31} & l_{32} & l_{33} \end{bmatrix} \equiv \begin{bmatrix} l_1 & m_1 & n_1 \\ l_2 & m_2 & n_2 \\ l_3 & m_3 & n_3 \end{bmatrix}$$

1.1.1 Direct Stress

Direct stress σ measures the intensity of a reaction to externally applied loading. In fact, σ refers to the internal force acting perpendicular to a unit of area within a material. For example, when a uniaxial external force is either tensile(+) or compressive(−), σ is simply

$$\sigma = \pm\, W / A \tag{1.1}$$

where W is the magnitude of the externally applied force and A is the original normal area (see Fig. 1.1a). The elastic reduction in a section area under stress is negligibly small and hence it is unnecessary to distinguish between initial and current areas within eq(1.1). Elasticity is clearly evident from the initial linear plot of stress versus strain in Fig. 1.1b.

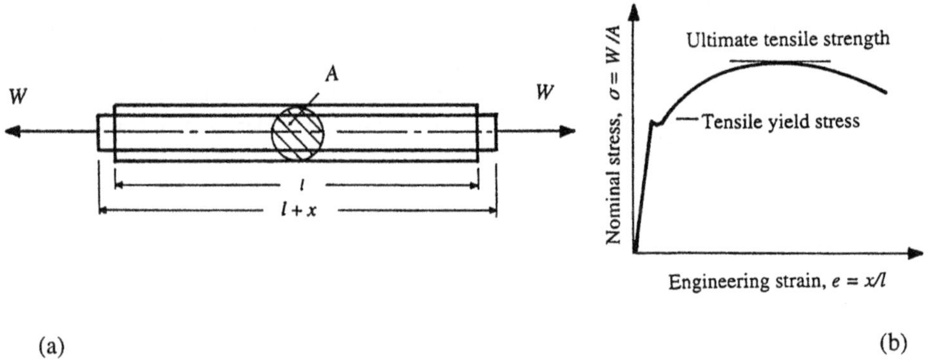

Figure 1.1 Direct tensile stress showing elastic and plastic strain responses

Note, from Fig. 1.1a, that the corresponding direct strain e is the amount by which the material extends per unit of its length as shown. For displacements under tension or compression,i.e. $\pm x$, occurring over a length l, the corresponding strains are:

$$e = \pm\, x / l \tag{1.2}$$

This engineering definition of strain applies to small, elastic displacements. With larger deformations in the plastic range a true stress is calculated from the current area and plastic strains are calculated from referring the displacement to the current length. The true stress and true strain are developed further in this and the following chapters.

1.1.2 Shear Stress

Let an applied shear force F act tangentially to the top area A, as shown in Fig. 1.2a.

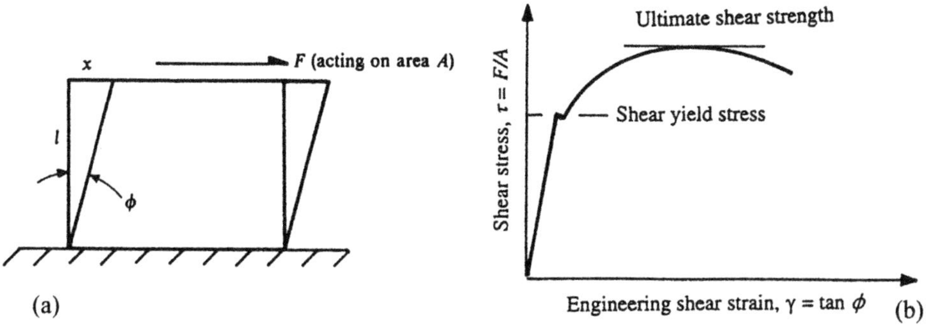

Figure 1.2 Shear distortion showing elastic and plastic strain responses

The shear stress intensity τ, sustained by the material as it maintains equilibrium with this force, is given by

$$\tau = F / A \tag{1.3}$$

The abscissa in the shear stress versus shear strain plot Fig. 1.2b refers to the angular distortion that a material suffers in shear. The shear strain is a dimensionless measure of distortion and is defined in Fig. 1.2a as

$$\gamma = \tan\phi = x / l \tag{1.4}$$

In eq(1.4) ϕ is the angular change in the right angle measured in radians. Within the elastic region the shear displacement x is small when it follows from eq(1.4) that, with a correspondingly small ϕ, the shear strain may be approximated as $\gamma \approx \phi$ (rad).

The original area A in eq(1.3) will depend upon the mode of shear. For example, consider the two plates, in Fig. 1.3a joined with a single rivet, subjected to tensile force F. Since the rivet is placed in single shear, A refers to its cross-sectional area and F to the transverse shear force. In a double shear lap joint in Fig. 1.3b the effective area resisting F is doubled and so τ is halved.

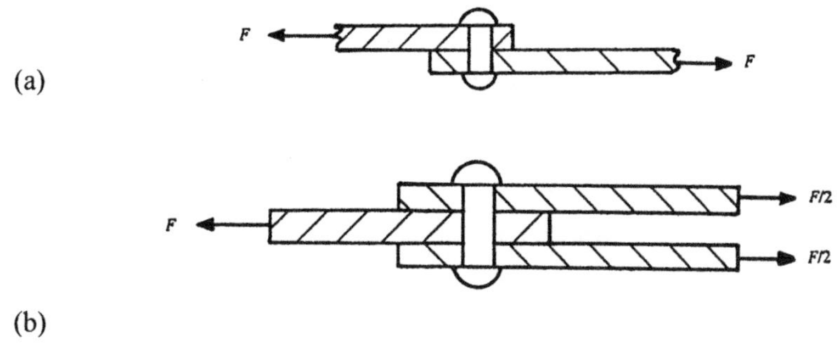

Figure 1.3 Riveted joints in single and double shear

1.2 Cauchy Definition of Stress

Consider an elemental area δa, on a plane B, that cuts through a loaded body in its deformed configuration (see Fig. 1.4).

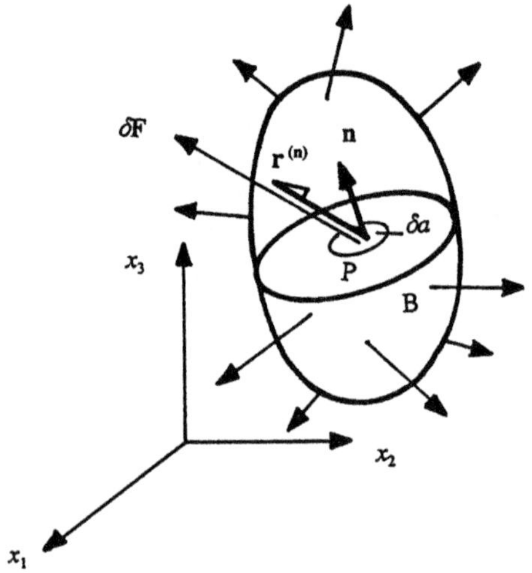

Figure 1.4 Force $\delta\mathbf{F}$ transmitted through area δa

Let a unit vector **n**, lying normal to δa at P, be directed outward from the positive side of B as shown. Due to the applied loading, an elemental resultant force vector $\delta\mathbf{F}$, acting in any direction on the positive side of δa, must also be transmitted to the negative side of B if the continuum is to remain in equilibrium. The traction acting across δa may be found from considering the lower half as a free body.

1.2.1 Stress Intensity

Let an average stress intensity, or traction vector $\mathbf{r}^{(\mathbf{n})}$, be the average force per unit area of δa, so that

$$\delta\mathbf{F} = \mathbf{r}^{(\mathbf{n})}\,\delta a \quad \text{or} \quad \delta F_i = r_i^{(\mathbf{n})}\,\delta a \tag{1.5a,b}$$

The alternative expression (1.5b) has employed the components $r_i^{(\mathbf{n})}$ of $\mathbf{r}^{(\mathbf{n})}$ in co-ordinates, x_i (where $i = 1, 2$ and 3). Equation (1.5a) shows that $\delta\mathbf{F}$ will depend upon the size and orientation of δa. The vector $\mathbf{r}^{(\mathbf{n})}$ emphasises this dependence upon the chosen area δa at P. For a given P, $\mathbf{r}^{(\mathbf{n})}$ is uniquely defined at the finite limit when δa tends to zero. This limit will further eliminate any moments of $\delta\mathbf{F}$ acting on δa. Thus, from eq(1.5a), the traction for any given normal direction **n**, through P, becomes

$$\mathbf{r}^{(\mathbf{n})} = \lim_{\delta a \to 0} \frac{\delta\mathbf{F}}{\delta a} = \frac{\mathrm{d}\mathbf{F}}{\mathrm{d}a} \quad \text{or} \quad r_i^{(\mathbf{n})} = \frac{\mathrm{d}F_i}{\mathrm{d}a} \tag{1.5c,d}$$

Equation (1.5d) reduces to the simple forms given in eqs(1.1) and (1.3) when a single force acts normal or parallel to a given surface. Where oblique forces act, the total stress vector $\mathbf{r}^{(\mathbf{n})}$ may be resolved into chosen co-ordinate directions, x_i. To define a general stress state completely, it is sufficient to resolve $\mathbf{r}^{(\mathbf{n})}$ into one normal and two shear stress components for the positive sides of orthogonal co-ordinate planes passing through point P. Such resolution reveals the tensorial nature of stress since it follows that there will be nine traction components when three orthogonal planes are considered. To show this, let $\mathbf{n}_j$ ($j = 1, 2, 3$) be unit vectors in the direction of the co-ordinates x_i so that $\mathbf{r}^{(\mathbf{n}_1)}$, $\mathbf{r}^{(\mathbf{n}_2)}$, $\mathbf{r}^{(\mathbf{n}_3)}$ become the traction vectors on the three faces shown in Fig. 1.5.

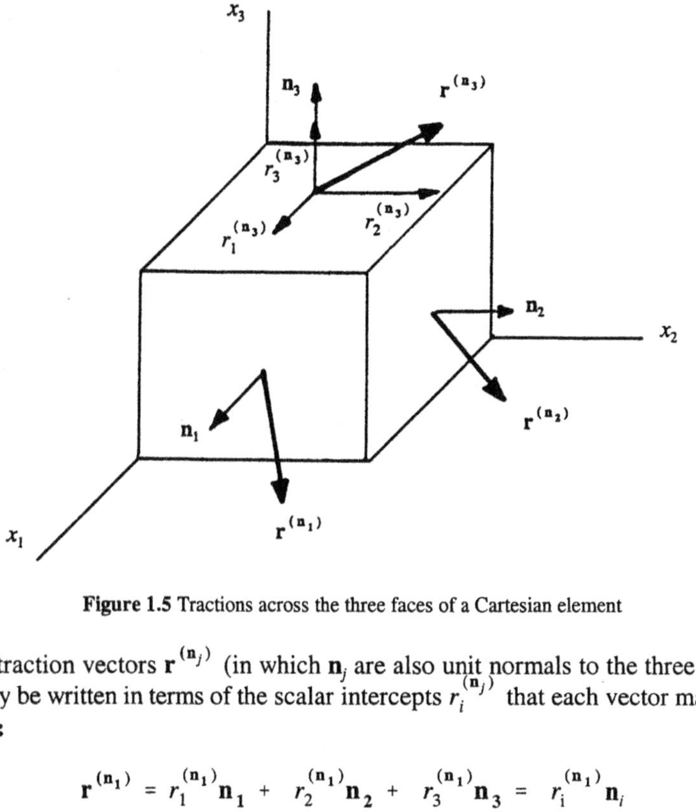

Figure 1.5 Tractions across the three faces of a Cartesian element

The three traction vectors $\mathbf{r}^{(\mathbf{n}_j)}$ (in which $\mathbf{n}_j$ are also unit normals to the three orthogonal planes) may be written in terms of the scalar intercepts $r_i^{(\mathbf{n}_j)}$ that each vector makes with x_i as follows:

$$\mathbf{r}^{(\mathbf{n}_1)} = r_1^{(\mathbf{n}_1)}\mathbf{n}_1 + r_2^{(\mathbf{n}_1)}\mathbf{n}_2 + r_3^{(\mathbf{n}_1)}\mathbf{n}_3 = r_i^{(\mathbf{n}_1)}\mathbf{n}_i$$

$$\mathbf{r}^{(\mathbf{n}_2)} = r_1^{(\mathbf{n}_2)}\mathbf{n}_1 + r_2^{(\mathbf{n}_2)}\mathbf{n}_2 + r_3^{(\mathbf{n}_2)}\mathbf{n}_3 = r_i^{(\mathbf{n}_2)}\mathbf{n}_i$$

$$\mathbf{r}^{(\mathbf{n}_3)} = r_1^{(\mathbf{n}_3)}\mathbf{n}_1 + r_2^{(\mathbf{n}_3)}\mathbf{n}_2 + r_3^{(\mathbf{n}_3)}\mathbf{n}_3 = r_i^{(\mathbf{n}_3)}\mathbf{n}_i$$

which, by the summation convention, may be contracted into a single equation:

$$\mathbf{r}^{(\mathbf{n}_j)} = r_i^{(\mathbf{n}_j)}\mathbf{n}_i \qquad (1.6a)$$

where $i, j = 1, 2$ and 3. The nine scalar components $r_i^{(\mathbf{n}_j)}$ form the components of a second order Cartesian stress tensor $\sigma_{ij} = r_i^{(\mathbf{n}_j)}$. Thus, the system of eqs(1.6a) becomes:

$$\mathbf{r}^{(\mathbf{n}_j)} = \sigma_{ij}\,\mathbf{n}_i \qquad (1.6b)$$

Equation(1.6b) satisfies force equilibrium parallel to each co-ordinate directions. This equilibrium condition will appear later with the alternative engineering stress notation (see eqs(1.11a,b,c)). The Cauchy stress tensor, **T**, with components σ_{ij} (where $i, j = 1, 2, 3$), is defined in from eq(1.6b) when the co-ordinates x_i are referred to the deformed configuration.

1.2.2 General Stress State

Within a general three-dimensional stress state both normal and shear stresses components comprise the tensor components σ_{ij} within eq(1.6b). Two conventions are employed to distinguish between these components and to identify the directions in which they act. In the *engineering notation*, σ denotes normal stress and τ denotes shear stress. Let these appear with Cartesian co-ordinates x, y and z, as shown in Fig. 1.6a.

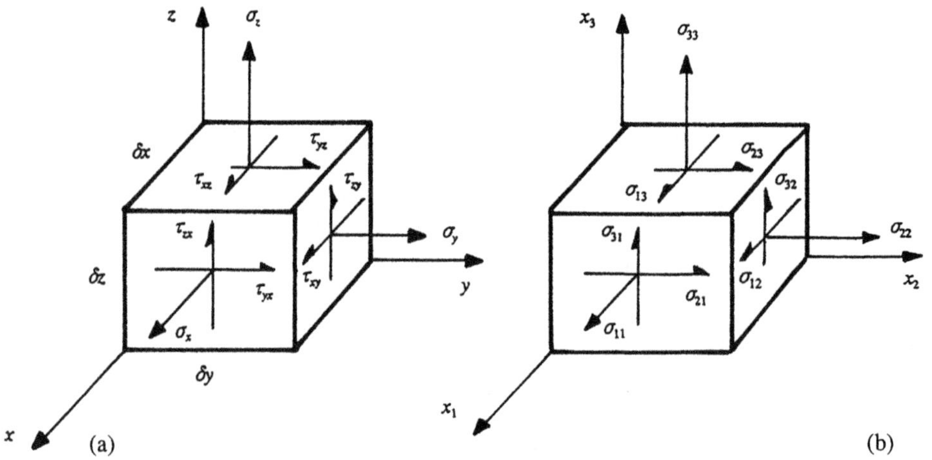

Figure 1.6 General stress states in (a) engineering and (b) mathematical notations

A single subscript on σ identifies the direction of the three normal stress components. The double subscript on τ distinguishes between the six shear components. The first subscript denotes the direction of the stress and the second the direction of the normal to the plane on which that stress acts, e.g. τ_{xy} is a shear stress aligned with the x-direction on the plane whose normal is aligned with the y-direction (Note: some texts interchange these subscripts by writing the normal direction first). Only three shear stresses components are independent. The complementary nature of the shear stresses: $\tau_{xy} = \tau_{yx}$, $\tau_{xz} = \tau_{zx}$ and $\tau_{yz} = \tau_{zy}$, ensures that moments produced by the force resultants about any point are in equilibrium. To show this, take moments on four faces in the x-y plane about a point along the z-axis in Fig. 1.6a:

$$\sigma_y(\delta x)^2(\delta z)/2+\sigma_x(\delta y)^2(\delta z)/2+\tau_{xy}(\delta x)(\delta z)(\delta y) = \sigma_x(\delta y)^2(\delta z)/2+\sigma_y(\delta x)^2(\delta z)/2+\tau_{yx}(\delta x)(\delta y)(\delta z)$$

which leads to $\tau_{xy} = \tau_{yx}$. In Fig. 1.6b, an alternative Cartesian frame x_i (x_1, x_2 and x_3) is employed to identify the stress components according to the *mathematical tensor notation*. Here, the single symbol σ is used for both normal and shear stress components. They are distinguished with double subscripts referring to directions and planes as before. Thus, σ_{11} is a normal stress aligned with the 1-direction and the normal to its plane is also in the x_1-direction. Normal stresses will always appear with two similar subscripts in this notation.

Different subscripts denote shear stresses, e.g. σ_{12} is aligned with the x_1-direction but acts on the plane whose normal is aligned with the x_2-direction. Great care must be taken not to confuse generalised co-ordinates: x_1, x_2 and x_3 with the system of co-ordinates 1, 2 and 3 used in the following section to identify principal stresses. Since shear stress is absent along principal directions we may employ a single subscript 1, 2 or 3 with σ to identify principal stresses unambiguously.

1.2.3 Stress Tensor

It is seen that six independent scalar components of stress are required to define the general state of stress at a point. This identifies stress as a Cartesian tensor of second order. The components appear in the tensor notation as $\sigma_{ij} = \sigma_{ji}$ (where $i = j = 1, 2$ and 3). Note that a vector is a tensor of the first order since it is defined from the three scalar intercepts the vector makes with its co-ordinate axes. The following section shows that the scalar components of the stress tensor may be transformed for any given rotation in the co-ordinate axes. These tensor components are often expressed in the form of a symmetrical 3 × 3 matrix **T**. The following matrices of stress tensor components are thus equivalent and we shall alternate between them throughout this and other chapters.

$$\sigma_{ij} = \begin{bmatrix} \sigma_x & \tau_{xy} & \tau_{xz} \\ \tau_{yx} & \sigma_y & \tau_{yz} \\ \tau_{zx} & \tau_{zy} & \sigma_z \end{bmatrix} \equiv \begin{bmatrix} \sigma_{11} & \sigma_{12} & \sigma_{13} \\ \sigma_{21} & \sigma_{22} & \sigma_{23} \\ \sigma_{31} & \sigma_{32} & \sigma_{33} \end{bmatrix} = \mathbf{T} \qquad (1.7a,b)$$

The matrices (1.7a and b) are symmetrical about a leading diagonal composed of the three independent direct stress components.

1.3 Three-Dimensional Stress Analysis

Let an oblique triangular plane ABC in Fig. 1.7a cut through the stressed Cartesian element in Fig. 1.6a to produce a tetrahedron OABC. The six known independent stress components: σ_x, σ_y, σ_z, $\tau_{xy} = \tau_{yx}$, $\tau_{xz} = \tau_{zx}$ and $\tau_{yz} = \tau_{zy}$ now act on the back three triangular faces OAB, OBC and OAC in the negative co-ordinate directions.

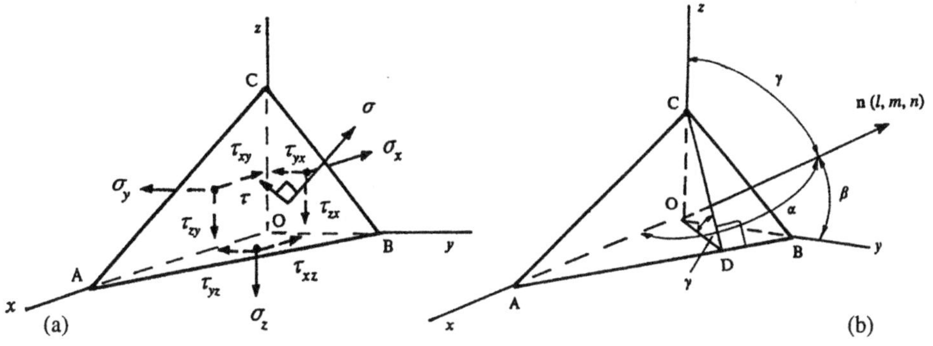

Figure 1.7 General stress state for a tetrahedron showing direction cosines for oblique plane ABC

Since the element must remain in equilibrium, the force resultants produced by the action of these stresses are equilibrated by a normal stress σ and a shear stress τ on the oblique plane ABC in Fig. 1.7a. The objective is to find this stress state (σ, τ) in both magnitude and direction, by the methods of force resolution and tensor transformation.

1.3.1 Direction Cosines

It is first necessary to find the areas of each back face. Let the area ABC in Fig. 1.7b be unity. Construct a perpendicular CD to AB and join OD. A normal vector **n** to plane ABC is defined by direction cosines l, m and n, measured relative to x, y and z respectively as follows:

$$l = \cos\alpha,\ m = \cos\beta \ \text{ and } \ n = \cos\gamma \tag{1.8a,b,c}$$

Then, as Area ABC = ½AB × CD and Area OAB = ½AB × OD:

$$(\text{Area OAB}) / (\text{Area ABC}) = \text{OD} / \text{CD} = \cos\gamma = n$$

Hence: Area OAB = n. Similarly: Area OBC = l and Area OAC = m. The direction cosines are not independent. Their relationship follows from the equation of vector **n**:

$$\mathbf{n} = n_x \mathbf{u}_x + n_y \mathbf{u}_y + n_z \mathbf{u}_z \tag{1.9a}$$

where $\mathbf{u}_x$, $\mathbf{u}_y$ and $\mathbf{u}_z$ are unit vectors and n_x, n_y and n_z are scalar intercepts with the co-ordinates x, y and z, as shown in Fig. 1.8a.

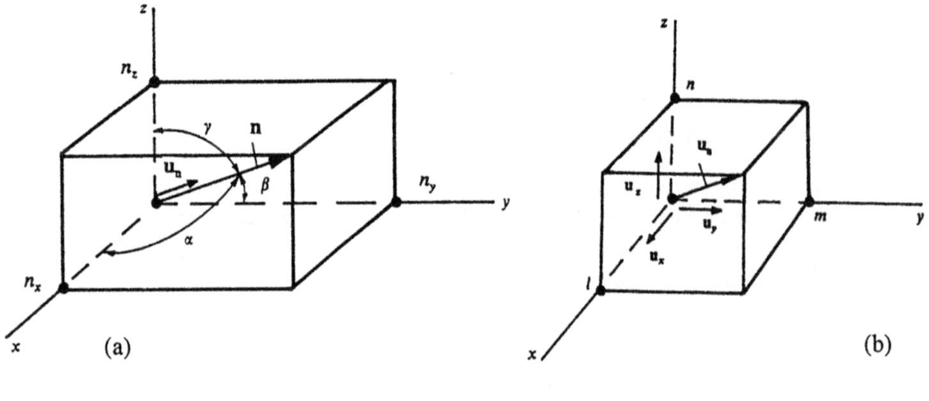

Figure 1.8 Scalar intercepts for (a) normal vector **n** and (b) unit normal $\mathbf{u}_n$

The unit vector $\mathbf{u}_n$, for the normal direction (see Fig. 1.8b), is found from dividing eq(1.9a) by the magnitude $|\mathbf{n}|$:

$$\mathbf{u}_n = (n_x / |\mathbf{n}|)\, \mathbf{u}_x + (n_y / |\mathbf{n}|)\, \mathbf{u}_y + (n_z / |\mathbf{n}|)\, \mathbf{u}_z \tag{1.9b}$$

Substituting from eqs(1.8a,b,c): $l = \cos\alpha = n_x/|\mathbf{n}|$, $m = \cos\beta = n_y/|\mathbf{n}|$ and $n = \cos\gamma = n_z/|\mathbf{n}|$, eq(1.9b) becomes

$$\mathbf{u}_n = l\, \mathbf{u}_x + m\, \mathbf{u}_y + n\, \mathbf{u}_z \tag{1.9c}$$

It follows that l, m and n are also the intercepts that the unit normal vector $\mathbf{u}_n$ makes with x, y and z (shown in Fig. 1.8b). Furthermore, since

$$(n_x)^2 + (n_y)^2 + (n_z)^2 = |\mathbf{n}|^2$$

$$(n_x / |\mathbf{n}|)^2 + (n_y / |\mathbf{n}|)^2 + (n_z / |\mathbf{n}|)^2 = 1$$

the direction cosines obey the relationship:

$$l^2 + m^2 + n^2 = 1 \tag{1.10}$$

1.3.2 Force Resolution

(a) *Magnitudes of σ and τ*

Let σ and τ be the normal and shear stress components of the resultant force or traction vector $\mathbf{r}$, acting upon plane ABC in Fig. 1.9a.

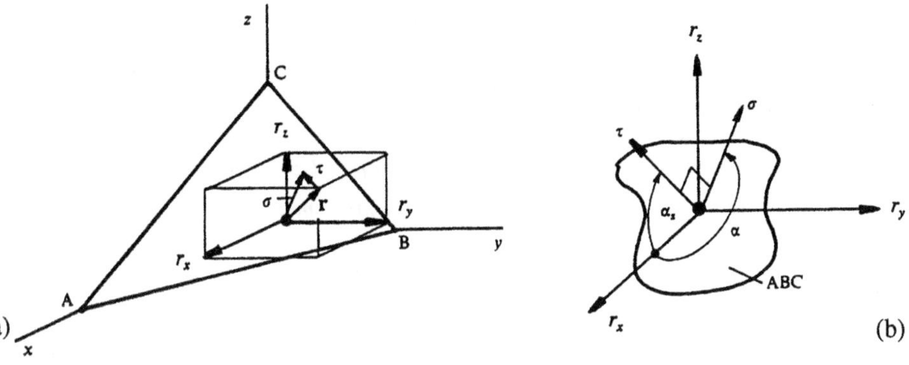

Figure 1.9 Stress state for the oblique plane ABC

The components of vector $\mathbf{r}$ are r_x, r_y and r_z as shown. Since $\mathbf{r}$ must equilibrate the forces due to stress components applied to the back faces (see Fig. 1.7a), it follows that

$$r_x = l\sigma_x + m\tau_{xy} + n\tau_{xz} \tag{1.11a}$$
$$r_y = m\sigma_y + n\tau_{yz} + l\tau_{yx} \tag{1.11b}$$
$$r_z = n\sigma_z + l\tau_{zx} + m\tau_{zy} \tag{1.11c}$$

Writing eqs(1.11a,b,c) in the contracted form: $r_i = \sigma_{ij} n_j$, it is seen that these become a re-statement of eq(1.6) in which $\sigma_{ij} = \sigma_{ji}$. Using the engineering notation, the corresponding matrix equation, $\mathbf{r} = \mathbf{T}\mathbf{n}$, gives

$$\begin{bmatrix} r_x \\ r_y \\ r_z \end{bmatrix} = \begin{bmatrix} \sigma_x & \tau_{xy} & \tau_{xz} \\ \tau_{yx} & \sigma_y & \tau_{yz} \\ \tau_{zx} & \tau_{zy} & \sigma_z \end{bmatrix} \begin{bmatrix} l \\ m \\ n \end{bmatrix}$$

Now as the area of ABC is unity, σ is the sum of the r_x, r_y and r_z force components resolved into the normal direction. This gives

$$\sigma = r_x \cos\alpha + r_y \cos\beta + r_z \cos\gamma = r_x l + r_y m + r_z n \tag{1.12a}$$

where, from eqs(1.11a,b,c)

$$\sigma = \sigma_x l^2 + \sigma_y m^2 + \sigma_z n^2 + 2\,(lm\tau_{xy} + mn\tau_{yz} + ln\tau_{zx}) \tag{1.12b}$$

The magnitude of the resultant force on ABC is expressed in two ways:

$$r^2 = r_x^2 + r_y^2 + r_z^2 = \sigma^2 + \tau^2$$

$$\therefore\ \tau^2 = r^2 - \sigma^2 = r_x^2 + r_y^2 + r_z^2 - \sigma^2 \tag{1.12c}$$

and substituting eqs(1.11a-c) into (1.12c), τ can be found.

(b) *Directions of σ and τ*

Since σ lies parallel to **n**, the direction of σ is also defined by l, m and n for the plane ABC. The direction of τ in the plane ABC is defined by the directions: $l_s = \cos\alpha_s$, $m_s = \cos\beta_s$ and $n_s = \cos\gamma_s$ (see Fig. 1.9b). Because r_x, r_y and r_z are the resultant forces for the x, y and z components of σ and τ, this gives

$$r_x = \sigma\cos\alpha + \tau\cos\alpha_s = l\sigma + l_s\tau$$
$$r_y = \sigma\cos\beta + \tau\cos\beta_s = m\sigma + m_s\tau$$
$$r_z = \sigma\cos\gamma + \tau\cos\gamma_s = n\sigma + n_s\tau$$

Re-arranging gives

$$l_s = (r_x - l\sigma)/\tau \tag{1.13a}$$
$$m_s = (r_y - m\sigma)/\tau \tag{1.13b}$$
$$n_s = (r_z - n\sigma)/\tau \tag{1.13c}$$

Example 1.1 A stress resultant of 140 MPa makes respective angles of 43°, 75° and 50°53' with the x, y and z-axes. Determine the normal and shear stresses, in magnitude and direction, on an oblique plane whose normal makes respective angles of 67°13', 30° and 71°34' with these axes.

Referring to Fig. 1.9a, first resolve r = 140 MPa in the x, y and z directions to give its components as

$$r_x = 140\cos 43^\circ = 102.39 \text{ MPa}$$
$$r_y = 140\cos 75^\circ = 36.24 \text{ MPa}$$
$$r_z = 140\cos 50^\circ 53' = 88.33 \text{ MPa}$$

The normal stress is found from eq(1.12a), in which l, m and n are the direction cosines for the normal:

$$\sigma = r_x l + r_y m + r_z n = r_x\cos\alpha + r_y\cos\beta + r_z\cos\gamma$$
$$= 102.39\cos 67^\circ 13' + 36.24\cos 30^\circ + 88.33\cos 71^\circ 34' = 98.96 \text{ MPa}$$

Equation (1.12c) supplies the shear stress on this plane as

$$\tau = \surd\,(r^2 - \sigma^2) = \surd\,(140^2 - 98.96^2) = 99.03 \text{ MPa}$$

and eqs(1.13a,b,c) gives its direction cosines as

$$l_s = (r_x - l\sigma)/\tau = (102.39 - 98.96 \cos 67°13')/ 99.03 = 0.647 \quad (\alpha_s = 49°41')$$
$$m_s = (r_y - m\sigma)/\tau = (36.24 - 98.96 \cos 30°)/ 99.03 = -0.500 \quad (\beta_s = 120°)$$
$$n_s = (r_z - n\sigma)/\tau = (88.33 - 98.96 \cos 71°34')/ 99.03 = 0.576 \quad (\gamma_s = 54°50')$$

from which it can be checked that: $l_s^2 + m_s^2 + n_s^2 = 1$.

1.3.3 Stress Transformations in Tensor and Matrix Notations

It is now shown that components σ and τ in eqs(1.12a,b) appear as components in a general tensor transformation law for stress. The general transformation law for a tensor follows from the dyadic product of two vectors. The equations for any pair of arbitary vectors **a** and **b** (see, for example, Fig. 1.10a) will appear in Cartesian co-ordinates: x_1, x_2 and x_3, as

$$\mathbf{a} = a_1\mathbf{u}_1 + a_2\mathbf{u}_2 + a_3\mathbf{u}_3 = a_i\mathbf{u}_i \qquad (1.14a)$$
$$\mathbf{b} = b_1\mathbf{u}_1 + b_2\mathbf{u}_2 + b_3\mathbf{u}_3 = b_i\mathbf{u}_i \qquad (1.14b)$$

where a_i and b_i are the scalar intercepts and $\mathbf{u}_i$ are unit co-ordinate vectors. Figure 1.10a shows each of these for the vector **a**.

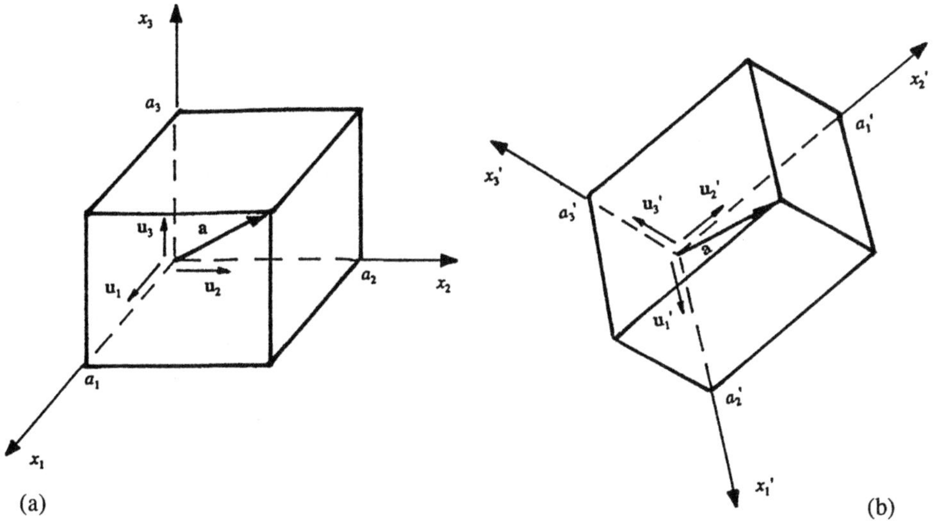

Figure 1.10 Components of a vector in co-ordinate frame x_1, x_2 and x_3 and x_1', x_2' and x_3'

Let the co-ordinate axes rotate about the origin to lie in the final orthogonal frame x_1', x_2' and x_3', as shown in Fig. 1.10b The equations of the stationary vectors **a** and **b** become

$$\mathbf{a} = a_1'\mathbf{u}_1' + a_2'\mathbf{u}_2' + a_3'\mathbf{u}_3' = a_j'\mathbf{u}_j' \qquad (1.15a)$$
$$\mathbf{b} = b_1'\mathbf{u}_1' + b_2'\mathbf{u}_2' + b_3'\mathbf{u}_3' = b_j'\mathbf{u}_j' \qquad (1.15b)$$

Next, consider the method for expressing this rotation. It has previously been shown that the components of a unit vector are the direction cosines (see eq(1.9c)). Thus, unit vectors $\mathbf{u}_1'$, $\mathbf{u}_2'$ and $\mathbf{u}_3'$ in the frame x_1', x_2' and x_3', may each be expressed in terms of unit vectors $\mathbf{u}_1$, $\mathbf{u}_2$ and $\mathbf{u}_3$, for the original frame x_1, x_2 and x_3, as follows:

$$\mathbf{u}_1' = l_{11}\,\mathbf{u}_1 + l_{12}\,\mathbf{u}_2 + l_{13}\,\mathbf{u}_3 \qquad (1.16a)$$
$$\mathbf{u}_2' = l_{21}\,\mathbf{u}_1 + l_{22}\,\mathbf{u}_2 + l_{23}\,\mathbf{u}_3 \qquad (1.16b)$$
$$\mathbf{u}_3' = l_{31}\,\mathbf{u}_1 + l_{32}\,\mathbf{u}_2 + l_{33}\,\mathbf{u}_3 \qquad (1.16c)$$

Using the summation convention eqs(1.16a,b,c) may be contracted to a single equation:

$$\mathbf{u}_i' = l_{ij}\,\mathbf{u}_j \qquad (1.16d)$$

To confirm this, sum eq(1.16d) over j = 1, 2 and 3, to give

$$\mathbf{u}_i' = l_{i1}\,\mathbf{u}_1 + l_{i2}\,\mathbf{u}_2 + l_{i3}\,\mathbf{u}_3$$

and, substituting i = 1, 2 and 3 provides the three relations in eqs(1.16a,b,c). The directions l_{ij} (i, j = 1, 2 and 3) in eq(1.16d), define each primed direction relative to the unprimed direction. That is: $l_{ij} = \cos(x_i', x_j)$. For example, $l_{11} = \cos(x_1', x_1)$, $l_{12} = \cos(x_1', x_2)$ and $l_{13} = \cos(x_1', x_3)$ define the directions of x_1' within the frame x_1, x_2 and x_3. It follows that the direction cosines l_{ij} are the components of the following rotation matrix **M**:

$$\mathbf{M} = \begin{bmatrix} l_{11} & l_{12} & l_{13} \\ l_{21} & l_{22} & l_{23} \\ l_{31} & l_{32} & l_{33} \end{bmatrix} \qquad (1.17a)$$

An orthogonal property of this matrix is that $l_{ij}\,l_{kj} = \delta_{ik}$ or $\mathbf{MM}^T = \mathbf{I}$. In full, this is:

$$\mathbf{MM}^T = \begin{bmatrix} l_{11} & l_{12} & l_{13} \\ l_{21} & l_{22} & l_{23} \\ l_{31} & l_{32} & l_{33} \end{bmatrix} \begin{bmatrix} l_{11} & l_{21} & l_{31} \\ l_{12} & l_{22} & l_{32} \\ l_{13} & l_{23} & l_{33} \end{bmatrix} \begin{bmatrix} 1 & 0 & 0 \\ 0 & 1 & 0 \\ 0 & 0 & 1 \end{bmatrix}$$

which contains the following relationships between cosines for each direction:

$$l_{11}^2 + l_{12}^2 + l_{13}^2 = 1 \;\; (\mathbf{u}_1' \bullet \mathbf{u}_1' = l_{1i}\,l_{1i} = \mathbf{u}_1^T\mathbf{u}_1 = 1) \text{ for } x_1'$$
$$l_{21}^2 + l_{22}^2 + l_{23}^2 = 1 \;\; (\mathbf{u}_2' \bullet \mathbf{u}_2' = l_{2i}\,l_{2i} = \mathbf{u}_2^T\mathbf{u}_2 = 1) \text{ for } x_2' \qquad (1.17b)$$
$$l_{31}^2 + l_{32}^2 + l_{33}^2 = 1 \;\; (\mathbf{u}_3' \bullet \mathbf{u}_3' = l_{3i}\,l_{3i} = \mathbf{u}_3^T\mathbf{u}_3 = 1) \text{ for } x_3'$$

Additional relationships apply to pairs of orthogonal directions:

$$l_{11}\,l_{21} + l_{12}\,l_{22} + l_{13}\,l_{23} = 0 \;\; (\mathbf{u}_1' \bullet \mathbf{u}_2' = l_{1i}\,l_{2i} = \mathbf{u}_1^T\mathbf{u}_2 = 0) \text{ for } x_1' \text{ and } x_2'$$
$$l_{21}\,l_{31} + l_{22}\,l_{32} + l_{23}\,l_{33} = 0 \;\; (\mathbf{u}_2' \bullet \mathbf{u}_3' = l_{2i}\,l_{3i} = \mathbf{u}_2^T\mathbf{u}_3 = 0) \text{ for } x_2' \text{ and } x_3' \qquad (1.17c)$$
$$l_{11}\,l_{31} + l_{12}\,l_{32} + l_{13}\,l_{33} = 0 \;\; (\mathbf{u}_1' \bullet \mathbf{u}_3' = l_{1i}\,l_{3i} = \mathbf{u}_1^T\mathbf{u}_3 = 0) \text{ for } x_1' \text{ and } x_3'$$

The abbreviated expressions in parentheses show the equivalent equations appear in the respective notions of a direct tensor (i.e. the dot product), indicial tensor components and a matrix. Since $\mathbf{u}_1 = \{ l_{11}\ l_{12}\ l_{13} \}^T$ etc, denote column matrices it follows that a row matrix is formed from the transpose: $\mathbf{u}_1^T = \{ l_{11}\ l_{12}\ l_{13} \}$. Apart from the dot and cross products of vectors, the direct tensor notation will not be adopted further. Instead, we shall alternate between the tensor component and matrix notations in our consideration of the stress and strain tensors and the relationships that exist between them.

Combining eqs(1.14) and (1.15), the vectors **a** and **b** may be expressed in both systems of co-ordinates as

$$\mathbf{a} = a_i \mathbf{u}_i = a_j' \mathbf{u}_j' = a_j' (l_{ji} \mathbf{u}_i)$$
$$\mathbf{b} = b_i \mathbf{u}_i = b_j' \mathbf{u}_j' = b_j' (l_{ji} \mathbf{u}_i)$$

from which the vector transfomation laws follow:

$$a_i = a_j' l_{ji} = l_j a_j' \text{ and } b_i = b_j' l_{ji} = l_{ji} b_j \qquad (1.18a,b)$$

In the matrix notation eqs(1.18a,b) become

$$\mathbf{a} = \mathbf{M}^T \mathbf{a}', \mathbf{b} = \mathbf{M}^T \mathbf{b}' \qquad (1.18c,d)$$

where **a**, **a**′, **b** and **b**′ are column matrices, e.g. $\mathbf{a} = \{a_1\ a_2\ a_3\}^T$. To invert eqs(1.18a,b), multiply both sides by l_{ki}:

$$l_{ki} a_i = l_{ki} l_{ji} a_j' = \delta_{kj} a_j' = a_k'$$

where from eq(1.17b,c) $\delta_{ij} = 1$ for $k = j$ and $\delta_{kj} = 0$ for $k \neq j$. Reverting to i, j subscripts:

$$a_i' = l_{ij} a_j \text{ and, similarly, } b_i' = l_{ij} b_j \qquad (1.19a,b)$$

Correspondingly, to invert the matrix eq(1.18c) pre-multiply both sides by $(\mathbf{M}^T)^{-1}$. This gives

$$(\mathbf{M}^T)^{-1} \mathbf{a} = (\mathbf{M}^T)^{-1} \mathbf{M}^T \mathbf{a}' = \mathbf{I}\, \mathbf{a}' = \mathbf{a}' \qquad (1.19c)$$

Since the inverse of the square matrix **M** will obey $\mathbf{M}\mathbf{M}^{-1} = \mathbf{I}$ and as $\mathbf{M}\mathbf{M}^T = \mathbf{I}$, a further orthogonal property of the rotation matrix is that $\mathbf{M}^{-1} = \mathbf{M}^T$. It then follows that

$$(\mathbf{M}^T)^{-1} = (\mathbf{M}^{-1})^{-1} = \mathbf{M}$$

and eq(1.19c) becomes

$$\mathbf{a}' = \mathbf{M}\mathbf{a} \text{ and, similarly, } \mathbf{b}' = \mathbf{M}\mathbf{b} \qquad (1.19d,e)$$

A second-order Cartesian tensor may be formed from the dyadic product of two vectors. Note that this differs from the cross product which results in another vector lying normal to the plane containing the two vectors. The tensor or *dyadic product* of two vectors, **a** and **b**, is written as

$$\mathbf{a} \otimes \mathbf{b} = (a_i \mathbf{u}_i) \otimes (b_j \mathbf{u}_j) = a_i b_j (\mathbf{u}_i \otimes \mathbf{u}_j)$$

The tensor so formed appears as

$$\mathbf{K} = K_{ij} (\mathbf{u}_i \otimes \mathbf{u}_j)$$

where $K_{ij} = a_i b_j$ ($\mathbf{K} = \mathbf{a}\mathbf{b}^{\mathrm{T}}$) are the components of a second order Cartesian tensor $\mathbf{K}$ for which the unit vectors (dyads) $\mathbf{u}_i$ and $\mathbf{u}_j$ appear in linear combination. The components K_{ij} may be referred to both sets of orthogonal axes x_i and x_i' through the vector transformation laws (eqs(1.18a,b)). These give

$$K_{ij} = a_i b_j = (\, l_{pi} a_p' \,)\,(\, l_{qj} b_q' \,)$$

and putting $K_{pq}' = a_p'\, b_q'$ leads to the transformation law

$$K_{ij} = l_{pi}\, l_{qj}\, K_{pq}' \tag{1.20a}$$

Equations(1.19a,b) provide the components of the the inverse transformation matrix K_{ij}' as

$$K_{ij}' = a_i'\, b_j' = (l_{ip}\, a_p)(l_{jq}\, b_q)$$

Setting $K_{pq} = a_p b_q$, the general transformation law for any second order tensor is obtained:

$$K_{ij}' = l_{ip}\, l_{jq}\, K_{pq} \tag{1.21a}$$

In converting eqs(1.20a) and (1.21a) to matrix equations, similar subscripts must appear adjacent within each term to become consistent with matrix multiplication. That is

$$K_{ij} = l_{pi}\, l_{qj}\, K_{pq}' = l_{pi}\, K_{pq}'\, l_{qj} \quad \text{or} \quad \mathbf{K} = \mathbf{M}^{\mathrm{T}}\mathbf{K}'\,\mathbf{M} \tag{1.20b}$$

$$K_{ij}' = l_{ip}\, l_{jq}\, K_{pq} = l_{ip}\, K_{pq}\, l_{jq} \quad \text{or} \quad \mathbf{K}' = \mathbf{M}\,\mathbf{K}\,\mathbf{M}^{\mathrm{T}} \tag{1.21b}$$

Alternatively, direct matrix derivations are given by eqs(1.18c,d) as

$$\mathbf{K} = \mathbf{a}\,\mathbf{b}^{\mathrm{T}} = (\mathbf{M}^{\mathrm{T}}\mathbf{a}')(\mathbf{M}^{\mathrm{T}}\mathbf{b}')^{\mathrm{T}} = \mathbf{M}^{\mathrm{T}}(\mathbf{a}'\,\mathbf{b}'^{\,\mathrm{T}})\,\mathbf{M} = \mathbf{M}^{\mathrm{T}}\mathbf{K}'\,\mathbf{M}$$

$$\mathbf{K}' = \mathbf{a}'\,\mathbf{b}'^{\mathrm{T}} = (\mathbf{M}\,\mathbf{a})\,(\mathbf{M}\,\mathbf{b})^{\mathrm{T}} = \mathbf{M}\,(\mathbf{a}\,\mathbf{b}^{\mathrm{T}})\,\mathbf{M}^{\mathrm{T}} = \mathbf{M}\,\mathbf{K}\,\mathbf{M}^{\mathrm{T}}$$

It has been previously established that the physical quantity called stress is a second order tensor. The stress tensor must therefore transform in the manner of eqs(1.20) and (1.21). Normally, it is required to transform the known components of the stress tensor σ_{pq} in axes x_1, x_2 and x_3 (Fig. 1.11a) to components σ_{ij}' in axes x_1', x_2' and x_3', as shown in Fig. 1.11b.

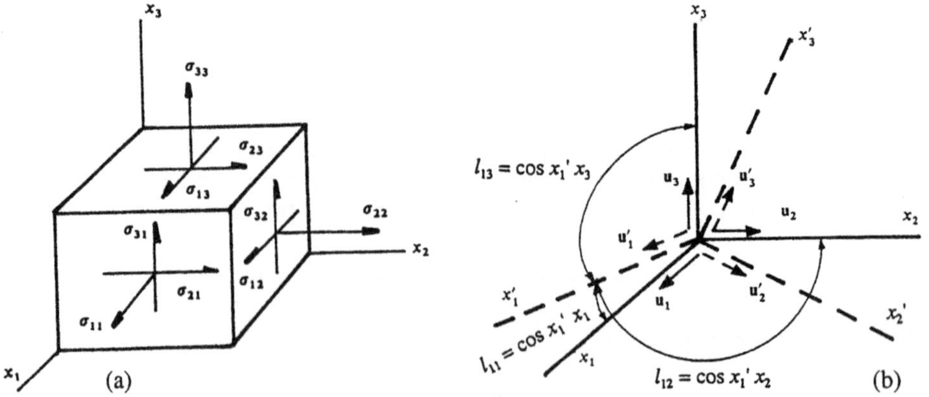

Figure 1.11 (a) Generalised stress components and (b) a rotation in orthogonal axes

It follows from eq(1.21a) that the law of transformation is

$$\sigma_{ij}' = l_{ip}\, l_{jq}\, \sigma_{pq} \quad \text{or} \quad \mathbf{T}' = \mathbf{M T M}^{\mathrm{T}} \tag{1.22a}$$

where $\mathbf{T} \equiv \sigma_{ij}$ and $\mathbf{T}' \equiv \sigma_{ij}'$ and $\mathbf{M} \equiv l_{ip}$. Writing the stress transformation law (1.22a) in full:

$$\begin{bmatrix} \sigma_{11}' & \sigma_{12}' & \sigma_{13}' \\ \sigma_{21}' & \sigma_{22}' & \sigma_{23}' \\ \sigma_{31}' & \sigma_{32}' & \sigma_{33}' \end{bmatrix} = \begin{bmatrix} l_{11} & l_{12} & l_{13} \\ l_{21} & l_{22} & l_{23} \\ l_{31} & l_{32} & l_{33} \end{bmatrix} \begin{bmatrix} \sigma_{11} & \sigma_{12} & \sigma_{13} \\ \sigma_{21} & \sigma_{22} & \sigma_{23} \\ \sigma_{31} & \sigma_{32} & \sigma_{33} \end{bmatrix} \begin{bmatrix} l_{11} & l_{21} & l_{31} \\ l_{12} & l_{22} & l_{32} \\ l_{13} & l_{23} & l_{33} \end{bmatrix} \tag{1.22b}$$

This gives one normal and two shear stresses for each of the three orthogonal planes in the x_i' ($i = 1, 2$ and 3) frame. In the analytical method, the stress state for a single oblique plane ABC (Fig. 1.7) was found. We can identify ABC with the plane lying normal to x_1' (say) with directions: l_{11}, l_{12} and l_{13}. The stress components for this plane (σ_{11}', σ_{21}' and σ_{31}'), are:

$$\begin{bmatrix} \sigma_{11}' \\ \sigma_{21}' \\ \sigma_{31}' \end{bmatrix} = \begin{bmatrix} l_{11} & l_{12} & l_{13} \end{bmatrix} \begin{bmatrix} \sigma_{11} & \sigma_{12} & \sigma_{13} \\ \sigma_{21} & \sigma_{22} & \sigma_{23} \\ \sigma_{31} & \sigma_{32} & \sigma_{33} \end{bmatrix} \begin{bmatrix} l_{11} \\ l_{12} \\ l_{13} \end{bmatrix} \tag{1.22c}$$

The normal and shear stress referred to in eqs(1.12b and c) now become $\sigma = \sigma_{11}'$ and $\tau = \surd[(\sigma_{21}')^2 + (\sigma_{31}')^2]$. Clearly, τ is the resultant shear stress acting on plane ABC and σ_{12}', σ_{13}' are its components aligned with the axes x_2' and x_3'.

It is important to note here that the prime on stress in eqs(1.22a-c) refers to the normal and shear stress components for the transformed axes x_i'. They are not to be confused with deviatoric stresses σ_{ij}' and $\mathbf{T}'$, shown in Fig. 3.3. The stress deviator has the hydrostatic part of the stress tensor removed, i.e. $\sigma_{ij}' = \sigma_{ij} - \frac{1}{3}\,\delta_{ij}\sigma_{kk}$, or, $\mathbf{T}' = \mathbf{T} - \frac{1}{3}\,\mathbf{I}\ \mathrm{tr}\ \mathbf{T}$ (see eqs(3.9a,b)) and retains the property of transformation, as in eq(1.22a).

1.4 Principal Stresses and Invariants

The three principal planes are orthogonal and free of shear stress. The three stresses normal to these planes are, by definition, principal stresses. Their magnitudes and orientation will now be derived from the known stress components for non-principal axes.

1.4.1 Magnitudes of Principal Stresses

For this let us employ the engineering notation, where the stress components shown in Fig. 1.6a correspond to the 3×3 matrix given in eq(1.7a). When the shear stress τ is absent for the plane ABC (l, m, n) in Fig. 1.7a then the normal stress σ becomes a principal stress. Force resolution in the x, y and z directions modifies eqs(1.11) to:

$$r_x = l\,\sigma = l\sigma_x + m\tau_{xy} + n\tau_{xz}$$
$$r_y = m\sigma = m\sigma_y + n\tau_{yz} + l\tau_{yx}$$
$$r_z = n\sigma = n\sigma_z + l\tau_{zx} + m\tau_{zy}$$

That is

$$l\,(\sigma_x - \sigma) + m\tau_{xy} + n\tau_{xz} = 0$$
$$l\tau_{yx} + m\,(\sigma_y - \sigma) + n\tau_{yz} = 0 \qquad (1.23a)$$
$$l\tau_{zx} + m\tau_{zy} + n\,(\sigma_z - \sigma) = 0$$

Writing the direction cosines in a column matrix $\mathbf{u} = \{l\ m\ n\}^{\mathrm{T}}$, the equivalent tensor component and matrix forms of eq(1.23a) will respectively appear as

$$(\sigma_{ij} - \sigma\,\delta_{ij})\,u_i = 0 \text{ or } (\mathbf{T} - \sigma\mathbf{I})\,\mathbf{u} = 0 \qquad (1.23b)$$

By Cramar's rule, the solution to σ is found from the determinant

$$\begin{vmatrix} (\sigma_x - \sigma) & \tau_{xy} & \tau_{xz} \\ \tau_{yx} & (\sigma_y - \sigma) & \tau_{yz} \\ \tau_{zx} & \tau_{zy} & (\sigma_z - \sigma) \end{vmatrix} = 0 \qquad (1.23c)$$

Contracted forms of eq(1.23c) appear, in the two alternative notations, as

$$\det\,(\sigma_{ij} - \sigma\,\delta_{ij}) = 0 \text{ or } \det\,(\mathbf{T} - \sigma\mathbf{I}) = 0$$

Expanding eq(1.23c), leads to a cubic (or characteristic) equation

$$(\sigma_x - \sigma)[(\sigma_y - \sigma)(\sigma_z - \sigma) - \tau_{yz}\tau_{zy}] - \tau_{xy}[\tau_{yx}(\sigma_z - \sigma) - \tau_{yz}\tau_{zx}] + \tau_{xz}[\tau_{yx}\tau_{zy}\ \ \tau_{zx}(\sigma_y - \sigma)] = 0$$
$$\sigma^3 - (\sigma_x + \sigma_y + \sigma_z)\sigma^2 + (\sigma_x\sigma_y + \sigma_y\sigma_z + \sigma_z\sigma_x - \tau_{xy}{}^2 - \tau_{yz}{}^2 - \tau_{zx}{}^2)\sigma$$
$$- (\sigma_x\sigma_y\sigma_z + 2\tau_{xy}\tau_{yz}\tau_{zx} - \sigma_x\tau_{yz}{}^2 - \sigma_y\tau_{zx}{}^2 - \sigma_z\tau_{xy}{}^2) = 0 \qquad (1.24a)$$

The three roots (the eigen values) to eq(1.24a) give the principal stress magnitudes σ_1, σ_2 and σ_3. Equation (1.24a) is usually written as

$$\sigma^3 - J_1\sigma^2 + J_2\sigma - J_3 = 0 \qquad (1.24b)$$

The principal stresses are unique for a given stress tensor. The coefficients J_1, J_2 and J_3 in eq(1.24b), are therefore independent of the co-ordinate frame, x, y, z, in Fig. 1.6a, chosen to define the stress tensor components. J_1, J_2 and J_3 are therefore called invariants of the stress tensor σ_{ij}. Equation (1.24a) must include an orientation where x, y and z coincide with the principal stress directions 1, 2 and 3. Thus the invariants may be expressed either in terms of general stress components (subscripts x, y and z) or in terms of principal stresses (subscripts 1, 2 and 3):

$$J_1 = \sigma_1 + \sigma_2 + \sigma_3 = \sigma_x + \sigma_y + \sigma_z = \sigma_{ii} = \operatorname{tr}\mathbf{T} \qquad (1.25a)$$
$$J_2 = \sigma_1\sigma_2 + \sigma_2\sigma_3 + \sigma_1\sigma_3 = \sigma_x\sigma_y + \sigma_y\sigma_z + \sigma_z\sigma_x - \tau_{xy}{}^2 - \tau_{yz}{}^2 - \tau_{zx}{}^2$$
$$= \tfrac{1}{2}(\sigma_{ii}\sigma_{jj} - \sigma_{ij}\sigma_{ji}) = \tfrac{1}{2}[(\operatorname{tr}\mathbf{T})^2 - \operatorname{tr}\mathbf{T}^2] \qquad (1.25b)$$
$$J_3 = \sigma_1\sigma_2\sigma_3 = \sigma_x\sigma_y\sigma_z + 2\tau_{xy}\tau_{yz}\tau_{zx} - \sigma_x\tau_{yz}{}^2 - \sigma_y\tau_{zx}{}^2 - \sigma_z\tau_{xy}{}^2$$
$$= \det\,(\sigma_{ij}) = \det\mathbf{T} \qquad (1.25c)$$

Also given in eqs(1.25a-c) are the contracted tensor and matrix expressions. The former is to be employed with tensor subscripts $i, j = 1, 2, 3$. Repeated subscripts on a single symbol, or within a term, denote summation.

Where there are exact roots to eq(1.24a), the principal stresses are more conveniently found from expanding the determinant (1.23a) following substitution of the numerical values of the stress components. Otherwise, the major σ_1, intermediate σ_2 and minor σ_3 principal stresses ($\sigma_1 > \sigma_2 > \sigma_3$) must be found from the solution to the cubic eq(1.24a). The Cayley-Hamilton theorem states that a square matrix will satisfy its own characteristic equation. For the 3 × 3 stress matrix, **T**, eq(1.24b) becomes

$$\mathbf{T}^3 - J_1\mathbf{T}^2 + J_2\mathbf{T} - \mathbf{I}\,J_3 = 0 \qquad (1.25d)$$

Substituting from eqs(1.25a-c), the theorem states that **T** must satisfy:

$$\mathbf{T}^3 - \mathbf{T}^2 \mathrm{tr}\,\mathbf{T} + \tfrac{1}{2}\mathbf{T}\,[\,(\mathrm{tr}\,\mathbf{T})^2 - \mathrm{tr}\,\mathbf{T}^2] - \mathbf{I}\,\mathrm{det}\mathbf{T} = 0 \qquad (1.25e)$$

Taking the trace of eq(1.25d) gives an alternative expression for J_3:

$$\mathrm{tr}\,\mathbf{T}^3 - J_1\,\mathrm{tr}\,\mathbf{T}^2 + J_2\,\mathrm{tr}\,\mathbf{T} - 3J_3 = 0$$

and substituting from eqs(1.25a and b) gives

$$J_3 = \frac{1}{6}\,(\,\mathrm{tr}\,\mathbf{T}\,)^3 - \frac{1}{2}\,\mathrm{tr}\,\mathbf{T}\,\mathrm{tr}\,\mathbf{T}^2 + \frac{1}{3}\,\mathrm{tr}\,\mathbf{T}^3$$

1.4.2 Principal Stress Directions

Let the direction cosines for σ_1 be l_1, m_1 and n_1 within a co-ordinate frame x, y, z. Substituting for the applied stresses into eq(1.23a), with $\sigma = \sigma_1$, leads to three simultaneous equations in l_1, m_1 and n_1. Only two of these are independent because of the relationship: $l_1^2 + m_1^2 + n_1^2 = 1$. (see eq(1.10)). A similar deduction can be made for further substitutions: $\sigma_2(l_2, m_2, n_2)$ and $\sigma_3(l_3, m_3, n_3)$ into eqs(1.23a). It follows from eq(1.9c) that the principal sets of direction cosines; (l_1, m_1, n_1), (l_2, m_2, n_2) and (l_3, m_3, n_3) define the unit vectors aligned with the principal directions (see Fig. 1.12).

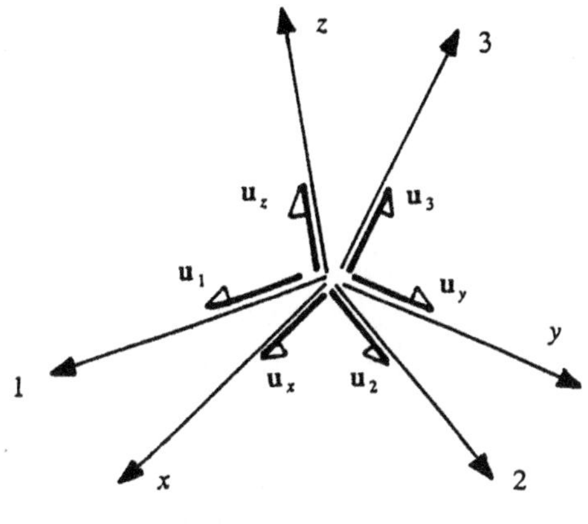

Figure 1.12 Principal directions

They are

$$\mathbf{u}_1 = l_1\mathbf{u}_x + m_1\mathbf{u}_y + n_1\mathbf{u}_z \quad (1.26a)$$
$$\mathbf{u}_2 = l_2\mathbf{u}_x + m_2\mathbf{u}_y + n_2\mathbf{u}_z \quad (1.26b)$$
$$\mathbf{u}_3 = l_3\mathbf{u}_x + m_3\mathbf{u}_y + n_3\mathbf{u}_z \quad (1.26c)$$

The three unit vectors are orthogonal when their dot products are zero. For the 1 and 2 directions:

$$\mathbf{u}_1 \bullet \mathbf{u}_2 = (l_1\,\mathbf{u}_x + m_1\,\mathbf{u}_y + n_1\,\mathbf{u}_z) \bullet (l_2\,\mathbf{u}_x + m_2\,\mathbf{u}_y + \mathrm{n}_2\,\mathbf{u}_z) = 0$$

Now $\mathbf{u}_x \bullet \mathbf{u}_x = \mathbf{u}_y \bullet \mathbf{u}_y = \mathbf{u}_z \bullet \mathbf{u}_z = 1$ and $\mathbf{u}_x \bullet \mathbf{u}_y = \mathbf{u}_x \bullet \mathbf{u}_z = \mathbf{u}_y \bullet \mathbf{u}_z = 0$. The further dot products, $\mathbf{u}_1 \bullet \mathbf{u}_3$ and $\mathbf{u}_2 \bullet \mathbf{u}_3$, determine the full orthogonality conditions:

$$l_1 l_2 + m_1 m_2 + n_1 n_2 = 0 \quad (1.27a)$$
$$l_2 l_3 + m_2 m_3 + n_2 n_3 = 0 \quad (1.27b)$$
$$l_1 l_3 + m_1 m_3 + n_1 n_3 = 0 \quad (1.27c)$$

The relationships (1.27a-c) are the only conditions that satisfy the simultaneous equations(1.26a-c), confirming that the principal stresses directions and their associated planes are orthogonal. In fact, $\mathbf{u}_\alpha$ ($\alpha = 1, 2, 3$) in eqs(1.26a,b,c), define the eigen vectors for the characteristic eq(1.24b). Since its roots are the eigen values σ_α ($\alpha = 1, 2, 3$), eq(1.23b) becomes

$$\sigma_{ij}\,l_{\alpha j} = \sigma_\alpha\,\delta_{ij}\,l_{\alpha j} = \sigma_\alpha\,l_{\alpha i} \quad \text{or} \quad (\mathbf{T} - \sigma_\alpha \mathbf{I})\mathbf{u}_\alpha = 0 \quad (1.28a,b)$$

To show that these forms are identical to eqs(1.23a), put $\alpha = 1$ in eq(1.28a) and expand for $i = 1$ over $j = 1, 2$ and 3:

$$\sigma_{1j}\,l_{1j} = \sigma_1\,l_{11} \;(\text{for } i = 1)$$
$$\sigma_{11}\,l_{11} + \sigma_{12}\,l_{12} + \sigma_{13}\,l_{13} = \sigma_1\,l_{11} \quad (\text{for } j = 1, 2, 3)$$
$$l_{11}(\sigma_{11} - \sigma_1) + l_{12}\,\sigma_{12} + l_{13}\,\sigma_{13} = 0$$

When converted to the engineering notation this becomes the first of eqs(1.23a), i.e. $l\,(\sigma_x - \sigma) + m\,\tau_{xy} + n\,\tau_{xz} = 0$. Post-multiplying eq(1.28a) by l_{pi} gives

$$\sigma_{ij}\,l_{\alpha j}\,l_{pi} = \sigma_\alpha l_{\alpha i}\,l_{pi} \quad (1.28c)$$

Then, for $\alpha = 1$ and $p = 2$ and $\alpha = 2$ and $p = 1$, eq(1.28c) gives

$$\sigma_{ij}\,l_{1j}\,l_{2i} = \sigma_1\,l_{1i}\,l_{2i}$$
$$\sigma_{ij}\,l_{2j}\,l_{1i} = \sigma_2\,l_{2i}\,l_{1i}$$

Subtracting these leads to

$$\sigma_{ij}\,l_{1j}\,l_{2i} - \sigma_{ij}\,l_{2j}\,l_{1i} = \sigma_1\,l_{1i}\,l_{2i} - \sigma_2\,l_{2i}\,l_{1i}$$
$$\sigma_{ij}\,l_{1j}\,l_{2i} - \sigma_{ij}\,l_{2i}\,l_{1j} = (\sigma_1 - \sigma_2)\,l_{1i}\,l_{2i}$$
$$(\sigma_{ij} - \sigma_{ji})\,l_{1j}\,l_{2i} = (\sigma_1 - \sigma_2)\,l_{1i}\,l_{2i}$$

but since $\sigma_{ij} = \sigma_{ji}$ and $\sigma_1 \neq \sigma_2$, it follows that

$$l_{1i}\,l_{2i} = l_{11}\,l_{21} + l_{12}\,l_{22} + l_{13}\,l_{23} = 0$$

In the engineering notation this is eq(1.27a): $l_1 l_2 + m_1 m_2 + n_1 n_2 = 0$, thereby confirming that directions 1 and 2 are orthogonal. Further pairs of substitutions: ($\alpha = 1, p = 3$; $\alpha = 3, p = 1$) and ($\alpha = 2, p = 3$; $\alpha = 3, p = 2$) will confirm eqs(1.27b,c). Within the three column matrices: $\mathbf{u}_1 = \{l_{11}\ l_{12}\ l_{13}\}^{\mathrm{T}}$, $\mathbf{u}_2 = \{l_{21}\ l_{22}\ l_{23}\}^{\mathrm{T}}$ and $\mathbf{u}_3 = \{l_{31}\ l_{32}\ l_{33}\}^{\mathrm{T}}$, which definite the 1, 2 and 3 directions respectively, the orthogonality conditions are written as

$$\mathbf{u}_1^{\mathrm{T}}\mathbf{u}_2 = \mathbf{u}_1\mathbf{u}_2^{\mathrm{T}} = 0$$
$$\mathbf{u}_1^{\mathrm{T}}\mathbf{u}_3 = \mathbf{u}_1\mathbf{u}_3^{\mathrm{T}} = 0$$
$$\mathbf{u}_2^{\mathrm{T}}\mathbf{u}_3 = \mathbf{u}_2\mathbf{u}_3^{\mathrm{T}} = 0$$

To show that a principal stress state exists when a rotation in the co-ordinates aligns them with the principal stress directions, eqs(1.22a) and (1.28b) are, respectively

$$\sigma_{i\alpha}' = l_{ip}\, l_{\alpha q}\, \sigma_{pq} \tag{1.29a}$$
$$\sigma_{pq}\, l_{\alpha q} = \sigma_\alpha\, l_{\alpha p} \tag{1.29b}$$

Pre-multiply eq(1.29b) by l_{ip} and substitute from eq(1.29a) gives

$$l_{ip}\, \sigma_{pq}\, l_{\alpha q} = l_{ip}\, \sigma_\alpha\, l_{\alpha p}$$
$$\sigma_{i\alpha}' = l_{ip}\, l_{\alpha p}\, \sigma_\alpha = \delta_{i\alpha}\sigma_\alpha \text{ or } \mathbf{T}' = \mathbf{I}\,\mathbf{t}$$

where $\mathbf{t} = \{\sigma_1\ \ \sigma_2\ \ \sigma_3\}^{\mathrm{T}}$ are the principal stresses.

Example 1.2 The stress components (in MPa) at a point within a loaded body are: $\sigma_x = 5$, $\sigma_y = 7$, $\sigma_z = 6$, $\tau_{xy} = 10$, $\tau_{xz} = 8$ and $\tau_{yz} = 12$. Find the magnitudes of the principal stresses and the maximum shear stress. Show that the principal stress directions are orthogonal.

It follows from eq(1.23c) that the principal stresses cubic (or characteristic equation) may be found either from (i) expanding the determinant:

$$\begin{vmatrix} 5-\sigma & 10 & 8 \\ 10 & 7-\sigma & 12 \\ 8 & 12 & 6-\sigma \end{vmatrix} = 0$$

or (ii) from direct substitution into eq(1.24a). These give

$$\sigma^3 - 18\sigma^2 - 201\sigma - 362 = 0$$

from which the invariants in eq(1.24b) are identified as: $J_1 = 18$ MPa, $J_2 = -201$ (MPa)^2 and $J_3 = 362$ (MPa)^3. The roots to this cubic are identified with the principal stresses according to $\sigma_1 > \sigma_2 > \sigma_3$ as follows: $\sigma_1 = 26.2$, $\sigma_2 = -2.37$ and $\sigma_3 = -5.83$ MPa. The three direction cosines for the 1-direction (l_1, m_1, n_1), are found from substituting $\sigma_1 = 26.2$ MPa into eq(1.23a):

$$-21.2\, l_1 + 10m_1 + 8n_1 = 0$$
$$10\, l_1 + 19.2\, m_1 + 12n_1 = 0$$
$$8\, l_1 + 12\, m_1 - 20.2\, n_1 = 0$$

of which only two equations are independent. To solve for l_1, m_1 and n_1 from these equations let a vector: $\mathbf{a} = a_1\mathbf{u}_1 + a_2\mathbf{u}_2 + a_3\mathbf{u}_3$ of arbitary magnitude $|\mathbf{a}| = \sqrt{(a_1^2 + a_2^2 + a_3^2)}$, lie along the 1-direction. Setting $l_1 = a_1/|\mathbf{a}|$, $m_1 = a_2/|\mathbf{a}|$ and $n_1 = a_3/|\mathbf{a}|$ enables a_1 to be set to unity (say) from which $a_2 = 1.222$ and $a_3 = 1.122$. Hence $|\mathbf{a}| = 1.937$ and $l_1 = 0.516$, $m_1 = 0.631$ and $n_1 = 0.579$. Thus a unit vector (eq(1.26a)) may be identified with the 1-direction as:

$$\mathbf{u}_1 = 0.516\mathbf{u}_x + 0.631\mathbf{u}_y + 0.579\mathbf{u}_z$$

Similarly setting $\sigma_2 = -\ 2.37$ MPa in eq(1.23a) leads to a unit vector for the 2-direction as

$$\mathbf{u}_2 = 0.815\mathbf{u}_x - 0.153\mathbf{u}_y - 0.560\mathbf{u}_z$$

Finally, setting $\sigma_3 = -\ 5.83$ MPa in eq(1.23a), leads to a unit vector for the 3-direction as

$$\mathbf{u}_3 = 0.265\mathbf{u}_x - 0.761\mathbf{u}_y + 0.592\mathbf{u}_z$$

Taking the dot products of $\mathbf{u}_1$, $\mathbf{u}_2$ and $\mathbf{u}_3$ shows that $\mathbf{u}_1 \bullet \mathbf{u}_2 = \mathbf{u}_1 \bullet \mathbf{u}_3 = \mathbf{u}_2 \bullet \mathbf{u}_3 = 0$, so confirming that the directions 1, 2 and 3 are orthogonal.

1.4.3 Reductions to Plane Stress

Consider the non-zero plane stress components σ_x, σ_y and τ_{xy}, shown in Fig. 1.13a. Direction cosines: $l = \cos\alpha$, $m = \cos\beta = \cos(90° - \alpha) = \sin\alpha$ and $n = 0$ define the direction normal to the oblique plane in x, y and z co-ordinates.

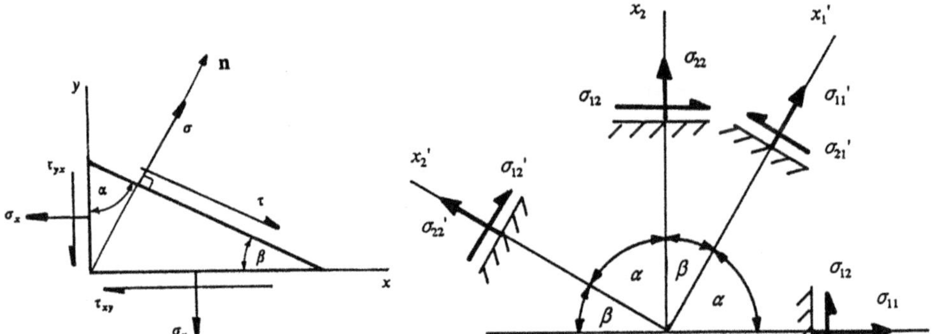

Figure 1.13 Plane stress in x - y and x_1 - x_2 co-ordinates

Substituting into eq(1.12a) gives the normal stress on the oblique plane

$$\begin{aligned}\sigma &= \sigma_x \cos^2\alpha + \sigma_y \sin^2\alpha + 2\tau_{xy}\cos\alpha\sin\alpha \\ &= \sigma_x\cos^2\alpha + \sigma_y\sin^2\alpha + \tau_{xy}\sin 2\alpha \\ \tau &= \tfrac{1}{2}(\sigma_x + \sigma_y) + \tfrac{1}{2}(\sigma_x - \sigma_y)\cos 2\alpha + \tau_{xy}\sin 2\alpha\end{aligned}$$

and from eqs(1.11) and (1.12b), the components of the traction vector $\mathbf{r}$ are:

$$r_x = \sigma_x \cos\alpha + \tau_{xy} \sin\alpha,\ r_y = \sigma_y \sin\alpha + \tau_{yx} \cos\alpha \text{ and } r_z = 0$$

The shear stress on this plane is found from the substitution of r_x, r_y and r_z into eq(1.12b) as

$$\tau^2 = (\sigma_x \cos\alpha + \tau_{xy} \sin\alpha)^2 + (\sigma_y \sin\alpha + \tau_{yx} \cos\alpha)^2 - (\sigma_x \cos^2\alpha + \sigma_y \sin^2\alpha + \tau_{xy} \sin 2\alpha)^2$$
$$= (\sigma_x^2 - \sigma_x\sigma_y + \sigma_y^2) \sin^2\alpha \cos^2\alpha + \tau_{xy} \sin 2\alpha\, [\sigma_x(1 - 2\cos^2\alpha) + \sigma_y(1 - 2\sin^2\alpha)]$$
$$+ \tau_{xy}^2 (\sin^2\alpha + \cos^2\alpha - \sin^2 2\alpha)$$
$$\tau^2 = \tfrac{1}{4} (\sigma_x - \sigma_y)^2 \sin^2 2\alpha - \tau_{xy}(\sigma_x - \sigma_y) \sin 2\alpha \cos 2\alpha + \tau_{xy}^2 \cos^2 2\alpha$$
$$= [\, \tfrac{1}{2} (\sigma_x - \sigma_y) \sin 2\alpha - \tau_{xy} \cos 2\alpha\,]^2$$
$$\tau = \tfrac{1}{2} (\sigma_x - \sigma_y) \sin 2\alpha - \tau_{xy} \cos 2\alpha$$

The respective matrix forms for this plane reduction is found from the transformation eq(1.22a): $\mathbf{T}' = \mathbf{M}\mathbf{T}\mathbf{M}^{\mathrm{T}}$ in which **T** and **M** are now 2 × 2 matrices:

$$\begin{bmatrix} \sigma_{11}' & \sigma_{12}' \\ \sigma_{21}' & \sigma_{22}' \end{bmatrix} = \begin{bmatrix} l_{11} & l_{12} \\ l_{21} & l_{22} \end{bmatrix} \begin{bmatrix} \sigma_{11} & \sigma_{12} \\ \sigma_{21} & \sigma_{22} \end{bmatrix} \begin{bmatrix} l_{11} & l_{21} \\ l_{12} & l_{22} \end{bmatrix}$$

where, $l_{11} = \cos\alpha$, $l_{12} = \cos(90 - \alpha) = \sin\alpha$, $l_{21} = \cos(90 + \alpha) = -\sin\alpha$ and $l_{22} = \cos\alpha$. Figure 1.13b shows that the stress state (σ_{11}', σ_{21}') applies to the x_1' - direction and thus the matrix equation reduces to

$$\begin{bmatrix} \sigma_{11}' \\ \sigma_{21}' \end{bmatrix} = \begin{bmatrix} l_{11} & l_{12} \\ l_{21} & l_{22} \end{bmatrix} \begin{bmatrix} \sigma_{11} & \sigma_{12} \\ \sigma_{21} & \sigma_{22} \end{bmatrix} \begin{bmatrix} l_{11} \\ l_{12} \end{bmatrix}$$

where the following associations are made with the engineering notation $\sigma_{11} = \sigma_x$, $\sigma_{22} = \sigma_y$ and $\sigma_{12} = \tau_{xy}$ in the co-ordinate directions and $\sigma_{11}' = \sigma$ and $\sigma_{21}' = \tau$ for the oblique plane.

In finding the plane principal stresses (σ_1 and σ_2), the invariants in eqs(1.25a-c) become:

$$J_1 = \sigma_x + \sigma_y,\ J_2 = \sigma_x\sigma_y - \tau_{xy}^2 \text{ and } J_3 = 0$$

and the principal stress cubic eq(1.24b) reduces to a quadratic

$$\sigma^2 - (\sigma_x + \sigma_y)\sigma + (\sigma_x\sigma_y - \tau_{xy}^2) = 0$$

The solution gives the two principal stresses as roots

$$\sigma_{1,2} = \tfrac{1}{2} (\sigma_x + \sigma_y) \pm \tfrac{1}{2} \sqrt{[\, (\sigma_x - \sigma_y)^2 + 4\tau_{xy}^2 \,]}$$

where the positive discriminant applies to σ_1.

1.5 Principal Stresses as Co-ordinates

Let the co-ordinate axes become aligned with the orthogonal principal directions 1, 2 and 3 so that the applied stresses are the principal stresses: $\sigma_1 > \sigma_2 > \sigma_3$, as shown in Fig. 1.14. Because there is no shear stress on faces ACO, ABO and BCO, the expressions for the normal and shear stresses (σ, τ) upon the oblique plane ABC are simplified.

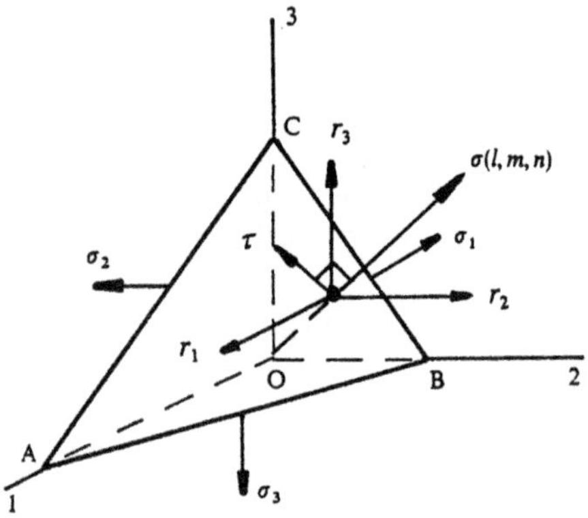

Figure 1.14 Oblique plane set in principal stress axes

Replacing x, y and z in eqs(1.11) and (1.12) with 1, 2 and 3 respectively and setting $\tau_{xy} = \tau_{yz} = \tau_{xz} = 0$ gives the following reduced forms:

$$r_1 = l\sigma_1,\ \ r_2 = m\sigma_2,\ \ r_3 = n\sigma_3 \tag{1.30a,b,c}$$

$$\begin{aligned}\sigma &= r_1 l + r_2 m + r_3 n \\ &= \sigma_1 l^2 + \sigma_2 m^2 + \sigma_3 n^2\end{aligned} \tag{1.31a}$$

The traction across ABC can be written as

$$r^2 = r_1^2 + r_2^2 + r_3^2 = \sigma^2 + \tau^2$$

from which

$$\begin{aligned}\tau^2 &= r^2 - \sigma^2 = (r_1^2 + r_2^2 + r_3^2) - \sigma^2 \\ &= (l\sigma_1)^2 + (m\sigma_2)^2 + (n\sigma_3)^2 - (\sigma_1 l^2 + \sigma_2 m^2 + \sigma_3 n^2)^2\end{aligned}$$

$$\tau = \surd[(l\sigma_1)^2 + (m\sigma_2)^2 + (n\sigma_3)^2 - (\sigma_1 l^2 + \sigma_2 m^2 + \sigma_3 n^2)^2] \tag{1.31b}$$

The direction cosines for τ are, from eqs(1.13),

$$l_s = (r_1 - l\,\sigma)/\tau = l\,(\sigma_1 - \sigma)/\tau \tag{1.32a}$$

$$m_s = (r_2 - m\sigma)/\tau = m(\sigma_2 - \sigma)/\tau \tag{1.32b}$$

$$n_s = (r_3 - n\sigma)/\tau = n\,(\sigma_3 - \sigma)/\tau \tag{1.32c}$$

In the mathematical notation, when x_1, x_2 and x_3 in Fig. 1.11a are aligned with the principal stress co-ordinates (1, 2 and 3), the transformation eq(1.22b) becomes

$$\begin{bmatrix}\sigma_{11}' & \sigma_{12}' & \sigma_{13}' \\ \sigma_{21}' & \sigma_{22}' & \sigma_{23}' \\ \sigma_{31}' & \sigma_{32}' & \sigma_{33}'\end{bmatrix} = \begin{bmatrix}l_{11} & l_{12} & l_{13} \\ l_{21} & l_{22} & l_{23} \\ l_{31} & l_{32} & l_{33}\end{bmatrix}\begin{bmatrix}\sigma_{11} & 0 & 0 \\ 0 & \sigma_{22} & 0 \\ 0 & 0 & \sigma_{33}\end{bmatrix}\begin{bmatrix}l_{11} & l_{21} & l_{31} \\ l_{12} & l_{22} & l_{32} \\ l_{13} & l_{23} & l_{33}\end{bmatrix}$$

This will contain the expressions (1.31a,b) for the normal and shear stresses on a single oblique plane ABC. Identifying direction cosines l_{11}, l_{12} and l_{13} for the normal to ABC gives its three stress components as

$$\begin{bmatrix} \sigma_{11}' \\ \sigma_{21}' \\ \sigma_{31}' \end{bmatrix} = \begin{bmatrix} l_{11} & l_{12} & l_{13} \end{bmatrix} \begin{bmatrix} \sigma_{11} & 0 & 0 \\ 0 & \sigma_{22} & 0 \\ 0 & 0 & \sigma_{33} \end{bmatrix} \begin{bmatrix} l_{11} \\ l_{12} \\ l_{13} \end{bmatrix}$$

in which $\sigma \equiv \sigma_{11}'$ and $\tau \equiv \sqrt{[(\sigma_{21}')^2 + (\sigma_{31}')^2]}$.

1.5.1 Maximum Shear Stress

It can be shown from eq(1.31b) that maximum shear stresses act on planes inclined at 45° to two principal planes and are perpendicular to the remaining plane. For the 1-2 plane in Fig. 1.15, for example, the normal **n** to the 45° plane shown, has directions $l = m = \cos 45° = 1/\sqrt{2}$ and $n = \cos 90° = 0$.

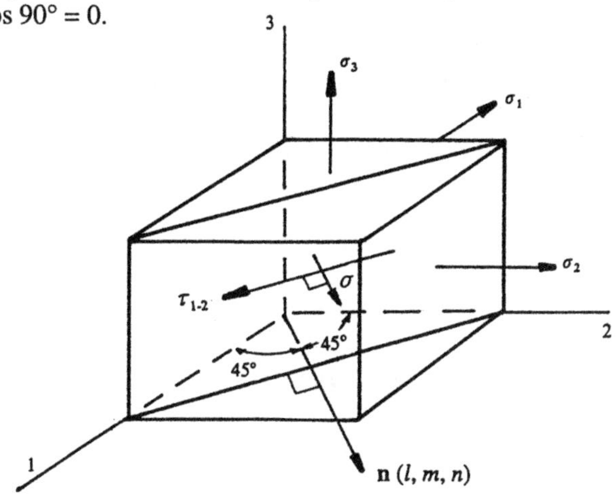

Figure 1.15 Maximum 45° shear plane

Substituting l, m and n into eq(1.31b), the maximum shear stress for this plane is

$$\tau_{1\text{-}2}^2 = \sigma_1^2/2 + \sigma_2^2/2 - (\sigma_1/2 + \sigma_2/2)^2 = \tfrac{1}{4}(\sigma_1^2 + \sigma_2^2 - 2\sigma_1\sigma_2) = \tfrac{1}{4}(\sigma_1 - \sigma_2)^2$$

$$\tau_{1\text{-}2} = \pm \tfrac{1}{2}(\sigma_1 - \sigma_2) \qquad (1.33a)$$

where the subscripts 1, 2 refer to those principal planes to which τ is equally inclined. Similarly, for the plane inclined at 45° to the 1 and 3 directions ($l = n = 1/\sqrt{2}$ and $m = 0$), the maximum shear stress is

$$\tau_{1\text{-}3} = \pm \tfrac{1}{2}(\sigma_1 - \sigma_3) \qquad (1.33b)$$

and, for the plane inclined at 45° to the 2 and 3 directions, where $n = m = 1/\sqrt{2}$ and $l = 0$, the maximum shear stress is

$$\tau_{2\text{-}3} = \pm \tfrac{1}{2}(\sigma_2 - \sigma_3) \qquad (1.33c)$$

The greatest of the three shear stresses, for a system in which $\sigma_1 > \sigma_2 > \sigma_3$, is $\tau_{max} = \tau_{1\text{-}3}$. When the 45° shear planes are considered in all four quadrants they form a *rhombic dodecahedron*. The normal stress acting upon the planes of maximum shear stress are found from eq(1.31a). For example, with $l = m = 1/\surd 2$ and $n = 0$ for the 45° plane shown in Fig. 1.13, the normal stress is

$$\sigma = \tfrac{1}{2}(\sigma_1 + \sigma_2) \tag{1.33d}$$

1.5.2 Octahedral Plane

The octahedral plane is equally inclined to the principal directions. It follows from eq(1.10) that the direction cosines of its normal are $l = m = n = 1/\surd 3$ ($\alpha = \beta = \gamma = 54.8°$). Substituting these into eq(1.31a), gives the octahedral normal stress

$$\sigma_o = \sigma_1 l^2 + \sigma_2 m^2 + \sigma_3 n^2 = \sigma_1 (1/\surd 3)^2 + \sigma_2 (1/\surd 3)^2 + \sigma_3 (1/\surd 3)^2$$
$$\sigma_o = \tfrac{1}{3}(\sigma_1 + \sigma_2 + \sigma_3) \tag{1.34a}$$

Since σ_o is the average of the principal stresses it is also called the mean or hydrostatic stress, σ_m. The octahedral shear stress is found from substituting $l = m = n = 1/\surd 3$ in eq(1.31b):

$$\tau_o^2 = (\sigma_1/\surd 3)^2 + (\sigma_2/\surd 3)^2 + (\sigma_3/\surd 3)^2 - [\sigma_1 (1/\surd 3)^2 + \sigma_2 (1/\surd 3)^2 + \sigma_3 (1/\surd 3)^2]^2$$
$$= (\sigma_1 + \sigma_2 + \sigma_3)/3 - [(\sigma_1 + \sigma_2 + \sigma_3)/3]^2 = (2/9)(\sigma_1^2 + \sigma_2^2 + \sigma_3^2 - \sigma_1\sigma_2 - \sigma_1\sigma_3 - \sigma_2\sigma_3)$$
$$= (1/9)[(\sigma_1 - \sigma_2)^2 + (\sigma_2 - \sigma_3)^2 + (\sigma_1 - \sigma_3)^2]$$
$$\tau_o = \tfrac{1}{3}\surd[(\sigma_1 - \sigma_2)^2 + (\sigma_2 - \sigma_3)^2 + (\sigma_1 - \sigma_3)^2] \tag{1.34b}$$

Combining eq(1.34b) with eqs(1.33a,b,c) gives σ_o in terms of the three maximum shear stresses:

$$\tau_o = (2/3)\surd(\tau_{1\text{-}2}^2 + \tau_{2\text{-}3}^2 + \tau_{1\text{-}3}^2) \tag{1.34c}$$

The direction of τ_o is found from eqs(1.32a,b,c) as

$$l_o = (\sigma_1 - \sigma_o)/(\surd 3\,\tau_o) \tag{1.35a}$$
$$m_o = (\sigma_2 - \sigma_o)/(\surd 3\,\tau_o) \tag{1.35b}$$
$$n_o = (\sigma_3 - \sigma_o)/(\surd 3\,\tau_o) \tag{1.35c}$$

When the eight octahedral planes in all four quadrants are considered they form the faces of the *regular octahedron*, shown in Fig. 1.16.

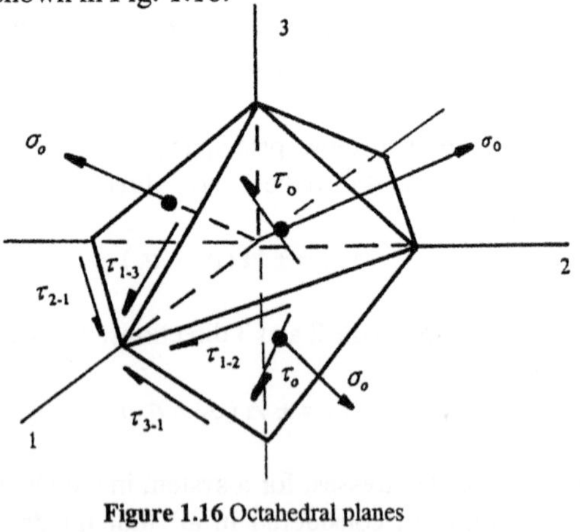

Figure 1.16 Octahedral planes

Here, σ_o and τ_o act on the eight planes while $\tau_{1\text{-}2}$, $\tau_{2\text{-}3}$ and $\tau_{1\text{-}3}$ act along their edges. The deformation that arises from a given stress state may be examined on an octahedral basis. Since σ_o acts with equal inclination and intensity it causes a recoverable elastic volume change irrespective of the principal stress magnitudes. Superimposed upon this is the distortion produced by τ_o. Since eq(1.34b) shows that the magnitude of τ_o depends upon differences between the principal stresses, a critical value of τ_o will determine whether the deformation will be elastic or elastic-plastic. In Chapter 3 it is shown that a yield criterion may be formulated on this basis.

Example 1.3 Given the principal stresses $\sigma_1 = 7.5$, $\sigma_2 = 3.1$ and $\sigma_3 = 1.4$ (MPa) find (i) the maximum shear stresses and their directions and (ii) the magnitude and direction of the normal and shear stresses for the octahedral plane. Confirm the answers using a Mohr's stress circle.

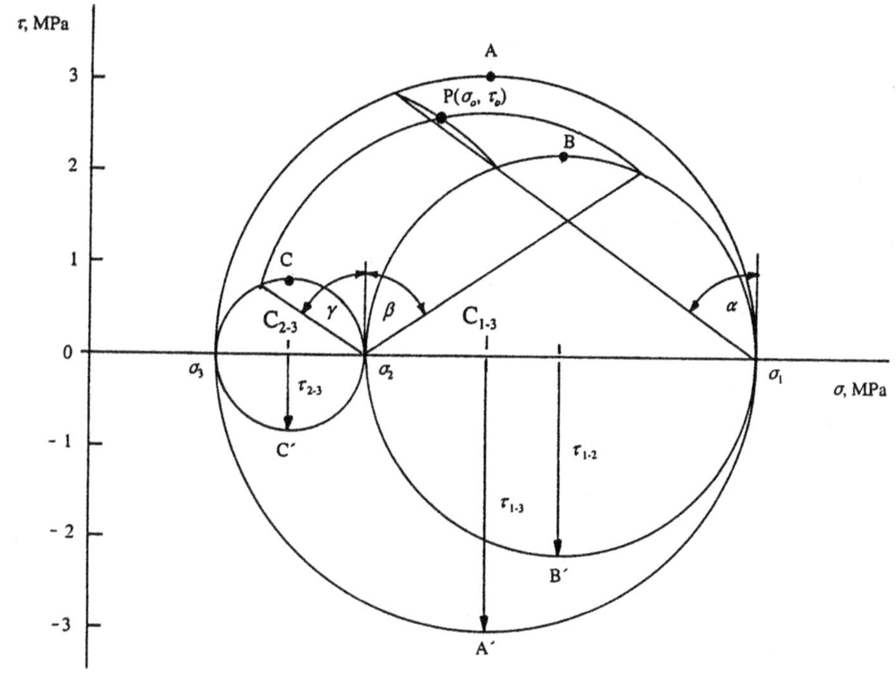

Figure 1.17 Mohr's circle showing max shear and octahedral shear planes

(i) From eqs(1.33a,b,c) the three shear stress maxima are:

$\tau_{1\text{-}2} = \pm \frac{1}{2}(7.5 - 3.1) = \pm 2.20$ MPa, at 45° to the directions 1 and 2 and perpendicular to 3
$\tau_{2\text{-}3} = \pm \frac{1}{2}(3.1 - 1.4) = \pm 0.85$ MPa, at 45° to the directions 2 and 3 and perpendicular to 1
$\tau_{1\text{-}3} = \pm \frac{1}{2}(7.5 - 1.4) = \pm 3.05$ MPa, at 45° to the directions 1 and 3 and perpendicular to 2

(ii) From eq(1.34a) the normal stress on the octahedral plane is:

$\sigma_o = \frac{1}{3}(7.5 + 3.1 + 1.4) = 4.0$ MPa equally inclined to directions 1, 2 and 3

The octahedral shear stress is, from eq(1.34b),

$$\tau_o = ⅓ \sqrt{[(7.5 - 3.1)^2 + (3.1 - 1.4)^2 + (7.5 - 1.4)^2]} = 2.57 \text{ MPa}$$

with direction cosines, from eq(1.35a,b,c):

$$l_o = (7.5 - 4.0) / (\sqrt{3} \times 2.57) = 0.786$$
$$m_o = (3.1 - 4.0) / (\sqrt{3} \times 2.57) = - 0.202$$
$$n_o = (1.4 - 4.0) / (\sqrt{3} \times 2.57) = - 0.584$$

Mohr's circle (see Fig. 1.17) confirms each answer. The maximum positive shear stresses are the vertical radii of the three circles at A, B and C. Complementary negative shear stresses are associated with the reflective points A′, B′ and C′. To locate the octahedral plane, mark off lines at $\alpha = \beta = 54.7°$ ($l = m = n = 1/\sqrt{3}$) from σ_1, σ_2 and σ_3 as shown. With centres $C_{1\text{-}3}$ and $C_{2\text{-}3}$ draw arcs to intersect at a point P whose co-ordinates are σ_o and τ_o. When P is coincident with A this locates a maximum shear plane where $\alpha = \beta = 45°$ and $\gamma = 90°$. Similarly with P at B and C the orientation of two further shear planes are confirmed.

1.5.3 Reductions to Plane Principal Stress

The plane stress transformation equations are particular cases of the foregoing 3D equations. In Fig. 1.18a, the applied principal stresses are σ_1 and σ_2 and the oblique plane is defined with respective directions: $l = \cos\alpha$, $m = \cos(90° - \alpha) = \sin\alpha$ and $n = 0$ for directions 1, 2 and 3.
and 3.

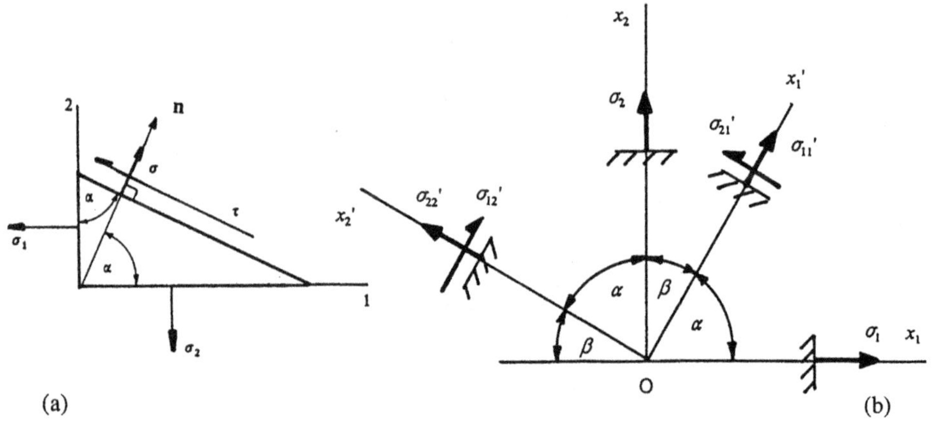

Figure 1.18 Reduction to plane stress

Substituting these into eq(1.31a,b) gives the normal and shear stresses for the oblique plane:

$$\sigma = \sigma_1 \cos^2\alpha + \sigma_2 \sin^2\alpha = \sigma_1 (1 + \cos 2\alpha)/2 + \sigma_2 (1 - \cos 2\alpha)/2$$
$$\sigma = ½(\sigma_1 + \sigma_2) + ½(\sigma_1 - \sigma_2) \cos 2\alpha$$

$$\tau^2 = \sigma_1^2 \cos^2\alpha + \sigma_2^2 \sin^2\alpha - (\sigma_1 \cos^2\alpha + \sigma_2 \sin^2\alpha)^2$$
$$= (\sigma_1 - \sigma_2)^2 \sin^2\alpha \cos^2\alpha$$
$$\tau = ½(\sigma_1 - \sigma_2) \sin 2\alpha$$

Alternatively, these can be found from the transformation matrix $\mathbf{T}' = \mathbf{MTM}^{\mathrm{T}}$. In full, this takes the plane form

$$\begin{bmatrix} \sigma_{11}' & \sigma_{12}' \\ \sigma_{21}' & \sigma_{22}' \end{bmatrix} = \begin{bmatrix} l_{11} & l_{12} \\ l_{21} & l_{22} \end{bmatrix} \begin{bmatrix} \sigma_1 & 0 \\ 0 & \sigma_2 \end{bmatrix} \begin{bmatrix} l_{11} & l_{21} \\ l_{12} & l_{22} \end{bmatrix}$$

Referring to Fig. 1.18b, the direction cosines for the x_1' axis become: $l_{11} = \cos\alpha$ and $l_{12} = \cos(90 - \alpha) = \sin\alpha$ and for the x_2' axis: $l_{21} = \cos(90 + \alpha) = -\sin\alpha$ and $l_{22} = \cos\alpha$. The required stress components σ_{11}' and σ_{21}' for the plane normal to x_1' are found from the reduced matrix multiplication:

$$\begin{bmatrix} \sigma_{11'} \\ \sigma_{21'} \end{bmatrix} = \begin{bmatrix} l_{11} & l_{12} \\ l_{21} & l_{22} \end{bmatrix} \begin{bmatrix} \sigma_1 & 0 \\ 0 & \sigma_2 \end{bmatrix} \begin{bmatrix} l_{11} \\ l_{12} \end{bmatrix}$$

where the correspondence is $\sigma_{11}' = \sigma$ and $\sigma_{12}' = \tau$. Full matrix multiplication will also supply those stress components σ_{12}', σ_{22}' on the plane normal to x_2', as shown in Fig. 1.18b.

1.6 Alternative Stress Definitions

The components σ_{ij}, of the Cauchy stress tensor **T**, appearing in eq(1.6), refer to the elemental deformed area δa of the surface on which the stress traction vector $\mathbf{r}^{(\mathbf{n})}$ acts. Thus, as the foregoing stress transformations apply to the geometry of the deformed material, the stress components are said to be true stresses. Provided deformations are less than 1%, as with the elasticity of metallic materials, it is unnecessary to distinguish between the initial and current areas. A nominal stress, calculated from the original area, will give the Cauchy stress with acceptable accuracy. For example, in a tensile test on a metal, it will only become necessary to convert nominal stresses to true stresses within the reduced cross-sectional area when deformation advances well into the plastic range. However, it is unacceptable to calculate Cauchy elastic stress components from the original area where large elastic deformations arise in certain non-metals. When the elastic extension in an incompressible rubber reaches 300% the cross-sectional area will have diminished by 67%. This means that the nominal stress is only 1/3 of the true stress! Clearly, the choice between initial and current areas can become critical to defining stress properly. The Cauchy definition may be regarded as a true stress. Alternative definitions of nominal stress, given by Piola and Kirchhoff, employ the original area for convenience.

In Fig. 1.19 the resultant force $\delta\mathbf{F}$ is shown for both the reference and current configurations, X_i and x_i, respectively. The traction vectors $\mathbf{r}^{(\mathbf{N})}$ and $\mathbf{r}^{(\mathbf{n})}$ are associated with the vectors **N** and **n**, lying normal to the reference and current areas δA and δa. The corresponding stress tensors are **T** (with Cauchy components σ_{ij} in x_i) and **S** (with nominal components S_{ij} in X_i). Since the force $\delta\mathbf{F}$ is common between the two configurations, it follows from eqs(1.5) and (1.6) that $\mathbf{r}^{(\mathbf{N})} = \mathbf{r}^{(\mathbf{n})}$, when:

$$\sigma_{ji} n_j \, \delta a = S_{ji} N_j \, \delta A \quad \text{or} \quad \mathbf{T}^{\mathrm{T}} \mathbf{n} \delta a = \mathbf{S}^{\mathrm{T}} \mathbf{N} \delta A \tag{1.36}$$

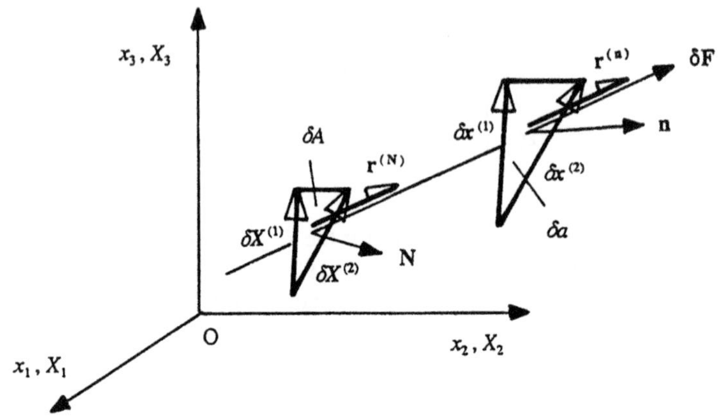

Figure 1.19 Tractions in reference and current configurations

To express the relationship between **n**, **N**, δa and δA it is convenient to take, without loss of generality, infinitesimal triangular areas. Let vectors $\delta\mathbf{x}^{(1)}$ and $\delta\mathbf{x}^{(2)}$ lie along the sides of δa and vectors $\delta\mathbf{X}^{(1)}$ and $\delta\mathbf{X}^{(2)}$ lie along the sides of δA as shown. The areas appear within their cross products

$$n_i\,\delta a = \tfrac{1}{2}\epsilon_{ijk}\delta x_j^{(1)}\delta x_k^{(2)} \quad \text{or} \quad \mathbf{n}\,\delta a = \tfrac{1}{2}\delta\mathbf{x}^{(1)}\times\delta\mathbf{x}^{(2)} \tag{1.37a}$$

$$N_i\delta A = \tfrac{1}{2}\epsilon_{ijk}\delta X_j^{(1)}\,\delta X_k^{(2)} \quad \text{or} \quad \mathbf{N}\delta A = \tfrac{1}{2}\delta\mathbf{X}^{(1)}\times\delta\mathbf{X}^{(2)} \tag{1.37b}$$

where ϵ_{ijk} is the alternating unit tensor. The latter gives: (i) unity for clockwise permutations of differently valued indices, i.e. $\epsilon_{123} = \epsilon_{231} = \epsilon_{312} = 1$, (ii) minus unity for anticlockwise differently valued indices, i.e. $\epsilon_{132} = \epsilon_{321} = \epsilon_{213} = -1$ and (iii) zero when any two or more indices are the same, e.g. $\epsilon_{112} = \epsilon_{323} = 0$ etc. Now, since $\mathbf{x} = \mathbf{x}\,(\mathbf{X})$ or $x_i = x_i\,(X_j)$, it follows:

$$\delta x_i = (\partial x_i / \partial X_j)\,\delta X_j \quad \text{or} \quad \delta\mathbf{x} = (\partial\mathbf{x} / \partial\mathbf{X})\,\delta\mathbf{X} \tag{1.38}$$

Substituting eq(1.38) into eq(1.37a) and making appropriate changes to the indices,

$$n_i\,\delta a = \tfrac{1}{2}\epsilon_{ijk}(\partial x_j / \partial X_q)\,\delta X_q^{(1)}(\partial x_k / \partial X_r)\,\delta X_r^{(2)}$$

Multiplying both sides by $\partial x_i / \partial X_p$ leads to

$$n_i(\partial x_i/\partial X_p)\,\delta a = \tfrac{1}{2}\epsilon_{ijk}(\partial x_i/\partial X_p)(\partial x_j/\partial X_q)(\partial x_k/\partial X_r)\,\delta X_q^{(1)}\,\delta X_r^{(2)} \tag{1.39}$$

Use is now made of the identity

$$\epsilon_{pqr}\det\mathbf{F} = \epsilon_{ijk}F_{ip}F_{jq}F_{kr} \tag{1.40a}$$

where $F_{ip} = \partial x_i / \partial X_p$ etc, are the deformation gradients and $J = \det\mathbf{F} = |\partial x_i / \partial X_j\,|$ is the Jacobian determinant:

$$J = \begin{vmatrix} \partial x_1/\partial X_1 & \partial x_1/\partial X_2 & \partial x_1/\partial X_3 \\ \partial x_2/\partial X_1 & \partial x_2/\partial X_2 & \partial x_2/\partial X_3 \\ \partial x_3/\partial X_1 & \partial x_3/\partial X_2 & \partial x_3/\partial X_3 \end{vmatrix}$$

A simple physical interpretation of J is as a ratio between material densities (or volume elements) in the undeformed and deformed configurations. That is, $J = \rho_o/\rho = \mathrm{d}V/\mathrm{d}V_o$. Combining eqs(1.39) and (1.40):

$$n_i F_{ip}\,\delta a = \tfrac{1}{2}\epsilon_{pqr}\,J\,\delta X_q^{(1)}\,\delta X_r^{(2)}$$

and substituting from eq(1.37b)

$$n_i F_{ip}\delta a = J\,N_p\delta A \quad \text{or} \quad \mathbf{n}^{\mathrm{T}}\mathbf{F}\,\delta a = J\,\mathbf{N}^{\mathrm{T}}\delta A$$

$$n_i\delta a = J\,(F_{pi}^{-1})N_p\delta A \quad \text{or} \quad \mathbf{n}^{\mathrm{T}}\delta a = J\,\mathbf{N}^{\mathrm{T}}\mathbf{F}^{-1}\delta A \tag{1.41}$$

The matrix form may further be expressed as

$$(\mathbf{n}^{\mathrm{T}})^{\mathrm{T}}\delta a = J\,(\mathbf{N}^{\mathrm{T}}\mathbf{F}^{-1})^{\mathrm{T}}\,\delta A$$

$$\mathbf{n}\,\delta a = J\,(\mathbf{F}^{-1})^{\mathrm{T}}\mathbf{N}\,\delta A$$

Substituting eq(1.41) into eq(1.36) gives

$$\sigma_{ji}J\,(F_{pj}^{-1})N_p\delta A = S_{ji}N_j\delta A$$

$$\sigma_{ji}J\,(F_{pj}^{-1})N_p\delta A = S_{pi}N_p\delta A$$

$$J\,\sigma_{ji}(F_{pj}^{-1}) = S_{pi} \quad \text{or} \quad J\,\mathbf{T}^{\mathrm{T}}(\mathbf{F}^{-1})^{\mathrm{T}} = J\,(\mathbf{F}^{-1}\mathbf{T})^{\mathrm{T}} = \mathbf{S}^{\mathrm{T}}$$

This transpose of the nominal stress tensor ($S_{ip} \equiv \mathbf{S}$) defines the first Piola-Kirchoff stress tensor. It is seen that the components S_{pi} are related to the Cauchy stress components σ_{ji} and the deformation gradients as follows:

$$S_{pi} = (\det\mathbf{F})(\partial X_p/\partial x_j)\,\sigma_{ji} \quad \text{or} \quad \mathbf{S}^{\mathrm{T}} = (\det\mathbf{F})\,(\mathbf{F}^{-1}\mathbf{T})^{\mathrm{T}} \tag{1.42a}$$

Note that the nominal stress tensor is

$$S_{ip} = (\det\mathbf{F})(\partial X_i/\partial x_j)\sigma_{jp} \quad \text{or} \quad \mathbf{S} = (\det\mathbf{F})\,\mathbf{F}^{-1}\mathbf{T} \tag{1.42b}$$

It follows from eqs(1.42a,b) that $S_{ip} \neq S_{pi}$ (or $\mathbf{S} \neq \mathbf{S}^{\mathrm{T}}$) so that the tensor is not symmetric. A second, symmetric Piola-Kirchoff stress tensor $\mathbf{S}'$ can be introduced within the relation

$$\mathbf{S} = \mathbf{S}'\,\mathbf{F}^{\mathrm{T}} \tag{1.43a}$$

Post-multiply eq(1.43a) by $(\mathbf{F}^{\mathrm{T}})^{-1}$ gives

$$\mathbf{S}' = \mathbf{S}\,(\mathbf{F}^{\mathrm{T}})^{-1} \tag{1.43b}$$

and also

$$(\mathbf{S}^{\mathrm{T}}) = (\mathbf{S}'\,\mathbf{F}^{\mathrm{T}})^{\mathrm{T}} = \mathbf{F}\mathbf{S}'^{\,\mathrm{T}}$$
$$\mathbf{S}'^{\mathrm{T}} = \mathbf{F}^{-1}(\mathbf{S}^{\mathrm{T}}) \qquad (1.43c)$$

Substituting eq(1.42a) into eq(1.43b), it follows that

$$\mathbf{S}' = (\det \mathbf{F})\,\mathbf{F}^{-1}\mathbf{T}\,(\mathbf{F}^{\mathrm{T}})^{-1} \qquad (1.44a)$$

from which the transpose is

$$\mathbf{S}'^{\mathrm{T}} = (\det \mathbf{F})\,\mathbf{F}^{-1}(\mathbf{F}^{-1}\mathbf{T}\,)^{\mathrm{T}} = (\det \mathbf{F})\,\mathbf{F}^{-1}\mathbf{T}^{\mathrm{T}}\,(\,\mathbf{F}^{-1})^{\mathrm{T}} \qquad (1.44b)$$

Equation (1.44b) can also be derived from substituting eq(1.42b) into eq(1.43c). With the following symmetries: $(\mathbf{F}^{\mathrm{T}})^{-1} = (\mathbf{F}^{-1})^{\mathrm{T}}$ and $\mathbf{T} = \mathbf{T}^{\mathrm{T}}$, it follows from eqs(1.44a,b) that $\mathbf{S}' = \mathbf{S}'^{\mathrm{T}}$ but this tensor has no physical interpretation. Finally, the relationship between the two Piola-Kirchoff stress tensors is found from eq(1.43a):

$$(\mathbf{S}^{\mathrm{T}}) = (\mathbf{S}'\,\mathbf{F}^{\mathrm{T}})^{\mathrm{T}} = \mathbf{F}\,\mathbf{S}'^{\mathrm{T}} = \mathbf{F}\,\mathbf{S}' \quad \text{or} \quad \mathbf{S}' = \mathbf{F}^{-1}(\mathbf{S}^{\mathrm{T}})$$

Now, from eq(1.44a), the tensor $\mathbf{S}'$, with components S_{ip}', is written in terms of the Cauchy stress tensor as

$$S_{ip}' = (\det \mathbf{F})\,(\partial X_i/\partial x_j)\,(\partial X_j/\partial x_i)\,\sigma_{jp} \qquad (1.44c)$$

This form of nominal stress tensor is suitable for the formulation of certain finite constitutive relations. Note that when the deformation gradients remain small during infinitesimal straining the distinction between the three stress definitions disappears, with the reductions to eqs(1.44a,b,c) showing that $\mathbf{T} = \mathbf{S}'$ and $\mathbf{S}^{\mathrm{T}} = \mathbf{I}\,\mathbf{S}'$.

Example 1.4 Express the Cauchy stress tensor $\mathbf{T}$ in terms of deformation gradients $\mathbf{F}$, the density ratio ρ/ρ_o and (i) the nominal stress tensor $\mathbf{S}$, (ii) the first Piola-Kirchoff stress tensor $\mathbf{S}^{\mathrm{T}}$ and (iii) the second Piola-Kirchoff stress tensor $\mathbf{S}'$.

(i) Writing eq(1.42a) as

$$\mathbf{S}^{\mathrm{T}} = J\,(\mathbf{F}^{-1}\mathbf{T})^{\mathrm{T}}$$

the transpose gives $\mathbf{S}$ as

$$\mathbf{S} = J\,\mathbf{F}^{-1}\mathbf{T}$$

Pre-multiplying by $\mathbf{F}$ gives the Cauchy stress $\mathbf{T}$ as

$$\mathbf{T} = (1/J)\mathbf{F}\mathbf{S} = (\rho/\rho_o)\,\mathbf{F}\mathbf{S} \quad \text{or} \quad \sigma_{ij} = (\rho/\rho_o)\,F_{ik}S_{kj}$$

(ii) Again, from eq(1.42a) we write

$$\mathbf{S}^{\mathrm{T}} = J\,\mathbf{T}^{\mathrm{T}}(\,\mathbf{F}^{-1})^{\mathrm{T}}$$

Post-multiplying by $\mathbf{F}^{\mathrm{T}}$

$$\mathbf{S}^{\mathrm{T}}\mathbf{F}^{\mathrm{T}} = J\,\mathbf{T}^{\mathrm{T}}(\,\mathbf{F}^{-1})^{\mathrm{T}}\mathbf{F}^{\mathrm{T}} = J\,\mathbf{T}^{\mathrm{T}}(\,\mathbf{F}\,\mathbf{F}^{-1})^{\mathrm{T}} = J\,\mathbf{T}^{\mathrm{T}}$$

from which

$$\mathbf{T}^{\mathrm{T}} = (1/J\,)\,\mathbf{S}^{\mathrm{T}}\mathbf{F}^{\mathrm{T}} = (\rho/\rho_o)\,\mathbf{S}^{\mathrm{T}}\mathbf{F}^{\mathrm{T}} = (\rho/\rho_o)\,(\,\mathbf{F}\,\mathbf{S})^{\mathrm{T}} \quad \text{or} \quad \sigma_{ji} = (\rho/\rho_o)\,S_{jk}F_{ki}$$

(iii) Writing eq(1.44a) as

$$\mathbf{S}' = J\,\mathbf{F}^{-1}\mathbf{T}\,(\mathbf{F}^{\mathrm{T}})^{-1}$$

Post-multiply by $\mathbf{F}^{\mathrm{T}}$:

$$\mathbf{S}'\mathbf{F}^{\mathrm{T}} = J\,\mathbf{F}^{-1}\mathbf{T}$$

Pre-multiply by $\mathbf{F}$:

$$\mathbf{T} = (1/J)\,\mathbf{F}\,\mathbf{S}'\mathbf{F}^{\mathrm{T}} = (\rho/\rho_o)\,\mathbf{F}\,\mathbf{S}'\mathbf{F}^{\mathrm{T}} \quad \text{or} \quad \sigma_{ij} = (\rho/\rho_o)\,F_{ip}F_{jq}S_{pq}'$$

The final result in (i) and (ii) shows that since $\mathbf{T} = \mathbf{T}^{\mathrm{T}}$ then $\mathbf{FS} = (\,\mathbf{FS}\,)^{\mathrm{T}}$. That is, the matrix product **FS** is symmetrical. The stress tensors, most often employed for finite studies, are those of Cauchy (eq 1.6) and Piola-Kirchoff (eqs (1.42a and 1.44c)), depending upon whether they are to be found from the current or initial areas.

Bibliography

Arridge R. G. C. *Mechanics of Polymers*, 1975, Clarendon, Oxford.
Arridge R. G. C. *An Introduction to Polymer Mechanics*, 1985, Taylor and Francis.
Backofen W. A. *Deformation Processing*, 1972, Addison-Wesley.
Barnes H. A, Hutton J.F and Walters K *An Introduction to Rheology*, Ser. 3, 1989, Elsevier.
Billington E. W. *Introduction to the Mechanics and Physics of Solids*, 1986, Adam Hilger.
Billington E. W. and Tate A. *The Physics of Deformation and Flow*, 1981, McGraw-Hill.
Boresi A. P. and Sidebottom O.M. *Advanced Mechanics of Materials*, 1985, Wiley and Son.
Calcote L. R. *Introduction to Continuum Mechanics*, 1968, van Nostrand.
Ford H. and Alexander J. M. *Advanced Mechanics of Materials*, 1963, Longman.
Malvern L. E. *Introduction to the Mechanics of a Continuous Medium*, 1969, Prentice-Hall.
Rees D. W. A. *Mechanics of Solids and Structures*, 2000, IC Press, World Scientific.
Spencer A. J. M. *Continuum Mechanics*, 1980, Longman.
Tanner R. I. *Engineering Rheology*, 1985, Clarendon, Oxford.
Williams J. G. *Stress Analysis of Polymers*, 1980, Ellis Horwood.
Zyczkowski M. *Combined Loadings in the Theory of Plasticity*, 1981, PWN.

Exercises

1.1 Given the following stress tensor components (in MPa): $\sigma_{11} = 2$, $\sigma_{22} = 2$, $\sigma_{12} = 3$, $\sigma_{13} = 1$, $\sigma_{23} = 3$ and the first invariant $J_1 = 5$ MPa, determine σ_{33}, the principal stresses and the maximum shear stress.
[Answers: 1, 6.4, 0.6, − 2, 4.2 (MPa)]

1.2 Find the principal stresses for a stress tensor with the components: $\sigma_{11} = 1$, $\sigma_{22} = 4$, $\sigma_{33} = 1$, $\sigma_{12} = \sigma_{21} = 2$, $\sigma_{13} = \sigma_{31} = 3$ and $\sigma_{23} = \sigma_{32} = 6$ (MPa). Show that their directions are orthogonal.
[Answer: $\sigma_1 = 10$, $\sigma_2 = 0$ and $\sigma_3 = -4$ (MPa)]

1.3 The principal stresses: $\sigma_1 = 15.4$, $\sigma_2 = 12.65$ and $\sigma_3 = 6.8$ (MPa) act at a point. Determine the normal and resultant shear stresses acting upon an oblique plane whose normal is defined by the unit vector: $\mathbf{u}_n = 0.732\,\mathbf{u}_1 + 0.521\mathbf{u}_2 + 0.439\mathbf{u}_3$. What is the state of stress upon the octahedral plane? Check your answers graphically.
[Answers: $\sigma = 12.63$, $\tau = 3.38$, $\tau_o = 11.17$, $\tau_o = 3.51$(MPa)]

1.4 The stress state at a point is described by the stress components: $\sigma_{11} = 14$, $\sigma_{22} = 10$, $\sigma_{33} = 35$, $\sigma_{12} = \sigma_{21} = 7$, $\sigma_{13} = \sigma_{31} = -7$ and $\sigma_{23} = \sigma_{32} = 0$ (MPa). Find the normal and resultant shear stresses acting

upon a plane whose normal is defined by the direction cosines: $l = 2/\surd 14$ $m = -1/\surd 14$ and $n = 3/\surd 14$.
[Answer: $\sigma = 19.21$, $\tau = 14.95$ (MPa)]

1.5 The stress state at a point is described by the stress components: $\sigma_{11} = 6$, $\sigma_{22} = 0$, $\sigma_{33} = 0$, $\sigma_{12} = \sigma_{21} = 2$, $\sigma_{13} = \sigma_{31} = 2$ and $\sigma_{23} = \sigma_{32} = 4$ (MPa). Find the principal stresses, the greatest shear stress and the state of stress upon the octahedral plane.
[Answer: $\sigma_1 = 8$, $\sigma_2 = 2$ and $\sigma_3 = -4$, $\tau_o = 4.9$, $\sigma_o = 2$(MPa)]

1.6 Determine the invariants J_1, J_2 and J_3 for the stress tensor components: $\sigma_{11} = 6$, $\sigma_{22} = 6$, $\sigma_{33} = 8$, $\sigma_{12} = \sigma_{21} = -3$, $\sigma_{13} = \sigma_{23} = 0$ (MPa). Show that the same invariants values apply to the principal stresses for this system. Determine the octahedral shear stress and the maximum shear stress and the planes on which they act.
[Answers: $J_1 = 20$, $J_2 = 123$, $J_3 = 216$, $\tau_o = 2.625$, $\tau_{max} = 3$ (MPa)]

1.7 Transform the following non-zero components of a stress matrix: $\sigma_{11} = 15$, $\sigma_{22} = 5$, $\sigma_{33} = 20$, $\sigma_{12} = \sigma_{21} = -10$ and $\sigma_{13} = \sigma_{23} = 0$ (MPa) using the following non-zero components of a rotation matrix: $l_{11} = 0.6$, $l_{22} = 1$, $l_{33} = 0.6$, $l_{13} = -l_{31} = -0.8$.
[Answers: $\sigma_{11}' = 18.2$, $\sigma_{22}' = 5$, $\sigma_{33}' = 16.8$, $\sigma_{12}' = \sigma_{21}' = -6$, $\sigma_{13}' = \sigma_{31}' = -2.4$, $\sigma_{23}' = \sigma_{32}' = -8$ (MPa)]

1.8 What are the deviatoric stress components corresponding to each of the stress tensors given in Exercises 1.1-1.4 ?
[Answers: $\sigma_{11}' = 1/3$, $\sigma_{22}' = 1/3$, $\sigma_{33}' = -2/3$; $\sigma_{11}' = -1$, $\sigma_{22}' = 2$, $\sigma_{33}' = -1$; $\sigma_{11}' = 3.78$, $\sigma_{22}' = 1.03$, $\sigma_{33}' = -4.81$; $\sigma_{11}' = -5.67$, $\sigma_{22}' = -9.67$, $\sigma_{33}' = 15.34$ (MPa)]

1.9 Show, from the general transformation law $\sigma_{ij}' = l_{ip} l_{jq} \sigma_{pq}$, that the relationship between the shear stress τ expression (1.12c) for any oblique plane is the resultant of its two shear traction components.

1.10 The following stress tensor components (in MPa): $\sigma_{11} = 1$, $\sigma_{12} = 5$, $\sigma_{13} = -5$, $\sigma_{22} = 0$, $\sigma_{23} = 0$ and $\sigma_{33} = -1$ apply to Cartesian axes x_1, x_2 and x_3. Determine the stress components for a rotation to new axes x_1', x_2' and x_3' defined by the respective orthogonal vectors:

$$\mathbf{a} = \mathbf{u}_1 + 2\mathbf{u}_2 + 3\mathbf{u}_3,\ \mathbf{b} = \mathbf{u}_1 + \mathbf{u}_2 - \mathbf{u}_3 \text{ and } \mathbf{c} = -5\mathbf{u}_1 + 4\mathbf{u}_2 - \mathbf{u}_3.$$

[Answers: $\sigma_{11}' = -1.286$, $\sigma_{12}' = 1.389$, $\sigma_{13}' = 1.980$, $\sigma_{22}' = 6.667$, $\sigma_{23}' = -2.762$ and $\sigma_{33}' = -5.381$]

CHAPTER 2

STRAIN ANALYSIS

2.1 Introduction

The amount of stretch, compression and distortion within a deforming solid is defined in terms of its strain. Various definitions have been used so it is pertinent to begin with explanations as to how strain is defined. As with stress, it is essential that the tensorial nature of strain is understood. Where deformation is small (infinitesimal) we examine: (i) the representations of strain in the tensor component and matrix notations and (ii) the manner in which distortion is decomposed into strain and rigid body rotation. It will be recognised from this that stress and strain are identical second rank Cartesian tensors with similar transformation properties. In fact, displacement derivatives will define the strain tensor components when metallic materials suffer infinitesimal elastic deformations. The strain at the limit of elasticity may be exceeded one hundred times by plastic deformation in a metal. Comparitavely large elastic strains in non-metals can arise from rapid rates of straining. Since the infinitesimal definition of strain is inappropriate under these conditions, alternative finite strain measures are discussed. Among these are the natural or logarithmic strain, the deformation gradient and the extension ratio.

In Chapter 1, direct and shear strain components were shown to accompany stress for respective uniaxial and shear loadings (see Figs 1.1 and 1.2). The two strains can co-exist in various combinations under combined loadings. In general, six independent strain components appear within the general strain tensor. As with stress, the following strain analyses will alternate between the engineering and mathematical notations (see Table 1.1) to enable the reader to recognise their equivalence.

2.2 Infinitesimal Strain Tensor

A line element in an unloaded body may distort in a number of ways when forces are applied to it. In general, the element can translate, rotate and change its length. Length changes and angular distortions are normally associated with direct and shear strain components respectively. Rotations and translations, arising from rigid body motion, do not strain a body. When infinitesimal displacements of line elements arise from a combination of these motions, it becomes necessary to separate the physical deformation from rigid body motion.

2.2.1 Distortion and Rotation

Here it will be necessary to specify the deformation that arises from applying the six independent components of the stress tensor σ_{ij} to a body element. In the engineering notation, Fig. 1.6a identifies these components as σ_x, σ_y, σ_z, τ_{xy}, τ_{xz} and τ_{yz}. Two types of distortion arise: (i) direct strains ε_x, ε_y and ε_z from σ_x, σ_y and σ_z and (ii) angular distortions e_{xy}, e_{xz} and e_{yz} from τ_{xy}, τ_{xz} and τ_{yz}. Let one corner of the body element, $\delta x \times \delta y \times \delta z$, originally lie at the origin O in the x-y plane, as shown in Fig. 2.1a.

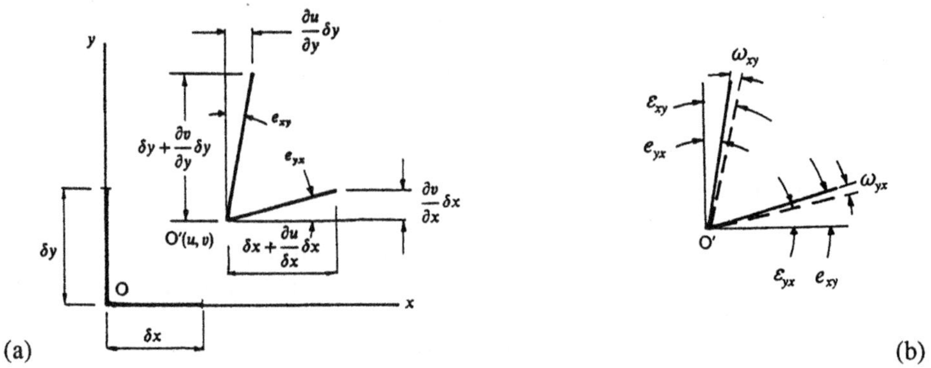

Figure 2.1 Distortion of a corner in the x-y plane

The angular change to the corner is composed of pure shear strains, due to τ_{xy}, τ_{xz} and τ_{yz}, with superimposed rigid body rotations. By subtracting rotations from the angular change, the shear strain associated with shape change is found. In Fig. 2.1b, the angular changes, e_{xy} and e_{yx}, to the corner are shown. With superimposed direct strains, ε_x and ε_y, due to σ_x and σ_y, the corner is also displaced from O to O' in the x and y directions. The respective displacements, u and v, are each functions of the Cartesian co-ordinates: $u = u(x, y, z)$ and $v = v(x, y, z)$. Thus, line elements δx and δy will change their lengths in proportion to the displacement gradients. The latter define the two direct strains in the x-y plane as

$$\varepsilon_x = \partial u/\partial x \quad \text{and} \quad \varepsilon_y = \partial v/\partial y \qquad (2.1a,b)$$

In general, $\delta x \neq \delta y$ and $\varepsilon_x \neq \varepsilon_y$ and so the angular changes are unequal. That is: $e_{xy} \neq e_{yx}$ where $e_{xy} = \partial u/\partial y$ and $e_{yx} = \partial v/\partial x$. The engineering shear strain γ_{xy} is defined as the total angular distortion to the right angle:

$$\gamma_{xy} = \partial u/\partial y + \partial v/\partial x \qquad (2.1c)$$

In Fig. 2.1b the distorted corner is rotated so that it becomes equally inclined to the x- and y - directions. This configuration defines two tensor shear strain components as

$$\varepsilon_{xy} = \varepsilon_{yx} = \tfrac{1}{2}\gamma_{xy} = \tfrac{1}{2}(\partial u/\partial y + \partial v/\partial x)$$

with their associated rotations

$$\omega_{xy} = e_{xy} - \varepsilon_{xy} = \partial u/\partial y - \tfrac{1}{2}(\partial u/\partial y + \partial v/\partial x) = \tfrac{1}{2}(\partial u/\partial y - \partial v/\partial x)$$
$$\omega_{yx} = e_{yx} - \varepsilon_{yx} = \partial v/\partial x - \tfrac{1}{2}(\partial u/\partial y + \partial v/\partial x) = -\tfrac{1}{2}(\partial u/\partial y - \partial v/\partial x)$$

obeying skew-symmetry, $\omega_{xy} = -\omega_{yx}$. Further strain-displacement relationships, similar in form to eqs(2.1a,b,c), will apply to normal and shear distortions in the x-z and y-z planes. These require a further displacement function: $w = w(x, y, z)$, for the z - direction.

Consequently, the complete distortion of a volume element $\delta x \times \delta y \times \delta z$ may be expressed as the sum of corresponding strains and rotations in the matrix form

$$\begin{bmatrix} e_x & e_{xy} & e_{xz} \\ e_{yx} & e_y & e_{yz} \\ e_{zx} & e_{zy} & e_z \end{bmatrix} = \begin{bmatrix} \varepsilon_x & \varepsilon_{xy} & \varepsilon_{xz} \\ \varepsilon_{yx} & \varepsilon_y & \varepsilon_{yz} \\ \varepsilon_{zx} & \varepsilon_{zy} & \varepsilon_z \end{bmatrix} + \begin{bmatrix} \omega_x & \omega_{xy} & \omega_{xz} \\ \omega_{yx} & \omega_y & \omega_{yz} \\ \omega_{zx} & \omega_{zy} & \omega_z \end{bmatrix} \tag{2.2a}$$

where $\omega_x = \omega_y = \omega_z = 0$. In the tensor component notation, eq(2.2a) appears as

$$e_{ij} = \varepsilon_{ij} + \omega_{ij} \quad \text{(for } i \text{ and } j = 1, 2 \text{ and } 3) \tag{2.2b}$$

The strain and rotation components for each tensor in eq(2.2b) may be expressed in terms of their respective displacement gradients. Firstly, the angular changes appear as

$$e_{ij} = \begin{bmatrix} \dfrac{\partial u}{\partial x} & \dfrac{\partial u}{\partial y} & \dfrac{\partial u}{\partial z} \\[2ex] \dfrac{\partial v}{\partial x} & \dfrac{\partial v}{\partial y} & \dfrac{\partial v}{\partial z} \\[2ex] \dfrac{\partial w}{\partial x} & \dfrac{\partial w}{\partial y} & \dfrac{\partial w}{\partial z} \end{bmatrix}$$

It follows that the strain matrix is

$$\varepsilon_{ij} = \begin{bmatrix} \dfrac{\partial u}{\partial x} & \dfrac{1}{2}\left(\dfrac{\partial u}{\partial y} + \dfrac{\partial v}{\partial x}\right) & \dfrac{1}{2}\left(\dfrac{\partial u}{\partial z} + \dfrac{\partial w}{\partial x}\right) \\[2ex] \dfrac{1}{2}\left(\dfrac{\partial u}{\partial y} + \dfrac{\partial v}{\partial x}\right) & \dfrac{\partial v}{\partial y} & \dfrac{1}{2}\left(\dfrac{\partial v}{\partial z} + \dfrac{\partial w}{\partial y}\right) \\[2ex] \dfrac{1}{2}\left(\dfrac{\partial u}{\partial z} + \dfrac{\partial w}{\partial x}\right) & \dfrac{1}{2}\left(\dfrac{\partial v}{\partial z} + \dfrac{\partial w}{\partial y}\right) & \dfrac{\partial w}{\partial z} \end{bmatrix}$$

and the rotation matrix is

$$\omega_{ij} = \begin{bmatrix} 0 & \dfrac{1}{2}\left(\dfrac{\partial u}{\partial y} - \dfrac{\partial v}{\partial x}\right) & \dfrac{1}{2}\left(\dfrac{\partial u}{\partial z} - \dfrac{\partial w}{\partial x}\right) \\[2ex] -\dfrac{1}{2}\left(\dfrac{\partial u}{\partial y} - \dfrac{\partial v}{\partial x}\right) & 0 & \dfrac{1}{2}\left(\dfrac{\partial v}{\partial z} - \dfrac{\partial w}{\partial y}\right) \\[2ex] -\dfrac{1}{2}\left(\dfrac{\partial u}{\partial z} - \dfrac{\partial w}{\partial x}\right) & -\dfrac{1}{2}\left(\dfrac{\partial v}{\partial z} - \dfrac{\partial w}{\partial y}\right) & 0 \end{bmatrix}$$

All direct and shear strain components appear within a linear combination of displacement derivatives, i.e. the *Eulerian infinitesimal strain tensor*:

$$\varepsilon_{ij} = \frac{1}{2}\left(\frac{\partial u_i}{\partial x_j} + \frac{\partial u_j}{\partial x_i}\right) \qquad \text{or} \qquad \mathbf{E} = \tfrac{1}{2}\,(\mathbf{u}^{\mathrm{T}}\nabla + \nabla^{\mathrm{T}}\mathbf{u}) \tag{2.3}$$

The *Eulerian rotation tensor* is defined as

$$\omega_{ij} = \frac{1}{2}\left(\frac{\partial u_i}{\partial x_j} - \frac{\partial u_j}{\partial x_i}\right) \qquad \text{or} \qquad \boldsymbol{\Omega} = \tfrac{1}{2}\,(\mathbf{u}^{\mathrm{T}}\nabla - \nabla^{\mathrm{T}}\mathbf{u}) \tag{2.4}$$

in whch x_i are the spacial, or current, co-ordinates of the deformed body (see Fig. 1.19). When the corresponding engineering notation is required, the subscripts i and j on the symbols u and x are to be read as: $u_1 = u$, $u_2 = v$, $u_3 = w$, $x_1 = x$, $x_2 = y$ and $x_3 = z$. Note that $\mathbf{u}$ and ∇ in eqs(2.3) and (2.4) define column matrices of their components: $\mathbf{u} = \{u_1\ u_2\ u_3\}^{\mathrm{T}}$ and $\mathbf{V} = \{\ \partial/\partial x_1\ \ \partial/\partial x_2\ \ \partial/\partial x_3\}^{\mathrm{T}}$. Hence $\mathbf{u}^{\mathrm{T}}$ and ∇^{T} define their row matrices. Equation (2.2b) shows that the 3 × 3 strain matrix $\mathbf{E}$ is symmetric (i.e. $\varepsilon_{ij} = \varepsilon_{ji}$) and that the 3 × 3 rotation matrix $\boldsymbol{\Omega}$ is skew symmetric (i.e. $\omega_{ij} = -\ \omega_{ji}$). Under applied principal stresses, when all the rotations $\omega_{ij} = 0$, the deformation is said to be irrotational.

2.2.2 *Strain Transformations in Tensor and Matrix Notations*

It may be deduced that the transformation properties of the symmetric 3 × 3 strain matrix ε_{ij} are identical to those of Cauchy stress, the latter being a symmetric 3 × 3 matrix ($\sigma_{ij} = \sigma_{ji}$). Thus, the transformation of an Eulerian strain tensor, following a rotation in the orthogonal axes from x_i to x_i', will obey a law equivalent in form to eq(1.22a):

$$\varepsilon_{ij}' = l_{ip}\,l_{jq}\,\varepsilon_{pq}, \quad \text{or} \quad \mathbf{E}' = \mathbf{MEM}^{\mathrm{T}} \tag{2.5a}$$

In full, eq(2.5a) becomes

$$\begin{bmatrix} \varepsilon_{11}' & \varepsilon_{12}' & \varepsilon_{13}' \\ \varepsilon_{21}' & \varepsilon_{22}' & \varepsilon_{23}' \\ \varepsilon_{31}' & \varepsilon_{32}' & \varepsilon_{33}' \end{bmatrix} = \begin{bmatrix} l_{11} & l_{12} & l_{13} \\ l_{21} & l_{22} & l_{23} \\ l_{31} & l_{32} & l_{33} \end{bmatrix} \begin{bmatrix} \varepsilon_{11} & \varepsilon_{12} & \varepsilon_{13} \\ \varepsilon_{21} & \varepsilon_{22} & \varepsilon_{23} \\ \varepsilon_{31} & \varepsilon_{32} & \varepsilon_{33} \end{bmatrix} \begin{bmatrix} l_{11} & l_{21} & l_{31} \\ l_{12} & l_{22} & l_{32} \\ l_{13} & l_{23} & l_{33} \end{bmatrix} \tag{2.5b}$$

It follows that there is a correspondence between the stress and strain transformation expressions. For this to be achieved, τ must be interchanged with the tensor definition of shear strain ($\gamma/2$). For example, to find the normal strain on an oblique plane in terms of engineering co-ordinate strains, eq(1.12b) is converted from stress to strain as follows:

$$\varepsilon = \varepsilon_x\, l^2 + \varepsilon_y\, m^2 + \varepsilon_z\, n^2 + 2\,(\,lm\varepsilon_{xy} + mn\varepsilon_{yz} + ln\varepsilon_{zx}) \tag{2.6a}$$

$$= \varepsilon_x\, l^2 + \varepsilon_y\, m^2 + \varepsilon_z\, n^2 + lm\,\gamma_{xy} + mn\,\gamma_{yz} + ln\,\gamma_{zx} \tag{2.6b}$$

The principal strain cubic is similarly deduced from eq(1.24b) as

$$\varepsilon^3 - I_1\,\varepsilon^2 + I_2\,\varepsilon - I_3 = 0 \tag{2.7a}$$

where the strain invariants are

$$I_1 = \varepsilon_x + \varepsilon_y + \varepsilon_z = \varepsilon_{ii} = \text{tr}\,\mathbf{E} \tag{2.7b}$$

$$I_2 = \varepsilon_x\varepsilon_y + \varepsilon_y\varepsilon_z + \varepsilon_z\varepsilon_x - \varepsilon_{xy}^{\ 2} - \varepsilon_{yz}^{\ 2} - \varepsilon_{zx}^{\ 2} = \tfrac{1}{2}\,(\varepsilon_{ii}\varepsilon_{jj} - \varepsilon_{ij}\varepsilon_{ji}) = \tfrac{1}{2}[(\text{tr}\,\mathbf{E})^2 - \text{tr}\,\mathbf{E}^2] \tag{2.7c}$$

$$\begin{aligned} I_3 &= \varepsilon_x\varepsilon_y\varepsilon_z + 2\varepsilon_{xy}\varepsilon_{zx}\varepsilon_{yz} - \varepsilon_x\varepsilon_{yz}^{\ 2} - \varepsilon_y\varepsilon_{zx}^{\ 2} - \varepsilon_z\varepsilon_{xy}^{\ 2} \\ &= \varepsilon_x\varepsilon_y\varepsilon_z + \tfrac{1}{4}\gamma_{xy}\gamma_{zx}\gamma_{yz} - \tfrac{1}{4}\varepsilon_x\gamma_{yz}^{\ 2} - \tfrac{1}{4}\varepsilon_y\gamma_{zx}^{\ 2} - \tfrac{1}{4}\varepsilon_z\gamma_{xy}^{\ 2} = \det(\varepsilon_{ij}) = \det\mathbf{E} \end{aligned} \tag{2.7d}$$

Because of the identical nature of stress and strain transformations, the directions of the principal stress and strain axes are coincident. It is not possible to convert the general shear stress expression (in eq(1.12c)) directly to shear strain without firstly specifying the initial perpendicular directions to which the shear strain applies. Performing the matrix multiplication (2.5b) for any one tensor shear strain (i.e. with $i \neq j$) provides a method for finding the shear strain between a pair of perpendicular lines. For example, with $i = 1$ and $j = 2$, eq(2.5b) becomes $\varepsilon_{12}' = l_{1p}\,l_{2q}\,\varepsilon_{pq}$. In the engineering notation this is equivalent to an abbreviated matrix multiplication:

$$\frac{\gamma}{2} = \begin{bmatrix} l_1 & m_1 & n_1 \end{bmatrix} \begin{bmatrix} \varepsilon_x & \varepsilon_{xy} & \varepsilon_{xz} \\ \varepsilon_{xy} & \varepsilon_y & \varepsilon_{yz} \\ \varepsilon_{xz} & \varepsilon_{yz} & \varepsilon_z \end{bmatrix} \begin{bmatrix} l_2 \\ m_2 \\ n_2 \end{bmatrix} \tag{2.8a}$$

$$= \begin{bmatrix} (l_1\,\varepsilon_x + m_1\,\varepsilon_{xy} + n_1\,\varepsilon_{xz}) & (l_1\,\varepsilon_{xy} + m_1\,\varepsilon_y + n_1\,\varepsilon_{yz}) & (l_1\,\varepsilon_{xz} + m_1\,\varepsilon_{yz} + n_1\,\varepsilon_z) \end{bmatrix} \begin{bmatrix} l_2 \\ m_2 \\ n_2 \end{bmatrix}$$

$$\begin{aligned} &= (l_1\varepsilon_x + m_1\varepsilon_{xy} + n_1\varepsilon_{xz})l_2 + (l_1\varepsilon_{xy} + m_1\varepsilon_y + n_1\varepsilon_{yz})m_2 + (l_1\varepsilon_{xz} + m_1\varepsilon_{yz} + n_1\varepsilon_z)n_2 \\ &= l_1l_2\varepsilon_x + m_1m_2\varepsilon_y + n_1n_2\varepsilon_z + (l_1m_2 + l_2m_1)\varepsilon_{xy} + (m_1n_2 + m_2n_1)\varepsilon_{yz} + (l_1n_2 + l_2n_1)\varepsilon_{xz} \end{aligned} \tag{2.8b}$$

Alternatively, the generalised shear strain (eq(2.8b)) will appear within the tensor and matrix notations as

$$\tfrac{1}{2}\,\gamma = l_{1i}\,\varepsilon_{ij}\,l_{2j} = \mathbf{u}_1'^{\mathrm{T}}\,\mathbf{E}\,\mathbf{u}_2' \tag{2.8c}$$

where column matrices $\mathbf{u}_1' = \{\, l_{11}\ l_{12}\ l_{13} \,\}^{\mathrm{T}}$ and $\mathbf{u}_2' = \{\, l_{21}\ l_{22}\ l_{23} \,\}^{\mathrm{T}}$ express the direction cosines of unit vectors aligned with a given pair of perpendicular directions: x_1' and x_2'. It is possible to identify the correspondence between τ in the stress eq(1.12c) and $\gamma/2$ in eq(2.8b). Referring to the plane ABC Fig. 1.7a, the perpendicular directions are those defined by the direction cosines (1.13a,b,c) for the shear stress τ and those (1.8a,b,c) for the normal stress σ. That is, in eq(2.8b) l_2, m_2 and n_2 are identified with l_s, m_s and n_s in eqs(1.13a-c) and l_1, m_1 and n_1 with l, m and n in eqs(1.8a-c). Equation (2.8b) will then supply the shear strain for original perpendicular directions: one parallel to the normal stress σ and the other aligned with the direction of the resultant shear stress τ in Fig. 1.7a.

Example 2.1 The state of strain at a point is specified, in the engineering notation, by the following components of its strain tensor ($\times 10^{-4}$): $\varepsilon_x = 1$, $\varepsilon_y = 1$, $\varepsilon_z = 4$, $\varepsilon_{xy} = -3$, $\varepsilon_{xz} = \sqrt{2}$ and $\varepsilon_{yz} = -\sqrt{2}$. Determine: (i) the normal strain component in a direction defined by the unit vector: $\mathbf{u} = (1/2)\mathbf{u}_x - (1/2)\mathbf{u}_y + (1/\sqrt{2})\mathbf{u}_z$, (ii) the shear strain between the normal and a perpendicular direction whose unit vector equation is $\mathbf{u} = -(1/2)\mathbf{u}_x + (1/2)\mathbf{u}_y + (1/\sqrt{2})\mathbf{u}_z$, (iii) the invariants and (iv) the principal strains.

(i) Substituting $l = 1/2$, $m = -1/2$ and $n = 1/\sqrt{2}$ together with the component strains into eq(2.6a) gives the normal strain

$$\varepsilon/10^{-4} = [(1/2)^2 \times 1] + [(-1/2)^2 \times 1] + [(1/\sqrt{2})^2 \times 4]$$
$$+ 2[(1/2)(-1/2)(-3) + (-1/2)(1/\sqrt{2})(-\sqrt{2}) + (1/2)(1/\sqrt{2})(\sqrt{2})] = 6$$

(ii) The engineering shear strain γ is found from eq(2.8b). The direction cosines: $l_1 = 1/2$, $m_1 = -1/2$ and $n_1 = 1/\sqrt{2}$ apply to the normal direction and $l_2 = -1/2$, $m_2 = 1/2$ and $n_2 = 1/\sqrt{2}$ apply to the perpendicular direction. (Note: $l_1 l_2 + m_1 m_2 + n_1 n_2 = 0$.)

$$\gamma/2 \times 10^{-4} = [(1/2)(-1/2) \times 1] + [(-1/2)(1/2) \times 1] + [(1/\sqrt{2})(1/\sqrt{2}) \times 4]$$
$$+ [(1/2)(1/2) + (-1/2)(-1/2)](-3) + [(-1/2)(1/\sqrt{2}) + (1/2)(1/\sqrt{2})](-\sqrt{2})$$
$$+ [(1/2)(1/\sqrt{2}) + (-1/2)(1/\sqrt{2})](\sqrt{2}) = 0$$

This zero value for γ shows that there is no change in the right angle between the two perpendicular directions, i.e. they are principal directions and therefore the normal strain found in (i) is one of the principal strains.

(iii) The invariants I_1, I_2 and I_3 are found from eqs(2.7b,c,d):

$$I_1/10^{-4} = (1 + 1 + 4)10^{-4} = 6$$
$$I_2/10^{-8} = [(1)(1) + (1)(4) + (4)(1) - (-3)2 - (-\sqrt{2})2 - (\sqrt{2})2] = -4$$
$$I_3/10^{-12} = \det(\varepsilon_{ij}) = 1(4 - 2) + 3(-12 + 2) + \sqrt{2}(3/2 - \sqrt{2}) = -24$$

(iv) The invariants give a principal strain cubic, from eq(2.7a)

$$\varepsilon^3 - (6 \times 10^{-4})\varepsilon^2 - (4 \times 10^{-8})\varepsilon + (24 \times 10^{-12}) = 0$$

for which one root is 6×10^{-4}. The two remaining principal strains may be found from the coefficients a, b and c within the quadratic equation $(a\varepsilon^2 + b\varepsilon + c)$ as follows:

$$[\varepsilon - (6 \times 10^{-4})](a\varepsilon^2 + b\varepsilon + c) = \varepsilon^3 - (6 \times 10^{-4})\varepsilon^2 - (4 \times 10^{-8})\varepsilon + (24 \times 10^{-12}) = 0$$

Equating coefficients gives $a = 1$, $b = 0$ and $c = -4 \times 10^{-8}$. The quadratic equation simplifies to $\varepsilon^2 - 4 \times 10^{-8} = 0$, for which the roots are: $\varepsilon = \pm 2 \times 10^{-4}$. Thus, the principal strains are identified as: $\varepsilon_1 = 6 \times 10^{-4}$, $\varepsilon_2 = 2 \times 10^{-4}$ and $\varepsilon_3 = -2 \times 10^{-4}$.

2.2.3 Reduction to Plane Strain

Putting $l = \cos\alpha$, $m = \cos(90° - \alpha) = \sin\alpha$ and $n = \cos 90° = 0$ in eq(2.6b), leads to the normal strain in the direction of x' in Fig. 2.2a.

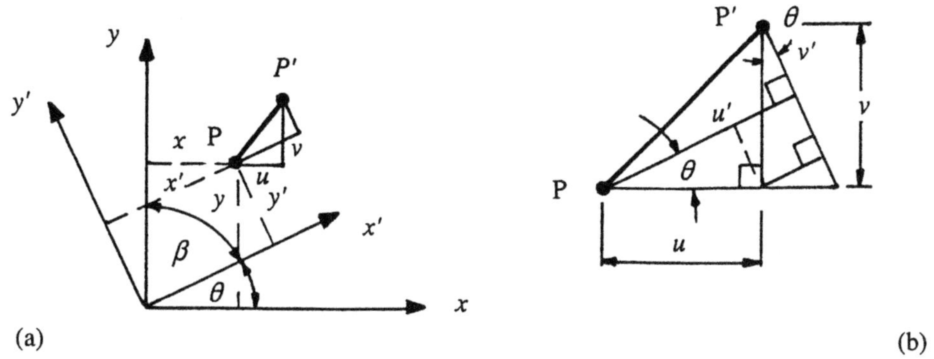

Figure 2.2 (a) Plane co-ordinate and (b) displacement transformations

$$\begin{aligned}\varepsilon_x' &= \varepsilon_x l^2 + \varepsilon_y m^2 + \varepsilon_z n^2 + \gamma_{xy} lm + \gamma_{yz} mn + \gamma_{zx} ln \\ &= \varepsilon_x \cos^2\alpha + \varepsilon_y \sin^2\alpha + \gamma_{xy} \sin\alpha \cos\alpha \\ &= ½(\varepsilon_x + \varepsilon_y) + ½(\varepsilon_x - \varepsilon_y) \cos 2\alpha + (½\gamma_{xy}) \sin 2\alpha \qquad (2.9a)\end{aligned}$$

The normal strain in the y' - direction is found from setting $l = \cos(90 + \alpha) = -\sin\alpha$, $m = \cos\alpha$ and $n = 0$ in eq(2.9a). This gives

$$\begin{aligned}\varepsilon_y' &= \varepsilon_x \sin^2\alpha + \varepsilon_y \cos^2\alpha - \gamma_{xy} \sin\alpha \cos\alpha \\ &= ½(\varepsilon_x + \varepsilon_y) - ½(\varepsilon_x - \varepsilon_y) \cos 2\alpha - (½\gamma_{xy}) \sin 2\alpha \qquad (2.9b)\end{aligned}$$

The shear strain between the primed directions x' and y' is found from eq(2.8b). For the x'-direction, we set: $l_1 = \cos\alpha$, $m_1 = \cos(90° - \alpha) = \sin\alpha$ and $n_1 = \cos 90° = 0$ and for the y'-direction, set $l_2 = \cos(90 + \alpha) = -\sin\alpha$, $m_2 = \cos\alpha$ and $n_2 = 0$:

$$\begin{aligned}½\gamma_{xy}' &= -\varepsilon_x \sin\alpha \cos\alpha + \varepsilon_y \sin\alpha \cos\alpha + \gamma_{xy}(\cos^2\alpha - \sin^2\alpha) \\ &= -½\varepsilon_x \sin 2\alpha + ½\varepsilon_y \sin 2\alpha + ½\gamma_{xy} \cos 2\alpha \\ &= -½(\varepsilon_x - \varepsilon_y) \sin 2\alpha + ½\gamma_{xy} \cos 2\alpha \qquad (2.9c)\end{aligned}$$

The strains given in eqs(2.9a-c) may be confirmed from direct differentiation of the displacements according to the infinitesimal strain definitions (2.1a-c). Figure 2.2a provides the geometrical relationships between the co-ordinates:

$$\begin{aligned}x &= x' \cos\alpha - y' \sin\alpha \\ y &= x' \sin\alpha + y' \cos\alpha\end{aligned}$$

The displacements of a point P as it moves to P′ (see Fig. 2.2b), are given by

$$\begin{aligned}u' &= u \cos\alpha + v \sin\alpha \\ v' &= -u \sin\alpha + v \cos\alpha\end{aligned}$$

where u and v are the displacements of P′ along x and y and u' and v' are the displacements of P′ along x' and y', as shown. The normal strain and shear strains in the x'-y' plane are found from eqs(2.1a-c):

$$\begin{aligned}
\varepsilon_x' &= \partial u'/\partial x' = (\partial u'/\partial x)(\partial x/\partial x') + (\partial u'/\partial y)(\partial y/\partial x')\\
&= [\,(\partial u/\partial x)(\partial u'/\partial u) + (\partial v/\partial x)(\partial u'/\partial v)](\partial x/\partial x') + [(\partial u/\partial y)(\partial u'/\partial u) + (\partial v/\partial y)(\partial u'/\partial v)](\partial y/\partial x')\\
&= [(\partial u/\partial x)\cos\alpha + (\partial v/\partial x)\sin\alpha\,]\cos\alpha + [(\partial u/\partial y)\cos\alpha + (\partial v/\partial y)\sin\alpha\,]\sin\alpha\\
&= \varepsilon_x\cos^2\alpha + \varepsilon_y\sin^2\alpha + \gamma_{xy}\sin\alpha\cos\alpha\\
&= ½\,(\varepsilon_x + \varepsilon_y) + ½\,(\varepsilon_x - \varepsilon_y)\cos 2\alpha + ½\,\gamma_{xy}\sin 2\alpha
\end{aligned}$$

$$\begin{aligned}
\varepsilon_y' &= \partial v'/\partial y' = (\partial v'/\partial y)(\partial y/\partial y') + (\partial v'/\partial x)(\partial x/\partial y')\\
&= [(\partial u/\partial y)(\partial v'/\partial u) + (\partial v/\partial y)(\partial v'/\partial v)](\partial y/\partial y') + [(\partial u/\partial x)(\partial v'/\partial u) + (\partial v/\partial x)(\partial v'/\partial v)](\partial x/\partial y')\\
&= [\,-\,(\partial u/\partial y)\sin\alpha + (\partial v/\partial y)\cos\alpha\,]\cos\alpha + [\,-\,(\partial u/\partial x)\sin\alpha + (\partial v/\partial x)\cos\alpha\,](-\sin\alpha)\\
&= \varepsilon_x\sin^2\alpha + \varepsilon_y\cos^2\alpha - \gamma_{xy}\sin\alpha\cos\alpha\\
&= ½\,(\varepsilon_x + \varepsilon_y) - ½\,(\varepsilon_x - \varepsilon_y)\cos 2\alpha - ½\gamma_{xy}\sin 2\alpha
\end{aligned}$$

$$\begin{aligned}
\gamma_{xy}' &= \partial v'/\partial x' + \partial u'/\partial y'\\
&= (\partial v'/\partial x)(\partial x/\partial x') + (\partial v'/\partial y)(\partial y/\partial x') + (\partial u'/\partial x)(\partial x/\partial y') + (\partial u'/\partial y)(\partial y/\partial y')\\
&= [(\partial u/\partial x)(\partial v'/\partial u)+(\partial v/\partial x)(\partial v'/\partial v)](\partial x/\partial x') + [(\partial u/\partial y)(\partial v'/\partial u) + (\partial v/\partial y)(\partial v'/\partial v)](\partial y/\partial x')\\
&\quad + [(\partial u/\partial x)(\partial u'/\partial u) + (\partial v/\partial x)(\partial u'/\partial v)](\partial x/\partial y') + [(\partial u/\partial y)(\partial u'/\partial u) + (\partial v/\partial y)(\partial u'/\partial v)](\partial y/\partial y')\\
&= [\,-\,(\partial u/\partial x)\sin\alpha + (\partial v/\partial x)\cos\alpha\,]\cos\alpha + [\,-\,(\partial u/\partial y)\sin\alpha + (\partial v/\partial y)\cos\alpha\,]\sin\alpha\\
&\quad + [(\partial u/\partial x)\cos\alpha + (\partial v/\partial x)\sin\alpha\,](\,-\sin\alpha) + [(\partial u/\partial y)\cos\alpha + (\partial v/\partial y)\sin\alpha\,]\cos\alpha\\
&= -\,2(\partial u/\partial x)\sin\alpha\cos\alpha + 2(\partial v/\partial y)\sin\alpha\cos\alpha + [(\partial v/\partial x) + (\partial u/\partial y)](\cos^2\alpha - \sin^2\alpha)
\end{aligned}$$

$$½\gamma_{xy}' = -\,½\,(\varepsilon_x - \varepsilon_y)\sin 2\alpha + ½\gamma_{xy}\cos 2\alpha$$

It is shown in Section 2.4 that infinitesimal strain transformation equations refer to line elements in the current configuration (the deformed solid). However, with the small elastic strains involved here, there is negligible error in choosing line elements in the original reference configuration (the undeformed solid). Thus, the strain components ε_x, ε_y and γ_{xy} may be identified with the nominal (engineering) strains in which displacements are more conveniently referred to the initial geometry.

2.3 Large Strain Definitions

When strains exceed 1%, we cannot employ the same approximations used previously for deriving the infinitesimal strain-displacement relations. A number of large (finite) strain definitions are available and these are reviewed here. They can be applied wherever large deformations arise, say, in the stretching of rubbery materials and in deforming metallic materials at very rapid rates under impact loading.

2.3.1 Natural, True or Logarithmic Strain

The theory of metal plasticity employs an incremental strain definition. This was first employed by Ludwik [1] and Hencky [2], who referred the change in length δl of a given line element to its current length, l. This gives an incremental strain $\delta\varepsilon = \delta l/l$, for which the true, or natural, strain is characterised by any one of the integrated quantities:

$$\varepsilon = \int \delta l / l \qquad (2.10a)$$
$$= \ln (l / l_o) = \ln [(l_o + \Delta l) / l_o]$$
$$= \ln (1 + \Delta l / l_o) = \ln (1 + e) \qquad (2.10b)$$
$$\approx e - e^2/2 + e^3/3 - ...$$

where the nominal (engineering) strain $e = \delta l/l_o$ refers δl to its original length l_o. These two measures of strain, e and ε, are approximately equal for nominal strain values less than 2%. The logarithmic strain provides the correct measure of the final strain when deformation takes place in a series of increments. For example, let an initial length l_o increase to a final length l_f, under an incremental tensile loading for which there are six intermediate lengths: l_1, l_2, l_3, l_4, l_5 and l_6. It follows from eq(2.10a) that the final, true strain $\varepsilon = \ln (l_f / l_o)$, will be given by the sum of the increments:

$$\varepsilon = \ln (l_1 / l_o) + \ln (l_2 / l_1) + \ln (l_3 / l_2) + \ln (l_4 / l_3) + \ln (l_5 / l_4) + \ln (l_6 / l_5) + \ln (l_f / l_6)$$
$$= \ln [(l_1 / l_o)(l_2 / l_1)(l_3 / l_2)(l_4 / l_3)(l_5 / l_4)(l_6 / l_5)(l_f / l_6)] = \ln (l_f / l_o)$$

Here, as with all proportional loading paths, the final strain does not depend upon the intermediate strains, i.e. the same strain would be achieved if a single continuous loading were to produce l_f. Applying the engineering strain definition to each stage of the deformation shows that successive strains are not additive under uniaxial loading. That is,

$$(l_1 - l_o)/l_o + (l_2 - l_1)/l_1 + (l_3 - l_2)/l_2 + (l_4 - l_3)/l_3 + (l_5 - l_4)/l_4 + (l_6 - l_5)/l_5 \neq (l_f - l_o)/l_o$$

The logarithmic strain measure will therefore account for the influence of a strain path more reliably than engineering strain. This particularly applies when the load path changes direction under a non-proportional loading, giving a final strain that depends upon the history of strain. The essential tensorial character of normal strain is retained within this logarithmic definition but there is not a corresponding shear strain. Large shear strain is expressed as $\gamma = \tan\phi = x/l$ (see Fig. 1.2) but this will only approximate to the engineering shear strain i.e. $\gamma = \tan\phi \approx \phi$ rad, for distortion angles $\phi < 10°$.

2.3.2 Extension Ratio

The extension ratio is a measure of large deformation, often used in rubber and polymer mechanics. It is, simply, the ratio between the final length l and the initial length l_o:

$$\lambda_{(\mathbf{u})} = l / l_o \qquad (2.11a)$$

where the unit vector **u** subscript denotes the original direction of l_o. Elastomeric fibres and polymers can stretch by multiples of their original lengths to give typical engineering strains of between 500% and 600%. The extension ratio bears a simple relationship to the engineering strain e. When l_o is aligned with a unit vector $\mathbf{u}_x$ in the x - co-ordinate direction:

$$e_x = \frac{l - l_o}{l_o} = \lambda_{(\mathbf{u}_x)} - 1 \qquad (2.11b)$$

If an incompressible (constant volume) deformation of a unit cube occurs withour shear, eq(2.11b) supplies a relation between the principal extension ratios:

$$(1 + e_1)(1 + e_2)(1 + e_3) - 1 = 0 \quad \text{or} \quad \lambda_1 \lambda_2 \lambda_3 - 1 = 0 \qquad (2.12a,b)$$

Note that it is only when the strains are small can the left-hand side of eq(2.12a) be approximated to

$$e_1 + e_2 + e_3 = 0 \quad \text{giving} \quad \lambda_1 + \lambda_2 + \lambda_3 = 3$$

2.3.3 Finite Homogenous Strains

(a) *Direct Strain*

Consider uni-directional straining of a line element PQ, aligned with the x - direction, as shown in Fig. 2.3. Let end points P and Q, with co-ordinates P(x, 0, 0) and Q($x + \delta x$, 0, 0), move to new positions P′ and Q′.

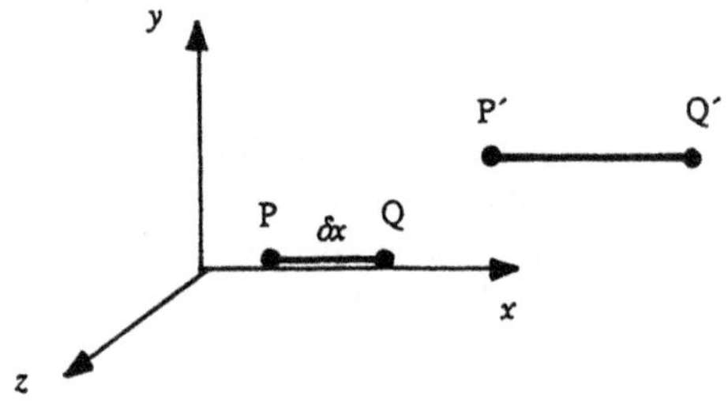

Figure 2.3 Finite extension of a line element aligned with the x - direction

Given that the displacements of point P to P′ are u, v and w in the x, y and z - directions respectively, the co-ordinates of P′ become P′($x + u$, v, w). Now, these displacements are functions of the co-ordinates: $u = u(x, y, z)$, $v = v(x, y, z)$ and $w = w(x, y, z)$, and so it follows that the co-ordinates of point Q′ are

$$Q' \, [x + u + \delta(x + u), \; v + \delta v, \; w + \delta w \,]$$

and, since both δy and δz are zero in the unstrained configuration,

$$\delta(x + u) = \delta x + (\partial u/\partial x)\delta x, \; \delta v = (\partial v/\partial x)\delta x \; \text{ and } \; \delta w = (\partial w/\partial x)\delta x \qquad (2.13a)$$

The finite normal strain for P′Q′ is expressed as

$$\{(P'Q')^2 - (PQ)^2\}/(PQ)^2 = \{[\delta x + (\partial u/\partial x)\delta x]^2 + [(\partial v/\partial x)\delta x]^2 + [(\partial w/\partial x)\delta x]^2 - (\delta x)^2\}/(\delta x)^2$$

$$= [1 + (\partial u/\partial x)]^2 + (\partial v/\partial x)^2 + (\partial w/\partial x)^2 - 1$$

$$2\varepsilon_x^G = 2\{(\partial u/\partial x) + \tfrac{1}{2}[(\partial u/\partial x)^2 + (\partial v/\partial x)^2 + (\partial w/\partial x)^2]\} \qquad (2.13b)$$

The quantity { } in eq(2.13b) was proposed by Green and Adkins [3] as a finite strain measure. This is written as ε^G and bears the following relationship with the extension ratio $\lambda = \lambda_{(u_x)} = P'Q'/PQ$:

$$\lambda^2 - 1 = 2\varepsilon^G \quad \text{or} \quad \varepsilon^G = \tfrac{1}{2}(\lambda^2 - 1) \qquad (2.13c)$$

Also, ε^G relates to the engineering strain $e = \mathrm{P'Q'/PQ} - 1$:

$$e = \sqrt{\{[1 + (\partial u/\partial x)]^2 + (\partial v/\partial x)^2 + (\partial w/\partial x)^2\}} - 1$$
$$= \sqrt{[1 + 2\{(\partial u/\partial x) + \tfrac{1}{2}[(\partial u/\partial x)^2 + (\partial v/\partial x)^2 + (\partial w/\partial x)^2]\}]} - 1$$
$$= \sqrt{(1 + 2\varepsilon^G)} - 1 = \lambda - 1 \qquad (2.13d)$$

where

$$\varepsilon^G = \tfrac{1}{2}\,[(1 + e)^2 - 1] = e + \tfrac{1}{2}\,e^2 \qquad (2.13e)$$

(b) *Volumetric Strain*

The strained volume V, accompanying finite deformation, appears in terms of the three principal extension ratios λ_1, λ_2 and λ_3 and the original volume V_o, as

$$(V / V_o)^2 = (\lambda_1 \lambda_2 \lambda_3)^2 \qquad (2.14a)$$

Substituting eq(2.13c) into eq(2.14c):

$$(V / V_o)^2 = (1 + 2\varepsilon_1{}^G)(1 + 2\varepsilon_2{}^G)(1 + 2\varepsilon_3{}^G)$$

from which we can write

$$\tfrac{1}{2}[\,(V / V_o)^2 - 1] = (\varepsilon_1{}^G + \varepsilon_2{}^G + \varepsilon_3{}^G) + 2(\varepsilon_1{}^G\varepsilon_2{}^G + \varepsilon_1{}^G\varepsilon_3{}^G + \varepsilon_2{}^G\varepsilon_3{}^G) + 4\varepsilon_1{}^G\varepsilon_2{}^G\varepsilon_3{}^G$$
$$= I_1{}^G + 2\,I_2{}^G + 4\,I_3{}^G \qquad (2.14b)$$

Note that $I_1{}^G$, $I_2{}^G$ and $I_3{}^G$ are the invariants of a principal *Green strain cubic*

$$(\varepsilon^G)^3 - I_1{}^G(\varepsilon^G)^2 + I_2{}^G\varepsilon^G - I_3{}^G = 0$$

which is identical in form to eq(2.7a).

A relationship between the three deformation measures is found from substituting the engineering volumetric strain $\Delta = (V - V_o) / V_o$ into eqs(2.14a and b):

$$\Delta + \Delta^2/2 = I_1{}^G + 2\,I_2{}^G + 4\,I_3{}^G = \tfrac{1}{2}\,[\,(\lambda_1\lambda_2\lambda_3)^2 - 1] \qquad (2.14c)$$

Equation (2.14c) reduces to a simple approximation for infinitesimal straining:

$$\Delta \approx e_1 + e_2 + e_3$$

(c) *Shear Strain*

Finite shear strain refers to an angular distortion similar to the infinitesimal shear strain expression (2.1c). Consider the shear deformation between two originally perpendicular line elements PQ and PR, lying in the *X-Y* plane (Fig. 2.4a). Finite shear deformation is identified with the cosine of the angle ϕ between P′Q′ and P′R′ in the deformed configuration (see Fig. 2.4b). Local co-ordinates *x*, *y* and *z*, with origin at P′, are aligned parallel to the original axes *X*, *Y* and *Z*, whose origin lies at point P. Note that points R′ and Q′ do not remain in the *x-y* plane. Green's finite shear strain is defined as: $2\varepsilon_{xy}{}^G = \cos\phi$, where $\cos\phi$ is the dot product of unit vectors $\mathbf{n}_1$ and $\mathbf{n}_2$ aligned with P′Q′ and P′R′. The scalar intercepts made by $\mathbf{n}_1$ and $\mathbf{n}_2$ with *x*, *y* and *z*, are the direction cosines for each vector. Hence, using eq(2.13a),

$$\mathbf{n}_1 = (1 + \partial u/\partial x)\mathbf{u}_x + (\partial v/\partial x)\mathbf{u}_y + (\partial w/\partial x)\mathbf{u}_z \qquad (2.15a)$$
$$\mathbf{n}_2 = (\partial u/\partial y)\mathbf{u}_x + (1 + \partial v/\partial y)\mathbf{u}_y + (\partial w/\partial y)\mathbf{u}_z \qquad (2.15b)$$

where $\mathbf{u}_x$, $\mathbf{u}_y$ and $\mathbf{u}_z$ are unit vectors in the *x*, *y* and *z* - directions.

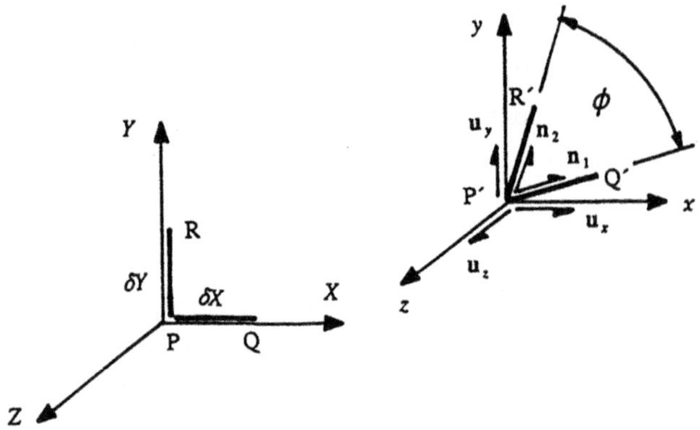

(a) (b)

Figure 2.4 Finite shear deformation referred to an x, y, z frame

The finite shear strain becomes

$$\mathbf{n}_1 \bullet \mathbf{n}_2 = \cos\phi = 2\varepsilon_{xy}{}^G \tag{2.16a}$$

Substituting eqs(2.15a,b) into eq(2.16a) gives

$$\begin{aligned} 2\varepsilon_{xy}{}^G = \cos\phi &= (\partial u/\partial y)(1 + \partial u/\partial x) + (\partial v/\partial x)(1 + \partial v/\partial y) + (\partial w/\partial x)(\partial w/\partial y) \\ &= \{\partial u/\partial y + \partial v/\partial x + (\partial u/\partial y)(\partial u/\partial x) + (\partial v/\partial x)(\partial v/\partial y) + (\partial w/\partial y)(\partial w/\partial x)\} \end{aligned} \tag{2.16b}$$

In the small strain theory, product terms in eq(2.16b) are ignored. Taking the result with the identity $\cos\phi = \sin(\pi/2 - \phi) \approx (\pi/2 - \phi)$ rad, the infinitesimal shear strain is recovered:

$$\gamma_{xy} = 2\varepsilon_{xy} = (\pi/2 - \phi) = \partial u/\partial y + \partial v/\partial x$$

In finite deformation, the engineering shear strain γ_{xy} is the angular change to the right angle, originally between PR and PQ. The relationship between the finite shear strain $2\varepsilon_{xy}{}^G$ and γ_{xy} follows from an alternative expression for the dot product of the vectors $\mathbf{P'R'}$ and $\mathbf{P'Q'}$:

$$\mathbf{P'R'} \bullet \mathbf{P'Q'} = |\mathbf{P'R'}|\,|\mathbf{P'Q'}| \cos(\pi/2 - \gamma_{xy}) = |\mathbf{P'R'}|\,|\mathbf{P'Q'}| \sin\gamma_{xy}$$

$$\sin\gamma_{xy} = \frac{\mathbf{P'R'} \bullet \mathbf{P'Q'}}{\mathrm{PR}\,\lambda_{(\mathbf{u}_y)}\,\mathrm{PQ}\,\lambda_{(\mathbf{u}_x)}} = \frac{\mathbf{n}_1 \bullet \mathbf{n}_2}{\lambda_{(\mathbf{u}_x)}\lambda_{(\mathbf{u}_y)}}$$

where PR = $|\mathbf{P'R'}|$ and PQ = $|\mathbf{P'Q'}|$ under pure shear. Then, from eqs(2.13c) and (2.16a),

$$\sin\gamma_{xy} = \frac{2\,\varepsilon_{xy}^G}{\sqrt{[(1 + 2\,\varepsilon_x^G)(1 + 2\,\varepsilon_y^G)]}} \tag{2.16c}$$

and again, when the shear strains are small, $\sin\gamma_{xy} \approx \gamma_{xy} \approx 2\varepsilon_{xy}$.

(d) *Combined Normal and Shear Strain*
Figure 2.5 shows the line element PQ with unstrained co-ordinates P(x, y, z) and Q(x + δx, y + δy, z + δz). In general, PQ will displace to P′Q′ with both extension and rotation in the manner shown.

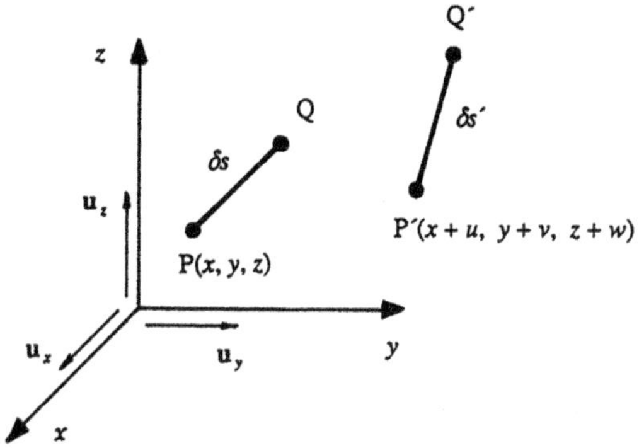

Figure 2.5 Finite extension and rotation of a line element PQ

Let the displacements of point P to P′ be u, v and w, so that the co-ordinates of P′ become P′($x + u$, $y + v$, $z + w$). The co-ordinates of Q′ become $[x + u + \delta(x + u), y + v + \delta(y + v), z + w + \delta(z + w)]$, where

$$\delta(x + u) = \delta x + (\partial u/\partial x)\delta x + (\partial u/\partial y)\delta y + (\partial u/\partial z)\delta z \tag{2.17a}$$
$$\delta(y + v) = \delta y + (\partial v/\partial x)\delta x + (\partial v/\partial y)\delta y + (\partial v/\delta z)\delta z \tag{2.17b}$$
$$\delta(z + w) = \delta z + (\partial w/\partial x)\delta x + (\partial w/\partial y)\delta y + (\partial w/\partial z)\delta z \tag{2.17c}$$

Set the infinitesimal lengths of PQ and P′Q′ in Fig. 2.5 to δs and $\delta s'$ respectively. The direction cosine l' for P′Q′ is found from dividing eqs(2.17a) by $\delta s'$. This gives

$$l' = \delta(x + u)/\delta s' = \delta x/\delta s' + (\partial u/\partial x)\delta x/\delta s' + (\partial u/\partial y)\delta y/\delta s' + (\partial u/\partial z)\delta z/\delta s'$$
$$= (\delta s/\delta s')[(\delta x/\delta s) + (\partial u/\partial x)(\delta x/\delta s) + (\partial u/\partial y)(\delta y/\delta s) + (\partial u/\partial z)(\delta z/\delta s)]$$

Substituting for the direction cosines of PQ in the undeformed configuration $l = \delta x/\delta s$, $m = \delta y/\delta s$ and $n = \delta z/\delta s$, with $\lambda = \delta s'/\delta s$, gives

$$\lambda l' = (1 + \partial u/\partial x)l + (\partial u/\partial y)m + (\partial u/\partial z)n \tag{2.18a}$$

Similarly, from eqs(2.17b and c), the two remaining direction cosines for P′Q′ appear as

$$\lambda m' = (\partial v/\partial x)l + (1 + \partial v/\partial y)m + (\partial v/\partial z)n \tag{2.18b}$$
$$\lambda n' = (\partial w/\partial x)l + (\partial w/\partial y)m + (1 + \partial w/\partial z)n \tag{2.18c}$$

Squaring and adding eqs(2.18a-c), and noting that $(l')^2 + (m')^2 + (n')^2 = 1$, leads to an expression in the extension ratio

$$\lambda^2 - 1 = 2\{l^2\varepsilon_x^G + m^2\varepsilon_y^G + n^2\varepsilon_z^G + 2\,(lm\varepsilon_{xy}^G + mn\varepsilon_{yz}^G + ln\varepsilon_{xz}^G)\}$$

from which Green's finite strain is defined as

$$\varepsilon^G = \tfrac{1}{2}(\lambda^2 - 1) = \{l^2\varepsilon_x^G + m^2\varepsilon_y^G + n^2\varepsilon_z^G + 2(lm\varepsilon_{xy}^G + mn\varepsilon_{yz}^G + ln\varepsilon_{xz}^G)\} \quad (2.19a)$$

Within the right-hand side of eq(2.19a) are the components of ε^G, aligned with the co-ordinate axes x, y and z. These are of the form ε_x^G and $2\varepsilon_{xy}^G$ as in eqs(2.13b) and (2.16b). These are, in full,

$$\varepsilon_x^G = (\partial u/\partial x) + \tfrac{1}{2}[(\partial u/\partial x)^2 + (\partial v/\partial x)^2 + (\partial w/\partial x)^2] = \tfrac{1}{2}[(\lambda_x)^2 - 1] \quad (2.19b)$$
$$\varepsilon_y^G = (\partial v/\partial y) + \tfrac{1}{2}[(\partial u/\partial y)^2 + (\partial v/\partial y)^2 + (\partial w/\partial y)^2] = \tfrac{1}{2}[(\lambda_y)^2 - 1] \quad (2.19c)$$
$$\varepsilon_z^G = (\partial w/\partial z) + \tfrac{1}{2}[(\partial u/\partial z)^2 + (\partial v/\partial z)^2 + (\partial w/\partial z)^2] = \tfrac{1}{2}[(\lambda_z)^2 - 1] \quad (2.19d)$$
$$2\varepsilon_{xy}^G = \{\partial u/\partial y + \partial v/\partial x + (\partial u/\partial y)(\partial u/\partial x) + (\partial v/\partial x)(\partial v/\partial y) + (\partial w/\partial y)(\partial w/\partial x)\} = \lambda_{xy}^2 \quad (2.19e)$$
$$2\varepsilon_{xz}^G = \{\partial w/\partial x + \partial u/\partial z + (\partial u/\partial z)(\partial u/\partial x) + (\partial v/\partial x)(\partial v/\partial z) + (\partial w/\partial z)(\partial w/\partial x)\} = \lambda_{xz}^2 \quad (2.19f)$$
$$2\varepsilon_{yz}^G = \{\partial w/\partial y + \partial v/\partial z + (\partial u/\partial y)(\partial u/\partial z) + (\partial v/\partial z)(\partial v/\partial y) + (\partial w/\partial y)(\partial w/\partial z)\} = \lambda_{yz}^2 \quad (2.19g)$$

where eqs(2.19b and e) correspond to expressions (2.13b) and (2.16b), derived previously for finite uniaxial strain and finite shear strain. An alternative derivation to eq(2.19a) employs $2\varepsilon^G = [(P'Q')^2 - (PQ)^2] / (PQ)^2$. Substituting from eqs(2.17a,b,c), gives the finite normal strain measure ($2\varepsilon^G$) for P′Q′ as

$$\{[(x + u) + \delta x + (\partial u/\partial x)\delta x + (\partial u/\partial y)\delta y + (\partial u/\partial z)\delta z - (x + u)]^2$$
$$+ [(y + v) + \delta y + (\partial v/\partial x)\delta x + (\partial v/\partial y)\delta y + (\partial v/\partial z)\delta z - (y + v)]^2$$
$$+ [(z + w) + \delta z + (\partial w/\partial x)\delta x + (\partial w/\partial y)\delta y + (\partial w/\partial z)\delta z - (z + w)]^2\} \div [(\delta x)^2 + (\delta y)^2 + (\delta z)^2] - 1$$

It is seen that the normal strain expression (2.19a) is identical to its infinitesimal counterpart (eq 2.6a), previously associated with a tensor transformation for strain. It now becomes apparent that finite strain components are defined to preserve a similar tensor transformation. That is: $\varepsilon_{ij}^{G\prime} = l_{ip}\, l_{jq}\varepsilon_{pq}^G$, where the prime refers to the rotated axes. Note, from eqs(2.19e-g), how the *shear extension ratios* λ_{xy}, λ_{xz} and λ_{yz}, relate to Green's shear strain. Substituting eqs(2.19e-g) into eq(2.19a), gives $2\varepsilon^G$ in terms of extension ratios

$$(\lambda^2 - 1) = l^2(\lambda_x^2 - 1) + m^2(\lambda_y^2 - 1) + n^2(\lambda_z^2 - 1) + 2(lm\lambda_{xy}^2 + mn\lambda_{yz}^2 + ln\lambda_{xz}^2) \quad (2.20a)$$
$$= l^2\lambda_x^2 + m^2\lambda_y^2 + n^2\lambda_z^2 - (l^2 + m^2 + n^2) + 2(lm\lambda_{xy}^2 + mn\lambda_{yz}^2 + ln\lambda_{xz}^2)$$

from which

$$\lambda^2 = l^2\lambda_x^2 + m^2\lambda_y^2 + n^2\lambda_z^2 + 2(lm\lambda_{xy}^2 + mn\lambda_{yz}^2 + ln\lambda_{xz}^2) \quad (2.20b)$$

The right-hand side of eq(2.20b) is identical in form to the right-hand sides of eqs(2.6b) and (2.19a). It follows that the finite shear strain definition will also ensure that extension ratios conform to a transformation equation $\lambda_{ij}^{2\prime} = l_{ip}\, l_{jq}\, \lambda_{pq}^2$.

Example 2.2 Convert the following matrix of Green's strain ε_{pq}^G to an appropriate matrix of extension ratios that will allow their transformation.

$$\varepsilon_{pq}^G = \begin{bmatrix} 1 & 3 & -2 \\ 3 & 1 & -2 \\ -2 & -2 & 6 \end{bmatrix}$$

Then perform the transformation $\lambda_{ij}^{2\prime} = l_{ip}\, l_{jq}\, \lambda_{pq}^{2}$, given rotations in the co-ordinates from x to x' and y to y'. Unit vectors, aligned with the directions x' and y', are given by

$$\mathbf{u}_x' = \sqrt{(2/3)}\mathbf{u}_x + (1/\sqrt{6})\mathbf{u}_y + (1/\sqrt{6})\mathbf{u}_z \quad \text{and} \quad \mathbf{u}_y' = (1/\sqrt{3})\mathbf{u}_x - (1/\sqrt{3})\mathbf{u}_y - (1/\sqrt{3})\mathbf{u}_z$$

respectively. Confirm the result from the corresponding transformation: $\varepsilon_{ij}^{G\prime} = l_{ip}\, l_{jq}\, \varepsilon_{pq}^{G}$.

Firstly, the unit vector $\mathbf{u}_z$, describing the rotation from z to z', can be found from the cross product:

$$\mathbf{u}_z = \mathbf{u}_x \times \mathbf{u}_y = \begin{bmatrix} \mathbf{u}_x & \mathbf{u}_y & \mathbf{u}_z \\ \sqrt{\dfrac{2}{3}} & \dfrac{1}{\sqrt{6}} & \dfrac{1}{\sqrt{6}} \\ \dfrac{1}{\sqrt{3}} & -\dfrac{1}{\sqrt{3}} & -\dfrac{1}{\sqrt{3}} \end{bmatrix} = \frac{1}{\sqrt{2}}\mathbf{u}_y - \frac{1}{\sqrt{2}}\mathbf{u}_z$$

Note: $\mathbf{u}_x' \bullet \mathbf{u}_y' = \mathbf{u}_x' \bullet \mathbf{u}_z' = \mathbf{u}_y' \bullet \mathbf{u}_z' = 0$. Converting ε_{pq}^{G} to ε_{pq}^{2} from eqs(2.19b - g), the required transformation $\lambda_{ij}^{2\prime} = l_{ip}\, l_{jq}\, \lambda_{pq}^{2}$, becomes

$$\lambda_{ij}^{2\,\prime} = \begin{vmatrix} \sqrt{(2/3)} & 1/\sqrt{6} & 1/\sqrt{6} \\ 1/\sqrt{3} & -1/\sqrt{3} & -1/\sqrt{3} \\ 0 & 1/\sqrt{2} & -1/\sqrt{2} \end{vmatrix} \begin{vmatrix} 3 & 6 & -4 \\ 6 & 3 & -4 \\ -4 & -4 & 13 \end{vmatrix} \begin{vmatrix} \sqrt{(2/3)} & 1/\sqrt{3} & 0 \\ 1/\sqrt{6} & -1/\sqrt{3} & 1/\sqrt{2} \\ 1/\sqrt{6} & -1/\sqrt{3} & -1/\sqrt{2} \end{vmatrix} = \begin{vmatrix} 4.667 & -0.943 & 2.888 \\ -0.943 & 2.333 & 8.166 \\ 2.888 & 8.166 & 12.002 \end{vmatrix}$$

The result may be checked from $\varepsilon_{ij}^{G\prime} = l_{ip}\, l_{jq}\, \varepsilon_{pq}^{G}$

$$\varepsilon_{ij}^{G\,\prime} = \begin{bmatrix} \sqrt{(2/3)} & 1/\sqrt{6} & 1/\sqrt{6} \\ 1/\sqrt{3} & -1/\sqrt{3} & -1/\sqrt{3} \\ 0 & 1/\sqrt{2} & -1/\sqrt{2} \end{bmatrix} \begin{bmatrix} 1 & 3 & -2 \\ 3 & 1 & -2 \\ -2 & -2 & 6 \end{bmatrix} \begin{bmatrix} \sqrt{(2/3)} & 1/\sqrt{3} & 0 \\ 1/\sqrt{6} & -1/\sqrt{3} & 1/\sqrt{2} \\ 1/\sqrt{6} & -1/\sqrt{3} & -1/\sqrt{2} \end{bmatrix} = \begin{bmatrix} 1.883 & -0.4715 & 1.444 \\ -0.4715 & 0.6666 & 4.083 \\ 1.444 & 4.083 & 5.501 \end{bmatrix}$$

from which $\lambda_{ij}^{2\prime}$ are found from eqs(2.19b - g). For example, $\lambda_x^{2\prime} = 1 + (2 \times 1.833) = 4.667$ and $\lambda_{xy}^{2\prime} = 2 \times (-0.4715) = -0.943$.

2.4 Finite Strain Tensors

In Fig. 2.6, two sets of independent Cartesian co-ordinates are superimposed. *Lagrangian material co-ordinates* $X_i = X_1, X_2$ and X_3 define the unstrained line element PQ in the material at time $t = 0$. *Eulerian spacial co-ordinates* $x_i = x_1, x_2$ and x_3, define the strained line element pq after time t.

The co-ordinates X_i and x_i allow two mappings of the deformation. In the first mapping the present position x_i is a function of the original position X_i:

$$x_i = x_i\,(X_j, t) = x_i(X_1, X_2, X_3, t) \quad \text{or} \quad \mathbf{x} = \mathbf{x}\,(\mathbf{X}, t) \tag{2.21a}$$

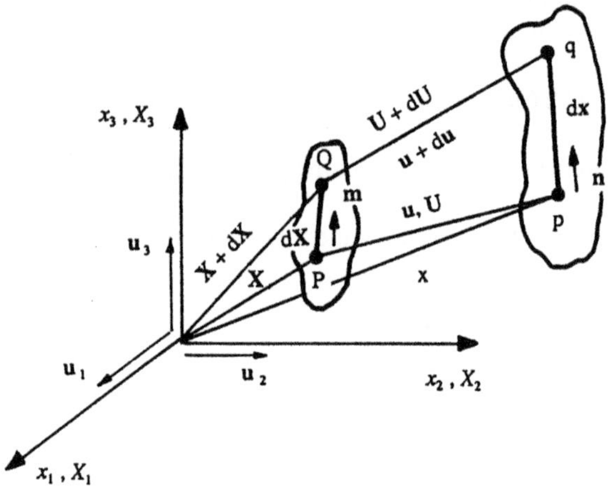

Figure 2.6 Finite deformation within material X_i and spacial x_i co-ordinates

The components $\mathrm{d}x_i$, of the current differential line element pq, are those of vector **dx**:

$$\mathrm{d}x_i = (\partial x_i / \partial X_j)\, \mathrm{d}X_j \quad \text{or} \quad \mathbf{dx} = \mathbf{F}\, \mathbf{dX} \tag{2.21b}$$

where **F** is a matrix of *material deformation gradients* components:

$$F_{ij} = \partial x_i / \partial X_j \quad \text{or} \quad \mathbf{F} = \mathbf{x}\, \forall^{\mathrm{T}} \tag{2.21c}$$

where $\forall = \{\partial/\partial X_1 \;\; \partial/\partial X_2 \;\; \partial/\partial X_3\}^{\mathrm{T}}$. The second mapping traces the original position from the current position. This appears in the function

$$X_i = X_i\,(x_j\,,\,t) \;= X_i\,(x_1, x_2\,, x_3\,, t\,) \quad \text{or} \quad \mathbf{X} = \mathbf{X}\,(\mathbf{x}, t\,) \tag{2.22a}$$

This inverse function will exist when the determinant of the Jacobian eq(1.40b) does not vanish. The components $\mathrm{d}X_i$ of the vector **dX**, for the original differential line element PQ, are then

$$\mathrm{d}X_i = (\partial X_i \,/\, \partial x_j)\, \mathrm{d}x_j \quad \text{or} \quad \mathbf{dX} = \mathbf{H}\, \mathbf{dx} \tag{2.22b}$$

where $\mathbf{H} = \mathbf{F}^{-1}$ is the matrix of *spacial deformation gradients*

$$H_{ij} = \partial X_i / \partial x_j \quad \text{or} \quad \mathbf{H} = \mathbf{X}\, \nabla^{\mathrm{T}} \tag{2.22c}$$

where $\nabla = \{\partial/\partial x_1 \;\; \partial/\partial x_2 \;\; \partial/\partial x_3\}^{\mathrm{T}}$. The square of the current differential length pq is given as

$$\mathrm{d}x_i \mathrm{d}x_i = \delta_{ij}\, \mathrm{d}x_i \mathrm{d}x_j \quad \text{or} \quad |\mathbf{dx}|^2 = \mathbf{dx}^{\mathrm{T}}\, \mathbf{dx} \tag{2.23a}$$

Substituting eq(2.21b) into eq(2.23a) with k and l as dummy subscripts

$$\begin{aligned} |\mathbf{dx}|^2 &= \delta_{ik}\, (\partial x_i / \partial X_j)\, \mathrm{d}X_j\, (\partial x_k / \partial X_l\,)\, \mathrm{d}X_l = (\partial x_k / \partial X_j\,)\, \mathrm{d}X_j\, (\partial x_k / \partial X_l\,)\, \mathrm{d}X_l \\ &= (\partial x_k / \partial X_i)(\partial x_k / \partial X_j\,)\, \mathrm{d}X_i \mathrm{d}X_j = G_{ij} \mathrm{d}X_i \mathrm{d}X_j \end{aligned} \tag{2.23b}$$

or

$$|\mathbf{dx}|^2 = \mathbf{dX}^{\mathrm{T}}\mathbf{G}\,\mathbf{dX} \tag{2.23c}$$

The components G_{ij} of the right Cauchy Green deformation tensor **G** are defined from eq(2.23b) as follows:

$$G_{ij} = (\partial x_k/\partial X_i)(\partial x_k/\partial X_j) = F_{ki}F_{kj} \quad \text{or} \quad \mathbf{G} = \mathbf{F}^{\mathrm{T}}\mathbf{F} \tag{2.23d}$$

Now the square of the current differential length PQ, is

$$\mathrm{d}X_i\mathrm{d}X_i = \delta_{ij}\mathrm{d}X_i\mathrm{d}X_j \quad \text{or} \quad |\mathbf{dX}|^2 = \mathbf{dX}^{\mathrm{T}}\mathbf{dX} \tag{2.24a}$$

Substituting eq(2.22b) into eq(2.24a) leads to

$$|\mathbf{dX}|^2 = \delta_{ik}(\partial X_i/\partial x_j)\,\mathrm{d}x_j(\partial X_k/\partial x_l)\,\mathrm{d}x_l = (\partial X_k/\partial x_j)\,\mathrm{d}x_j(\partial X_k/\partial x_l)\,\mathrm{d}x_l$$
$$= (\partial X_k/\partial x_i)(\partial X_k/\partial x_j)\,\mathrm{d}x_i\mathrm{d}x_j = C_{ij}\mathrm{d}x_i\mathrm{d}x_j$$

or

$$|\mathbf{dX}|^2 = \mathbf{dx}^{\mathrm{T}}\mathbf{C}\,\mathbf{dx} \tag{2.24b}$$

where **C** is the Cauchy deformation tensor with components C_{ij}:

$$C_{ij} = (\partial X_k/\partial x_i)(\partial X_k/\partial x_j) = H_{ki}H_{kj} \quad \text{or} \quad \mathbf{C} = \mathbf{H}^{\mathrm{T}}\mathbf{H} \tag{2.24c}$$

The motion of point P to p may be expressed in terms of either the material or the spacial components of the respective displacement vectors, **U** or **u**, in Fig. 2.6. Working with the material displacement vector:

$$x_i = U_i + X_i \quad \text{or} \quad \mathbf{x} = \mathbf{U} + \mathbf{X} \tag{2.25a}$$

for which its derivative is

$$\partial x_i/\partial X_j = \partial U_i/\partial X_j + \partial X_i/\partial X_j \tag{2.25b}$$

Substituting from eq(2.21c) into eq(2.25b) gives

$$F_{ij} = \partial U_i/\partial X_j + \delta_{ij} \tag{2.25c}$$

where $\partial U_i/\partial X_j$ are the *material displacement gradients*. When working with the spacial displacement vector,

$$x_i = u_i + X_i \quad \text{or} \quad \mathbf{x} = \mathbf{u} + \mathbf{X} \tag{2.26a}$$

the derivative is

$$\partial x_i/\partial x_j = \partial u_i/\partial x_j + \partial X_i/\partial x_j \tag{2.26b}$$

and substituting eq(2.22c) into eq(2.26b) gives

$$H_{ij} = \delta_{ij} - \partial u_i/\partial x_j \tag{2.26c}$$

where $\partial u_i/\partial x_j$ are the *spacial displacement gradients*.

Example 2.3 Find the components of **F**, **H**, **G** and **C** given that Lagrangian components of a particle motion $x_i = x_i(X_i, t)$ are as follows: $x_1 = X_1 + X_3(\mathrm{e}^t - 1)$, $x_2 = X_2 + X_3(\mathrm{e}^t - \mathrm{e}^{-t})$ and $x_3 = X_3\mathrm{e}^t$, where e is a constant.

Note that when time $t = 0$, $x_1 = X_1$, $x_2 = X_2$ and $x_3 = X_3$, the spacial x_i and material X_i co-ordinates coincide. The components of the deformation gradient are, from eq(2.21c),

$$F_{11} = \partial x_1/\partial X_1 = 1,\ F_{12} = \partial x_1/\partial X_2 = 0,\ F_{13} = \partial x_1/\partial X_3 = e^{t} - 1$$
$$F_{21} = \partial x_2/\partial X_1 = 0,\ F_{22} = \partial x_2/\partial X_2 = 1,\ F_{23} = \partial x_2/\partial X_3 = e^{t} - e^{-t}$$
$$F_{31} = \partial x_3/\partial X_3 = 0,\ F_{32} = \partial x_3/\partial X_2 = 0 \text{ and } F_{33} = \partial x_3/\partial X_3 = e^{t}$$

These define the material deformation gradient matrix:

$$\mathbf{F} = \begin{bmatrix} 1 & 0 & (e^{t} - 1) \\ 0 & 1 & (e^{t} - e^{-t}) \\ 0 & 0 & e^{t} \end{bmatrix}$$

Since $J = \det \mathbf{F} = e^{t} \neq 0$, the inverse function $X_i = X_i(x_i, t)$ exists. Its components are

$$X_1 = x_1 - x_3(1 - e^{-t}),\quad X_2 = x_2 - x_3(1 - e^{-2t}) \text{ and } X_3 = x_3 e^{-t}$$

Hence, using eq(2.22c), the spacial deformation gradient matrix $\mathbf{H}$ is formed from its components as follows:

$$H_{11} = \partial X_1/\partial x_1 = 1,\ H_{12} = \partial X_1/\partial x_2 = 0,\ H_{13} = \partial X_1/\partial x_3 = -(1 - e^{-t})$$
$$H_{21} = \partial X_2/\partial x_1 = 0,\ H_{22} = \partial X_2/\partial x_2 = 1,\ H_{23} = \partial X_2/\partial x_3 = -(1 - e^{-2t})$$
$$H_{31} = \partial X_3/\partial x_3 = 0,\ H_{32} = \partial x_3/\partial X_2 = 0,\ H_{33} = \partial x_3/\partial X_3 = e^{-t}$$

$$\mathbf{H} = \begin{bmatrix} 1 & 0 & -(1 - e^{-t}) \\ 0 & 1 & -(1 - e^{-2t}) \\ 0 & 0 & e^{-t} \end{bmatrix}$$

which may be confirmed from $\mathbf{H} = \mathbf{F}^{-1}$.

The right Cauchy-Green deformation tensor $\mathbf{G}$ is found from eqs(2.23c) as

$$\mathbf{G} = \begin{bmatrix} 1 & 0 & 0 \\ 0 & 1 & 0 \\ (e^{t} - 1) & (e^{t} - e^{-t}) & e^{t} \end{bmatrix} \begin{bmatrix} 1 & 0 & (e^{t} - 1) \\ 0 & 1 & (e^{t} - e^{-t}) \\ 0 & 0 & e^{t} \end{bmatrix} = \begin{bmatrix} 1 & 0 & (e^{t} - 1) \\ 0 & 1 & (e^{t} - e^{-t}) \\ (e^{t} - 1) & (e^{t} - e^{-t}) & 3e^{2t} - 2e^{t} + e^{-2t} - 1 \end{bmatrix}$$

The Cauchy deformation tensor $\mathbf{C}$ is found from $\mathbf{H}$ in eq(2.24c):

$$\mathbf{C} = \begin{bmatrix} 1 & 0 & 0 \\ 0 & 1 & 0 \\ -(e^{t} - 1) & -(1 - e^{-2t}) & e^{-t} \end{bmatrix} \begin{bmatrix} 1 & 0 & -(1 - e^{-t}) \\ 0 & 1 & -(1 - e^{-2t}) \\ 0 & 0 & e^{-t} \end{bmatrix} = \begin{bmatrix} 1 & 0 & -(1 - e^{t}) \\ 0 & 1 & -(1 - e^{-2t}) \\ -(1 - e^{-t}) & -(1 - e^{-2t}) & 2 - 2e^{-t} + e^{-4t} \end{bmatrix}$$

Alternatively **H** and **F** can be found from the displacement gradients in eqs(2.25c) and (2.26c). The material displacement components follow from eq(2.25a) as

$$U_1 = x_1 - X_1 = X_3(\mathrm{e}^t - 1),\ \ U_2 = x_2 - X_2 = X_3(\mathrm{e}^t - \mathrm{e}^{-t})\ \text{ and }\ U_3 = x_3 - X_3 = X_3(\mathrm{e}^t - 1)$$

These give the material displacement gradients as

$$\partial U_1/\partial X_1 = \partial U_1/\partial X_2 = \partial U_2/\partial X_1 = \partial U_2/\partial X_2 = \partial U_3/\partial X_1 = \partial U_3/\partial X_2 = 0$$

$$\partial U_1/\partial X_3 = (\mathrm{e}^t - 1),\ \ \partial U_2/\partial X_3 = (\mathrm{e}^t - \mathrm{e}^{-t})\ \text{ and }\ \partial U_3/\partial X_3 = (\mathrm{e}^t - 1)$$

when, from eq(2.25c):

$$\mathbf{F} = \begin{bmatrix} 0 & 0 & (\mathrm{e}^t - 1) \\ 0 & 0 & (\mathrm{e}^t - \mathrm{e}^{-t}) \\ 0 & 0 & (\mathrm{e}^t - 1) \end{bmatrix} + \begin{bmatrix} 1 & 0 & 0 \\ 0 & 1 & 0 \\ 0 & 0 & 1 \end{bmatrix} = \begin{bmatrix} 1 & 0 & (\mathrm{e}^t - 1) \\ 0 & 1 & (\mathrm{e}^t - \mathrm{e}^{-t}) \\ 0 & 0 & \mathrm{e}^t \end{bmatrix}$$

The spacial displacement components follow from eq(2.26a) as

$$u_1 = x_1 - X_1 = x_3(1 - \mathrm{e}^{-t}),\ \ u_2 = x_2 - X_2 = x_3(1 - \mathrm{e}^{-2t})\ \text{ and }\ u_3 = x_3 - X_3 = x_3(1 - \mathrm{e}^{-t})$$

These give the spacial displacement gradients as

$$\partial u_1/\partial x_1 = \partial u_1/\partial x_2 = \partial u_2/\partial x_1 = \partial u_2/\partial x_2 = \partial u_3/\partial x_1 = \partial u_3/\partial x_2 = 0$$

$$\partial u_1/\partial x_3 = (1 - \mathrm{e}^{-t}),\ \partial u_2/\partial x_3 = (1 - \mathrm{e}^{-2t})\ \text{and}\ \partial u_3/\partial x_3 = (1 - \mathrm{e}^{-t})$$

when, from eq(2.26c)

$$\mathbf{H} = \begin{bmatrix} 1 & 0 & 0 \\ 0 & 1 & 0 \\ 0 & 0 & 1 \end{bmatrix} - \begin{bmatrix} 0 & 0 & (1 - \mathrm{e}^{-t}) \\ 0 & 0 & (1 - \mathrm{e}^{-2t}) \\ 0 & 0 & (1 - \mathrm{e}^{-t}) \end{bmatrix} = \begin{bmatrix} 1 & 0 & -(1 - \mathrm{e}^{-t}) \\ 0 & 1 & -(1 - \mathrm{e}^{-2t}) \\ 0 & 0 & \mathrm{e}^{-t} \end{bmatrix}$$

Note that in this example **F, H, G** and **C** are time and not co-ordinate dependent. In general, **F** and **G** will depend upon t and X_i while **H** and **C** will depend upon t and x_i.

2.4.1 The Lagrangian Finite Strain Tensor

Two finite strain tensors arise from taking the difference between the squares of the current and original differential lengths. They are the Lagrangian and Eulerian measures, which we shall now describe separately.

To find the Lagrangian strain take the difference between eqs(2.23b) and (2.24a) and substitute from eq(2.25b):

$$
\begin{aligned}
|\mathbf{dx}|^2 - |\mathbf{dX}|^2 &= [\,(\partial x_k/\partial X_i)(\partial x_k/\partial X_j) - \delta_{ij}\,]\,\mathrm{d}X_i\,\mathrm{d}X_j \\
&= [\,(\partial U_k/\partial X_i + \delta_{ki})(\partial U_k/\partial X_j + \delta_{kj}) - \delta_{ij}\,]\,\mathrm{d}X_i\,\mathrm{d}X_j \\
&= [\,\partial U_j/\partial X_i + \partial U_i/\partial X_j + (\partial U_k/\partial X_i)(\partial U_k/\partial X_j)]\,\mathrm{d}X_i\,\mathrm{d}X_j \\
&= 2L_{ij}\,\mathrm{d}X_i\,\mathrm{d}X_j
\end{aligned}
\tag{2.27a}
$$

L_{ij} is the *Lagrangian or Green finite strain tensor*, defined from eqs(2.23c) and (2.27a) as

$$L_{ij} = ½[(\partial x_k/\partial X_i)(\partial x_k/\partial X_j) - \delta_{ij}] = ½[\,\partial U_j/\partial X_i + \partial U_i/\partial X_j + (\partial U_k/\partial X_i)(\partial U_k/\partial X_j)] \tag{2.27b}$$

which is also written as

$$\mathbf{L} = ½\,(\mathbf{G} - \mathbf{I}) = ½\,(\mathbf{F}^{\mathrm{T}}\mathbf{F} - \mathbf{I}) \tag{2.27c}$$

Note, the L_{ij} tensor components (2.27b) were those identified previously within eqs(2.19b-g) as Green's strains, using the engineering notation. These correspond to a Lagrangian strain measure since they were referred to the original material co-ordinates X_i.

2.4.2 The Eulerian Finite Strain Tensor

Taking the difference between the squares of the current and original differential lengths, from eqs(2.23a) and (2.24b), gives

$$
\begin{aligned}
|\mathbf{dx}|^2 - |\mathbf{dX}|^2 &= [\delta_{ij} - (\partial X_k/\partial x_i)(\partial X_k/\partial x_j)]\mathrm{d}x_i\,\mathrm{d}x_j \\
&= [\delta_{ij} - (\delta_{ki} - \partial u_k/\partial x_i)(\delta_{kj} - \partial u_k/\partial x_j)]\mathrm{d}x_i\,\mathrm{d}x_j \\
&= [\partial u_j/\partial x_i + \partial u_i/\partial x_j - (\partial u_k/\partial x_i)(\partial u_k/\partial x_j)]\mathrm{d}x_i\,\mathrm{d}x_j \\
&= 2E_{ij}\,\mathrm{d}x_i\,\mathrm{d}x_j
\end{aligned}
\tag{2.28a}
$$

E_{ij} is the *Eulerian or Almansi finite strain tensor* defined as:

$$E_{ij} = ½[\,\delta_{ij} - (\partial X_k/\partial x_i)(\partial X_k/\partial x_j)] = ½[\partial u_j/\partial x_i + \partial u_i/\partial x_j - (\partial u_k/\partial x_i)(\partial u_k/\partial x_j)] \tag{2.28b}$$

which is written as

$$\mathbf{E} = ½\,(\mathbf{I} - \mathbf{C}) = ½\,(\mathbf{I} - \mathbf{H}^{\mathrm{T}}\mathbf{H}) = ½\,(\mathbf{I} - \mathbf{B}^{-1}) \tag{2.28c}$$

where $\mathbf{C} = \mathbf{B}^{-1}$ in which $\mathbf{B}$ is the *left Cauchy Green deformation tensor*:

$$B_{ij} = (\partial x_i/\partial X_k)(\partial x_j/\partial X_k) \quad \text{or} \quad \mathbf{B} = \mathbf{F}\mathbf{F}^{\mathrm{T}} \tag{2.28d}$$

Note, that the Eulerian tensor **E** refers to spacial co-ordinates x_i and is less useful when it is required to refer finite deformation to the original material co-ordinates X_i. However, in the case of infinitesimal straining the product term in eqs(2.27b) and (2.28b) may be neglected. This results in the Lagrangian and Eulerian infinitesimal strain tensors respectively. Since the reference configuration is unimportant to small strain measures, it is traditional to refer the Eulerian tensor (2.3) to the material co-ordinates.

Example 2.4 Determine the Lagrangian, Eulerian and Cauchy finite strain tensors for the Lagrangian description of motion given in Example 2.3.

Using the matrices: **G**, **C** and **F** previously found, eq(2.27c) gives the Lagrangian strain tensor, **L** = ½(**G** − **I**):

$$\mathbf{L} = \frac{1}{2}\begin{bmatrix} 1 & 0 & (e^t - 1) \\ 0 & 1 & (e^t - e^{-t}) \\ (e^t - 1) & (e^t - e^{-t}) & (3e^{2t} - 2e^t + e^{-2t} - 1) \end{bmatrix} - \frac{1}{2}\begin{bmatrix} 1 & 0 & 0 \\ 0 & 1 & 0 \\ 0 & 0 & 1 \end{bmatrix} = \frac{1}{2}\begin{bmatrix} 0 & 0 & (e^t - 1) \\ 0 & 0 & (e^t - e^{-t}) \\ (e^t - 1) & (e^t - e^{-t}) & (3e^{2t} - 2e^t + e^{-2t} - 2) \end{bmatrix}$$

From eq(2.28c), the Eulerian tensor, **E** = ½(**I** − **C**), becomes

$$\mathbf{E} = \frac{1}{2}\begin{bmatrix} 1 & 0 & 0 \\ 0 & 1 & 0 \\ 0 & 0 & 1 \end{bmatrix} - \frac{1}{2}\begin{bmatrix} 1 & 0 & -(1 - e^{-t}) \\ 0 & 1 & -(1 - e^{-2t}) \\ -(1 - e^{-t}) & -(1 - e^{-2t}) & (2 - 2e^{-t} + e^{-4t}) \end{bmatrix} = \frac{1}{2}\begin{bmatrix} 0 & 0 & (1 - e^{-t}) \\ 0 & 0 & (1 - e^{-2t}) \\ (1 - e^{-t}) & (1 - e^{-2t}) & (2e^{-t} - e^{-4t} - 1) \end{bmatrix}$$

From eq(2.28d) the Cauchy-Green deformation tensor is $\mathbf{B} = \mathbf{F}\mathbf{F}^{\mathrm{T}}$:

$$\mathbf{B} = \begin{bmatrix} 1 & 0 & (e^t - 1) \\ 0 & 1 & (e^t - e^{-t}) \\ 0 & 0 & e^t \end{bmatrix}\begin{bmatrix} 1 & 0 & 0 \\ 0 & 1 & 0 \\ (e^t - 1) & (e^t - e^{-t}) & e^t \end{bmatrix} = \begin{bmatrix} 1 + (e^t - 1)^2 & (e^t - 1)(e^t - e^{-t}) & e^t(e^t - 1) \\ (e^t - e^{-t})(e^t - 1) & 1 + (e^t - e^{-t})^2 & e^t(e^t - e^{-t}) \\ e^t(e^t - 1) & e^t(e^t - e^{-t}) & e^{2t} \end{bmatrix}$$

Each component of **L**, **E** and **B** can be evaluated for a given time, *t*.

2.4.2 Extension (or Stretch) Ratios

The infinitesimal definition of the extension ratio is $\lambda = |\mathrm{d}\mathbf{x}|/|\mathrm{d}\mathbf{X}|$. This may be expressed in terms of both the material and spacial co-ordinates. With unit vectors **m** and **n**, aligned with line elements PQ and pq respectively, in Fig. 2.6, it follows that

$$\mathbf{m} = \mathrm{d}\mathbf{X} / |\mathrm{d}\mathbf{X}|, \quad \mathbf{n} = \mathrm{d}\mathbf{x} / |\mathrm{d}\mathbf{x}| \qquad (2.29a,b)$$

and, from eqs(2.23c) and (2.29a)

$$\lambda_{(\mathbf{m})}^2 = |\mathrm{d}\mathbf{x}|^2 / |\mathrm{d}\mathbf{X}|^2 = G_{ij}\,(\,\mathrm{d}X_i / |\mathrm{d}\mathbf{X}|\,)\,(\,\mathrm{d}X_j / |\mathrm{d}\mathbf{X}|\,)$$

$$\lambda_{(\mathbf{m})}^2 = G_{ij}\, m_i\, m_j \quad \text{or} \quad \lambda_{(\mathbf{m})}^2 = \mathbf{m}^{\mathrm{T}}\mathbf{G}\mathbf{m} \qquad (2.30a)$$

where $\mathbf{m} = \{m_1\, m_2\, m_3\}^{\mathrm{T}}$ is a column matrix of the direction cosines m_i. Now, from eq(2.27a)

$$|\mathrm{d}\mathbf{x}|^2 / |\mathrm{d}\mathbf{X}|^2 - 1 = 2L_{ij}\,(\,\mathrm{d}X_i / |\mathrm{d}\mathbf{X}|\,)\,(\,\mathrm{d}X_j / |\mathrm{d}\mathbf{X}|\,)$$

and substituting from eq(2.30a)

$$\lambda_{(\mathbf{m})}^2 - 1 = 2L_{ij}\, m_i\, m_j \quad \text{or} \quad \lambda_{(\mathbf{m})}^2 - 1 = \mathbf{m}^{\mathrm{T}} 2\mathbf{L}\mathbf{m} \qquad (2.30b)$$

Combining eqs(2.30a and b) connects the material finite strain tensors **L** and **G** with the extension ratio

$$\lambda_{(\mathrm{m})}^{\,2} = \mathbf{m}^{\mathrm{T}} 2\mathbf{L}\mathbf{m} + 1 = \mathbf{m}^{\mathrm{T}} \mathbf{G}\mathbf{m} \tag{2.30c}$$

Also, from eqs(2.24c) and (2.29b)

$$1/\lambda_{(\mathrm{n})}^{\,2} = |\mathrm{d}\mathbf{X}|^2 / |\mathrm{d}\mathbf{x}|^2 = C_{ij}\,(\mathrm{d}x_i/|\mathrm{d}\mathbf{x}|)\,(\mathrm{d}x_j/|\mathrm{d}\mathbf{x}|)$$

$$1/\lambda_{(\mathrm{n})}^{\,2} = C_{ij} n_i n_j \quad \text{or} \quad 1/\lambda_{(\mathrm{n})}^{\,2} = \mathbf{n}^{\mathrm{T}}\mathbf{C}\mathbf{n} \tag{2.31a}$$

where $\mathbf{n} = \{n_1\, n_2\, n_3\}^{\mathrm{T}}$ is a column matrix of the direction cosines n_i. Then, from eq(2.28b),

$$1 - |\mathrm{d}\mathbf{X}|^2 / |\mathrm{d}\mathbf{x}|^2 = 2E_{ij}\,(\mathrm{d}x_i/|\mathrm{d}\mathbf{x}|)\,(\mathrm{d}x_j/|\mathrm{d}\mathbf{x}|)$$

Substituting from eq(2.29b)

$$1 - 1/\lambda_{(\mathrm{n})}^{\,2} = 2E_{ij} n_i n_j \quad \text{or} \quad 1 - 1/\lambda_{(\mathrm{n})}^{\,2} = \mathbf{n}^{\mathrm{T}} 2\mathbf{E}\mathbf{n} \tag{2.31b}$$

Combining eqs(2.31a and b) connects the spacial finite strain tensors, **E** and **C**, to the extension ratio:

$$1/\lambda_{(\mathrm{n})}^{\,2} = 1 - \mathbf{n}^{\mathrm{T}} 2\mathbf{E}\mathbf{n} = \mathbf{n}^{\mathrm{T}}\mathbf{C}\mathbf{n} \tag{2.31c}$$

In general, $\lambda_{(\mathrm{m})} \neq \lambda_{(\mathrm{n})}$, from eqs(2.30c) and (2.31c), unless **n** remains in the same direction as **m**. One example where $\lambda_{(\mathrm{m})} = \lambda_{(\mathrm{n})}$ applies is to a stretch inclined at 45° to a pure shear deformation. This stretch occurs without rotation so that **m** and **n** remain parallel at 45°.

Referring to Fig. 2.7, the angular change between the two line elements PQ and PR is now required when these elements both stretch and rotate to their deformed positions pq and pr. Figure 2.7 represents this deformation, both within our material co-ordinates X_i and spacial co-ordinates x_i, in which the vectors $\mathbf{dX}^{(1)}$ and $\mathbf{dX}^{(2)}$ map into $\mathbf{dx}^{(1)}$ and $\mathbf{dx}^{(2)}$.

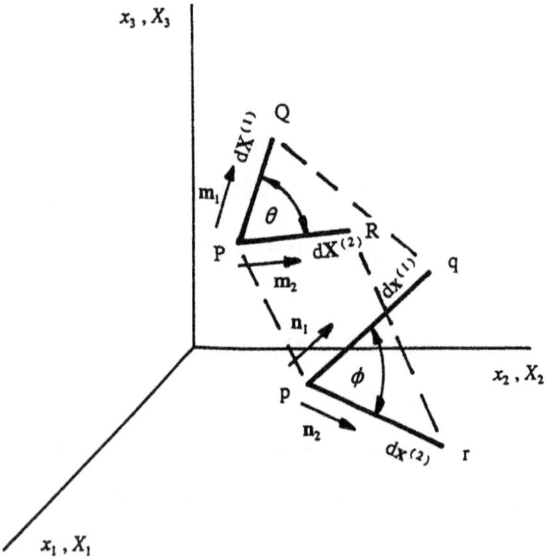

Figure 2.7 Shear deformation in material and spacial co-ordinates

From the dot product: $\mathbf{dx}^{(1)} \bullet \mathbf{dx}^{(2)} = |\mathbf{dx}^{(1)}|\,|\mathbf{dx}^{(2)}| \cos\phi$:

$$\cos\phi = \frac{\mathbf{dx}^{(1)}}{dx^{(1)}} \bullet \frac{\mathbf{dx}^{(2)}}{dx^{(2)}} = \frac{(\mathbf{dx}^{(1)})^{\mathrm{T}}}{dx^{(1)}} \frac{\mathbf{dx}^{(2)}}{dx^{(2)}} \qquad (2.32a)$$

where $dx = |\mathbf{dx}|$. Substituting eqs(2.21b) and (2.23b) into eq(2.32a)

$$\cos\phi = \frac{[\mathbf{F}\,\mathbf{dX}^{(1)}]^{\mathrm{T}}}{\sqrt{\mathbf{dX}^{(1)\mathrm{T}}\,\mathbf{G}\,\mathbf{dX}^{(1)}}} \times \frac{[\mathbf{F}\,\mathbf{dX}^{(2)}]}{\sqrt{\mathbf{dX}^{(2)\mathrm{T}}\,\mathbf{G}\,\mathbf{dX}^{(2)}}} \qquad (2.32b)$$

The numerator and denominator of eq(2.32b) are divided by the product $dX^{(1)}\,dX^{(2)}$ to allow $\cos\phi$ to be referred to unit vectors $\mathbf{m}_1 = \mathbf{dX}^{(1)}/dX^{(1)}$ and $\mathbf{m}_2 = \mathbf{dX}^{(2)}/dX^{(2)}$, aligned with the unstrained line elements PQ and PR:

$$\cos\phi = \frac{(\mathbf{F}\,\mathbf{m}_1)^{\mathrm{T}}}{\sqrt{\mathbf{m}_1^{\mathrm{T}}\mathbf{G}\,\mathbf{m}_1}} \times \frac{(\mathbf{F}\,\mathbf{m}_2)}{\sqrt{\mathbf{m}_2^{\mathrm{T}}\mathbf{G}\,\mathbf{m}_2}} = \frac{\mathbf{m}_1^{\mathrm{T}}(\mathbf{F}^{\mathrm{T}}\mathbf{F})\mathbf{m}_2}{\sqrt{\mathbf{m}_1^{\mathrm{T}}\mathbf{G}\,\mathbf{m}_1}\,\sqrt{\mathbf{m}_2^{\mathrm{T}}\mathbf{G}\,\mathbf{m}_2}} \qquad (2.32c)$$

Substituting eqs(2.23c) and (2.30a) into eq(2.32c) gives

$$\cos\phi = \frac{\mathbf{m}_1^{\mathrm{T}}\,\mathbf{G}\,\mathbf{m}_2}{\lambda_{(\mathbf{m}_1)}\,\lambda_{(\mathbf{m}_2)}} \qquad (2.32d)$$

The original included angle θ, between PQ and PR in Fig. 2.7, is found from the dot product $\mathbf{dX}^{(1)} \bullet \mathbf{dX}^{(2)} = |\mathbf{dX}^{(1)}|\,|\mathbf{dX}^{(2)}| \cos\theta$, in the spacial frame:

$$\cos\theta = \frac{\mathbf{dX}^{(1)}}{dX^{(1)}} \bullet \frac{\mathbf{dX}^{(2)}}{dX^{(2)}} = \frac{(\mathbf{dX}^{(1)})^{\mathrm{T}}}{dX^{(1)}} \frac{\mathbf{dX}^{(2)}}{dX^{(2)}} \qquad (2.33a)$$

Substituting eqs(2.22b) and (2.24b) into eq(2.33a)

$$\cos\theta = \frac{[\mathbf{H}\,\mathbf{dx}^{(1)}]^{\mathrm{T}}}{\sqrt{\mathbf{dx}^{(1)\mathrm{T}}\,\mathbf{C}\,\mathbf{dx}^{(1)}}} \times \frac{[\mathbf{H}\,\mathbf{dx}^{(2)}]}{\sqrt{\mathbf{dx}^{(2)\mathrm{T}}\,\mathbf{C}\,\mathbf{dx}^{(2)}}} \qquad (2.33b)$$

Dividing eq(2.33b) by the product $dx^{(1)}\,dx^{(2)}$ and introducing the unit vectors $\mathbf{n}_1 = \mathbf{dx}^{(1)}/dx^{(1)}$ and $\mathbf{n}_2 = \mathbf{dx}^{(2)}/dx^{(2)}$, aligned with strained line elements, gives

$$\cos\theta = \frac{(\mathbf{H}\,\mathbf{n}_1)^{\mathrm{T}}}{\sqrt{\mathbf{n}_1^{\mathrm{T}}\mathbf{C}\,\mathbf{n}_1}} \times \frac{(\mathbf{H}\,\mathbf{n}_2)}{\sqrt{\mathbf{n}_2^{\mathrm{T}}\mathbf{C}\,\mathbf{n}_2}} = \frac{\mathbf{n}_1^{\mathrm{T}}(\mathbf{H}^{\mathrm{T}}\mathbf{H})\,\mathbf{n}_2}{\sqrt{\mathbf{n}_1^{\mathrm{T}}\mathbf{C}\,\mathbf{n}_1}\,\sqrt{\mathbf{n}_2^{\mathrm{T}}\mathbf{C}\,\mathbf{n}_2}} \qquad (2.33c)$$

Substituting eqs(2.24c) and (2.31a) into eq(2.33c) gives

$$\cos\theta = \lambda_{(\mathbf{n}_1)}\lambda_{(\mathbf{n}_2)}\left(\mathbf{n}_1^{\mathrm{T}}\mathbf{C}\mathbf{n}_2\right) \tag{2.33d}$$

which requires the spacial unit vectors $\mathbf{n}_1$ and $\mathbf{n}_2$ for the elements pq and pr.

Example 2.5 A unit square lies in the X_2 - X_3 plane with one corner at the origin, as shown in Fig. 2.8a. Sketch the deformed shape when the Lagrangian description of motion for a given time is $x_1 = X_1 + BX_2X_3$, $x_2 = X_2 + BX_3^2$ and $x_3 = X_3 + BX_1^2$. Determine: (i) the extension ratio and the engineering strain for the diagonal joining the origin (0,0,0) to the corner Q(0,1,1) and (ii) the change in the right angle at the corner point Q when $B = 1$.

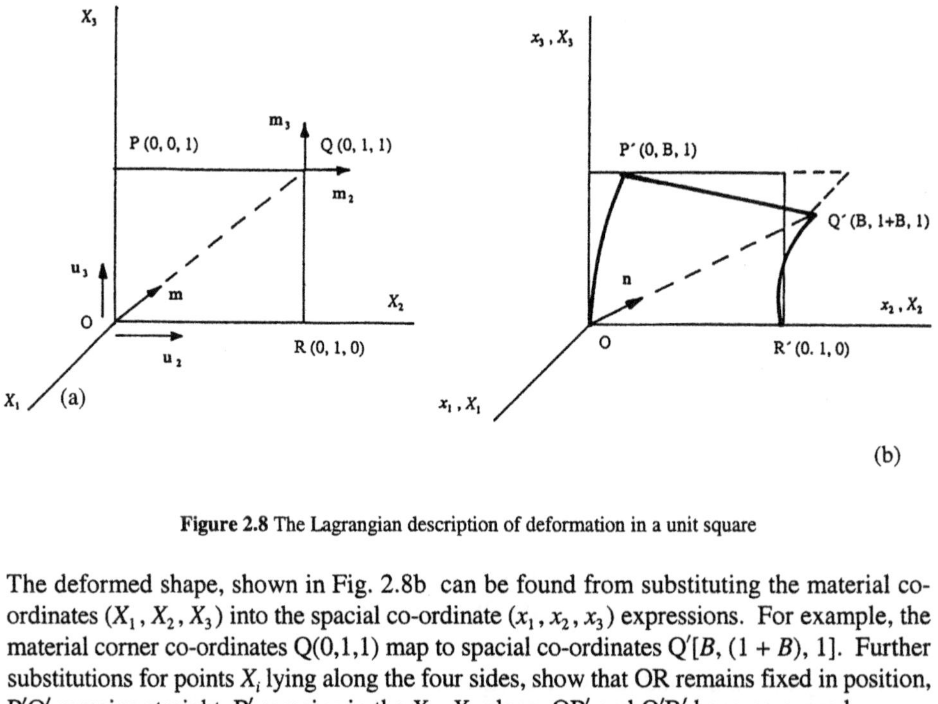

Figure 2.8 The Lagrangian description of deformation in a unit square

The deformed shape, shown in Fig. 2.8b can be found from substituting the material co-ordinates (X_1, X_2, X_3) into the spacial co-ordinate (x_1, x_2, x_3) expressions. For example, the material corner co-ordinates Q(0,1,1) map to spacial co-ordinates Q′[B, (1 + B), 1]. Further substitutions for points X_i lying along the four sides, show that OR remains fixed in position, P′Q′ remains straight, P′ remains in the X_2-X_3 plane, OP′ and Q′R′ become curved.

Equation (2.30a) provides an extension ratio where the unit vector **m** is aligned with the unstrained diagonal OQ. That is: $\mathbf{m} = (1/\sqrt{2})\mathbf{u}_2 + (1/\sqrt{2})\mathbf{u}_3$, from which the column matrix is $\mathbf{m} = \{0 \;\; 1/\sqrt{2} \;\; 1/\sqrt{2}\}^{\mathrm{T}}$. Now, **F** is the matrix of derivates $\partial x_i/\partial X_j$. Hence, $\mathbf{G} = \mathbf{F}^{\mathrm{T}}\mathbf{F}$ is

$$\mathbf{G} = \begin{bmatrix} 1 & 0 & 2BX_1 \\ BX_3 & 1 & 0 \\ BX_2 & 2BX_3 & 1 \end{bmatrix}\begin{bmatrix} 1 & BX_3 & BX_2 \\ 0 & 1 & 2BX_3 \\ 2BX_1 & 0 & 1 \end{bmatrix} = \begin{bmatrix} [1+(2BX_1)^2] & BX_3 & (BX_2+2BX_1) \\ BX_3 & [(BX_3)^2+1] & [(BX_2)(BX_3)+2BX_3] \\ (BX_2+2BX_1) & [(BX_2)(BX_3)+2BX_3] & [(BX_2)^2+(2BX_3)^2+1] \end{bmatrix}$$

Substituting **m** and **G** into eq(2.30c), with $X_1 = 0$, $X_2 = X_3 = 1$, gives

$$\lambda_{(\mathrm{m})}^2 = \begin{bmatrix} 0 & \frac{1}{\sqrt{2}} & \frac{1}{\sqrt{2}} \end{bmatrix} \begin{bmatrix} 1 & B & B \\ B & 1+B^2 & B(2+B) \\ B & B(2+B) & (1+5B^2) \end{bmatrix} \begin{bmatrix} 0 \\ \frac{1}{\sqrt{2}} \\ \frac{1}{\sqrt{2}} \end{bmatrix}$$

$$= \begin{bmatrix} \sqrt{2}B & \left(\frac{1}{\sqrt{2}} + \sqrt{2}B + \sqrt{2}B^2\right) & \left(\frac{1}{\sqrt{2}} + \sqrt{2}B + 3\sqrt{2}B^2\right) \end{bmatrix} \begin{bmatrix} 0 \\ \frac{1}{\sqrt{2}} \\ \frac{1}{\sqrt{2}} \end{bmatrix} = 1 + 2B + 4B^2$$

The engineering strain is from eq(2.11b):

$$e_{(\mathrm{m})} = \lambda_{(\mathrm{m})} - 1 = \sqrt{(1 + 2B + 4B^2)} - 1$$

From the displacement of Q to Q′ shown, it follows that

$$\mathrm{OQ'} = \sqrt{[B^2 + (1+B)^2 + 1^2]} = \sqrt{[2(1 + B + B^2)]}$$

and

$$\mathrm{OQ'/\,OQ} = \sqrt{(1 + B + B^2)}$$

This is not the same as $\lambda_{(\mathrm{m})}$ because the deformation in OQ is non-linear. In fact, $\lambda_{(\mathrm{m})}$ would describe the ratio between the length of a curved trajectory OQ′ to its original length. Moreover, we should not expect the same stretch ratio $\lambda_{(\mathrm{n})}$ from eq(2.31c) since the directions of **n** and **m** differ.

Unit vector components, which describe the unstrained sides PQ and RQ, are respectively: $\mathbf{m}_2 = \{0\ \ 1\ \ 0\}^{\mathrm{T}}$ and $\mathbf{m}_3 = \{0\ \ 0\ \ 1\}^{\mathrm{T}}$. Thus, the numerator in eq(2.32d) becomes

$$\mathbf{m}_2^{\mathrm{T}}\mathbf{G}\,\mathbf{m}_3 = \begin{bmatrix} 0 & 1 & 0 \end{bmatrix} \begin{bmatrix} 1 & B & B \\ B & (1+B^2) & B(2+B) \\ B & B(2+B) & (1+5B^2) \end{bmatrix} \begin{bmatrix} 0 \\ 0 \\ 1 \end{bmatrix} = \begin{bmatrix} B & (1+B^2) & B(2+B) \end{bmatrix} \begin{bmatrix} 0 \\ 0 \\ 1 \end{bmatrix} = B(2+B)$$

In the denominator of eq(2.32d), the stretch ratios for PQ and RQ are, respectively,

$$\lambda_{(\mathrm{m}_2)}^2 = \mathbf{m}_2^{\mathrm{T}}\mathbf{G}\,\mathbf{m}_2 = \begin{bmatrix} 0 & 1 & 0 \end{bmatrix} \begin{bmatrix} 1 & B & B \\ B & (1+B^2) & B(2+B) \\ B & B(2+B) & (1+5B^2) \end{bmatrix} \begin{bmatrix} 0 \\ 1 \\ 0 \end{bmatrix} = \begin{bmatrix} B & (1+B^2) & B(2+B) \end{bmatrix} \begin{bmatrix} 0 \\ 1 \\ 0 \end{bmatrix} = 1 + B^2$$

$$\lambda_{(\mathrm{m}_3)}^2 = \mathbf{m}_3^{\mathrm{T}}\mathbf{G}\,\mathbf{m}_3 = \begin{bmatrix} 0 & 0 & 1 \end{bmatrix} \begin{bmatrix} 1 & B & B \\ B & (1+B^2) & B(2+B) \\ B & B(2+B) & (1+5B^2) \end{bmatrix} \begin{bmatrix} 0 \\ 1 \\ 0 \end{bmatrix} = \begin{bmatrix} B & B(2+B) & (1+5B^2) \end{bmatrix} \begin{bmatrix} 0 \\ 0 \\ 1 \end{bmatrix} = 1 + 5B^2$$

Substituting into eq(2.32d):

$$\cos\theta = \frac{B(2+B)}{\sqrt{(1+B^2)(1+5B^2)}}$$

which, for $B = 1$, gives $\cos\theta = \sqrt{3}/2$. That is, $\theta = 30°$ amd so the change to the original right angle PQR is 60°.

2.5 Polar Decomposition

Equations (2.26a,b,c) show that an infinitesimal displacement gradient is composed of the sum of strain and rotation tensors. In the case of finite straining, a similar decomposition can be made to the deformation gradient **F**. Let some intermediate position vector **dx**′ lie between the mapping of line element vector **dX** in its initial position to its current position **dx**, in Fig. 2.6. The current position may be traced with two alternative local sequential mappings: (i) a stretch **dx**′ = **U dX**, followed by a rigid rotation **dx** = **R dx**′, or (ii) a rigid rotation **dx**′ = **R dX**, followed by a stretch **dx** = **Vdx**′. Thus, the final position **dx** is

$$\mathbf{dx} = (\mathbf{RU})\,\mathbf{dX} = (\mathbf{VR})\,\mathbf{dX} \tag{2.34a}$$

Comparing eqs(2.21b) and (2.34a) it follows that **F** is

$$\mathbf{F} = \mathbf{RU} = \mathbf{VR} \quad \text{or} \quad F_{ij} = \partial x_i / \partial X_j = R_{ik}U_{kj} = V_{ik}R_{kj} \tag{2.34b}$$

where **U** and **V** are the right and left stretch tensors, being positive, symmetric and possessing the same principal stretch values. Since det **F** = $\rho_o/\rho > 0$, **R** is an orthogonal rotation matrix, obeying

$$\mathbf{RR}^{\mathrm{T}} = \mathbf{R}^{\mathrm{T}}\mathbf{R} = \mathbf{I} \quad \text{or} \quad R_{ik}R_{jk} = R_{ki}R_{kj} = \delta_{ij} \tag{2.35a}$$

F, R, U and **V** will all depend upon position unless the deformation is homogenous, where **x** = **FX** and the components of **F, R, U** and **V** become constants. Combining eqs(2.34b) and (2.35a) gives relationships between **U** and **V**:

$$\mathbf{U} = \mathbf{R}^{\mathrm{T}}\mathbf{VR} \quad \text{and} \quad \mathbf{V} = \mathbf{RUR}^{\mathrm{T}} \tag{2.35b,c}$$

Now, from eqs(2.34b) and (2.23c), **G** and its components G_{ij}, may be written as

$$\begin{aligned}
\mathbf{G} &= \mathbf{F}^{\mathrm{T}}\mathbf{F} = (\mathbf{RU})^{\mathrm{T}}(\mathbf{RU}) \\
&= \mathbf{U}^{\mathrm{T}}(\mathbf{R}^{\mathrm{T}}\mathbf{R})\mathbf{U} = \mathbf{U}^{\mathrm{T}}\mathbf{I}\,\mathbf{U}
\end{aligned}$$

But as $\mathbf{U}^{\mathrm{T}} = \mathbf{U}$ and $\mathbf{IU} = \mathbf{U}$

$$\mathbf{G} = \mathbf{U}^2$$

$$\begin{aligned}
G_{ij} &= F_{ik}{}^{\mathrm{T}}F_{kj} = F_{ki}F_{kj} \\
&= (R_{kp}U_{pi})(R_{kq}U_{qj}) \\
&= (R_{kp}R_{kq})U_{pi}U_{qj} \\
&= \delta_{pq}U_{pi}U_{qj} = U_{qi}U_{qj}
\end{aligned} \tag{2.36}$$

Let $\mathbf{U}^*$ contain the principal stretches α_1, α_2 and α_3 of **U**. Substituting eq(2.36) into eq(2.30a), gives

$$\lambda_{(\mathbf{m})}{}^2 = \mathbf{m}^{\mathrm{T}}(\mathbf{U}^{*2})\,\mathbf{m} \tag{2.37a}$$

Taking the principal stretch directions of $\mathbf{U}^*$ as co-ordinates, so that $\mathbf{m}_1$ is a unit vector aligned with the major principal stretch direction, eq(2.37a) gives

$$\lambda^2_{(\mathbf{m}_1)} = [1 \;\; 0 \;\; 0] \begin{bmatrix} \alpha_1^2 & 0 & 0 \\ 0 & \alpha_2^2 & 0 \\ 0 & 0 & \alpha_3^2 \end{bmatrix} \begin{bmatrix} 1 \\ 0 \\ 0 \end{bmatrix} = \alpha_1^2 \qquad (2.37b)$$

Thus, from eq(2.37b), $\lambda_{(\mathbf{m}_1)} = \alpha_1$. Similarly, when further unit vectors $\mathbf{m}_2 = \{0 \;\; 1 \;\; 0\}^{\mathrm{T}}$ and $\mathbf{m}_3 = \{0 \;\; 0 \;\; 1\}^{\mathrm{T}}$ are aligned with their respective principal directions: $\lambda_{(\mathbf{m}_2)} = \alpha_2$ and $\lambda_{(\mathbf{m}_3)} = \alpha_3$. This means that the stretch ratios for the principal directions are the principal values of **U**.

From eqs (2.34b) and (2.28d), **B** and its components B_{ij}, may each be written as

$$\mathbf{B} = \mathbf{F}\mathbf{F}^{\mathrm{T}} = (\mathbf{V}\mathbf{R})(\mathbf{V}\mathbf{R})^{\mathrm{T}} \qquad B_{ij} = F_{ik}F_{kj}^{\mathrm{T}} = F_{ik}F_{jk}$$

$$= \mathbf{V}(\mathbf{R}\mathbf{R}^{\mathrm{T}})\mathbf{V}^{\mathrm{T}} = \mathbf{V}\mathbf{I}\mathbf{V}^{\mathrm{T}} \qquad = (V_{ip}R_{pk})(V_{jq}R_{qk})$$

But as $\mathbf{V}^{\mathrm{T}} = \mathbf{V}$ and $\mathbf{V}\mathbf{I} = \mathbf{V}$: $\qquad = V_{ip}V_{jq}(R_{pk}R_{qk})$

$$\mathbf{B} = \mathbf{V}^2 \qquad = V_{ip}V_{jq}\delta_{pq} = V_{iq}V_{jq} \qquad (2.38)$$

Let the principal directions of **U** be given by the three unit vectors $\mathbf{u}_i^*$ ($i = 1, 2, 3$), with their components in the X_i frame. They will satisfy an equation similar to eq(1.23b):

$$(\mathbf{U} - \alpha_i \mathbf{I})\,\mathbf{u}_i^* = 0 \qquad (2.39a)$$

Since $\mathbf{R}^{\mathrm{T}}\mathbf{R}\mathbf{u}^* = \mathbf{I}\,\mathbf{u}^* = \mathbf{u}^*$, eq(2.39a) can be re-written as

$$\mathbf{R}\,(\mathbf{U} - \alpha_i\mathbf{I})(\mathbf{R}^{\mathrm{T}}\mathbf{R})\mathbf{u}_i^* = [\,\mathbf{R}\mathbf{U}\mathbf{R}^{\mathrm{T}} - \mathbf{R}\,(\alpha_i\mathbf{I})\mathbf{R}^{\mathrm{T}}\,]\,(\mathbf{R}\mathbf{u}_i^*) = 0$$

Substituting from eq(2.35c)

$$(\mathbf{V} - \alpha_i\mathbf{I})(\mathbf{R}\mathbf{u}_i^*) = 0 \qquad (2.39b)$$

When the principal directions of **V** are defined by the unit vectors $\mathbf{v}^*_i$ ($i = 1, 2, 3$) they will satisfy the equation

$$(\mathbf{V} - \beta_i\mathbf{I})\,\mathbf{v}_i^* = 0 \qquad (2.39c)$$

Comparing eqs(2.39b and c) shows that: (i) the principal values of **U** and **V** are the same since $\alpha_i = \beta_i$ and (ii) **R** connects the principal directions of **U** and **V** within: $\mathbf{v}_i^* = \mathbf{R}\mathbf{u}_i^*$ or $\mathbf{u}_i^* = \mathbf{R}^{\mathrm{T}}\,\mathbf{v}_i^*$. Equations (2.36) and (2.38) show that **U** and **V** provide an equivalent physical interpretation of **G** and **B** though **G** and **B** remain the more convenient measures of strain.

Principal stretch tensors $\mathbf{U}^*$ and $\mathbf{V}^*$ are calculated from the square roots of the eigen values of **G** and **B** respectively. The 3×3 transformation matrix **M** of eigen vector components (for **G** and **B**) will not, in general, equal the components of the rotation matrix **R** as the following example shows. **M** appears in tensor transformation laws for **U** and **V**, similar to eq(1.21b):

$$\mathbf{U}^* = \mathbf{M}\mathbf{U}\mathbf{M}^{\mathrm{T}} \text{ and } \mathbf{V}^* = \mathbf{M}\mathbf{V}\mathbf{M}^{\mathrm{T}} \tag{2.40a,b}$$

Reversing eqs(2.40a,b) gives (see eq 1.20b)

$$\mathbf{U} = \mathbf{M}^{\mathrm{T}}\mathbf{U}^*\mathbf{M} \text{ and } \mathbf{V} = \mathbf{M}^{\mathrm{T}}\mathbf{V}^*\mathbf{M} \tag{2.40c,d}$$

Alternatively, eqs(2.35b,c) allow the calculation of **U** from **V** and vice-versa, without first finding their principal values.

Example 2.6 Find **U**, **V** and **R** for a Lagrangian description of the following motion: $x_1 = aX_1 + bX_2$, $x_2 = -aX_1 + bX_2$ and $x_3 = cX_3$ (where $a > b > c$). Show that the principal values of **U** and **V** are equal.

Having found the components $F_{ij} = \partial x_i / \partial X_j$, **G** is found from eq(2.23c) as

$$\mathbf{G} = \mathbf{F}^{\mathrm{T}}\mathbf{F} = \begin{bmatrix} a & -a & 0 \\ b & b & 0 \\ 0 & 0 & c \end{bmatrix} \begin{bmatrix} a & b & 0 \\ -a & b & 0 \\ 0 & 0 & c \end{bmatrix} = \begin{vmatrix} 2a^2 & 0 & 0 \\ 0 & 2b^2 & 0 \\ 0 & 0 & c^2 \end{vmatrix}$$

from which **U** is obtained:

$$\mathbf{U} = \sqrt{\mathbf{G}} = \begin{bmatrix} \sqrt{2}a & 0 & 0 \\ 0 & \sqrt{2}b & 0 \\ 0 & 0 & c \end{bmatrix} = \mathbf{U}^*$$

This result reveals that the directions of the principal, right stretches are aligned with the co-ordinates X_i. The rotation **R** then follows from eq(2.34b) as

$$\mathbf{R} = \mathbf{F}\mathbf{U}^{-1} = \begin{bmatrix} a & b & 0 \\ -a & b & 0 \\ 0 & 0 & c \end{bmatrix} \begin{bmatrix} \frac{1}{\sqrt{2}a} & 0 & 0 \\ 0 & \frac{1}{\sqrt{2}b} & 0 \\ 0 & 0 & 1/c \end{bmatrix} = \begin{bmatrix} \frac{1}{\sqrt{2}} & \frac{1}{\sqrt{2}} & 0 \\ -\frac{1}{\sqrt{2}} & \frac{1}{\sqrt{2}} & 0 \\ 0 & 0 & 1 \end{bmatrix}$$

These components are direction cosines corresponding to rotating X_1 and X_2 by 45° about the fixed X_3 axis. **B** is found from eq(2.28d) as

$$\mathbf{B} = \begin{bmatrix} a & b & 0 \\ -a & b & 0 \\ 0 & 0 & c \end{bmatrix} \begin{bmatrix} a & -a & 0 \\ b & b & 0 \\ 0 & 0 & c \end{bmatrix} = \begin{bmatrix} (a^2 + b^2) & (-a^2 + b^2) & 0 \\ (-a^2 + b^2) & (a^2 + b^2) & 0 \\ 0 & 0 & c^2 \end{bmatrix}$$

The stretch tensor **V** is more conveniently found from $\mathbf{V}^* = \sqrt{\mathbf{B}^*}$ when $\mathbf{B}^*$ is referred to its principal axes. The principal values of $\mathbf{B}^*$ (i.e. eigen values) are the three real and positive roots β_i ($i = 1, 2, 3$) of the characteristic equation for **B**. They follow from the expansion to the determinant $(\mathbf{B} - \beta_i \mathbf{I}) = 0$:

$$\begin{vmatrix} (a^2+b^2)-\beta & (-a^2+b^2) & 0 \\ (-a^2+b^2) & (a^2+b^2)-\beta & 0 \\ 0 & 0 & c-\beta \end{vmatrix} = 0$$

which gives $\beta_1 = 2a^2$, $\beta_2 = 2b^2$ and $\beta_3 = c^2$. Hence, we have

$$\mathbf{B} = \begin{bmatrix} 2a^2 & 0 & 0 \\ 0 & 2b^2 & 0 \\ 0 & 0 & c^2 \end{bmatrix}, \quad \mathbf{V} = \sqrt{\mathbf{B}} = \begin{bmatrix} \sqrt{2}a & 0 & 0 \\ 0 & \sqrt{2}b & 0 \\ 0 & 0 & c \end{bmatrix}$$

showing that the principal values of **U** and **V** (i.e. the components of $\mathbf{U}^*$ and $\mathbf{V}^*$) are equal.

The eigen vectors $\mathbf{v}_i^*$ ($i = 1, 2, 3$), which define the orthogonal, principal axes of $\mathbf{V}^*$, are found from the solution to:

$$(\mathbf{B} - \beta_i \mathbf{I})\, \mathbf{v}_i^* = 0 \qquad \text{(i)}$$

For $i = 1$, let $\mathbf{v}_1^* = l_1\mathbf{u}_1 + m_1\mathbf{u}_2 + n_1\mathbf{u}_3$ where $\mathbf{u}_1$, $\mathbf{u}_2$ and $\mathbf{u}_3$ are unit vectors aligned with the co-ordinates X_i (see Fig. 2.6) and l_1, m_1 and n_1 are the intercepts (direction cosines) with X_i. In eq(i) we write vector components $\mathbf{v}_1^* = \{l_1 \;\; m_1 \;\; n_1\}^{\mathrm{T}}$ and substitute for matrix **B** to give:

$$(a^2 + b^2 - \beta_1)\, l_1 + (-a^2 + b^2)\, m_1 = 0 \qquad \text{(ii)}$$
$$(-a^2 + b^2)\, l_1 + (a^2 + b^2 - \beta_1)\, m_1 = 0 \qquad \text{(iii)}$$
$$(c^2 - \beta_1)\, n_1 = 0 \qquad \text{(iv)}$$

Substituting $\beta_1 = 2a^2$ in eqs(ii)-(iv) we find $l_1 = 1/\sqrt{2}$, $m_1 = -1/\sqrt{2}$ and $n_1 = 0$ giving:

$$\mathbf{v}_1^* = (1/\sqrt{2})\mathbf{u}_1 - (1/\sqrt{2})\mathbf{u}_2$$

Similarly, for $i = 2$ and 3, the eigen vectors are:

$$\mathbf{v}_2^* = (1/\sqrt{2})\mathbf{u}_1 + (1/\sqrt{2})\mathbf{u}_2 \;\text{ and }\; \mathbf{v}_3^* = \mathbf{u}_3$$

in which $\mathbf{v}_1^* \bullet \mathbf{v}_2^* = \mathbf{v}_2^* \bullet \mathbf{v}_3^* = \mathbf{v}_3^* \bullet \mathbf{v}_1^* = 0$ confirms orthogonality in these vectors. The components of these vectors define the transformation matrix **M** in eq(2.40a-d) so that:

$$\mathbf{V} = \begin{bmatrix} 1/\sqrt{2} & 1/\sqrt{2} & 0 \\ -1/\sqrt{2} & 1/\sqrt{2} & 0 \\ 0 & 0 & 1 \end{bmatrix} \begin{bmatrix} \sqrt{2}a & 0 & 0 \\ 0 & \sqrt{2}b & 0 \\ 0 & 0 & c \end{bmatrix} \begin{bmatrix} 1/\sqrt{2} & -1/\sqrt{2} & 0 \\ 1/\sqrt{2} & 1/\sqrt{2} & 0 \\ 0 & 0 & 1 \end{bmatrix} = \begin{bmatrix} (a+b)/\sqrt{2} & -(a-b)/\sqrt{2} & 0 \\ (a-b)/\sqrt{2} & (a+b)/\sqrt{2} & 0 \\ 0 & 0 & c \end{bmatrix}$$

Finally, we may check this result from $\mathbf{F} = \mathbf{VR}$:

$$\mathbf{F} = \begin{bmatrix} (a+b)/\sqrt{2} & -(a-b)/\sqrt{2} & 0 \\ (a-b)/\sqrt{2} & (a+b)/\sqrt{2} & 0 \\ 0 & 0 & c \end{bmatrix} \begin{bmatrix} 1/\sqrt{2} & 1/\sqrt{2} & 0 \\ -1/\sqrt{2} & 1/\sqrt{2} & 0 \\ 0 & 0 & 1 \end{bmatrix} = \begin{bmatrix} a & b & 0 \\ -a & b & 0 \\ 0 & 0 & c \end{bmatrix}$$

2.6 Strain Definitions

It is seen that there is no single fundamental definition of strain. Indeed, the expressions for the natural, Lagrangian, Eulerian strains and the extension ratio will all differ, even for a simple uniaxial, finite deformation. Thus, the flow behaviour of a material will depend upon the chosen definition of strain. The Lagrangian finite strain tensor, **L** in eq(2.27b,c), is more commonly used for referring finite deformation in a solid to its original shape.

For infinitesimal strains, the differential products in eqs(2.27b) and (2.28b) may be ignored. The Eulerian and Langrangian finite strain tensors **E** and **L**, will then reduce to eq(2.3), implying that there is negligible displacement between the material and spacial co-ordinates. Seth [4] related the three uniaxial strains to the extension ratio within a single formula. Given initial and final lengths l_o and l respectively, these strains are expressed from eqs(2.11b), (2.30c) and (2.31c) as

$$\varepsilon^{(a)} = (1/a)[1 - (l_o/l)^a] = (1/a)[1 - (1/\lambda)^a]$$

where $a = -1$, $a = -2$ and $a = +2$ according to the engineering definition and those of Lagrange and Euler respectively. Other values of a may lead to simpler constitutive relations, e.g. when a is chosen to provide linear stress-strain behaviour, but the corresponding stress definition becomes uncertain and the tensorial nature of strain is lost.

The reader will find further discusion of strain tensors and their transformation properties in books on continuum mechanics [5-8].

References

1. Ludwik P. *Elemente der Technologischen Mechanik*, 1909, Springer, Berlin.
2. Hencky H. "Uber die form des elastizitatsgesetzes bei ideal elastischen stoffen", *Z.T.P.* 1928, **9**, 214-223.
3. Green A. E. and Adkins J. E. *Large Elastic Deformations*, 1960, Oxford University Press, London.
4. Seth B. R. "Generalised strain measure with applications to physical problems", Proc: *Second Order Effects*, IUTAM, Pergamon Press, 1964.
5. Farrashkhalaut M. and Miles J. P. *Tensor Methods for Engineers and Scientists*, 1990, Ellis-Horwood.
6. Lai W. M., Rubin D. and Krempl E. *Introduction to Continuum Mechanics*, Pergamon Engineering Series, Vol.17, 1978, Pergamon Press.
7. Mase G. *Continuum Mechanics*, 1970, Schaum Outline Series, McGraw-Hill.
8. Mase G. E. and Mase G. T. *Continuum Mechanics for Engineers*, 1992, CRC Press.

Exercises

2.1 A displacement vector **u** is described by its three components u_i in spacial co-ordinates x_i, as

$$u_1 = (k_1x_1 - k_3x_3)^2, \quad u_2 = (k_2x_2 + k_3x_3)^2 \text{ and } \quad u_3 = (k_1x_1)(k_2x_2)$$

Determine the components of infinitesimal strain and rotation matrices, ε_{ij} and ω_{ij}, for a point P(0,1,2), when $k_1 = k_2 = 1$ and $k_3 = 2$. What then is the change in the right angle between a pair of perpendicular lines passing through P with direction cosines: (1/√3, 1/√3, - 1/√3) and (1/√14, 2/√14, 3/√14) with respect to x_1, x_2 and x_3?

2.2 The following infinitesimal, micro-strain components: $\varepsilon_{11} = 600$, $\varepsilon_{22} = \varepsilon_{33} = 0$, $\varepsilon_{12} = \varepsilon_{21} = 200$, $\varepsilon_{23} = \varepsilon_{32} = 400$ and $\varepsilon_{13} = \varepsilon_{31} = 200$ (all multipied by 10^{-6}) define a strain tensor. Determine the magnitude and direction of the principal strains, the maximum and octahedral shear strains. What is the state of strain for a plane with normal direction cosines $\alpha = 0.53$, $\beta = 0.35$ and $\gamma = 0.77$?

[Answers: 800, 200, - 400, 600, 9.8, 622 ($\times 10^{-6}$)]

2.3 The components of a micro-strain tensor are $\varepsilon_{11} = 100$, $\varepsilon_{22} = \varepsilon_{23} = 0$, $\varepsilon_{33} = -100$, $\varepsilon_{12} = \varepsilon_{21} = 500$ and $\varepsilon_{13} = \varepsilon_{31} = -500$ in x_1, x_2 and x_3 co-ordinates. Transform these to x_1', x_2' and x_3' co-ordinates given that the direction cosines for the x_1'- direction are: $l_{11} = 1/\sqrt{14}$, $l_{12} = 2/\sqrt{14}$, $l_{13} = 3/\sqrt{14}$ and those for the x_2'- direction are: $l_{21} = 1$, $l_{22} = 1$, $l_{23} = -1$.

[Answers: $\varepsilon_{11}' = -128.6$, $\varepsilon_{22}' = 666.7$, $\varepsilon_{33}' = 538.1$, $\varepsilon_{12}' = 138.9$, $\varepsilon_{13}' = 198$ and $\varepsilon_{23}' = -276.2$]

2.4 Under plane-strain deformation, a line element increases in length by 17% in the X_1- direction and by 22% in the X_2- direction. What is the percentage increase in length and the rotation of a line element originally at 45° to the reference configuration X_1 and X_2?

[Answers: 27%, 2.4° to X_1]

2.5 Given the three components of a spacial co-ordinate function $\mathbf{x} = \mathbf{x}(\mathbf{X}, t)$ are $x_1 = A_1(X_1 + BX_2)$, $x_2 = A_2X_2$ and $x_3 = A_3X_3$, where A and B are constants at time t, determine: (i) the components of the deformation gradients **F** and $\mathbf{F}^{-1}$, (ii) the left and right Cauchy Green deformation tensors **G**, $\mathbf{G}^{-1}$, **B** and $\mathbf{B}^{-1}$ and (iii) the Lagrangian and Eulierian strain tensors, **L** and **E** respectively.

2.6 The spacial co-ordinates x_i, in Exercise 2.5, describe homogenous deformation at a given time t. Find the new direction and the stretch of a line element originally equally inclined to a reference configuration X_i ($i = 1, 2$ and 3) at time $t = 0$.

2.7 Given that the *spacial* components of motion, $\mathbf{x} = \mathbf{x}(\mathbf{X}, t)$, are:

$$x_1 = X_1 + X_2(e^t - K),$$

$$x_2 = X_1(e^{-t} - K) + X_2$$

$$x_3 = X_3,$$

show that the *material* components of motion $\mathbf{X} = \mathbf{X}(\mathbf{x}, t)$, are:

$$X_1 = [-x_1 + x_2(e^t - K)] / [K(K - e^t - e^{-t})]$$

$$X_2 = [x_1(e^{-t} - K) - x_2] / [K(K - e^t - e^{-t})]$$

$$X_3 = x_3$$

2.8 The spacial position vector **x** has components: $x_1 = X_1$, $x_2 = X_2 + KX_3$ and $x_3 = X_3 + KX_2$, where K is a constant. Determine the components of the displacement vector $\mathbf{u} = \mathbf{x} - \mathbf{X}$ (see Fig. 2.6) in both material and spacial co-ordinates.

[Answers: $u_1 = 0$, $u_2 = KX_3 = k(x_3 - Kx_2)/(1 - K^2)$ and $u_3 = KX_2 = k(x_2 - Kx_3)/(1 - K^2)$]

2.9 Show that the components of **G** and **L** for x_i in Exercise 2.8 are given as:

$$G_{ij} = \begin{bmatrix} 1 & 0 & 0 \\ 0 & 1+K^2 & 2K \\ 0 & 2K & 1+K^2 \end{bmatrix}, \quad L_{ij} = \begin{bmatrix} 0 & 0 & 0 \\ 0 & K^2 & 2K \\ 0 & 2K & K^2 \end{bmatrix}$$

Taking $K = 1$, determine the magnitude and direction of the principal Lagrangian strains.

2.10 Given that the material components of the displacement vector **U** (see Fig. 2.6) are: $U_1 = X_1X_3^2$, $U_2 = X_1^2X_2$ and $U_3 = X_2^2X_3$, determine the components of the material displacement gradient $\partial U_i/\partial X_j$ and the material deformation gradient $F_{ij} = \partial x_i/\partial X_j$. Hence confirm the derivative $F_{ij} = \partial U_i/\partial X_j + \delta_{ij}$.

2.11 Calculate the stretch ratio for a line element vector aligned with the X_1 - axis and another with direction cosines (0, 1/√2, 1/√2) w.r.t. X_i, following a shear deformation whose components x_i, are given in Exercise 2.8. Hint, use eq(2.13c) with the corresponding right Cauchy-Green components G_{ij} from Exercise 2.9. [Answer: 1, (1 + K)]

2.12 Show from eq(2.28d) that the extension ratio 1 + K, found in Exercise 2.11, applies when inverting the components from x_i in **x** = **x**(**X**) to X_i in **X** = **X**(**x**) upon forming the inverse of the left Cauchy-Green components B_{ij}. Explain why this is.

2.13 Determine the extension ratios λ_X and λ_x from eqs(2.30a) and (2.31a) for a line element originally aligned with the X_2 - axis, following a motion with x_i components given in Exercise 2.8. Explain why these differ.

2.14 Determine, for the unit square shown in Fig. 2.8a, the extension ratio for the diagonal PR and the change in the angle OPQ, when it is subjected to the same Lagrangian description **x** = **x**(**X**) of motion (see Example 2.5).

2.15 Determine the deformation gradient for the Lagrangian motion $x_1 = X_1 + 2X_3$, $x_2 = X_2 - 2X_3$ and $x_3 = -2X_1 + 2X_2 + X_3$. Find **R**, **U** and **V** by polar decomposition and show that the principal values of **U** and **V** are equal.

CHAPTER 3

YIELD CRITERIA

3.1 Introduction

Experiment has shown that an assumption of initial isotropy is usually a good approximation when quoting a yield stress value for a polycrystalline material. One exception is where the yield stress becomes direction-dependent following a history of severe deformation processing, as with the cold-rolling of sheet metals. There are two approaches, microscopic and macroscopic, used in the analysis of this form of anisotropy. The first approach examines the manner in which various slip systems control anisotropy within a given structure. The second approach, which will be adopted here, quantifies the effect that anisotropy has upon the initial yield surface. Typically, thie yield surface can display different initial yield stresses in tension and compression, so that it appears asymmetric relative to its stress co-ordinate axes. As with an annealed, isotropic material, it is convenient to express the inherent, or residual, anisotropy, within the structure of a worked material, with an appropriate yield function. Suitable functions are given for conditions of isotropy, transverse isotropy and orthotropy. A second form of anisotropy, which accentuates yield surface asymmetry, is induced by subsequent straining of a hardening material beyond its initial yield point. This is a deformation-induced anisotropy, of which the *Bauschinger effect* is one manifestation. To describe this, the initial yield function requires a modification using a *hardening rule* (see Chapter 10).

Similar functions are often employed to describe the initial *yield surface* in metals and the *failure surface* for brittle non-metals, despite their different responses to loading. In a non-hardening metal, the yield and fracture surfaces are coincident and a single function applies. The term *elastic, perfectly-plastic* is used to describe this. The material response is elastic within the surface's interior region. Along its boundary, plastic flow occurs under a constant yield stress. In contrast, a purely brittle material remains elastic to the point of fracture, where the critical stress combination lies on its failure surface. However, as both yield and failure surfaces bound an elastic interior they appear similar mathematically.

A flow rule describes the manner in which plastic strain depends upon a given combination of stress. It will be shown in Chapter 4 that there are two types of flow rule for elastic-perfect plasticity. These rules differ between their incremental and total strain formulations, the former with and the latter without an association to the yield surface. The incremental strain approach will be outlined here.

3.2 Yielding of Ductile Isotropic Materials

In Fig. 1.1b it is seen that a ductile metallic material will begin to deform and harden plastically under a uniaxial stress when the yield stress is exceeded. In practice, the stress state is biaxial or triaxial and the question arises as to what magnitudes of combined stress will cause the onset of yielding? Those who first attempted to answer this question sought a suitable criterion based upon stress, strain or strain energy for the complex system that could be related to the corresponding quantity at the uniaxial yield point. Where a material yields isotropically, the critical value of the chosen parameter at yield is independent of the orientation of the stress system. All criteria may then be related to a constant uniaxial yield stress Y, which is most conveniently measured in a simple tension test (see Fig. 3.1a). Isotropic yield criteria are normally expressed in principal stress form. The latter refers to the stress system σ_1, σ_2 and σ_3 in Fig. 3.1b, where, conventionally, $\sigma_1 > \sigma_2 > \sigma_3$. The stress transformation equations, given in Chapter 1, will enable these yield criteria to be expressed in any two- or three-dimensional combination of applied direct and shear stresses.

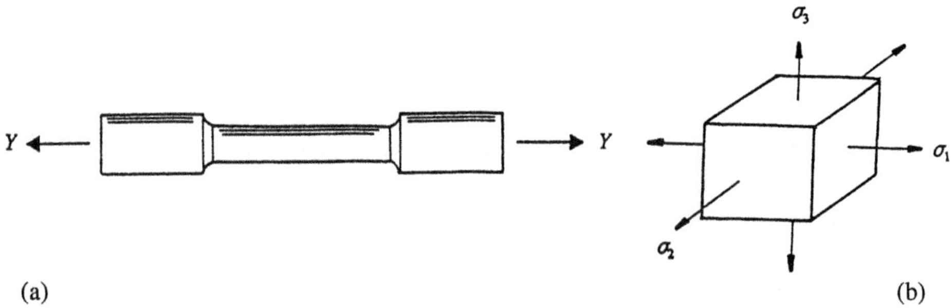

Figure 3.1 Uniaxial and triaxial principal stress states

A number of yield criteria have been proposed over the past two centuries [1], though it is now accepted that the criteria commonly attributed to von Mises [2] and Tresca [3] are most representative of initial yielding in isotropic, metallic materials. It is instructive to consider the derivation and verification of these two criteria using appropriate experimental data available in the literature.

3.2.1 Maximum Shear Stress Theory

Attributed jointly to Tresca, Coloumb, and Guest, the maximum shear stress theory assumes that yielding, under the principal stress system in Fig. 3.1b, begins when the maximum shear stress reaches a critical value. The latter is taken as the maximum shear stress k at the point of yielding under simple tension or compression. That is, $k = Y/2$, which acts along planes at 45° to the tensile stress axis, shown in Fig. 3.2a.

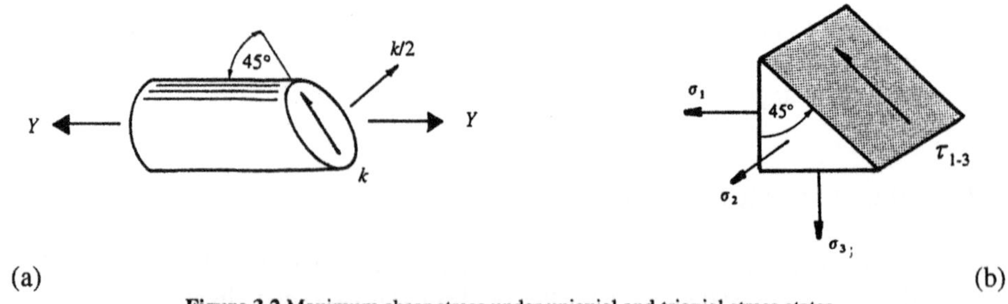

Figure 3.2 Maximum shear stress under uniaxial and triaxial stress states

Under a principal, triaxial stress state (Fig. 3.1b), the greatest shear stress is, from eq(1.33b), $\tau_{1\text{-}3} = ½(\sigma_1 - \sigma_3)$. This acts along the plane inclined at 45° to the 1 and 3 directions, as shown in Fig. 3.2b. Equating $\tau_{1\text{-}3}$ to the shear stress $k = Y/2$, for the uniaxial case, leads to the Tresca yield criterion

$$\sigma_1 - \sigma_3 = Y \tag{3.1a}$$

Numerical values of σ_1 and σ_3 must be substituted into the left-hand side of eq(3.1a) with signs denoting tension or compression, e.g. if $\sigma_1 = 30$, $\sigma_2 = -15$ and $\sigma_3 = -20$ MPa, then the left side value is 30 - (- 20) = 50 MPa, showing that the intermediate stress value is irrelevant. Often, eq(3.1a) appears in the alternative descriptive form

Greatest principal stress - Least principal stress = Tensile yield stress (3.1b)

In the case of plane stress, where σ_x, σ_y, τ_{xy} are non-zero, the principal stresses have been derived in Section 1.4.3 (p.20):

$$\sigma_1, \sigma_2 = ½\,(\sigma_x + \sigma_y) \pm ½\,\surd[(\sigma_x - \sigma_y)^2 + 4\tau_{xy}^{\;2}] \text{ and } \sigma_3 = 0 \tag{3.2}$$

where σ_1 is tensile (greatest positive value) and σ_2 is compressive (least negative value). Setting $\sigma_y = 0$ in eq(3.2) and substituting into eq(3.1b) gives a simplified Tresca criterion

$$\sigma_x^2 + 4\tau_{xy}^{\;2} = Y^2 \tag{3.3}$$

which defines the equation of a two-dimensional elliptical yield locus in axes of σ_x and τ_{xy}.

3.2.2 Shear Strain Energy Theory

This is the most commonly used yield criterion. It is named after Richard von Mises but the crterion is also associated with others: Maxwell, Huber and Hencky. Huber proposed that the total strain energy was composed of a dilatational (volumetric) component and a distortional (shear) component. The former depends upon the mean, or hydrostatic, component of the applied stress and the latter upon the remaining reduced, or deviatoric, component of stress, as shown in Figs 3.3a-c.

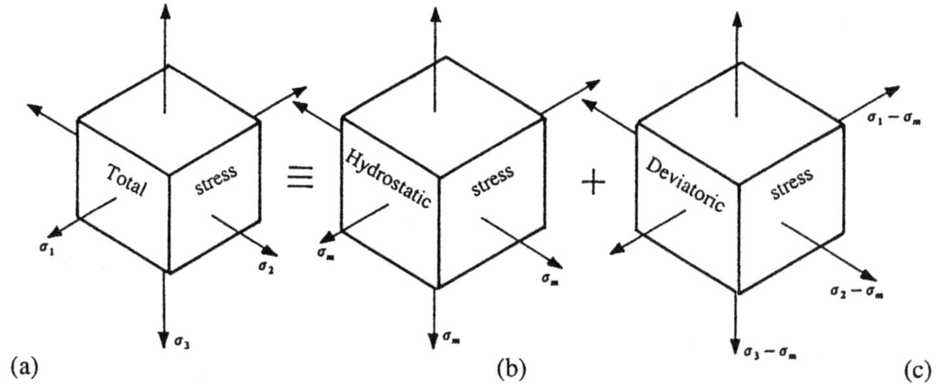

Figure 3.3 Components of a triaxial stress state

Maxwell believed that hydrostatic stress played no part upon yielding. He proposed that yielding occurred when the shear strain energy component of the total energy reached a critical value. This value is taken to be the shear strain energy at the tensile yield point. Alternative approaches, adopted by von Mises and Hencky, which lead to the same yield criterion, are outlined below. It will be shown in the following section, that there is considerable experimental evidence supporting a von Mises criterion for ductile, initially isotropic, metallic materials.

(a) *Shear Strain Energy* U_s

The total energy density for Fig. 3.3a is given by

$$U = \int_{\varepsilon} \sigma \, d\varepsilon = \int_{\varepsilon} (\sigma_1 \, d\varepsilon_1 + \sigma_2 \, d\varepsilon_2 + \sigma_3 \, d\varepsilon_3)$$

Substituting from the elastic constitutive relations:

$$\varepsilon_1 = (1/E)[\sigma_1 - \nu(\sigma_2 + \sigma_3)]$$
$$\varepsilon_2 = (1/E)[\sigma_2 - \nu(\sigma_1 + \sigma_3)]$$
$$\varepsilon_3 = (1/E)[\sigma_3 - \nu(\sigma_1 + \sigma_2)]$$

where E is Young's modulus and ν is Poisson's ratio. This leads to

$$U = (\sigma_1^2 + \sigma_2^2 + \sigma_3^2)/(2E) - (\nu/E)(\sigma_1 \sigma_2 + \sigma_2 \sigma_3 + \sigma_1 \sigma_3) \tag{3.4}$$

The volumetric strain energy density arises from the hydrostatic component of stress (see Fig. 3.3b):

$$U_v = \int_{\varepsilon} \sigma \, d\varepsilon = \int_{\varepsilon} \sigma_m \, d\varepsilon + \sigma_m \, d\varepsilon + \sigma_m \, d\varepsilon$$

Substituting $\varepsilon = \sigma_m/(3K)$ for the linear strain along an edge, where K is the bulk modulus:

$$U_v = \sigma_m^2/(2K) = 3(1 - 2\nu)\sigma_m^2/(2E)$$
$$= (1 - 2\nu)(\sigma_1 + \sigma_2 + \sigma_3)^2/(6E) \tag{3.5}$$

where $\sigma_m = \sigma_{kk}/3$. Here a relationship between the three elastic constants has been used: $E = 3K(1 - 2\nu)$. Subtracting eq(3.5) from eq(3.4) leads to the shear strain energy associated with the deviatoric stress, in Fig. 3.3c:

$$U_S = U - U_v$$
$$= (\sigma_1^2 + \sigma_2^2 + \sigma_3^2)/(2E) - \nu(\sigma_1 \sigma_2 + \sigma_2 \sigma_3 + \sigma_1 \sigma_3)/E - (1 - 2\nu)(\sigma_1 + \sigma_2 + \sigma_3)^2/(6E)$$
$$= (1 + \nu)[(\sigma_1 - \sigma_2)^2 + (\sigma_2 - \sigma_3)^2 + (\sigma_1 - \sigma_3)^2]/(6E) \tag{3.6a}$$

The value of U_S at the tensile yield point is found from setting $\sigma_1 = Y$, $\sigma_2 = \sigma_3 = 0$ in eq(3.6a):

$$U_S = (1 + \nu)Y^2/(3E) \tag{3.6b}$$

Equating (3.6a) and (3.6b) provides the principal stress form of the von Mises criterion:

$$(\sigma_1 - \sigma_2)^2 + (\sigma_2 - \sigma_3)^2 + (\sigma_1 - \sigma_3)^2 = 2Y^2 \tag{3.7a}$$

If one principal stress, say σ_3, is zero the biaxial stress form of eq(3.7a) becomes

$$\sigma_1^2 - \sigma_1\sigma_2 + \sigma_2^2 = Y^2 \tag{3.7b}$$

Substitutions from eqs(3.2) for σ_1 (tensile) and σ_2 (compressive) will provide a general plane stress form in terms of σ_x, σ_y and τ_{xy}. Setting $\sigma_y = 0$ within this reveals the more common plane stress form of the Mises criterion:

$$\sigma_x^2 + 3\tau_{xy}^2 = Y^2 \tag{3.8}$$

Subscripts x and y are often omitted from eq(3.8) to account for yielding under a given combination of direct and shear stress, set within either Cartesian or polar co-ordinates.

(b) *Octahedral Shear Stress* τ_o
This approach proposes that yielding, under the triaxial stress system in Fig. 3.1b, commences when τ_o in eq(1.34b) reaches its critical value at the tensile yield point, for which $\sigma_1 = Y$ and $\sigma_2 = \sigma_3 = 0$. This gives

$$\frac{1}{3}\sqrt{[(\sigma_1 - \sigma_2)^2 + (\sigma_2 - \sigma_3)^2 + (\sigma_1 - \sigma_3)^2]} = \frac{1}{3}\sqrt{(2Y^2)}$$

which again leads to eq(3.7a). This is because the normal stress for the octahedral planes, i.e. σ_o in eq(1.34a), is numerically equal to the mean stress $\sigma_m = ⅓\sigma_{kk}$, which produces dilatation only. Thus, the distortion which occurs under τ_o, leads to yielding under a critical combination of principal stress differences. Clearly, energy considerations are not required when formulating the yield criterion in this way.

(c) *Deviatoric Stress Invariants* J_2' and J_3'
Because the mean or hydrostatic stress plays no part in yielding, it follows that the yield criterion should be a function of the deviatoric, or reduced, stress in Fig. 3.3c. This shows that the deviatoric stress tensor σ_{ij}' is the remaining part of the absolute tensor σ_{ij} after the mean or hydrostatic stress σ_m has been subtracted. This reduction applies only to the normal stress components σ_{11}, σ_{22} and σ_{33}, since shear stresses σ_{12}, σ_{13} and σ_{23} will cause no dilatation. The introduction of the unit matrix **I**, equivalent to Kronecker's delta δ_{ij}, ensures the correct reduction:

$$\sigma_{ij}' = \sigma_{ij} - \delta_{ij}\sigma_m \quad \text{or} \quad \mathbf{T}' = \mathbf{T} - \frac{1}{3}\mathbf{I}\,\mathrm{tr}\,\mathbf{T} \tag{3.9a,b}$$

where δ_{ij} is unity for $i = j$ and zero for $i \neq j$. For example, with $i = 1$ and $j = 1, 2$ and 3, eq(3.9a) gives

$$\sigma_{11}' = \sigma_{11} - \sigma_m = \sigma_{11} - ⅓(\sigma_{11} + \sigma_{22} + \sigma_{33})$$
$$\sigma_{12}' = \sigma_{12} \text{ (no change)}$$

With further reductions, it follows that the resulting deviatoric stress matrix **T** ', in eq(3.9b), is composed of σ_{11}', σ_{22}' and σ_{33}' and the original shear stresses σ_{12}, σ_{13} and σ_{23}:

$$\begin{bmatrix} \sigma_{11}' & \sigma_{12} & \sigma_{13} \\ \sigma_{21} & \sigma_{22}' & \sigma_{23} \\ \sigma_{31} & \sigma_{32} & \sigma_{33}' \end{bmatrix} = \begin{bmatrix} \sigma_{11} & \sigma_{12} & \sigma_{13} \\ \sigma_{21} & \sigma_{22} & \sigma_{23} \\ \sigma_{31} & \sigma_{32} & \sigma_{33} \end{bmatrix} - \frac{\sigma_{11} + \sigma_{22} + \sigma_{33}}{3} \begin{bmatrix} 1 & 0 & 0 \\ 0 & 1 & 0 \\ 0 & 0 & 1 \end{bmatrix} \tag{3.9c}$$

When σ_{ij}' is expressed in the form $f(\sigma_{ij}')$ = constant, this function is known as a yield criterion and it predicts the onset of plasticity under a critical deviatoric stress state. Moreover, as yielding is a property of the material itself f must be independent of the co-ordinates used to define σ_{ij}'. Yielding must, therefore, be a function of the deviatoric stress invariants J_1', J_2' and J_3' since they will remain unaltered by co-ordinate transformation. The deviatoric invariants may be obtained by subtracting σ_m from the absolute stress invariants, J_1, J_2 and J_3, in eq(1.25a-c). In terms of principal stresses the deviatoric invariants become

$$\begin{aligned} J_1' &= \sigma_1' + \sigma_2' + \sigma_3' = (\sigma_1 - \sigma_m) + (\sigma_2 - \sigma_m) + (\sigma_3 - \sigma_m) \\ &= (\sigma_1 + \sigma_2 + \sigma_3) - 3\sigma_m = 0 \end{aligned} \tag{3.10a}$$

$$\begin{aligned} -J_2' &= \sigma_1'\sigma_2' + \sigma_2'\sigma_3' + \sigma_1'\sigma_3' = (\sigma_1 - \sigma_m)(\sigma_2 - \sigma_m) + (\sigma_2 - \sigma_m)(\sigma_3 - \sigma_m) + (\sigma_1 - \sigma_m)(\sigma_3 - \sigma_m) \\ &= (\sigma_1\sigma_2 + \sigma_2\sigma_3 + \sigma_1\sigma_3) - 3\sigma_m^2 \\ &= ⅓\,[(\sigma_1\sigma_2 + \sigma_2\sigma_3 + \sigma_1\sigma_3) - (\sigma_1^2 + \sigma_2^2 + \sigma_3^2)] \\ &= -1/6\,[(\sigma_1 - \sigma_2)^2 + (\sigma_2 - \sigma_3)^2 + (\sigma_1 - \sigma_3)^2] \end{aligned} \tag{3.10b}$$

$$\begin{aligned} J_3' &= \sigma_1'\,\sigma_2'\,\sigma_3' = (\sigma_1 - \sigma_m)(\sigma_2 - \sigma_m)(\sigma_3 - \sigma_m) \\ &= \sigma_1\sigma_2\sigma_3 - \sigma_m(\sigma_1\sigma_2 + \sigma_2\sigma_3 + \sigma_1\sigma_3) + 2\sigma_m^3 \\ &= \sigma_1\sigma_2\sigma_3 - ⅓(\sigma_1 + \sigma_2 + \sigma_3)(\sigma_1\sigma_2 + \sigma_2\sigma_3 + \sigma_1\sigma_3) + 2/27(\sigma_1 + \sigma_2 + \sigma_3)^3 \\ &= (⅓)^4\,[(2\sigma_1 - \sigma_2 - \sigma_3)^3 + (2\sigma_2 - \sigma_1 - \sigma_3)^3 + (2\sigma_3 - \sigma_1 - \sigma_2)^3] \end{aligned} \tag{3.10c}$$

The sign of the J_2' expression (3.10b) is positive within its corresponding deviatoric stress cubic as follows:

$$(\sigma')^3 - J_2'(\sigma') - J_3' = 0$$

Note, for general stress deviator σ_{ij}', the invariants in eqs(3.10a,b,c) appear in the tensor and matrix notations as

$$J_1' = \sigma_{ii}' = \mathrm{tr}\,\mathbf{T}' = 0 \tag{3.11a}$$

$$J_2' = ½\sigma_{ij}'\,\sigma_{ji}' = ½\,\mathrm{tr}\,(\mathbf{T}')^2 \tag{3.11b}$$

$$J_3' = ⅓\,\sigma_{ij}'\,\sigma_{jk}'\,\sigma_{ki}' = ⅓\,\mathrm{tr}\,(\mathbf{T}')^3 \tag{3.11c}$$

Yielding begins when a function of the two, non-zero, deviatoric invariants J_2' and J_3', attains a critical, constant value, C :

$$f(J_2', J_3') = C \tag{3.12}$$

C is normally defined from reducing eq(3.12) to yielding in simple tension or torsion. For example, when J_3' is omitted from the function in eq(3.12), f is simply equated to J_2' to give the von Mises yield function:

$$J_2' = k^2 \tag{3.13a}$$

where $k^2 = Y^2/3$, is found from setting $\sigma_1 = Y$ (the tensile yield stress) with $\sigma_2 = \sigma_3 = 0$ in eq(3.10b). It follows from eqs(3.10b and 3.13a) that

$$[(\sigma_1 - \sigma_2)^2 + (\sigma_2 - \sigma_3)^2 + (\sigma_1 - \sigma_3)^2]/\,6 = Y^2/3 \tag{3.13b}$$

which again gives the yield criterion in eq(3.8). It is this derivation, in which the second invariant of deviatoric stress attains a critical value at the yield point, which has become the yield criterion normally associated with von Mises.

3.3 Experimental Verification

3.3.1 Determination of the Initial Yield Point

Where a material displays a sharp yield point, as shown in Fig. 3.4a, the division between the elastic and elastic-plastic regions is clearly defined and a yield stress is easily found.

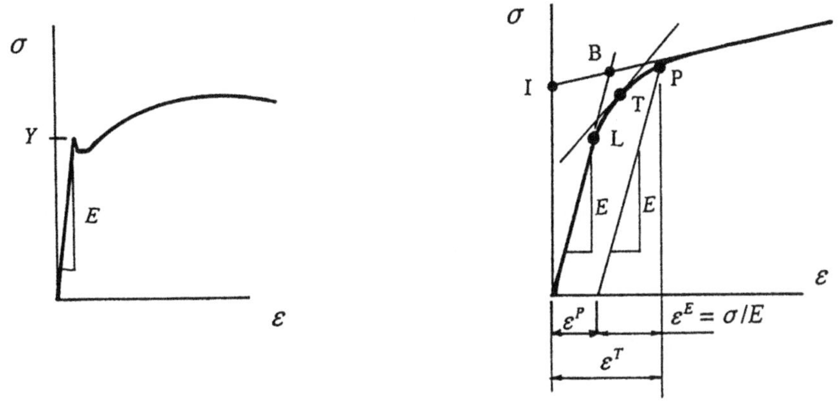

(a) (b)

Figure 3.4 Definitions of yielding

Key: B-Back extrapolation, I-Intercept stress, L-Limit of Proportionality, P-Proof stress, T-Tangent stress

There may be some doubt as to what the inital yield stress value is for the many metallic materials that display a gradual transition between their elastic and elastic-plastic regions. Various definitions of yield have been employed to overcome such uncertainty in the determination of yield stress. The most commonly used definition is the *proof stress*, i.e. the stress at P in Fig. 3.4b. The proof stress corresponds to a plastic strain ε^P, offset by a given small amount, usually taken from the range: 0.001% - 0.01%. Under combined stresses, a similar equivalent plastic strain value is chosen to determine the amount that the component strains are to be offset. This method is often employed to determine the combined yield stresses necessary to construct a yield locus. One difficulty arising with this is that the usual Mises form of equivalent plastic strain expression,

$$\bar{\varepsilon}^P = \sqrt{\frac{2}{3}\left(\varepsilon_{ij}^P \varepsilon_{ij}^P\right)}$$

pre-supposes a Mises yield surface. If the yield points, so determined, do not lie on a Mises locus the implication is that another $\bar{\varepsilon}^P$ definition is required, corresponding to a locus that would contain the yield points. Unfortunately, we have no way of knowing in advance what the true yield function can be. All that can be said is that the von Mises offset strain method checks the validity of the Mises function. Many investigations on the shape of the yield surface, reviewed in [4], have attempted to avoid this problem by employing low $\bar{\varepsilon}^P$ Mises strain values. This ensures that yield stresses, found from offsetting strains by other definitions of $\bar{\varepsilon}^P$, would all lie within a narrow range, just beyond the limit of elastic proportionality L, in Fig. 3.4b.

An alternative, *back-extrapolation* yield point, B in Fig. 3.4b, is much simpler to locate but its application requires large excursions into the plastic range. Since B is the junction in the bi-linear approximation to a stress-strain curve, it estimates of the yield stress at large plastic strain. The extrapolation is suited to materials, e.g. alloy steels, that harden in an approximately linear manner. It is less applicable to annealed materials with well-rounded plastic regions, e.g. copper and aluminium. The figure shows that the proof stress P, found from increasing the offset strain, will approach the back-extrapolation point B. Here the determination of both P and B involves considerable plastic strain and this requires a new testpiece for every stress probe. With repeated probing upon a single testpiece, the plastic strains accumulate to confuse the initial reference condition for the material.

Of all the yield definitions available, that which derives a yield stress from the *limit of proportionality* L, is the only one with physical significance. It divides the regions of elastic lattice distortion from inelastic slip. It is because the stress value at the limit of proportionality is sensitive to individual judgement that alternative definitions have been employed. Michno and Findley [5] have discussed instances in Fig. 3.4b, where the stress intercept I and a tangent point T, of predetermined slope, have defined yield. More recently, point L has been found consistently from the stepped change in temperature that occurs with the transition from elastic to plastic deformation [6].

3.3.2 Comparisons Between Tresca and von Mises

Experimental determination of yield loci are usually conducted within the simplest two-dimensional stress states. These allow comparisons to be made between experimental loci and Tresca, von Mises and other predictions. An ideal comparison should employ a metallic material with a well-defined yield point, such as low carbon steel, as in Fig. 3.4a. It is also instructive to examine the influence of yield point definition for other materials with less distinct yield points. For these, initial yield stresses can be determined at the limit of proportionality, by the proof strain and extrapolation methods. The comments made previously upon this issue are pertinent to the following comparisons.

(a) *The σ_1 versus σ_2 plane*
A principal, biaxial stress state is achieved in the wall of a thin-walled tube when it is subjected to combined internal pressure and axial load. The radial stress σ_3 may be ignored provided the diameter to thickness ratio of the tube exceeds 15. Yield loci have either been determined from proportional or step-wise loading paths, these producing tensile stresses lying within the first (positive) quadrant. In the remaining quadrants of the yield locus, compressive buckling can precede yielding, particularly with yield definitions involving larger amounts of plastic strain.

Table 3.1 summarises the test conditions for published data [7-12] used in the construction of Fig. 3.5. The materials investigated had received a processing method, involving heat treatment, to attain an initially isotropic condition. There was an obvious yield point (y.p.) in En 24 steel [7]. In the remaining investigations [8-12], the yield stress was identified with the limit of proportionality (l.p.) or with the proof stress at the indicated offset strain value ($\overline{\varepsilon}^P$). All stress paths were radial, with one exception [9], in which it was shown that it was possible to determine the full yield locus in alloy steels from applying a sequence of non-radial stress probes to a single testpiece. The choice of yield stress at the l.p. minimised the accumulation of plastic strain arising from the repeated probing. To avoid strain history effects from radial probing well into the plastic range, a new testpiece should be employed for each stress ratios (σ_2/σ_1). This also applies when defining a yield point at a large offset strain or by back extrapolation.

Table 3.1 Experimental yield point investigations in σ_1, σ_2 space

Material	Heat Treatment	Yield Definition	Symbol	Reference
En 24 carbon steel	annealed	y.p.	×	7
SAE 1045 carbon steel	hot rolled	$\bar{\varepsilon}^P = 0.7\%$	■	8
2¼% Cr, 1% Mo steel	stress relieved	l.p.	○	9
X-60 alloy steel	normalised	l.p.	▽	9
304 stainless steel	stress relieved	l.p.	Δ	9
306 stainless steel	solution treated	l.p.	□	9
M-63 brass	annealed	l.p.	◥	10
14S-T4 Al alloy	hot rolled	$\bar{\varepsilon}^P = 0.2\%$	⬓	11
Ni-Cr-Mo steel	annealed	$\bar{\varepsilon}^P = 0.2\%$	●	12

The theoretical loci, shown in Fig. 3.5, are plotted with their axes normalised by the tensile yield stress Y.

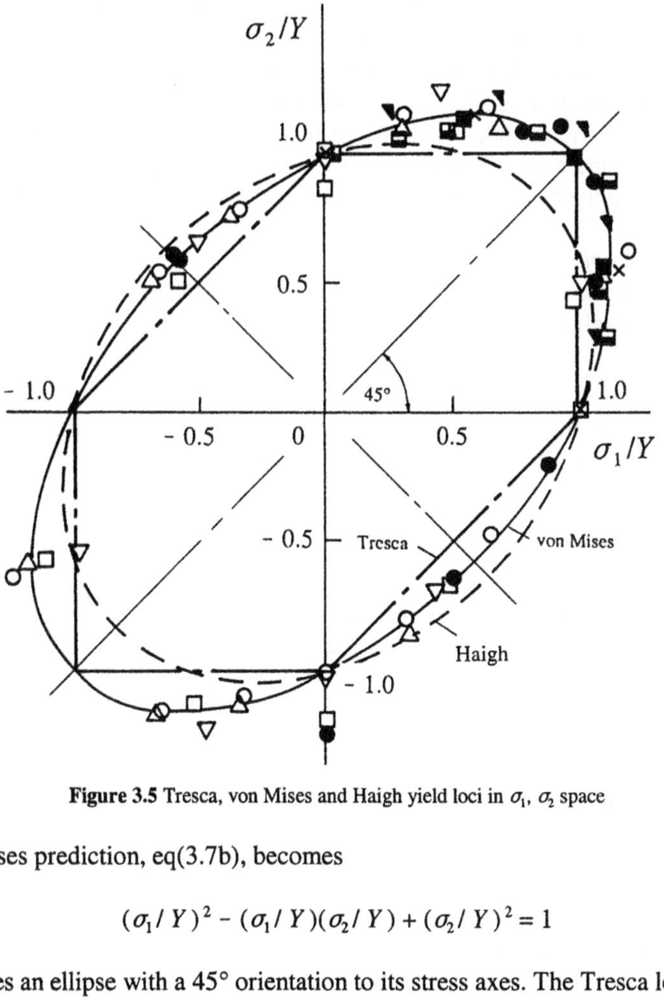

Figure 3.5 Tresca, von Mises and Haigh yield loci in σ_1, σ_2 space

The von Mises prediction, eq(3.7b), becomes

$$(\sigma_1/Y)^2 - (\sigma_1/Y)(\sigma_2/Y) + (\sigma_2/Y)^2 = 1$$

which defines an ellipse with a 45° orientation to its stress axes. The Tresca locus is found from applying eq(3.1b) to each quadrant. For example, in quadrant 1, where $0 \le \sigma_1/Y < 1$, the greatest principal stress is $\sigma_2/Y = 1$ (constant) and the least is $\sigma_3/Y = 0$. Also, in this quadrant, where $0 \le \sigma_2/Y < 1$, the greatest principal stress is $\sigma_1/Y = 1$ and the least value is $\sigma_3/Y = 0$. The respective yield criteria, σ_2/Y and $\sigma_1/Y = 1$, thus describe the horizontal and vertical sides of the Tresca hexagon as shown. In quadrant 2, σ_2/Y is greatest in tension while σ_1/Y is least in compression. Hence, the left sloping side of the hexagon conforms to $\sigma_2/Y - \sigma_1/Y = 1$. The completed Tresca hexagon is inscribed within the Mises ellipse. The Haigh total energy yield criterion, as derived from eq(3.4), is also shown in this figure. The yield locus associated with this criterion has an elliptical equation involving Poisson's ratio:

$$(\sigma_1/Y)^2 - 2\nu(\sigma_1/Y)(\sigma_2/Y) + (\sigma_2/Y)^2 = 1$$

but this is not often used since it assumes an influence of hydrostatic stress upon yield. Overall, the superimposed experimental data in Fig. 3.5 lie closer to the Mises prediction and appear independent of the test conditions. The conclusion to be drawn from these is that an initially isotropic von Mises yield condition applies to ductile polycrystalline materials. Most data lies outside the hexagon, confirming that a Tresca prediction of yielding for these materials is conservative. It is shown later that where the yield points for an individual material lie between or outside the two loci they may be represented by the function in eq(3.12). The uncertainties in yield point determination may however cast doubt on the requirement for an initial yield function to fit the test data precisely. This author's view is that a von Mises function is adequate provided the material has been heat-treated to an approximately isotropic condition. An examination of plastic strain paths under proportional loading may provide a more definitive test of an appropriate yield function, given that normality between the plastic strain path and the yield locus is accepted.

(b) *The σ versus τ plane*

The respective Tresca and Mises yield criteria are given in eqs(3.3) and (3.8). The total energy theory (Haigh) follows from eq(3.4) as

$$(\sigma/Y)^2 + 2(1 + \nu)(\tau/Y)^2 = 1$$

Here we shall omit the subscripts x and y when applying yield criteria to this simple, combined stress state, i.e σ and τ. The experiments, detailed in Table 3.2, were all conducted on thin-walled tubes, subjected to torsion combined with either a circumferential tension [7] or, an axial tension/compression [13-19].

Table 3.2 Experimental yield point investigations in σ, τ space

Material	Heat Treatment	Yield Definition	Symbol	Reference
En 24 carbon steel	annealed	y.p.	×	7
En 25 carbon steel	annealed	y.p.	○	13
19S Al alloy	stress relieved	l.p.	▽	14
1100-0 Al	annealed	l.p.	■	15
PA6 Al	as-received	$\bar{\varepsilon}^P = 50\mu\varepsilon$	+	16
Brass	as-received	$\bar{\varepsilon}^P = 200\mu\varepsilon$	◣	17
Ti-50A Ti-alloy	stress relieved	$\bar{\varepsilon}^P = 20\mu\varepsilon$	●	18
Copper (99.8%)	annealed	b.e.	Δ	19
Aluminium (99.7%)	annealed	b.e.	□	19

These tension-torsion combinations include the original experiments of Taylor and Quinney [19] who, in 1931, pioneered the backward extrapolation (b.e.) technique. With the exception of the stepped-stress probes, employed by Ivey [14] and Phillips and Tang [15], all other investigations, reported in Table 3.2, employed radial loading. Ellyin and Grass [18] applied multiple, radial probes, each emanating from the stress origin, to a single testpiece. The final probe, which duplicated their initial probe, revealed little difference between their yield stresses. Consequently, the effect of accumulated plastic strain, arising from intermediate probing was avoided by employing a small, offset strain ($\overline{\varepsilon}^P = 20\mu\varepsilon$) to define yield.

The comparison between theory and experiment in Fig. 3.6 reveals again that most of the experimental data in quadrants 2 and 4 lie closer to the yield locus of Mises than to Tresca, irrespective of the chosen yield definition. There is an obvious difference between the von Mises and Tresca predicted shear yield stress. That is, the lengths of the semi-minor axis are $k = Y/\sqrt{3}$ and $Y/2$ respectively.

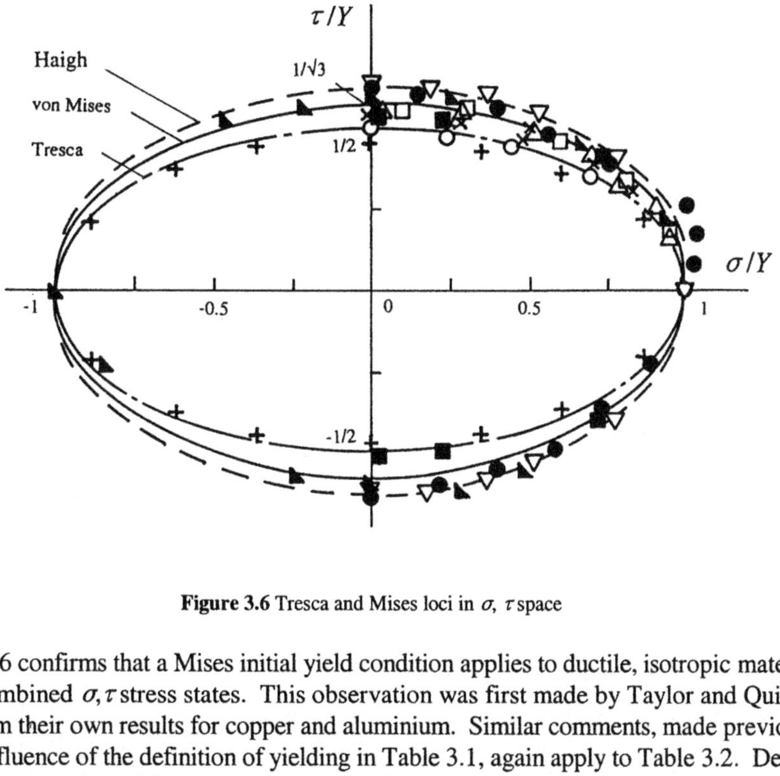

Figure 3.6 Tresca and Mises loci in σ, τ space

Figure 3.6 confirms that a Mises initial yield condition applies to ductile, isotropic materials under combined σ, τ stress states. This observation was first made by Taylor and Quinney [19], from their own results for copper and aluminium. Similar comments, made previously on the influence of the definition of yielding in Table 3.1, again apply to Table 3.2. Despite the uncertainty in yield point determination, it appears, from Fig. 3.6, that Tresca will continue to provide a conservative estimate whatever the chosen definition of yield.

Note that hot-rolled [8,11] and as-received material [16,17], given in Tables 3.1 and 3.2, did not receive any further heat treatment prior to test. It is possible that their departure from a Mises initial condition is due to the influence of initial anisotropy. The latter is evident when initial tensile yield stresses for orthogonal directions are dissimilar, but this behaviour was not reported in these investigations.

There are many practical instances where direct and shear stress states are combined. They arise from loading a shaft axially in tension or compression in combination with a transverse shear force or axial torsion. Combining a bending moment with a torque produces a similar stress state. A combination of direct stress and shear stress exists at points in the cross-section of a transversely loaded beam, away from its neutral axis. Equations (3.3) and (3.8) may also apply to yielding under these superimposed loadings.

3.3.3 *Influence of the Third Invariant*

(a) *Symmetrical Functions*

The fact that some initial yield points for isotropic material do not lie on the Mises loci, in Figs 3.5 and 3.6, suggests that initial yielding may conform to a more general function containing both deviatoric invariants. Firstly, note that Tresca's yield criterion may be expressed as the following complex function [20] of these invariants:

$$f = 4 J_2'^3 - 27 J_3'^2 - 36k^2 J_2'^2 + 96k^4 J_2' = 64\, k^6 \tag{3.14a}$$

where k, the shear yield stress, bears a simple relationship to the tensile yield stress Y, i.e. $k = Y/2$. Substituting eqs(3.10b,c) into (3.14a), and factorising gives

$$[(\sigma_1' - \sigma_2')^2 - 4k^2][(\sigma_2' - \sigma_3')^2 - 4k^2][(\sigma_1' - \sigma_3')^2 - 4k^2] = 0 \tag{3.14b}$$

The familiar Tresca criterion follows from the last of the following solutions to eq(3.14b):

$$(\sigma_1 - \sigma_2) = 2k, \quad (\sigma_2 - \sigma_3) = 2k \quad \text{and} \quad (\sigma_1 - \sigma_3) = 2k \tag{3.14c}$$

in which $(\sigma_1 - \sigma_3) = (\sigma_1' - \sigma_3')$. There are many other isotropic functions containing both invariants. Among these are the following homogenous stress functions for f in eq(3.12):

$$f = J_2'^3 - c\, J_3'^2 = k^6 \tag{3.15a}$$

$$f = J_2' - b\,(J_3'/J_2')^2 = k^2 \tag{3.15b}$$

$$f = (J_2')^{3/2} - d\, J_3' = k^3 \tag{3.15c}$$

where c, b, and d are material constants and k is the shear yield stress. Unlike eq(3.14a), it is now possible to select a value for the constants in eqs(3.15a-c) to fit the initial yield behaviour of most metallic materials [21]. Of these three yield functions, the most well-known is eq(3.15a), proposed by Drucker [22]. This is a homogenous function in stress of the sixth degree, for which c must lie in the range $-27/8 \le c \le 9/4$, to ensure convexity of the yield surface. These limits are shown in Fig. 3.7 where bounding loci, corresponding to the positive and negative limiting values $c_{crit} = 9/4$ and $-27/8$, ensure a closed surface [23]. It is seen that corners begin to appear at the negative limit for c_{crit}

Figure 3.7 also demonstrates how closely the data given for stainless steel [9] can be represented by taking $c = -2$ in eq(3.15a). The corresponding locus f_o represents the data better than would be found from either Mises or Tresca. The following chapter will show, by including J_3' in a homogenous yield function, that the associated stress-plastic strain relations become somewhat cumbersome. Consequently, the simplified relations associated with the Mises yield function $f = J_2'$ are often preferred.

The subsequent yield locus f, shown in Fig. 3.7, is almost a rigid translation of the initial yield locus f_o. Locus f applies to a plastic pre-stress P reached under a radial path having a constant stress ratio $R = \sigma_2/\sigma_1$. Comparing f to f_o reveals the manner in which the limits of

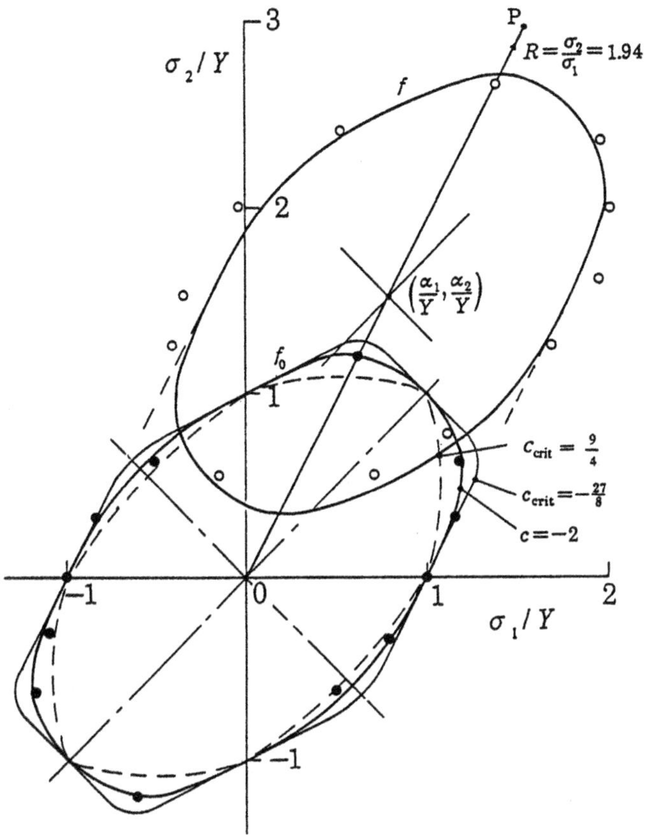

Figure 3.7 Drucker function applied to initial and subsequent yielding in stainless steel

subsequent elastic proportionality have been altered by work hardening. In fact, when the centre-co-ordinates (α_1/Y, α_2/Y) re-define the origin, eq(3.15a) may again be used as the subsequent function f. This is the principle of *kinematic hardening*, to which we shall return in Chapter 10.

(b) *Non-Symmetrical Functions*

Consider a linear combination of the two deviatoric invariants, J_2' and J_3', in the following form [21]:

$$f = J'_2{}^n + (p / \sigma_t)^{(2n-1)} J_3' \tag{3.16}$$

in which n and p are constants. The tensile yield stress σ_t is employed within the denominator of the second terrm in eq(3.16) to ensure homogeneity in stress. Functions of this form account for so-called *second order* effects arising from incompressible plasticity [24]. Among these effects are (i) non-linear plastic strain paths under radial loading [21], (ii) the accumulation of axial strain under pure torsion [25] and (iii) a difference between the initial tensile and compressive (σ_c) yield stresses. Explicit forms of eq(3.16) are found from substituting J_2' and J_3' from eq(3.10b and c). For example, taking $n = 1$ in eq(3.16), within a principal biaxial stress state (i.e. $\sigma_3 = 0$), gives an initial yield function

$$f = ⅓\,(\sigma_1^2 - \sigma_1\sigma_2 + \sigma_2^2) + (⅓)^4\,[(2\sigma_1 - \sigma_2)^3 + (2\sigma_2 - \sigma_1)^3 - (\sigma_1 + \sigma_2)^3] \tag{3.17}$$

The effect of (iii) above is revealed from respective substitutions $\sigma_1 = \sigma_t$ and $\sigma_1 = -\sigma_c$, each with $\sigma_2 = 0$. This gives

$$f = \sigma_t^2\,(1/3 + 2p/27) = \sigma_c^2/3 - 2p\sigma_c^3/(27\sigma_t) \tag{3.18a}$$

Employing the ratio $\rho = \sigma_c/\sigma_t$ in eq(3.18a) leads to a relationship between p and ρ:

$$(2p/9)\rho^3 - \rho^2 + (2p/9 + 1) = 0 \tag{3.18b}$$

We may normalise the function f with σ_t from equating (3.17) and (3.18a). This gives

$$(\sigma_1/\sigma_t)^2 - (\sigma_1/\sigma_t)(\sigma_2/\sigma_t) + (\sigma_2/\sigma_t)^2 + (p/27)[(2\sigma_1/\sigma_t - \sigma_2/\sigma_t)^3 + (2\sigma_2/\sigma_t - \sigma_1/\sigma_t)^3 - (\sigma_1/\sigma_t + \sigma_2/\sigma_t)^3] = 1 + 2p/9 \tag{3.19}$$

Figure 3.8 presents yield loci, from eq(3.19), with $\rho = ½, ⅔, 1, 2$ and 3. Equation (3.18b) shows that these correspond to $p = -3, -27/14, 0, 3/2$ and 9/7 respectively.

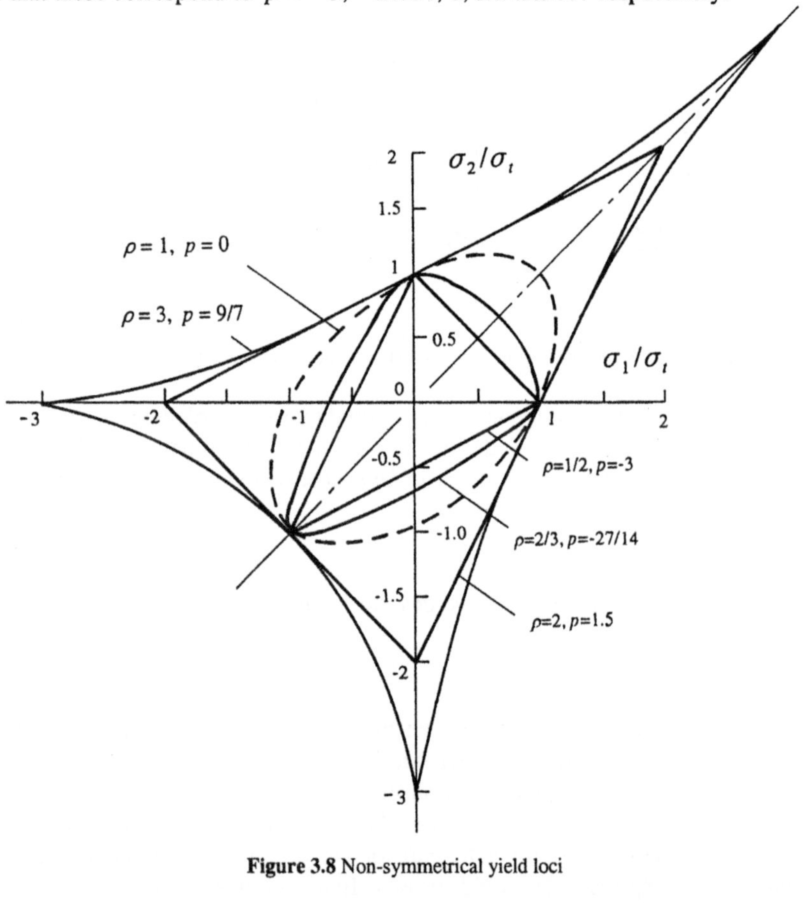

Figure 3.8 Non-symmetrical yield loci

When $p = 0$ ($\rho = 1$), the isotropic von Mises locus is recovered (broken line). For $\rho = ½$ and $\rho = 2$, the loci are bounded by three straight sides, which contrasts with Tresca's six-sided isotropic locus in Fig. 3.5. Convexity in these loci is ensured when p lies in the range $-3 < p < 3/2$. This means that eq(3.19) is restricted to where one uniaxial yield stress is not more than 50% of the other. For example, taking an intermediate value $p = -27/14$ at $\rho = 2/3$, we find σ_t exceeds σ_c by 33%. Because stress deviator invariants appear in eq(3.16), the function preserves plastic incompressibility while accounting for any second-order phenomena that may appear in quasi-isotropic, ductile polycrystals.

Uncertainty in yield point determination can cloud the influence that J_3' has upon the initial yield stress. It is also possible that initial anisotropy has a greater influence than J_3', in promoting deviations from the Mises and Tresca loci, shown in Figs 3.5 and 3.6. This is more likely for the non-heat treated materials in Table 3.2. It will, therefore, be appropriate to consider later (see p. 83), a number of anisotropic yield functions.

3.3.5 Superimposed Hydrostatic Stress

We have seen that when a yield function is formulated from the stress deviator invariants it assumes that initial yielding is unaffected by the magnitude of hydrostatic stress. This simplifying assumption has enabled the development of various isotropic yield functions for polycrystalline materials. The literature reveals experimental evidence, in support of this assumption. High magnitudes of hydrostatic pressure, up to 3 kbar, when superimposed upon torsion [26-28], tension [29, 30] and compression [31], have not altered significantly the yield stresses for mild steel, copper, aluminium and brass. To show this theoretically, we assume a yield function of the general isotropic form $f\,(J_2', J_3')$. With incremental changes to J_2' and J_3', the change in f becomes:

$$\mathrm{d}f = [(\partial f/\partial J_2')\,\mathrm{d}J_2' + (\partial f/\partial J_3')\,\mathrm{d}J_3'\,] \tag{3.20a}$$

Since f, J_2' and J_3' are functions of σ_{ij}', the total differential is

$$\mathrm{d}f/\,\mathrm{d}\sigma_{ij}' = [(\partial f/\partial J_2')(\partial J_2'/\partial \sigma_{ij}') + (\partial f/\partial J_3')(\partial J_3'/\partial \sigma_{ij}')] \tag{3.20b}$$

Under torsion, with $J_2' = \tau^2$ and $J_3' = 0$:

$$\mathrm{d}f/\,\mathrm{d}\tau = (\,\partial f/\,\partial J_2'\,) \times 2\tau$$

Under tension, the second invariant is

$$J_2' = \tfrac{1}{2}\,(\sigma'^{\,2}_{11} + \sigma'^{\,2}_{22} + \sigma'^{\,2}_{33})$$

so that: $\partial J_2'/\partial \sigma'_{11} = \sigma'_{11}$, $\partial J_2'/\partial \sigma'_{22} = \sigma'_{22}$ and $\partial J_2'/\partial \sigma'_{33} = \sigma'_{33}$. The third invariant is

$$J_3' = \tfrac{1}{3}\,(\sigma'^{\,3}_{11} + \sigma'^{\,3}_{22} + \sigma'^{\,3}_{33})$$

from which $\partial J_3'/\partial \sigma'_{11} = \sigma'^{\,2}_{11}$, $\partial J_3'/\partial \sigma'_{22} = \sigma'^{\,2}_{22}$ and $\partial J_3'/\partial \sigma'_{33} = \sigma'^{\,2}_{33}$. Substituting into eq(3.20b) gives:

$$\mathrm{d}f/\mathrm{d}\sigma'_{11} = [(\partial f/\partial J_2')\,\sigma'_{11} + (\partial f/\partial J_3')\,\sigma'^{\,2}_{11}] \tag{3.21a}$$
$$\mathrm{d}f/\mathrm{d}\sigma'_{22} = [(\partial f/\partial J_2')\,\sigma'_{22} + (\partial f/\partial J_3')\,\sigma'^{\,2}_{22}] \tag{3.21b}$$
$$\mathrm{d}f/\mathrm{d}\sigma'_{33} = [(\partial f/\partial J_2')\,\sigma'_{33} + (\partial f/\partial J_3')\,\sigma'^{\,2}_{33}] \tag{3.21c}$$

It is apparent from eqs(3.21a-c) that a change $\mathrm{d}f > 0$ depends upon deviatoric stress and the changes to the stress deviators. Thus, the flow stress remains unaffected by a superimposed mean, or hydrostatic, stress σ_m of any intensity. This can also be shown from the total differential, with respect to σ_{kk}:

$$df/d\sigma_{kk} = [(\partial f/\partial J_2')(\partial J_2'/\partial \sigma_{kk}) + (\partial f/\partial J_3')(\partial J_3'/\partial \sigma_{kk})] = 0$$

where, for a hydrostatic pressure $-p$, imposed upon σ_{ij}:

$$\sigma_{kk} = 3\sigma_m = [(\sigma_{11} - p) + (\sigma_{22} - p) + (\sigma_{33} - p)]$$

It follows that plastic flow is associated with the partial derivatives of f with respect to the deviatoric stress. However, since σ_{ii} has been removed from within the deviatoric invariant function f, plasticity will depend directly upon the partial derivitives $\partial f/\partial \sigma_{ij}$. Thus, incremental plastic strains appear within an *associated flow rule*:

$$d\varepsilon_{ij}^{P} = d\lambda\, \partial f/\partial \sigma_{ij} \tag{3.22}$$

where $d\lambda$ is a scalar multiplier, which derives from the unique, equivalent stress-strain relation for a given material (see Chapter 4). Note, however, that a superimposed hydrostatic pressure can increase the strain to fracture by inhibiting void formation prior to onset of fracture [32]. In the absence of a superimposed pressure, the mean component of the applied stress $\sigma_m = ⅓\,(\sigma_{11} + \sigma_{22} + \sigma_{33})$ is of comparatively moderate intensity and lies within the region where yield stress is unaffected by σ_m. Few experimental data are availabe that show an influence of mean stress upon yielding of anisotropic polycrystals. Consequently, given an appropriate yield function, the associated flow rule should ensure that mean stress σ_m does not influence anisotropic yielding and subsequent plastic flow remains incompressible.

Under certain conditions, an allowance for the influence of a superimposed hydrostatic pressure may be required. For example, the upper shear yield stress $k = 200$ MPa for mild steel was reduced by 5-6% under $p = 3$ kbar [26]. For pre-strained brass, with k lying in the range 110-190 MPa, variations to k between $-$ 6% and 3% were observed when superimposing pressure between 1 and 4 kbar [30]. To account for these, Hu [27] separated the influence of σ_m upon yielding by including $J_1 = \sigma_{kk}$ in the yield function $f(J_1, J_2', J_3')$. Functions of this kind are also required where the magnitude of σ_m is critical to the failure in brittle, non-metals and to plastic flow in porous compacts. The following section develops one such failure criterion for cast iron.

3.3.4 Influence of the First Invariant

Equation (3.16) does not allow the mean, or hydrostatic, stress to promote a difference between σ_t and σ_c. Here, a yield, or fracture, function should contain the mean stress. Conveniently, σ_m is proportional to the first invariant of absolute stress: $J_1 = \sigma_{ii} = 3\sigma_m$. Thus, for brittle materials, which are influenced by hydrostatic stress, J_1 is combined with J_2' to formulate the function. The latter will define a fracture surface in the absence of plasticity. One notable example is cast iron, for which a fracture surface follows from a simple linear sum of J_1' and J_2':

$$J_2' + \alpha J_1 = c \tag{3.23a}$$

Substituting from eqs(3.10a,b),

$$(1/6)[(\sigma_1 - \sigma_2)^2 + (\sigma_2 - \sigma_3)^2 + (\sigma_1 - \sigma_3)^2] + \alpha(\sigma_1 + \sigma_2 + \sigma_3) = c \tag{3.23b}$$

Constant c is found from substituting the uniaxial condition $(\sigma_1, 0, 0)$ in eq(3.23b):

$$1/6\,(2\sigma_1^{\,2}) + \alpha\,\sigma_1 = c \tag{3.24a}$$

Since $\sigma_1 = \sigma_t$ and $\sigma_1 = -\,\sigma_c$ are the roots of eq(3.24a), it follows that

$$(\sigma_1 - \sigma_t)(\sigma_1 + \sigma_c) = \sigma_1^{\,2} + 3\alpha\,\sigma_1 - 3\,c$$
$$\sigma_1^{\,2} + \sigma_1\,(\sigma_c - \sigma_t) - \sigma_c\,\sigma_t = \sigma_1^2 + 3\alpha\,\sigma_1 - 3\,c \tag{3.24b}$$

Coefficients in eq(3.24b) give $\alpha = ⅓\,(\sigma_c - \sigma_t)$ and $c = ⅓\,\sigma_c\,\sigma_t$. Substituting into eq(3.23b),

$$[(\sigma_1 - \sigma_2)^2 + (\sigma_2 - \sigma_3)^2 + (\sigma_1 - \sigma_3)^2\,] + 2\sigma_t(\sigma_c/\sigma_t - 1)(\sigma_1 + \sigma_2 + \sigma_3) = 2\sigma_c\,\sigma_t \tag{3.25a}$$

Putting $\rho = \sigma_c/\sigma_t$ in eq(3.25a) leads to the Stassi fracture criterion [33]:

$$[(\sigma_1 - \sigma_2)^2 + (\sigma_2 - \sigma_3)^2 + (\sigma_1 - \sigma_3)^2] + 2\sigma_t(\rho - 1)(\sigma_1 + \sigma_2 + \sigma_3) = 2\rho\,\sigma_t^2 \tag{3.25b}$$

Setting $\sigma_3 = 0$ in eq(3.25b), provides a normalised, biaxial criterion used to predict the Stassi fracture loci shown in Fig. 3.9.

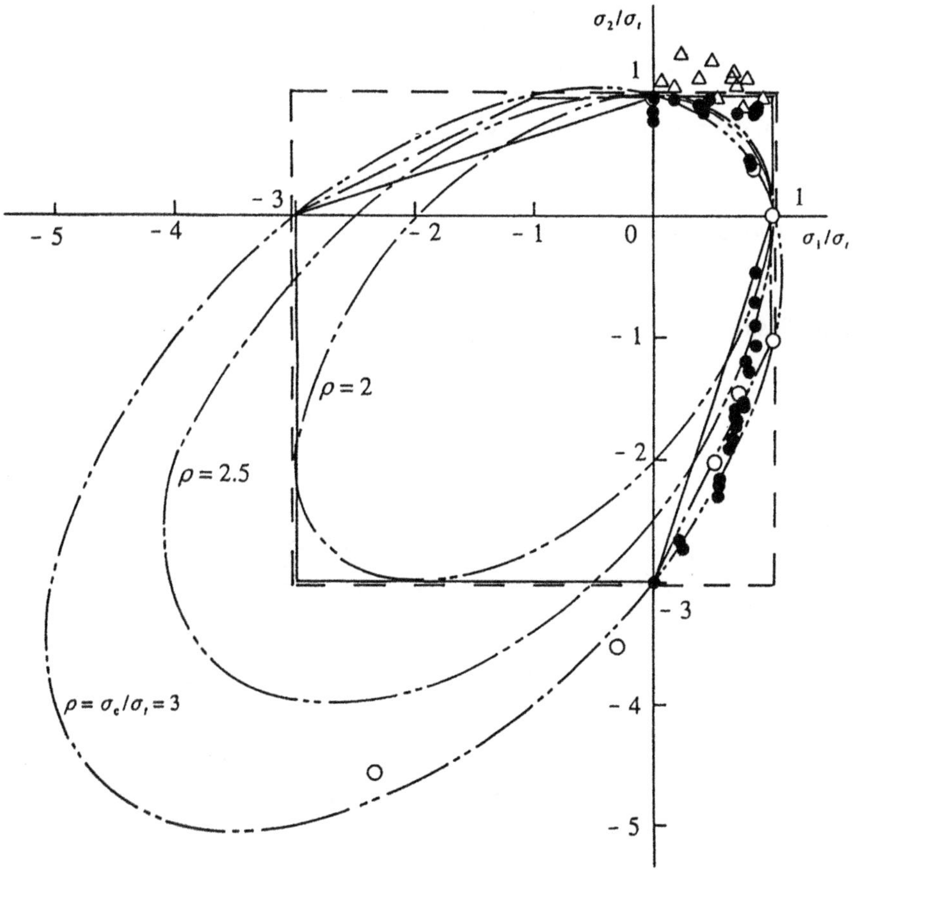

Figure 3.9 Brittle fracture criteria applied to cast iron

Key: ———— — — ———— Stassi; ———— Coloumb-Mohr; — — — Rankine; ———— — ———— Modified-Mohr

That is:

$$(\sigma_1/\sigma_t)^2 - (\sigma_1/\sigma_t)(\sigma_2/\sigma_t) + (\sigma_2/\sigma_t)^2 + (\rho - 1)(\sigma_1/\sigma_t + \sigma_2/\sigma_t) = \rho \tag{3.25c}$$

for which $\rho = \sigma_c/\sigma_t = 2$, 2.5 and 3 when principal stresses are normalised with σ_t. Stassi did not employ invariants for his derivation of eq(3.25b) but it is clear that hydrostatic stress accounts for the difference between tensile and compressive strengths in this material. In Fig. 3.9, a comparison is made between various strength criteria, including eq(3.25c), and experimental data for grey cast iron. A detailed description of the alternative predictions: Rankine, Coloumb, Mohr etc, has been given elsewhere [1]. It appears, from Fig. 3.9, that experiment is unable to discriminate between the various predictions since test data falls mainly within quadrants 1 and 4. This is largely due to the difficulty in acquiring reliable data under biaxial compression. Under biaxial tension, the low-stress combinations at failure are not biased to any one theory. However, few data lying within quadrant 3 in Fig. 3.9, suggest that Stassi's criterion will predict enhanced compressive strengths most reliably by using the appropriate ρ value.

When the derivation is extended to include shear stress, J_1 remains unchanged but additional terms in σ_{12}, σ_{23} and σ_{13} appear within J_2':

$$(\sigma_{11} - \sigma_{22})^2 + (\sigma_{22} - \sigma_{33})^2 + (\sigma_{11} - \sigma_{33})^2 + 6(\sigma_{12}^2 + \sigma_{13}^2 + \sigma_{23}^2) + 2\sigma_t(\rho - 1)(\sigma_{11} + \sigma_{22} + \sigma_{33}) = 2\rho\,\sigma_t^2 \tag{3.26a}$$

Equation (3.26a) defines a locus in a reduced space σ_{11}, σ_{12}, as

$$\sigma_{11}^2 + 3\sigma_{12}^2 + \sigma_t(\rho - 1)\sigma_{11} = \rho\,\sigma_t^2 \tag{3.26b}$$

for which the normalised form is

$$(\sigma_{11}/\sigma_t)^2 + 3(\sigma_{12}/\sigma_t)^2 + (\rho - 1)(\sigma_{11}/\sigma_t) = \rho \tag{3.26c}$$

Figure 3.10 illustrates the fracture loci from eq(3.26c) for $\rho = 1.25$, 2, 2.5 and 3.

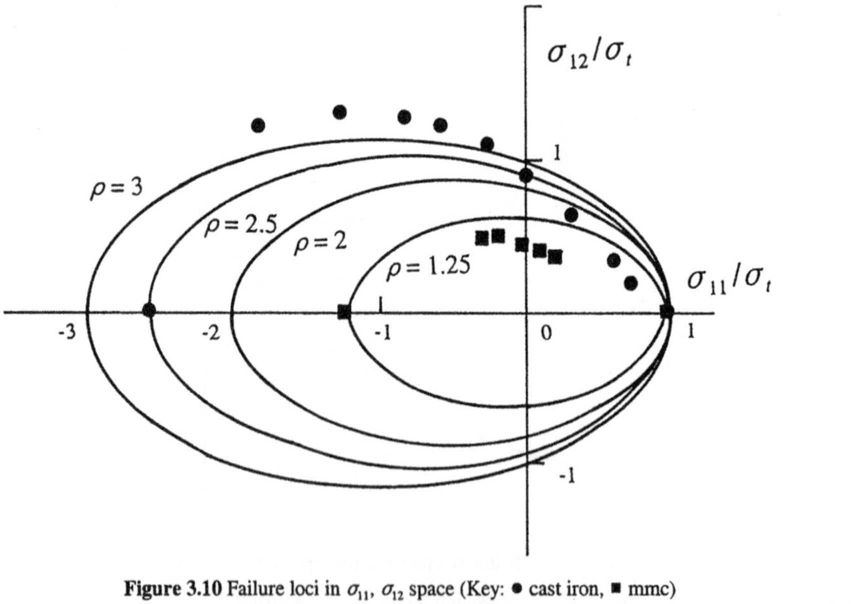

Figure 3.10 Failure loci in σ_{11}, σ_{12} space (Key: ● cast iron, ■ mmc)

A few of this author's previously unpublished data for cast iron (5% ferrite with 6-12 mm graphite flakes) and a metal matrix composite (17% volume of 3μ, Si-C particles in a 2124 Cu-Al metal matrix) are shown in Fig. 3.10. Dissimilar, uniaxial strengths in each material and the combined stress fracture values are predictable within the range of ρ shown.

We see that the major advantage of an invariant function lies in its reduction to any given stress state. When $\sigma_c = \sigma_t$, i.e. $\rho = 1$, the von Mises function is recovered and hydrostatic stress plays no part in failure.

3.4 Anisotropic Yielding in Polycrystalls

In metallic materials, plastic-anisotropy is associated with a directional variation in yield stress. None of the foregoing isotropic functions can display similar behaviour. Also, with the exception of eq(3.16), they show yield stresses of equal magnitude irrespective of the sense of stress. For example, from the symmetry of Figs 3.5 - 3.7, the magnitude of the uniaxial tensile and compressive yield stresses are equal and the positive and negative shear yield stresses are also equal. This assumption is reasonably consistent with the observed, initial yield behaviour of annealed, or stress-relieved, metals and alloys. In the absence of heat treatment, the orthotropic form of initial anisotropy in polycrystals is most common, arising from the processing method, e.g. rolling, drawing and extrusion. An orthotropic material has different initial yield stresses for axes lying parallel and normal to the direction in which it has been worked. Cold-rolled sheet and extruded bar will often continue to display distinctly different tensile stress-strain behaviour along these orthotropic axes, well into the plastic range. This is apparent from a divergence in these stress-strain curves when overlaid.

3.4.1 Initial Yielding

In his pivotal paper of 1948, Hill [34] generalised the von Mises isotropic yield criterion to account for anisotropic yielding in an orthotropic material. Hill's yield function describes a yield surface whose stress axes align with the three principal axes of orthotropy in the material. To establish the yield condition an applied stress σ_{ij} must, therefore, be resolved along the orthotropic axes. Writing the latter as *x*, *y* and *z*, the yield function is written as

$$2f(\sigma_{ij}) = F(\sigma_{yy} - \sigma_{zz})^2 + G(\sigma_{xx} - \sigma_{zz})^2 + H(\sigma_{xx} - \sigma_{yy})^2 + 2L\tau_{yz}^2 + 2M\tau_{xz}^2 + 2N\tau_{xy}^2 = 1 \quad (3.27)$$

where *F*, *G*, *H*, *L*, *M* and *N* are six coefficients that characterise an orthotropic symmetry in the yield stresses. In fact, eq(3.27) is a restricted form of a general quadratic yield function given later by Edelman and Drucker [35]:

$$f(\sigma_{ij}) = \tfrac{1}{2} C_{ijkl}\, \sigma_{ij}\, \sigma_{kl} \quad (3.28)$$

The fourth rank tensor C_{ijkl} contains 81 components, similar to that used for defining a strain energy function for anisotropic elasticity [36]. The number of independent components reduces to 21 with imposed symmetry conditions: (i) $\sigma_{ij} = \sigma_{ji}$ in the Cauchy stress tensor, giving $C_{ijkl} = C_{jikl} = C_{ijlk}$ and (ii) coincidence between the axes of stress and orthotropy, which gives $C_{ijkl} = C_{klij}$. With these reduced coefficients, the function can account for an influence of hydrostatic stress upon anisotropic yielding. When the condition of incompressibility is imposed upon eq(3.28), the number of independent coefficients is further reduced to 15. The corresponding relation is found from substituting eq(3.28) into the flow rule, eq(3.22). The incremental plastic strains become

$$\delta\varepsilon_{ij}^{P} = \delta\lambda\, \partial f(\sigma_{ij})/\partial\sigma_{ij} = \delta\lambda \times \tfrac{1}{2} C_{ijkl}\, \sigma_{kl}$$

Multiplying by δ_{ij} gives

$$\delta\varepsilon_{ij}^{P}\delta_{ij} = \delta\lambda \times \tfrac{1}{2} C_{ijkl}\, \sigma_{kl}\, \delta_{ij}$$

the left-hand side is $\delta\varepsilon_{ii}^{P} = 0$, when it follows that $C_{iikl} = 0$. This condition applies to normal stress components, for which $k = l$, so that

$$C_{ii11} = C_{ii22} = C_{ii33} = 0$$

Alternatively, the 15 independent components will appear directly from the expansion of a Mises generalised quadratic potential, written in stress deviator form:

$$f(\sigma_{ij}') = \tfrac{1}{2} H_{ijkl}\, \sigma_{ij}'\, \sigma_{kl}' = 1 \tag{3.29}$$

In general, the elastic-plastic deformation of an anisotropic solid is characterised by two, fourth-order tensors with 21 elastic components [36] and 15 plastic components. These tensors will ensure elastic compressibility and plastic incompressibility, in common with metals. With further reductions to the number of independent components H_{ijkl}, eq(3.29) reveals particular material symmetries. Among these are: orthotropy (six coefficients), transverse isotropy (3 coefficients) and a cubic form of anisotropy characterised by 2 coefficients [37]. The first of these corresponds to Hill's eq(3.27). The remaining two symmetries follow from eq(3.27) by putting: (i) $G = H$ with $M = N$ and (ii) $F = G = H$ with $L = M = N$ respectively. Wider application of the general forms of eqs(3.28) and (3.29) have been demonstrated in the absence of such symmetry [38, 39]. Further applications and developments to Hill's quadratic function have been made under particular conditions of plane stress and plane strain [40-42]. Alternative yield criteria [43-50] for anisotropic metals are summarised in Table 3.3 for various stress states (general, principal and plane).

Table 3.3 Yield criteria for various initially anisotropic conditions in metals

	Reference	Anisotropic Yield Function
(i)	[43]	$f = C_{11}\sigma_{11}^2 + C_{12}\sigma_{11}\sigma_{22} + C_{13}\sigma_{11}\sigma_{12} + C_{22}\sigma_{22}^2 + C_{23}\sigma_{22}\sigma_{12} + C_{33}\sigma_{12}^2 = 1$
(ii)	[44]	$f = 3/2[(\sigma_1/\sigma_{1y})^2 - (\sigma_1/\sigma_{1y})(\sigma_2/\sigma_{2y}) + (\sigma_2/\sigma_{2y})^2]^3$ $- \tfrac{1}{8}(\sigma_1/\sigma_{1y} + \sigma_2/\sigma_{2y})^2(\sigma_1/\sigma_{1y} - 2\sigma_2/\sigma_{2y})^2(\sigma_2/\sigma_{2y} - 2\sigma_1/\sigma_{1y})^2 = 1$
(iii)	[45]	$f = J_2'^2(J_2' + A_{ijkl}\sigma_{ij}'\sigma_{kl}') - CJ_3'^2 = k^6$
(iv)	[46]	$f = f(\sigma_2 - \sigma_3)^m + g(\sigma_1 - \sigma_3)^m + h(\sigma_1 - \sigma_2)^m$ $+ a(2\sigma_1 - \sigma_2 - \sigma_3)^m + b(2\sigma_2 - \sigma_3 - \sigma_1)^m + c(2\sigma_3 - \sigma_1 - \sigma_2)^m = \sigma^m$
(v)	[47]	$f = 3\sigma_x^3 - 6\sigma_x^2\sigma_y - 6\sigma_x\sigma_y^2 + 4\sigma_y^3 + (4\sigma_x + 21\sigma_y)\tau_{xy}^2$
(vi)	[48]	$f = \sum_{ijk} A_{ijk}\sigma_x^i\sigma_y^j\tau_{xy}^{2k}$, for $i + j + 2k \le 4$
(vii)	[49]	$f = a_1(\sigma_x - \sigma_y)^2 + a_2(\sigma_x - \sigma_z)^2 + a_3(\sigma_y - \sigma_z)^2 + a_4\tau_{yz}^2 + a_5\tau_{xy}^2 + a_6\tau_{xz}^2$ $+ a_7(\sigma_x - \sigma_y) + a_8(\sigma_x - \sigma_z) + a_9(\sigma_y - \sigma_z) + a_{10}\tau_{yz} + a_{11}\tau_{xy} + a_{12}\tau_{xz} = 1$
(viii)	[50]	$f = A\sigma_1^2 + B\sigma_2^2 + C\sigma_3^2 - D\sigma_1\sigma_2 - E\sigma_2\sigma_3 - F\sigma_1\sigma_3 + L\sigma_1 + M\sigma_2 + N\sigma_3 = 1$

In the plane, quadratic stress function (i) in Table 3.3, the cross-product terms in $\sigma\tau$ introduce two additional constants (Hill has a total of four). This form was used to predict the initial yield stresses in anisotropic sheet metal. The function (ii) is a normalised form of Drucker's eq(3.11a) for a principal, biaxial stress state. The stresses, σ_1 and σ_2, are divided by their respective yield stresses, σ_{1y} and σ_{2y}, which differ between the 1- and 2-directions. In eq(iii), Drucker's isotropic function is modified with an orthotropic quadratic term. This form accounted for deviations from a Mises isotropic condition due to the combined influences of J_3' and initial anisotropy. A non-quadratic function (iv) was proposed by Hill [46] for principal stresses aligned with a material's orthotropic axes. The constants σ^m, a, b, c, f, g and h are all positive. It can be seen how this function has modified the two isotropic, deviatoric invariants:

$$J_2' = (1/6)[(\sigma_1 - \sigma_2)^2 + (\sigma_2 - \sigma_3)^2 + (\sigma_1 - \sigma_3)^2]$$
$$J_3' = (1/3)^4[(2\sigma_1 - \sigma_2 - \sigma_3)^3 + (2\sigma_2 - \sigma_1 - \sigma_3)^3 + (2\sigma_3 - \sigma_1 - \sigma_2)^3]$$
$$= (1/3)^3(2\sigma_1 - \sigma_2 - \sigma_3)(2\sigma_2 - \sigma_1 - \sigma_3)(2\sigma_3 - \sigma_1 - \sigma_2)$$

It is also apparent from the first equation how Hill's orthotropic yield function (3.21) modifies J_2'. Using a non-integer exponent for eq(iv) in the range: $1 \le m \le 2$, extends its application beyond the quadratic form. In particular, by taking $a = b = 0$ and $f = g = 0$, this equation has been shown to predict the plastic strain ratios arising from applying uniaxial and equi-biaxial tension to rolled, polycrystalline sheets with planar-isotropy [51, 52]. Interestingly, in 1935, Bailey [53] predicted multiaxial, secondary creep rates observed in steam piping at 850°C. The anisotroic flow potential, which was given later by Davis [54], becomes similar to eq(iv) when a, b and c are set to zero. Note, that it is permissible to compare the creep potential to a yield function in this way when creep rates are time-dependent plastic strains. However, Hill [46] has shown that the terms containing the constants f, g, h and m, when taken alone, cannot admit all forms of anisotropy. For example, the four or more ears, which have appeared from cupping rolled sheet metal, are unpredictable. The quadratic function, eq(3.27), can provide for two and, at most, four ears in certain cases. An account of more ears requires a higher order polynomial yield function. Bourne and Hill [47] showed that with five anisotropy coefficients, their cubic stress function (v), matched the six ears observed in brass. A quartic function (vi), proposed by Gotoh [48], represents up to a maximum of eight ears observed in soft aluminium and its alloy. The explicit, plane stress, form of eq(vi) is

$$f = A_1\,\sigma_x^4 + A_2\,\sigma_x^3\,\sigma_y + A_3\,\sigma_x^2\,\sigma_y^2 + A_4\,\sigma_x\sigma_y^3 + A_5\,\sigma_y^4 + (A_6\,\sigma_x^2 + A_7\,\sigma_x\sigma_y + A_8\,\sigma_y^2)\,\tau_{xy}^2 + A_9\,\tau_{xy}^4$$

where eight coefficients $A_2 \ldots\, A_9$, together with $A_1 = 1$, provide a description of eight ears. A further term: $A_0\,(\sigma_x + \sigma_y)^2$, is added where an account of compressibility is required. These coefficients are found from the yield stresses under uniaxial tension and either through-thickness plane strain compression, or in-plane equi-biaxial tension. Further consideration of the yield criteria employed for forming rolled sheet metals is given in Chapter 11 (see Section 11.8, p. 365).

3.4.2 The Bauschinger Effect

The quadratic yield functions, in eqs(3.27) and (3.28), assume that the tensile and compressive yield stresses are equal for each principal direction in the material. This also applies to all other homogenous stress functions in Table 3.3. It should be noted that Hill had originally intended his 1948 function for backward-extrapolated yield points, which do show similar stresses in tension and compression. The unequal tensile and compressive yield stresses which appear within non-homogenous yield functions are consistent with other yield point definitions. We shall use the term *Bauschinger effect* here to denote this form of asymmetry in the initial yield surface, though the effect is more often associated with a subsequent yield surface (as with f in Fig. 3.7). A simple account of different positive and negative yield stress values appears with linear stress terms in the yield function. Two such functions are given in Table 3.3: eq(vii) in plane principal space and eq(viii) in general stress space. Stassy-D'Alia [49] examined both forms after placing restrictions on certain constants. We can arrive at a more general formulation of linear plus quadratic terms when yielding remains independent of a superimposed hydrostatic stress. Equation (3.29) is modified to

$$f(\sigma_{ij}') = H_{ij}\sigma_{ij}' + \tfrac{1}{2}H_{ijkl}\,\sigma_{ij}'\,\sigma_{kl}' = 1 \tag{3.30}$$

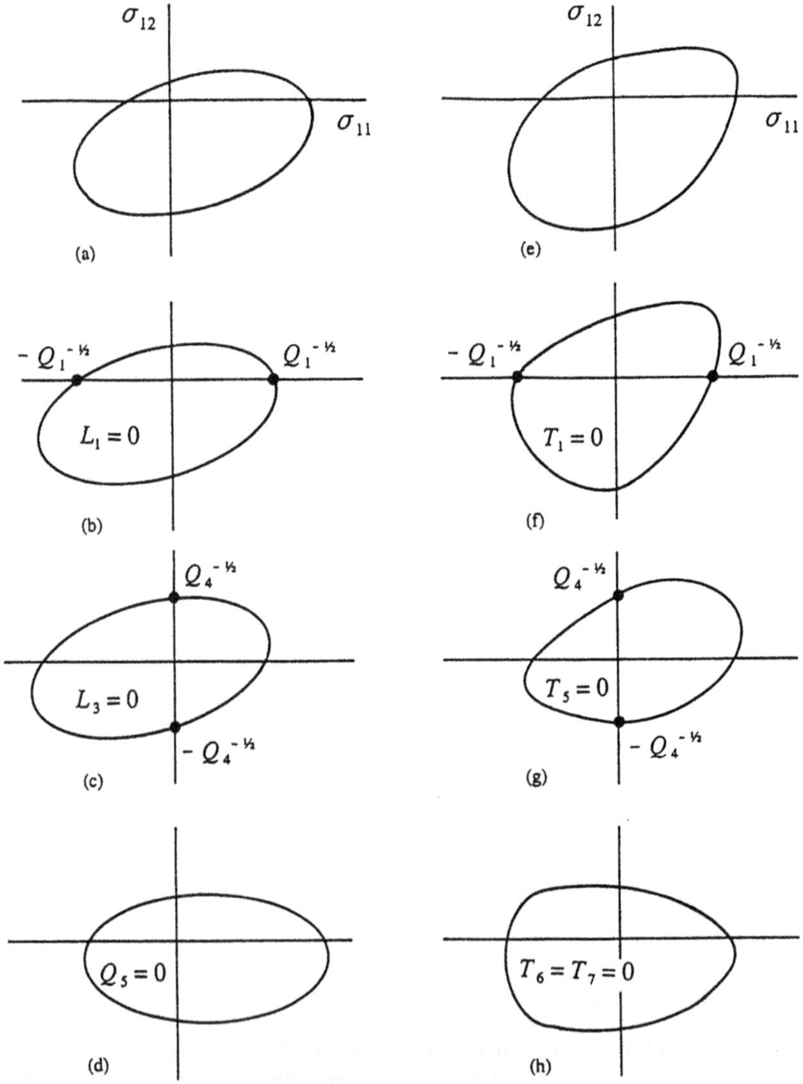

Figure 3.11 Yield loci in combined σ_{11}, σ_{12} space

When the 1-and 2-axes of orthotropy coincide with the axes of a plane stress state, σ_{11}, σ_{22} and σ_{12}, eq(3.30) expands into

$$f = L_1 \sigma_{11} + L_2 \sigma_{22} + L_3 \sigma_{12} + Q_1 \sigma_{11}^2 + Q_2 \sigma_{22}^2 + Q_3 \sigma_{11} \sigma_{22} + Q_4 \sigma_{12}^2 + Q_5 \sigma_{11} \sigma_{12} + Q_6 \sigma_{22} \sigma_{12} = 1 \tag{3.31a}$$

The coefficients L and Q are derived from H_{ij} and H_{ijkl} using plastic incompressibility [55]. In a reduced stress space, σ_{11} and σ_{12}, eq(3.30) becomes

$$f = L_1 \sigma_{11} + L_3 \sigma_{12} + Q_1 \sigma_{11}^2 + Q_4 \sigma_{12}^2 + Q_5 \sigma_{11} \sigma_{12} = 1 \tag{3.31b}$$

Figures 3.11a-d show four cases of anisotropy provided by eq(3.31b). The coefficients L and Q provide equal and unequal yield stress intercepts as shown. When tensile and compressive yield stresses differ, they become the roots to $Q_1 \sigma_{11}^2 + L_1 \sigma_{11} - 1 = 0$, in Figs 3.11a, c and d. When forward and reversed shear yield stresses differ, they become the roots to $Q_4 \sigma_{12}^2 + L_3 \sigma_{12} - 1 = 0$, in Fig. 3.11a, b and d. The product term $Q_5 \sigma_{11} \sigma_{12}$, in eq(3.31b), provides an inclination to the yield locus, as shown in Figs 3.11a, b and c. A simple check for the inclination is whether axial strain is produced from torsion. This requires an examination of the direction of the outward normal at a point of intersection between the yield locus and the the σ_{12}- axis. An inclined normal has a component of axial strain aligned with the σ_{11}- axis, in addition to a shear strain component, aligned with the σ_{12}- axis. This effect is due to anisotropy and is not the second-order effect, previously associated with J_3'. Where no axial strain is observed, Q_5 is set to zero, resulting in the locus of Fig. 3.11d.

When shear stress is absent along a material's orthotropic directions 1 and 2, eq(3.31a) shows that the principal, biaxial stresses σ_1 and σ_2, appear in dimensionless form:

$$f = L_1(\sigma_1/\sigma_{1t}) + L_2(\sigma_2/\sigma_{1t}) + Q_1(\sigma_1/\sigma_{1t})^2 + Q_2(\sigma_2/\sigma_{1t})^2 + Q_3(\sigma_1/\sigma_{1t})(\sigma_2/\sigma_{1t}) = 1 \tag{3.31c}$$

in which σ_{1t} is the tensile yield stress in the 1 - direction. Applications of eq(3.31c) have shown good agreement with initial yield loci for anisotropic sheets of Ti-Al alloy and a zircaloy [55]. An example of the former material [56] is shown in Fig. 3.12.

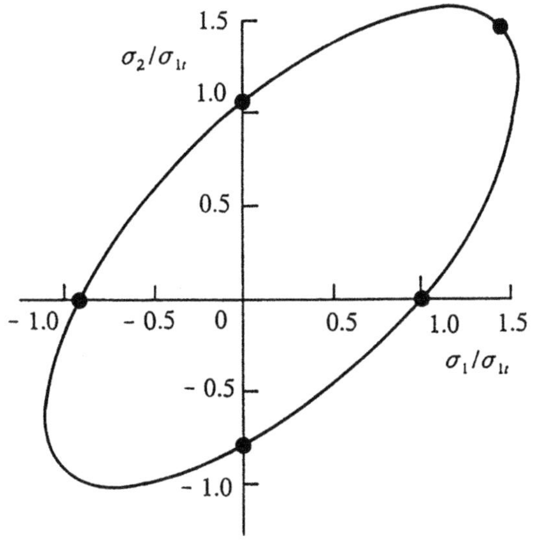

Figure 3.12 Initial yield locus for anisotropic Ti-Al alloy

It is apparent that, within this elliptical function, the five yield points may be fitted exactly with an equal number of coefficients in L and Q. Comparisons of this kind are best made at the limit of proportionality since an equivalent, anisotropic, offset strain cannot be defined. Moreover, if yield were to be so defined, anisotropy, initially present in the yield locus, would diminish with increasing offset strain.

To recover an initial Mises condition from eq(3.30), we identify H_{ij} and H_{ijkl} with the corresponding isotropic tensors:

$$H_{ij} = \delta_{ij} \tag{3.32}$$

and, for the general fourth-order isotropic tensor [57]:

$$H_{ijkl} = I_{ijkl} = \lambda \delta_{ij}\delta_{kl} + \mu \delta_{ik}\delta_{jl} + \nu\, \delta_{il}\delta_{kj} \tag{3.33a}$$

Substituting eqs(3.32) and (3.33a) into eq(3.30) gives

$$\begin{aligned} f(\sigma_{ij}') &= \delta_{ij}\sigma_{ij}' + ½\,(\lambda\delta_{ij}\delta_{kl} + \mu\,\delta_{ik}\delta_{jl} + \nu\,\delta_{il}\delta_{kj})\,\sigma_{ij}'\,\sigma_{kl}' \\ &= \sigma_{ii}' + ½\,(\lambda\sigma_{ii}'\,\sigma_{kk}' + \mu\,\sigma_{kj}'\,\sigma_{kj}' + \nu\,\sigma_{lj}'\,\sigma_{jl}') \\ &= ½\,(\mu\,\sigma_{kj}'\,\sigma_{kj}' + \nu\,\sigma_{ij}'\,\sigma_{ij}') \end{aligned}$$

where $\sigma_{ii}' = \sigma_{kk}' = 0$. Putting $\mu = \nu = ½$ leads to the von Mises form

$$f(\sigma_{ij}') = ½\,\sigma_{ij}'\,\sigma_{ij}'$$

It follows that, by setting $\lambda = 0$ and $\mu = \nu = ½$ in eq(3.33a), a reduced form of fourth order, isotropic tensor is suited to stress deviators:

$$I_{ijkl} = ½\,(\delta_{ik}\,\delta_{jl} + \delta_{il}\,\delta_{jk}) \tag{3.33b}$$

since this will lead directly to the von Mises condition. Note that eq(3.30) describes an ellipsoidal yield surface and will not account for any distortion that may initially be present.

3.4.3 Distorted Yield Loci

The experiments described in Tables 3.3 and 3.4 have often revealed distorted yield loci. Distortion, which appears to be a dominant feature in certain alloys, can be modelled when cubic stress terms appear in the yield function. Consider, for example, an anisotropic yield function formed from the sum of quadratic and cubic stress deviator terms [58]:

$$f(\sigma_{ij}') = ½\,H_{ijkl}\sigma_{ij}'\,\sigma_{kl}' + ⅓\,H_{ijklmn}\,\sigma_{ij}'\,\sigma_{kl}'\,\sigma_{mn}' = 1 \tag{3.34}$$

Restrictions upon the number of sixth-order tensor components H_{ijklmn} follow from symmetry in the stress tensor $\sigma_{ij}' = \sigma_{ji}'$, $\sigma_{kj}' = \sigma_{jk}'$ and $\sigma_{mn}' = \sigma_{nm}'$. This gives $H_{ijklmn} = H_{jiklmn} = H_{ijlkmn} = H_{ijklnm}$. With the axes of stress coincident with material orthotropy axes, we further have $H_{ijklmn} = H_{klmnij} = H_{mnijkl} = H_{ijmnkl} = H_{mnklij} = H_{klijmn}$. Moreover, a condition of incompressibility is implied in using stress deviators in eq(3.34) and this reduces the number of coefficients to 35. Hill's modified orthotropic function, eq(iv) in Table 3.3, is one form of eq(3.34), where $m = 2$ in the first line of eq(iv) and $m = 3$ in the second line. The reduction in eq(3.34) to a direct stress combined with a shear stress becomes:

$$f = Q_1 \sigma_{11}^2 + Q_4 \sigma_{12}^2 + Q_5 \sigma_{11} \sigma_{12} + T_1 \sigma_{11}^3 + T_5 \sigma_{12}^3 + T_6 \sigma_{11} \sigma_{12}^2 + T_7 \sigma_{11}^2 \sigma_{12} = 1 \quad (3.35a)$$

Figures 3.11e-h apply to eq(3.35a). These contrast with the quadratic predictions (Fig. 3.11a-d) for four similar cases of anisotropy described previously. Clearly distortion is now a consisitent feature within each case. Equation (3.34) can be written in a dimensionless form when principal, biaxial stress directions are aligned with orthotropic axes 1 and 2:

$$Q_1 (\sigma_1 / \sigma_{1t})^2 + Q_2 (\sigma_2 / \sigma_{1t})^2 + Q_3 (\sigma_1 / \sigma_{1t})(\sigma_2 / \sigma_{1t})$$
$$+ T_1 (\sigma_1 / \sigma_{1t})^3 + T_2 (\sigma_2 / \sigma_{1t})^3 + T_3 (\sigma_1 / \sigma_{1t})^2 (\sigma_2 / \sigma_{1t}) + T_4 (\sigma_1 / \sigma_{1t})(\sigma_2 / \sigma_{1t})^2 = 1 \quad (3.35b)$$

The coefficients Q and T in eq(3.35a,b) are derived from H_{ijkl} and H_{ijklmn} in [48]. Equations (3.30) and (3.35b) are applied to an initial yield locus for extruded magnesium in Fig. 3.13.

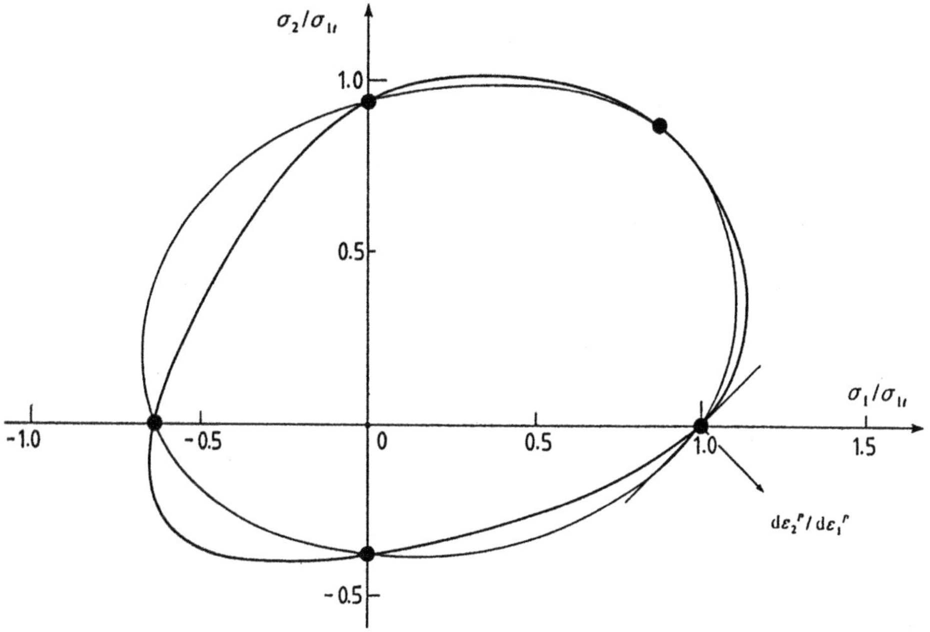

Figure 3.13 Yield locus for an orthotropic magnesium extrusion

The 5 constants, L and Q in eq(3.30), are determined directly from the five experimental yield points shown. The determination of seven constants in eq(3.35b) requires an additional yield point (interpolated) and a strain vector direction as indicated. It is seen that both predictions represent the pronounced Bauschinger effect along each axis of orthotropy. Both quadratic and cubic yield functions appear to describe the measured initial yield points equally well. However, eq(3.35b) provides the better account of distortion appearing in a locus connecting these yield points. Such distortion is linked to the direction of the plastic strain increment vector through the normality rule, i.e. this vector is aligned with the direction of the exterior normal to the yield locus. It appears from Fig. 3.13 that additional yield points would be required to test an initial distortion prediction more precisely. In general, the literature reveals far stronger evidence for a strain-induced form of distortion within a subsequent yield locus than there is for distortion within the initial yield surface.

When it is required to recover a condition of isotropy from eq(3.34), H_{ijkl} is defined from eq(3.33b) and H_{ijklmn} appears as a general sixth order isotropic tensor [23]:

$$H_{ijklmn} = I_{ijklmn} = a\delta_{ij}\delta_{kl}\delta_{mn} + b(\delta_{ij}\delta_{km}\delta_{ln} + \delta_{ij}\delta_{kn}\delta_{ml} + \delta_{kl}\delta_{im}\delta_{jn} + \delta_{kl}\delta_{in}\delta_{jm} + \delta_{mn}\delta_{ik}\delta_{jl} + \delta_{mn}\delta_{il}\delta_{jk})$$
$$+ c(\delta_{ik}\delta_{lm}\delta_{nj} + \delta_{ik}\delta_{jm}\delta_{nl} + \delta_{il}\delta_{jm}\delta_{nk} + \delta_{il}\delta_{jn}\delta_{mk} + \delta_{im}\delta_{jk}\delta_{ln} + \delta_{im}\delta_{jl}\delta_{kn} + \delta_{in}\delta_{jk}\delta_{lm} + \delta_{in}\delta_{jl}\delta_{km}) \quad (3.36)$$

Substituting from eq(3.36), reduces the second term in eq(3.34) to the third deviatoric invariant: $J_3' = \frac{1}{3}\sigma_{ij}'\,\sigma_{jk}'\,\sigma_{ki}'$, when $a = b = 0$ and $c = \frac{1}{8}$. It follows that the first two terms in eq(3.36) are unnecessary for the recovery of this isotropic stress deviator.

3.5 Choice of Yield Function

It has been seen that the classical, von Mises and Tresca theories of yielding for metals may each be formulated as a function of the stress deviator invariants. Such macroscopic predictions to the initial yield surface provide for all possible stress combinations. However, these do not offer information about the microstructural mechanisms of yielding, discussed in Chapter 8. In consideration of subsequent yielding, where plastic strain paths, calculated from $\varepsilon^P = \varepsilon^T - \sigma/E$ (see Fig. 1.1), remain linear, this indicates that a simple rule of *isotropic hardening* may be applied for loading within the plastic range. This employs the concept of an expanding yield surface which retains its initial shape and orientation, to which we shall return in Chapter 9 and 10.

It has been seen that the departure from the isotropic function $f(J_2')$ may either be due to the influence of J_3' (for an isotropic material), or to the presence of initial anisotropy. Clearly, when selecting an appropriate yield function, checks are necessary to establish precisely the initial condition of the material. Anisotropic yielding and flow behaviour may be identified with distinct differences in the stress-strain curves obtained from testpieces machined from different directions in the material. For rolled sheets, the off-axis tensile test will reveal directional differences between yield stresses when anisotropy is present, but the more usual measure of anisotropy for sheet metals is the r value. This is the ratio between the plastic components of an incremental width and thickness strain in a tension test. Anisotropy is revealed when $r \neq 1$. The formability of sheet metal is enhanced when $r > 1$. Bramley and Mellor [59] showed, from eq(3.27), how it was possible to describe the effects of initial anisotropy in transversely isotropic sheets using the constant gradients of linear plastic strain paths (see Chapter 11).

Alternatively, when isotropy is assumed in a tension test the material may be taken to conform to the general isotropic function $f(J_2', J_3')$. Substituting this for f into eq(3.22) gives the plastic strain increment tensor:

$$d\varepsilon_{ij}^P = d\lambda\,[\,(\partial f/\partial J_2')(\partial J_2'/\partial\sigma_{ij}) + (\partial f/\partial J_3')(\partial J_3'/\partial\sigma_{ij})] \quad (3.37)$$

If σ_1 is the non-zero tensile stress, it follows from eqs(3.10b and c) that $\partial J_2'/\partial\sigma_1 = 2\sigma_1/3$, $\partial J_2'/\partial\sigma_2 = -\sigma_1/3$, $\partial J_3'/\partial\sigma_1 = 2\sigma_1^2/9$ and $\partial J_3'/\partial\sigma_2 = -\sigma_1^2/9$. Then, from eq(3.37), the axial and lateral plastic strain increments become

$$d\varepsilon_1^P = d\lambda\,[(\partial f/\partial J_2')(2\sigma_1/3) + (\partial f/\partial J_3')(2\sigma_1^2/9)] \quad (3.38a)$$
$$d\varepsilon_2^P = d\lambda\,[(\partial f/\partial J_2')(-\sigma_1/3) + (\partial f/\partial J_3')(-\sigma_1^2/9)] \quad (3.38b)$$

Dividing eqs(3.38a and b) gives the constant ratio $d\varepsilon_2^P/d\varepsilon_1^P = -\frac{1}{2}$, irrespective of the yield function $f(J_2', J_3')$. That is, the gradient of the lateral versus axial plastic strain plot remains linear, with a gradient of $-\frac{1}{2}$. Figures 3.14a-e compares this isotropic prediction with experimental plastic strain paths for aluminium, steel, copper and brass.

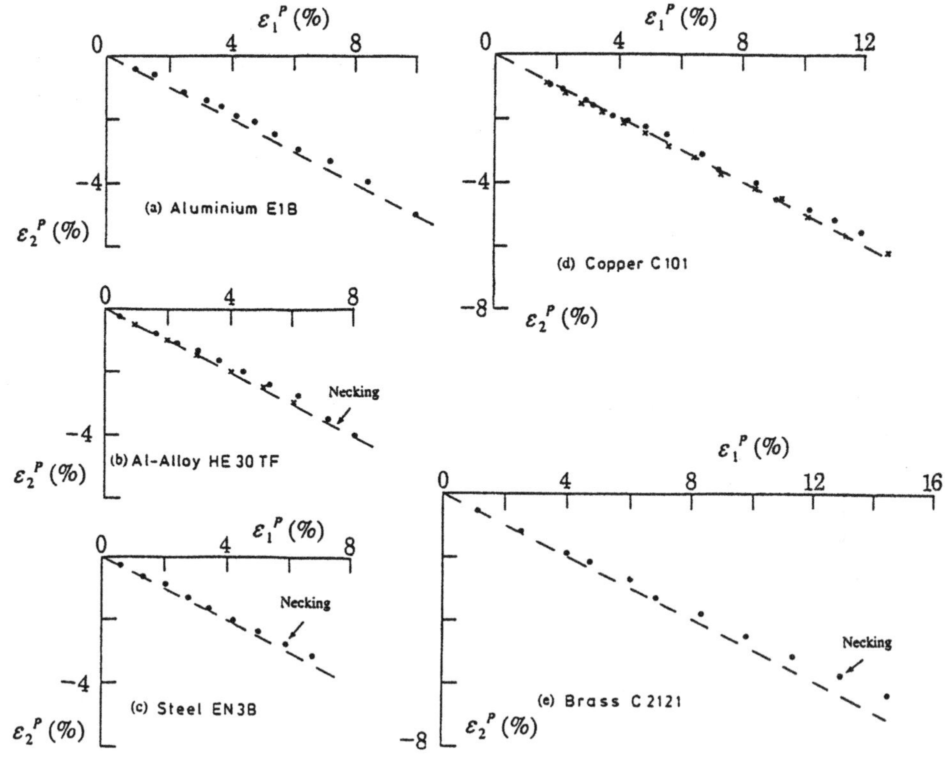

Figure 3.14 Lateral versus axial plastic strain paths under tension

The agreement found between the gradient of each plot and the theoretical gradients value of - ½ confirms that testpieces machined longitudinally from extruded bars become almost isotropic following heat treatment. In contrast, the interstage anneals employed for rolling sheet metal may not leave the material in an isotropic condition. It will be seen later that it is desirable to retain anisotropy arising from rolling when it enhances formability.

References

1. Rees D. W. A. *The Mechanics of Solids and Structures*, 2000, IC Press, U.K.
2. von Mises R. *Nachr. Ges. Wiss.* Gottingen, 1913, 582. (also *ZAMM* 1928, **8**, 161)
3. Tresca H. "Memoire sur l'ecoulement des corps solides", *Memoirs Par Divers Savants*, Paris, 1968, **18**, 733 and 1972, **20**, 75.
4. Ikegami K. J. *Soc. Mat. Sci.* 1975, **24**(261), 491 and **24**(263), 709 (see also BISI translation 14420, The Metals Soc., Sept 1976)
5. Michno M. J. and Findley W. N. *Int. J. Non-Linear Mech.*, 1976, **11**, 59.
6. Banabic D. et al, Proceedings: *4th ESAFORM Conf on Material Forming*, (Ed. Habraken A. M.) Vol. 1, p. 297, University of Liége, Belgium, April 2001.
7. Shahabi S. N. and Shelton A. J. *Mech. Eng. Sci.*, 1975, **17**, 82.
8. Johnson K. R. and Sidebottom O. M. *Expl Mech.*, 1972, **12**, 264.
9. Moreton D. N., Moffat, D. G. and Parkinson, D. B. *J. Strain Anal.*, 1981, **16**, 127.

10. Miastkowski J and Szczepinski W. *Int. J. Solids and Structures*, 1965, **1**, 189.
11. Marin J., Hu L. W. and Hamburg J. F. *Proc. A.S.M.*, 1953, **45**, 686.
12. Lessels J. M. and MacGregor, C. W. *Jl Franklin Inst.*, 1940, **230**, 163.
13. Rogan J. and Shelton A. *J. Strain Anal.*, 1969, **4**, 127.
14. Ivey H. J. *J. Mech. Eng. Sci.*, 1961, **3**, 15.
15. Phillips A. and Tang J-L. *Int. J. Solids and Struct.*, 1972, **8**, 463.
16. Kowalewski Z. L., Dietrich L. and Socha G. Proceedings: *Third Int Congress on Thermal Stress*, 1999, 181-184, University of Kracow.
17. Shiratori E., Ikegami K. and Kaneko K. *Trans Japan Soc. Mech. Engrs*, 1973, **39**, 458.
18. Ellyin F. and Grass J-P. *Trans Can. Soc for Mech, Engrs*, 1975, **3**, 156.
19. Taylor G. I. and Quinney, H. *Phil Trans Roy Soc A*, 1931, **230**, 323.
20. Shrivastava H. P., Mroz, Z. and Dubey, R. N. *ZAMM*, 1973, **53**, 625.
21. Rees D. W. A. *Proc. R. Soc. Lond.*, 1982, **A383**, 333.
22. Drucker D. C. *J. Appl. Mech.*, 1949, **16**, 349.
23. Betten J. *Acta Mechanica*, 1976, **25**, 79.
24. Freudenthall A. M. and Gou P. F. *Acta Mechanica*, 1969, **8**, 34.
25. Rees D. W. A. *J. Applied Mech.*, 1982, **49**, 663.
26. Crossland B. *Proc. I. Mech. E.*, 1954, **168**, 935.
27. Hu L. W. Proc: *Naval Structures*, (eds Lee E. H. and Symonds P. S.) 1960, 924, Pergamon, Oxford, U.K.
28. Pugh H. Ll. D. and Green D. *Proc. I. Mech. E.*, 1964, **179**, 1.
29. Ratner S. I. *Zeitschrift fur Technische Physik*, 1949, **19**, 408.
30. Fung P. K., Burns D. J. and Lind N. C. Proc: *Foundations of Plasticity*, (ed A.Sawczuk) 1973, 287, Noordhoff, Warsaw.
31. Ros M. and Eichinger A. *Eidgenoess, Materialpruef. Versuchanstalt Ind. Bauw.* Gewerbe, Zurich, 1929, **34**, 1.
32. Brandes M. in: *The Mechanical Behaviour of Materials Under Pressure*, (ed Pugh H. Ll.) 1971, Ch. 6, 236, Applied Science, London.
33. Stassi-D'Alia F. *Meccanica*, 1967, **2**, 178.
34. Hill R. *Proc. Roy. Soc.*, 1948, **A193**, 281.
35. Edelman F. and Drucker D. C. *J. Franklin Inst.*, 1951, **251**, 581.
36. Hearmon R. F. S. *Applied Anisotropic Elasticity*, Clarendon Press, Oxford 1961.
37. Olsak W. and Urbanowski W. *Arch of Mech.*, 1956, **8**, 671.
38. Sobotka, Z. *ZAMM*, 1969, **49**, 25.
39. Troost A. and Betten J. *Mech. Res. Comms*, 1974, **1**, 73.
40. Fava F. *J. Annals of the C.I.R.P*, 1967, **15**, 411.
41. Hu L. W. *J. Appl. Mech.*, 1956, **23**, 444.
42. Hazlett T. H., Robinson A. T. and Dorn J. E. *Trans ASME*, 1950, **42**, 1326.
43. Jones S. E. and Gillis P. P. *Met Trans A*, 1984, **15A**, 129.
44. Hu L. W. and Marin J. *J. Appl. Mech.*, 1955, **22**, 77.
45. Takeda T. and Nasu Y. *J. Strain Analysis*, 1991, **26**, 47.
46. Hill R. *Math. Proc. Camb. Phil. Soc.*, 1979, **85**, 179.
47. Bourne L. and Hill R. *Phil. Mag.*, 1950, **41**, 671.
48. Gotoh M. *Int. J. Mech. Sci.*, 1977, **19**, 505.
49. Stassi-D'Alia F. *Meccanica*, 1969, **4**, 349.
50. Harvey S. J., Adkin P. and Jeans P. J. *Fatigue of Eng Mats and Struct.*, 1983, **6**, 89.
51. Dodd B. *Int. J. Mech. Sci.*, 1984, **12**, 587.
52. Parmar A. and Mellor P. B. *Int. J. Mech. Sci.*, 1978, **20**, 385.
53. Bailey R. W. *Proc. I. Mech. E.*, 1935, **131**, 131.

54. Davis E. A. *Trans ASME*, 1961, **28E**, 310.
55. Rees D. W. A. *Acta Mechanica*, 1982, **43**, 223.
56. Lee D. and Backofen W. A. *Trans Met. Soc. AIME*, 1966, **236**, 1077.
57. Spencer A. J. M. *Continuum Mechanics*, 1980, Longman, London.
58. Rees D. W. A. *Acta Mechanica*, 1984, **52**, 15.
59. Mellor P. B. in: *Mechanics of Solids*, (eds Hopkins H. G. and Sewell M. J.) 1982, 383, Pergamon Press, Oxford, U.K.

Exercises

3.1 Show that the constants c, b and d in eqs(3.15a,b and c) are defined from the relationship between Y and k as follows: (i) $Y = k\,(1/27 - 4c/9^3)^{1/6}$, (ii) $Y = 9k\,/\,(27 - 4b)^{1/2}$ and (iii) $Y = k\,/\,(1/3^{3/2} - 2d/27)^{1/3}$.

3.2 Normalise the Stassi fracture criterion (3.25b) with the compressive fracture stress σ_c and plot the family of loci in σ_1, σ_2 space for $\rho = \sigma_c/\sigma_t = 2$, 3 and ∞.

3.3 Construct the family of yield loci in σ, τ space from the unsymmetrical yield function:

$$f = J_2' + (p\,/\,Y)\,J_3'$$

within the range $1.5 \le p \le -3$.

3.4 Construct a family of yield loci for the unsymmetrical yield function (3.16) for $n = 2$. Determine the range of p values that will ensure a closed yield surface.

3.5 Establish the right-hand sides of the expressions (iii) and (iv) given in Table 3.3 in terms of the known yield stress σ_{1y} along the principal, 1- axis, of orthotropy.

3.6 Compare the yield criteria (i) - (viii), listed in Table 3.3, when they are reduced to a principal biaxial stress space, i.e. for $\sigma_3 = 0$ and $\tau_{xy} = \tau_{xz} = \tau_{yz} = 0$, as appropriate.

3.7 Derive from eq(3.22) the principal, plastic constitutive relations corresponding to each of the isotropic yield functions given in eqs(3.15a -c).

3.8 The coefficients L_1, L_2, Q_1, Q_2 and Q_3 in eq(3.31c) can account for four cases in which the tensile and compressive yield stresses in the 1 and 2 directions may be the same or different under a principal biaxial stress state σ_1, σ_2. Show each case with a plot similar to Fig. 3.11a-d.

3.9 Show how the coefficients T_1, T_2, T_3, T_4, Q_1, Q_2 and Q_3 in eq(3.35b) can account for several conditions of anisotropy under a principal biaxial stress state σ_1, σ_2. Show each case with a plot similar to Fig. 3.11e-h.

3.10 Reduce eq(3.27) to criteria of yielding under a principal, biaxial stress state (σ_1, σ_2) for material: (i) orthotropy and (ii) transverse isotropy, where yield stresses are the same in the 2- and 3-directions.

3.11 Show that Hill's yield criterion (3.27) reduces to a plane stress form:

$$(G + H)\sigma_x^2 - 2H\,\sigma_x\,\sigma_x + (H + F)\sigma_x + 2N\tau_{xy}^{\,2} = 1$$

when plane stresses σ_x, σ_y and τ_{xy} are applied within the plane of a thin sheet and aligned with the orthotropic axes x and y (z is through the thickness). Use this equation to show that the variation in yield stress Y with orientation θ to x within the plane of the sheet, is given by:

$$Y(\theta) = \frac{1}{\sqrt{F\sin^2\theta + G\cos^2\theta + H + \frac{1}{4}(2N - F - G - 4H)\sin^2 2\theta}}$$

3.12 If, in exercise 3.11, the yield stresses in the three orthogonal directions x, y and z are denoted by X, Y and Z respectively and P is the shear yield stress in the x-y plane of the sheet, show that the coefficients F, G, H and N are given by:

$$F = \frac{1}{2}\left(\frac{1}{Y^2} + \frac{1}{Z^2} + \frac{1}{X^2}\right), \quad G = \frac{1}{2}\left(\frac{1}{Z^2} + \frac{1}{X^2} + \frac{1}{Y^2}\right),$$

$$H = \frac{1}{2}\left(\frac{1}{X^2} + \frac{1}{Y^2} + \frac{1}{Z^2}\right), \quad N = \frac{1}{2P^2}$$

3.13 Using the orthotropic yield function (3.27), examine how you would predict the onset of yielding in an orthotropic sheet metal when a direct stress σ is combined with a shear stress τ in each of the following cases: (i) when both σ and τ are aligned with the in-plane orthotropic axes x and y and (ii) when they are inclined at θ to x and y, as shown in Fig. 3.15. Note, the presence of complementary shear in each case.

[Answer (i): $\sigma_x^2 + \frac{2N}{H + G}\tau_{xy}^2 = X^2$ where X is the yield stress in the x-direction]

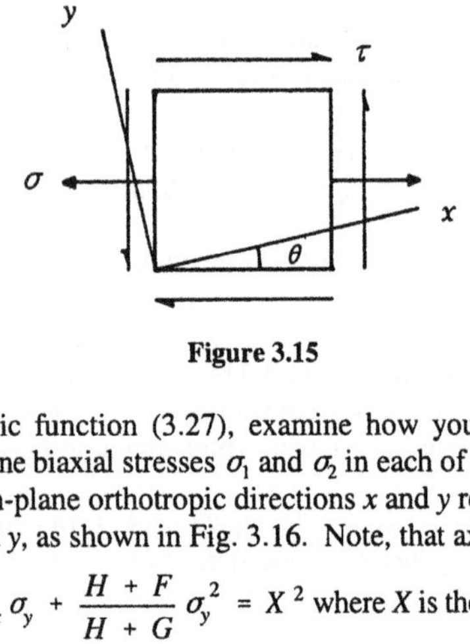

Figure 3.15

3.14 Using Hill's orthotropic function (3.27), examine how you would predict yielding in an orthotropic sheet under in-plane biaxial stresses σ_1 and σ_2 in each of the following cases: (i) when σ_1 and σ_2 are aligned with the in-plane orthotropic directions x and y respectively and (ii) when σ_1 and σ_2 are inclined at θ to x and y, as shown in Fig. 3.16. Note, that axis z is through the thickness.

[Answer (i): $\sigma_x^2 - \frac{2H}{H + G}\sigma_x\sigma_y + \frac{H + F}{H + G}\sigma_y^2 = X^2$ where X is the yield stress in the x-direction]

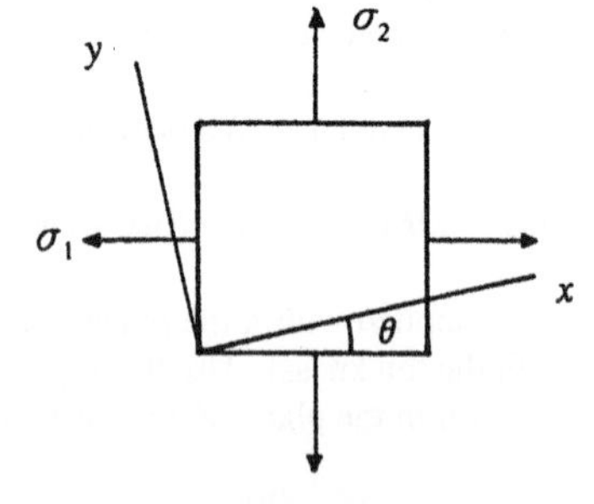

Figure 3.16

CHAPTER 4

NON-HARDENING PLASTICITY

4.1 Introduction

Non-hardening theories of plasticity are very useful for providing solutions to practical problems with static indeterminancy, i.e. where satisfying force equilibrium alone is insufficient. A solution is reached from combinining force equilibrium with the yield condition and strain compatibility. Here it is convenient to distinguish between uniform and non-uniform stress states. The two classical theories available are normally referred to as Hencky-Ilyushin and Prandtl-Reuss. Though formulated differently, each theory can provide the total strain arising from realistic, non-uniform, multiaxial, stress states. The question arises as to which theory is the more reliable? To answer this, a number of comparisons are made between their predictions to the deformation in solid and hollow bars subjected to combinations of axial load and torque, for which experimental results are available.

4.2 Classical Theories of Plasticity

Two classical theories of plasticity are the total strain (or deformation) theory of Hencky [1], Illyushin [2] (also Nadai [3]) and the incremental (or flow) theory of Prandtl-Reuss [4, 5]. It is instructive to examine the ease with which they can solve for the plasticity of a non-hardening material subjected to both uniform and non-uniform, multiaxial, stress states. It will be seen that, although each theory is constructed from different assumptions, they do provide comparable solutions to the load-deformation behaviour and the internal stress distributions under simple loading paths. Firstly, an outline of each theory is given.

4.2.1 Hencky-Ilyushin

In the formulation of the deformation theory of non-hardening plasticity, total strains are used. The strains that arise from the application of normal stresses, within the plastic range, are the sum of deviatoric and volumetric strain components. The latter is associated with the mean normal stress. Shear deformation occurs independently under the application of shear stress. In this way the total strain tensor must appear, in the respective tensor subscript and matrix notations, as

$$\varepsilon_{ij}^{t} = \phi\, \sigma_{ij}' + (\delta_{ij}\, \sigma_{kk}) / (9K) \quad \text{or} \quad \mathbf{E} = \phi\, \mathbf{T}' + (\, \mathbf{I}\, \mathrm{tr}\, \mathbf{T}\,) / (9K) \qquad (4.1\text{a,b})$$

where the scalar $\phi > 0$ during loading and $\phi = 0$ with unloading. The use of the Kronecker delta δ_{ij} in eq(4.1a) ensures that the final volumetric strain term does not contribute to the components of shear strain, since $\delta_{ij} = 0$ for $i \neq j$. The absolute stress appears with the substitution of the deviatoric (or reduced) stress expression $\sigma_{ij}' = \sigma_{ij} - \delta_{ij}\sigma_{kk}/3$, into eq(4.1a):

$$\varepsilon_{ij}^{t} = \phi\,\sigma_{ij} + [(1-2\nu)/E - \phi]\delta_{ij}\sigma_{kk}/3 \qquad (4.2a)$$

or

$$\mathbf{E} = \phi\,\mathbf{T} + [1/(3K) - \phi](\mathbf{I}\,\mathrm{tr}\,\mathbf{T})/3 \qquad (4.2b)$$

where $E = 3K(1-2\nu)$ connects the three elastic constants. Equation (4.2a) indicates a considerable simplification from assuming $\nu = ½$, i.e. an elastically incompressible material. This provides a good approximation to the total strain tensor:

$$\varepsilon_{ij}^{t} = \phi\,\sigma_{ij} - \phi\,\delta_{ij}\,\sigma_{kk}/3 \qquad (4.2b)$$

This total strain theory has been used mostly for the determination of limit loads in structures of elastically-perfectly plastic material [6, 7]. Such loading involves combinations of tension, torsion, bending, pressure and shear force. The theory provides for the resulting stress and strain distributions of structures in both the elastic-plastic and fully-plastic conditions [6]. Equations (4.2a,b) are not linked to a particular yield function and therefore may be combined with any suitable function, including those of Tresca and von Mises. The plastic strain supplied by eq(4.2b) is probably the earliest example of a *non-associated flow rule.*

4.2.2 Prandtl-Reuss

In the derivation of the flow theory of Prandtl-Reuss, the total incremental strain tensor $d\varepsilon_{ij}^{t}$ is the sum of elastic and plastic incremental strain tensors. This gives

$$d\varepsilon_{ij}^{t} = d\varepsilon_{ij}^{P} + d\varepsilon_{ij}^{e} = d\lambda\,\sigma_{ij}' + \left[\frac{(1-2\nu)\,\delta_{ij}\,d\sigma_{kk}}{3E} + \frac{d\sigma_{ij}'}{2G}\right] \qquad (4.3a)$$

In matrix notation eq(4.3a) becomes

$$d\mathbf{E} = d\lambda\,\mathbf{T}' + \frac{(1-2\nu)}{3E}\,\mathbf{I}\,\mathrm{tr}\,(d\mathbf{T}) + \frac{1}{2G}\,d\mathbf{T}' \qquad (4.3b)$$

Components of the plastic strain tensor $d\varepsilon_{ij}^{P}$ are identified with the flow rule of Levy-Mises. That is, f in eq(3.22) is associated with the von Mises criterion of yielding $f = ½\sigma_{ij}'\sigma_{ij}'$, in which the stress deviator σ_{ij}', ensures plastic incompressibility. The incremental, recoverable, elastic strain $d\varepsilon_{ij}^{e}$ appears within [ ] in eq(4.3a). This tensor component form of Hooke's law reveals that elastic strain is the sum of hydrostatic and deviatoric stress contributions to elasticity. There is no contribution from the former under shear stress, since $\delta_{ij} = 0$. Under normal stress ($\delta_{ij} = 1$) this component will be small without a superimposed pressure. Here, an assumption of elastic incompressibility ($\nu = ½$) will reduce eq(4.3a) to an acceptable approximation:

$$d\varepsilon_{ij}^{t} = d\lambda\,\sigma_{ij}' + d\sigma_{ij}'/(2G) \qquad (4.3c)$$

In eqs(4.3a,c), the scalar multiplier $d\lambda$ will characterise the changing nature of the plastic hardening behaviour for the material. This is normally defined from uniaxial tension where the equivalent stress and the equivalent, incremental plastic strain are connected from within the first term in eq(4.3c):

$$d\bar{\varepsilon}^{P} = d\lambda\left(\bar{\sigma} - \frac{1}{3}\bar{\sigma}\right) \quad \Rightarrow \quad d\lambda = \frac{3\,d\bar{\varepsilon}^{P}}{2\bar{\sigma}}$$

in which $\bar{\sigma} = \bar{\sigma}(\bar{\varepsilon}^P)$. In the case of a non-hardening, perfectly-plastic material $\bar{\sigma} = Y =$ constant. Note, that it is then possible to integrate eq(4.3a) to give the components of a total strain tensor as

$$\varepsilon_{ij}^{\,t} = \int_0^{\bar{\varepsilon}^P} \frac{3\sigma_{ij}'}{2Y}\,\mathrm{d}\bar{\varepsilon}^P + \int_0^{\sigma_{kk}} \frac{(1-2\nu)}{3E}\,\delta_{ij}\,\mathrm{d}\sigma_{kk} + \int_0^{\sigma_{ij}'} \frac{1}{2G}\,\mathrm{d}\sigma_{ij}'$$

$$= \frac{3\bar{\varepsilon}^P}{2Y}\sigma_{ij}' + \frac{(1-2\nu)}{3E}\delta_{ij}\,\sigma_{kk} + \frac{1}{2G}\sigma_{ij}' \qquad (4.3d)$$

The integration assumes that the components of the stress deviator tensor σ_{ij}' in eq(4.3d) increase: (i) proportionately, or (ii) in a stepwise manner. This allows stress components to be separated within the integral. The following example will illustrate how condition (ii) applies to a combination of a single shear stress and a single direct stress.

4.2.3 Theoretical Comparison Between Hencky and Prandtl-Reuss

We can further examine the differences in formulation between the compressible forms of the deformation and flow theories from making a theoretical comparison between eqs(4.1a) and (4.3a). Take, for example, a stress system composed of combined tension σ_{11} and one shear stress component σ_{12}. The Hencky eq(4.1a) gives the total direct and shear strains as

$$\begin{aligned}\varepsilon_{11}^{\,t} &= \phi\,\sigma_{11}' + \delta_{11}\,\sigma_{11}/(9K)\\ &= \sigma_{11}[\,2\phi/3 + 1/(9K)\,]\end{aligned} \qquad (4.4a)$$

$$\begin{aligned}\varepsilon_{22}^{\,t} = \varepsilon_{33}^{\,t} &= -\,\phi\,\sigma_{11}/3 + (1-2\nu)\sigma_{11}/(3E)\\ &= \sigma_{11}[-\,\phi/3 + 1/(9K)\,] = \varepsilon_{33}^{\,t}\end{aligned} \qquad (4.4b)$$

$$\begin{aligned}\varepsilon_{12}^{\,t} &= \phi\,\sigma_{12} + \delta_{12}\,\sigma_{12}/(9K)\\ \gamma_{12}^{\,t} &= 2\varepsilon_{12}^{\,t} = 2\phi\,\sigma_{12}\end{aligned} \qquad (4.4c)$$

where $\delta_{11} = 1$ and $\delta_{12} = 0$. From the Prandtl-Reuss eq(4.3d), we have

$$\varepsilon_{11}^{\,t} = \int_0^{\bar{\varepsilon}^P} \frac{3\mathrm{d}\bar{\varepsilon}^P}{2Y}\left(\frac{2}{3}\sigma_{11}\right) + \int_0^{\sigma_{11}}\left[\frac{(1-2\nu)}{3E} + \frac{1}{3G}\right]\mathrm{d}\sigma_{11}$$

$$= \frac{\bar{\varepsilon}^P}{Y}\sigma_{11} + \left[\frac{(1-2\nu)}{3E} + \frac{1}{3G}\right]\sigma_{11} \qquad (4.5a)$$

$$\varepsilon_{22}^{\,t} = \int_0^{\bar{\varepsilon}^P} \frac{3\mathrm{d}\bar{\varepsilon}^P}{2Y}\sigma_{22}' + \int_0^{\sigma_{11}} \frac{(1-2\nu)}{3E}\delta_{22}\,\mathrm{d}\sigma_{11} + \int_0^{\sigma_{22}'} \frac{1}{2G}\,\mathrm{d}\sigma_{22}'$$

$$= -\int_0^{\bar{\varepsilon}^P} \frac{\mathrm{d}\bar{\varepsilon}^P}{2Y}\sigma_{11} + \int_0^{\sigma_{11}}\left[\frac{(1-2\nu)}{3E} - \frac{1}{6G}\right]\mathrm{d}\sigma_{11}$$

$$= -\frac{\bar{\varepsilon}^P}{2Y}\sigma_{11} + \left[\frac{(1-2\nu)}{3E} - \frac{1}{6G}\right]\sigma_{11} = \varepsilon_{33}^{\,t} \qquad (4.5b)$$

$$\varepsilon_{12}{}' = \int_0^{\bar{\varepsilon}^P} \frac{3\mathrm{d}\bar{\varepsilon}^P}{2Y}\sigma_{12} + \int_0^{\sigma_{11}} \frac{(1-2\nu)}{3E}\delta_{12}\,\mathrm{d}\sigma_{11} + \int_0^{\sigma_{12}} \frac{1}{2G}\,\mathrm{d}\sigma_{12}$$

$$= \frac{3\bar{\varepsilon}^P}{2Y}\sigma_{12} + \frac{1}{2G}\sigma_{12}$$

$$\mathrm{d}\gamma_{12}{}' = 2\varepsilon_{12}{}' = \frac{3\bar{\varepsilon}^P}{Y}\sigma_{12} + \frac{1}{G}\sigma_{12} \tag{4.5c}$$

It will be seen that the total strain predictions are different except for the case where the stresses increase proportionately, i.e. σ_{12}/σ_{11} = constant. The corresponding total strain ratio is, from the Hencky eqs(4.4a and c),

$$\frac{\gamma_{12}^t}{\varepsilon_{11}^t} = \left(\frac{2\phi}{2\phi/3 + 1/(9K)}\right)\frac{\sigma_{12}}{\sigma_{11}} \tag{4.6a}$$

and, from the Prandtl-Reuss eqs(4.5a and c), this ratio becomes

$$\frac{\gamma_{12}^t}{\varepsilon_{11}^t} = \left(\frac{3\bar{\varepsilon}^P/Y + 1/G}{\bar{\varepsilon}^P/Y + 1/(9K) + 1/(3G)}\right)\frac{\sigma_{12}}{\sigma_{11}} \tag{4.6b}$$

The total strain ratios in eqs(4.6a and b) are approximately equal when we identify $\phi = (3\bar{\varepsilon}^P)/(2Y)$ and take the reciprocals of the elastic constants to be negligibly small. The ratios in eqs(4.6a,b) are exactly equal when the deviatoric stress contribution to plasticity is separated from within each ratio. That is, both give $\gamma_{12}{}^P/\varepsilon_{11}{}^P = 3(\sigma_{12}/\sigma_{11})$. In general, the plastic response from the deformation and flow theories of plasticity are identical under proportional (radial) loading conditions.

Note that, in contrast to the Prandtl-Reuss flow theory, the Hencky eq(4.2a) makes no specific identity with the elastic and plastic components of the total strain. Nor does Hencky depend upon the strain path as it rests solely upon the final strain condition. Consequently, Hencky's load limit predictions may be in error for certain non-proportional stress paths in which the history of deformation is influential. In contrast, the flow theory traces the path through an incremental strain summation. We next examine this area of uncertainty from making further comparisons between theory and experiment. Also, a consideration is given to the influence of the elastic component of strain upon compressibility and total strain.

4.3 Application of Classical Theory to Uniform Stress States

Near-uniform stress states are achieved from the application of various loadings to thin-walled tubes and plates. Experiments that combine tension, torsion and internal pressure in thin-walled tubes have often been used to provide data for the appraisal of a plasticity theory. The nature of hardening assumed should reflect that in a given material. Here, we examine variations in stress with strain from two theories of non-hardening plasticity without a simplifying assumption of incompressible elasticity. A comparison is made with the results from four different experiments in which one component of total strain is held constant.

4.3.1 Combined Tension-Torsion with Constant Shear Strain

Hohensemer [8] conducted the original experiment in which a thin-walled tube of non-hardening, low-carbon steel was extended longitudinally while the shear strain γ_o at yield was held constant. This gave

$$\gamma_o = k/\,G = 2Y\,(1 + \nu)/\,(\sqrt{3}E) = \text{constant} \tag{4.7}$$

where $k = Y/\sqrt{3}$ and $E = 2G(1 + \nu)$. It is required to determine the variations in the ensuing axial σ and shear τ stresses with total axial strain ε^t, according to the two theories of plasticity. The von Mises yield criterion (3.8) is used, in which the shear stress through the wall of the tube is taken to be uniform, and the influence of radial stress and hoop stress are small enough to be neglected. Stress subscripts will be omitted for simplicity when we identify $\sigma = \sigma_{11}$ and $\tau = \sigma_{12}$. Equation (3.8) is written as

$$\sigma^2 + 3\tau^2 = Y^2 \quad \Rightarrow \quad S^2 + 3T^2 = 1 \tag{4.8a,b}$$

for $S = \sigma/Y$ and $T = \tau/Y$.

(a) *Hencky*
The deformation theory (4.2a) provides the following total axial and shear strain components under the stress combination σ, τ:

$$\varepsilon^t = \frac{2}{3}\phi\sigma + \frac{(1 - 2\nu)}{3E}\sigma \tag{4.9a}$$

$$\gamma^t = 2\phi\,\tau \tag{4.9b}$$

Putting $\gamma^t = \gamma_o$ = constant in eq(4.9b) defines the scalar multiplier for the ensuing deformation as $\phi = \gamma_o/(2\tau)$. Substituting into eq(4.9a) and expressing τ, from within eq(4.8a), leads to

$$\varepsilon^t = \frac{\sigma\gamma_o}{\sqrt{3(Y^2 - \sigma^2)}} + \frac{(1 - 2\nu)\sigma}{3E} \tag{4.10a}$$

Substituting from eq(4.7), gives the normalised form of eq(4.10a):

$$\frac{\varepsilon^t E}{Y} = \frac{S}{3}\left[(1 - 2\nu) + \frac{2(1 + \nu)}{\sqrt{1 - S^2}}\right] \tag{4.10b}$$

which is to be used with the normalised yield criterion, eq(4.8b).

(b) *Prandtl-Reuss*
The application of the flow theory (4.3a) supplies total incremental strains, each as the sum of elastic and plastic components:

$$\mathrm{d}\varepsilon^t = 2\,\mathrm{d}\lambda\,\sigma/3 + \mathrm{d}\sigma/E \tag{4.11a}$$

$$\mathrm{d}\gamma^t = 2\,\mathrm{d}\lambda\tau + \mathrm{d}\tau/G \tag{4.11b}$$

Now, putting $d\gamma' = 0$ in eq(4.11b) gives $d\lambda = -d\tau/(2\tau G)$. Substituting this into eq(4.11a), we find

$$d\varepsilon^t = -\frac{\sigma\, d\tau}{3\tau G} + \frac{d\sigma}{E} \tag{4.12a}$$

where from eq(4.8a), $d\tau = -\sigma d\sigma/(3\tau)$. Substituting this into eq(4.12a) gives

$$d\varepsilon^t = \frac{\sigma^2 d\sigma}{9\tau^2 G} + \frac{d\sigma}{E} = \frac{\sigma^2 d\sigma}{3G(Y^2 - \sigma^2)} + \frac{d\sigma}{E} \tag{4.12b}$$

Integrating eq(4.12b) by partial fractions

$$\varepsilon^t = \frac{1}{3G}\int_0^\sigma \left(\frac{Y/2}{Y+\sigma} + \frac{Y/2}{Y-\sigma} - 1\right) d\sigma + \frac{1}{E}\int_0^\sigma d\sigma$$

$$= \frac{Y}{6G}\ln\left(\frac{Y+\sigma}{Y-\sigma}\right) - \frac{\sigma}{3G} + \frac{\sigma}{E} \tag{4.12c}$$

Taken with $E = 2G(1+\nu)$, eq(4.12c) reduces to the normalised form

$$\frac{\varepsilon^t E}{Y} = \frac{(1-2\nu)S}{3} + \frac{(1+\nu)}{3}\ln\left(\frac{1+S}{1-S}\right) \tag{4.12d}$$

Figure 4.1a-c compares the predictions supplied by eqs(4.10b) and (4.12d) with Hohensemer's experimental results.

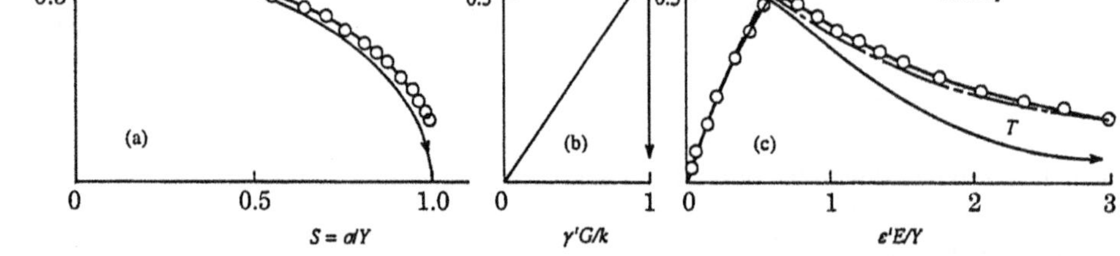

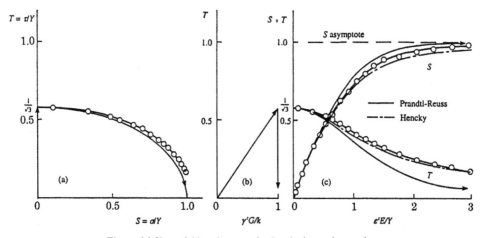

Figure 4.1 Shear yield strain constrained under increasing tension

Both theories appear to be in reasonable agreement with observed experimental trends. Figure 4.1c shows that the total axial strain increases with S initially at a gradient of E. Thereafter, $\varepsilon' E/Y$ becomes infinite as S approaches unity asymptotically. Also, $T = \tau/Y$

approaches zero simultaneously as γ' becomes a permanent strain. The fact that Hencky appears to be more representative of this data led early workers [9, 10] to favour a total strain formulation. Now, however, the generally held view is that the Prandtl-Reuss theory is more reliable in the presence of stress gradients and non-radial loading. The deviation found between the flow theory and this experiment is likely to be due to the occurrence of some hardening and a disparity with the von Mises yield criterion for the steel in question. The Hencky theory retains more than an historical interest since as we see here and later, for other uniform stress states, it is easier to apply and provides perfectly acceptable predictions.

The Hencky eq(4.10a) can also be applied where an elastic shear strain ($\gamma_o = \tau/G$, in which $\tau_o < k$), is held constant under increasing tension. According to the Prandtl-Reuss theory, the material first yields under a stress state σ_o, τ_o, according to

$$\sigma_o^{\,2} + 3\tau_o^{\,2} = Y^2 \tag{4.13}$$

With Y constant and $\sigma > \sigma_o$, it follows from eq(4.13) that $\tau < \tau_o$. The ensuing, total axial strain again follows from eq(4.12c) but with new integration limits:

$$\varepsilon^t = \frac{1}{3G}\int_{\sigma_o}^{\sigma}\left(\frac{Y/2}{Y+\sigma} + \frac{Y/2}{Y-\sigma} - 1\right)\mathrm{d}\sigma + \frac{1}{E}\int_{\sigma_o}^{\sigma}\mathrm{d}\sigma \tag{4.14a}$$

Equation(4.14a) integrates to

$$\varepsilon^t = \frac{Y}{6G}\ln\left[\frac{(Y+\sigma)(Y-\sigma_o)}{(Y+\sigma_o)(Y-\sigma)}\right] + \left(\frac{1}{E} - \frac{1}{3G}\right)(\sigma - \sigma_o) \tag{4.14b}$$

Equation (4.14b) relates normalised stress to total strain in the manner of Figs 4.2a-c.

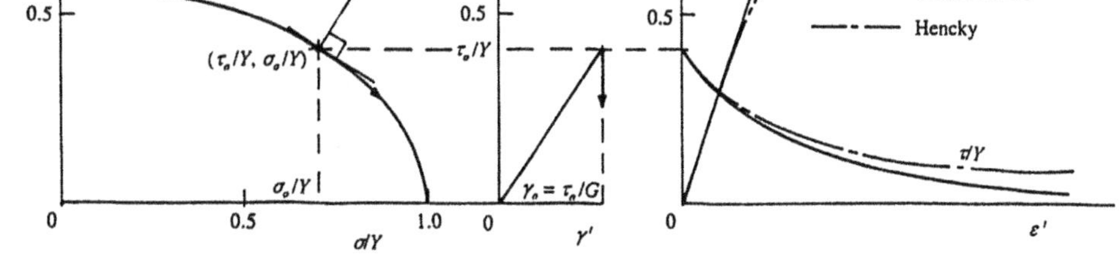

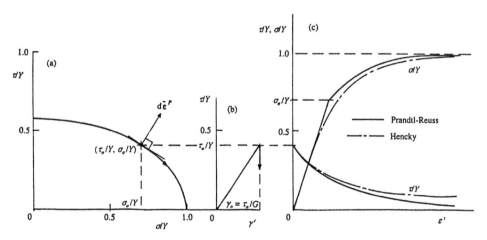

Figure 4.2 Elastic shear strain constrained under increasing tension

Initial loading to σ_o is elastic with slope E. Thereafter, with $\sigma > \sigma_o$, σ and τ vary with ε^t in a similar asymptotic manner to Fig. 4.1c. As the stress state follows the boundary of the yield locus, the plastic strain increment vector $d\bar{\varepsilon}^P$ aligns with its exterior normal (Fig. 4.2a). The two components of this vector identify the axial and plastic shear strains. Figure 4.2b shows that restraining the initial elastic shear strain results in its gradual conversion to plastic strain.

(c) *Other theories*
The incompressible forms of Prandtl-Reuss and Hencky (or Nadai) are found from putting $\nu = ½$. In Hohensemer's experiment, for example, substituting $\nu = ½$ in eqs(4.10a) and (4.12d) reduces them to simplified forms, originally given by Prager [11] and Neuber [12]:

$$S = \tanh(\varepsilon E/Y) \text{ and } S = (\varepsilon E/Y)/[1 + (E\varepsilon/Y)] \qquad (4.15a,b)$$

These predictions to the S variations would be less satisfactory than those of compressible theories shown in Fig. 4.1. Betten [13] generalised eqs(4.15a,b) to become

$$S = [\tanh(\varepsilon E/Y)^n]^{1/n} \text{ and } S = (\varepsilon E/Y)/[1 + (E\varepsilon/Y)^n]^{1/n} \qquad (4.16a,b)$$

He demonstrated that with an integer value $n > 1$, eqs(4.16a,b) ensured an improved fit to this data. However, it is doubtful whether Hencky's incompressible theory could allow a similar modification with its application to other, reduced stress states.

4.3.2 Combined Tension-Torsion with Constant Axial Strain

Consider a thin tube subjected to constrained deformation in which the tensile yield strain is held constant under increasing torsion. The following analyses provide predictions to the growth in shear strain.

(a) *Hencky*
The axial strain at yield is

$$\varepsilon_o = \frac{Y}{E} = \frac{\sqrt{3}k}{2G(1+\nu)} \qquad (4.17a)$$

Putting $\varepsilon' = \varepsilon_o$ = constant, in eq(4.9a), defines the scalar multiplier for the ensuing deformation as

$$\phi = \frac{3\,\varepsilon_o}{2\,\sigma} - \frac{(1-2\nu)}{2E} \qquad (4.17b)$$

Substituting eq(4.17b) into eq(4.9b) and eliminating σ from eq(4.8a), leads to

$$\gamma^t = \frac{3\,\tau\,\varepsilon_o}{\sqrt{Y^2 - 3\tau^2}} - \frac{(1-2\nu)\,\tau}{E} \qquad (4.18a)$$

Employing eq(4.17a), and setting $Y = \sqrt{3}k = E\varepsilon_o$, gives a normalised form of eq(4.18a)

$$\frac{\gamma^t G}{\sqrt{3}k} = \frac{3T}{2(1+\nu)\sqrt{1-3T^2}} - \frac{(1-2\nu)T}{2(1+\nu)} \qquad (4.18b)$$

(b) *Prandtl-Reuss*
Here we put $\mathrm{d}\varepsilon' = 0$ in eq(4.11a), to give

$$\mathrm{d}\lambda = -\frac{3\,\mathrm{d}\sigma}{2E\sigma} \tag{4.19a}$$

where, from eq(4.8a),

$$\frac{\mathrm{d}\sigma}{\sigma} = \frac{-3\tau\,\mathrm{d}\tau}{\sigma^2} = \frac{3\tau\,\mathrm{d}\tau}{(3\tau^2 - Y^2)} \tag{4.19b}$$

Substituting into eq(4.11b) gives

$$\mathrm{d}\gamma' = \frac{-9\tau^2\,\mathrm{d}\tau}{E(3\tau^2 - Y^2)} + \frac{\mathrm{d}\tau}{G} \tag{4.20a}$$

Integrating by partial fractions gives

$$\gamma' = \frac{3}{E}\int_0^{\tau}\left(-1 + \frac{Y/2}{Y - \sqrt{3}\tau} + \frac{Y/2}{Y + \sqrt{3}\tau}\right)\mathrm{d}\tau + \int_0^{\tau}\frac{\mathrm{d}\tau}{G}$$

$$= \frac{3}{E}\left[-\tau + \frac{Y}{\sqrt{3}}\ln\left(\frac{Y + \sqrt{3}\tau}{Y - \sqrt{3}\tau}\right)\right] + \frac{\tau}{G} \tag{4.20b}$$

The corresponding dimensionless form of eq(4.20b) is

$$\frac{G\gamma'}{\sqrt{3}k} = \frac{\sqrt{3}}{4(1+\nu)}\ln\left(\frac{1 + \sqrt{3}T}{1 - \sqrt{3}T}\right) - \frac{(1 - 2\nu)T}{2(1+\nu)} \tag{4.20c}$$

The stress variations, predicted from eqs(4.18a) and (4.20b), are shown in Figs 4.3a-c.

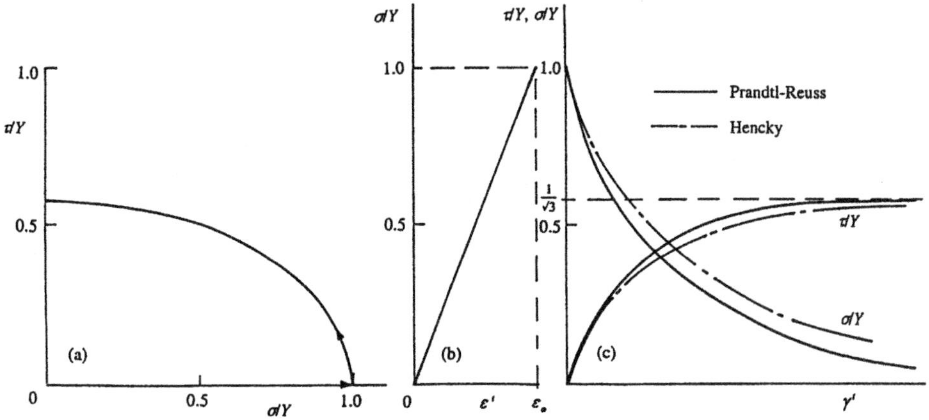

Figure 4.3 Tensile yield strain constrained under increasing torsion

In this case the axial strain (see Fig. 4.3b) becomes permanent as the corresponding stress reduces to zero.

When an initially elastic, axial strain ε_o^e is held constant under increasing torsion, the lower limit of the integral in eq(4.20b) is altered to τ_o. This provides the total shear strain for the ensuing deformation under $\tau > \tau_o$, as

$$\gamma^t = \frac{3}{E}\int_{\tau_o}^{\tau}\left(-1 + \frac{Y/2}{Y - \sqrt{3}\,\tau} + \frac{Y/2}{Y + \sqrt{3}\,\tau}\right)\mathrm{d}\tau + \int_{\tau_o}^{\tau}\frac{\mathrm{d}\tau}{G}$$

$$= \frac{3}{E}\left(-(\tau - \tau_o) + \frac{Y}{2\sqrt{3}}\ln\left[\frac{(Y + \sqrt{3}\,\tau)(Y - \sqrt{3}\,\tau_o)}{(Y - \sqrt{3}\,\tau)(Y + \sqrt{3}\,\tau_o)}\right]\right) + \frac{(\tau - \tau_o)}{G}$$

$$= \left(\frac{1}{G} - \frac{3}{E}\right)(\tau - \tau_o) + \frac{\sqrt{3}\,Y}{2E}\ln\left[\frac{(Y+\sqrt{3}\,\tau)(Y-\sqrt{3}\,\tau_o)}{(Y-\sqrt{3}\,\tau_o)(Y+\sqrt{3}\,\tau_o)}\right] \qquad (4.21)$$

Equation (4.21) will also apply to a stress path in which an initially elastic, axial stress σ_o, is allowed to alter under an increasing shear stress. This is because σ_o can only alter in a non-hardening material by maintaining the total axial strain constant. That is, $\mathrm{d}\varepsilon^t = 0$ for the ensuing deformation, when from eq(4.11a), the axial plastic strain will alter according to:

$$\mathrm{d}\varepsilon^P = (2\,\mathrm{d}\lambda/3)\sigma = -\,\mathrm{d}\sigma/E \qquad (4.22a)$$

Integration of eq(4.22a) reveals a conversion from initial elastic strain $\varepsilon_o^e = \sigma_o/E$, to plastic strain in the form

$$\varepsilon^P = (\sigma_o - \sigma)/E \qquad (4.22b)$$

where $\sigma_o > \sigma$. Clearly, $\varepsilon^P = \varepsilon_o^e$ when σ in eq(4.22b), falls to zero. The incremental plastic shear strain component $\mathrm{d}\gamma^P$ is associated with the first term in eq(4.11b). Dividing this by eq(4.22a) gives the ratio between the incremental plastic strains as

$$\mathrm{d}\gamma^P/\mathrm{d}\varepsilon^P = 3\tau/\sigma \qquad (4.23a)$$

All tension-torsion stress paths conform to the relationship given in eq(4.23a), showing that the plastic strain increment vector is aligned with the exterior normal to the yield surface. We can also show this from eq(4.8a), where the gradient of a tangent to this surface is given by $\mathrm{d}\sigma/\mathrm{d}\tau = -\,\sigma/(3\tau)$. It follows that the gradient to the normal equals the right-hand side of eq(4.23a) since the product of these gradients is -1. If we wish to separate the plastic component from eq(4.21), then eqs(4.22a) and (4.23a) are combined with eq(4.8a) to give

$$\mathrm{d}\gamma^P = \frac{-\sqrt{3}}{E}\left[\frac{\sqrt{Y^2 - \sigma^2}\,\mathrm{d}\sigma}{\sigma}\right]$$

$$\gamma^P = \frac{-\sqrt{3}\,Y}{E}\int_0^S \frac{\sqrt{1 - S^2}\,\mathrm{d}S}{S}$$

Making a trigonometric substitution, $S = \sin\theta$, leads to

$$\gamma^P = \frac{\sqrt{3}Y}{E}\left[\left(\frac{1}{S} + \sqrt{\sin^2 S - 1}\right)\sin^{-1} S + \sqrt{1 - S^2}\right] \qquad (4.23b)$$

The difference $\gamma' - \gamma^P$, found from eqs(4.21) and (4.23b), is identified with the elastic component of the total shear strain.

4.3.3 Neutral Loading Under Internal Pressure Combined with Torsion

Cylindrical co-ordinates will be employed for the plane stress state arising in the wall of a tube under combined torsion and internal pressure. That is, the non-zero stress components are written as σ_θ, σ_z and $\tau_{z\theta}$. The small magnitude of radial stress can be ignored when the wall is thin. Gill [14] conducted a neutral loading experiment on a tube of non-hardening, pre-strained, 70/30 brass. A total shear-prestrain of 0.44% was sufficient to ensure that the material did not harden further when combined stress components were adjusted to follow the contour of the yield locus. The latter is derived from eq(3.13a) as

$$\sigma_z^2 - \sigma_z\sigma_\theta + \sigma_\theta^2 + 3\tau_{z\theta}^{\ 2} = Y^2 \qquad (4.24a)$$

Note that under the internal pressure, where $\sigma_\theta = R\sigma_z$ ($R = 2.2$), two independent stress components are required to define the yield criterion. That is, eq(4.24a) appears in terms of σ_z and $\tau_{z\theta}$ as

$$Q\,\sigma_z^2 + 3\tau_{z\theta}^{\ 2} = Y^2 \qquad (4.24b)$$

where $Q = 1 - R + R^2$. It follows from our analysis of Hohensemer's experiment that neutral loading arises when the shear yield strain for the ensuing deformation is constrained:

$$\gamma_o^{\,t} = k/G = 2Y(1 + \nu)/(\sqrt{3}E) = \text{constant} \qquad (4.25)$$

The following classical theoretical solutions apply.

(a) *Hencky*
The total strains are, from eq(4.2a),

$$\begin{aligned}\varepsilon_z^{\,t} &= \phi\,\sigma_z + [(1 - 2\nu)/E - \phi](\sigma_z + \sigma_\theta)/3\\ &= \sigma_z\{\phi + 1/3(1 + R)[(1 - 2\nu)/E - \phi]\}\end{aligned} \qquad (4.26a)$$

$$\begin{aligned}\varepsilon_\theta^{\,t} &= \phi\,\sigma_\theta + [(1 - 2\nu)/E - \phi](\sigma_z + \sigma_\theta)/3\\ &= \sigma_z\{R\phi + 1/3(1 + R)[(1 - 2\phi)/E - \phi]\}\end{aligned} \qquad (4.26b)$$

$$\gamma_{z\theta}^{\,t} = 2\phi\,\tau_{z\theta} \qquad (4.26c)$$

where $R = \sigma_\theta/\sigma_z$. Putting $\gamma_{z\theta}^{\,t} = \gamma_o^{\,t} =$ constant in eq(4.26c) gives the scalar $\phi = \gamma_o^{\,t}/(2\tau_{z\theta})$. Substituting ϕ into eqs(4.26a,b) and eliminating $\tau_{z\theta}$ from eq(4.24b) gives the respective total axial and circumferential strains:

$$\varepsilon_z^t = \frac{(2 - R)\, \gamma_o^t\, \sigma_z}{2\sqrt{3\,(Y^2 - Q\,\sigma_z^2)}} + \frac{(1 + R)(1 - 2\,\nu)\,\sigma_z}{3E} \tag{4.27a}$$

$$\varepsilon_\theta^t = \frac{(2R - 1)\, \gamma_o^t\, \sigma_z}{2\sqrt{3\,(Y^2 - Q\,\sigma_z^2)}} + \frac{(1 + R)(1 - 2\,\nu)\,\sigma_z}{3E} \tag{4.27b}$$

From eq(4.25) and setting $S_z = \sigma_z / Y$, the dimensionless forms of eqs(4.27a,b) become

$$\frac{\varepsilon_z^t E}{Y} = \frac{(2 - R)(1 + \nu)\,S_z}{3\sqrt{(1 - Q\,S_z^{\,2})}} + \frac{(1 + R)(1 - 2\,\nu)\,S_z}{3} \tag{4.27c}$$

$$\frac{\varepsilon_\theta^t E}{Y} = \frac{(2R - 1)(1 + \nu)\,S_z}{3\sqrt{(1 - Q\,S_z^{\,2})}} + \frac{(1 + R)(1 - 2\,\nu)\,S_z}{3} \tag{4.27d}$$

(b) *Prandtl-Reuss*

The total incremental strains are, from eq(4.3b),

$$\mathrm{d}\varepsilon_z^{\,t} = (2\,\mathrm{d}\lambda/3)(\sigma_z - \sigma_\theta/2) + (1/E)(\mathrm{d}\sigma_z - \nu\,\mathrm{d}\sigma_\theta) \tag{4.28a}$$

$$\mathrm{d}\varepsilon_\theta^{\,t} = (2\,\mathrm{d}\lambda/3)(\sigma_\theta - \sigma_z/2) + (1/E)(\mathrm{d}\sigma_\theta - \nu\,\mathrm{d}\sigma_z) \tag{4.28b}$$

$$\mathrm{d}\gamma_{z\theta}^{\,t} = 2\,\mathrm{d}\lambda\,\tau_{z\theta} + \mathrm{d}\tau_{z\theta}/G \tag{4.28c}$$

Putting $\mathrm{d}\gamma_{z\theta}^{\,t} = 0$ in eq(4.28c)

$$\mathrm{d}\lambda = -\,\mathrm{d}\tau_{z\theta}/\,(2\,\tau_{z\theta}G) \tag{4.29a}$$

Differentiating eq(4.24b)

$$3\,\tau_{z\theta}\,\mathrm{d}\tau_{z\theta} = -\,Q\,\sigma_z\mathrm{d}\sigma_z \tag{4.29b}$$

Substituting eqs(4.29a,b) into eqs(4.28a,b,c) provides the total, axial and circumferential strain increments:

$$\mathrm{d}\varepsilon_z^t = \frac{Q(2 - R)}{6G} \times \frac{\sigma_z^2\,\mathrm{d}\sigma_z}{(Y^2 - Q\,\sigma_z^2)} + \frac{(1 - R\,\nu)\,\mathrm{d}\sigma_z}{E} \tag{4.30a}$$

$$\mathrm{d}\varepsilon_\theta^t = \frac{Q(2R - 1)}{6G} \times \frac{\sigma_z^2\,\mathrm{d}\sigma_z}{(Y^2 - Q\,\sigma_z^2)} + \frac{(R - \nu)\,\mathrm{d}\sigma_z}{E} \tag{4.30b}$$

Integrating eqs(4.30a,b) by partial fractions leads to dimensionless total strains:

$$\frac{E\varepsilon_z^t}{Y} = \frac{(1-2\nu)(1+R)S_z}{3} + \frac{(1+\nu)(2-R)}{6\sqrt{Q}}\ln\left(\frac{1+\sqrt{Q}S_z}{1-\sqrt{Q}S_z}\right) \tag{4.30c}$$

$$\frac{E\varepsilon_\theta^t}{Y} = \frac{(1-2\nu)(1+R)S_z}{3} + \frac{(1+\nu)(2R-1)}{6\sqrt{Q}}\ln\left(\frac{1+\sqrt{Q}S_z}{1-\sqrt{Q}S_z}\right) \tag{4.30d}$$

The predictions from eqs(4.27c,d) and (4.30c,d) are compared with Gill's results in Figs 4.4a-d, for $\nu = ⅓$.

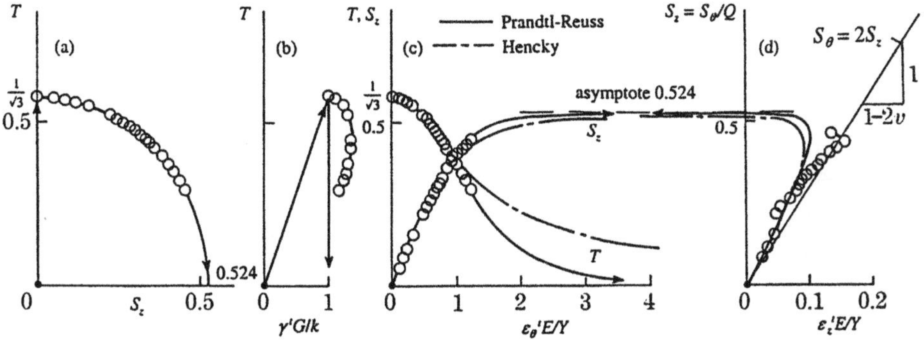

Figure 4.4 Neutral loading under combined internal pressure and axial torsion

It is seen from Fig. 4.4a that the relation between the stress components is in good agreement with the yield function (4.24b). Figure 4.4b confirms that the shear strain remains approximately constrained. Both predictions, given in Figs 4.4c,d, show that $T\ (= \tau/Y)$ diminishes to zero as $S_z\ (= \sigma_z/Y)$ becomes asymptotic to $1/\sqrt{Q}\ (= 0.524)$. Correspondingly, the predicted strains, $\varepsilon_\theta{}^t$ and $\varepsilon_z{}^t$, become infinite in tension and compression respectively. Because the experiment was terminated well before large values of these strains were achieved, it does not reveal which theory is the better. It is seen that the axial strain is extremely sensitive to the stress ratio. With $\sigma_\theta = R\sigma_z$ $(R = 2)$, both eqs(4.27c) and (4.30c) give an elastic axial strain response according to

$$\varepsilon_z{}^e E/Y = (1 - 2\nu)S_z \tag{4.31}$$

Equation (4.31) defines the line of slope 3 in Fig. 4.4d, which is apparently in closer agreement with the experimental data. Gill was surprised to find plastic deformation from his 'neutral' path but we should expect this path to result in plasticity for a non-hardening material. When a neutral path coincides with the initial yield condition in a hardening material, the stress state will remain at the yield point without producing plasticity.

4.3.4 Internal Pressure Combined with Axial Tension

When a thin walled, closed tube is pressurised the radial stress is negligible compared to the axial and circumferential stresses $(\sigma_z, \sigma_\theta)$. This allows a comparison to be made between the total and incremental theories within a biaxial, principal stress space. In Schlafer and Sidebottom's experiments [15], the axial and circumferential strain was constrained in separate experiments on a thin-walled cylinder of non-hardening, annealed SAE 1035 steel. In their first experiment, the cylinder was pre-strained to the uniaxial yield strain ε_o^t, under circumferential tension, then extended under axial tension with the pre-strain held constant. Here, the von Mises yield criterion (4.24a) reduces to a common form for the two theories:

$$\sigma_z^2 - \sigma_z\sigma_\theta + \sigma_\theta^2 = Y^2 \tag{4.32a}$$

The normalised form of eq(4.32a) is

$$S_z^2 - S_z S_\theta + S_\theta^2 = 1 \tag{4.32b}$$

where $S_z = \sigma_z/Y$ and $S_\theta = \sigma_\theta/Y$. If we wish to eliminate one stress component, say σ_θ, within the first quadrant of stress σ_θ, σ_z, eq(4.32a) gives

$$\sigma_\theta = \tfrac{1}{2}\sigma_z \pm \tfrac{1}{2}\sqrt{(4Y^2 - 3\sigma_z^2)} \tag{4.32c}$$

in which the positive discriminant applies.

(a) *Hencky*
Using cylindrical co-ordinates, the total strains are, from eq(4.2a),

$$\varepsilon_\theta^t = \phi\,\sigma_\theta + [(1-2\nu)/E - \phi](\sigma_z + \sigma_\theta)/3 = \varepsilon_o^t \tag{4.33a}$$
$$\varepsilon_z^t = \phi\,\sigma_z + [(1-2\nu)/E - \phi](\sigma_z + \sigma_\theta)/3 \tag{4.33b}$$

Equation (4.33a) supplies the scalar ϕ as

$$\phi = \frac{3E\varepsilon_o^t - (1-2\nu)(\sigma_z + \sigma_\theta)}{E(2\sigma_\theta - \sigma_z)} \tag{4.34}$$

Substituting eq(4.34) into eq(4.33b) leads to

$$\varepsilon_z^t = \frac{\varepsilon_o^t(2\sigma_z - \sigma_\theta)}{(2\sigma_\theta - \sigma_z)} + \frac{3(1-2\nu)(\sigma_z^2 - \sigma_\theta^2)}{3E(2\sigma_\theta - \sigma_z)} \tag{4.35a}$$

We may apply eq(4.35a) to the deformation ensuing from the yield strain $\varepsilon_o^t = Y/E$, under stress components connected through the yield criterion. The following normalised form of eq(4.35a) is found when σ_θ is eliminated from eq(4.32c):

$$\frac{E\varepsilon_z^t}{Y} = \frac{3\sigma_z/2 - 2\sqrt{Y^2 - 3\sigma_z^2/4} + [(1-2\nu)/Y]\left(Y^2 + \sigma_z\sqrt{Y^2 - 3\sigma_z^2/4} - 3\sigma_z^2/2\right)}{2\sqrt{Y^2 - 3\sigma_z^2/4}}$$

$$= \frac{3S_z/2 - 2\sqrt{1 - 3S_z^2/4} + (1-2\nu)\left(1 + S_z\sqrt{1 - 3S_z^2/4} - 3S_z^2/2\right)}{2\sqrt{1 - 3S_z^2/4}} \tag{4.35b}$$

where $\nu = 0.285$ and $E/Y = 1345.3$. The variation in $E\varepsilon_z^t/Y$ with S_θ follows from the simultaneous eqs(4.32b) and (4.35b). Equation (4.35b) gives the origin of normalised strain as $E\varepsilon_z^t/Y = -0.785$ (see Fig. 4.5) from substituting $S_z = 0$.

(b) *Prandtl-Reuss*

For the circumferential pre-straining, the volume of the cylinder material remains constant, giving

$$\mathrm{d}\varepsilon_r^P + \mathrm{d}\varepsilon_z^P + \mathrm{d}\varepsilon_\theta^P = 0$$

$$\therefore\ \mathrm{d}\varepsilon_r^P = \mathrm{d}\varepsilon_z^P = -\tfrac{1}{2}\,\mathrm{d}\varepsilon_\theta^P \qquad (4.36)$$

Now, from eqs(4.3a) and (4.36), $\mathrm{d}\varepsilon_z^t$ bears the following relationship to $\mathrm{d}\varepsilon_\theta^P$:

$$\mathrm{d}\varepsilon_z^t = \mathrm{d}\varepsilon_z^P + \mathrm{d}\varepsilon_z^e = -\tfrac{1}{2}\,\mathrm{d}\varepsilon_\theta^P - \nu\,\mathrm{d}\sigma_\theta/E \qquad (4.37a)$$

Integrating eq(4.37a), with limits of σ_θ from 0 to Y, defines a normalised axial strain origin:

$$\frac{E\,\varepsilon_z^t}{Y} = -\frac{E\,\varepsilon_\theta^P}{2Y} - \nu \qquad (4.37b)$$

That is, with $\varepsilon_\theta^P = Y/E$ and $\nu = 0.285$, eq(4.37b) gives $E\varepsilon_z^t/Y = -0.785$, which agrees with Hencky. Under constrained deformation, the total incremental strains are found from the Prandtl-Reuss eq(4.3b):

$$\mathrm{d}\varepsilon_z^t = (2\,\mathrm{d}\lambda/3)(\sigma_z - \sigma_\theta/2) + (1/E)(\mathrm{d}\sigma_z - \nu\,\mathrm{d}\sigma_\theta) \qquad (4.38a)$$

$$\mathrm{d}\varepsilon_\theta^t = (2\,\mathrm{d}\lambda/3)(\sigma_\theta - \sigma_z/2) + (1/E)(\mathrm{d}\sigma_\theta - \nu\,\mathrm{d}\sigma_z) \qquad (4.38b)$$

where σ_z is the total axial stress. Stress variations are found from putting $\mathrm{d}\varepsilon_\theta^t = 0$ in eq(4.38b). This gives

$$\mathrm{d}\lambda = -3(\mathrm{d}\sigma_\theta - \nu\,\mathrm{d}\sigma_z)/[E(2\sigma_\theta - \sigma_z)] \qquad (4.39a)$$

where from eq(4.32a)

$$\mathrm{d}\sigma_\theta = -(2\sigma_z - \sigma_\theta)\mathrm{d}\sigma_z/(2\sigma_\theta - \sigma_z) \qquad (4.39b)$$

Substituting eqs(4.39a,b) into eq(4.38a) leads to

$$\mathrm{d}\varepsilon_z^t = \frac{[2Y^2(2-\nu) + (2\nu - 1)(3\sigma_z\sigma_\theta - Y^2)]\mathrm{d}\sigma_z}{E(2\sigma_\theta - \sigma_z)^2} \qquad (4.40a)$$

Substituting eq(4.32c) into eq(4.40a) and integrating provides the total axial strain:

$$\varepsilon_z^t = \frac{Y^2(5-4\nu)}{E}\int_0^{\sigma_z}\frac{\mathrm{d}\sigma_z}{(4Y^2 - 3\sigma_z^2)} + \frac{3(2\nu-1)}{2E}\int_0^{\sigma_z}\frac{\sigma_z^2\,\mathrm{d}\sigma_z}{(4Y^2 - 3\sigma_z^2)} \pm \frac{3(2\nu-1)}{2E}\int_0^{\sigma_z}\frac{\sigma_z\,\mathrm{d}\sigma_z}{\sqrt{4Y^2 - 3\sigma_z^2}}$$

$$= \frac{Y(5-4\nu)}{4\sqrt{3}E}\ln\left(\frac{Y+\sqrt{3}\sigma_z/2}{Y-\sqrt{3}\sigma_z/2}\right) + \frac{Y(2\nu-1)}{2\sqrt{3}E}\left[\ln\left(\frac{Y+\sqrt{3}\sigma_z/2}{Y-\sqrt{3}\sigma_z/2}\right) - \frac{\sqrt{3}\sigma_z}{Y}\right] + \frac{(2\nu-1)}{E}\left[Y - \sqrt{Y^2 - \left(\frac{\sqrt{3}\sigma_z}{2}\right)^2}\right]$$

$$= \frac{\sqrt{3}Y}{4E}\ln\left(\frac{Y+\sqrt{3}\sigma_z/2}{Y-\sqrt{3}\sigma_z/2}\right) + \frac{(2\nu-1)}{E}\left[Y - \frac{\sigma_z}{2} - Y\sqrt{1 - \left(\frac{\sqrt{3}\sigma_z}{2Y}\right)^2}\right] \qquad (4.40b)$$

The corresponding dimensionless form of eq(4.40b) is

$$\frac{E\varepsilon_z^t}{Y} = \frac{\sqrt{3}}{4}\ln\left(\frac{1+\sqrt{3}S_z/2}{1-\sqrt{3}S_z/2}\right) - (1-2\nu)\left(1 - S_z/2 - \sqrt{1-3S_z^2/4}\right) \qquad (4.40c)$$

Figure 4.5a-c shows the Hencky and Prandtl-Reuss predictions, from eqs(4.35b) and (4.40c) respectively. These are consistent with the observation of Schlafer and Sidebottom, in that S_z increases continuously while S_θ increases initially before it decreases.

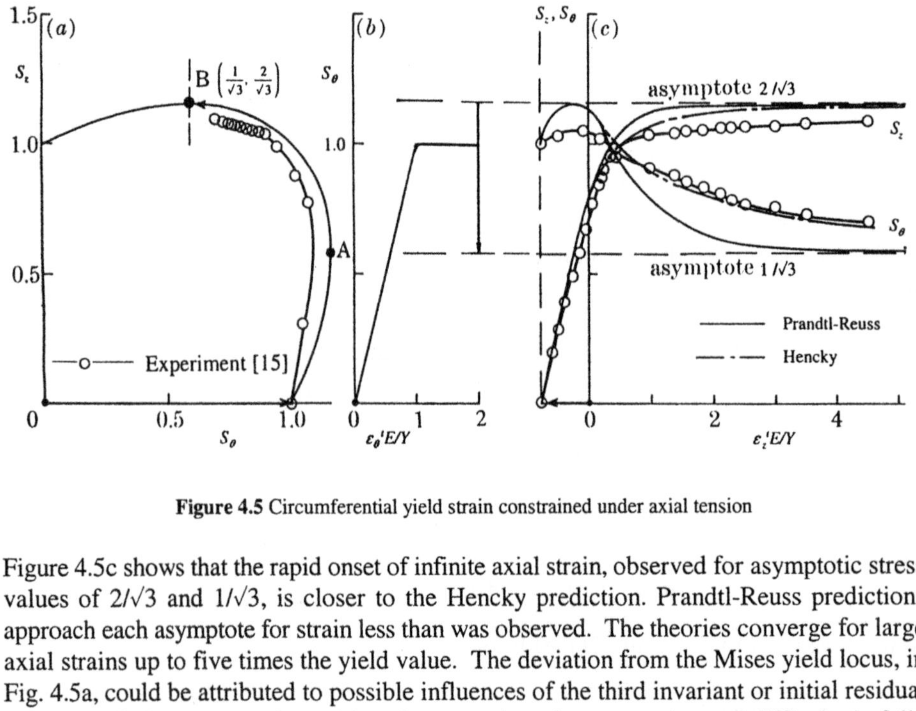

Figure 4.5 Circumferential yield strain constrained under axial tension

Figure 4.5c shows that the rapid onset of infinite axial strain, observed for asymptotic stress values of $2/\sqrt{3}$ and $1/\sqrt{3}$, is closer to the Hencky prediction. Prandtl-Reuss predictions approach each asymptote for strain less than was observed. The theories converge for large axial strains up to five times the yield value. The deviation from the Mises yield locus, in Fig. 4.5a, could be attributed to possible influences of the third invariant or initial residual anisotropy following annealing. Also, there may have been experimental difficulty in fully constraining the circumferential pre-strain, as it should appear from Fig. 4.5b.

Figure 4.6a-c shows that similar conclusions apply to the magnitude of the circumferential strain in a second test, on the same material, when an axial pre-strain was held constant. The deformation under increasing circumferential tensile stress may be deduced from replacing θ with z in eqs(4.35b) and (4.40c). This gives respective Hencky and Prandtl-Reuss predictions:

$$\frac{E\varepsilon_\theta^t}{Y} = \frac{3S_\theta/2 - 2\sqrt{1-3S_\theta^2/4} + (1-2\nu)\left(1 + S_\theta\sqrt{1-3S_\theta^2/4} - 3S_\theta^2/2\right)}{2\sqrt{1-3S_\theta^2/4}}$$

$$\frac{E\varepsilon_\theta^t}{Y} = \frac{\sqrt{3}}{4}\ln\left(\frac{1+\sqrt{3}S_\theta/2}{1-\sqrt{3}S_\theta/2}\right) - (1-2\nu)\left(1 - \frac{S_\theta}{2} - \sqrt{1-\frac{3S_\theta^2}{4}}\right)$$

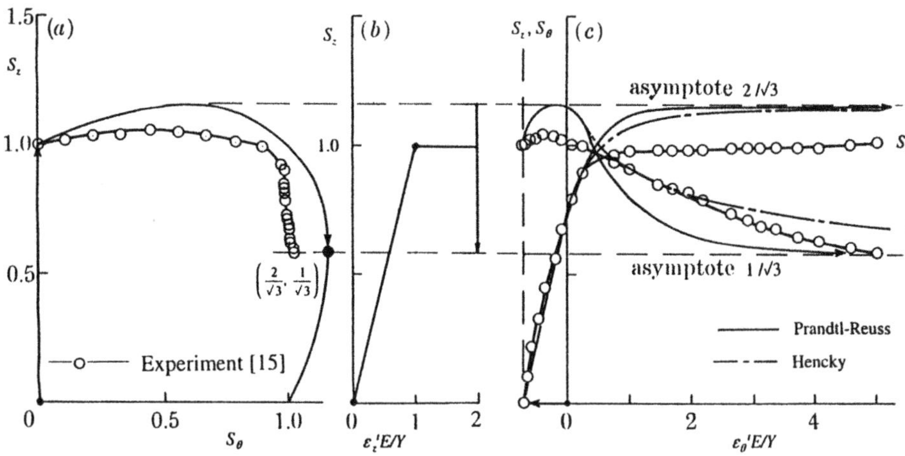

Figure 4.6 Axial yield strain constrained under circumferential tension

The corresponding origin in Fig. 4.6c is found from interchanging θ with z in eq(4.37b):

$$E\varepsilon_\theta'/Y = -(E\varepsilon_z^P)/(2Y) - \nu = -0.785$$

where $\varepsilon_z^P = Y/E$. The theoretical asymptotes in Figs 4.5c and 4.6c limit the stresses with greater accuracy than has been found from a Tresca yield criterion [15]. However, in practice, before the infinite theoretical strains ε_z' and ε_θ' in Figs 4.5c and 4.6c could be reached, the respective tube would become unstable from local necking and bulging.

4.4 Application of Classical Theory to Non-Uniform Stress States

Here we shall examine the ability with which the two classical theories of plasticity can provide solutions to elastic, perfect-plastic deformation under non-uniform stress states. The solutions to similar problems for hardening materials will be considered later in Chapter 10. For the purpose of making an experimental appraisal, results will be used for solid circular bars, subjected to different combinations of axial tension and torsion. It will be seen that, though constructed differently, the two theories supply comparable solutions to the deformation response and to the internal stress distributions from loading bars under relatively simple loading paths. The Hencky theory supplies closed-form equations that may be solved by trial. The Prandtl-Reuss equations, on the other hand, require numerical solutions, unless, as we have seen, the stresses are uniformly distributed. A further assumption of elastic incompressibility (i.e. $\nu = ½$), which lead to eqs(4.2b) and (4.3c), will simplify the application of each theory without incurring unacceptable error. Applying the 'incompressible' theory to a given stress state provides the loading corresponding to elastic-plastic and fully-plastic bars. These solutions also employ a criterion of yielding with the force and torque equilibrium equations.

In the case of a tensile force P, combined with a torque C, the equilibrium equations for a fully-plastic, solid bar of outer radius r_o, are

$$P = 2\pi \int_0^{r_o} \sigma_x r\,dr, \quad C = 2\pi \int_0^{r_o} \tau_x r^2\,dr \tag{4.41a,b}$$

For a thick-walled tube, the inner radius r_i becomes the lower limit of integration in eqs(4.41a,b). It is convenient to employ the following dimensionless forms of P and C for a solid bar :

$$n = \frac{P}{P_y} = \frac{P}{\pi Y r_o^2} \quad \text{and} \quad m = \frac{C}{C_y} = \frac{2\sqrt{3}C}{\pi Y r_o^3} \tag{4.42a,b}$$

where P_y and C_y, found from eqs(4.41a,b), apply to when the bar first becomes plastic under the separate action of tension and torsion. In a similar manner, the combined strains are normalised with respect to yield strains $\varepsilon_y = Y/E$ and $\gamma_y = Y/(\sqrt{3}G)$:

$$e = \varepsilon^t / \varepsilon_y = \varepsilon^t E / Y \tag{4.43a}$$

$$g = \gamma^t / \gamma_y = \sqrt{3}G\gamma^t / Y = \sqrt{3}G\,(\chi r_o)/Y \tag{4.43b}$$

where $\chi = \theta / l$ is thetwist/unit length giving χr_o as the outer diameter shear strain. The following loading paths facilitate comparison between theory with experiment.

4.4.1 Radial Loading

Let P and C increase proportionately (i.e. n/m = constant) from zero in a solid circular bar. The theoretical analyses each employ the von Mises yield criterion (4.8).

(a) *Hencky*

The total strain components are again expressed from eqs(4.9a,b) by assuming that any radial and hoop stress arising from P and C are negligibly small. Eliminating ϕ between eqs(4.8) and (4.9), the normalised axial stress $S = \sigma/Y$, for the plastic region, is found from the solution to

$$(1-2\nu)^2 S^4 - 6(1-2\nu)eS^3 + [9e^2 - (1-2\nu)^2 + 4(1+\nu)2g^2\rho^2]S^2 + 6(1-2\nu)eS - 9e^2 = 0 \tag{4.44}$$

Equation (4.44) applies to given values of $\rho = r/r_o$, e, and g from eqs(4.43a,b). The corresponding normalised shear stress $T' = \tau/k$, is found from eqs(4.8a) and (4.9b):

$$T' = \frac{2g\rho(1+\nu)}{\sqrt{[3e - S(1-2\nu)]^2 + 4(1+\nu)^2 g^2 \rho^2}} \tag{4.45}$$

Note that in eq(4.8b), $T = T'/\sqrt{3}$. Now, consider the elastic-plastic bar, where the plastic zone has penetrated from the outside radius to a radius r_{ep} in the cross-section. This interface radius is normalised $\rho_{ep} = r_{ep}/r_o$ so that, for the elastic core $0 \le \rho \le \rho_{ep}$, the normalised stresses are simply

$$S = e, \quad T' = g\rho \tag{4.46a,b}$$

The equilibrium equations (4.41a,b) for P and C, involve contributions from the elastic and plastic zones. The axial force becomes

$$P = 2\pi \int_0^{r_{ep}} \sigma r\,\mathrm{d}r + 2 \int_{r_{ep}}^{r_o} \sigma r\,\mathrm{d}r \tag{4.47a}$$

where, from eqs(4.42), (4.43) and (4.46), the dimensionless form becomes

$$n = 2\int_0^{\rho_{ep}} S(\rho)\rho\,d\rho + 2\int_{\rho_{ep}}^{1} S(\rho)\rho\,d\rho = e\rho_{ep}^2 + 2\int_{\rho_{ep}}^{1} S(\rho)\rho\,d\rho \qquad (4.47b)$$

The torque C is similarly normalised to give m:

$$C = 2\pi\int_0^{r_{ep}} \tau r^2\,dr + 2\pi\int_{r_{ep}}^{r_o} \tau r^2\,dr \qquad (4.48a)$$

$$m = 4\int_0^{\rho_{ep}} T'(\rho)\rho^2\,d\rho + 4\int_{\rho_{ep}}^{1} T'(\rho)\rho^2\,d\rho = g\,\rho_{ep}^4 + 4\int_{\rho_{ep}}^{1} T'(\rho)\rho^2\,d\rho \qquad (4.48b)$$

The second integral in eqs(4.47b) and (4.48b) may be evaluated numerically, e.g. by Simpson's rule. Here $S(\rho)$ and $T'(\rho)$ are the plastic zone stresses supplied by eqs(4.44) and (4.45). The condition that S and T' are singular on the elastic-plastic boundary is ensured from equating either (4.45) and (4.46b) or (4.44) and (4.46a). This gives

$$e^2 + g^2\rho_{ep}^2 = 1 \qquad (4.49)$$

When $\rho_{ep} = 1$, the cylinder is fully elastic (f.e.) and eqs(4.47)-(4.49) supply relationships between the applied loading:

$$n = e \text{ and } m = g \qquad (4.50a)$$

$$n^2 + m^2 = 1 \qquad (4.50b)$$

For a given plastic penetration ρ_{ep}, the solution to $S(\rho)$, $T'(\rho)$, n and m is found as follows. Firstly, assume a value for e and from eq(4.49), find g. At a given normalised radius ρ, where $\rho_{ep} \le \rho \le 1$ in the plastic zone, S and T' are found from eqs(4.44) and (4.45). A value for n/m follows from the integration of eqs(4.47b) and (4.48b). If this does not equal the true applied load ratio $n/m = (P/C)[r_o/2\sqrt{3})]$, found from eqs(4.42a,b), then another e value must be assumed and the procedure repeated until agreement is found. This method will supply the stress distributions rapidly when programmed for a computer. The original limit load theory [6] was simplified upon setting $\nu = ½$, from the assumption of elastic incompressibility. Equations (4.44) and (4.45) then provide the following closed solutions within the plastic region: $\rho_{ep} \le \rho \le 1$, for an elastic-plastic cylinder:

$$S = \frac{e}{\sqrt{e^2 + g^2\rho^2}}, \qquad T' = \frac{g\rho}{\sqrt{e^2 + g^2\rho^2}} \qquad (4.51a,b)$$

Since eqs(4.46a,b) remain valid for the elastic core, $0 \le \rho \le \rho_{ep}$, their substitutions, together with those of eqs(4.51a,b), into (4.47b) and (4.48b), lead to

$$n = e\rho_{ep}^2 + (2e/g^2)(e^2 + g^2)^{1/2} - (2e/g^2)(e^2 + g^2\rho_{ep}^2)^{1/2} \qquad (4.52a)$$

$$m = g\rho_{ep}^4 + (4/g)(e^2 + g^2)^{1/2} - [8/(3g^3)](e^2 + g^2)^{3/2}$$
$$- (4\rho_{ep}^2/g)(e^2 + g^2\rho_{ep}^2)^{1/2} + [8/(3g^3)](e^2 + g^2\rho_{ep}^2)^{3/2} \qquad (4.52b)$$

With a given n/m value, e and g are found from the simultaneous solution to eqs(4.49) and (4.52a,b). Take, for example, radial loading of a solid circular bar under $n/m = 0.8$ to an elastic-plastic interface radius ρ_{ep} = 0.4. We have from eq(4.49): $e^2 = 1 - 0.16g^2$. Substituting this into eqs(4.52a,b) leads to the following expression in g:

$$0.0204\, g + (3.18/g)(1 + 0.84g^2)^{1/2} - (2.12/g^3)(1 + 0.84g^2)^{3/2} - 0.509/g + 2.12/g^3$$
$$= 0.16\,(1 - 0.16g^2)^{1/2} + (2/g^2)(1 - 0.16g^2)^{1/2}(1 + 0.84g^2)^{1/2} - (2/g^2)(1 - 0.16g^2)^{1/2}$$

A trial solution gives $g = 1.19$ and hence $e = 0.879$, $m = 0.926$ and $n = 0.741$. Equations (4.46a,b) supply the elastic-zone stresses as: $S = e = 0.879$ and $T' = g\rho = 1.19\rho$. Equations (4.51a,b) supply the plastic-zone stresses as:

$$S = 1/\,(1 + 1.835\rho^2)^{1/2} \text{ and } T' = 1.355\rho/\,(1 + 1.835\rho^2)^{1/2}$$

In the fully plastic (f.p.) case, where $\rho_{ep} = 0$, eq(4.49) gives $e = 1$. Equations (4.52a,b) reduce to

$$n = (2e/g^2)(e^2 + g^2)^{1/2} - 2e^2/g^2 \tag{4.53a}$$

$$m = (4/g)(e^2 + g^2)^{1/2} - [8/(3g^3)](e^2 + g^2)^{3/2} + 8e^3/(3g^3) \tag{4.53b}$$

Equations(4.53a,b) give the following expression in g for the ratio $n/m = 0.8$.

$$(3.18/g)(1 + g^2)^{1/2} - (2.12/g^3)(1 + g^2)^{3/2} + 2.12/g^3 = (2/g^2)(1 + g^2)^{1/2} - 2/g^2$$

from which $g = 1.38$, $m = 0.93$ and $n = 0.744$. Here, the fully plastic stresses follow from eqs(4.51a,b):

$$S = 1/\,(1 + 1.896\rho^2)^{1/2} \text{ and } T' = 1.377\rho/\,(1 + 1.896\rho^2)^{1/2}$$

The strains corresponding to each ρ_{ep} value may be found from eqs(4.43a,b), given Y and E for the material. With increasing strain beyond full plasticity, i.e. $e > 1$ and $g > 1.38$, eqs(4.53a,b) show that n and m remain constant. Zyczkowski [6] has shown that when e and g are eliminated between eqs(4.53a,b), three relations between m and n apply:

$$m = (2/3)(1 - n)^{1/2}(2 + n) \tag{4.54a}$$
$$m = (2/3)(2 - 3n)/\,(1 - n)^{3/2} \tag{4.54b}$$
$$m = (2/3)(1 + n)^{1/2}(2 - n) \tag{4.54c}$$

It is seen from Fig. 4.7 that only eq(4.54a) expresses the true interaction between m and n at the f.p. load-limit carrying capacity.

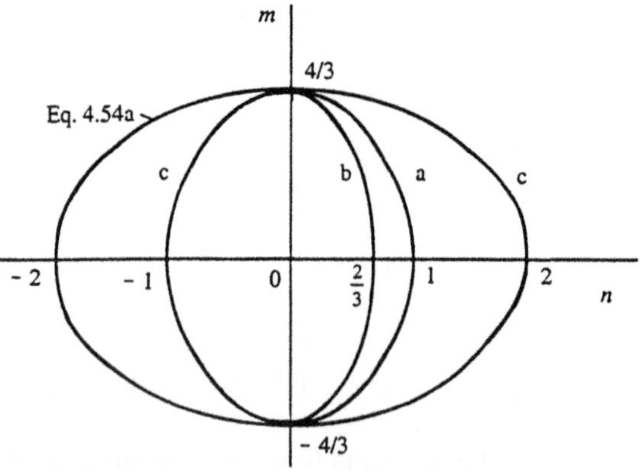

Figure 4.7 Interactions between n and m at limit loading

Thus, for the example cited above, eq(4.54a) is satisfied by $m = 0.93$ and $n = 0.744$. The inner curve b must be rejected because it does not give $n = 1$ under a fully-plastic tensile load. Figures 4.8a,b give the normalised stress distributions with incompressible elasticity. These are found from eq(4.51a,b) for 0.2 intervals in ρ_{ep} from the fully elastic (f.e.) to the fully plastic (f.p.) condition.

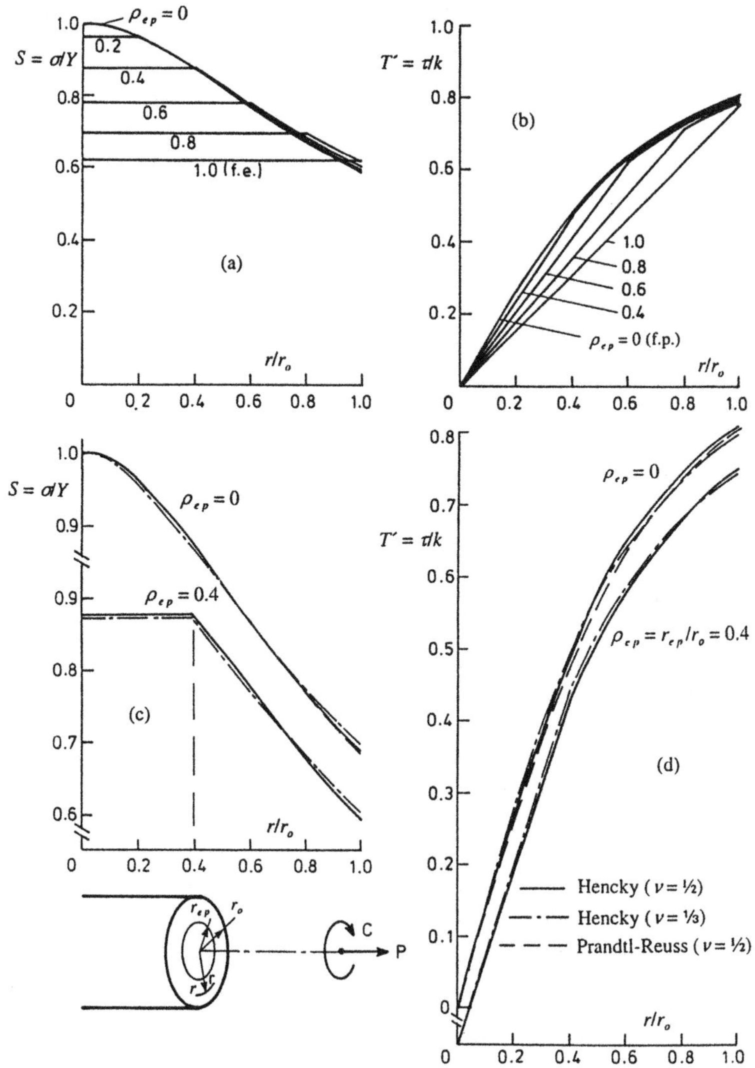

Figure 4.8 Stress distributions in a solid bar under combined tension and torsion

With full elasticity, the axial stress is uniform and the shear stress varies linearly with the radius. The penetration of a plastic zone increasingly destroys stress uniformity and linearity as shown. The compressible ($\nu = ⅓$) and incompressible ($\nu = ½$) Hencky solutions are compared within the enlarged scales of Figs 4.8c and d. These apply to an elastic-plastic condition ($\rho_{ep} = 0.4$) and to a fully plastic condition ($\rho_{ep} = 0$). The comparison shows negligible error in the stress magnitudes for $\nu = ½$. It follows that the incompressible theory provides realistic distributions of stress as the bar deforms from an initially elastic condition, through the elastic-plastic region, to attain full plasticity.

The strains increase rapidly to infinity under the limit loading in the manner of Fig. 4.9. Experimental data [16] is compared with the plot between n and m, predicted from eqs(4.42) and with a plot between the outer diameter axial and shear strain, ε and γ, from eqs(4.43).

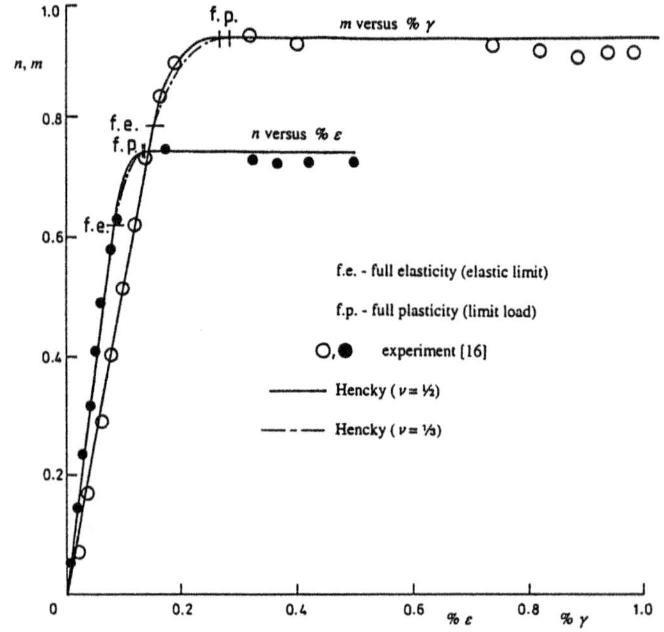

Figure 4.9 Load-deformation behaviour in a solid bar under combined tension-torsion

Clearly, limit loads from the incompressible, total-strain theory are accurate. Slight differences observed between total strain predictions, with and without incompressibility, are confined only to the region between the f.e. and f.p. conditions.

Strictly, a triaxial stress state will arise from the Hencky theory for a tension-torsion member of material with $\nu \neq ½$ [17]. The perturbation method of solution showed uniform radial and tangential compressive stresses within the elastic zone of a partially plastic bar. These stresses will depend upon the penetration depth, but do not amount to more than 10% of the yield stress within the plastic zone.

(b) *Prandtl-Reuss*

Gaydon [18] derived a closed-form, non-hardening, Prandtl-Reuss solution to this problem from assuming that $\nu = ½$ in eq(4.3a). Within the plastic region ($\rho_{ep} \leq \rho \leq \rho_o$), the total axial and shear strain increments, eqs(4.11a,b), were written as

$$d\varepsilon^t = dl/l = (2d\lambda/3)\sigma + d\sigma/(3G) \tag{4.55a}$$

$$d\gamma^t = r\,d\theta/l = 2d\lambda\tau + d\tau/G \tag{4.55b}$$

Eliminating $d\lambda$ between eqs(4.55a,b) gives

$$\frac{1}{l}\frac{dl}{d\theta} = \frac{\sigma}{3\tau}\frac{r}{l} - \left(\frac{\sigma}{3G\tau}\right)\frac{d\tau}{d\theta} + \frac{1}{3G}\frac{d\sigma}{d\theta} \tag{4.56a}$$

Since the stresses must satisfy the von Mises yield criterion (4.8a), σ in eq(4.56a) may be eliminated to give

$$\frac{3}{l} \times \frac{dl}{d\theta} = \frac{r}{\tau l}\sqrt{Y^2 - 3\tau^2} - \frac{Y^2}{G\tau\sqrt{Y^2 - 3\tau^2}} \times \frac{d\tau}{d\theta} \tag{4.56b}$$

With the substitutions $T' = \tau/k = \sqrt{3}\tau/Y$ and $g = \sqrt{3}(G/Y)(r_o/l_o)\theta$, where l_o is the initial length of the bar, eq(4.56b) becomes

$$\frac{dT'}{dg} = \frac{r\,l_o}{r_o l}\left(1 - T'^2\right) - \frac{\alpha T'}{\sqrt{3}}\sqrt{1 - T'^2} \tag{4.56c}$$

In eq(4.56c), $\alpha = (\sqrt{3}G/Yl)(dl/dg)$ is a variable which integrates to

$$\int \alpha\, dg = (\sqrt{3}G/Y)\ln(l/l_o) \tag{4.57a}$$

$$l/l_o = e^{\frac{Y}{\sqrt{3}G}\int \alpha\, dg} \tag{4.57b}$$

The axial strain is defined as

$$\varepsilon' = \ln(l/l_o) \tag{4.58a}$$

for which the normalised form is

$$e = \varepsilon'/(Y/E) = \sqrt{3}\int \alpha\, dg \tag{4.58b}$$

where $E = 3G$. Now $S = e$ and $T' = g\rho$ for the elastic core: $0 \le \rho \le \rho_{ep}$. Then, from eq(4.8), $S^2 + T'^2 = 1$ at the elastic-plastic interface. Substituting from eqs(4.49) and (4.58) gives

$$3\left(\int_0^g \alpha\, dg\right)^2 + (\rho_{ep}\, g)^2 = 1 \tag{4.59}$$

The integral term ( ) in eq(4.59) must take the constant value e at full elasticity ($\rho_{ep} = 1$). This condition is given by $e^2 + g^2 = 1$, where $e = n$ and $g = m$. Under radial loading, e is found from combining eqs(4.42a,b) and (4.50a,b) to give

$$[P/(\pi Y r_o^2)]^2 + [2\sqrt{3}C/(\pi Y r_o^3)]^2 = 1 \tag{4.60a}$$

$$e = n = P/(\pi Y r_o^2) = [1 + (12/r_o^2)(C/P)^2]^{-1/2} \tag{4.60b}$$

As the normalised load and torque are again found from eqs(4.47a) and (4.48a), the ratio $n/m = (Pr_o)/(2\sqrt{3}C)$ is a constant, since P/C = constant. The three equations (4.56c), (4.59) and (4.60b) are then solved by trial from assuming an initial value for α.

In the case of a fully plastic bar, putting $dT'/dg = 0$ and $l/l_o \approx 1$ in eq(4.56c), leads to

$$T'(\rho) = \rho/(\rho^2 + \alpha^2/3)^{1/2} \tag{4.61a}$$

$$S(\rho) = (1 - T'^2)^{1/2} = \alpha/(\alpha^2 + 3\rho^2)^{1/2} \tag{4.61b}$$

Also, with $\rho_{ep} = 0$, we have from eqs(4.47b) and (4.48b)

$$n/m = (Pr_o)/(2\sqrt{3}C) = [S(\rho)\rho\, d\rho]/\{2[T'(\rho)\rho^2 d\rho]\} \tag{4.62a}$$

Substituting eqs(4.61a,b) into eq(4.62a) leads to

$$\frac{Pr_o}{C} = \frac{\int_0^1 \alpha\rho\,\mathrm{d}\rho/(\alpha^2 + 3\rho^2)^{1/2}}{\int_0^1 \rho^3\,\mathrm{d}\rho/(\alpha^2 + 3\rho^2)^{1/2}} \tag{4.62b}$$

from which α may be solved. In the radial loading test employed by Shammamy and Sidebottom [16] the test conditions were $P/C = 0.374\ \mathrm{mm}^{-1}$ ($9.5\ \mathrm{in}^{-1}$) and $r_o = 7.37$ mm (0.29 in). The properties of the non-hardening, SAE 1035 steel used were: $Y = 283.3$ MPa (41.1×10^3 lbf/in^2) and $E = 207$ GPa (30×10^6 lbf/in^2). We find by trial that eq(4.62b) is satisfied by $\alpha = 1.261$. The stresses are then found from eqs(4.61a,b). The comparison between the Hencky and Prandtl-Reuss stress distributions at full plasticity in Fig. 4.8 reveals that these are almost coincident. A further check can be made on the assumption that $l / l_o \approx 1$ at full plasticity, from eqs(4.57b). Correspondingly, when $\rho_{ep} = 0$, eq(4.59) gives $\int \alpha \mathrm{d}g = 0.5773$. Substituting this, together with Y and $E = 3G$ into eq(4.57b), gives $l / l_o =$ 1.0014. The assumption is valid because, when the bar first becomes fully plastic, the strains are of an elastic order. It may be deduced from Fig. 4.8d that the distributions of stress, found from the 'incompressible' Prandtl-Reuss theory, are again realistic as the bar deforms in the elastic-plastic region.

Sved and Brooks [19] have given the governing equations for a Prandtl-Reuss solution to this problem when accounting for compressible elasticity. When $\nu \neq ½$, it becomes necessary to admit the tangential and radial stresses in addition to the axial and shear stresses. They showed that this leads to the requirement for a numerical solution to a system of four, simultaneous, partial differential equations. The additional stresses, so found, were of similar small magnitude to those found from the compressible Hencky theory [17]. Tangential and radial stresses are absent when $\nu = ½$ but, as we have seen, a different ν does not alter the more dominant axial and shear stresses by more than a few percent.

4.4.2 Constrained Deformation

Consider a deformation path where the tensile yield strain ε_y is held constant while increasing the shear strain γ, within a thick-walled tube of non-hardening material. The theories predict the following variations in both P and C with γ.

(a) *Hencky*

According to eqs(4.11), when $\nu = ½$ and $\varepsilon_y = Y/E$, the stresses during the ensuing deformation are given by

$$\sigma = 3\varepsilon_y / (2\phi), \quad \tau = \gamma / (2\phi) \tag{4.63a,b}$$

Substituting eqs(4.63a,b) into eq(4.8a), supplies the scalar multiplier ϕ

$$\phi = \surd(9\varepsilon_y^{\,2} + 3\gamma^2) / (2Y) \tag{4.64}$$

Equations (4.63a,b) and (4.64) provide the following dimensionless stress components:

$$S = (1 + \rho^2 g^2)^{-1/2} \quad \text{and} \quad T' = \rho g\,(1 + \rho^2 g^2)^{-1/2} \tag{4.65a,b}$$

where

$$S^2 + T'^2 = 1 \tag{4.65c}$$

Note that $T' = \tau/k$, which is not to be confused with $T = \tau/Y$, is used for a uniform stress analysis given in Section 4.3. The relationship $E = 3G$ is implied with assumed incompressibility. Since the tube is fully plastic, it follows from the corresponding forms of eqs(4.41a,b) and (4.42a,b) that

$$n = \frac{2}{(1 - D^2)} \int_0^1 \rho S \, \mathrm{d}\rho \,, \quad m = \frac{4}{(1 - D^4)} \int_0^1 \rho^2 T' \mathrm{d}\rho \tag{4.66a,b}$$

where $D = r_i / r_o < 1$, is the tube diameter ratio. Combining eqs(4.65a,b) and (4.66a,b), leads to the integrated forms for n and m

$$n = \frac{2\left[(1 + g^2)^{\frac{1}{2}} - (1 + D^2 g^2)^{\frac{1}{2}}\right]}{g^2(1 - D^2)} \tag{4.67a}$$

$$m = \frac{4\left[(1 + g^2)^{\frac{1}{2}} - \frac{2}{3g^2}(1 + g^2)^{\frac{3}{2}} - D^2(1 + D^2 g^2)^{\frac{1}{2}} + \frac{2}{3g^2}(1 + D^2 g^2)^{\frac{3}{2}}\right]}{g(1 - D^4)} \tag{4.67b}$$

With a given value of g for the ensuing deformation, the distribution of stress through the wall of the cylinder is supplied by eq(4.65a,b). Equations (4.67a,b) express the loading necessary to achieve that stress state.

(b) *Prandtl-Reuss*

The total, incremental strains are given by the Prandtl-Reuss eqs(4.11a,b). Since $\mathrm{d}\varepsilon' = 0$, it follows from eqs(4.8a) and (4.11a) that the scalar $\mathrm{d}\lambda$ is

$$\mathrm{d}\lambda = -\,3\mathrm{d}\sigma/\,(2E\sigma) = 9\tau\,\mathrm{d}\tau/\,(Y^2 - 3\tau^2) \tag{4.68}$$

Substituting eq(4.68) into eq(4.11b) and integrating between 0 and τ leads to the γ^t prediction, given by Smith and Sidebottom [20]:

$$\gamma^t = r\chi = (2\nu - 1)\tau/E + (\sqrt{3}Y/E)\tanh^{-1}(\sqrt{3}\tau/Y) \tag{4.69a}$$

Employing the present dimensionless parameters, eq(4.69a) becomes

$$(1 + \nu)g\rho = (\nu - ½)T' + ¾ \ln\left[(1 + T')/\,(1 - T')\right] \tag{4.69b}$$

Thus, T' can be found from eq(4.69b) for a given g over the range $D \le \rho \le 1$. S follows from eq(4.65c) and n and m from the numerical integration of eqs(4.66a,b). The incompressible solution is found from putting $\nu = ½$, giving a relationship between the elastic constants as $1/E = 1/(9K) + 1/(3G)$. Equations (4.65c) and (4.69b) then supply dimensionless stresses in closed form:

$$S = \mathrm{sech}\,(\rho g\,), \quad T' = \tanh\,(\rho g\,) \tag{4.70a,b}$$

Equations (4.70a,b) originally appeared in the work of Gaydon [18], Prager and Hodge [21] and Nadai [22]. Gaydon also solved the case of loading where the axial strain was initially elastic. Combining eqs(4.66) and (4.70), the corresponding loading parameters become

$$n = \frac{2}{(1 - D^2)} \int_D^1 \rho \operatorname{sech}(\rho g)\, d\rho\,, \quad m = \frac{4}{(1 - D^4)} \int_D^1 \rho^2 \tanh(\rho g)\, d\rho \qquad (4.71a,b)$$

The few published results, which enable comparison with this theory, are again due to Shammamy and Sidebottom [16]. These apply to the same grade of non-hardening SAE 1035 steel in a tube with $D = 0.93$. Figures 4.10a and b compare the theoretical stress distributions and, in Fig. 4.10c, a comparison is made between experiment and the predictions from eqs(4.66a,b), (4.67a,b) and (4.71a,b).

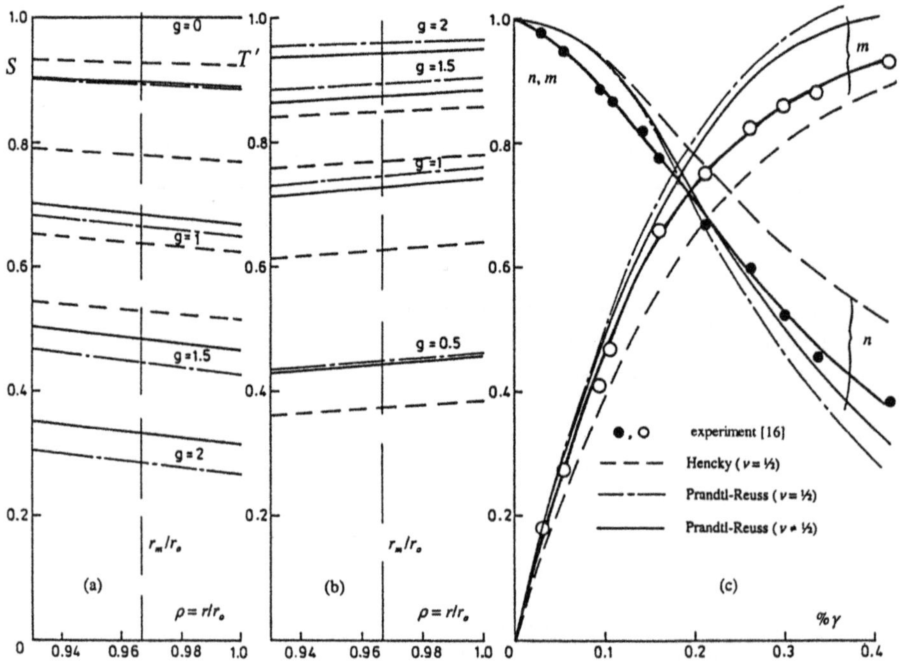

Figure 4.10 Constrained deformation in a thick-walled tube under combined tension and torsion

It is seen from Fig. 4.10c that deviations between the three predictions increase with increasing shear deformation. The greatest deviation in the stress distributions, for a given g value in Figs 4.10a and b, occurs between the incremental and total strain theories when $\nu = ½$. This further applies in respect of the loading parameters n and m in (c), although experiment is unable to distinguish clearly between the various theoretical predictions. Again, the $\nu = ½$ assumption leads to acceptable simplified solutions. A similar test, conducted on a solid bar of non-hardening material, would be more conclusive for making a comparative study of this nature. Because it is evident from Figs 4.10a,b that the wall stress gradients for this tube are not severe, a thin-walled theory would approximate its singular wall stress value to the indicated mid-wall radius.

As the material in Fig. 4.10 is non-hardening, the theory need not account for the influence of the initial 0.2% plastic loading strain. In fact, a general feature of the Hencky theory is that it applies only to the current stress state, i.e. it does not refer to how that state was achieved. This deficiency is corrected with an incremental strain summation and explains why the Prandtl-Reuss theory is generally believed to be more representative of path dependent plasticity under non-proportional loading.

Sved and Brooks [19] applied the compressible Prandtl-Reuss solution to the converse problem in which the twist was held constant at its yield value, in a solid bar, under increasing axial strain. They showed that radial and hoop stresses also arose with maximum values in compression at the bar centre. These were never greater than 5% of the shear yield stress, even at full plasticity and low values $\nu \approx 0.1$. Their work confirmed that the variation in Poisson's ratio is of sufficiently small influence, justifying Gaydon's [17] application on an incompressible Prandtl-Reuss theory.

4.4.3 Stepped Loading

Here an elastic torque C_o is held constant under an increasing axial force P (Fig. 4.11a). This gives $m < 1$, and with $\nu = ½$, it is first necessary to find n from eq(4.50a) for the fully-elastic condition ($\rho_{ep} = 1$). For an increasing n within the elastic-plastic region ($0 \le \rho_{ep} \le 1$), eqs(4.46a,b), (4.49), (4.51a,b) and (4.52a,b) are applied to determine e, g and n for a given $\rho_{ep} = r_{ep}/r_o$ and m.

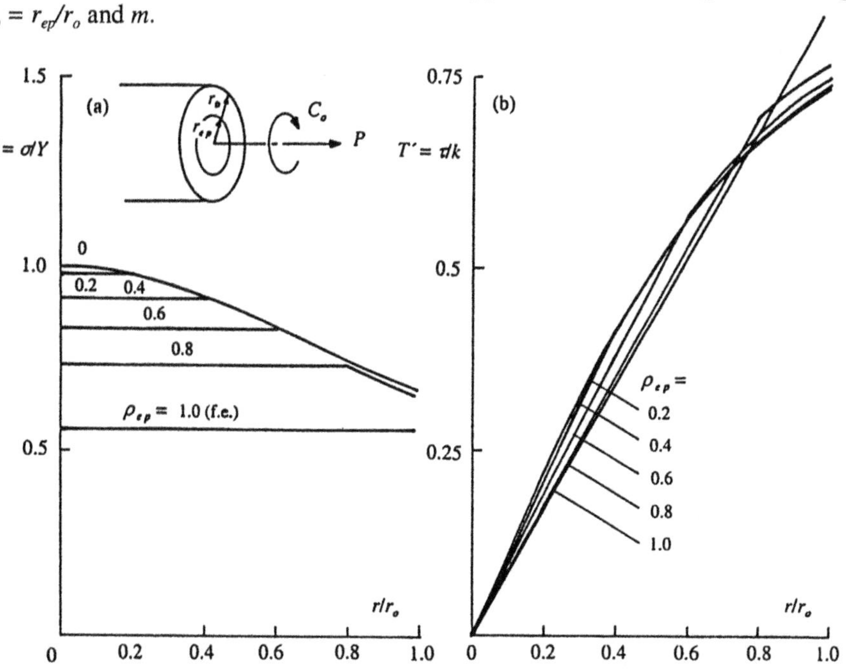

Figure 4.11 Normalised stress distributions for a stepped loading of increasing P under constant C_o

The following procedure applies to the simplified 'incompressible' Hencky theory. Say we wish to establish the stress distributions for $m = 0.83$ = constant, where the force P has produced: (i) a fully elastic bar ($\rho_{ep} = 1$) and (ii) an elastic-plastic bar with $\rho_{ep} = 0.6$. For (i), it follows from eqs(4.46a,b) that $T' = m\rho = 0.83\rho$ (i.e. linear in ρ) in which $0 \le \rho \le 1$ and $S = n = 0.56$. For condition (ii), eq(4.49) gives $e^2 = 1 - 0.36g^2$. Substituting into eq(4.52b):

$$0.83 = 0.13g + (4/g)(1 + 0.64g^2)^{1/2} - [8/(3g^3)](1 + 0.64g^2)^{3/2} - 1.44/g + 8/(3g^3) \quad (4.72)$$

A trial solution to eq(4.72) gives $g = 0.934$ and $e = 0.828$. Substituting the appropriate e, g and ρ_{ep} values into eq(4.52a), provides $n = 0.77$ as a dimensionless measure of the increased force necessary for partial yielding of the section. Within the elastic zone ($0 \le \rho \le 0.6$), the stresses are found from eqs(4.46a,b) to be $S = e = 0.828$ and $T' = g\rho = 0.934\rho$. In the plastic zone ($0.6 \le \rho \le 1$), eqs(4.51a,b) give

$$S = 1/(1 + 1.272\rho^2)^{1/2} \quad \text{and} \quad T' = 1.128\rho/(1 + 1.272\rho^2)^{1/2} \qquad (4.73\text{a,b})$$

The normalised stresses, from eq(4.73a,b), are distributed according to Figs 4.11a and b. Other distributions shown are calculated in a similar manner for interface radii: $\rho_{ep} = 0$ (i.e. fully plastic), 0.2, 0.4, 0.8 and 1.0 (i.e. fully elastic). For each ρ_{ep} value, a zero shear stress occurs at the bar centre, corresponding to maximum axial stress. Note that the elastic and elastic-plastic stresses must agree at the interface radius. Experiments [18] show that even in mildly hardening materials, n and m continue to increase beyond their fully plastic values. The rule of *isotropic hardening* can account for this by allowing the yield surface to expand (see p. 313). This rule has enabled a corresponding extension to the Hencky theory to account for the greater load carrying capacity which most metallic materials display [23].

4.4.4 Residual Stress Distributions

Self-equilibrating residual stresses result from unloading an elastic-plastic bar. They are found from subtracting elastic stresses, S_e and T_e', that recover with unloading, from elastic-plastic stresses that arise when the loading is applied. For example, with radial loading of incompressible elastic material in Fig. 4.8, the Hencky eqs(4.46a,b), (4.50a,b) and (4.51a,b), supply the following normalised, residual stress expressions

$$S_R = S - S_e \qquad (4.74\text{a})$$
$$= (1 + g^2\rho^2/e^2)^{-1/2} - n \qquad (4.74\text{b})$$

$$T_R' = T' - T_e' \qquad (4.75\text{a})$$
$$= (g/e)\rho\,(1 + g^2\rho^2/e^2)^{-1/2} - \rho m \qquad (4.75\text{b})$$

where e, g, m and n are found from eqs(4.49) and (4.52a,b) for a given n/m and ρ_{ep}. The distributions, found from eqs(4.74b) and (4.75b), are shown in the upper two diagrams of Fig. 4.12a. These correspond to an elastic-plastic cylinder with $\rho_{ep} = 0.6$ and a fully plastic cylinder (i.e. $\rho_{ep} = 0$). The latter condition increases the magnitude of the residual stresses following radial loading of a non-hardening material.

The residual stress distributions, S_R and T_R', resulting from the release of a non-radial loading in Fig. 4.10 need separate consideration. When eqs(4.74a) and (4.75a) are applied to the constrained test conditions, S_R and T_R' are distributed in the manner of Fig. 4.12b. Clearly, for constrained deformation, the residuals show a greater dependency upon the particular theory chosen to determine S and T'. That is, differences between the Hencky and Prandtl-Reuss predictions, shown here for $g = 2$, are a consequence of variations between the theoretical loading stresses given in Figs. 4.10a,b. In Fig. 4.12c, the residuals are shown for $\rho_{ep} = 0$ and 0.6, based upon the Hencky applied stress distributions (Fig. 4.11) under stepped loading. These residuals are similar to those following the release of a radial loading (Fig. 4.12a) with comparable elastic-plastic interface radii. It has long been known that residual stresses can have the beneficial effect of enhancing fatigue life. Here, it becomes important to establish the residual stresses with reasonable accuracy, particularly in the case of non-radial loading, where large errors can arise from simplifying the solution to the resulting stress [24, 25].

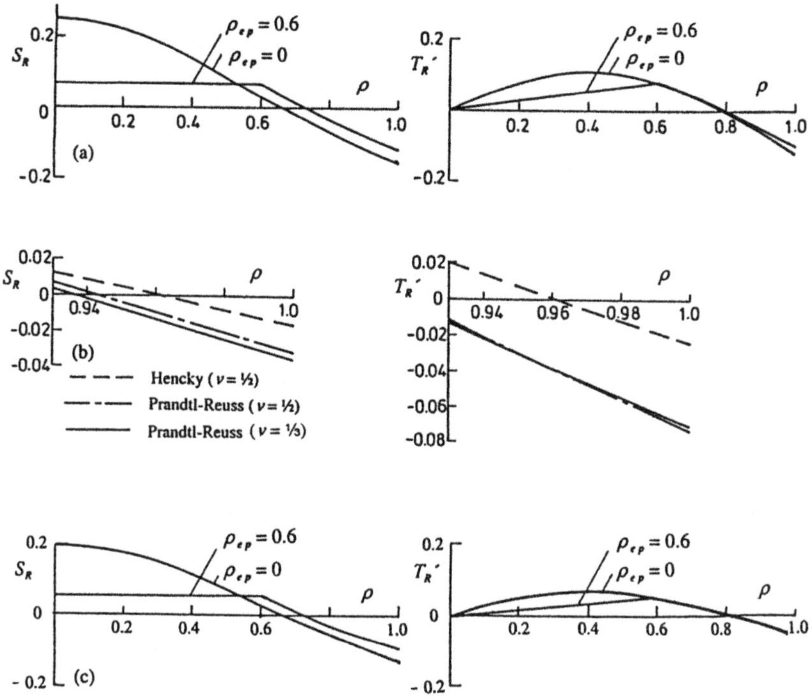

Figure 4.12 Residual stress resulting from various loadings

4.5 Hencky versus Prandtl-Reuss

Choosing between the two classical descriptions of deformation arising from a combined stress state will depend upon the manner in which stress is distributed. When the latter is uniform, there is little difference between predictions from the Hencky deformation theory and the Prandtl-Reuss flow theory. In fact, experiment appears to agree better with the simpler, closed-form Hencky predictions to deformation under non-radial loading paths. In the analysis of the deformation behaviour of structures under non-uniform stress states, experiment shows that the effects of history are insignificant with proportional loadings from zero stress. The Hencky total-strain deformation theory will represent, within closed solutions of acceptable accuracy, the distribution of stress and the load-deformation response with the transition from elastic to fully plastic behaviour. Deviations arise between the total and incremental theories under non-proportional loadings, including stepped-stress paths or paths in which one component of strain is constrained. Although the Hencky theory appears to remain conservative in its prediction of strain, the generally held view, partly substantiated here, is that the incremental Prandtl-Reuss theory (with elastic compressibility) provides more realistic predictions whenever history effects are present. However, it is known that errors can arise from Prandtl-Reuss. We have seen in Chapter 3 that it does not account for plasticity in anisotropic material. Other deficiencies include the neglect of softening resulting from stress reversal and cold creep, i.e. low-temperature, time-dependent strain.

References

1. Hencky H. *Zeit Angew Math Mech*, 1924, **4**, 323.
2. Ilyushin A. A. *Prik Matem Mekh*, AN SSSR, 1946, **10**, 347.
3. Nadai A. *Plasticity: A Mechanics of the Plastic State of Matter*, 1931, McGraw-Hill, New York.
4. Prandtl L. *Proc. 1st Int Congr Appl Mech. Delft*, 1924, p.43 (also *Zeit Angew Math Mech* 1928, **8**, 85.)
5. Reuss A. *Zeit Angew Math Mech*, 1930, **10**, 266.
6. Zyczkowski M. *Combined Loadings in the Theory of Plasticity*, 1981, PWN, Warsaw.
7. Massonnet C. *Plasticity in Structural Engineering, Fundamentals and Applications*, 1979, Ch. 1, Springer Verlag.
8. Hohensemer K. *Zeit Angew Math Mech*, 1951, **11**, 515.
9. Prager W. *J. Appl. Phys*, 1944, **15**, 65.
10. Hohensemer K. and Prager W. *Zeit Angew Math Mech*, 1932, **12**, 1.
11. Prager W. Proc: *5th Int Congr Appl Mech*, Cambridge, 1938, S.234.
12. Neuber H. *Kerbspannungslehre, 2 Aufl*, Berlin Gottingen, Heidelberg 1958.
13. Betten J. *Zeit Angew Math Mech*, 1975, **55**, 119.
14. Gill S. S. *J. Appl. Mech*, 1956, **23**, 497.
15. Schlafer J. L. and Sidebottom O. M. *Expl Mech*, 1969, **9**, 500.
16. Shammamy M. R. and Sidebottom O. M. *Expl Mech*, 1967, **7**, 497.
17. Zyczkowski M. *Rozpr Inz*, Polska Akademia, 1955, **3**, 285.
18. Gaydon F. A. *Quart. J. Mech. and Appl Math*, 1952, **5**, 29.
19. Sved G. and Brooks D. S. *Acta Techn. Hung*, 1965, **50**, 337.
20. Smith J. O. and Sidebottom O. M. *Inelastic Behaviour of Load-Carrying Members*, 1965, J. Wiley and Son.
21. Prager W. and Hodge P. *Theory of Perfectly Plastic Solids*, 1951, J. Wiley and Son, New York.
22. Nadai A. *Theory of Flow and Fracture of Solids*, 1950, McGraw-Hill, London.
23. Rees D. W. A. *Applied Solid Mechanics* 1988, Chapter 17, 321-346 (Eds Tooth A. S. and Spence J.) , Elsevier.
24. Hoskin B. C. *Austral. J. Appl. Sci*, 1961, **12**, 255.
25. Mii H. J. *Japan Soc Appl Mech*, Tokyo, 1950, **3**, 195.

Exercises

Uniform Stress States

4.1 A thin-walled cylinder is stressed to the point of yield under an axial-shear stress ratio $\sigma/\tau = 2$. The shear strain is then held constant while the cylinder is subjected to increasing tension. Derive expressions for the subsequent variation in axial strain with the axial and shear stresses (σ, τ) according to the Hencky and Prandtl-Reuss theories. Compare, graphically showing the normalised stress asymptotes σ/Y and τ/Y. Take $G = 77$ GPa, $\nu = 0.3$ and $Y = 230$ MPa.

4.2 A thin-walled cylinder is stressed to the point of yield under an axial-shear stress ratio $\sigma/\tau = 2$. The axial strain is then held constant while the cylinder is subjected to increasing torsion. Derive expressions for the subsequent variation in shear strain with the axial and shear stresses (σ, τ) according to the Hencky and Prandtl-Reuss theories. Compare graphically showing the normalised stress asymptotes σ/Y and τ/Y. Take $G = 77$ GPa, $\nu = 0.3$ and $Y = 230$ MPa.

4.3 The internal pressure in a closed-end, thin tube is increased to produce yielding of the tube material under a ratio $\sigma_\theta / \sigma_z = 2$. The circumferential strain is then held constant while the tube is progressively twisted. Show, according to the Prandtl-Reuss theory, that τ varies with total shear strain γ^t as

$$\gamma^t = \frac{k(2 - \nu)}{2G(1 + \nu)} \ln\left(\frac{1 + T'}{1 - T'}\right) - \frac{(1 - 2\nu)T'}{(1 + \nu)}$$

where $T' = \tau/k$. Why is axial plastic strain absent when the radial stress is ignored?

4.4 In a thin tube of non-hardening material the axial strain at yield is held constant. Show, with progressively increasing circumferential tension, that the Prandtl-Reuss prediction to the ensuing hoop strain is given by

$$\varepsilon_\theta^t = \frac{\sqrt{3}Y}{4E} \ln\left(\frac{1 + \sqrt{3}S_\theta/2}{1 - \sqrt{3}S_\theta/2}\right) + \frac{Y(2\nu - 1)}{E}\left[\left(1 - \frac{3S_\theta}{2}\right) - \sqrt{1 - \frac{3S_\theta^2}{4}}\right]$$

where $S_\theta = \sigma_\theta/E$. Hence, show that the normalised stress asymptotes are $\sigma_\theta/Y = 2/\sqrt{3}$ and $\sigma_z/Y = 1/\sqrt{3}$.

4.5 A thin-walled tube is twisted under torsion to the point of yield. The total shear strain is then held constant while the tube is subjected to an increasing internal pressure for which $\sigma_\theta = 2\sigma_z$. Show, for the ensuing deformation, that the respective total circumferential and axial strains are given by:

$$\varepsilon_\theta^t = \frac{Y}{E}\left[(1 - 2\nu)S_z + \frac{(1 + \nu)}{2\sqrt{2}} \ln\left(\frac{1 + \sqrt{2}S_z}{1 - \sqrt{2}S_z}\right)\right]$$

$$\varepsilon_z^t = \frac{Y}{E}(1 - 2\nu)S_z$$

where $S_z = \sigma_z/Y$. Plot S_z versus $\varepsilon_\theta{}^t E/Y$ and $\varepsilon_z{}^t E/Y$ showing the S_z asymptotes.

4.6 A non-hardening, thin-walled cylinder, with inner and outer radii 7.1 and 7.6 mm respectively, is strained to the tensile yield point. Thereafter the cylinder is subjected to an increasing torque whilst the simultaneous tension is adjusted to maintain the tensile yield strain Y/E constant. Compare predictions to the deformation that ensues from the Hencky incompressible theory ($\nu = ½$) and the Prandtl-Reuss theory with $\nu = ¼$. Take $E = 207$ GPa and $Y = 283$ MPa.

Non-Uniform Stress States

4.7 A solid cylindrical bar, 14.75 mm diameter, is made from a non-hardening steel with a yield stress of 283 MPa. Given that the torque C and the axial load P increase proportionately in the ratio $P/C = 0.375\ \text{mm}^{-1}$, find P and C: (i) at the limit of elastic deformation, (ii) at the elastic-plastic mean radius and (iii) for a fully plastic condition. Assume elastic incompressibility with $E = 207$ GPa. [Answer: (i) 80.5 Nm, 30.1 kN, (ii) 94.12 Nm, 35.3 kN, (iii) 95.7 Nm, 35.8 kN]

4.8 Plot for the cylinder in exercise 4.7, the distribution of tensile and shear stress with radius for each of the conditions (i), (ii) and (iii). Assuming elastic unloading, plot the distribution of residual stress corresponding to (ii) and (iii). Take $E = 207$ GPa and assume elastic incompressibility.

4.9 A solid cylindrical bar, 12.8 mm diameter, is made from a non-hardening steel with a yield stress of 305 MPa. The bar is subjected to a constant torque of 60.7 Nm with a steadily increasing axial force. Calculate the value of the force corresponding to: (i) the limit of elasticity, (ii) an elastic-plastic interface coincident with the mean radius and (iii) full plasticity. Take $E = 207$ GPa and $\nu = ½$.
[Answer: 21.9 kN, 31.05 kN, 31.6 kN]

4.10 Plot the distributions of axial and shear stress with radius for the cylinder of exercise 4.9 in each of (i), (ii) and (iii). Determine which condition (ii) and (iii) has the greatest extent of the residual compression following unloading.

4.11 Apply the Prandtl-Reuss solution, with compressible elasticity, to the deformation in a solid bar subjected to torque and axial load which increase proportionately within the plastic range.

4.12 Derive expressions for the normalised stress components $S = \sigma_z / Y$ and $T' = \tau / k$ from the incompressible Prandtl-Reuss theory when the shear strain in a solid circular bar is increased while the elastic axial strain is held constant.

4.13 A solid circular bar supports a constant, initially elastic-plastic torque while being subjected to increasing tension. Determine the Hencky stress distributions for suitable stages of the ensuing deformation to full plasticity.

4.14 Derive the Prandtl-Reuss solutions, for $\nu = ½$, to the stress distributions in a solid circular bar under increasing axial force combined with a constant, initially elastic torque.

4.15 Establish the stress distributions from exercise 4.14 for $m = 0.83$, when the force has produced a fully elastic bar ($\rho_{ep} = 1$) and an elastic-plastic bar with $\rho_{ep} = 0.6$. Compare graphically with the Hencky solution (see Fig. 4.11) for the same conditions.

4.15 Establish the governing equations from the Prandtl-Reuss theory, with compressible elasticity, for the deformation which ensues when an initially elastic torque C_o is held constant under an increasing force P in a solid bar.

CHAPTER 5

ELASTIC-PERFECT PLASTICITY

5.1 Introduction

With ideal, elastic-perfectly plastic behaviour, elastic strain accompanies plastic strain in the absence of hardening. This simplification will often provide closed-form solutions to the stress distribution in loaded structures, revealing a clear distinction between zones of elasticity and plasticity. We examine, using this model, the non-hardening elastic-plastic behaviour of beams in bending and solid shafts under torsion. The use of a yield criterion becomes necessary when the stress state is multi-axial. This is illustrated with the determination of principal stress distributions for axially symmetric structures, including a thick-walled pressurised cylinder, with different end conditions, an annular disc and a rotating disc. All solutions to the plastic-zone stresses employ the equilibrium condition and the yield criterion to ensure stress continuity at the elastic-plastic interface. Both elastic and plastic zone stress distributions must satisfy the conditions of internal force equilibrium and strain compatibility. From applying these, the limiting plastic loads may be found for beams, torsion bars, thick cylinders and discs. Such load limit predictions are, of course, restricted to where the material within these structures is non-hardening.

5.2 Elastic-Plastic Bending of Beams

Let an applied moment M_Y lead to yielding within the most highly stressed fibres furthest from the neutral axis (NA). All other fibres in the cross-section remain elastic. As the moment is increased beyond M_Y, the cross-section becomes partially plastic as successive interior fibres, approaching the NA, reach the yield stress Y of the beam material. This state constitutes an elastic-plastic beam under an applied moment M_{ep}. When the plastic zone has penetrated through the whole cross-section on the tensile and compressive sides of the NA, this condition will determine the ultimate moment M_{ult} a given beam can withstand. It is assumed that a beam of any section collapses under M_{ult} and that the beam material behaves in an elastic-perfect plastic manner, where no increase in stress beyond Y occurs during the inward plastic penetration.

Consider a beam with a rectangular section, breadth b, depth d under a positive (hogging) bending moment M (see Fig.5.1a). As M increases, the material of the beam responds elastically initially but thereafter attains its yield point as plastic penetration begins. This results in an elastic-plastic cross-section, prior to the beam collapsing as the cross-section becomes fully plastic under M_{ult}. It is first necessary to derive the bending moment at each stage of the deformation. This analysis employs the stress distributions involved in the transition from elastic to fully plastic behaviour, as shown in Figs 5.1b-d.

5.2.1 Full Elasticity

Under M_Y in Fig. 5.1b, the outer fibres are stressed to the yield point but the interior cross-section remains elastic. Therefore, the engineer's theory of bending applies to the outer fibres where $\sigma = Y$ for $y = d/2$.

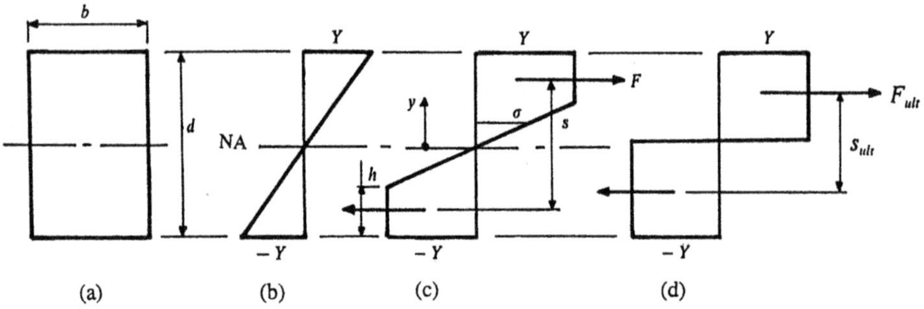

Figure 5.1 Penetration of a plastic zone in a rectangular section beam

This theory gives the applied momemt expression for initial yielding as

$$M_Y = \frac{YI}{y} = \frac{Y(bd^3/12)}{(d/2)} = \frac{bd^2Y}{6} \tag{5.1}$$

The radius of curvature is

$$R = \frac{Ey}{\sigma} = \frac{Ed}{2Y} \tag{5.2}$$

5.2.2 Elasto-Plasticity

Under an intermediate elastic-plastic moment M_{ep} in Fig. 5.1c, the plastic zone has penetrated by the amount h, from the top and bottom surfaces. The cross-section's moment of resistance is composed of elastic and plastic components, such that:

$$M_{ep} = M_e + M_p = YI_e/y + Fs$$

where $s = d - h$ is the distance between the net force $F = Ybh$, in each plastic region. Dimension $y = (d - 2h)/2$ defines the distance of the elastic-plastic interface from the NA. The second moment of area for the inner elastic core is

$$I_e = b\,(d - 2h)^3 / 12$$

Then M_{ep} may be expressed in a dimensionless form as follows:

$$M_{ep} = \frac{2Yb(d - 2h)^3}{12(d - 2h)} + (Ybh)(d - h)$$

$$= \frac{bd^2Y}{6}\left[1 + \frac{2h}{d}\left(1 - \frac{h}{d}\right)\right] \tag{5.3a}$$

$$\therefore \quad \frac{M_{ep}}{M_Y} = 1 + \frac{2h}{d}\left(1 - \frac{h}{d}\right) \tag{5.3b}$$

The radius of curvature R_{ep} of the NA, within an elastic-plastic section, is found from applying the elastic bending theory to the inner elastic core:

$$\frac{M_e}{I_e} = \frac{E}{R_{ep}} = \frac{Y}{y}$$

and with $y = (d - 2h)/2$, this gives

$$R_{ep} = \frac{Ey}{Y} = \frac{E(d - 2h)}{2Y} \tag{5.4a}$$

or

$$R_{ep} = \frac{EI_e}{M_e} = \frac{EI_e\, y}{YI_e} = \frac{E(d - 2h)}{2Y} \tag{5.4b}$$

5.2.3 Full Plasticity

Under the fully-plastic moment M_{ult} in Fig. 5.1d, the plastic zone has fully penetrated both the tensile and compressive sides of the NA. The resistive moment of the fully-plastic section is simply

$$M_{ult} = F_{ult} \times s_{ult} = \frac{bdY}{2} \times \frac{d}{2} = \frac{bd^2Y}{4} \tag{5.5}$$

(a) *Shape Factor*
The shape factor Q is defined as the ratio between the ultimate and initial yield moments:

$$Q = \frac{M_{ult}}{M_Y} > 1 \tag{5.6a}$$

Substituting eqs(5.1) and (5.5) into eq(5.6a), the shape factor becomes

$$Q = \frac{M_{ult}}{M_Y} = \frac{bd^2Y/4}{bd^2Y/6} = \frac{3}{2} \tag{5.6b}$$

Equation (5.6b) shows that the constant Q value depends solely upon the beam cross-section, i.e. Q is independent of the applied loading.

(b) *Load Factor*
This is defined as the ratio between the corresponding collapse load W_{ult} and the safe elastic working load W:

$$L = W_{ult} / W \tag{5.7}$$

Hence L will depend upon the section, the applied loading and the manner of its support. Consider a simply supported beam that collapses from the formation of a single plastic hinge, as shown in Fig. 5.2a.

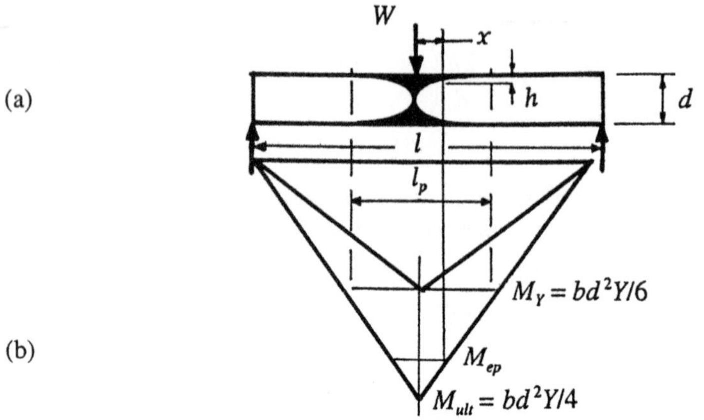

Figure 5.2 Plastic collapse of a simply-supported beam

The elastic safety factor S is

$$S = \sigma_a / \sigma_w \tag{5.8a}$$

where σ_a is the allowable working stress and σ_w is the safe working stress. Putting $\sigma_a = Y$ in eq(5.8a) gives:

$$\sigma_w = Y/S \tag{5.8b}$$

Now, from bending theory, the corresponding safe working moment is

$$M = \frac{\sigma_w I}{y} = \frac{YI}{Sy} \tag{5.9}$$

The maximum bending moment $M = Wl/4$ occurs beneath the load. It follows from eq(5.9) that the safe working load is

$$W = \frac{4M}{l} = \frac{4YI}{lSy} = \frac{2Ybd^2}{3lS} \tag{5.10}$$

Substituting from eq(5.5) gives the ultimate load for this beam as

$$W_{ult} = \frac{4M_{ult}}{l} = \frac{bd^2Y}{l} \tag{5.11}$$

Employing eqs(5.6), (5.7), (5.10) and (5.11), the load factor L may be combined with the shape factor Q and the safety factor S in a simple general form

$$L = \frac{(bd^2Y/l)}{(2Ybd^2)/(3lS)} = QS \tag{5.12}$$

In taking $S = 1$ in eq(5.10), the safe load becomes the initial yield load. We then have equal load and shape factors, $L = Q = 3/2$, for a simply supported beam with rectangular section.

(c) *Plastic Hinge*

The plastic hinge length l_p is determined from the ultimate moment diagram at the point of collapse (see Fig. 5.2). The moments are: $M = M_{ult}$, beneath the load and $M = M_Y$, at the extremities of the hinge. Now, from the geometry of the M - diagram in Fig. 5.2b, we have

$$2\,(M_{ult} - M_Y)\,/\,l_p = 2\,M_{ult}/l \tag{5.13a}$$

from which

$$l_p = l\,(1 - M_Y/M_{ult}) = l\,(1 - 1/Q) \tag{5.13b}$$

Substituting Q = 3/2 from eqs(5.6b) into eq(5.13a) shows that the hinge length of a rectangular cross-section is $l_p = l/3$. That is, the central hinge extends over one third of the beam's length. For other axisymmetric beam sections, Q in eq(5.13b) is a numerical value defining the cross-section. For example, it can be shown that Q = 1.7 for a solid circular section and Q = 2 for a square section when oriented with one diagonal as the horizontal axis of symmetry.

The shape of the hinge follows from the similar triangles within Fig. 5.2b. Also, employing eqs(5.3b) and (5.6b), we find for the rectangular section

$$x\,/\,l = \tfrac{1}{2}\,(1 - M_{ep}/\,M_{ult})$$

$$= \tfrac{1}{2}\{1 - \tfrac{2}{3}[\,1 + 2h/d - 2(h/d)^2]\} \tag{5.13c}$$

$$x = \frac{l}{6}\left(1 - \frac{2h}{d}\right)^2$$

which reveals that the hinge has the parabolic profile in x versus h co-ordinates as shown.

Where more than one hinge is necessary for failure, they form elsewhere instantaneously by stress redistribution. For the encastre-beam in Fig. 5.3a, the collapse mechanism requires that M_{ult} is reached beneath the load and at each fixed-end simultaneously.

(a)

(b)

(c)

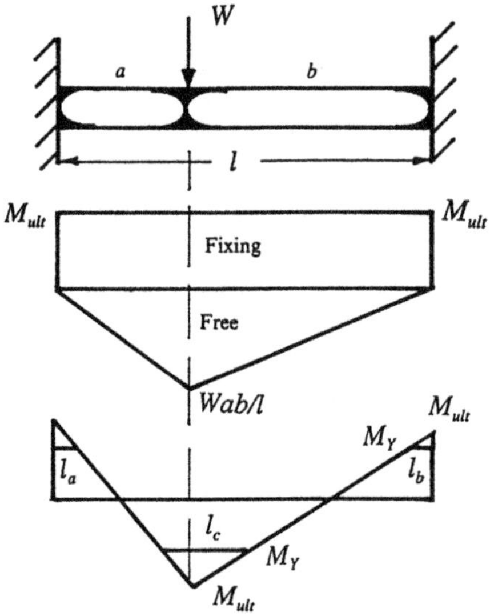

Figure 5.3 Collapse of an encastre beam

Equating the moment at the end to the net moment at the centre in Fig. 5.3b

$$M_{ult} = Wab\,/\,l - M_{ult} \tag{5.14a}$$

Thus, the collapse load is $W = 2M_{ult}\,l/ab$ and the plastic hinge lengths l_a, l_b and l_c follow from the geometry of similar triangles within Fig. 5.3c:

$$\frac{M_{ult} - M_Y}{l_a} = \frac{2M_{ult}}{a}$$

$$l_a = \frac{a}{2}\left(1 - \frac{M_Y}{M_{ult}}\right) = \frac{a}{2}\left(1 - \frac{1}{Q}\right) \tag{5.14b}$$

$$\frac{M_{ult} - M_Y}{l_b} = \frac{2M_{ult}}{b}$$

$$l_b = \frac{b}{2}\left(1 - \frac{M_Y}{M_{ult}}\right) = \frac{b}{2}\left(1 - \frac{1}{Q}\right) \tag{5.14c}$$

$$l_c = (a + b)\left(1 - \frac{M_Y}{M_{ult}}\right) = \frac{l}{2}\left(1 - \frac{1}{Q}\right) \tag{5.14d}$$

where Q in eqs(5.14a-c) will again depend upon the beam cross-section.

5.2.4 Residual Bending Stresses

When an elastic-plastic beam is fully unloaded from the moment M_{ep}, a state of residual stress σ_R will remain in the cross-section following recovery of elastic bending stresses σ_E. It is assumed that purely elastic unloading occurs from the elastic-plastic stress state σ that exists under M_{ep}. This gives

$$\sigma_R = \sigma - \sigma_E \tag{5.15}$$

For the elastic-plastic rectangular section in Fig. 5.1c, the elastic stress recovers over the whole depth $-d/2 \le y \le d/2$, following unloading from M_{ep}. Elastic bending theory gives the recovered elastic stress as

$$\sigma_E = M_{ep}\,y / I = (M_{ep} / M_Y)(M_Y / I)\,y$$

where $I = bd^3/12$. Substituting from eqs(5.1) and (5.3b),

$$\sigma_E = (M_{ep}/M_Y)(2Y/d)\,y$$

$$= \frac{2Y}{d}\left[1 + \frac{2h}{d}\left(1 - \frac{h}{d}\right)\right]y \tag{5.16}$$

Within the plastic zone $\sigma = Y$, for y lying in the outer regions $\pm(d/2 - h) \le y \le \pm d/2$. The residual stresses follow from eqs(5.15) and (5.16) as

$$\sigma_R = Y - \frac{2Y}{d}\left[1 + \frac{2h}{d}\left(1 - \frac{h}{d}\right)\right]y \tag{5.17}$$

Within the elastic zone, y lies in the inner regions $0 \le y \le \pm(d/2 - h)$. Bending theory supplies the applied stress as

$$\sigma = \frac{M_e\, y}{I} = \frac{Yb(d - 2h)^2 y}{6} \times \frac{12}{b(d - 2h)^3} = \frac{2Yy}{(d - 2h)} \tag{5.18}$$

which can readily be checked from similar triangles in Fig. 5.1c. The residual stress for this zone is found from subtracting eq(5.16) from eq(5.18):

$$\begin{aligned}\sigma_R &= \frac{2Yy}{(d - 2h)} - \frac{2Y}{d}\left[1 + \frac{2h}{d}\left(1 - \frac{h}{d}\right)\right]y \\ &= Yy\left\{\frac{2}{(d - 2h)} - \frac{2}{d}\left[1 + \frac{2h}{d}\left(1 - \frac{h}{d}\right)\right]\right\}\end{aligned} \tag{5.19}$$

Equations (5.17 and 5.19) show that σ_R varies linearly with y within each zone. Figure 5.4a illustrates σ_R corresponding to an elastic-plastic condition. In Fig. 5.4b, σ_R for the fully plastic condition is given where, with $h = d/2$ in eq(5.17), $\sigma_R = Y(1 - 3y/d)$. This shows that the maximum residual stress possible, $\sigma_R = Y$, occurs at the beam centre where $y = 0$. At the top and botom edges, where $y = \pm\, d/2$, $\sigma_R = \pm\, Y/2$.

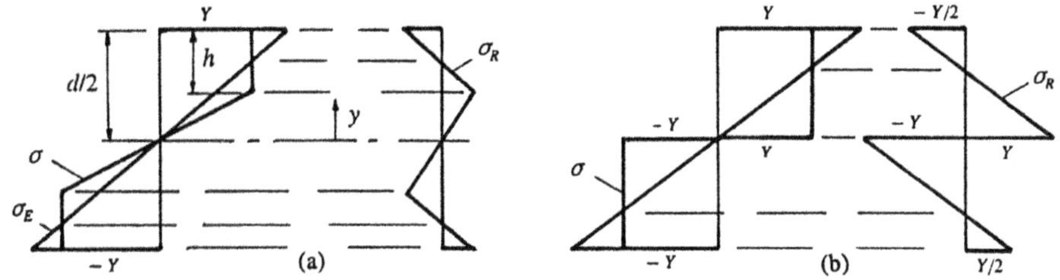

Figure 5.4 Residual stress distributions

Residual stresses have an important influence upon fatigue strength of a structure, where it is desirable that they serve to prolong life. Here, the sense of the residual stress must oppose the mean stress within a superimposed fatigue cycle arising from external loading.

5.2.5 Residual Strains and Curvature

The elastic bending theory also provides the relationship between the longitudinal bending strain ε and the radius of curvature R of the neutral axis:

$$\varepsilon = \sigma / E = y / R \tag{5.20a}$$

where y is the distance from the NA to any point within the elastic core. When the outer fibres reach the yield stress, eq(5.20a) gives the radius of the beam curvature as

$$R_Y = Ey / \sigma = (Ed) / (2Y) \tag{5.20b}$$

and therefore the longitudinal strain at position y is

$$\varepsilon_Y = y \,/\, R_Y = (2Yy) \,/\, (Ed) \tag{5.20c}$$

Now, from eqs(5.4a) and (5.20b), the normalised curvature of an elastic-plastic section is defined from its elastic core:

$$\frac{R_{ep}}{R_Y} = \frac{E(d - 2h)}{2Y} \times \frac{2Y}{Ed} = 1 - \frac{2h}{d} \tag{5.21a}$$

Equation (5.21a) shows that the ratio between these elastic curvatures diminishes from unity to zero as h penetrates to the full depth. Inverting the ratio in eq(5.21a) defines the normalised elastic-plastic strain when, from eq(5.20a),

$$\frac{\varepsilon_{ep}}{\varepsilon_Y} = \frac{y/R_{ep}}{y/R_Y} = \frac{R_Y}{R_{ep}} = \frac{1}{1 - 2h/d} \tag{5.21b}$$

Equation (5.21b) applies to all y within the elastic core $0 \le y \le (d/2 - h)$. Figure 5.5 shows that the strain ratio $\varepsilon_{ep}/\varepsilon_Y$ becomes infinite as the ratio M_{ep}/M_Y approaches its asymptote 3/2.

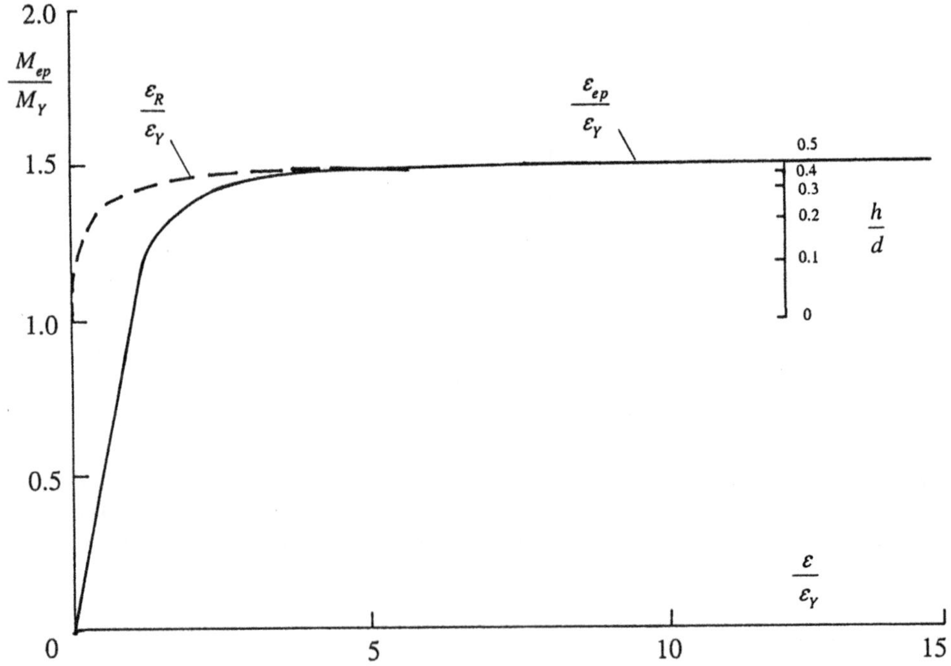

Figure 5.5 Relationship between longitudinal strain and the elastic-plastic moment

Following the removal of M_{ep}, the residual strain is found from

$$\varepsilon_R = \varepsilon_{ep} - \varepsilon_E \tag{5.22a}$$

where the elastic recovered strain is $\varepsilon_E = (M_{ep}y)/(IE)$. Normalising eq(5.22a) with the elastic core strain ε_Y from eq(5.20c) and substituting from eq(5.1), leads to

$$\frac{\varepsilon_R}{\varepsilon_Y} = \frac{\varepsilon_{ep}}{\varepsilon_Y} - \frac{M_{ep}\, y}{I E \varepsilon_Y} = \frac{\varepsilon_{ep}}{\varepsilon_Y} - \frac{M_{ep}}{M_Y} \tag{5.23a}$$

where from eq(5.1) $M_Y = 2IY/d$. Substituting eqs(5.3b) and (5.21b) into eq(5.23a) gives residual strain for $0 \le y \le (d/2 - h)$:

$$\frac{\varepsilon_R}{\varepsilon_Y} = \frac{1}{(1 - 2h/d)} - \left[1 + \frac{2h}{d}\left(1 - \frac{h}{d}\right)\right] \tag{5.23b}$$

$$= \frac{2(h/d)^2(3 - 2h/d)}{(1 - 2h/d)} \tag{5.23c}$$

Equation (5.23c) shows that $\varepsilon_R/\varepsilon_Y$ increases from zero, at $h/d = 0$, to infinity when $h/d = 0.5$ as M_{ult} is reached (see Fig. 5.5). Comparing eqs(5.23b) and (5.19) shows that residual stress and strains in the elastic core obey a simple elastic relation $\varepsilon_R = \sigma_R/E$. The same applies to the outer plastic zones in non-hardening material, i.e. where σ_R is given by eq(5.17). It follows that the residual strain distribution is identical to Fig.5.4b but with a scaling factor $1/E$. This fact implies that originally plane cross-sections of the rectangular section beam remain plane. Because eq(5.23b) applies within the elastic core, $0 \le y \le (d/2 - h)$, which exists under M_{ep}, it will also provide a residual curvature R_R as

$$R_R = y/\varepsilon_R = (y/\varepsilon_Y)(\varepsilon_Y/\varepsilon_R) \tag{5.24a}$$

Substituting eqs(5.20b) and (5.23b) into eq(5.24a) gives

$$R_R = R_Y\left(\frac{\varepsilon_Y}{\varepsilon_R}\right) = \frac{Ed}{2Y}\left(\frac{\varepsilon_Y}{\varepsilon_R}\right) = \frac{Ed^3(1 - 2h/d)}{4Yh^2(3 - 2h/d)}$$

5.2.6 Non-Axisymmetric Section

In the T-section of Fig. 5.6a, the unstressed neutral does not lie at the central depth but originally passes through the centroid of its area. Yielding under M_Y will commence at the web bottom since this is the more highly stressed surface. As the bending moment is increased to M_{ep}, the plastic zone penetrates inwards from the bottom surface, as shown in Fig. 5.6b. During penetration the horizontal tensile and compressive forces, T and C respectively, remain balanced, i.e. $T = C$. For this to occur, the stress re-distributes so that Fig. 5.6b accommodates a shift in the NA to the position y_{ep} shown, giving $M_{ep} = Cs = Ts$. With increasing M_{ep}, the section stress continues to redistribute, maintaining $C = T$, until full plasticity is reached under M_{ult} in Fig. 5.6c. The NA then divides the section area equally.

5.2.7 Influence of Hardening

It is possible to account for the effect of linear hardening upon the moment carrying capacity of the section. The shaded areas in Fig. 5.6c represent this increased capacity when the gradient of the stress-strain curve $K = \mathrm{d}\sigma/\mathrm{d}\varepsilon^P$ is identified with the gradient of the stress distribution [1]. This is illustrated in the following example.

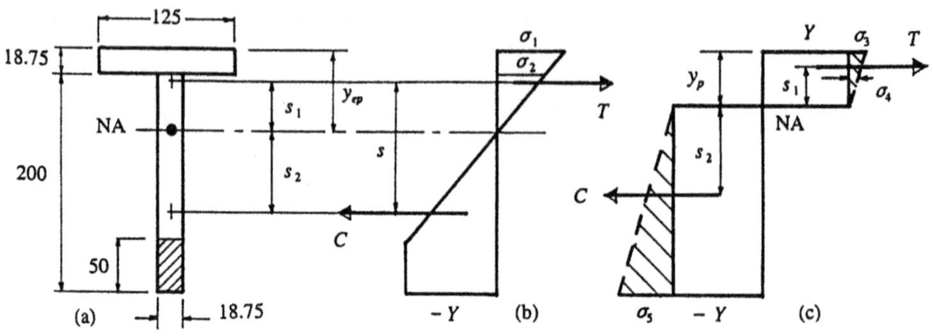

Figure 5.6 Elastic-plastic bending stress in a T-section

Example 5.2 In the T-section of Fig. 5.6a, yielding has occurred 50 mm from the web bottom. If the yield stress is constant at $Y = 278$ MPa, find the applied moment, the stress at the flange top and the position of the neutral axis. Estimate the fully plastic moment when (i) the beam material is non-hardening and (ii) when it hardens according to the linear law $\sigma = Y + K\varepsilon^P$, in which $K = 1/10$.

In Fig. 5.6b, the condition $T = C$ gives

$$(125 \times 18.75)(\sigma_1 + \sigma_2)/2 + (y_{ep} - 18.75)18.75\sigma_2/2 = (18.75 \times 50)Y + (168.75 - y_{ep})18.75Y/2 \quad \text{(i)}$$

where the stress ordinates appear in terms of y_{ep}. From the geometry of Fig. 5.6b:

$$\sigma_1/Y = y_{ep}/\,(168.75 - y_{ep}) \quad \text{and} \quad \sigma_2/Y = (y_{ep} - 18.75)\,/\,(168.75 - y_{ep}) \qquad \text{(ii,iii)}$$

Substituting eqs(ii) and (iii) into eq(i), leads to an equation which locates the position of the neutral axis at $y_{ep} = 72.84$ mm. Equation (ii) then gives the elastic stress at the flange top as $\sigma_1 = 211.1$ MPa. The forces $T = C = 510.6$ kN follow from the left- and right-hand sides of eq(i). These act at the respective centroids for the areas above and below the neutral axis. That is: $s_1 = \sum(A_i y_i)/A = 52.465$ mm and $s_2 = 72.955$ mm. The moment M_{ep} is found from

$$M_{ep} = Cs = C\,(s_1 + s_2) = 510.6\,(52.465 + 72.955)10^{-3} = 64.04 \text{ kNm}$$

When the fully plastic moment is reached (see Fig. 5.6c), the condition $T = C$ becomes

$$[(125 \times 18.75) + (y_p - 18.75) \times 18.75]Y = (218.75 - y_p) \times 18.75Y \qquad \text{(iv)}$$

The neutral axis now divides the section into two equal areas to give $y_p = 56.25$ mm. The centroids of these areas are $s_1 = \sum(A_i y_i)/A = 40.385$ mm and $s_2 = 81.25$ mm. Equation (iv) gives $T = C = 847.04$ kN and M_{ult} is found from

$$\begin{aligned} M_{ult} &= C\,s = (218.75 - y_p)18.75Y\,(s_1 + s_2) \\ &= 162.5 \times 18.75 \times 278\,(40.385 + 81.25)10^{-6} = 103.03 \text{ kNm} \end{aligned}$$

To account for hardening, the force within each shaded area in Fig. 5.6c must be added to eq(iv). This modifies the position of the neutral axis to y_p'. Using the condition $T' = C'$:

$$\begin{aligned} &[\,(125 \times 18.75) + (y_p' - 18.75)18.75]Y + (y_p' - 18.75)\sigma_4/2 + (18.75 \times 125)(\sigma_3 + \sigma_4)/2 \\ &= (218.75 - y_p')18.75Y + (218.75 - y_p')18.75\sigma_5/2 \end{aligned} \qquad \text{(v)}$$

in which the stress ordinates (see Fig. 5.6c) are:

$$\sigma_3 = K\,y_p', \quad \sigma_4 = K\,(y_p' - 18.75) \quad \text{and} \quad \sigma_5 = K\,(218.75 - y_p') \qquad \text{(vi-viii)}$$

Substituting eqs(vi - viii) into eq(v), with $K = 1/10$, leads to an equation which re-locates the position of the neutral axis to $y_p' = 57.38$ mm. From this, we find from eq(v), $T' = C' = 865.55$ kN. The centroid for the upper area is modified to $s_1' = \sum(A_i y_i)/A = 41.23$ mm and for the lower web area $s_2 = 80.69$ mm. Hence the collapse moment becomes

$$M_{ult}' = C'\,(s_1' + s_2') = 865.55\,(41.23 + 80.68)10^{-3} = 105.52 \text{ kNm}$$

which shows a slight increase over the non-hardening value.

5.3 Elastic-Plastic Torsion

Elastic-plastic torsion of a circular shaft is similar to elastic-plastic bending of beams in that three conditions apply to the shaft: fully elastic, elastic-plastic and fully plastic. The corresponding torques are derived for both solid and hollow circular section shafts.

5.3.1 Solid Cylinder

Consider a solid circular shaft of outer radius r_o, subjected to an increasing torque C. Provided the cross-section is elastic, torsion theory will describe the twist, shear stress and strain. This includes the case where the outer fibres have reached the shear yield stress k. Here, an initial yield torque C_Y for the limiting, fully-elastic section is

$$C_Y = \frac{Jk}{r_o} = \frac{\pi r_o^4 k}{2 r_o} = \frac{\pi r_o^3 k}{2} \qquad (5.25)$$

Increasing the torque beyond C_Y results in the penetration of a plastic zone inwards from the outer radius. This produces an elastic-plastic section, shown in Fig. 5.7a, where an elastic core is surrounded by a plastic annulus, interfaced at a common elastic-plastic radius r_{ep}.

(a)

(b)

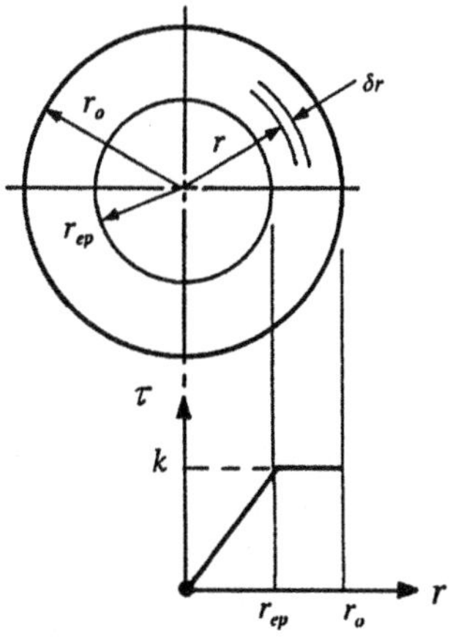

Figure 5.7 Elastic-plastic bar in torsion

Since the material does not work-harden (see Fig. 5.7b) k is constant for $r_{ep} \leq r \leq r_o$. It follows that there is a linear variation in elastic shear stress τ within the core, $0 \leq r \leq r_{ep}$:

$$\tau = k\,(r / r_{ep}) \tag{5.26}$$

The corresponding elastic-plastic torque C_{ep} is found from the following equilibrium condition:

$$C_{ep} = 2\pi \int_{r_{ep}}^{r_o} k r^2 \mathrm{d}r + 2\pi \int_0^{r_{ep}} \tau r^2 \mathrm{d}r \tag{5.27a}$$

$$= 2\pi k \int_{r_{ep}}^{r_o} r^2 \mathrm{d}r + \frac{2\pi k}{r_{ep}} \int_0^{r_{ep}} r^3 \mathrm{d}r$$

$$= \frac{2\pi k}{3} \left| r^3 \right|_{r_{ep}}^{r_o} + \frac{2\pi k}{4 r_{ep}} \left| r^4 \right|_0^{r_{ep}}$$

$$= \frac{\pi k r_o^3}{2} \left\{ \frac{4}{3}\left[1 - \left(\frac{r_{ep}}{r_o} \right)^3 \right] + \left(\frac{r_{ep}}{r_o} \right)^3 \right\} \tag{5.27b}$$

Dividing eq(5.27b) by C_Y from eq(5.25) gives the normalised torque:

$$\frac{C_{ep}}{C_Y} = \frac{4}{3}\left[1 - \frac{1}{4}\left(\frac{r_{ep}}{r_o} \right)^3 \right] \tag{5.28}$$

The angle of twist θ_{ep} is found from applying elastic torsion theory to the core. At the elastic-plastic interface, where $\tau = k$ for $r = r_{ep}$ (Fig. 5.7b), we find, for a shaft of length l,

$$\theta_{ep} = \frac{kl}{G r_{ep}} = \frac{C_e l}{G J_e} \tag{5.29}$$

in which $J_e = \pi r_{ep}^{\,4}/2$ and C_e is the torque carried by the elastic core, i.e. the second integral in eq(5.27a). C_e is not to be confused with the fully elastic torque eq(5.25). The section becomes fully plastic through to the bar centre under its ultimate torque C_{ult}. Putting $r_{ep} = 0$ in eq(5.27b), gives

$$C_{ult} = (2/3)(\pi r_o^{\,3} k) \tag{5.30a}$$

and substituting from eq(5.25) defines the shape factor for the solid circular section:

$$C_{ult} / C_Y = 4/3 \tag{5.30b}$$

(a) *Residual Shear Stress*

When the elastic-plastic bar is fully unloaded, elastic stresses recover to leave a residual shear stress distribution τ_R. This distribution is found from

$$\tau_R = \tau - \tau_E \tag{5.31}$$

where τ is the shear stress in either zone under the elastic-plastic torque C_{ep}. Torsion theory supplies the elastic shear stress τ_E that recovers upon removal of C_{ep}:

$$\tau_E = C_{ep}\, r / J \tag{5.32a}$$

where from eq(5.25)

$$C_{ep} / J = (C_{ep} / C_Y)(C_Y / J) = (C_{ep} / C_Y)(k / r_o) \tag{5.32b}$$

It follows from eqs(5.28), (5.31) and (5.32b) and with $J = \pi r_o^4 / 2$, that the residual shear stress distribution in the plastic zone, $r_{ep} \le r \le r_o$, where $\tau = k$ (see Fig. 5.8a) is given by

$$\tau_R = k - \frac{C_{ep} r}{J} = k - \left(\frac{C_{ep}}{C_Y}\right)\left(\frac{k r}{r_o}\right)$$

$$= k\left\{1 - \frac{4}{3}\left(\frac{r}{r_o}\right)\left[1 - \frac{1}{4}\left(\frac{r_{ep}}{r_o}\right)^3\right]\right\} \tag{5.33a}$$

and, from eqs(5.26), (5.28), (5.31) and (5.32b), the residual shear stress distribution within the elastic zone, $0 \le r \le r_{ep}$, is

$$\tau_R = k\left(\frac{r}{r_{ep}}\right) - \frac{4k}{3}\left(\frac{r}{r_o}\right)\left[1 - \frac{1}{4}\left(\frac{r_{ep}}{r_o}\right)^3\right]$$

$$= k\left(\frac{r}{r_{ep}}\right)\left\{1 - \frac{4}{3}\left(\frac{r_{ep}}{r_o}\right)\left[1 - \frac{1}{4}\left(\frac{r_{ep}}{r_o}\right)^3\right]\right\} \tag{5.33b}$$

The τ, τ_E and τ_R distributions are represented graphically in Fig. 5.8b. The residual stresses supplied by eqs(5.33a,b), are simply the difference between stress ordinates τ and τ_E within each zone, their signs determined from eq(5.31). It is seen that τ_R has its two largest values at the outer and interface radii, the greater value depending upon the depth of penetration.

(a)

(b)

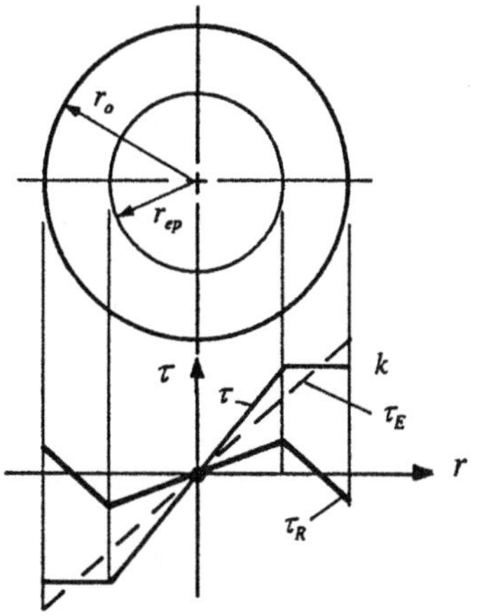

Figure 5.8 Residual stress in a partially-plastic solid bar

(b) *Residual Angular Twist*

When the bar's outer fibres reach the shear yield stress k, the fully-elastic angular twist θ_Y is supplied by torsion theory:

$$\theta_Y = \frac{kl}{Gr_o} \tag{5.34}$$

Combining eqs(5.29) and (5.34), gives a twist ratio for the elastic-plastic bar

$$\frac{\theta_{ep}}{\theta_Y} = \frac{(kl)/(Gr_{ep})}{(kl)/(Gr_o)} = \frac{r_o}{r_{ep}} \tag{5.35}$$

The following normalised, torque-twist relationship is found from combining eq(5.28) with eq(5.35):

$$\frac{C_{ep}}{C_Y} = \frac{4}{3}\left[1 - \frac{1}{4}\left(\frac{\theta_Y}{\theta_{ep}}\right)^3\right] \tag{5.36}$$

The graph of eq(5.36), given in Fig. 5.9, shows that the twist ratio becomes infinite as the normalised torque increases beyond unity to its ultimate value of 4/3.

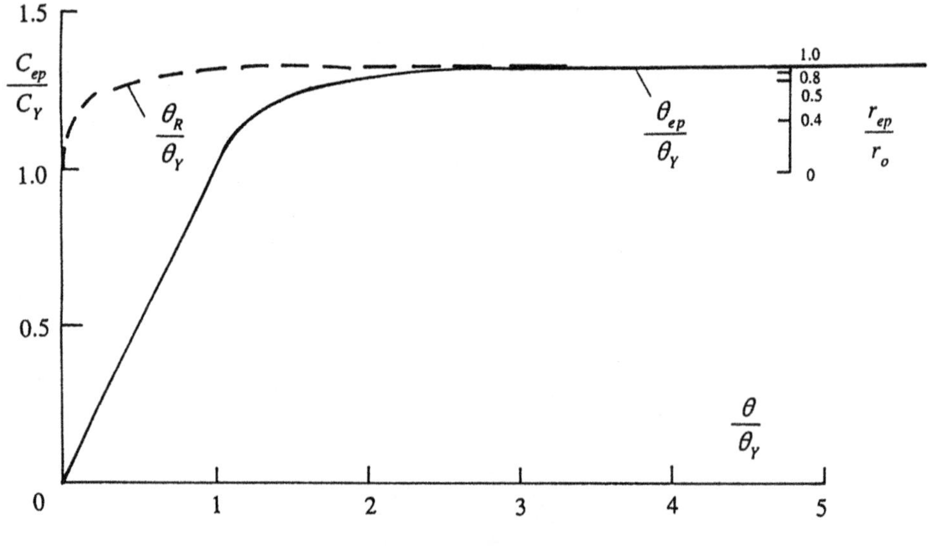

Figure 5.9 Dependence of twist upon applied torque in an elastic-plastic bar

Since the ordinate C_{ult}/C_Y is the asymptote 4/3, the implication is that C_{ult} could not be attained in practice if the material was truly elastic-perfectly plastic. However, most metallic materials do strain-harden to some degree, which increases the torque-carrying capacity above C_{ult}.

Following unloading from any intermediate torque value $C_{ep} < C_{ult}$, the state of residual twist θ_R is found from

$$\theta_R = \theta_{ep} - \theta_E \tag{5.37a}$$

where θ_E is the twist which recovers elastically from the removal of C_{ep}.

Substituting $\theta_E = (C_{ep}l)/(GJ)$, together with eq(5.29), into eq(5.37a) gives

$$\theta_R = \frac{kl}{Gr_{ep}} - \frac{C_{ep}l}{GJ} \tag{5.37b}$$

Normalising with θ_Y from eq(5.34):

$$\frac{\theta_R}{\theta_Y} = \frac{(kl)/(Gr_{ep})}{(kl)/(Gr_o)} - \frac{(C_{ep}l)/(GJ)}{(kl)/(Gr_o)} \tag{5.38a}$$

Substituting eqs(5.25), (5.35) and (5.36) into eq(5.38a), gives this twist ratio as

$$\frac{\theta_R}{\theta_Y} = \frac{r_o}{r_{ep}} - \frac{4}{3}\left[1 - \frac{1}{4}\left(\frac{r_{ep}}{r_o}\right)^3\right] \tag{5.38b}$$

$$= \frac{\theta_{ep}}{\theta_Y} - \frac{4}{3}\left[1 - \frac{1}{4}\left(\frac{\theta_Y}{\theta_{ep}}\right)^3\right] \tag{5.38c}$$

Equations (5.36) and (5.38c) establish a dependence between C_{ep}/C_Y, θ_{ep}/θ_Y and θ_R/θ_Y, shown in Fig. 5.9. A residual, angular twist exists following unloading from a torque which exceeds C_Y. The ratio θ_R/θ_Y increases to infinity under C_{ult}. When $C \le C_Y$, elastic torsion theory describes the torque versus twist line of unit slope, since $C/C_Y = \theta/\theta_Y = (Cr_o)/(Jk)$.

5.3.2 Hollow Cylinder

Figure 5.10a shows a partially yielded hollow cylinder, with inner and outer radii r_i and r_o respectively. The plastic zone has penetrated to intermediate radius r_{ep} under a torque C_{ep}. The distribution of shear stress is given in Fig. 5.10b. Again, an elastic-perfectly plastic material is assumed in which the plastic zone spreads under a constant yield stress value k.

(a)

(b)

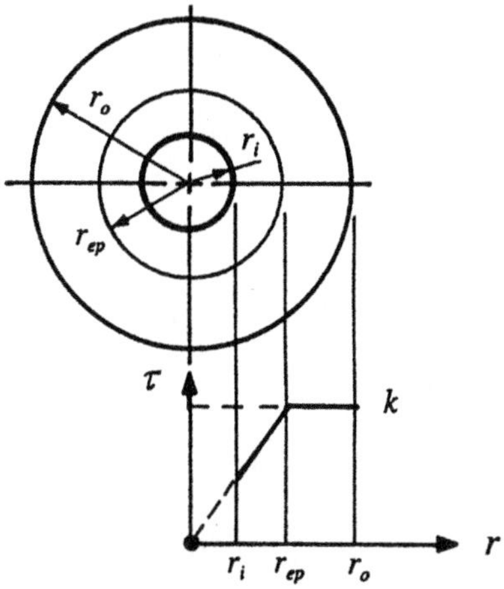

Figure 5.10 Elastic-plastic hollow cylinder under torsion

Torque equilibrium expresses the sum of elastic and plastic torques from each zone:

$$C_{ep} = C_e + C_p = \frac{2\pi k}{r_{ep}}\int_{r_i}^{r_{ep}} r^3 \mathrm{d}r + 2\pi k\int_{r_{ep}}^{r_o} r^2 \mathrm{d}r \tag{5.39a}$$

$$= \frac{\pi k}{2r_{ep}}\left(r_{ep}^4 - r_i^4\right) + \frac{2\pi k}{3}\left(r_o^3 - r_{ep}^3\right) \tag{5.39b}$$

Putting $r_{ep} = r_o$ in eq(5.39b) gives the initial yield torque C_Y as

$$C_Y = \frac{\pi k}{2r_o}(r_o^4 - r_i^4) \tag{5.40}$$

Putting $r_{ep} = r_i$ in eq(5.39b) gives the ultimate plastic torque C_{ult}. When C_{ult} is normalised with C_Y, from eq(5.40):

$$\frac{C_{ult}}{C_Y} = \frac{4[1 - (r_i/r_o)^3]}{3[1 - (r_i/r_o)^4]} \tag{5.41}$$

Combining eqs(5.39b) and (5.40) gives a dimensionless elastic-plastic torque C_{ep} for interface radii in the range $r_i \le r_{ep} \le r_o$:

$$\frac{C_{ep}}{C_Y} = \frac{4r_o(r_o^3 - r_{ep}^3)}{3(r_o^4 - r_i^4)} + \frac{r_o(r_{ep}^4 - r_i^4)}{r_{ep}(r_o^4 - r_i^4)}$$

$$= \frac{\frac{4}{3}\left[1 - \frac{1}{4}\left(\frac{r_{ep}}{r_o}\right)^3\right] - \frac{r_i}{r_{ep}}\left(\frac{r_i}{r_o}\right)^3}{1 - \left(\frac{r_i}{r_o}\right)^4} \tag{5.42}$$

Equation (5.42) reduces to eq(5.28) with $r_i = 0$ for a solid cylinder. The angular twist θ is found from the inner elastic core, $r_i \le r \le r_{ep}$, where the elastic torsion theory applies:

$$\frac{C_e}{J_e} = \frac{G\theta}{l} = \frac{k}{r_{ep}} \tag{5.43}$$

The elastic twist is, from eq(5.43),

$$\theta = \frac{C_e l}{GJ_e} = \frac{kl}{Gr_{ep}} \tag{5.44a}$$

where J_e applies to the elastic annulus:

$$J_e = \frac{\pi}{2}\left(r_{ep}^4 - r_i^4\right) \tag{5.44b}$$

and C_e is supplied from the first term in eq(5.39b):

$$C_e = \frac{\pi k}{2r_{ep}}(r_{ep}^4 - r_i^4) \tag{5.44c}$$

A residual angle of twist θ_R will remain when C_{ep} is released. Since the recovered twist θ_E is assumed to be elastic, eqs(5.37a,b) again apply but with $J = \pi(r_o^4 - r_i^4)/2$, for the full cross-section. Substituting eqs(5.39b) into eq(5.37b):

$$\theta_R = \frac{kl}{Gr_{ep}} - \frac{\pi k l r_o^3}{2JG}\left\{\frac{4}{3}\left[1 - \frac{1}{4}\left(\frac{r_{ep}}{r_o}\right)^3\right] - \left(\frac{r_i}{r_{ep}}\right)\left(\frac{r_i}{r_o}\right)^3\right\} \tag{5.45a}$$

which is constant for a given r_{ep}. We may normalise eq(5.45a) with the twist at the yield point for the outer radius, $\theta_Y = kl/Gr_o$. This gives

$$\frac{\theta_R}{\theta_Y} = \frac{r_o}{r_{ep}} - \frac{\pi r_o^4}{2J}\left\{\frac{4}{3}\left[1 - \frac{1}{4}\left(\frac{r_{ep}}{r_o}\right)^3\right] - \left(\frac{r_i}{r_{ep}}\right)\left(\frac{r_i}{r_o}\right)^3\right\} \tag{5.45b}$$

Putting $r_i = 0$ in eq(5.45b), with $J = \pi r_o^4/2$, leads to the residual twist for a solid shaft, given previously in eq(5.38b).

The residual shear stress in the outer, plastic zone, $r_{ep} \le r \le r_o$, is found from substituting eqs(5.32a) and (5.39b) into eq(5.31):

$$\tau_R = \tau - \frac{C_{ep}\,r}{J} = k - k\left(\frac{C_{ep}}{C_Y}\right)\left(\frac{r}{r_o}\right)$$

$$= k\left[1 - \left(\frac{C_{ep}}{C_Y}\right)\left(\frac{r}{r_o}\right)\right]$$

$$= k\left\{1 - \frac{4r}{3}\left[\frac{(r_o^3 - r_{ep}^3)}{(r_o^4 - r_i^4)} + \frac{3(r_{ep}^4 - r_i^4)}{4r_{ep}(r_o^4 - r_i^4)}\right]\right\} \tag{5.46a}$$

The residual stress in the inner, elastic zone $r_i \le r \le r_{ep}$, is found from substituting eqs(5.26), (5.32b) and (5.39b) into eq(5.31):

$$\tau_R = \tau - \frac{C_{ep}\,r}{J} = k\left(\frac{r}{r_{ep}}\right) - k\left(\frac{C_{ep}}{C_Y}\right)\left(\frac{r}{r_o}\right)$$

$$= k\left[\frac{r}{r_{ep}} - \left(\frac{C_{ep}}{C_Y}\right)\left(\frac{r}{r_o}\right)\right]$$

$$= k\left\{\frac{r}{r_{ep}} - \frac{4r}{3}\left[\frac{(r_o^3 - r_{ep}^3)}{(r_o^4 - r_i^4)} + \frac{3(r_{ep}^4 - r_i^4)}{4r_{ep}(r_o^4 - r_i^4)}\right]\right\} \tag{5.46b}$$

Alternatively, we could employ eq(5.42) within the second line of each expression (5.46a,b). It is seen that when $r_i = 0$ in eqs(5.46a,b) they reduce to the solid bar residual stresses, given previously in eqs(5.33a,b).

5.3.3 *Sandhill Analogy*

A simple analogy can be employed to find the fully plastic (ultimate) torque of solid, non-circular sections [2]. To show this, compare the expression for the volume V of a cone of sand that would rest on the end of an upright circular bar (see Table 5.1), with eq(5.30a). This shows that $C_{ult} = 2V$, where $k = h/r_o$ defines the sloping side of the cone. The ultimate torque C_{ult} for the rectangular and equilateral triangular sections can be derived in a similar manner, i.e. from knowing the volume of a prism of sand each section would support and that each side gradient is k.

Table 5.1 Fully plastic torques C_{ult} for solid sections

Section	V	Slope	$C_{ult} = 2V$
circle (radius r_o)	$\pi r_o^2 h/3$	$k = h/r_o$	$2\pi r_o^2 h/3 = 2\pi r_o^3 k/3$
rectangle ($b \times t$)	$(th/6)(3b - t)$	$k = 2h/t$	$kt^2(3b - t)/6$
triangle (side a)	$a^2 h/(4\sqrt{3})$	$k = 2\sqrt{3}h/a$	$ka^3/12$

With I, T and U sections under torsion, we take the sum of C_{ult} for each rectanglar web and flange. The C_{ult} expression for a rectangle will also apply to thin-walled, curved sections, where b is the perimeter length and t is the constant wall thickness. For example, $b = 2\pi r_m$ in a split, thin-walled circular tube, of mean wall radius r_m, giving $C_{ult} \approx \pi k t^2 r_m$. For closed tubes we may use the difference in C_{ult} between the inner and outer 'solid' sections.

5.4 Thick-Walled, Pressurised Cylinder with Closed-Ends

Thick-walled cylinders may be employed safely even when an internal pressure induces a partially plastic state within its wall. This pressure is intermediate to those pressures associated with full elasticity and full plasticity. As with the elastic-perfectly plastic beam and the torsion bar, the non-hardening cylinder is also a statically determinate problem. That is, the critical pressures may be found from combining an equilibrium equation with a yield criterion. These are sufficient to ensure that a condition of strain compatibility exists at the elastic-plastic interface.

5.4.1 *Initial Yielding*

Let a thick-walled cylinder of non-hardening material, with inner and outer radii r_i and r_o respectively, be subjected to a steadily increasing internal (gauge) pressure. The external pressure remains atmospheric at zero gauge pressure. Under elastic conditions, a principal, triaxial stress state is distributed through the wall. This state consists of radial, hoop and axial stresses σ_r, σ_θ and σ_z respectively, these being expressed from the Lamé theory [2] as

$$\sigma_r = \frac{p_i r_i^2 (1 - r_o^2/r^2)}{(r_o^2 - r_i^2)} = \frac{p_i (1 - r_o^2/r^2)}{(K^2 - 1)} \quad (5.47a)$$

$$\sigma_\theta = \frac{p_i r_i^2 (1 + r_o^2/r^2)}{(r_o^2 - r_i^2)} = \frac{p_i (1 + r_o^2/r^2)}{(K^2 - 1)} \quad (5.47b)$$

$$\sigma_z = \frac{(\sigma_r + \sigma_\theta)}{2} = \frac{p_i}{(K^2 - 1)} \quad (5.47c)$$

where $K = r_o/r_i$. As the pressure increases, the bore fibres are the first to reach the yield stress Y. Let us assume, firstly, that the material conforms to the von Mises yield criterion. Equation (3.7a) appears in its principal, polar co-ordinate form

$$(\sigma_\theta - \sigma_z)^2 + (\sigma_\theta - \sigma_r)^2 + (\sigma_r - \sigma_z)^2 = 2Y^2 \quad (5.48)$$

Substituting eqs(5.47a-c) into eq(5.48), with $r = r_i$, gives the initial yield pressure

$$p_Y = \frac{Y(K^2 - 1)}{\surd 3K^2} \quad (5.49)$$

In Tresca's prediction to the yield pressure, eq(3.1b) is written as

$$\sigma_\theta - \sigma_r = Y \quad (5.50)$$

From eqs(5.47a,b) and (5.50), with $r = r_i$

$$p_Y = \frac{Y(K^2 - 1)}{2K^2} \quad (5.51)$$

Comparing eqs(5.49) and (5.51), the cylinder first yields under Tresca's pressure. Von Mises requires the internal pressure be increased by a further 15% before yielding commences.

5.4.2 Elastic-Plastic Cylinder

Here we wish to determine the internal pressure which penetrates a plastic zone to a radial depth r_{ep}, where $r_i \le r_{ep} \le r_o$. Since the outer annulus $r_{ep} \le r \le r_o$ is elastic, the radial and hoop stresses are again supplied by the Lamé eqs(5.47a,b,c). The boundary conditions are: (i) at the interface radius r_{ep}, $\sigma_r = -p_{ep}$ and (ii) at outer radius r_o, $\sigma_r = 0$. These give

$$\sigma_r = \frac{p_{ep} r_{ep}^2 (1 - r_o^2/r^2)}{(r_o^2 - r_{ep}^2)} \quad (5.52a)$$

$$\sigma_\theta = \frac{p_{ep} r_{ep}^2 (1 + r_o^2/r^2)}{(r_o^2 - r_{ep}^2)} \quad (5.52b)$$

$$\sigma_z = \frac{\sigma_\theta + \sigma_r}{2} = \frac{p_{ep} r_{ep}^2}{(r_o^2 - r_{ep}^2)} \quad (5.52c)$$

At the interface radius, the material is at its initial yield point. Substituting eqs(5.52a-c) into eq(5.48), with $r = r_{ep}$, leads to the interface pressure

$$p_{ep} = \frac{Y}{\surd 3}\left(1 - \frac{r_{ep}^2}{r_o^2}\right) \tag{5.53}$$

Substituting eq(5.53) into eqs(5.52a,b,c) provides the triaxial stresses in the elastic zone:

$$\sigma_r = \frac{Y}{\surd 3}\left(\frac{r_{ep}}{r_o}\right)^2\left(1 - \frac{r_o^2}{r^2}\right) \tag{5.54a}$$

$$\sigma_\theta = \frac{Y}{\surd 3}\left(\frac{r_{ep}}{r_o}\right)^2\left(1 + \frac{r_o^2}{r^2}\right) \tag{5.54b}$$

$$\sigma_z = \frac{Y}{\surd 3}\left(\frac{r_{ep}}{r_o}\right)^2 \tag{5.54c}$$

A radial equilibrium equation applies to the axial symmetry [2]:

$$\sigma_\theta - \sigma_r = r\frac{\mathrm{d}\sigma_r}{\mathrm{d}r} \tag{5.55}$$

When the LHS of eq(5.55) is constant it becomes easily to apply. Within the plastic annulus $r_i \le r \le r_{ep}$, axial plastic strain is ignored when applying internal pressure to a closed-end cylinder. Substituting $\mathrm{d}\varepsilon_z^P = 0$ within the first term of eq(4.3c) leads to

$$\mathrm{d}\varepsilon_z^P = 0 = \mathrm{d}\lambda\,[\sigma_z - \tfrac{1}{2}(\sigma_r + \sigma_\theta)]$$

$$\therefore \quad \sigma_z = \tfrac{1}{2}(\sigma_r + \sigma_\theta) \tag{5.56}$$

Equation (5.56) holds consistently with the axial stress in elastic region, given by eq(5.52c). Substituting eq(5.56) into eq(5.48), the von Mises criterion is, simply

$$\sigma_\theta - \sigma_r = 2Y/\surd 3 \tag{5.57}$$

Combining eqs(5.55) and (5.57) allows integration:

$$\sigma_r = \frac{2Y}{\surd 3}\ln r + A \tag{5.58a}$$

in which A is found from the condition that $\sigma_r = -\,p_{ep}$ for $r = r_{ep}$. Substituting from eq(5.53):

$$A = -\,p_{ep} - \frac{2Y}{\surd 3}\ln r_{ep} = -\frac{Y}{\surd 3}\left(1 - \frac{r_{ep}^2}{r_o^2}\right) - \frac{2Y}{\surd 3}\ln r_{ep} \tag{5.58b}$$

Combining eqs(5.58a and b) with eqs(5.56) and (5.57), leads to expressions for the triaxial stress state in the plastic zone:

$$\sigma_r = \frac{2Y}{\sqrt{3}} \ln\left(\frac{r}{r_{ep}}\right) - \frac{Y}{\sqrt{3}}\left(1 - \frac{r_{ep}^2}{r_o^2}\right) \tag{5.59a}$$

$$\sigma_\theta = \frac{2Y}{\sqrt{3}} \ln\left(\frac{r}{r_{ep}}\right) + \frac{Y}{\sqrt{3}}\left(1 + \frac{r_{ep}^2}{r_o^2}\right) \tag{5.59b}$$

$$\sigma_z = \frac{2Y}{\sqrt{3}} \ln\left(\frac{r}{r_{ep}}\right) + \frac{Y}{\sqrt{3}}\left(\frac{r_{ep}^2}{r_o^2}\right) \tag{5.59c}$$

Now as $\sigma_r = -p_i$ for $r = r_i$, eq(5.59a) gives the corresponding internal pressure:

$$p_i = \frac{2Y}{\sqrt{3}} \ln\left(\frac{r_{ep}}{r_i}\right) + \frac{Y}{\sqrt{3}}\left(1 - \frac{r_{ep}^2}{r_o^2}\right) \tag{5.60a}$$

The pressure p_{ult}, required to produce a fully plastic cylinder, is found from putting $r_{ep} = r_o$ in eq(5.60a):

$$p_{ult} = \frac{2Y}{\sqrt{3}} \ln\left(\frac{r_o}{r_i}\right) \tag{5.60b}$$

which depends upon the yield stress of the cylinder material and its radius ratio. Crossland et al. [3,4] showed that eq(5.60b) was in reasonable agreement with their burst pressures for pressure vessel steel cylinders, when Y was identified with the upper yield point. However, an empirical modification to eq(5.60a) was made for when material within an elastic-plastic cylinder displayed both upper and lower yield points. They used

$$p_i = \frac{Y_2}{\sqrt{3}} \ln\left(\frac{r_{ep}}{r_i}\right)^2 + \frac{Y_1}{\sqrt{3}}\left[1 - \left(\frac{r_{ep}}{r_o}\right)^2\right] \tag{5.60c}$$

where $(Y_1/\sqrt{3})$ and $(Y_2/\sqrt{3})$ were identified with the respective upper and lower shear yield stresses in a torsion test. The elastic and plastic eqs(5.54a-c) and (5.59a-c) respectively, give radial, hoop and axial stress under pressure p_i. They are repesented graphically in Fig. 5.11a for when the interface radius co-incides with the mean wall radius, i.e. $r_{ep} = ½ (r_i + r_o)$.

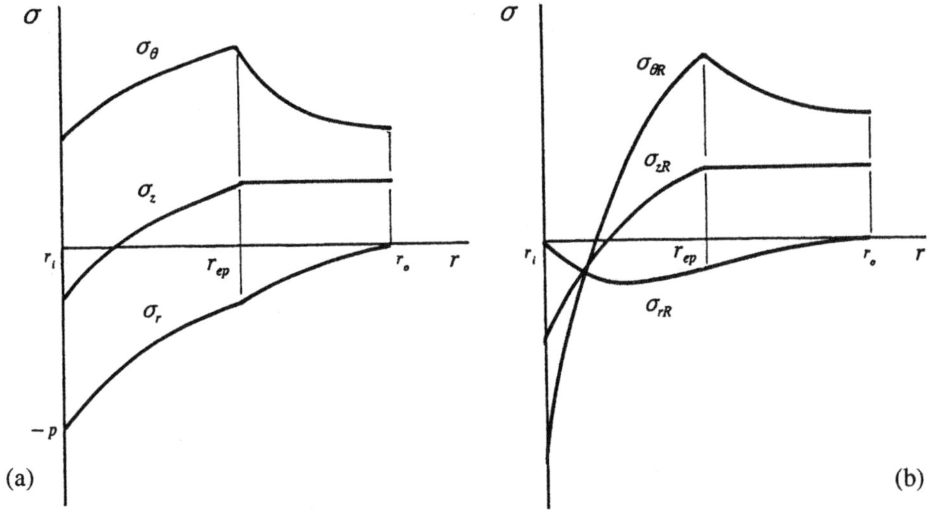

Figure 5.11 (a) Elastic-plastic and (b) residual stress distributions within cylinder wall

Figure 5.11a shows that the hoop stress is greatest in tension at the interface radius. At the bore, the magnitude of the maximum, compressive radial stress equals the internally applied pressure. Also, the axial stress averages the radial and hoop streses across both zones.

5.4.3 Residual Stresses

A triaxial state of residual stress will remain distributed throughout the wall of the cylinder when an elastic-plastic pressure is released. That is, when the bore pressure p_i in eq(5.60a) is released, elastic stresses σ_E recover to leave a residual stress distribution σ_R. The latter is found from applying $\sigma_R = \sigma - \sigma_E$, where σ are the elastic-plastic, applied stresses appearing in eqs(5.54a-c) and (5.59a-c). An elastic recovery occurs over the whole section $r_i \le r \le r_o$, upon releasing p_i. Thus, σ_E assumes the usual Lamé forms, given in eq(5.47a-c). Subtracting σ_E from the corresponding stresses σ under pressure in each zone leads to the requires residuals. For the inner, plastic zone , where $r_i \le r \le r_o$,

$$\sigma_{\theta R} = \sigma_\theta - \sigma_{\theta E}$$

$$= \frac{2Y}{\sqrt{3}} \ln\left(\frac{r}{r_{ep}}\right) + \frac{Y}{\sqrt{3}}\left(1 + \frac{r_{ep}^2}{r_o^2}\right) - \frac{p_i r_i^2 (1 + r_o^2/r^2)}{(r_o^2 - r_i^2)} \tag{5.61a}$$

$$\sigma_{rR} = \sigma_r - \sigma_{rE}$$

$$= \frac{2Y}{\sqrt{3}} \ln\left(\frac{r}{r_{ep}}\right) - \frac{Y}{\sqrt{3}}\left(1 - \frac{r_{ep}^2}{r_o^2}\right) - \frac{p_i r_i^2 (1 - r_o^2/r^2)}{(r_o^2 - r_i^2)} \tag{5.61b}$$

$$\sigma_{zR} = \sigma_z - \sigma_{zE}$$

$$= \frac{2Y}{\sqrt{3}} \ln\left(\frac{r}{r_{ep}}\right) + \frac{Y}{\sqrt{3}}\left(\frac{r_{ep}^2}{r_o^2}\right) - \frac{p_i r_i^2}{(r_o^2 - r_i^2)} \tag{5.61c}$$

For the outer, elastic zone where $r_{ep} \le r \le r_o$,

$$\sigma_{\theta R} = \sigma_\theta - \sigma_{\theta E}$$

$$= \frac{Y}{\sqrt{3}}\left(\frac{r_{ep}}{r_o}\right)^2 \left(1 + \frac{r_o^2}{r^2}\right) - \frac{p_i r_i^2 (1 + r_o^2/r^2)}{(r_o^2 - r_i^2)} \tag{5.61d}$$

$$\sigma_{rR} = \sigma_r - \sigma_{rE}$$

$$= \frac{Y}{\sqrt{3}}\left(\frac{r_{ep}}{r_o}\right)^2 \left(1 - \frac{r_o^2}{r^2}\right) - \frac{p_i r_i^2 (1 - r_o^2/r^2)}{(r_o^2 - r_i^2)} \tag{5.61e}$$

$$\sigma_{zR} = \tfrac{1}{2}(\sigma_{\theta R} + \sigma_{rR})$$

$$= \frac{Y}{\sqrt{3}}\left(\frac{r_{ep}}{r_o}\right)^2 - \frac{p_i r_i^2}{(r_o^2 - r_i^2)} \tag{5.61f}$$

where p_i is given by eq(5.60a). Equations (5.61a-f) are distributed in the manner of Fig. 5.11b. Compressive stresses left at the bore are particularly beneficial to improving fatigue strength. They oppose and reduce the magnitudes of peak, cyclic tensile stresses due to fluctuating, internal service pressure. The process of pre-pressurising thick cylinders to produce compressive residual stresses is known as *autofrettage*. This process has long been used for strengthening cylindrical pressure vessels and cannon tubing against fatigue failure.

A simpler, alternative analysis employs the Tresca yield criterion in eq(5.50) instead of the von Mises eq(5.57). Since no assumption needs to be made about axial strain, eq(5.50) is readily combined with the equilibrium eq(5.55), to give the radial stress $\sigma_r = Y \ln r + C$. It follows from eq(5.58a) that we can arrive at the Tresca solutions from multiplying, with the constant factor $\surd 3/2$, all the forgoing (von Mises) pressures and stresses, including the residuals in eqs(5.61a-f).

5.5 Open-Ended Cylinder and Thin Disc Under Pressure

In an open-ended, thick-walled cylinder, bore pistons produce and contain an internal fluid pressure. When the force exerted by each piston upon the fluid is reacted externally, e.g. by hydraulic jacks, the cylinder wall remains unstressed axially, giving $\sigma_z = 0$. The latter also applies during the radial, pressurised expansion of a hole in a disc. Here, the z-dimension is small but the diameter ratio is comparable with that of a thick cylinder, i.e. a plane stress condition applies. In both the disc and the open-ended cylinder the axial stress σ_z remains the intermediate (i.e. absent) stress $\sigma_\theta > \sigma_z > \sigma_r$. It follows, therefore, that Tresca's hoop and radial stresses in a closed-end cylinder will also apply to a cylinder with open-ends. Again, these are found from multiplying the σ_θ and σ_r expressions (5.54a,b), (5.59a,b) and (5.61a,b,d,e) by a factor of $\surd 3/2$. The von Mises yield criterion is more difficult to combine with the equilibrium condition for an open-ended cylinder. Nadai's parametric approach [1] is particularly useful for providing a solution to the radial and hoop stresses in an elastic-plastic, open-ended cylinder and a thin disc of non-hardening material. In contrast to Tresca's common solution, it will now be shown that a von Mises, open-end solution differs from its closed-ended solution, given above.

5.5.1 Initial Yield Pressure

When $\sigma_z = 0$, the von Mises criterion, eq(5.48), reduces to its biaxial form

$$\sigma_\theta^2 - \sigma_\theta \sigma_r + \sigma_r^2 = Y^2 \tag{5.62a}$$

With an initial yielding of the bore fibres, we put $r = r_i$ and $p_i = p_Y$ in the Lamé elastic stress eqs(5.47a,b), to give the bore stresses σ_θ and σ_r. Substituting these into eq(5.62a) provides the condition for yielding within the bore

$$p_Y^2 [\, (K^2 + 1)^2 + (K^2 - 1)(K^2 + 1) + (K^2 - 1)^2] \, / \, (K^2 - 1)^2 = Y^2 \tag{5.62b}$$

where $K = r_o / r_i$. Re-arranging eq(5.62b) supplies the yield pressure as

$$p_Y = \frac{Y(K^2 - 1)}{\surd(3K^4 + 1)} \tag{5.63}$$

Clearly, eq(5.63) differs from the closed-end yield pressure in eq(5.49).

5.5.2 Full Plasticity

We first derive the bore pressure that will produce full plasticity in a thin disc or open cylinder of non-hardening, von Mises material. At any radius in the wall, the yield criterion (5.62a) and the radial equilibrium eq(5.55) applies. Making the following substitutions

$$\sigma = \frac{1}{\sqrt{2}}(\sigma_\theta + \sigma_r), \qquad \sigma' = \frac{1}{\sqrt{2}}(\sigma_\theta - \sigma_r) \tag{5.64a,b}$$

into eq(5.62a), reduces it to

$$\frac{\sigma^2}{a^2} + \frac{\sigma'^2}{b^2} = 1 \tag{5.65}$$

where $a = \sqrt{2}Y$ and $b = \sqrt{(2/3)}Y$. Equation (5.65) describes the ellipse, shown in Fig. 5.12.

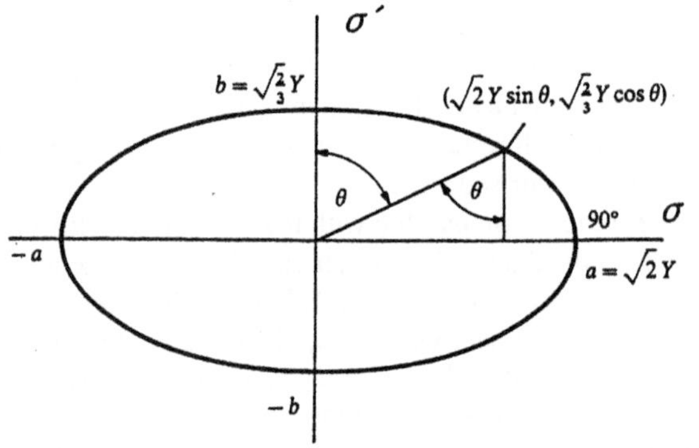

Figure 5.12 Nadai's parametric representation of von Mises yield criterion

The respective lengths, a and b, of the semi-major and -minor ellipse axes, re-appear within Nadai's parameters [1]:

$$\sigma = a\sin\theta = \sqrt{2}Y\sin\theta \tag{5.66a}$$
$$\sigma' = b\cos\theta = \sqrt{(2/3)}\,Y\cos\theta \tag{5.66b}$$

Solving eqs(5.64a,b) for σ_θ and σ_r and combining with eqs(5.66a,b) gives

$$\sigma_\theta = \frac{1}{\sqrt{2}}(\sigma + \sigma') = Y\left[\sin\theta + \frac{1}{\sqrt{3}}\cos\theta\right] = \frac{2Y}{\sqrt{3}}\sin\left(\theta + \frac{\pi}{6}\right) \tag{5.67a}$$
$$\sigma_r = \frac{1}{\sqrt{2}}(\sigma - \sigma') = Y\left[\sin\theta - \frac{1}{\sqrt{3}}\cos\theta\right] = \frac{2Y}{\sqrt{3}}\sin\left(\theta - \frac{\pi}{6}\right) \tag{5.67b}$$

Substituting eqs(5.67a,b) into eq(5.55) leads to a differential equation:

$$r\frac{\mathrm{d}}{\mathrm{d}r}\left[\sin\left(\theta - \frac{\pi}{6}\right)\right] = \left[\sin\left(\theta + \frac{\pi}{6}\right) - \sin\left(\theta - \frac{\pi}{6}\right)\right] = \cos\theta$$

Putting $y = \cos\theta$ enables the variables to be separated into the form

$$\frac{\mathrm{d}y}{\sqrt{1 - y^2}} + \frac{\mathrm{d}y}{\sqrt{3}} = -\frac{2}{\sqrt{3}}\frac{\mathrm{d}r}{r}$$

Integrating this leads to:

$$A^2/r^2 = y\,\mathrm{e}^{-\sqrt{3}\cos^{-1}y} = \cos\theta \times \mathrm{e}^{-\sqrt{3}\theta} \tag{5.68a}$$

where A is an integration constant, found from the condition that $\sigma_r = 0$ for $r = r_o$. Correspondingly, eq(5.67b) gives $\theta_o = \pi/6$ and eq(5.68a) becomes

$$A^2 = \frac{\sqrt{3}}{2} r_o^2\, \mathrm{e}^{-\sqrt{3}\pi/6}$$

$$\frac{r_o^2}{r^2} = \frac{2}{\sqrt{3}} \cos\theta\; \mathrm{e}^{\sqrt{3}(\pi/6 - \theta)} \tag{5.68b}$$

Equation (5.68b) enables θ to be found at any radius $r_i \le r \le r_o$, when σ_θ and σ_r will follow from eqs(5.67a,b). For example, with $r_o = 62.5$ and $r_i = 25$ mm the stress distributions follow the broken lines in Fig. 5.13a.

(a)

(b)

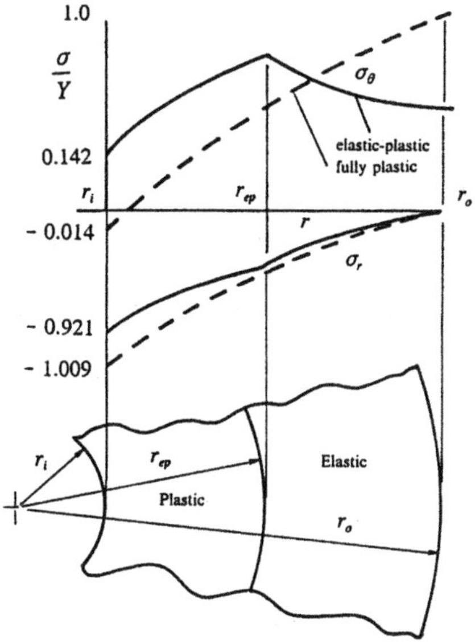

Figure 5.13 Yielding in a thin pressurised disc

The internal pressure to produce full plasticity identifies with the radial stress at the bore: $\sigma_r = -\,p_i$ for $r = r_i$ in eq(5.67b). This gives

$$p_i = \frac{2Y}{\sqrt{3}} \sin\left(\frac{\pi}{6} - \theta_i\right) \tag{5.68c}$$

where θ_i at r_i is found from eq(5.68b)

$$\left(\frac{r_o}{r_i}\right)^2 = \frac{2}{\sqrt{3}} \cos\theta_i\; \mathrm{e}^{\sqrt{3}(\pi/6 - \theta_i)}$$

The radial stress remains compressive for valid values of θ in the range $-\pi/3 < \theta < \pi/6$. For example, with $r_o/r_i = 2.5$, the disc becomes fully plastic (the lower broken line in Fig. 5.13a), when from eq(5.68b),

$$\left(\frac{62.5}{25}\right)^2 = \frac{2}{\sqrt{3}}\cos\theta_i \; e^{\sqrt{3}(\pi/6 - \theta_i)}$$

A trial solution gives $\theta_i = -30.9° = -0.539$ rad. The corresponding normalised, internal pressure is, from eq(5.68c),

$$\frac{p_i}{Y} = \frac{2}{\sqrt{3}}\sin\left(\frac{\pi}{6} + 0.539\right) = 1.009$$

Figure 5.13a shows that the maximum radial compressive stress is $\sigma_r/Y = -p_i/Y = -1.009$ within the bore, where the hoop stress is $\sigma_\theta/Y = -0.014$. Substituting these stresses into the yield criterion (5.62a) confirms that the bore fibres are at the point of yield. The maximum hoop stress is tensile for $r = r_o$. Since $\sigma_r = 0$ for $r = r_o$, eq(5.62a) confirms a maximum value $\sigma_\theta = Y$ (i.e. $\sigma_\theta/Y = 1$) for a fully plastic disc.

5.5.3 Elastic-Plastic Deformation

We can now determine the internal pressure p_i to yield the disc to any intermediate radius r_{ep}, where $r_i \le r_{ep} \le r_o$ (see Fig. 5.13b). Lamé stresses in the outer, elastic zone $r_{ep} \le r \le r_o$, follow from eqs(5.47a,b):

$$\sigma_\theta = \frac{p_{ep}\, r_{ep}^2\left(1 + r_o^2/r^2\right)}{\left(r_o^2 - r_{ep}^2\right)} \tag{5.69a}$$

$$\sigma_r = \frac{p_{ep}\, r_{ep}^2\left(1 - r_o^2/r^2\right)}{\left(r_o^2 - r_{ep}^2\right)} \tag{5.69b}$$

with $\sigma_z = 0$. These are shown as the continuous lines in Fig. 5.13a. The interface yield pressure p_{ep} is found from eq(5.63), where $K = r_o/r_{ep}$. This gives

$$p_{ep} = \frac{Y[(r_o/r_{ep})^2 - 1]}{\sqrt{3(r_o/r_{ep})^4 + 1}} \tag{5.70}$$

The stresses within the inner, plastic zone $r_i \le r \le r_{ep}$ are again supplied by eqs(5.67a,b). The radial and hoop stress plastic distributions must each match the corresponding elastic values at the elastic-plastic, interface radius. The condition that σ_r is common to both zones at $r = r_{ep}$ is used to re-determine the constant A in eq(5.68a) for an elastic-plastic disc. That is, from eqs(5.67b) and (5.70), $\sigma_r = -p_{ep}$ for $\theta = \theta_{ep}$:

$$\frac{Y[1 - (r_o/r_{ep})^2]}{\sqrt{3(r_o/r_{ep})^4 + 1}} = \frac{2Y}{\sqrt{3}}\sin\left(\theta_{ep} - \frac{\pi}{6}\right)$$

$$\frac{1 - (r_o/r_{ep})^2}{\sqrt{(r_o/r_{ep})^4 + 1/3}} = 2\sin\left(\theta_{ep} - \frac{\pi}{6}\right) \tag{5.71}$$

where θ_{ep} must be found by trial. It follows from eq(5.68a) that

$$A^2 = r_{ep}^2 \cos\theta_{ep}\, e^{-\sqrt{3}\,\theta_{ep}}$$

$$\frac{r_{ep}^2}{r^2} = \frac{\cos\theta}{\cos\theta_{ep}}\, e^{\sqrt{3}(\theta_{ep} - \theta)} \qquad (5.72)$$

Equation (5.72) enables θ to be found at any plastic radius $r_i \le r \le r_{ep}$. In particular, the solution $\theta = \theta_i$ for $r = r_i$, when substituted into eq(5.68c), supplies the internal pressure p_i to produce an elastic-plastic disc. For example, with $r_i = 25$ mm and $r_o = 62.5$ mm, we can find the pressure to yield the disc to its mean radius $r_{ep} = ½(25 + 62.5) = 43.75$ mm, as follows:

$$r_{ep} / r_i = 43.75 / 25 = 1.75$$
$$r_o / r_{ep} = (r_o / r_i) \times (r_i / r_{ep}) = 2.5 / 1.75 = 1.4286$$

Substituting into eq(5.71) gives

$$-\,0.4908 = 2 \sin(\theta_{ep} - \pi/6)$$

A trial solution gives $\theta_{ep} = 15.8° = 0.2758$ rad, which is a constant for a given depth of penetration. From eq(5.72), with $r = r_i$ and $\theta = \theta_i$,

$$1.75^2 = 1.0393 \cos\theta_i \times \exp[\sqrt{3}\,(0.2758 - \theta_i)]$$

A trial solution yields $\theta_i \simeq -\,22.9° = -\,0.3997$ rad. Hence, from eq(5.68c)

$$p_i/Y = (2/\sqrt{3}) \sin(\pi/6 + 0.3997) = 0.921$$

and from eq(5.67a), $\sigma_\theta/Y = 0.142$. The stress distribution within the plastic zone in Fig. 5.13 is found for further solutions to θ from eq(5.72) at given radii r within the range $r_i \le r \le r_{ep}$. Equations (5.69a,b) and (5.70) supply the elastic zone stress distributions for $r_{ep} \le r \le r_o$.

5.5.4 Residual Stresses

When the elastic-plastic pressure p_i in eq(5.68c) is released, it is assumed that elastic stress σ_E is recovered for all radii according to the Lamé eqs(5.47a,b). Applying $\sigma_R = \sigma - \sigma_E$ to the plastic zone $r_i \le r \le r_{ep}$, we subtract eqs(5.47a,b) from eqs(5.67a,b) to give the hoop and radial, residual stresses as:

$$\sigma_{\theta R} = \frac{2Y}{\sqrt{3}} \sin\left(\theta + \frac{\pi}{6}\right) - \frac{p_i(1 + r_o^2/r^2)}{r_o^2/r_i^2 - 1} \qquad (5.73a)$$

$$\sigma_{rR} = \frac{2Y}{\sqrt{3}} \sin\left(\theta - \frac{\pi}{6}\right) - \frac{p_i(1 - r_o^2/r^2)}{r_o^2/r_i^2 - 1} \qquad (5.73b)$$

In the elastic zone $r_{ep} \le r \le r_o$, the applied stresses σ are given by eqs(5.69a,b) so that the residuals within this zone become

$$\sigma_{\theta R} = \frac{p_{ep}(1 + r_o^2/r^2)}{r_o^2/r_{ep}^2 - 1} - \frac{p_i(1 + r_o^2/r^2)}{r_o^2/r_i^2 - 1} \tag{5.73c}$$

$$\sigma_{rR} = \frac{p_{ep}(1 - r_o^2/r^2)}{r_o^2/r_{ep}^2 - 1} - \frac{p_i(1 - r_o^2/r^2)}{r_o^2/r_i^2 - 1} \tag{5.73d}$$

where p_{ep} is given by eq(5.70). The residual stress distributions $\sigma_{\theta R}$ and σ_{rR}, according to eqs(5.73a-d), are similar to those of $\sigma_{\theta R}$ and σ_{rR} in Fig. 5.11b, for a closed-end cylinder. The maximum residual hoop stress is compressive within the bore fibres. Note that it is not permissible for $\sigma_{\theta R}$ to exceed Y for $r = r_i$ in a cylinder of non-hardening material. If eq(5.73a) predicts $\sigma_{\theta R} > Y$, the implication is that reversed yielding of the bore has occurred, so invalidating our assumption of an elastic recovery following pressure release. The experimental determination of residual stresses, by the hole boring technique [5], confirms that this assumption will overestimate the compressive residual hoop stresses in the the bore fibres whenever reversed yielding occurs.

5.6 Rotating Disc

Here we wish to determine the angular velocities of a spinning disc required to: (i) initiate yielding, (ii) penetrate a plastic zone into the wall and (iii) produce full plasticity. Consider a uniformly thin, solid disc of outer radius r_o and let the respective velocities be ω_Y, ω_{ep} and ω_{ult}. Firstly, a Tresca yield criterion is employed for simplicity. Later, Nadai's parameters will be employed to define the plane-stress state for a similar disc of von Mises material.

5.6.1 Initial Yielding

The elastic, radial and hoop stresses σ_r, σ_θ, at a radius $r \le r_o$ in a solid disc, when rotating at a speed ω rad/s, are [1]

$$\sigma_r = \frac{\rho\omega^2}{8}(3 + \nu)(r_o^2 - r^2) \tag{5.74a}$$

$$\sigma_\theta = \frac{\rho\omega^2}{8}\left[(3 + \nu)r_o^2 - (1 + 3\nu)r^2\right] \tag{5.74b}$$

where ρ is the density of the disc material. Equations (5.74a,b) are distributed in the manner shown in Fig. 5.14a.

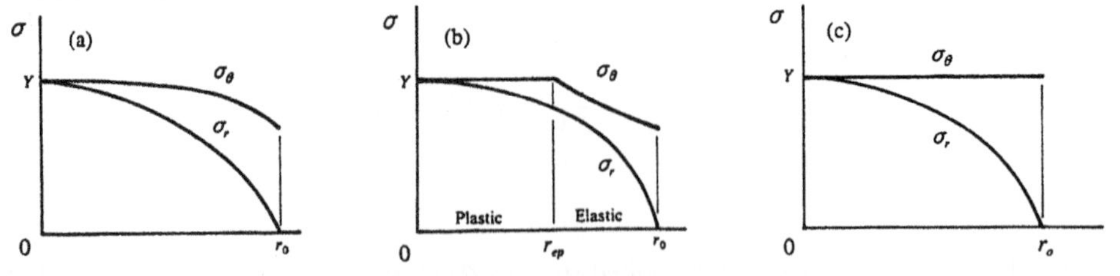

Figure 5.14 Radial and hoop stress distributions

Since both σ_θ and σ_r are tensile, the Tresca criterion (5.50) employs the magnitude of the greater tensile stress together with the least stress (i.e. zero axial stress σ_z). This gives

$$\sigma_\theta = Y \qquad (5.75)$$

Clearly, yielding will commence at the centre where σ_θ is a maximum. Substituting eq(5.74b) into (5.75) for $r = 0$ supplies the speed ω_Y to initiate yielding:

$$\frac{1}{8}\rho\omega_Y^2(3 + \nu)r_o^2 = Y$$

$$\therefore \omega_Y^2 = \frac{8Y}{\rho(3 + \nu)r_o^2} \qquad (5.76)$$

5.6.2 Elastic-Plastic Disc

The equilibrium eq(5.55) for an elastic-plastic disc is modified for the centrifugal force within the inner plastic core $0 \le r \le r_{ep}$. This gives

$$\sigma_r + r\frac{d\sigma_r}{dr} - \sigma_\theta + \rho\omega^2 r^2 = 0 \qquad (5.77)$$

Combining eq(5.75) with eq(5.77), provides a solution to σ_r as follows

$$\frac{d}{dr}(r\sigma_r) = Y - \rho\omega_{ep}^2 r^2$$

$$r\sigma_r = Yr - \frac{\rho\omega_{ep}^2 r^3}{3} + A$$

$$\sigma_r = Y - \frac{\rho\omega_{ep}^2 r^2}{3} + \frac{A}{r} \qquad (5.78a)$$

where ω_{ep} is the speed to produce partial yielding, yet to be determined. From Fig. 5.14b, since $\sigma_\theta = \sigma_r = Y$ for $r = 0$, the integration constant $A = 0$ in eq(5.78a). Thus, at the elastic-plastic radius $r = r_{ep}$, from within the plastic zone, eq(5.78a) becomes

$$\sigma_r = Y - \frac{\rho\omega_{ep}^2 r_{ep}^2}{3} \qquad (5.78b)$$

For the outer, elastic annulus $r_{ep} \le r \le r_o$, Lamé's equations are modified with a centrifugal stress term [1]:

$$\sigma_r = a - \frac{b}{r^2} - (3 + \nu)\frac{\rho\omega_{ep}^2 r^2}{8} \qquad (5.79a)$$

$$\sigma_\theta = a + \frac{b}{r^2} - (1 + 3\nu)\frac{\rho\omega_{ep}^2 r^2}{8} \qquad (5.79b)$$

Note, eqs(5.79a,b) reduce to eqs(5.74a,b) when, for a solid disc, the constants are $b = 0$ and $a = (3 + \nu)\rho\omega^2 r_o^2/8$, apply to the respective conditions: (i) that the stresses cannot be infinite at the disc centre and (ii) $\sigma_r = 0$ for $r = r_o$. Applying new boundary conditions to eqs(5.79a,b) enable the constants a and b to be found. They are (i) $\sigma_r = 0$ for $r = r_o$,

$$0 = a - \frac{b}{r_o^2} - (3 + \nu)\frac{\rho\omega_{ep}^2 r_o^2}{8}$$

and (ii) $\sigma_\theta = Y$ at $r = r_{ep}$

$$Y = a + \frac{b}{r_{ep}^2} - (1 + 3\nu)\frac{\rho\omega_{ep}^2 r_{ep}^2}{8}$$

Solving these simultaneously for a and b gives

$$a = \left\{ Y - \frac{\rho\omega_{ep}^2}{8}\left[(3 + \nu)r_o^2 - (1 + 3\nu)r_{ep}^2\right]\right\}\frac{r_{ep}^2}{(r_o^2 + r_{ep}^2)} + (3 + \nu)\frac{\rho\omega_{ep}^2 r_o^2}{8} \tag{5.80a}$$

$$b = \left\{ Y - \frac{\rho\omega_{ep}^2}{8}\left[(3 + \nu)r_o^2 - (1 + 3\nu)r_{ep}^2\right]\right\}\frac{r_o^2 r_{ep}^2}{(r_o^2 + r_{ep}^2)} \tag{5.80b}$$

The angular speed ω_{ep} is found from the condition that σ_r is common to both zones at the interface radius r_{ep}. Within the elastic annulus, eq(5.79a) gives, for $r = r_{ep}$,

$$\sigma_r = a - \frac{b}{r_{ep}^2} - (3 + \nu)\frac{\rho\omega_{ep}^2 r_{ep}^2}{8} \tag{5.81a}$$

Substituting a and b from eqs(5.80a,b) into eq(5.81a) leads to

$$\sigma_r = (3 + \nu)\frac{\rho\omega_{ep}^2}{8}(r_o^2 - r_{ep}^2) - \frac{Y(r_o^2 - r_{ep}^2)}{(r_o^2 + r_{ep}^2)}$$

$$+ \frac{\rho\omega_{ep}^2}{8}\left[(3 + \nu)r_o^2 - (1 + 3\nu)r_{ep}^2\right]\frac{(r_o^2 - r_{ep}^2)}{(r_o^2 + r_{ep}^2)} \tag{5.81b}$$

Equating (5.78b) and (5.81b)

$$2Yr_o^2 = \frac{\rho\omega_{ep}^2}{8}\left\{(3 + \nu)(r_o^2 - r_{ep}^2)(2r_o^2 + r_{ep}^2) + r_{ep}^2\left[\frac{8}{3}(r_o^2 + r_{ep}^2) - (1 + 3\nu)(r_o^2 - r_{ep}^2)\right]\right\}$$

leads to the speed of the elastic-plastic disc

$$\omega_{ep}^2 = \frac{16\,Y}{\rho r_o^2\left\{(3 + \nu)\left(1 - \frac{r_{ep}^2}{r_o^2}\right)\left(2 + \frac{r_{ep}^2}{r_o^2}\right) + \left(\frac{r_{ep}^2}{r_o^2}\right)\left[\frac{8}{3}\left(1 + \frac{r_{ep}^2}{r_o^2}\right) - (1 + 3\nu)\left(1 - \frac{r_{ep}^2}{r_o^2}\right)\right]\right\}} \tag{5.82}$$

Equation (5.82) supplies a speed $\omega_{ep}{}^2 = (2.671Y)/(\rho r_o{}^2)$, for when the elastic-plastic interface radius coincides with the mean radius $r_{ep}/r_o = ½$, in a steel disc with $\nu = ¼$. Dividing by ω_Y from eq(5.76), gives $\omega_{ep}/\omega_Y = 1.042$. This shows that only a 4.2% increase in the initial yield speed is required to produce this amount of plastic penetration.

5.6.3 Fully Plastic Disc

When the disc becomes fully plastic, eq(5.78a) applies to $0 \le r \le r_o$. Again, $\sigma_r = Y$ for $r = 0$, giving $A = 0$. Also, because $\sigma_r = 0$ for $r = r_o$, eq(5.78b) gives

$$0 = Y - \frac{\rho\,\omega_{ult}^2\, r_o^2}{3}$$

The fully plastic speed ω_{ult} is, therefore

$$\omega_{ult}^2 = \frac{3Y}{\rho\, r_o^2} \qquad (5.83)$$

Dividing eq(5.83) with ω_Y, from eq(5.76), gives the angular speed ratio:

$$\frac{\omega_{ult}}{\omega_Y} = \sqrt{\frac{3(3+\nu)}{8}}$$

This gives $\omega_{ult}/\omega_Y = 1.104$ for $\nu = ¼$, showing that a 10.4% increase in the initial yield speed is required to attain full plasticity in a solid disc. Distributions in σ_r and σ_θ for the fully plastic disc are shown in Fig. 5.14c. Clearly, the use of the Tresca criterion, where $\sigma_\theta = Y$ is constant throughout the region of plasticity, greatly simplifies this solution.

5.6.4 Tresca Versus von Mises

A numerical, von Mises solution to the stress distribution within a non-hardening, elastic-plastic, solid disc is also possible. The elastic zone stresses are again given by eq(5.79a,b) for the same boundary condition: $\sigma_r = 0$ for $r = r_o$. Moreover, eqs(5.79a,b) must satisfy the von Mises yield condition at the interface radius, i.e. where σ_θ and σ_r must equal the corresponding plastic zone stresses ar r_{ep}. Within the inner plastic core, these two stresses must satisfy the von Mises yield criterion and the equilibrium equation simultaneously:

$$\sigma_\theta^2 - \sigma_\theta\sigma_r + \sigma_r^2 = Y^2 \qquad (5.84a)$$

$$\sigma_r + r\frac{d\sigma_r}{dr} = \sigma_\theta - \rho\omega^2 r^2 \qquad (5.84b)$$

Using Nadai's eqs(5.67a,b) to separate σ_θ and σ_r in eq(5.84a) and then substituting these into eq(5.84b) leads to a first order differential equation in r and θ [6]:

$$\frac{d\theta}{dr} = \frac{2Y\cos\theta - \sqrt{3}\rho\,\omega^2 r^2}{Yr(\sqrt{3}\cos\theta + \sin\theta)} \qquad (5.85)$$

We may then solve eq(5.85) using a Runga-Kutta method, taking a single starting value for θ, depending upon whether the disc is solid or hollow (i.e. $\theta_o = \pi/2$ or $\pi/6$ respectively). The von Mises solid disc solution shows that both σ_r and σ_θ must remain very nearly equal to Y in this zone, in contrast to the Tresca solution (see Figs 5.15a,b).

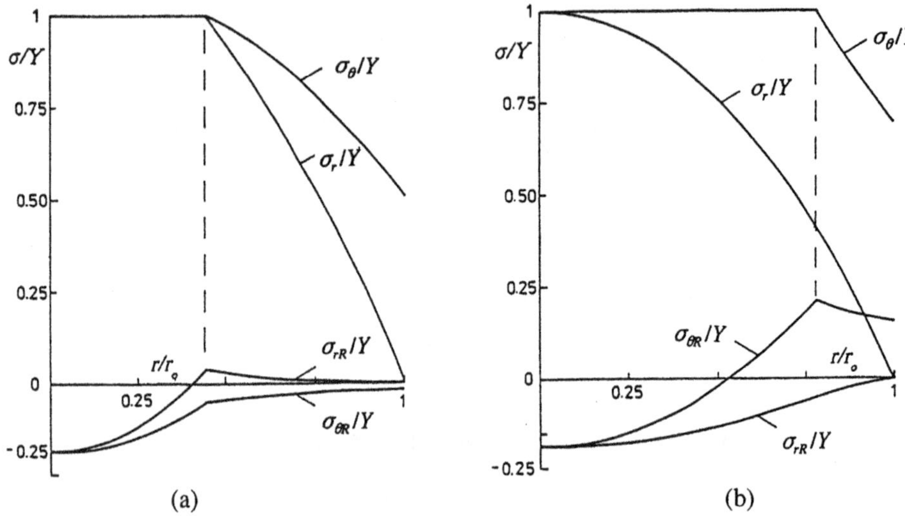

Figure 5.15 (a) von Mises and (b) Tresca stresses for solid disc at similar speeds

The hollow disc solution reveals a similar σ_r distributions in Figs 5.16a and b but the von Mises σ_θ is not constant, increasing above Y in the plastic zone interior $r_i < r < r_{ep}$.

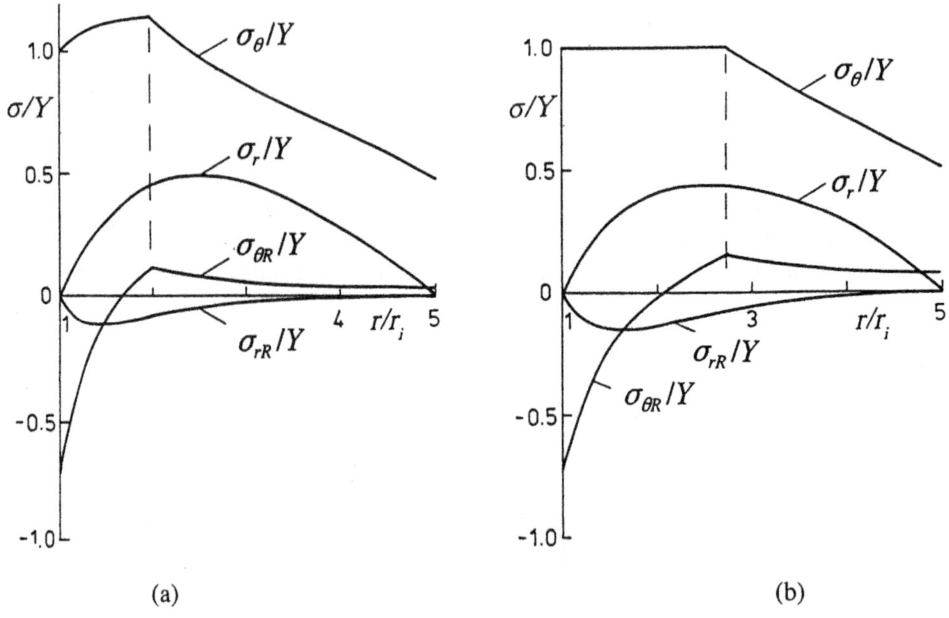

Figure 5.16 (a) von Mises, (b) Tresca stress distributions for a hollow rotating disc

The greater differences arising in σ_θ and σ_r, from applying different yield criteria to a solid disc, are reflected in their residual stress estimates, $\sigma_{\theta R}$ and σ_{rR}, as shown. For example, on applying $\sigma_R = \sigma - \sigma_E$ to each zone in Fig. 5.15b, the Tresca residuals follow from eqs(5.74a,b), (5.78a) and (5.79a,b) as:

$0 \le r \le r_{ep}$

$$\sigma_{\theta R} = Y - \frac{\rho\omega_{ep}^2}{8}\left[(3 + \nu)r_o^2 - (1 + 3\nu)r^2\right] \quad (5.86a)$$

$$\sigma_{rR} = Y - \frac{\rho\omega_{ep}^2 r^2}{3} - \frac{\rho\omega_{ep}^2}{8}(3 + \nu)(r_o^2 - r^2) \quad (5.86b)$$

$r_{ep} \le r \le r_o$

$$\sigma_{\theta R} = a + \frac{b}{r^2} - (1 + 3\nu)\frac{\rho\omega_{ep}^2 r^2}{8} - \frac{\rho\omega_{ep}^2}{8}\left[(3 + \nu)r_o^2 - (1 + 3\nu)r^2\right] \quad (5.86c)$$

$$\sigma_{rR} = a - \frac{b}{r^2} - (3 + \nu)\frac{\rho\omega_{ep}^2 r^2}{8} - \frac{\rho\omega_{ep}^2}{8}(3 + \nu)(r_o^2 - r^2) \quad (5.86d)$$

where a, b and ω_{ep} are given by eqs(5.80a,b) and (5.82). Equations (5.86a-d) were applied to a solid, steel disc with $Y = 310$ MPa, $\nu = 0.28$, $\rho = 7750$ kg/m^3 and $r_o = 250$ mm, following a speed $N = 13000$ rev/min. Figures 5.15a,b show these Tresca's residuals are greater and with different sign in the outer zone, compared to von Mises residuals [6]. This contrasts with the conservative nature of Tresca, which predicts a greater spread of plasticity at similar rotational speeds.

References

1. Nadai A. *Theory of Flow and Fracture of Solids*, 1950, McGraw-Hill, London.
2. Timoshenko S. and Goodier J. N. *Theory of Elasticity*, 1951, McGraw-Hill, New York.
3. Crossland B. and Bones J. A. *Proc. I. Mech. E*, 1958, **172**, 777.
4. Crossland B. *Proc. I. Mech. E*, 1954, **168**, 935.
5. Franklin G. J. and Morrison J. L. *Proc. I. Mech. E*, 1960, **174**, 947.
6. Rees D. W. A. *Zeit Angew Math Mech*, 1999, **79**, 281.

Exercises

Elastic-Perfectly Plastic Beams

5.1 A simply supported beam of length l, with a rectangular section breadth b and depth d, carries a uniformly distributed load w/unit length. Determine the ultimate moment, the collapse load and the plastic hinge length, given that the yield stress Y is constant,.
[Answer: $M_{ult} = bd^2Y/4$, $w = 2bd^2Y/l^2$, $l_p = l/\sqrt{3}$]

5.2 Determine the collapse load and the length of the plastic hinges when an encastre beam of length l, with rectangular section $b \times d$, carries a single concentrated load W, that divides this length into p and q, i.e. $p + q = l$.
[Answer: $M_{ult} = (bd^2lY)/(2pq)$, $p/6$, $q/6$, $l/6$]

5.3 Show that the collapse moment for a ⊥ section, made from equal rectangles each of length a and thickness t, is given by $M_{ult} = Yah(a + h)/2$.

5.4 Examine the manner in which a cantilever of length l will collapse when carrying a uniformly distributed load w/unit length, with a prop to prevent deflection at its free end. Show that the initial yield loading is $w_Y = 8M_Y/l^2$ and the collapse loading is $w_P = 11.73M_P/l^2$. Hence fimd the ratio between these loads for a rectangular section $b \times d$. Hint: The prop reaction is given by $3wl/8$.

Elastic-Perfectly Plastic Torsion Bars

5.5 A bar of diameter d and length l is bored to diameter $d/2$ over half its length. If the outer diameter of the solid shaft reaches its yield point under an applied torque, show that the diameter of the elastic-plastic interface d_{ep}, within the hollow shaft section, may be found from the solution to the quartic equation: $d_{ep}^4 - d^3 d_{ep} + 3d/16 = 0$. Show that the ratio between the angular twists for the hollow and solid shafts is given by d/d_{ep}.

5.6 Derive and plot the normalised torque-twist relationship for a hollow bar. Show the accumulation of normalised residual strain on this plot in a similar manner to that given in Fig. 5.9 for a solid bar.

5.7 What value of torque is required for an elastic-plastic interface to lie at the mean radius in a tube of inner and outer diameters 25 and 100 mm respectively? Determine the residual stress distribution and the residual twist when this torque is subsequently removed. Take $k = 230$ MPa and $G = 78$ GPa.

Elastic-Perfectly Plastic Cylinders and Discs

5.8 What is the maximum, limiting diameter ratio of an annular disc beyond which it is not possible to achieve a full spread of plasticity when radial pressure is applied to the inner diameter? At what pressure does this occur? [Answer: 2.963, $2Y/\sqrt{3}$]

5.9 An annular disc, with inner and outer radii 25 mm and 62.5 mm respectively, is machined from an alloy steel with a yield stress 500 MPa. Determine the internal pressures necessary to: (a) initiate yielding, (b) produce a fully plastic disc and (c) produce partial plasticity to the mean radius. [Answer: 2.415 kbar, 5.045 kbar, 4.584 kbar]

5.10 Compare the residual stress distributions in open-end and closed-end thick walled cylinders resulting from an autofrettage pressure sufficient to penetrate an elastic-plastic interface to coincide with the mean wall radius. Employ a von Mises yield criterion and normalise the stresses with the yield stress Y for a cylinder of diameter ratio 3.

5.11 Show that the speeds necessary to initate yielding within solid and hollow discs are each independent of the yield criterion and are respectively:

$$\omega_Y = \sqrt{\frac{4Y}{\rho[r_i^2(1 - \nu) + r_o^2(3 + \nu)]}}, \qquad \omega_Y = \sqrt{\frac{8Y}{(3 + \nu)\rho r_o^2}}$$

5.12 Using the Nadai's parameter approach, determine the von Mises stress distributions in an elastic-plastic, hollow disc, $r_i = 50$ mm and $r_o = 250$ mm, rotating at a speed of 11000 rev/min. Derive from these the residual stresses distribution for when the disc is brought to rest. The following conditions apply to the disc material: $Y = 310$ MPa, $\nu = 0.28$, and $\rho = 7750$ kg/m^3. Compare and contrast each distribution with the corresponding Tresca solution. [Answer is given in Figs 5.16a,b]

CHAPTER 6

SLIP LINE FIELDS

6.1 Introduction

Rigid plasticity theory ignores elastic strain and assumes that plastic deformation occurs without hardening under a constant yield stress. This material model is often justified for the description of a forming process involving large plastic strains. When extruding carbon steel, for example, the maximum elastic and plastic strains are of the typical contrasting magnitudes: 0.1% and 50% respectively. In general, a plastic-rigid model is an appropriate choice for the analyses of large scale deformation processes. These processes may be framed within co-ordinates of plane strain and axial symmetry, depending upon the shape that they produce. In this chapter it is shown how the slip-line field (SLF) method can represent the flow behaviour of plane strain processes with a plastic-rigid material model. Slip-lines can be constructed only where plane strain conditions are upheld. Consequently, this method is restricted to processes that include: extrusion of rectangular stock, indentation under a narrow, parallel die, rolling of thin sheet and orthogonal machining. Where plane strain conditions do not exist, as with the extrusion of circular bar, alternative analyses are required. Among these are the limit and slab analyses. A limit analysis provides simple, upper and lower bound approximations to the forming forces. A slab analysis is based upon the force equilibrium of a deforming element. In principle, these techniques place no restriction upon the deformation mode but are most often employed where they lead to convenient closed solutions. In particular, analyses are made in Chapters 7 and 14 of the plastic collapse of beams and circular plates under lateral loadings, hot and cold forging, rolling, wire and strip drawing.

Note, that in the absence of elasticity and hardening, SLF takes no account of the effects of creep, strain rate, temperature generation and thermal gradients. A more serious effect is likely to arise when plane cross sections distort out of their original planes. This can occur under extrusion conditions for the very severe deformations in the die region. Its extent has been examined from the distortion that occurs to an orthogonal grid [1].

6.2 Slip Line Field Theory

A plane strain condition refers to a process where strain in a direction perpendicular to the plane of deformation is zero or a constant. This applies to rolling, extrusion and indentation processes, where lateral flow is negligible. The slip-line field will approximate the forming loads required for each process. In an extrusion and indentation process, the boundary conditions for the contacting surfaces may vary from a constant coefficient of friction lying between zero and a sticking value.

We shall present here a graphical interpretation of equations, due to Hencky [2] and Geiringer [3], which govern plane strain processes. Three diagrams are: (i) the physical plane, defining the geometry of the process, (ii) the stress plane, showing the variation in normal stress on a slip-line and (iii) the hodograph of velocity streamlines.

6.2.1 Yield and Flow

When $f = J_2'$, given by eq(3.10b), is substituted into eq(3.22) it results in the *Levy-Mises flow rule*. The differentiation gives the principal plastic strain increments:

$$d\varepsilon_1 = \tfrac{2}{3}\, d\lambda\, [\sigma_1 - \tfrac{1}{2}(\sigma_2 + \sigma_3)] \quad (6.1a)$$
$$d\varepsilon_2 = \tfrac{2}{3}\, d\lambda\, [\sigma_2 - \tfrac{1}{2}(\sigma_1 + \sigma_3)] \quad (6.1b)$$
$$d\varepsilon_3 = \tfrac{2}{3}\, d\lambda\, [\sigma_3 - \tfrac{1}{2}(\sigma_1 + \sigma_2)] \quad (6.1c)$$

In eqs(6.1a,b,c) the strain superscript P is omitted since $d\varepsilon$ become total strain increments in the absence of elasticity. Taking 1 and 3 to lie in the plane of deformation, with $d\varepsilon_2 = 0$ for plane strain, eq(6.1b) leads to

$$\sigma_2 = \tfrac{1}{2}(\sigma_1 + \sigma_3) \quad (6.2)$$

Thus, σ_2 is the intermediate stress $\sigma_1 \geq \sigma_2 \geq \sigma_3$, giving Tresca's yield criterion (3.1a) as

$$\sigma_1 - \sigma_3 = 2k \quad (6.3)$$

where k is the shear yield stress. Equation (6.3) remains unchanged when eq(6.2) is substituted into the von Mises criterion (eq 3.7a). However, the relation between k and Y differ between the two criteria. For Tresca $k = Y/2$ and for von Mises $k = Y/\surd 3$. With plane-strain yielding under the general stress components σ_x, σ_y and τ_{xy}, the in-plane principal stresses are

$$\sigma_{1,3} = \tfrac{1}{2}(\sigma_x + \sigma_y) \pm \tfrac{1}{2} \surd[\,(\sigma_x - \sigma_y)^2 + 4\tau_{xy}^{\ 2}\,] \quad (6.4a)$$

Substituting eq(6.4a) into eq(6.3), the yield condition becomes

$$\tfrac{1}{2}(\sigma_1 - \sigma_3) = \tfrac{1}{2} \surd[\,(\sigma_x - \sigma_y)^2 + 4\tau_{xy}^{\ 2}\,] = k \quad (6.4b)$$

Equation (6.4b) is the limiting radius of a Mohr's circle (see Fig. 6.1a) describing a general stress state at the point of yield.

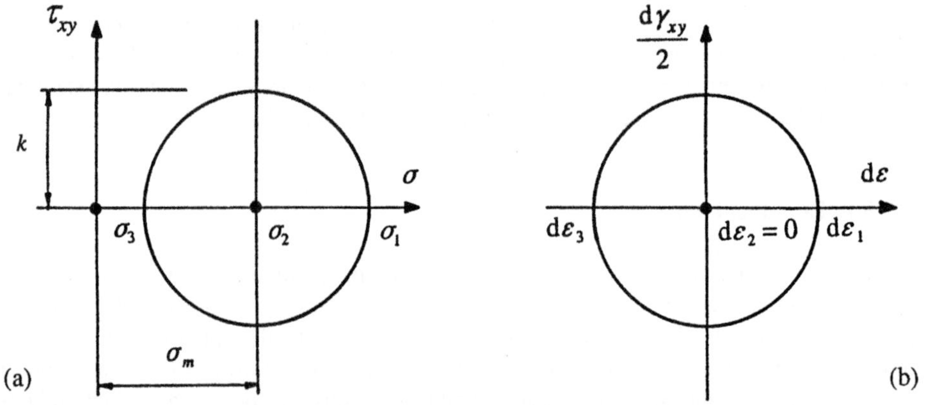

Figure 6.1 Mohr's circles of stress and strain

The centre of Mohr's stress circle follows from eqs(6.2) and (6.4a):

$$\tfrac{1}{2}(\sigma_1 + \sigma_3) = \sigma_2 = \tfrac{1}{2}(\sigma_x + \sigma_y) \quad (6.4c)$$

This is also the mean, or hydrostatic, stress $\sigma_m = ⅓\, \sigma_{kk}$:

$$\sigma_m = ⅓\,(\sigma_1 + \sigma_2 + \sigma_3)$$
$$= ⅓\,[\sigma_1 + ½(\sigma_1 + \sigma_3) + \sigma_3] = ½\,(\sigma_1 + \sigma_3)$$

since it equals the intermediate stress σ_2 in eq(6.2). Now, as the volume of material is conserved $dV/V \simeq d\varepsilon_{kk} = 0$, when it follows from eqs(6.1a,c) that $d\varepsilon_1 = -\, d\varepsilon_3$. The relationship between $d\varepsilon_1$ and $d\varepsilon_3$ and the component strains $d\varepsilon_x$, $d\varepsilon_y$ and $d\gamma_{xy}$, where also $d\varepsilon_x = -\, d\varepsilon_y$, is given by

$$d\varepsilon_{1,3} = ½\,(d\varepsilon_x + d\varepsilon_y) \pm ½\,\sqrt{[\,(d\varepsilon_x - d\varepsilon_y)^2 + (d\gamma_{xy})^2\,]}$$

The radius of the strain circle (see Fig. 6.1b) becomes

$$½\,(d\varepsilon_1 - d\varepsilon_3) = ½\,\sqrt{[\,(d\varepsilon_x - d\varepsilon_y)^2 + (d\gamma_{xy})^2]} \tag{6.5a}$$

with centre co-ordinate

$$½\,(d\varepsilon_1 + d\varepsilon_3) = d\varepsilon_2 = ½\,(d\varepsilon_x + d\varepsilon_y) = 0 \tag{6.5b}$$

It follows, from eqs(6.4a) and (6.5a), that the stress and strain circles are geometrically similar (i) for the shear strain ordinate $½\gamma_{xy}$, given in Fig. 6.1b and (ii) when the mean component of normal stress is subtracted from the stress abscissa in Fig. 6.1a. This does not effect the radius $½[\sigma_1 - \sigma_m) - (\sigma_3 - \sigma_m)]$ of the stress circle, given in eq(6.4a).

The Mohr's circles describe the stress and strain states at a given point. The top and bottom of a vertical diameter represent two maximum shear planes, lying at right angles to each other. Upon these planes of maximum shear, an association is made between: (i) the maximum shear stress aligned with the direction of these planes and (ii) a hydrostatic stress lying normally to the planes, for which there is no direct strain. In the theory the maximum shear stress τ_{max} has attained the shear yield stress k so that maximum shear planes become known as *slip lines* or *shear lines*. The stress state along a slip line will vary from point to point since the hydrostatic stress σ (subscript m omitted) is a variable. This variation can be observed within a slip-line field construction, shown in Fig. 6.2.

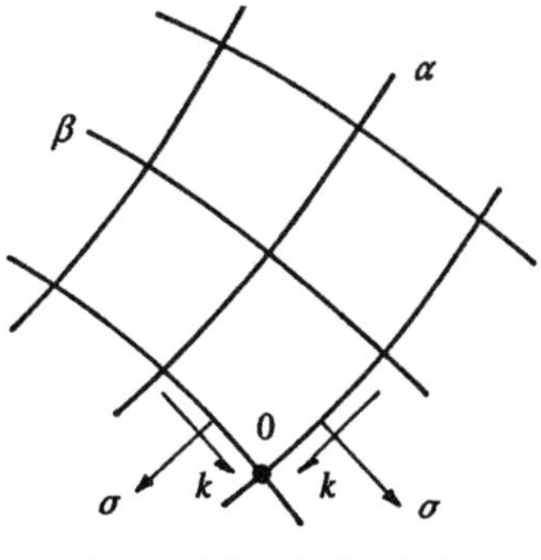

Figure 6.2 A slip line field

The SLF is an orthogonal, curvilinear network of α and β lines. The convention for constructing Mohr's circle for an intersecting pair of α, β lines is that tensile mean stress is positive and a clockwise shear stress is positive, i.e. to the right of the slip line normal.

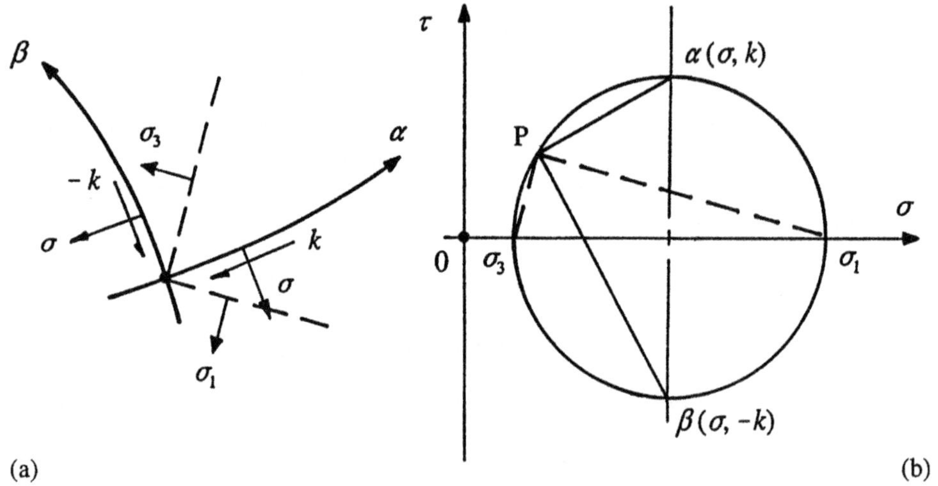

Figure 6.3 Stress state for a pair of intersecting slip lines

It follows from Fig. 6.3a that α - lines are all associated with a positive shear stress (+ k) and the β - lines with a negative shear stress (− k). The corresponding Mohr's circle construction is given in Fig. 6.3b. Firstly, the co-ordinates: $\alpha(\sigma, k)$, $\beta(\sigma, -k)$ are located on the circle according to our convention. The circle is drawn with points α and β at opposite ends of a vertical diameter. The directions of α and β are projected through points α and β to intersect the circle in a common pole point, P. Actually, one such projection is sufficient to locate the pole. It follows that the planes on which principal stresses σ_1 and σ_3 act, are parallel to the broken lines joining P to σ_1 and P to σ_3. Transferring principal planes to Fig. 6.3a, we see that the direction of the major principal stress σ_1 lies between α and β.

6.2.2 Hencky's Theorems

A unique relationship will hold between σ and k for each line. This is found from applying force equilibrium to Cartesian axes x and y, inclined at ϕ to any pair of slip lines α and β in Fig. 6.4.

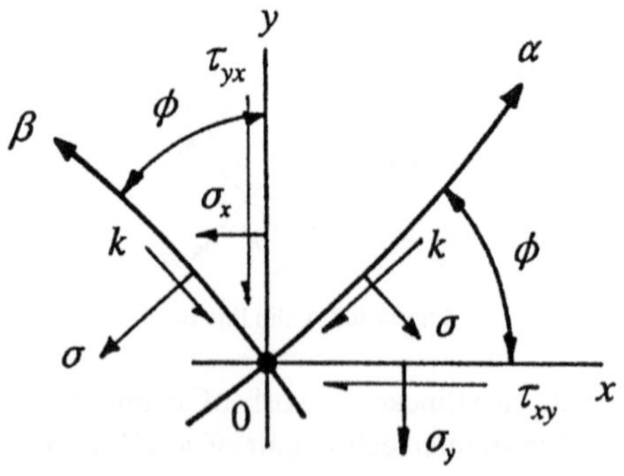

Figure 6.4 Equilibrium in Cartesian axes

The plane equilibrium conditions are referred to x and y:

$$\frac{\partial \sigma_x}{\partial x} + \frac{\partial \tau_{xy}}{\partial y} = 0 \tag{6.6a}$$

$$\frac{\partial \sigma_y}{\partial y} + \frac{\partial \tau_{yx}}{\partial x} = 0 \tag{6.6b}$$

The Cartesian stress components σ_x, σ_y and τ_{xy} are derived from σ and k by inverting the stress transformation law in eq(1.22a). Firstly, note that α and β are the primed axes along which the stress components (σ, k) are known. Pre- and post-multiply $\mathbf{T}'$ by $\mathbf{M}^T$ and $\mathbf{M}$ respectively:

$$\mathbf{M}^T\mathbf{T}'\mathbf{M} = \mathbf{M}^T(\mathbf{M}\mathbf{T}\mathbf{M}^T)\mathbf{M} = (\mathbf{M}^T\mathbf{M})\mathbf{T}(\mathbf{M}^T\mathbf{M})$$

Since $\mathbf{I} = \mathbf{M}^T\mathbf{M} = \mathbf{M}\mathbf{M}^T$, we have

$$\mathbf{T} = \mathbf{M}^T\mathbf{T}'\mathbf{M}$$

$$\begin{bmatrix} \sigma_x & \tau_{xy} \\ \tau_{yx} & \sigma_y \end{bmatrix} = \begin{bmatrix} \cos\phi & -\sin\phi \\ \sin\phi & \cos\phi \end{bmatrix} \begin{bmatrix} \sigma & k \\ k & \sigma \end{bmatrix} \begin{bmatrix} \cos\phi & \sin\phi \\ -\sin\phi & \cos\phi \end{bmatrix}$$

The components of the rotation matrix $\mathbf{M}$ are: $l_{11} = \cos(\alpha 0x) = \cos\phi$, $l_{12} = \cos(\alpha 0y) = \cos(90 - \phi) = \sin\phi$, $l_{21} = \cos(\beta 0x) = \cos(90 + \phi) = -\sin\phi$ and $l_{22} = \cos(\beta 0y) = \cos\phi$. Matrix multiplication gives

$$\sigma_x = \sigma - k\sin 2\phi \tag{6.7a}$$

$$\sigma_y = \sigma + k\sin 2\phi \tag{6.7b}$$

$$\tau_{xy} = \tau_{yx} = k\cos 2\phi \tag{6.7c}$$

Substituting eqs(6.7a-c) into eqs(6.6a,b), provides two relationships

$$\frac{\partial}{\partial x}(\sigma - k\sin 2\phi) + \frac{\partial}{\partial y}(k\cos 2\phi) = 0$$

$$\frac{\partial}{\partial y}(\sigma + k\sin 2\phi) + \frac{\partial}{\partial x}(k\cos 2\phi) = 0$$

In the limit of these equations $\sin 2\phi \to 0$ and $\cos 2\phi \to 1$ as $\phi \to 0$. We find, for the α - line

$$\frac{\partial \sigma}{\partial x} - 2k\frac{\partial \phi}{\partial x} = 0 \tag{6.8a}$$

and for the β - line,

$$\frac{\partial \sigma}{\partial y} + 2k\frac{\partial \phi}{\partial y} = 0 \tag{6.8b}$$

Integrating eqs(6.8a,b), leads to the Hencky equations for any pair of orthogonal slip lines

$$\sigma - 2k\phi = c_1 \quad \text{and} \quad \sigma + 2k\phi = c_2$$

where c_1 and c_2 are constants. Usually, σ is replaced by the acting mean compressive mean stress $-p$ to give

$$p + 2k\phi = c_1 \quad \text{and} \quad p - 2k\phi = c_2 \tag{6.9a,b}$$

where ϕ is positive ACW. Equations (6.9a,b) lead to Hencky's *first theorem,* which states that when two α - slip lines are cut by β - slip lines (see Fig. 6.5), then the angle subtended by tangents to the α - line at the intersection points A, B, P and Q is constant along the length of α.

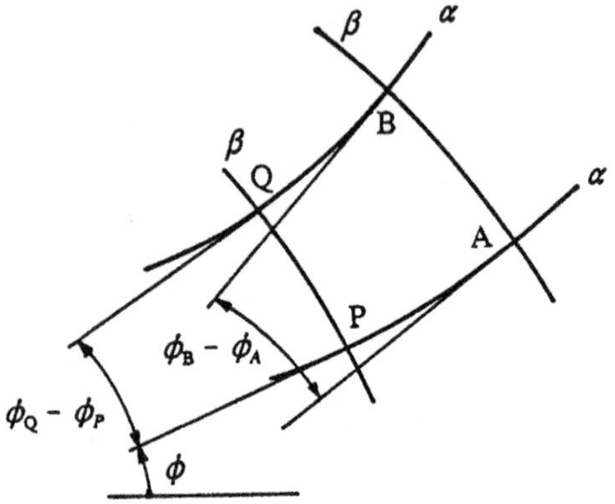

Figure 6.5 Hencky's theorem

That is, $(\phi_Q - \phi_P) = (\phi_B - \phi_A)$. The proof uses eqs(6.9a,b) with a fixed datum for ϕ as shown. Along α, eq(6.9a) gives

$$p_Q + 2k\phi_Q = c_1 \quad \text{and} \quad p_B + 2k\phi_B = c_1$$
$$p_P + 2k\phi_P = c_1 \quad \text{and} \quad p_A + 2k\phi_A = c_1$$

and along β, eq(6.9b) gives

$$p_Q - 2k\phi_Q = c_2 \quad \text{and} \quad p_B - 2k\phi_B = c_2$$
$$p_P - 2k\phi_P = c_2 \quad \text{and} \quad p_A - 2k\phi_A = c_2.$$

Now

$$\begin{aligned} p_Q - p_A &= (p_Q - p_B)_\alpha + (p_B - p_A)_\beta \\ &= 2k\,(\phi_B - \phi_Q) + 2k\,(\phi_B - \phi_A) \end{aligned} \tag{6.10a}$$

and

$$\begin{aligned} p_Q - p_A &= (p_Q - p_P)_\beta + (p_P - p_A)_\alpha \\ &= 2k\,(\phi_Q - \phi_P) + 2k\,(\phi_A - \phi_P) \end{aligned} \tag{6.10b}$$

Equating (6.10a and b) we find

$$2k\,(\phi_B - \phi_Q) + 2k\,(\phi_B - \phi_A) = 2k\,(\phi_Q - \phi_P) + 2k\,(\phi_A - \phi_P)$$

$$\therefore \phi_Q - \phi_P = \phi_B - \phi_A$$

so confirming a constant angle of intersection.

In his *second theorem,* Hencky related the radius of curvature of a slip line to the distance along it. This has been found useful for numerical solutions to slip line fields as discussed by Hill [4], Prager and Hodge [5]. These are not employed here. Instead, we employ Prager's geometrical interpretation [6] of this theorem. This enables a convenient graphical construction, as outlined in the following section.

6.2.3 Stress Plane

To follow Prager's geometrical interpretation of Hencky's equations, consider firstly the variation in stress between points 1 and 2 along an α - line in Fig. 6.6a.

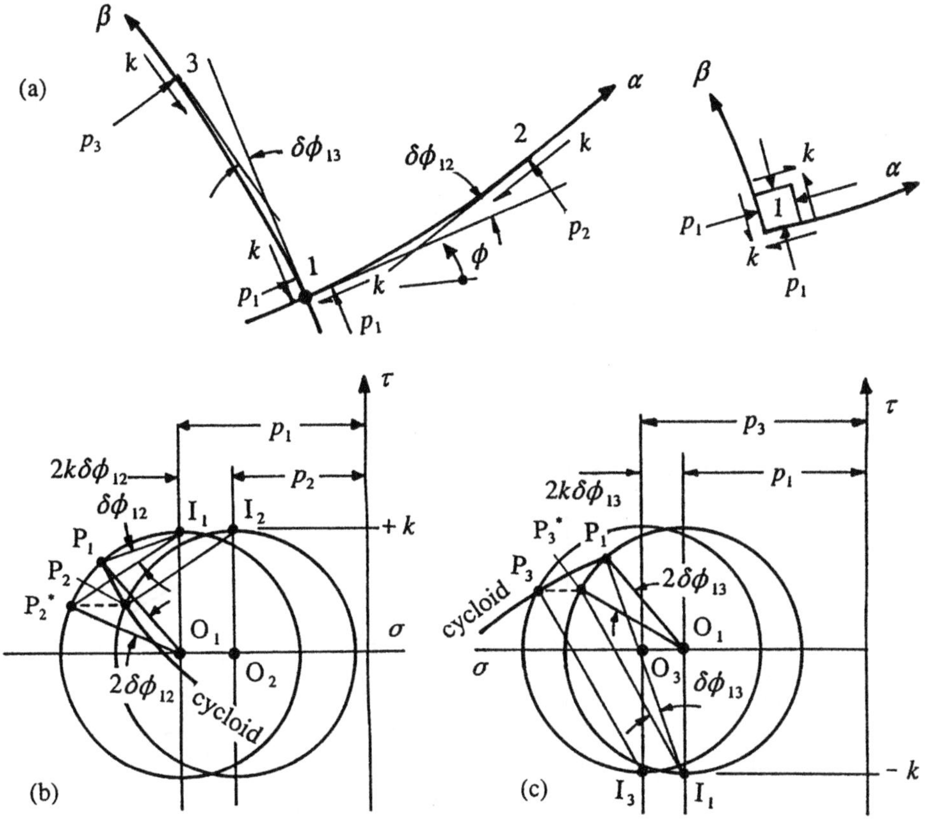

Figure 6.6 Prager's geometrical interpretation

If there is no variation in p, the constant stress state $(-p, k)$, appears as a point I_1 on a fixed circle with centre O_1. The pole point P_1 for this circle is the intersection point, found from projecting the direction of the plane on which $-p$ and k act, through I_1 (see Fig. 6.6b). We could also have arrived at P_1 from the constant stress state $(-p, -k)$ along the β - line, for which I_1 would lie at the opposite end of the vertical diameter. Now let the mean pressure vary along each slip line. Let the stress state at point 2 on the α - line be $(-p_2, k)$, where $p_1 \geq p_2$. This state locates a point I_2 on another circle, with centre O_2 and pole P_2. If the positive, anti-clockwise (ACW) difference between the slopes of the tangents at points 1 and 2 is $\delta\phi_{12}$, then the pressure difference is found from eq(6.9a) to be

$$p_2 - p_1 = -2k\delta\phi_{12} \tag{6.11a}$$

confirming that p_2 is less than p_1. As $\delta\phi_{12} = P_1 I_1 P_2^*$ in Fig. 6.6b, this locates the image, P_2 of P_2^*, on the original circle such that the angle subtended at the centre is $P_1 O_1 P_2^* = 2\delta\phi_{12}$. The image point P_2, reveals that a translation from P_1 to P_2 could be obtained from rolling the original circle without slip along the ordinate $\tau = + k$. The geometrical interpretation of eq(6.11a) is that the pole will trace a cycloid $P_1 P_2$... in the same direction as the changing slope of the α - line (ACW). A similar conclusion may be drawn from stress state $(p_3, - k)$ at point 3 on the β - line (Fig. 6.6a). The circle construction, in Fig. 6.6c, locates I_3 and the pole P_3. The pressure difference is found, from eq(6.9b), to be

$$p_3 - p_1 = 2k\delta\phi_{12} \tag{6.11b}$$

showing that p_3 is greater than p_1. A cycloid $P_1 P_3$... is generated by the pole as this circle rolls without slip along the ordinate $\tau = - k$, in the ACW direction of the changing slope of the β - line. Thus, slip lines in the physical plane will map into a system of congruent cycloids in Mohr's stress plane. In each of Figs 6.6b,c, I_1 is the instantaneous centre of rotation of point P_1 and $I_1 P_1$ is parallel to the respective α - and β - lines. Therefore, each element of a slip line is orthogonal to its corresponding element of cycloid.

6.2.4 *Stress Discontinuities*

Where, in Fig. 6.7a, the normal and shear stresses σ_N and τ, for a point A on an irregular free surface are known to produce a state of yielding, the α, β slip lines must pass through A.

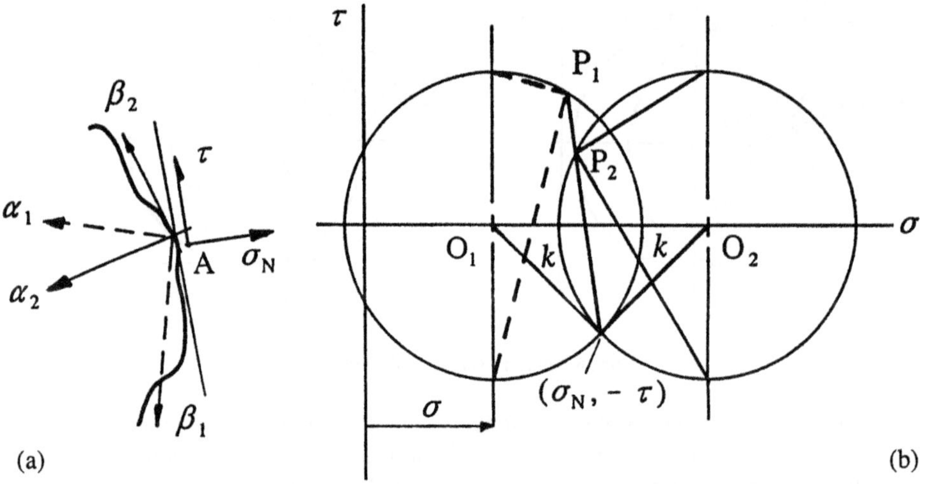

Figure 6.7 Strong and weak circles

However, the stress plane construction in Fig. 6.7b shows that α and β are not uniquely defined. Marking the radius k from the stress point $(\sigma_N, - \tau)$, in (b) locates the centres, O_1 and O_2, of two circles with respective poles P_1 and P_2. Joining the pole to points on the vertical diameter of each circle shows that there are two sets of posible slip lines, one pair being weak (broken lines), the other pair being strong. The latter contains the larger hydrostatic stress σ. Recall that the α - line is associated with $+ k$ and the β - line with $- k$. Thus, in the physical plane, the two α - lines are drawn parallel to the lines from P_1 and P_2 to the top of each circle. The two β - lines are parallel to the lines joining P_1 and P_2 to the bottom of each circle. It will be seen that it is usually possible to judge which of the two solutions is the appropriate one for a particular problem.

Next, consider the curved boundary between two plastic zones in which the adjacent material experiences different hoop tensions, i.e. $\sigma_x \neq \sigma_x'$ in Fig. 6.8a. Equilibrium is maintained through a continuity in both radial stress σ_y and in shear stress ($\tau_{xy} = \tau_{yx}$) across the boundary.

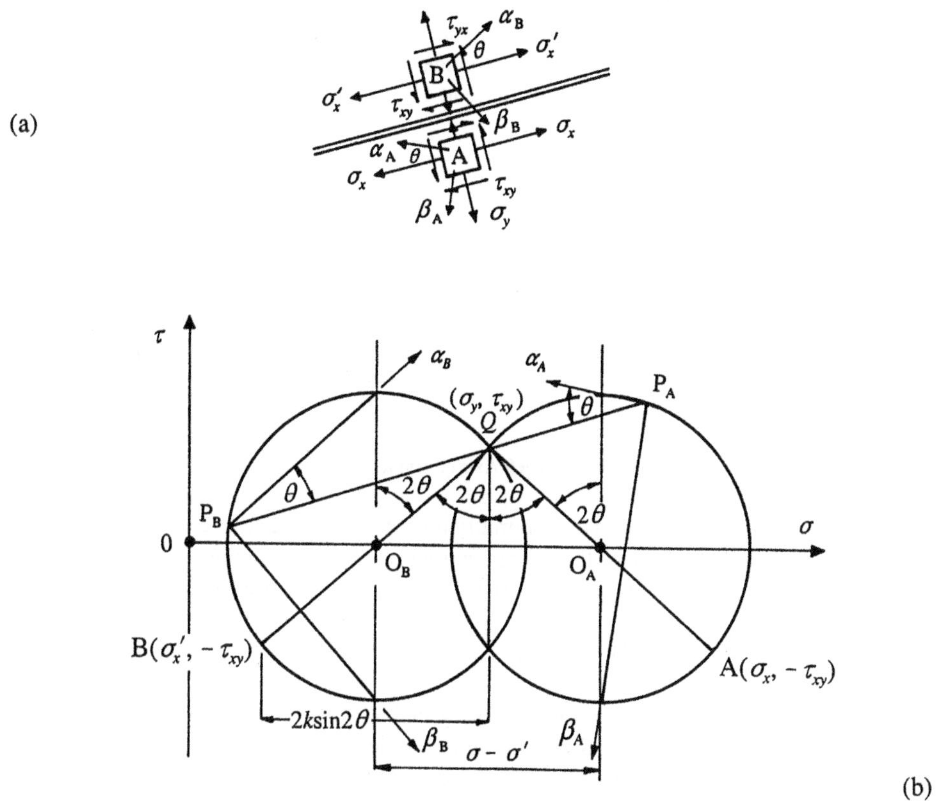

Figure 6.8 Tangential stress discontinuity

The boundary is called a *stress discontinuity*. Joining points $Q(\sigma_y, \tau_{xy})$, $A(\sigma_x, -\tau_{yx})$ and $B(\sigma_x', -\tau_{yx})$ within the stress plane in Fig. 6.8b locates the centres O_A and O_B, of circles, each of radius k. Projecting the tangent to the boundary through the point $Q(\sigma_y, \tau_{xy})$, locates pole points, P_A and P_B. Pole points establish the fact that the directions of α- and β- lines at points A and B in the physical plane are reflected across the discontinuity. That is, adjacent slip lines α_A and α_B for maximum positive shear stress lie with an equal inclination θ to the discontinuity. Also, the orthogonal slip lines β_A and β_B for the maximum negative shear stress will lie with an equal inclination $\theta + 90°$ to the discontinuity (see Fig. 6.8a). A stress discontinuity can neither coincide with a slip line nor, as it will be shown, with a velocity discontinuity. The magnitude of the stress discontinuity is, from the circle,

$$\sigma_x - \sigma_x' = 4k \sin 2\theta$$

where θ is the inclination of the reflection α_A to α_B across the discontinuity. When $\theta = 0°$, there is no discontinuity and the circles coincide. When $\theta = 45°$, the maximum difference arises, $\sigma_x' - \sigma_x = 4k$, as the circles become tangential at point Q lying on the σ- axis. Here, the greatest difference between the mean stress components is $\sigma - \sigma' = 2k$. Figure 6.8b refers to material that has yielded on either side of the discontinuity band while the interior of the band may remain elastic. As the band is penetrated from both sides σ_x' and σ_x will continuously change until they equalise. This must be so if circles, with diminishing radii less than k, pass through the common point Q, to ensure σ_y and τ_{xy} remain unchanged.

Where there is an axis of symmetry (a centre line) in the physical plane then this axis and another axis perpendicular to it, will coincide with the principal stress planes. Planes of maximum shear stress, i.e. the slip lines, lie at 45° to the principal planes (see Fig. 6.9).

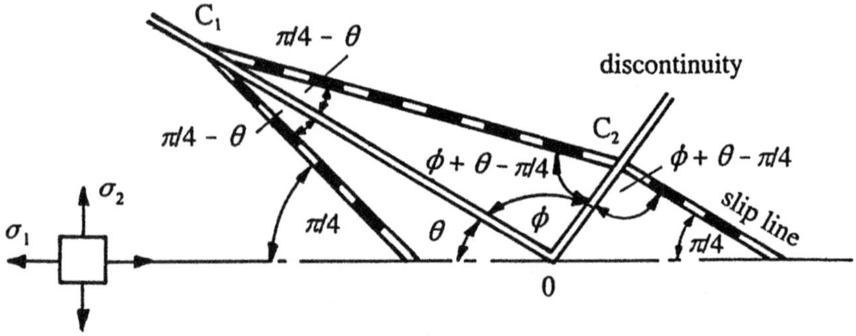

Figure 6.9 Axis of symmetry

Figure 6.9 also shows that a stress discontinuity must be reflected at an axis of symmetry if slip lines are to be reflected at their intersections with discontinuities. For the simple configuration of reflections at C_1 and C_2, the sum of included angles within OC_1C_2 give

$$(\pi/4 - \theta) + \phi + (\phi + \theta - \pi/4) = \pi$$

from which $\phi = 90°$, showing that OC_1 is orthogonal to OC_2. The full picture displays the mirror image of these slip lines and discontinuities beneath the symmetry axis. Note that it is possible for OC_1 and OC_2 to subtend angles of $\pi/4$ and $3\pi/4$ at the symmetry axis and so co-incide with the slip lines [7].

6.2.5 Geiringer's Equations

Shear strain, which is at its maximum between slip lines, is a radian measure of their angular distortion. A hydrostatic stress does not permanently strain the slip line. A further pair of compatibility relationships will ensure this inextensibility of α, β- slip lines (see Fig. 6.10).

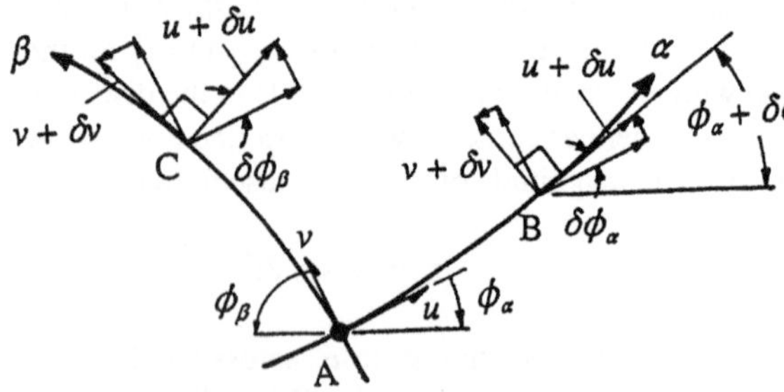

Figure 6.10 Velocity variation

Let u and v be the respective velocities in the directions of α and β at point A. These velocities increase by δu and δv at point B along the α - line as shown. Since the velocities $u + \delta u$ and $v + \delta v$ at B have rotated by $\delta\phi_\alpha$ they must be resolved in the directions of u and v at A. The fact that the α - line does not change its length is expressed in

$$(u + \delta u) \cos\delta\phi_\alpha - (v + \delta v) \sin\delta\phi_\alpha = u$$

Similarly at point C, inextensibility in the β - line is stated as

$$(v + \delta v) \cos\delta\phi_\beta + (u + \delta u) \sin\delta\phi_\beta = v$$

As $\delta\phi \to 0$, $\cos\delta\phi \to 1$ and $\sin\delta\phi \to \delta\phi$. Ignoring the products of infinitesimals leads to the *Geiringer equations* [3]:

$$\frac{\partial u}{\partial \phi_\alpha} - v = 0$$
$$\frac{\partial v}{\partial \phi_\beta} + u = 0 \qquad (6.12a,b)$$

Note that if velocities u and v are normalised over a given length they become strain rates.

6.2.6 Hodograph (Velocity Plane)

The hodograph is a graphical construction of those velocities which satisfy eqs(6.12a,b). Let resultant velocities, v_A and v_B, apply to respective points A and B along an α - line (see Fig. 6.11a). It is seen in Fig. 6.11b that their projections onto AB must be equal if the initially straight length of AB is not to change.

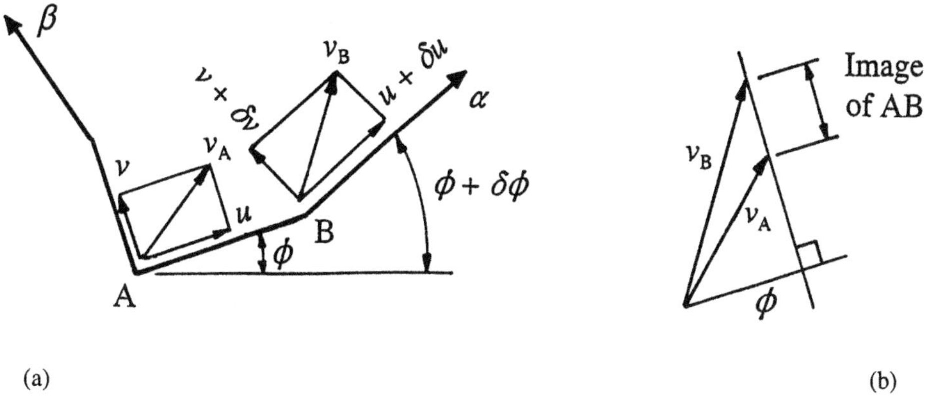

Figure 6.11 Hodograph for a portion AB of a straight slip line

The line joining v_A to v_B in Fig. 6.11b is the orthogonal image of the slip line between A and B in the physical plane (Fig. 6.11a). Because elements in the physical plane and stress plane are orthogonal, it follows that elements in the hodograph and the stress plane will be parallel. Where the α - slip line is curved between A and B (see Fig. 6.12a), it may be deduced that the curve joining v_A to v_B in Fig. 6.12b becomes the orthogonal image of the slip line.

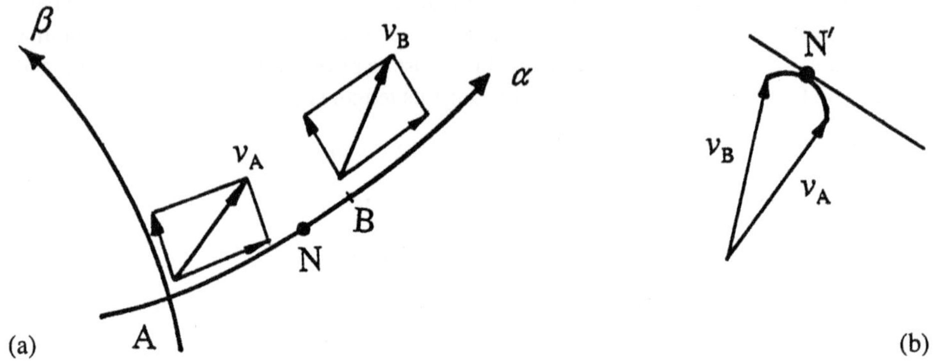

Figure 6.12 Hodograph for a curved slip line

The construction in which the tangent at point N in Fig. 6.12a is orthogonal to the tangent at point N′ in Fig. 6.12b, ensures that the initially curved length of AB will not change.

6.2.7 Velocity Discontinuities

A velocity discontinuity can arise when, in Fig. 6.13, the tangential velocities on either side of the discontinuity are unequal, i.e. $u \neq u'$.

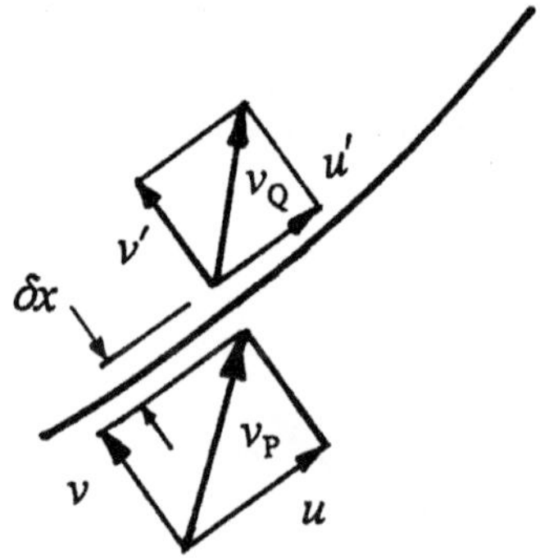

Figure 6.13 Tangential velocity discontinuity

Let the variation $\delta u = u - u'$, occur in a distance δx that straddles the discontinuity. It can be seen that as $\delta x \to 0$, the shear strain rate $\delta u/\delta x \to \infty$. We may conclude from this that lines of velocity discontinuity and slip lines are coincident. Applying Geiringer's eq(6.12a,b) to regions adjacent to the discontinuity gives

$$du = v\, d\phi \quad \text{and} \quad du' = v'\, d\phi \qquad (6.13a,b)$$

Normal velocities must remain equal, i.e. $v = v'$, if the material is to deform without volume change. Equations (6.13a,b) give

$$du = du' \quad \text{or} \quad u - u' = \text{constant} \qquad (6.14)$$

indicating that the velocity discontinuity has a constant magnitude. Since $u \neq u'$ the resultant velocities, v_P and v_Q in Fig. 6.13a will differ in both magnitude and direction as the slip line is crossed. Where u and u' vary along the slip line, the magnitudes and directions of the resultant velocities are not constant. Because $v = v'$, eq(6.14) implies that the vector sum of v_P and v_Q will be constant at all points along the slip line/velocity discontinuity. This is readily seen within a hodograph construction (Fig. 6.14b) for a length of α - slip line. Here the resultant velocities are v_{P1}, v_{Q1} at point 1 and v_{P2}, v_{Q2} at point 2 as shown in Fig. 6.14a.

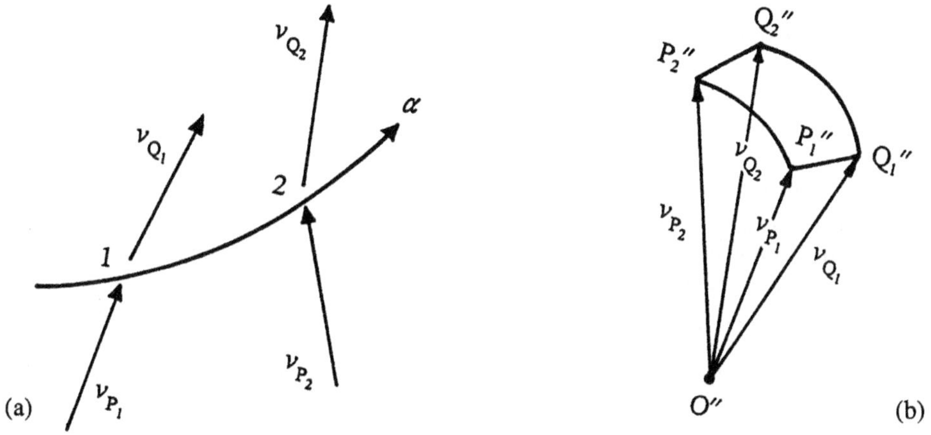

Figure 6.14 Equal velocity discontinuities

When constructing the hodograph in Fig. 6.14b, the velocities v_P and v_Q, are drawn through a common origin O″, parallel to velocity vectors in the physical plane. The magnitudes of the velocities are determined from two requirements:

(i) that $P_1''Q_1''$ and $P_2''Q_2''$ are equal and parallel to the respective tangents at points 1 and 2. Their magnitudes represent the jump in the tangential u - component of velocity from crossing the α - line at points 1 and 2 (see eq 6.14),

(ii) that the inextensibility of the α - line requires $P_1''P_2''$ and $Q_1''Q_2''$ to be orthogonal images of the slip line between points 1 and 2.

It is possible for u, as well as v, on one side of a slip line to be uniform when the material moves as a rigid body. The resultant velocity is then constant, as shown by v_P in Fig. 6.15a.

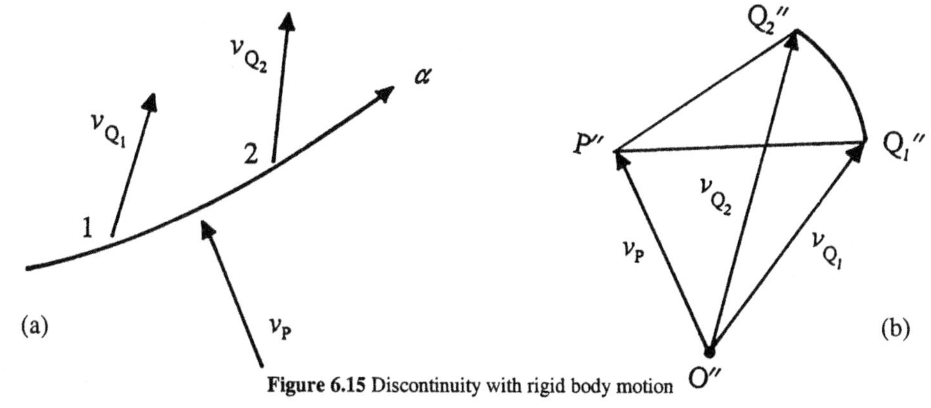

Figure 6.15 Discontinuity with rigid body motion

Here u' varies between points 1 and 2, on the other side of the slip line. The resultant velocities v_{Q1} and v_{Q2} will again differ in magnitude and in direction as shown. The simplified hodograph (see Fig. 6.15b) shows that $Q_1''Q_2''$ is the orthogonal image of slip line 12 and $P''Q_1'' = P''Q_2''$ is the jump in the tangential velocity at 1 and 2. This hodograph reveals that when v_P is constant then v_Q must vary. It is not possible, therefore, for the material on both sides of the slip line to move as rigid bodies between points 1 and 2.

6.2.8 Construction of Full Slip Line Fields

The general method employed for the determination of a slip line field (SLF) is based upon what is known about their 45° intersections with axes of symmetry and their reflections at stress discontinuities. The stress plane and hodograph follow from knowing that their own elements are parallel but perpendicular to those in the physical plane. If the hodograph closes then it is a possible solution, but is valid only when a positive rate of working is confirmed. Finally, it must be checked that the stress state within dead metal zones remains elastic. In this way, a unique solution can be established. The distortion of material lying within the slipped region is found from integrating slip-line velocities and the deforming forces are found from integrating slip line stress distributions. In constructing a SLF it is also necessary to satisfy the following three conditions prevailing at a boundary [1].

(a) *To determine slip lines for interior yielding given surface stress components.*

In the physical plane shown in Fig. 6.16a, the normal and shear stresses are known for points 1 and 2 on a curved boundary. It is possible to construct the stress plane in Fig. 6.16b where stress co-ordinates locate points 1′ and 2′ on circles each of radius k. Pole points P_1 and P_2, are found from the projections of tangents to the boundary at 1 and 2. The directions of the α - and β -lines lie parallel to lines connecting P_1 and P_2 to $\pm k$. In this case, the α - line is taken from circle 1 and the β - line from circle 2. The α - line may be extended from the cycloid generated by point P_1 as circle 1 rolls along $\tau = +k$ in Fig. 6.16b. Similarly P_2 traces a cycloid when circle 2 is rolled along $\tau = -k$. Cycloid intersection occurs at point a′. The corresponding intersection point a in the physical plane is constructed from the tangents at a′ and $P_{1,2}$ with adjustment for equal lengths aside the change in slip line direction (Fig. 6.16a). In the complete field, both the α - and β - lines intersect at the boundary points 1 and 2. The remaining slip lines (broken lines) connect pole points to points of maximum shear. At their intersections, the α - and β - lines are subjected to the same hydrostatic stress. The *strong* solution to this problem would correspond to the reflection of circles of radii k, through points 1 and 2. This corresponds to an α - line with the greatest possible hydrostatic stress at point 1. It is possible to select the appropriate solution to a given problem from the known deformation mode. Note that where surface tractions are given, these are interpreted as the resultant stresses σ_{R1} and σ_{R2} in Fig. 6.16a.

A given absolute velocity v of boundary material implies that points 1, 2 and c_A all move with velocity v. Thus, in the hodograph (Fig. 6.16c) the corresponding points, 1″ 2″ and c_A'', show a velocity v relative to a stationary origin, O″. The remainder of the hodograph construction employs the orthogonal relation between elements of the SLF and the corresponding elements of the hodograph. The additional points, a, b, c_B, d and e, assume the lower positions shown in Fig. 6.16a. This ensures that, when a slip line is crossed, the proper identity is made with a velocity discontinuity. For example, in crossing the lower β - line edb, equal magnitude discontinuities are found from joining points 2″e″, c_B''d″ and a″b″ in the hodograph.

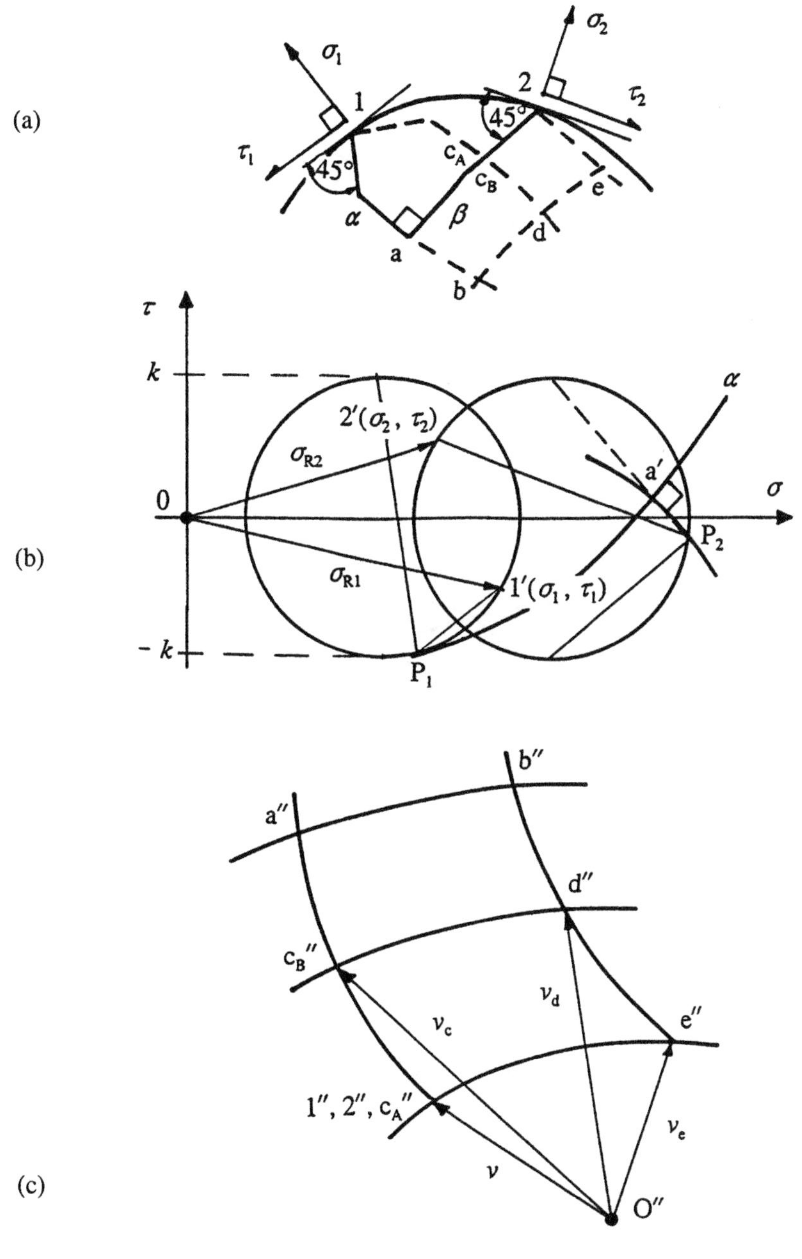

Figure 6.16 Known boundary stresses

In the crossing of the lower α - line ab, the corresponding, equal-magnitude, velocity discontinuities are: a″c_B″ and b″d″. Absolute velocities, v_c, v_d and v_e, are found from joining the respective points, c″, d″ and e″, to the origin O″.

(b) *To determine the slip-line field between two bounding, intersecting α, β - slip lines.*

Firstly, establish the circle for the stress state at the intersection point 1 in Fig. 6.17a. As in (a) the pole position 1' in Fig. 6.17b follows from projecting tangents to α and β, at 1.

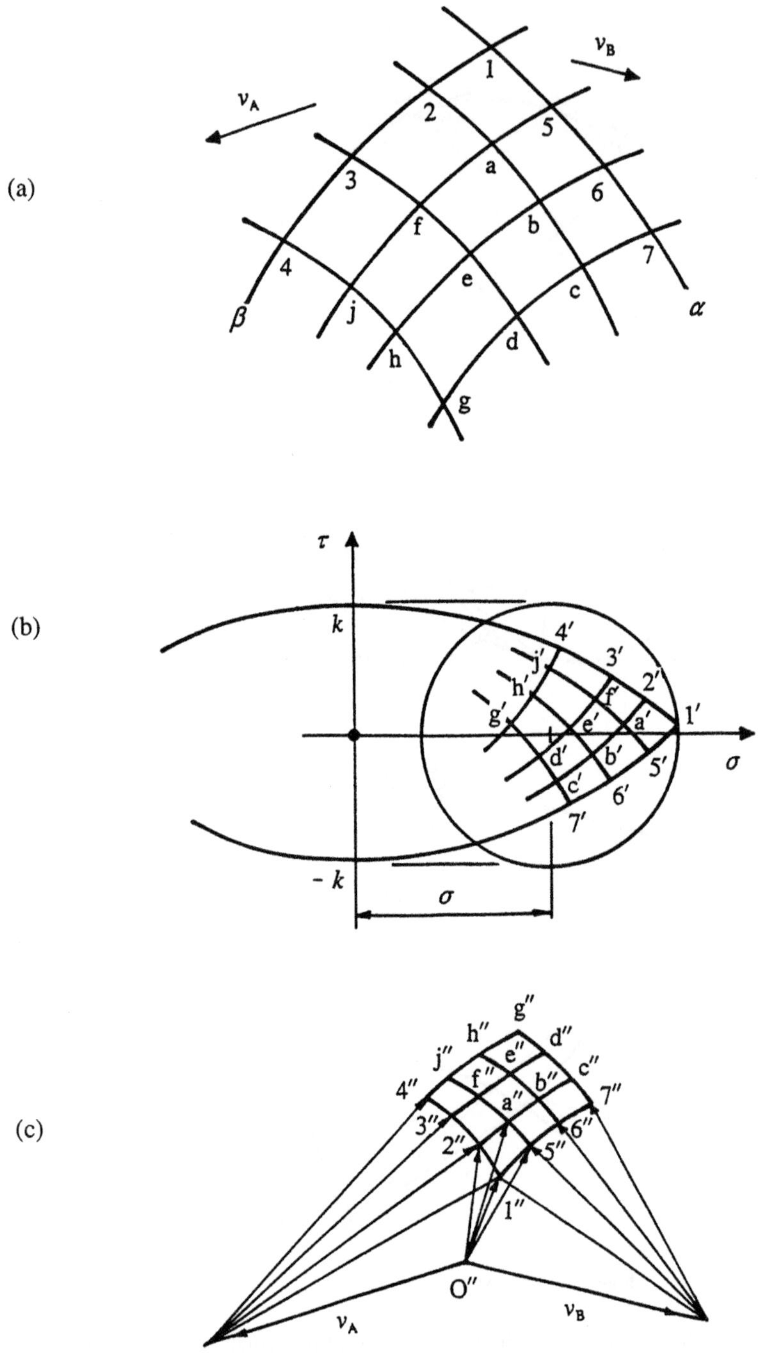

Figure 6.17 Bounding intersecting slip lines

Correspondence must apply between the normals at additional points 2, 3, 4, 5, 6 and 7 in the physical plane and the tangents to these points in the stress plane. The tangents to the slip lines rotate clockwise (CW) for the α - line and anti-clockwise (ACW) for the β - line.

It follows that the circle must roll ACW along $\tau = -k$ through β- points 2′, 3′ and 4′ and CW along $\tau = +k$ through α- points 5′, 6′ and 7′. Interior cycloid intersections at a′, b′, c′, d′, e′, f′, g′, h′ and j′ are found from reversing the direction of rotation at points 2′, 3′, 4′, 5′, 6′ and 7′ about the ordinate $\tau = -k$. Tangents to the orthogonal network of cycloids are then used to construct the SLF by extending slip lines, e.g. point a is found from points 2 and 5, b from a and 6, c from b and 7 etc.

Where rigid body velocities v_A and v_B, exterior to the boundary, are known and α- and β- lines are discontinuities, it is possible to construct the hodograph for material enclosed by the slip lines. In the physical plane, the SLF network above is positioned within lower regions where velocities are unknown. The absolute velocities v_A and v_B, are first drawn in Fig. 6.17c. Point 1″ is determined from the intersection between tangents to α and β at 1. Points 2″, 3″ and 4″ follow from the direction of the tangents to points 2, 3 and 4, their magnitudes being equal. Points 5″, 6″ and 7″ are similarly established. Point a″ is then derived from 2″ and 5″ using the tangents at corresponding points in the physical plane. The construction for the remaining intersection points within the hodograph is similar to that used for extending slip lines. This is because the interior SLF must be an orthogonal image of the hodograph. The resulting hodograph becomes a composite of the two discontinuities discussed separately in Figs 6.14 and 6.15. For example, the construction in Fig. 6.14 applies to crossing an interior slip line 5afj. Thus, in Fig. 6.17c, constant magnitude velocity discontinuities are found from joining 1″5″, 2″a″, 3″f″ and 4″j″. The absolute velocites on adjacent sides of this slip line are found from joining points 1″, 5″, 2″, a″, 3″, f″, 4″ and j″ to the origin 0″. Figure 6.15 applies to the absolute velocities in crossing the outer α- line at points 1, 5, 6 and 7, i.e. their magnitudes are found from joining 1″, 5″, 6″ and 7″ to O″ in the hodograph. Fig. 6.17c shows that equal tangential velocity discontinuities along α appear as a fan of vectors originating from the end of v_B.

(c) *To construct the SLF where a given slip line meets an axis of symmetry.*

In Fig. 6.18a, an inclination of 45° exists between the given β-line and the symmetry axis, the latter being a principal stress direction. The circle in Fig. 6.18b applies to the stress state $(\sigma, -k)$ at point 1. The pole 1′ is located from the tangent direction to β at point 1. As the circle rolls CW along $\tau = -k$, the pole traces the portion of cycloid 1′, 2′, 3′ shown. For the intermediate positions 2′ and 3′, the circle is rolled ACW along $\tau = k$ to trace portions of orthogonal cycloids which meet the axis of symmetry at a′ and c′. Lastly, a circle with pole a′ is rolled CW along $\tau = -k$ to give the image of the β-line a′ b′. Tangents to β at points 2′ and a′ in the stress plane locate the positions of 2 and a in Fig. 6.18a. The remaining points 3, b and c are found from the tangents at points 3′, b′ and c′ in a similar manner.

(a)

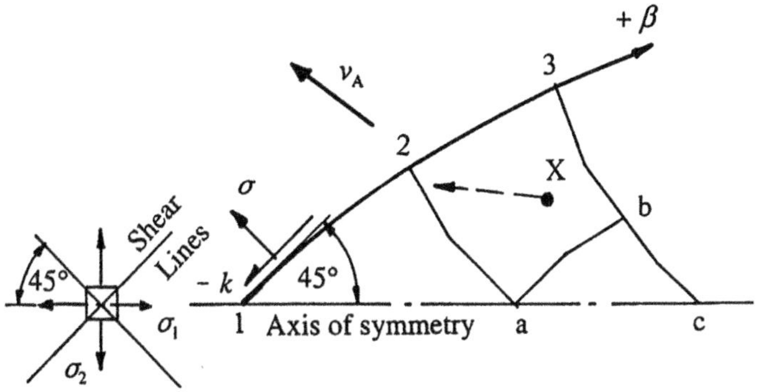

Figure 6.18 Intersection with an axis of symmetry

(b)

(c)

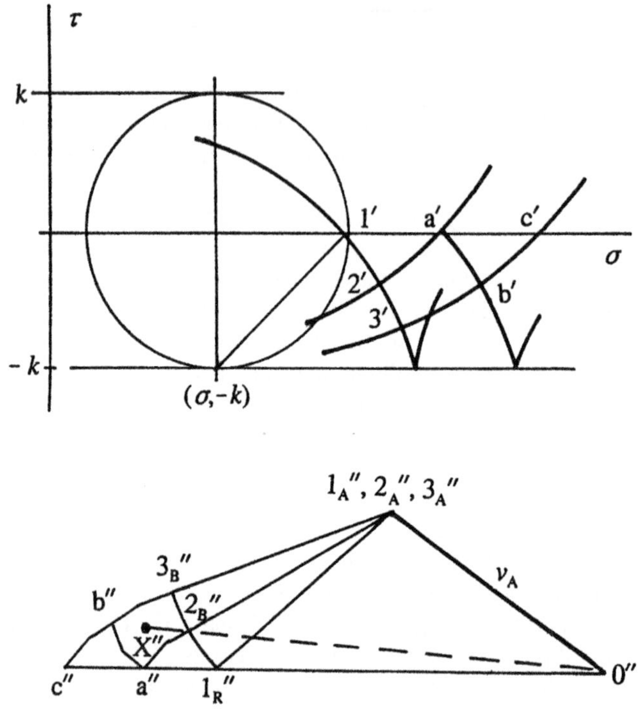

Figure 6.18 Intersection with an axis of symmetry (contd)

With a rigid body velocity v_A, exterior to the β-line in Fig. 6.18a, a fan of discontinuities ($1_A''\ 1_R''$, $2_A''\ 2_B''$ and $3_A''\ 3_B''$) in Fig. 6.18c originates from the end of $v_A(1_A'', 2_A'', 3_A'')$. Subscripts refer to positions above, right and below the given point. The hodograph is extended to points a″, b″ and c″ employing parallel elements from the stress plane as shown. The absolute velocity of any arbitary point X is found from joining the corresponding point X″ in the hodograph to its origin O″.

6.2.9 Resultant Force and Pressure on a Curved Slip Line

The stress plane supplies stress states for points along a curved slip line. Where horizontal forming forces are required, the stress distribution is integrated by one of two methods.

(a)

(b)

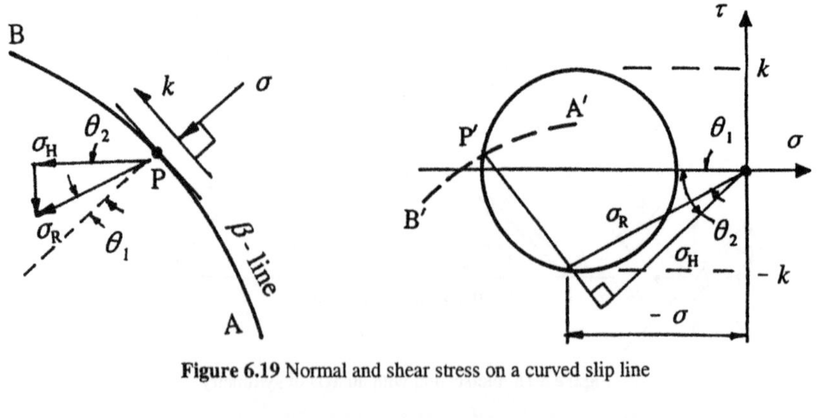

Figure 6.19 Normal and shear stress on a curved slip line

(a) *Resultant Stress*

Consider a point P on a β slip line AB in Fig. 6.19a. In the stress plane, the corresponding circle (Fig. 6.19b) has its pole at P′. The stress state at P ($-\sigma$, $-k$) in (a) follows from projecting the tangent to P through P' in Fig. 6.19b. This intersects the circle with these co-ordinates ($-\sigma$, $-k$), confirming AB as a β- line. Let the resultant stress (force on unit area) at P, i.e. $\sigma_R = \surd(\sigma^2 + k^2)$, have a horizontal component σ_H. We may regard σ_R and σ_H as forces/unit length of slip line when the thickness is unity. The construction for σ_R and σ_H shown applies to a given cycloidal image of the slip line within the stress plane. Because the cycloid is the locus of pole points, it is only necessary to connect the pole either to $+k$, for an α - line, or to $-k$, for a β - line, on the corresponding circle. The length of the perpendicular, dropped from the stress origin to the extension of these lines (see Fig. 6.19b), is the required component of horizontal stress σ_H. Taking other points between A′ and B′, we can repeat this construction to establish the variation of σ_H over AB, shown in Fig. 6.20.

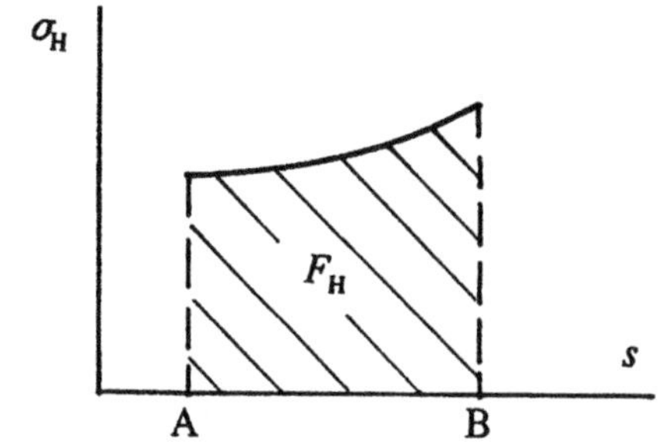

Figure 6.20 Resultant horizontal force on a slip line

It follows that the net horizontal force on a slip line is

$$F_H = \int_{s_A}^{s_B} \sigma_H \, ds \tag{6.15}$$

which is the area beneath Fig. 6.20, when drawn on the true, base length s of AB.

(b) *Cartesian Stresses*

In Fig. 6.21a, the portion AB of β - slip line is approximated with a series of vertical and horizontal planes to reveal Cartesian stress components σ_x, $\tau_{xy} = \tau_{xy}$ as shown.

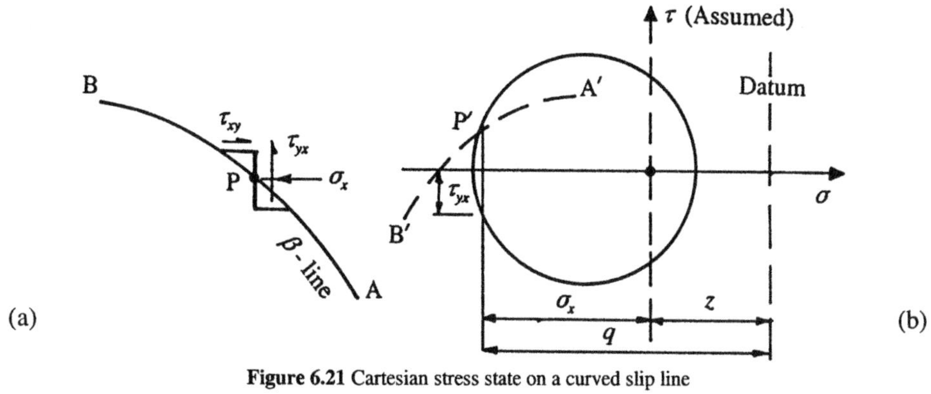

Figure 6.21 Cartesian stress state on a curved slip line

The plane stress state P(σ_x, τ_{yx}) upon the vertical plane lies at the intersection between a vertical line through pole P′ and the circle in Fig. 6.21b. Repeating this construction for a number of poles between A' and B' will establish the variations in σ_x and τ_{xy} with the projected lengths of AB, i.e. a and b in directions x and y respectively (see Fig. 6.22).

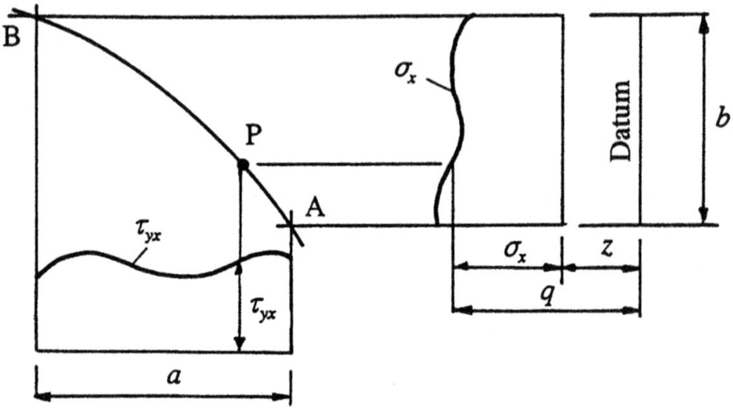

Figure 6.22 Stress component variations

Now τ_{xy}, upon the horizontal plane (Fig. 6.21a), is complementary to τ_{yx} and acts in opposition to σ_x. The net horizontal compressive force is found from

$$F_H = \int_0^b \sigma_x \, dy - \int_0^a \tau_{xy} \, dx \tag{6.16a}$$

Equation(6.16a) expresses the difference between the respective enclosed areas in Fig. 6.22. When the origin of stress is unknown, but F_H is known, we may work from an assumed origin for τ. Let $q = \sigma_x + z$ locate be the true datum for τ, so modifying eq(6.16a) to

$$F_H = \int_0^b \sigma_x \, dy + \int_0^b z \, dy - \int_0^a \tau_{xy} \, dx \tag{6.16b}$$

When $F_H = 0$ in eq(6.16b), the separation z between the true and assumed τ axes becomes

$$z = \frac{1}{b}\left[\int_0^a \tau_{xy} \, dx - \int_0^b \sigma_x \, dy\right] \tag{6.16c}$$

6.3 Frictionless Extrusion Through Parallel Dies

The following SLF solutions apply to a rectangular die of any width for the given reduction ratios $R = H/h$. The force analyses apply to unit width under a plane strain condition.

6.3.1 Extrusion Ratio R = 2

In a lubricated vessel (see Fig. 6.23a) there are no frictional forces parallel to the container walls and so interfacial shear stress is absent. Hence, the walls are principal planes and the slip lines, i.e. maximum shear planes, meet the container walls (points 0 and 5) at 45°.

(a)

(b)

(c)

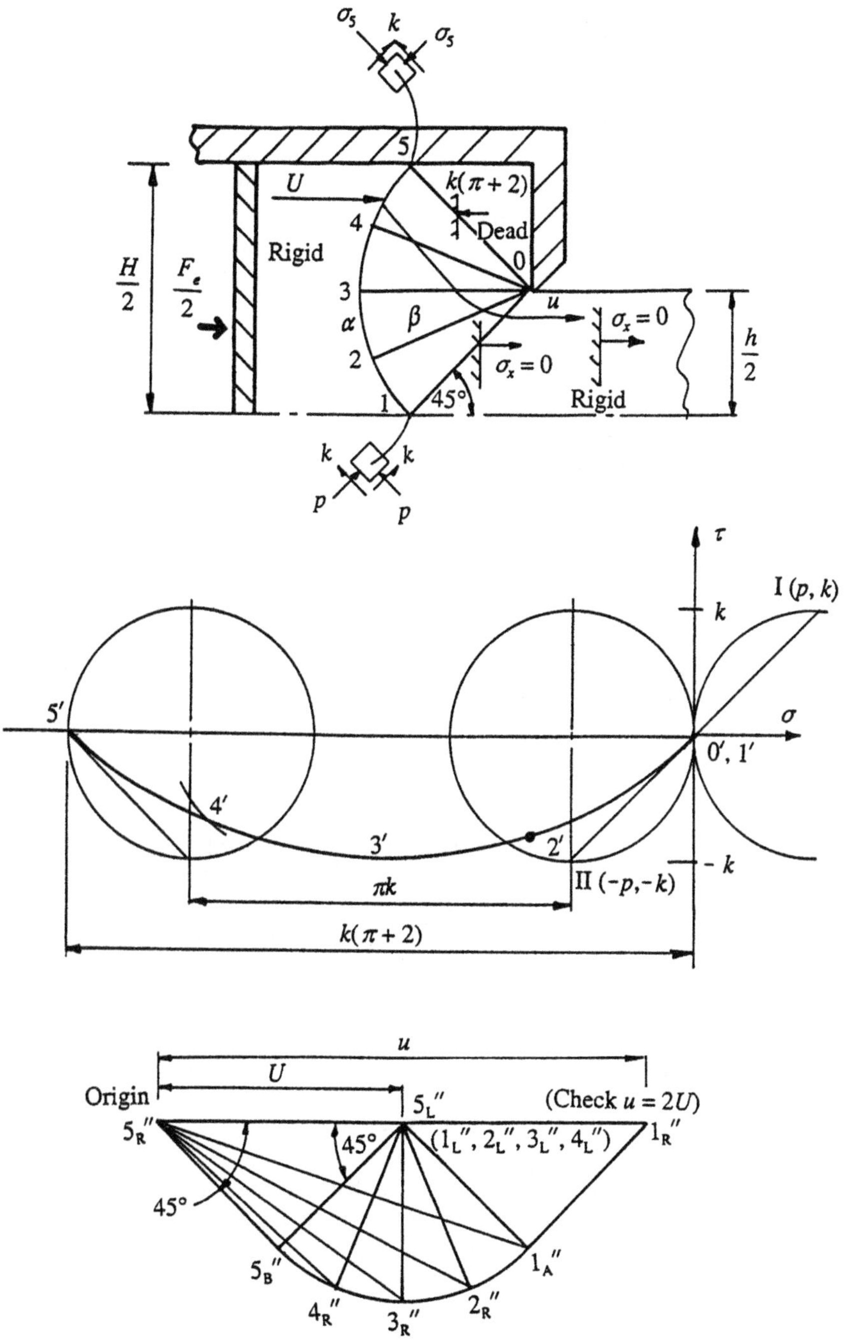

Figure 6.23 Frictionless extrusion $R = 2$

The slip line, joining points 0 and 5 in the physical plane (see Fig. 6.23a), is assumed to be straight. Another slip line 01 must be orthogonal to 05 at 0 and meet the centre line at 45°. The problem is then to construct the slip line connecting points 1 to 5. We begin with the fact that the major principal stress σ_x, normal to the free-end of the extruded billet, is zero.

This determines that the pole point $0'1'$ is coincident with the origin of the stress plane in Fig. 6.23b. In order to construct the initial circle, the complete stress state for a second plane is required. Because the shear stress along 01 is k, two circles, I and II, apply when the mean stress for this plane is assumed tensile and compressive respectively. Mean tension is impossible under the applied compressive force and so the weaker circle II applies to this problem. The true stress state for 01 is found by projecting a line parallel to 01 through the pole. This gives $p = -k$, $\tau = -k$, so establishing 01 as a β - line. It follows that 12345 will be an α - line. This line is established from the cycloid traced out by the pole as the circle rolls along the ordinate $+k$ without slip. Tangents drawn to this cycloid at intermediate points, $2'$, $3'$ and $4'$, are normals to the slip line at points 2, 3 and 5. Thus, the SLF is composed simply of a single α - line with two orthogonal β - lines, 01 and 05. Note that $5'$ is the pole of the circle for the stress state at 5. It follows from this that the compressive stress on a vertical plane through 5 is $k(\pi + 2)$. Since this acts perpendicular to the front face of the container, the die pressure is $p_d = k(\pi + 2)$. The mean extrusion pressure p_e is found from equating horizontal forces above and below the centre line. With unit width and an extrusion ratio $R = H/h = 2$:

$$p_e H = p_d (H - h)$$
$$p_e H = k (\pi + 2)(H - h)$$
$$p_e = k (1 + \pi /2) \tag{6.17a}$$

Multiplying p_e in eq(6.17a) by the ram area (or multiplying p_d by the die area) gives the extrusion force F_e/unit area:

$$F_e = p_e H = p_d (H - h)$$
$$F_e /H = k (1 + \pi /2) \tag{6.17b}$$

We could have found the die pressure from the application of eq(6.9a) to the α - slip line, 15. Because $\sigma_x = 0$ for 01, the constant stress state along 01 is $(-p, -k)$ where $p = k$. Thus an element at point 1 is stressed as shown in Fig. 6.23a. Let the datum for ϕ (increasing CW) be the tangent to α at 1. It follows from eq(6.9a) that at point 5, where $\phi = -\pi /2$,

$$\sigma_5 + 2k (- \pi /2) = c_1 \tag{6.18a}$$

The constant c_1 in eq(6.18a) is found from applying eq(6.9a) to point 1 on α, where $\phi = 0$. This gives

$$p + 2k (0) = c_1 = k \tag{6.18b}$$

It is seen from eqs(6.18a) that $\sigma_5 = k (1 + \pi)$, so that an element at point 5 in Fig. 6.23a is stressed as shown. Resolving forces along 05 within the dead zone, confirms that the horizontal die pressure is $p_d = k (\pi + 2)$. Equating the rate of external plastic work to the energy dissipated within a homogenous deformation zone allows estimates to p_e and F_e:

$$\dot{W} = \int p_e \dot{V} = \left[\int Y \mathrm{d}\varepsilon \right] \dot{V} \tag{6.19}$$

where $Y = 2k$ is the constant, plane strain yield stress, assumed within a deforming zone under an incremental strain $\mathrm{d}\varepsilon = \mathrm{d}l/l$. Dividing eq(6.19) by the extruded volume/unit time $\dot{V}$, gives the extrusion pressure as

$$p_e = \int Y \mathrm{d}\varepsilon = Y \int \mathrm{d}l/l = Y \ln (l_2 / l_1) = 2k \ln (l_2 / l_1) \tag{6.20a}$$

In preserving volume, the ratio between final and initial lengths is $l_2/l_1 = A_1/A_2 = H/h = R$. Then, from eq(6.20a),

$$p_e = F_e/H = 2k\ \ln R \tag{6.20b}$$

It follows that the limiting reduction occurs when $p_e = 2k$, giving $R = 2.72$. With $R = 2$, eq(6.20b) gives $p_e/2k = 0.693$, which is considerably less than the value 1.286 found from the SLF (eq 6.17a). As the work formula gives the minimum possible work for a forming process, the amount by which it underestimates the actual forming force may be attributed to the effects of frictional work and the redundant shear work.

In the construction of the hodograph in Fig. 6.23c, let the velocity within the container be U and that of the extruded material be u. An incompressible material obeys: $U\ (H \times 1) = u\ (h \times 1)$, so that $u = 2U$ for $H = 2h$. The origin, $5_R''$, refers to the stationary material within the dead-metal zone to the right of point 5 in Fig. 6.23a. To the left of point 5, the material in the body of the container has velocity U which appears as the horizontal vector joining points $5_R''$ and $5_L''$ in the hodograph. Crossing the slip lines from each of these regions to a point directly beneath point 5, establishes the vectors, $5_R''5_B''$ and $5_L''5_B''$. Their directions are parallel to the slip lines at point 5. Leftward velocity points $1_L''$, $2_L''$, $3_L''$ and $4_L''$ must be coincident with $5_L''$, since all move horizontally with velocity U. A fan of vectors $4_L''4_R''$, $3_L''3_R''$, $2_L''2_R''$ and $1_L''1_A''$ emanate from the singular point $5_L''$ in directions parallel to the tangents at points 4, 3, 2 and 1 in the physical plane. In order to ensure inextensibility of slip-line 1 2 3 4 5, the curve $5_B''\ 4_R''\ 3_R''\ 2_R''\ 1_A''$ becomes the parallel image of points $5'\ 4'\ 3'\ 2'\ 1'$ in the stress plane. Similarly, we can construct the remainder of the hodograph, $1_L''1_R''1_A''$, for velocities to the left, right and above point 1 in the physical plane. Knowing that vector $5_R''1_R''$ is the horizontal billet velocity u, the hodograph must terminate at the closing point $1_R''$ where $u = 3U$. Within the hodograph (see Fig. 6.23c), we can identify velocity discontinuities associated with rigid body motion adjacent to a slip line (refer to Fig. 6.15). They are:

(i) $5_R''5_B''$, along 05, with stationary dead-metal to the right,

(ii) $1_A''1_R''$, along 01, with a rigid-body velocity u to the right and

(iii) the fan of constant magnitude vectors $1_L''1_A''$, $2_L''2_R''$, $3_L''3_R''$ etc, at points 1, 2, 3 etc along the α - line when, to its left, the rigid body motion is U.

It follows that the slip lines are also velocity discontinuities bounding the deformation zone. Resultant velocities at points 1, 2, 3, 4 and 5 on the right side of α differ. Their magnitudes and directions are given by the vectors $5_R''1_A''$, $5_R''2_R''$, $5_R''3_R''$ and $5_R''4_R''$. Their directions enable the motion of particles to be traced from container to billet in the physical plane as shown. The solution applies only to an extrusion ratio of 2. Greater or lesser ratios need separate consideration.

6.3.2 Extrusion for $2 < R < 3$

In the SLF construction given in Fig. 6.24a, the weaker starting circle has its pole $0'1'$ at the origin of the stress plane in Fig. 6.24b. This corresponds to the known β - slip line, 01 as before. It is seen, however, that end point $5'$ for cycloid $1'\ 2'\ 3'\ 4'\ 5'$, which establishes an orthogonal α - line, cannot be fixed at the container wall without an extension to the field. The requirement is that another α - line, originating at point 6 and the extension to the β - line from 5 to 9, must both meet the wall at 45°.

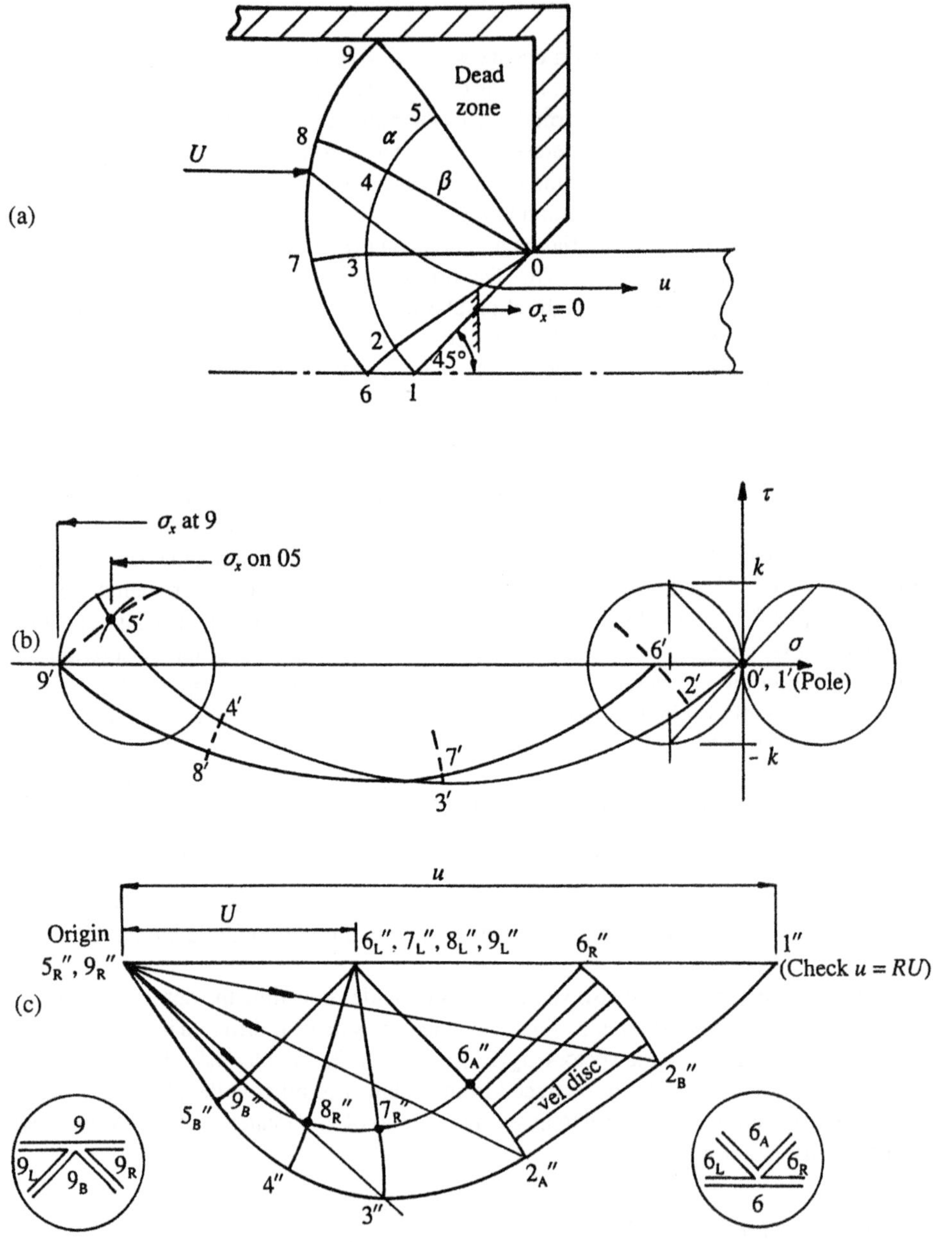

Figure 6.24 Frictionless extrusion for $2 < R < 3$

Thus, in the stress plane, both $6'$ and $9'$ lie on the $\tau = 0$ axis. Point $6'$ is derived from an orthogonal cycloid originating from the selected point, $2'$. Cycloid $2'6'$ is found from the rotation of a circle with pole point, $2'$ and radius k, which rolls along $\tau = -k$. Point 6 is found graphically from extending 02 to meet the centre line with a 45° inclination. This employs equal line elements whose directions are the normals to the corresponding cycloid at points 2′ and 6′. A similar constructions is used to establish the position of point 7 from the extension to the slip lines at points 3 and 6, using normals to points 2′, 3′, 6′ and 7′ in the stress plane. This construction is equivalent to replacing the extended portion of the slip line with circular arcs. In repeating this construction, point 8 is obtained from the extensions to slip lines at points 4 and 7 and finally, point 9 from the extensions at points 5 and 8. If 9 does not fall on the inner wall of the container, the position of point 2 must be adjusted and the construction repeated until this is achieved. Note, that the arc 1345 and radii 01, 03 and 04, together with the extensions 37 and 48, are constructions and not part of the final SLF solution. That is, the final SLF comprises an orthogonal network with a single α - line 6789 and two β - lines, 026 and 059, containing the deformation zone.

The construction of the hodograph in Fig. 6.24c for the α - line 6789 is similar to that given previously. Again, there is rigid body motion U to its left. Therefore, the coincident points, $6_L''$, $7_L''$, $8_L''$, $9_L''$, all move with velocity U relative to zero velocity in the dead metal zone $5_R''$. The fan of velocity discontinuity vectors, emanating from these coincident points, are equal in magnitude and lie parallel to tangents at points 9, 8, 7 and 6 along the slip line. They terminate at points $9_B''$, $8_R''$, $7_R''$ and $6_A''$, giving an orthogonal image of this slip line. Point $6_R''$ is found from drawing $6_A''6_R''$, parallel to the tangent at 6. The extension to the hodograph is achieved using a construction similar to that for extending slip lines. It begins with establishing point $5_B''$ from $5_R''$ and $9_B''$. To achieve this, the velocity directions are drawn orthogonal to the tangents at points 5 and 9 in the physical plane. Point 4″ is then found from $5_B''$ and $8_R''$, employing construction lines drawn perpendicular to the tangents at 5 and 8. Note that a single velocity applies at 4 and 3 since no slip lines are crossed. However, in crossing slip line 026 it is again necessary to distinguish between a point above ($2_A''$) and below ($2_B''$) point 2, where a velocity discontinuity arises. With points taken above ($6_A''$) and to the right ($6_R''$) of point 6, the hodograph shows that the magnitude of this discontinuity is constant at all points between 2 and 6 in the manner of Fig. 6.14. Here again all slip lines in (a) are velocity discontinuities. Having arrived at the destination point 1″, the magnitude must be checked from the fact that the final velocity $u = RU$. The absolute velocity vectors, originating from stationary points $5_R''$ and $9_R''$, enable the path of a particle through the deformation zone to be shown as a streamline.

If we needed to find the net horizontal force F_H from eq(6.16b) then the variations in σ_x and τ_{xy} are required either along 059 or 6789. The circles, with pole points 0′, 5′ and 9′ in (b), show that σ_x varies from zero at point 0 to the indicated values at points 5 and 9. The corresponding shear stress τ_{xy} varies from zero at point 0, through a maximum of $\tau = +k$ between 0 and 5, falling to zero at point 9. Clearly, it would be necessary to select more intermediate values when applying eq(6.16b) to this problem.

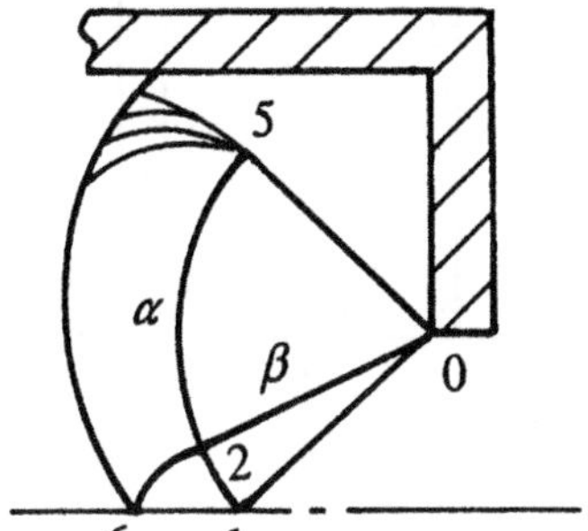

Figure 6.25 Alternative SLF for $2 < R < 3$

There is an alternative solution [1] for this extrusion ratio. In this, the single slip line 59 in Fig. 6.24a is replaced with a fan of slip lines, shown in Fig. 6.25. The envelope originates from 5 permitting a variation in the velocity discontinuity from the dead-metal zone across the fan. For this to be possible, the discontinuities cannot be coincident with each of the interior slip lines.

6.3.3 *Extrusion For 1 < R' < 2*

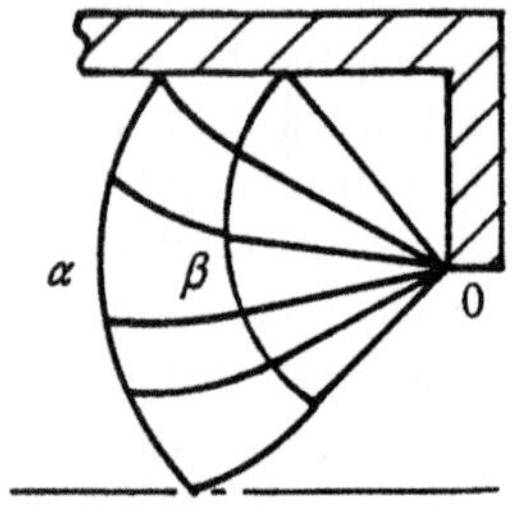

Figure 6.26 Frictionless extrusion, $1 < R < 2$

Figure 6.26 shows the SLF for an extrusion ratio $R' = H/h$, less than 2 [4]. This appears as an inversion of the SLF for $2 < R < 3$ about the singular point 0 in Fig. 6.24a. For an exact inversion to apply, the % increase in the extrusion ratio above 2 must equal the % decrease below 2. In Fig. 6.24 this gives $R' + R = 4$, so if $R = 2.4$, then $R' = 1.6$. Under the latter ratio the stress plane and the hodograph also become lateral inversions of Figs 6.24b,c about their origins [1]. However, where the indicated β - line is partly curved for $R' < 2$, the circle is not stationary and the true stress origin is unknown. Here, an origin for stress is assumed for completing the construction. The true origin is found thereafter from eq(6.16c), where the total horizontal force is zero to the right of this β - line. The origin for the inverted hodograph simply becomes the checkpoint 1″ in Fig. 6.24c.

6.3.4 *Extrusion For R = 3*

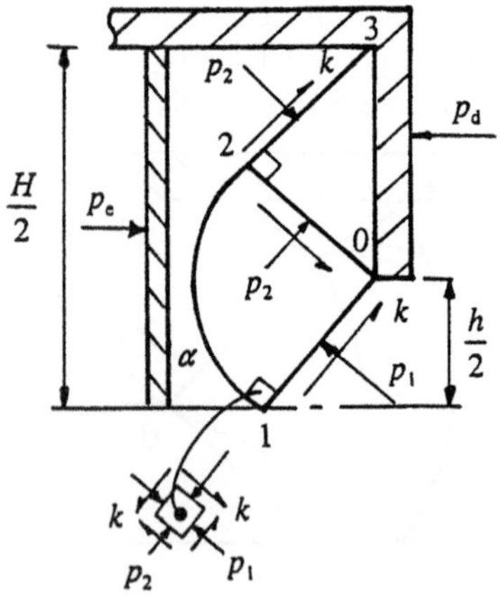

Figure 6.27 Frictionless extrusion for $R = 3$

Lee [8] and Green [9] proposed a SLF for $R = 3$, given in Fig. 6.27. In the absence of a dead zone, Lee [8] showed that this field provided the minimum possible extrusion force. Moreover, alternative solutions, discussed in the next section, were found to violate the yield criterion when applied to $R = 3$. As with $R = 2$, eqs(6.9a,b) may be applied to the SLF for $R = 3$. Again $\sigma_x = 0$ for 01 and the constant stress state along 01 is (p_1, k) where $p_1 = k$. Thus an element at point 1 is stressed as shown. To find p_2, let the datum for ϕ be the tangent to α - line at 1. Applying eq(6.9a) to point 2, where $\phi = -\pi/2$:

$$p_2 + 2k(-\pi/2) = c_1 \tag{6.21a}$$

in which $c_1 = k$ for $\phi = 0$. Equation (6.21a) gives

$$p_2 = k(1 + \pi) \tag{6.21b}$$

so that element 023 is stressed as shown. Resolving forces gives the die pressure as

$$p_d = k(2 + \pi) \tag{6.22a}$$

from which the extrusion pressure is

$$p_e = \left(1 - \frac{1}{R}\right) p_d = \frac{2p_d}{3} \tag{6.22b}$$

Combining eqs(6.22a and b) gives $p_e/2k = 1.714$.

6.3.5 *Extrusion For R > 3*

The SLF, given in Fig. 6.28a, will apply in the absence of a dead-metal zone. It is an extension to the SLF for $R = 3$ discussed previously.

(a)

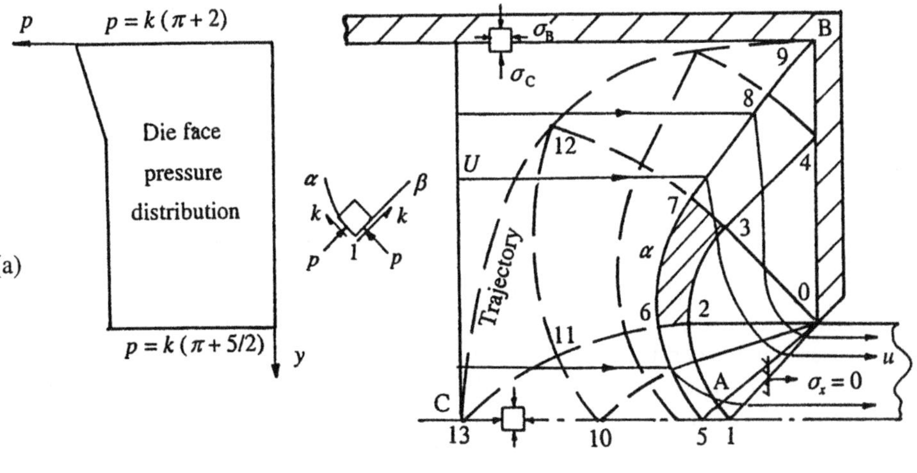

Figure 6.28 Frictionless extrusion $R > 3$

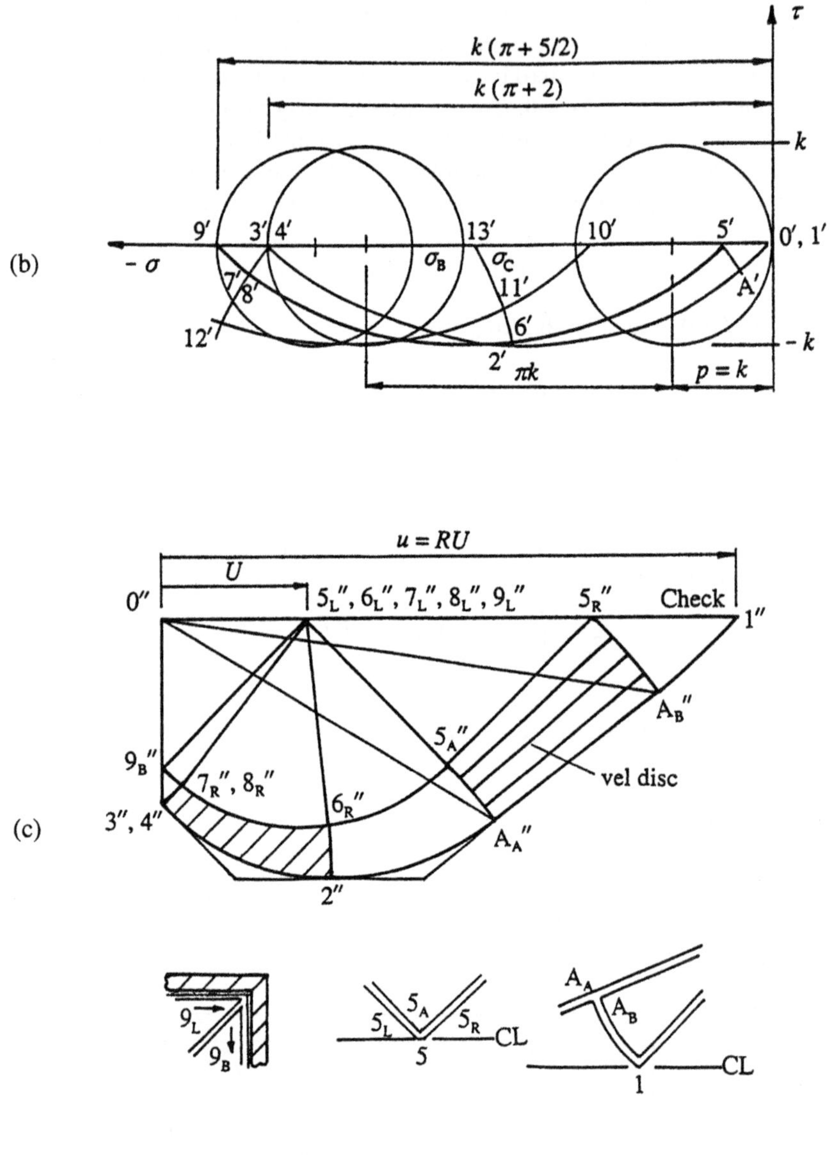

Figure 6.28 Frictionless extrusion $R > 3$ (continued)

It has been shown [10] that the extrusion pressure for Fig. 6.28a is less than would be found from extending either Fig. 6.24 or 6.25 to greater R. Moreover, when these were extended to give $R = 3$, they violated the yield criterion. It is seen from Fig. 6.28a that Lee's α - line 1234 and two β - lines, 01 and 03, would fit exactly within a vessel for $H/h = 3$. These lines are now used in the construction. To extend them to greater R values, we arbitarily select point A on the α - line and identify A′ in the stress plane of Fig. 6.28b. Next, we locate 5′ on the $\tau = 0$ axis from the orthogonal cycloid and thus its position 5 on the centre line. For the correct position of 5, end points 9′ and 9 will give zero shear stress at the corner of the container. Normals, drawn at the intersection points 2′3′4′ and 5′6′7′8′, within the orthogonal network of cycloids, extend the slip-line field in the usual way. The hodograph origin 0″ in (c), refers to any point on the stationary vessel. Thus, vector $0''\,9_B''$ is identified with the vertically downward motion of material beneath point 9 relative to the top corner point. Coincident points, $5_L''$, $6_L''$, $7_L''$, $8_L''$ and $9_L''$, in the region to the left of the α - line, move with rigid body velocity U. Velocity discontinuities arise in crossing this α - line, their magnitudes being given by velocity vectors: $9_L''9_B''$, $8_L''8_R''(\equiv 7_L''7_R'')$, $6_L''6_R''$ and $5_L''\,5_A''$. Their terminal points are joined by the orthogonal image of the α - line. The point $5_R''$ is then located from points $5_A''$ and $5_L''$. Tangents drawn at intersection points $2'3'4'$ and $5'6'7'8'$, within the orthogonal network of cycloids in the stress plane, are used to extend the hodograph. This gives points $3''(\equiv 4'')$, $2''$, A_A'' and A_B''. A fan of constant-magnitude, velocity discontinuities applies to all points between 5 and A in the manner of Fig. 6.14. Finally, tangents projected through points, A_B'' and $5_R''$, locate the checkpoint, $1''$, for which $u = RU$. Absolute velocity vectors, with common origin at $0''$, may be transferred to their appropriate slip lines to trace the motion of a particle through the deformation zone. Three such streamlines are illustrated in Fig. 6.28a.

The mean extrusion pressure is found from the known normal stresses at points 0, 4 and 9 along the die face. The stress plane shows that these are principal stresses (i.e. $\tau = 0$ along 049). Their magnitudes are $k\,(\pi + 2)$ = constant along 04, increasing to a maximum value of $k\,(\pi + 5/2)$ at point 9. Assuming the linear variation in normal stress between 4 and 9 (Fig. 6.28a) enables the mean extrusion pressure p_e to be estimated from

$$p_e H = k(H - h)(\pi + 2) + \frac{k}{2}\left[H - (R - 1)h\right]\left[\left(\pi + \frac{5}{2}\right) - (\pi + 2)\right]$$

$$p_e = k\left(1 - \frac{1}{R}\right)(\pi + 2) + \frac{k}{2}\left[1 - \frac{R - 1}{R}\right]\left[\left(\pi + \frac{5}{2}\right) - (\pi + 2)\right] \tag{6.23}$$

where $R = H/h$. Taking, for example, $R = 4$ in eq(6.23) gives $p_e = 3.92k$. The total force/unit width of die becomes $F_e = Hp_e$.

In the test for a complete solution, two conditions must be satisfied. Firstly, there is the requirement for positive plastic work: $\int \sigma_{ij}\, \mathrm{d}\varepsilon_{ij} > 0$. This involves a consistency check between relative directions in the hodograph and the deformation within a constant region of stress, e.g. the region 2376 bounded by slip lines in Fig. 6.28a. The pole positions in the stress plane reveal that the shear yield stresses $\pm\, k$ act in the directions shown in Fig. 6.29.

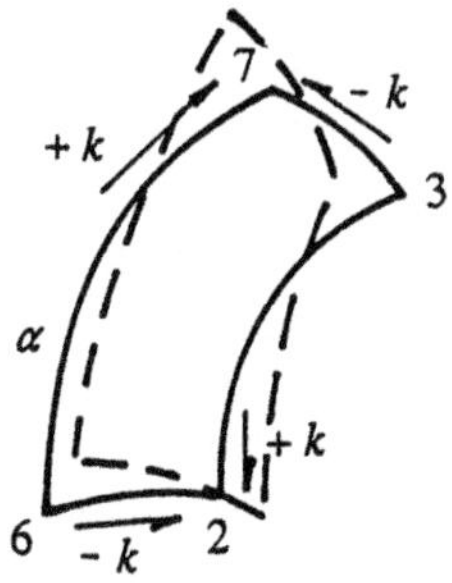

Figure 6.29 Positive plastic work check

The compressive hydrostatic stress acting on the α, β boundaries causes no deformation, the latter being due entirely to the shear stress k. The shear distortion is that indicated with broken lines. The element of the hodograph in Fig. 6.28c confirms the inward direction of point 3 relative to 6 along the line connecting 3 to 6. It follows that the product of a compressive stress and strain rate in the 36 direction will produce the required rate of positive plastic work. Four examples of inadmissible elements of hodograph are given in Fig. 6.30a-d for the element of SLF in Fig. 6.29.

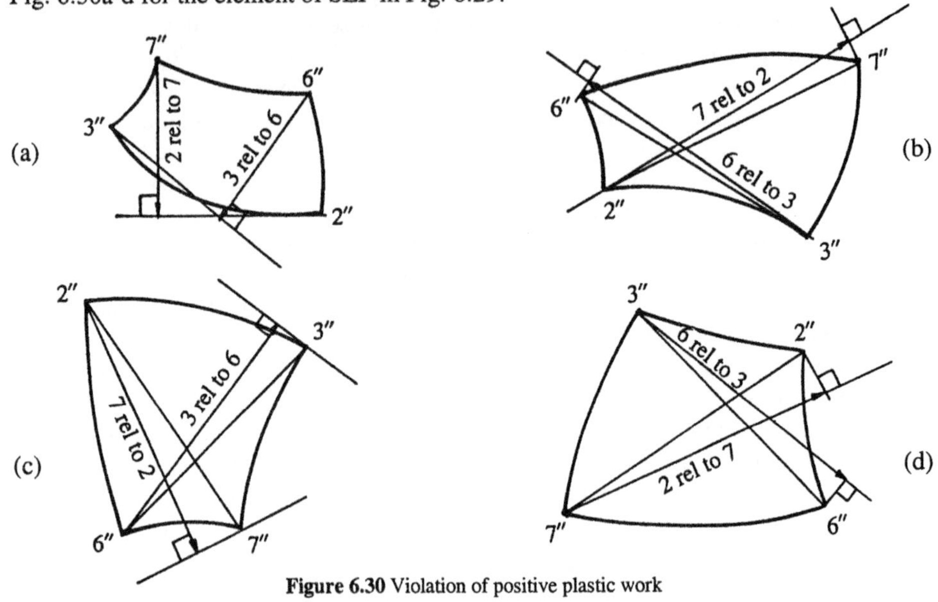

Figure 6.30 Violation of positive plastic work

These could arise when there is more than one hodograph fulfilling the required conditions. However, since these show relative motions between points 2, 3, 6 and 7 that are inconsistent with the known deformation from Fig. 6.29, they must be rejected in favour of the unique solution supplying positive plastic work.

Secondly, a check is made that the yield criterion is nowhere violated within the rigid material. This means that a rigid-plastic zone should not deform under the stresses imposed around its boundary. Bishop [11] showed that a slip-line field is complete where it can be extended into the rigid zone with pseudo slip-lines that become coincident with a stress-free boundary. The approach has been applied to frictionless extrusion with $R = 2$ [1] and $R = 3$ [10]. In applying Bishop's method to $R > 3$ in Fig. 6.28a, we need to ensure that the material to the right of 01 is non-deforming. The initial circle shows that the principal stresses for all points along 01 are $\sigma_y = 2k$ and $\sigma_x = 0$, acting in the vertical and horizontal directions respectively. The material to the right of the slip line suffers less vertical stress and therefore the application of eq(6.3) shows that this material has not yielded. In the rigid region to the left of 56789 an orthogonal network of cycloids are constructed within the stress plane to enable the extension of pseudo slip lines (broken lines). Their end-points are located on a principal stress trajectory 9-12-13 which lies at 45° to α and β and is tangential to a stress-free point 9 on the container wall (Fig. 6.28a). The principal stresses, normal to the trajectory, are found from projecting the tangent to the trajectory in (a) through the corresponding pole point in (b). In this manner σ_B and σ_C act normal to the trajectory at points B and C as shown. Equilibrium requires that σ_B and σ_C act with equal magnitudes in vertical and horizontal directions. Since absent shear stress is a consequence of frictionless conditions, σ_B and σ_C are transmitted into the remaining material to align with the container wall. The greatest principal stress difference, $\sigma_B - \sigma_C$, applies to the top left corner. The stress plane reveals that this difference is less than $2k$ and hence the yield criterion is nowhere violated.

6.4 Frictionless Extrusion Through Inclined Dies

The extrusion pressure can be reduced by spreading the extrusion force over a larger area but there is a practical limit to increasing the width of the ram. However, the die angle may be inclined to achieve the same effect.

6.4.1 Extrusion For R = 2

Taking a die angle of $\theta = 30°$, the simple field in Fig. 6.31a applies for $R = H/h = 2$. This is constructed from the requirement that slip lines meet the die face and the axis of symmetry at 45°.

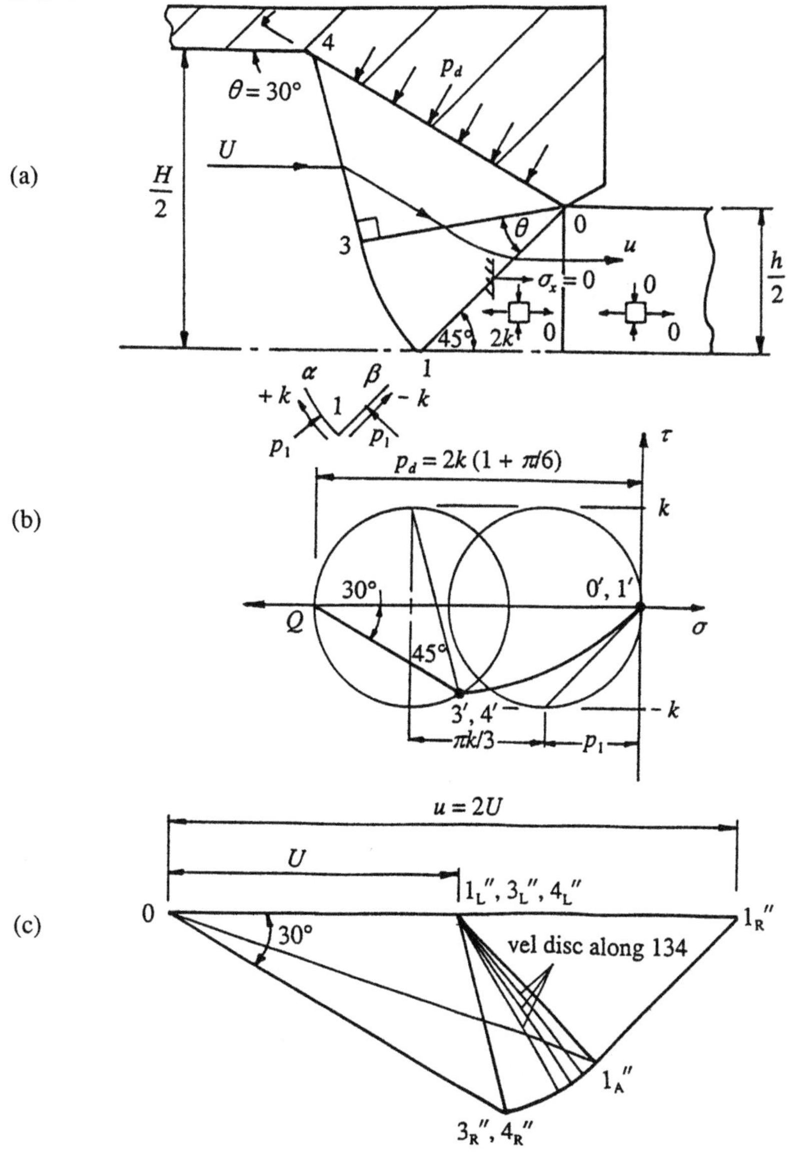

Figure 6.31 Frictionless extrusion through inclined dies with $R = 2$

The initial circle in Fig. 6.31b applies to the β - line 01, with its pole position coincident with the origin, so giving $\sigma_x = 0$. In rolling this circle along the ordinate $\tau = +k$, the variation in mean stress is established for that portion 13 of the α - line. The remainder 34 of this slip line must remain straight if it is to meet the die face at 45°, the circle being stationary with coincident pole points 3′ and 4′. Projecting a line parallel to the die face through pole point 4′, gives the stress state at Q in which the normal stress is identified with a constant die pressure

$$p_d = 2k\,(1 + \pi/6) \tag{6.24a}$$

The extrusion pressure p_e follows from equating horizontal forces/unit width. This gives

$$p_e H = \frac{p_d (H - h)\sin\theta}{\sin\theta} = 2k\left(1 + \frac{\pi}{6}\right)(H - h)$$

$$p_e = 2k(1 + \pi/6)(1 - 1/R) = k\,(1 + \pi/6) \tag{6.24b}$$

from which the horizontal extrusion force/unit width is

$$F_e = p_e \times H = H\,k\,(1 + \pi/6)$$

In general, for the inclined die geometry given in Fig. 6.31a:

$$R = H/h = 1 + 2\sin\theta \tag{6.25}$$

Equation (6.25) applies to the limit of the maximum reduction ratio that is possible in an inclined die [1]. For example, $R = 2$ for $\theta = 30°$ and $R = 2.414$ for $\theta = 45°$. The corresponding extrusion pressure is found from applying eq(6.9a) to the α - line 134. Referring to Fig. 6.31a, $p_1 = k$ at point 1. At point 3, where $\phi = -\theta$, substitution into eq(6.9a) gives

$$p_3 + 2k\,(-\theta) = c_1 \tag{6.26a}$$

where $c_1 = k$ for $\phi = 0$. Equation (6.26a) then gives

$$p_3 = k\,(1 + 2\theta) \tag{6.26b}$$

Resolving forces for the element 034 supplies the die face pressure

$$p_d = 2k\,(1 + \theta) \tag{6.27a}$$

from which the extrusion pressure is found as before in eq(6.24b)

$$\frac{p_e}{2k} = (1 + \theta)\left(1 - \frac{1}{R}\right) \tag{6.27b}$$

Equations (6.25) and (6.27b) supply an optimum die angle associated with the lowest extrusion pressure. A different SLF will apply to a given R value when θ is greater than the limiting value from eq(6.25). This is because the redundant work increases as frictional work decreases with increasing θ.

The hodograph construction in Fig. 6.31c shows rigid body horizontal velocity U to the left of points 1, 3 and 4. Material to the right of points 3 and 4 moves in the direction of the die face. Points $1_A''$, $3_R''$ and $4_R''$ are located from the projecting tangent directions at 1, 3 and 4 through $1_L''$, $3_L''$ and $4_L''$. The vectors $3_L''3_R''$ and $1_L''1_R''$, identify velocity discontinuities of the type in Fig. 6.15, in which the end points $3_R''(\equiv 4_R'')$ and $1_A''$ lie on a curve that is the orthogonal image of 13. Finally, the horizontal velocity u of a point to the right of 1 ($1_R''$) must give $u = 2U$. The particle flow direction, shown in Fig. 6.31a, is found from the directions of the absolute velocities on adjacent sides of the slip lines it crosses.

6.4.2 Extrusion For $R > 2$

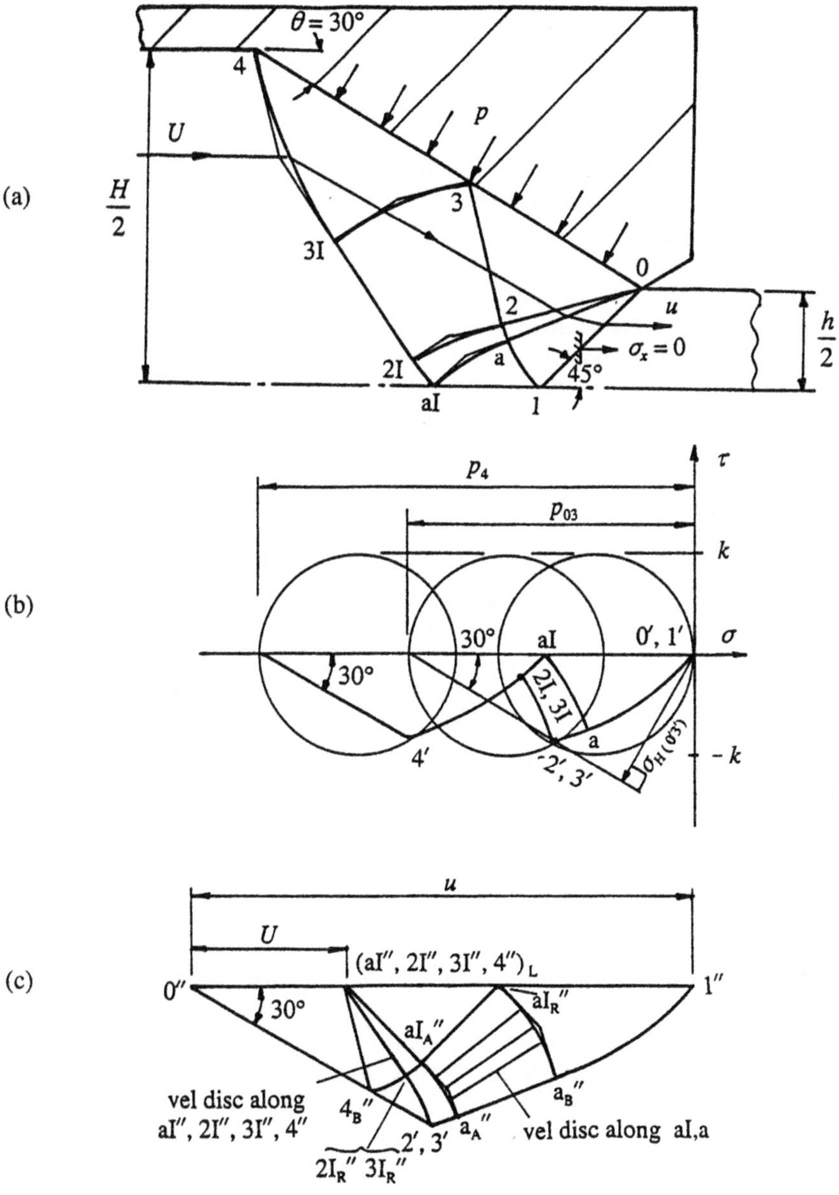

Figure 6.32 Frictionless extrusion through inclined dies with $R > 2$

The limiting solution, discussed in the previous section, may be employed to extend slip-line fields for extrusion ratios that are either greater or less than the ratio given by eq(6.25). Take the case of $R > 2$, with $\theta = 30°$, shown in Fig. 6.32a. A point a, between 1 and 2, must be found such that when the field is extended by the usual method, the point 4 lies in the container corner. Note the similarity between this slip-line field and that for square dies with $R \geq 3$ in Fig. 6.28a. The stress plane in Fig. 6.32b shows that the normal pressure p along the inclined die is constant at p_{03} along 03 but increases from 3 to a maximum value p_4 at 4. One method for obtaining the extrusion pressure is to integrate σ_x and τ_{xy} along the die face in the manner of eq(6.16b). Alternatively eq(6.15) could be used with σ_H typically as indicated. However, the stress plane can only supply stresses at points 0′, 3′ and 4′ as 034 is not a slip line. To improve accuracy, a mesh refinement would be necessary giving further intersections between pseudo slip-lines at intermediate points along 03 and 34. For this reason it is better to determine the horizontal extrusion force F_H by either method along the curved α - slip line aI, 2I, 3I, 4. The mean extrusion pressure is then $p_e = F_H /$ ram area.

The hodograph construction in Fig. 6.32c reveals two types of discontinuity: (i) that associated with rigid body motion to the left of 4, 3I, 2I, aI (refer to Fig. 6.15) and (ii) the fan of constant magnitude velocity vectors for adjacent points on either side of aI,a (refer to Fig. 6.14). For example, the relative velocity is constant for points above and below point, a (a_A'' and a_B'') and above and to the right of point, aI (aI_A'' and aI_R''). Their directions vary with the normals to the slip line at these points. It is seen that the path of a streamline now involves two changes of direction through the deformation zone. Finally the checkpoint 1″ may be used to ensure $U = Ru$ in obeyance with incompressible flow.

6.4.3 Extrusion For $R < 2$

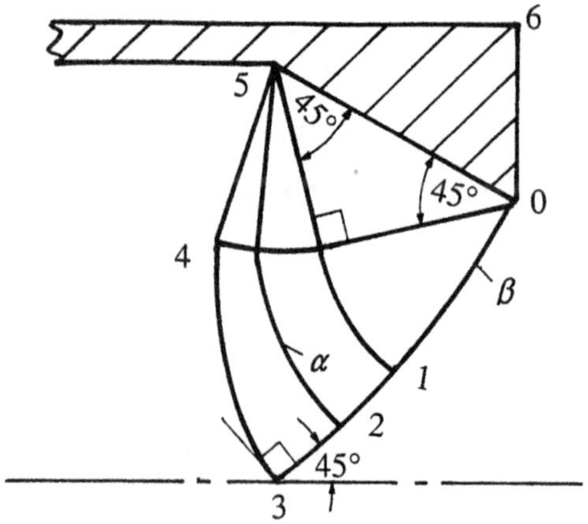

Figure 6.33 Frictionless extrusion through inclined dies with $R < 2$

Figure 6.33 gives a SLF for $R < 2$ in strip drawing [4]. Here, it is necessary to find point 1 on the β - line by trial. This enables the β-extension 23 to reach the axis of symmetry. Intersections at points 2 and 3, with orthogonal α - lines must meet at the top corner point 5 of the inclined die. As the α - lines cannot all meet the container at 45°, the stress state at point 4 is discontinuous. Because the circle is not stationary along points 1, 2 and 3, the stress plane is constructed for an assumed origin [1]. The true origin is found from applying eq(6.16c) in which the total horizontal force is zero for material to the right of the β - line.

6.5 Extrusion With Friction Through Parallel Dies

These solutions apply where the coefficient of friction is sufficiently high to induce frictional shear forces which yield the material in contact with the container walls. It follows that slip lines will meet the wall tangentially and normally within the deformation zone. The slip line field solution for a reduction ratio of 2 is given in Fig. 6.34. The requirement is that the α - line BD meets the axis of symmetry at 45° and the container wall at 90°. To achieve this, the position of 3, i.e. point 3′ in the stress plane (Fig. 6.34b) is varied until the pole position D′ lies on the $\tau = -k$ ordinate. A line drawn perpendicular to the wall through D′ confirms that BD is an α - line associated with $\tau = +k$.

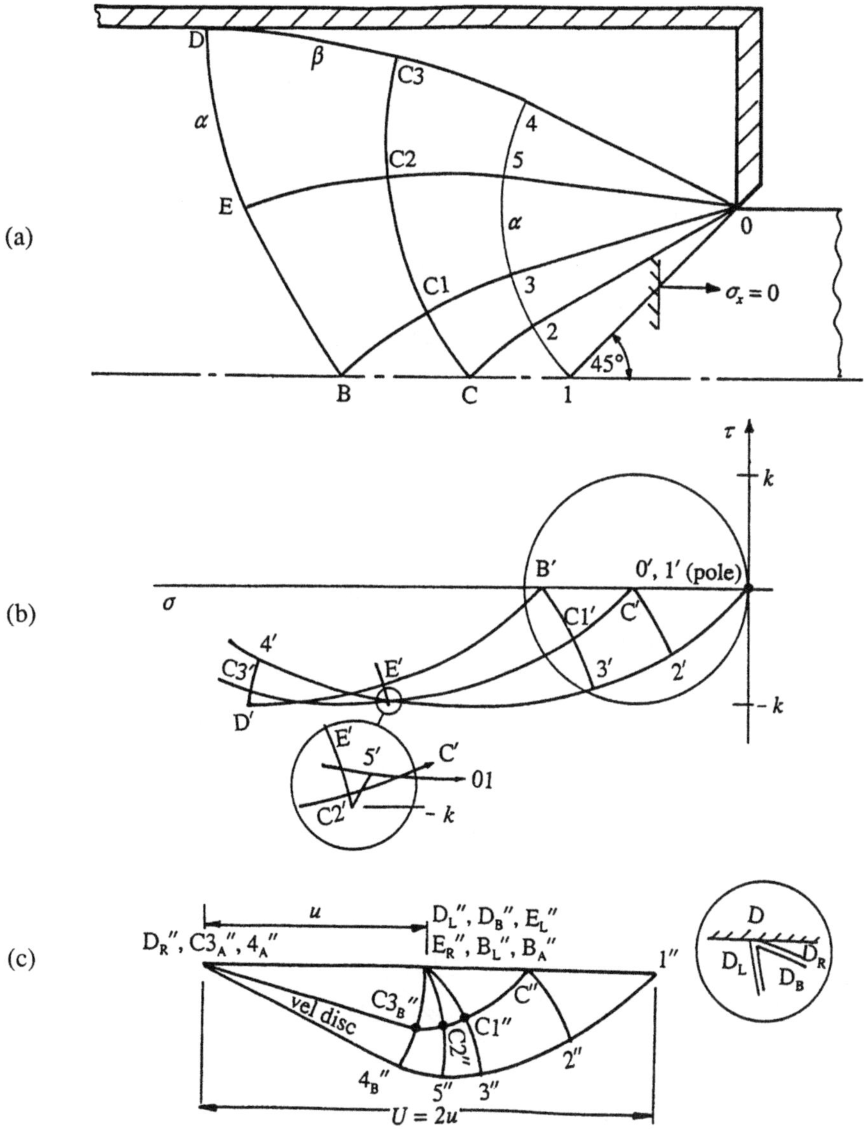

Figure 6.34 Extrusion with friction, $R = 2$

The tangential β - line at point D is determined from the orthogonal cycloid at D′ which terminates at point 4′. Intermediate slip lines are constructed from point C lying between 1 and B and from 5 between 3 and 4. Points to the right of D, and above C3 and 4, lie within the dead metal zone and, therefore, lie at the origin for the hodograph in Fig. 6.34c. Points beneath D, C3 and 4 are associated with the fan of velocity discontinuities. No discontinuity arises between points to the left and right of D, E and B, since these all move horizontally with a common velocity u. The checkpoint 1″ is found from extending the hodograph, with directions parallel to those in the stress plane, to give the final velocity $U = 2u$.

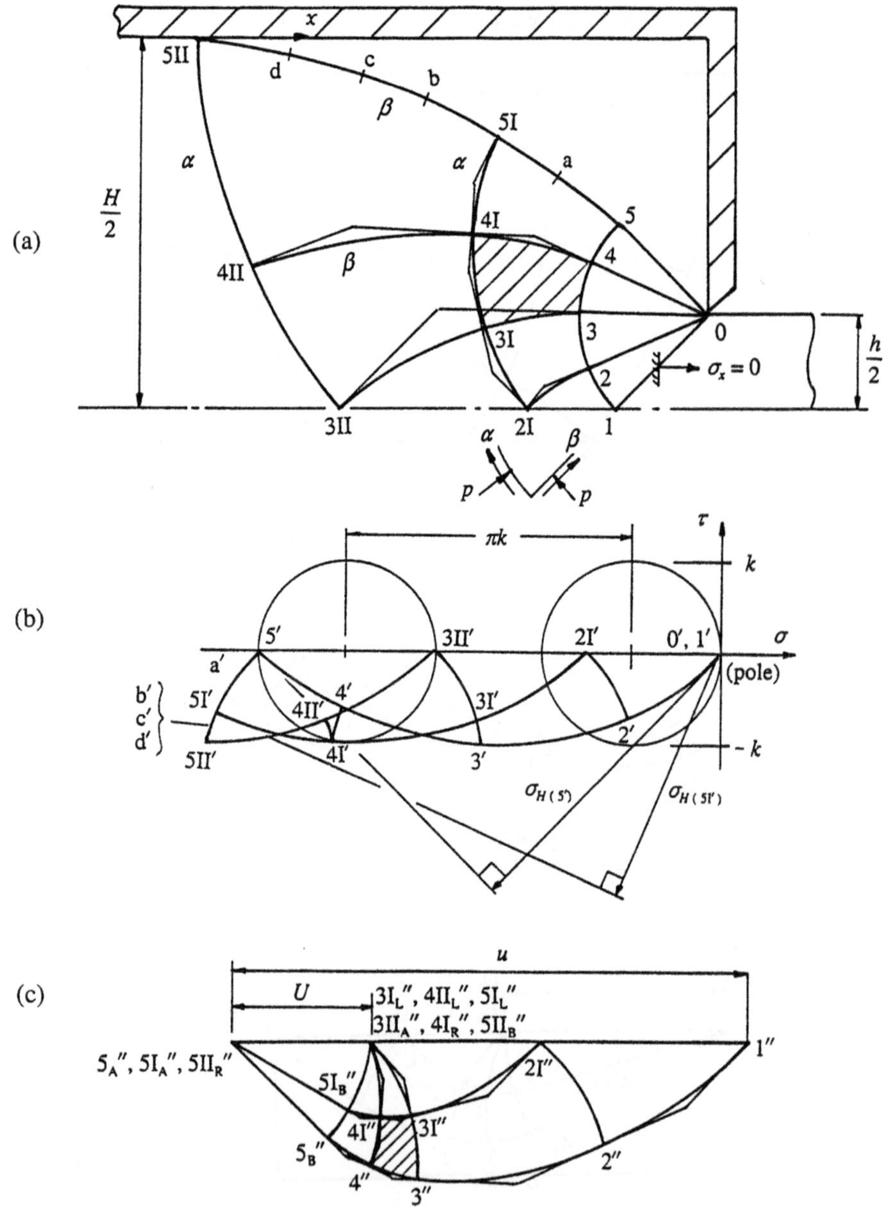

Figure 6.35 Extrusion with friction for $R = 4$

The solutions for different extrusion ratios and sticking friction are less distinct than those previously considered without friction. For example, the SLF solution for $R = 4$ (see Fig. 6.35a) is similar to that for $R = 2$ and, therefore, we need not detail the corresponding constructions for Figs 6.35b and c again. The additional construction lines in Fig. 6.35b, enable the determination of the total extrusion force/unit width by the method outlined in Fig. 6.19. This applies to the β - line 0, 5, 5I, 5II taking intermediate points a, b, c and d, to bound the dead metal zone. The corresponding poles, located on the orthogonal image in the stress plane, enable the centre and the ordinate position $\tau = -k$ to be located within each circle. Connecting the pole of each circle to its ordinate reveals the lengths of the perpendicular projections σ_H from the stress origin. These are expressed as a ratio of the circle's semi-circumferential length πk and plotted against the true length s of the β - line in Fig. 6.36.

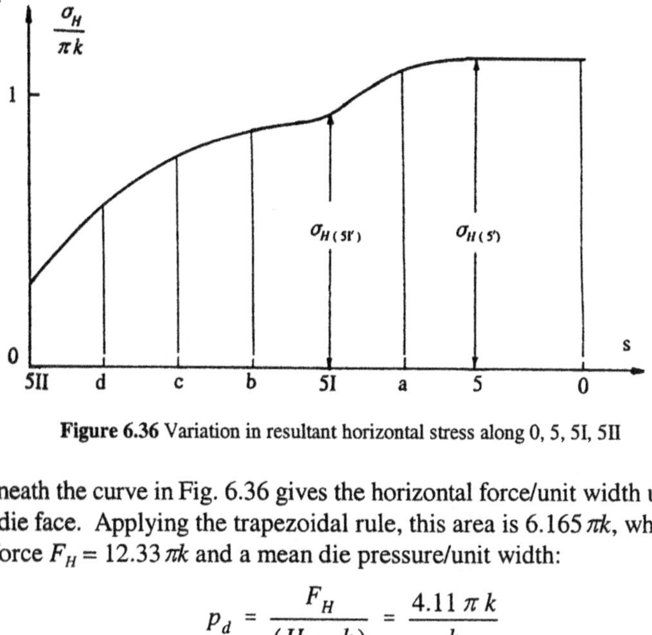

Figure 6.36 Variation in resultant horizontal stress along 0, 5, 5I, 5II

The area beneath the curve in Fig. 6.36 gives the horizontal force/unit width upon the upper rectangular die face. Applying the trapezoidal rule, this area is $6.165\,\pi k$, which gives a net horizontal force $F_H = 12.33\,\pi k$ and a mean die pressure/unit width:

$$p_d = \frac{F_H}{(H - h)} = \frac{4.11\,\pi k}{h}$$

The mean extrusion pressure p_e is found from

$$Hp_e = (H - h)\,p_d$$

$$p_e = 0.75\,p_d = \frac{3.083\,\pi k}{h}$$

Finally, a check is made that the rate of plastic work is positive. The shaded region within the hodograph in Fig. 6.35c shows that the motion of point 4I relative to 3 is consistent with the directions of k for the corresponding shaded region given in the stress plane (Fig. 6.35a).

6.6 Notched Bar in Tension

Deformation in a notched bar is in pseudo-steady state in which the bar size but not its shape changes with time. Fig. 6.37a shows one half of a symmetrical, double-notched tension bar. It will be shown that the semi-circular notch retains its shape when the through-thickness strain is zero under plane strain conditions.

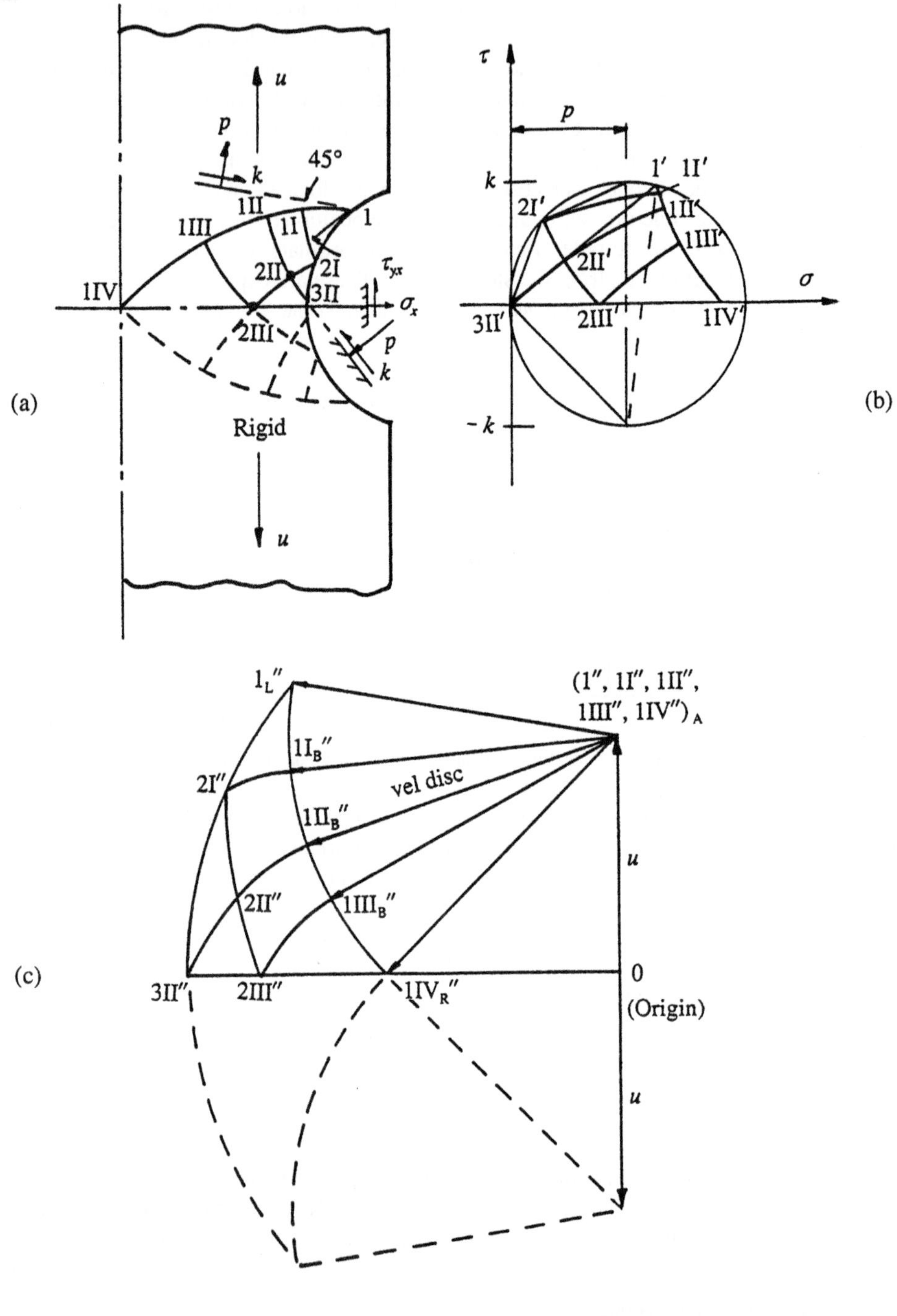

Figure 6.37 Notched tension under plane strain

The symmetry of the problem reduces the physical plane to ¼ of the bar only. Since both σ_x and τ_{yx} are zero at the notch root 3II, the circle in Fig. 6.37b must pass through a stress origin coincident with its pole point 3II′. The circle applies to the strong solution where the mean stress p is tensile. This circle will also represent the zero radial and tangential surface stress states for points 1 and 2I but with the different pole positions 1′ and 2I′ indicated. They are found from projecting tangents at 1 and 2I through the stress origin. The cycloid pattern shown originates from these pole positions when the circle rolls along $\tau = \pm k$. Points 2III′ and 1IV′ along the circle axis are associated with zero shear stress. Physically, they lie on the horizontal axis of symmetry, as this is a principal plane. The β - line: 3II, 2II, 1II, and the α - line: 2I, 2II, 2III, are derived from the corresponding cycloids in the stress plane. The problem lies in locating point 1. The requirement is that the α - line: 1, 1I, 1II, 1III, 1IV, is to terminate at the bar centre. Point 1II′ in the stress plane is determined by trial until this is achieved. We see that this is a particular type of boundary problem in which the surface tractions are zero. The applied force may be estimated from a stress distribution based upon the major principal (vertical) stresses at points 3II, 2III and 1IV.

In constructing the hodograph (Fig. 6.37c), the material outside the deformation zone moves with rigid body velocity u. This fixes coincident points: 1″, 1I″, 1II″, 1III″, 1IV″, above the bounding α - line. Within the deformation zone, the magnitude of the velocity discontinuities: $1''1_L''$, $1I''1I_B''$, $1II''1II_B''$, $1III''1III_B''$ and $1IV''1IV_R''$, are fixed since a point to the right of 1IV (i.e. $1IV_R''$) must also move horizontally relative to the origin. Remaining elements in the hodograph are established from orthogonality with the physical plane. It is seen that the deformed notch retains a semi-circular shape as points on its surface move radially inwards with uniform absolute velocity. That is, points $1_L''$, 2I″ and 3II″ lie on an arc with centre at the origin of the hodograph.

6.7 Die Indentation

With flat die indentation of a plate (see Fig. 6.38), the requirement is to develop a SLF for a given impression beneath the die. This leads to the die pressure and applied force F. Plasticity is localised for small b/t ratios, i.e. shallow impressions in very thick plates, but spreads through the thickness as b/t increases. Solutions are available for full friction, frictionless indentation and with an intermediate Coulomb friction between the contacting surfaces.

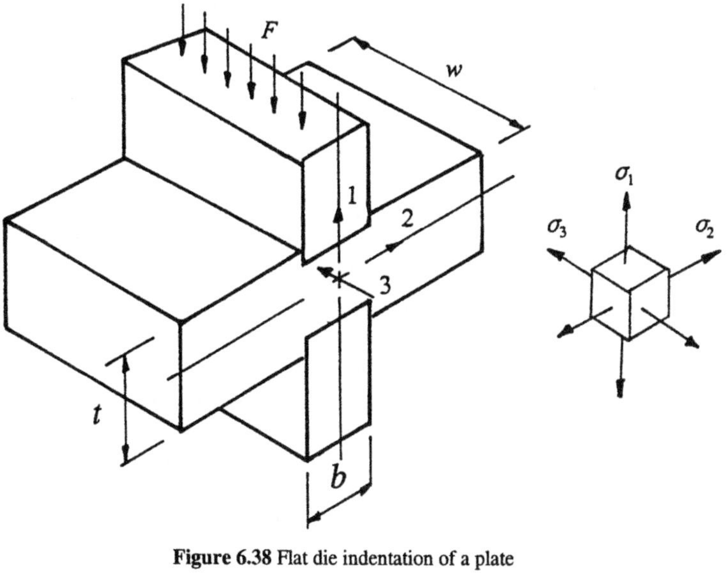

Figure 6.38 Flat die indentation of a plate

The Coloumb shear stress is the product of the frictional coefficient and the pressure normal to the die surface. Here, slip lines can meet the die surface at varying angles, in contrast to the respective tangential and 45° directions, with and without friction [7]. The SLF solutions for a blunt wedge indentor is surrounded by a dead metal zone similar to the SLF for flat die indentation of a thick plate. We will regard flat die indentation as a limiting case of the wedge indentor in which a plate of finite thickness t is compressed along a vertical axis between two flat platens each of breadth b (Fig. 6.38). Symmetry restricts the analysis necessary to ¼ of the deformation zone. Plane strain conditions apply when the plate width w is large compared to the die breadth. This means that plastic strain in the width direction is absent, this promoting an elastic constraint in each undeformed region adjacent to the deformation zone. The following geometries are considered from the interactions that arise between opposing slip line fields. The distinction between dies without friction (smooth) and with friction (rough) will need to be made for b/t ratios greater than unity.

6.7.1 *Indentation for $b/t < 0.115$*

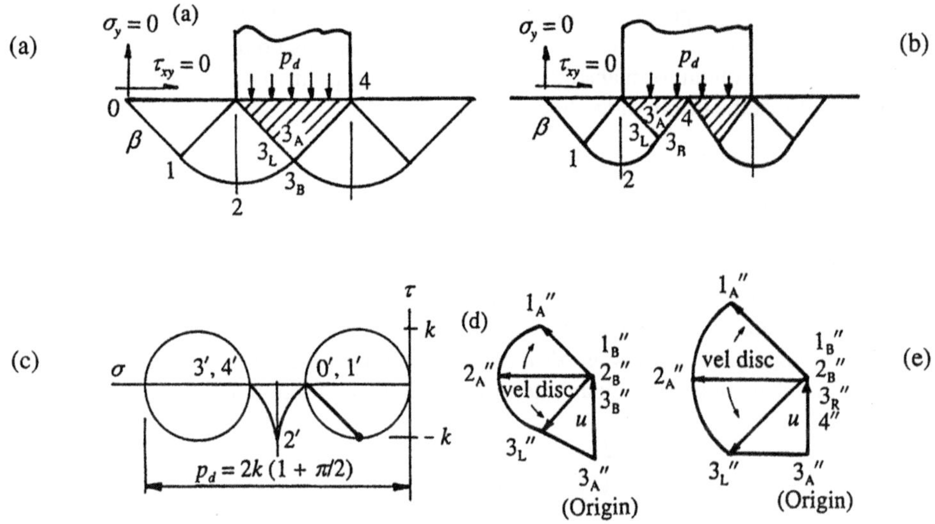

Figure 6.39 Flat die indentation for a thick plate

Two alternative SLF's for this b/t range (see Figs 6.39a,b) are due to the pioneering work of Prandtl [12] and Hill [4] respectively. Experiment has shown that Prandtl's SLF in (a) is more applicable to rough dies where there is less tendency for the dead zone (shaded) to slide on its contacting surface. However, should frictional forces along the narrow die face be zero, then surface roughness is unimportant. Hill's solution in (b) strictly applies to frictionless conditions in which slip lines intersect the die centre and edges at 45°. The same stress plane (Fig. 6.39c) applies to both slip-line fields. The initial, weak circle applies to the zero stress state (σ_y, τ_{xy}) at point 0 on the free surface. Projecting the horizontal plane through the stress origin locates the pole at 0′1′. Thus 01 is a β- line, which is extended to points 2 and 3 by rolling the circle along $\tau = -k$ to trace the cycloid image points 2′ and 3′. Beyond 3, the circle remains stationary with β adjacent to the dead zone. That is, pole points 3′ and 4′ coincide, corresponding to a straight line 34, bounding the dead zone.

Projecting a horizontal line, parallel to the die surface, through pole 3′4′ intersects this circle to give the stress normal to this surface. This is the die pressure

$$p_d = 2k\,(1 + \pi/2) \tag{6.28}$$

Prandtl and Hill's hodographs (Figs 6.39d,e respectively) show the downward velocity u of rigid wedges (points $3_A''$) relative to each origin. The latter are coincident points $1_B''$, $2_B''$, $3_{B,R}''$ and $4''$ for stationary material. A fan of velocity discontinuities appears where the outer β - line is crossed from stationary to moving material (points: $1_A''$, $2_A''$, $3_L''$). A further discontinuity $3_A''3_L''$ arises as the wedge is crossed from a point above and to the left of 3.

Both SLF solutions apply to regions beneath each die surface as they do not penetrate to the horizontal axis of symmetry. In fact, the upper field is not altered by the removal of the lower die (and vice versa).

6.7.2 *Indentation for 0.115 < b/t < 1*

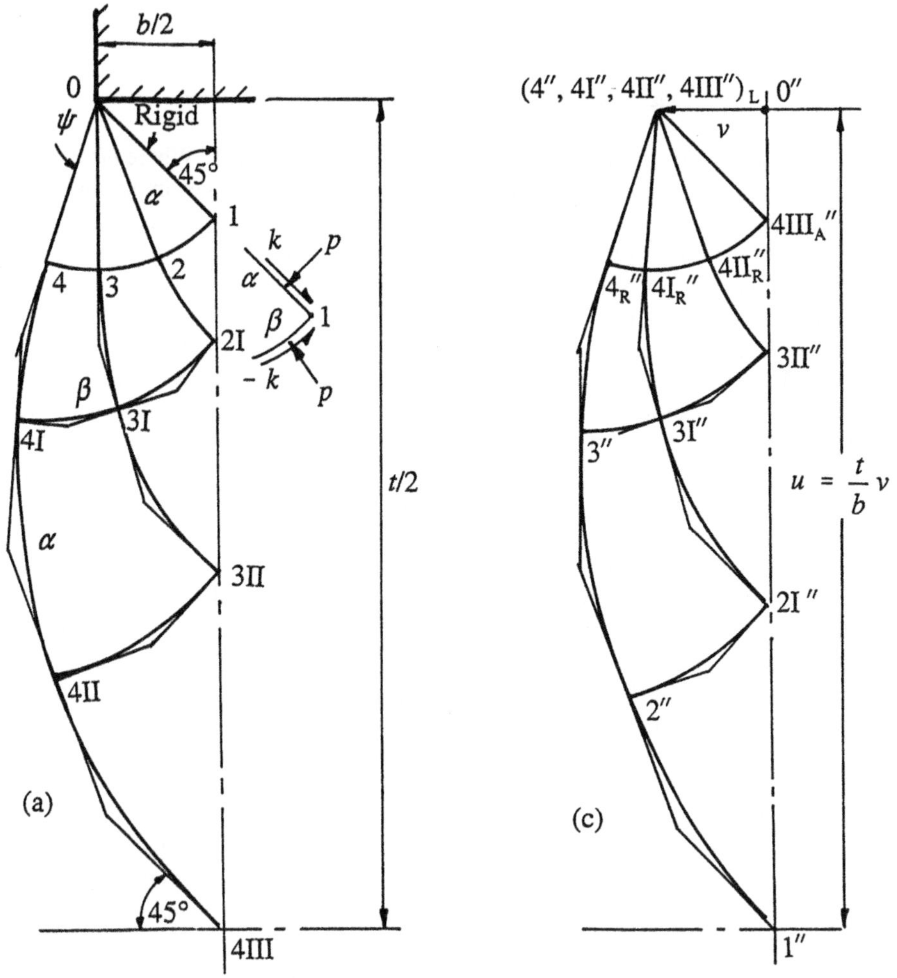

Figure 6.40 Flat die indentation, 0.115 < b/t < 1

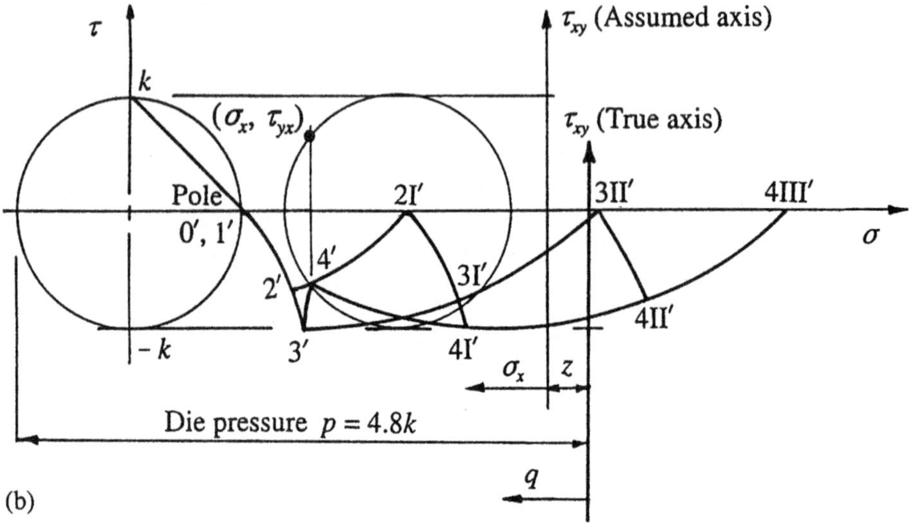

(b)

Figure 6.40 Flat die indentation, $0.115 < b/t < 1$ (continued)

Hill [13] derived a ratio $b/t = 0.115$, for opposing SLF to just meet. Where $b/t > 0.115$, an interaction between opposing SLF modifies their shape. A typical example for $b/t = 0.14$ is given in Figs 6.40a-c. As with Prandtl's solution (see Fig. 6.39a), the wedge of dead metal lying beneath the full width of the die in Fig. 6.40a implies that it is unimportant whether the die face is rough or smooth. Since the stress origin for Fig. 6.40b is not known, a stationary circle for the straight slip line 01, is chosen arbitrarily (left position). Assuming this an α - line, the stress state will lie at the top of the circle, from which the pole 0′1′ is located. The β - line 1234 is derived from the cycloid 1′2′3′4′, traced by this pole as the circle rolls in a clockwise direction along $\tau = -k$. Within the network of orthogonal cycloids point 4′ is selected to locate point 4III at the centre of the deformation zone.

The lateral velocity v of rigid material exterior to the deformation zone is found from an incompressibility condition: $b/t = v/u$, where u is the vertical die velocity. In the hodograph (Fig.7.40c) the coincidence of leftward points: $4_L''$, $4I_L''$, $4II_L''$ and $4III_L''$ is set at $v = 1$ unit from the origin 0″. The elements of the hodograph are orthogonal to those of the physical plane. These show discontinuities: $4_L''4_R''$, $4I_L''4I_R''$, $4II_L''4II_R''$ and $4III_L''4III_A''$, from crossing the outer α - line. Checkpoint 1″ reveals that the material within the dead zone moves with the correct downward velocity: $u = tv/b = 7.14v$.

The position of the stress origin is found by the second method outlined in Section 6.2.9 (p. 178). The requirement here is that the total horizontal force $F_H = 0$ for the outer, α slip line in Fig. 6.40a. Cartesian normal σ_x and shear stress τ_{xy} distributions are established from the points 0′, 4′, 4I′, 4II′ and 4III′ in the stress plane. These vary with the vertical and the horizontal projections of the slip line as shown in Fig. 6.41. Note that τ_{yx} is the full projection of a slip line that curves back on itself. For example, at point 4 the stress state (σ_x, τ_{yx}) is found from projecting a vertical line through 4′ in the stress plane. This shows that the sense of σ_x and τ_{xy} ($= \tau_{yx}$) oppose and so eq(6.16b) may be applied as in Figs 6.21 and 6.22. Here, q in Fig. 6.41 must be replaced with: (i) $\sigma_x + z$, for σ_x compressive in the region $c \le y \le b$ and (ii) $-(\sigma_x - z)$, for σ_x tensile in the region $0 \le y \le c$. Using ordinates in multiples of the circle radius k, the integrals in eq(6.16c) are evaluated by the trapezoidal rule. Setting $F_H = 0$ in eq(6.16b) determines z and the position of the true stress origin. From this the left point in the stress plane gives a normalised die pressure: $p_d/(2k) = 2.4$.

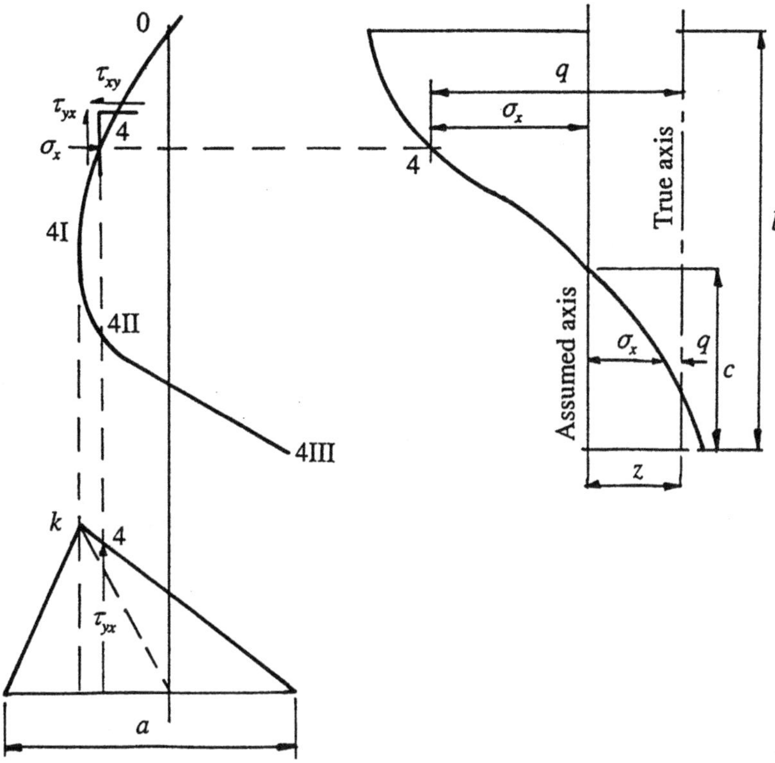

Figure 6.41 Variation in Cartesian stress components along α slip line

6.7.3 Indentation for b/t = 1

When $b/t = 1$, the SLF (Fig. 6.42a) consists of a single wedge of dead metal beneath the die face. As before, this implies the field is equally applicable to both rough and smooth dies.

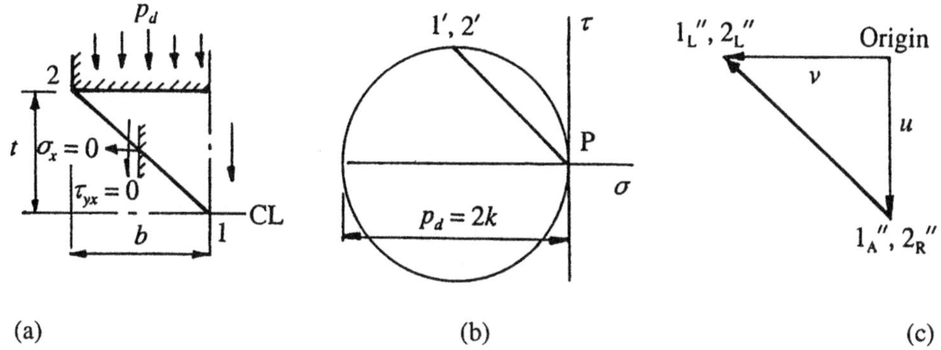

Figure 6.42 Flat die indentation, $b/t = 1$

With the pole P, at the origin of the stress plane in Fig. 6.42b, the constant die pressure is: $p_d/(2k) = 1$. In the hodograph (Fig. 6.42c), the lateral and vertical velocities are the same, i.e. $u = v$ and these are connected by the common discontinuity: $1_A''1_L''$, $2_R''2_L''$.

6.8 Rough Die Indentation

In the case of partially rough dies, SLF solutions are available for: (i) a constant shear stress at the boundary, less than k, and (ii) a constant coefficient of friction [1]. Condition (ii) allows the ratio between the shear force and the normal reaction to be matched more realistically. The following deals with an ideal case having completely rough dies.

6.8.1 Rough Indentation for 1 < b/t < 3.64

A detailed analysis [7] has shown that the slip line field in Fig. 6.43a does not meet the boundary tangentially when $b/t < 3.64$.

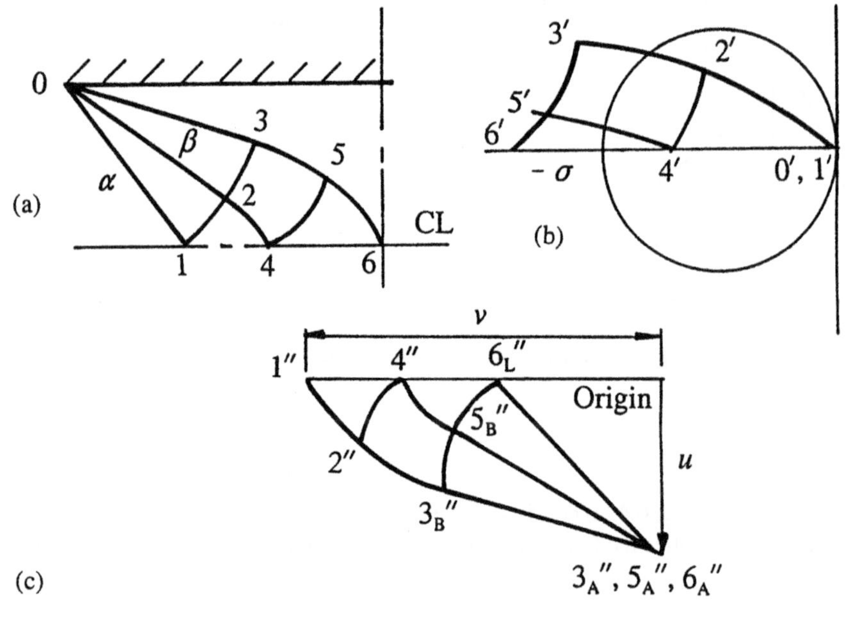

Figure 6.43 Rough die indentation, $1 < b/t < 3.64$

The 45° inclination of α - line 01 locates the pole of the weak circle at the origin of the stress plane in Fig. 6.43b. The cycloid 1′2′3′ is traced by the pole as the circle rotates ACW along $\tau = -k$ corresponding to β - line 123. Point 3 is adjusted until point 6 on α - line 356, which bounds the rigid zone, lies at the bar centre. This line can only meet the indentor edge tangentially for the limiting ratio: $b/t = 3.64$. For lower b/t, the edge inclinations of 0356 and another, intermediate α-line 024, lie between 0 and 45°. Slip lines meet axes of symmetry at 45° since these are principal stress planes. Elements of the hodograph in Fig. 6.43c are orthogonal to those in the physical plane (see Fig. 6.43a). The latter shows an array of discontinuity points, $3_B''$, $5_B''$ and $6_L''$, below the wedge. Points $3_A''$, $5_A''$ and $6_A''$ move rigidly downwards with velocity u. Checkpoint 1″ confirms the lateral velocity $v = (b/t)u$.

6.8.2 Rough Indentation for b/t > 3.64

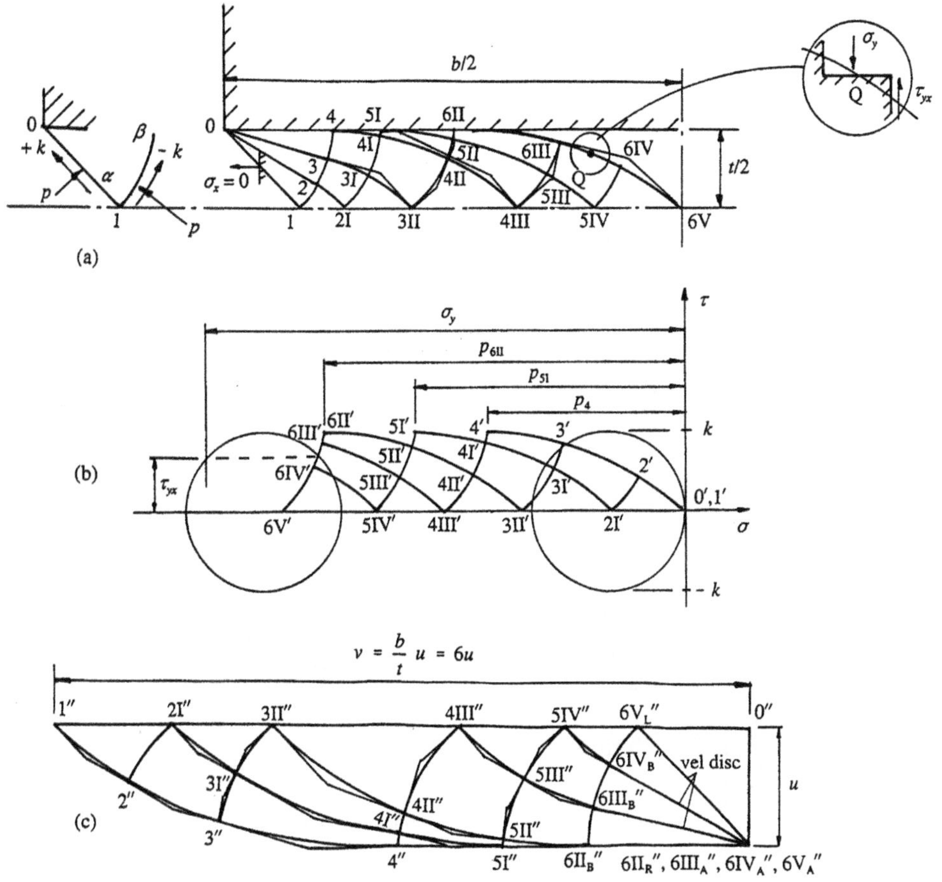

Figure 6.44 Rough die indentation $b/t = 6$

With $b/t > 3.64$ under full frictional conditions, the slip lines will meet a rough die surface tangentially to maintain a constant shear stress k at all points of contact. For example, a SLF solution for $b/t = 6$ is given in Fig. 6.44a. For the stress plane (Fig. 6.44b) the initial weak circle, with pole 0′1′ at the origin, applies to the stress state along α - line 01. The image of the β - line 1234 is a cycloid 1′2′3′4′, traced by the pole as this circle rolls along $\tau = -k$. The α - lines that intersect with points 2, 3 and 4 are constructed in the usual way. Point 3 must be adjusted so that point 6V lies at the bar centre. This shows that α - line: 6II .... 6V, contains the rigid wedge over half the die breadth. This SLF solution may be scaled to match any intermediate b/t ratio between 3.64 and 6.

Within the hodograph (Fig. 6.44c), the coincident points $6II_R''$, $6III_A''$, $6IV_A''$, and $6V_A''$, in the wedge zone, all move downwards with die velocity u. Crossing this α - line to points $6II_B''$ $6III_B''$, $6IV_B''$ and $6V_L''$, shows that this slip line is also a velocity discontinuity. The constant velocity is fixed by the horizontal motion of point $6V_L''$. The remaining elements of the hodograph are constructed from the orthogonal SLF elements, given the horizontal motion of points, 1″, 2I″, 3II″, 4III″, and 5IV″, along the axis of symmetry. The checkpoint 1″ confirms velocity $v = 6u$ for rigid material to the left of 01, by the constant volume condition. The hodograph also reveals how the horizontal velocity increases across any vertical section, i.e. by locating the intersections with slip lines within Fig. 6.44c.

The pressure distribution across the die face is not constant. A mean indentor pressure p_m can be found from the magnitudes of σ_y and τ_{yx} at points 0, 4, 5I, 6II, 6III, 6IV and 6V. These are provided by the stress plane (see Fig. 6.44b), from which the distribution is derived in Fig. 6.45.

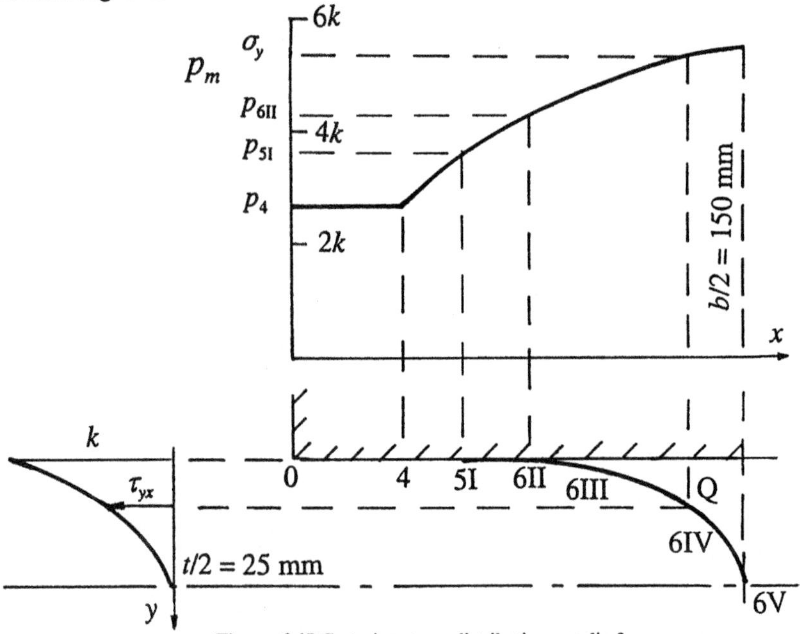

Figure 6.45 Cartesian stress distributions on die face

Using a similar derivation to eq(6.16a), the net compressive vertical force, per unit width, is given by

$$F_V = p_m b = \int \sigma_y \, dx - \int \tau_{yx} dy \qquad (6.29)$$

in which the stress component directions are opposed, as illustrated for point Q in Fig. 6.44a. Applying eq(6.29), with $b = 300$ mm and $t = 50$ mm, leads to: $F_V = 1400k$ per mm of width. Then, with k and p_m in MPa: $p_m/(2k) = 700/300 = 2.34$.

The forgoing solutions to each indentor enable a plot (see Fig. 6.46) of normalised pressure $p/(2k)$ versus b/t for the rough die indentation ($2k$ is the diameter of Mohr's circle). It is seen that a minimum appears in Figure 6.46 for the co-ordinates (1,1) and pressure is constant for $b/t < 0.115$. Hill [4] calculated the pressure across the die face for b/t ratios between 0 and 1. Experiments by Watts and Ford [14] confirmed Hill's theoretical prediction. With the exception of some integral ratios b/t, mean pressures must be employed for $b/t > 1$. Hill [15] estimated these from a linear relation

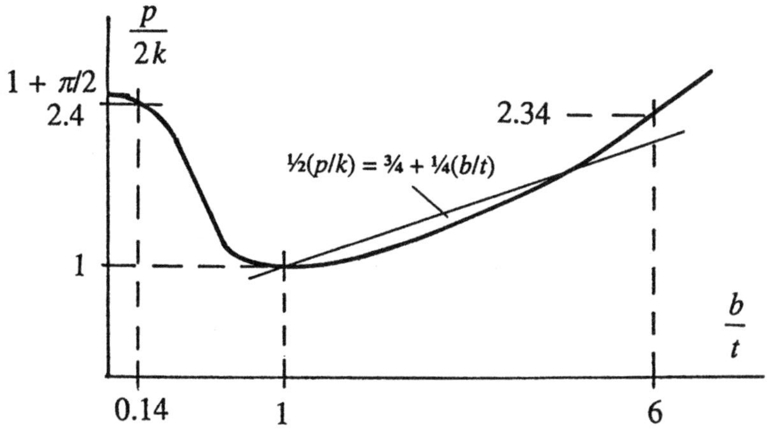

Figure 6.46 Pressure variation with b/t ratio for rough dies

$$p/2k = 3/4 + b/4t \qquad (6.30)$$

It is seen that eq (6.30) is in approximate agreement with the ratio $p/(2k) = 2.34$, as derived above for $b/t = 6$.

6.9 Lubricated Die Indentation

Complex, slip line fields for perfectly lubricated (smooth) dies, with $1 > b/t > 2$, were developed by Green [16]. Because slight differences arose for the intermediate critical ratio $b/t = \sqrt{2}$, a further subdivision of the b/t was necessary.

6.9.1 Smooth Indentation For $1 < b/t < \sqrt{2}$

The SLF, given in Fig. 6.47a, shows a single dead-metal zone lying at the centre of the die surface. Elsewhere slip lines intersect with the die face and both axes of symmetry at 45°.

(a)

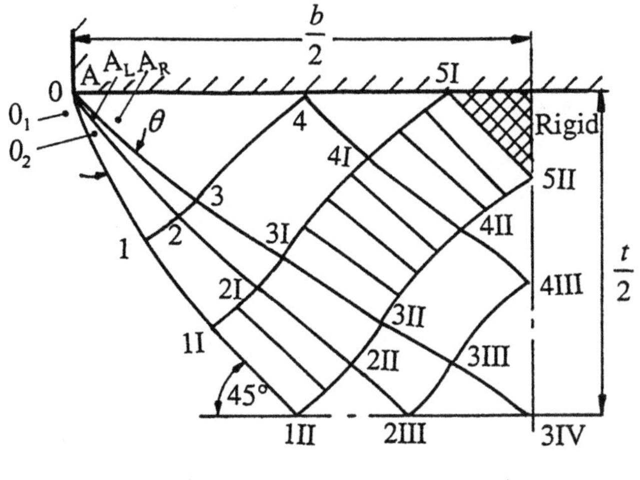

Figure 6.47 Smooth die indentation, $1 < b/t < \sqrt{2}$

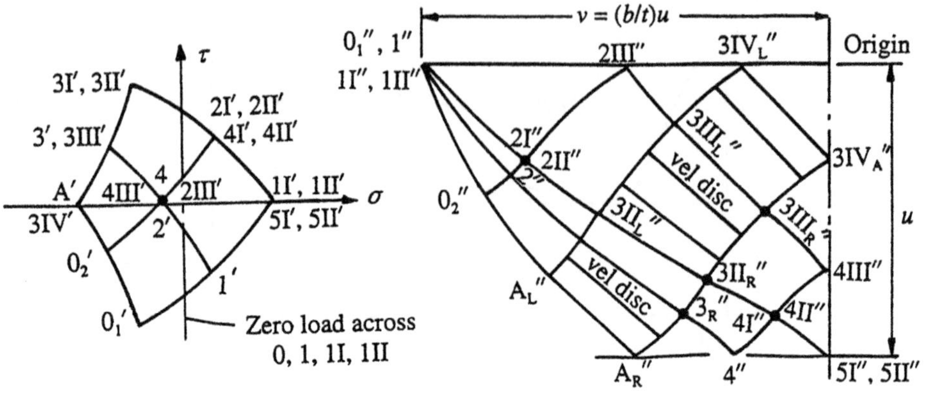

(b) (c)

Figure 6.47 Smooth die indentation, $1 < b/t < \sqrt{2}$ (continued)

The angle θ must be adjusted in Fig. 6.47a, along with the shape of the α - line: 0, 1, 1I, 1II, in order to locate point 3IV correctly on the vertical centre line. In Fig. 6.47b, the τ - axis is chosen to remove horizontal loading and, in Fig. 6.47c, the velocity of all leftward points $0_1''$, $1''$, $1I''$ and $1II''$ must conform to volume deformation.

6.9.2 Smooth Indentation For $\sqrt{2} < b/t < 2$

Within this b/t range, a lesser die pressure applies to a SLF with two dead-metal zones (see Fig. 6.48a). This also involves a more complex arrangement of velocity discontinuities as seen in Fig. 6.48c. Again, θ and the shape of the α - line 0, 1, 1I, 1II in Fig. 6.48a, must be adjusted to locate points 6II correctly at the centre. The stress origin is fixed from the condition that the net horizontal force is zero across this slip line. Outside the deformation zone the incompressible material flows laterally with velocity $v = (b/t)u$.

(a)

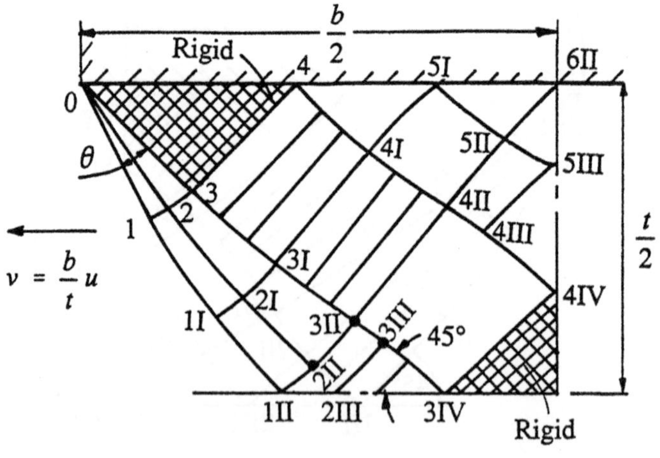

Figure 6.48 Smooth die indentation, $\sqrt{2} < b/t < 2$

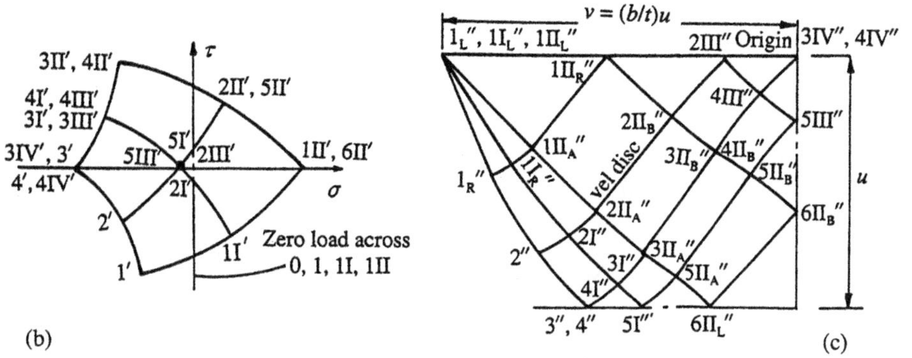

Figure 6.48 Smooth die indentation, $\surd 2 < b/t < 2$ (continued)

6.9.3 *Smooth Indentation For b/t ≥ 2*

Complete SLF solutions, available for smooth dies with integral ratios $b/t = 2, 3, 4$ etc, show rigid, block slip. For example, a simple SLF applies to $b/t = 2$, as shown in Fig. 6.49a.

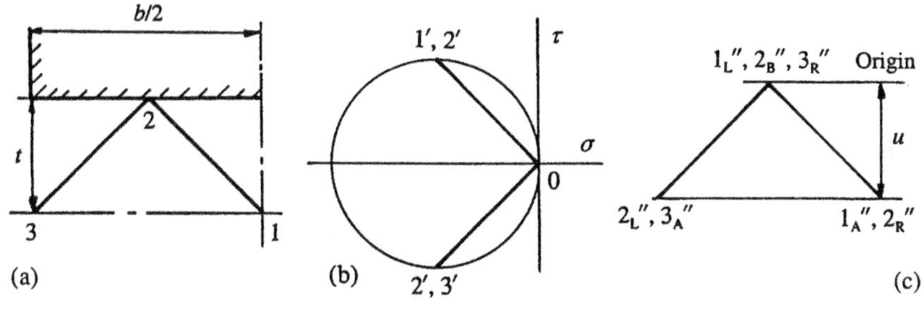

Figure 6.49 Smooth die indentation, $b/t = 2$

The weak, compressive circle in Fig. 6.49b will have its pole at the origin because $\sigma_x = 0$ in all horizontal directions. It follows that the mean stress on slip line is constant. That is, slip-lines remain straight, with every stress state represented within a single Mohr's circle. The maximum positive shear stress occurs at points 1′ and 2′, so identifying the α - line 12. Similarly, the β - line 23 is derived from points 2′, 3′ in the stress plane. The dead zone, above point 1 and to the right of 2, moves downwards with rigid velocity, u. This fixes points $1_A''$ and $2_R''$ within the hodograph (Fig. 6.49c). Velocity discontinuities coincide with slip lines, showing that the material points $1_L''$, $2_B''$ and $3_R''$ move leftward rigidly. Consideration of the horizontal and vertical components of velocity at points 2 and 3, reveals that the shape of elements within slip lines are retained with progressive deformation. With higher b/t ratios, a similar, extended pattern of slip lines and velocity discontinuities apply. These extensions introduce further 45° elements of alternating slope to reach the die edge.

Watts and Ford [14] recommended that b/t ratios within the range $2 \le b/t \le 4$ be employed for the plane strain compression test. Here t is the current thickness and the sheet width w is taken from the range $5 \le w/b \le 12$ (see Fig. 6.38). Their experiments confirmed the theoretical variation in $p/(2k)$ with b/t for frictionless compression in non-hardening material. Over a greater range of b/t the pressure variation is cyclic, as shown in Fig. 6.50.

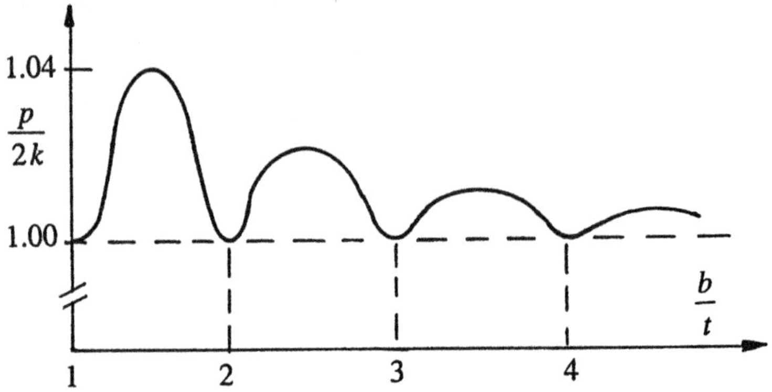

Figure 6.50 Cyclic variation in smooth die pressure with b/t ratio

Minimum values, $p/(2k) = 1$, occur for each integral ratio. With $b/t < 1$, the $p/(2k)$ variation will coincide with that found for rough dies (see Fig. 6.46). Testing within a limited b/t range allows for natural strain up to 200%. This is a far greater strain than can be reached in uniaxial compression upon a cylindrical testpiece where barrelling and buckling the impede uniform strain required (see Chapter 9).

Finally, the reader may wish to be challenged by noting that SLF's for smooth indentations with fractional b/t ratios beyond 2 are not known.

References

1. Ford H. and Alexander J. M. *Advanced Mechanics of Materials*, 1963, Longmans.
2. Hencky H. *Zeit Angew Math Mech*, 1923, **3**, 241.
3. Geiringer H. "Fondements mathematiques de la theorie des corps plastiques isotropes", *Mem Sci Math*, 1937, 86.
4. Hill R. *The Mathematical Theory of Plasticity*, 1950, Clarendon Press, Oxford.
5. Prager W. and Hodge P. G. *The Theory of Perfectly Plastic Solids*, 1951, Wiley, New York.
6. Prager W. "A geometrical discussion of the slip-line field in plane plastic flow", *Trans Roy Inst of Tech*, Stockholm, 1953, 65.
7. Ellis F. *Jl Strain Anal*, 1967, **2**, 52.
8. Lee E. H. "On stress discontinuities in plane plastic flow", Proc: *3rd Symposium in Applied Math*, 1950, 213, McGraw-Hill.
9. Green W. A. *Jl Mech. Phys Solids*, 1962, **10**, 225.
10. Alexander J. M. *Quart. Appl Maths*, 1961, **19**, 32.
11. Bishop J. F. W. *Jl Mech Phys Solids*, 1953, **2**, 43.
12. Prandtl L. *Nachr Ges Wiss Gottingen*, 1920, 74.
13. Hill R. *Jl Iron and Steel Inst.*, 1947, **156**, 513.
14. Watts A. B. and Ford H. *Proc. I. Mech. E*, 1952-3, **1**B, 448.
15. Hill R. in: *Strength of Materials*, 1950, HMSO, **6**(1), 87
16. Green A. P. *Phil Mag*, 1951, **42**(7), 900.

Exercises

6.1 Show that the slip line field and hodograph solutions for an extrusion ratio $R = 3$ produces a consistent positive rate of plastic work.

6.2 Construct extended, pseudo slip lines to show that, within the rigid body regions for $R = 2$ extrusion, the yield criterion is nowhere violated.

6.3 Determine the horizontal extrusion force F_H from the application of eq(6.16a) to the α - slip line aI, 2I, 3I, I4 in Fig. 6.32. Take the extrusion ratio $R = 3.3$ for a die inclination $\theta = 30°$.

6.4 In Fig. 6.51 the magnitude and direction of respective velocities v_A and v_B for points 1 and 2 on an α - line, together with v_C and v_D for points 1 and 3 on a β - line, are known. Construct the hodograph for the region enclosed within α and β.

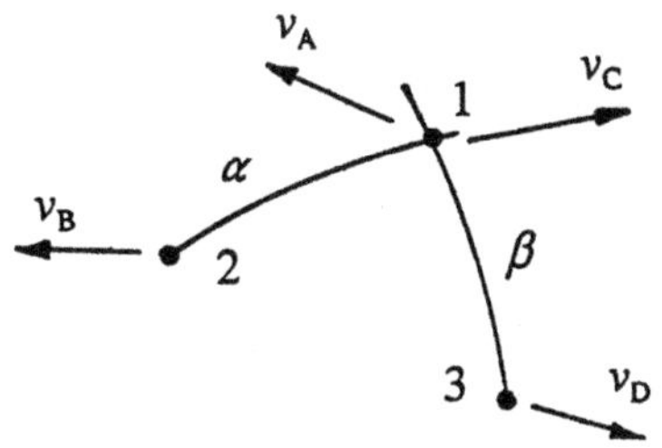

Figure 6.51

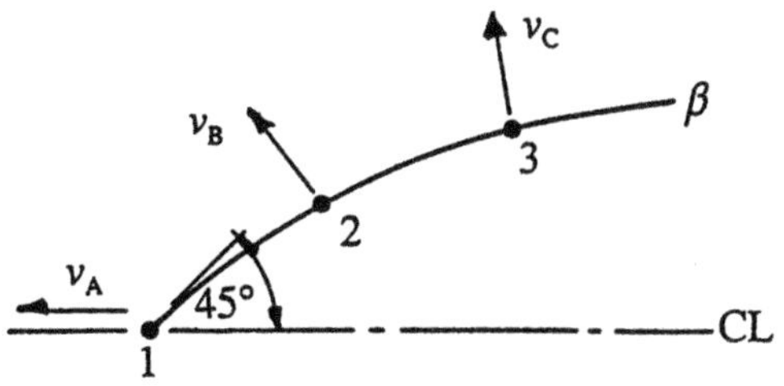

Figure 6.52

6.5 The magnitude and direction of absolute velocities v_A, v_B and v_C at points 1, 2 and 3 on a β - line are known. Construct the hodograph for the region enclosed within this β - line and an axis of symmetry (CL) when the intersection is as shown in Fig. 6.52.

6.6 The absolute velocities at two points 1 and 2 on a free boundary in Fig. 6.53 are known. Construct the hodograph for the lower, interior region.

Figure 6.53

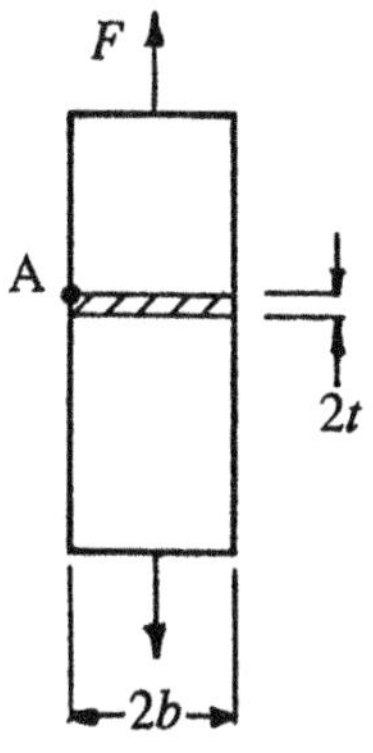

Figure 6.54

6.7 Figure 6.54 shows two rigid blocks connected with a layer of solder $2b \times 2t \times w$, where the width w is large compared to b and t. When carrying a tensile force F/unit width, the solder flows plastically under a shear yield stress k. Sketch the slip line field and hodograph and show that when $b/t = 11.5$, $\sigma = k(2.5\pi + 1)$ at point A, where the slip line bounding the central dead zone reaches the surface of the block. Estimate F and the mean normal stress along the interface for this condition.

6.8 In the side extrusion container in Fig. 6.55, the ram exerts a vertical force P to extrude horizontally a rectangular strip of thickness h. Assuming frictionless plane-strain conditions, determine appropriate upper bounds and associated hodographs for the extrusion pressure/unit width of container. How would these be modified by base friction?

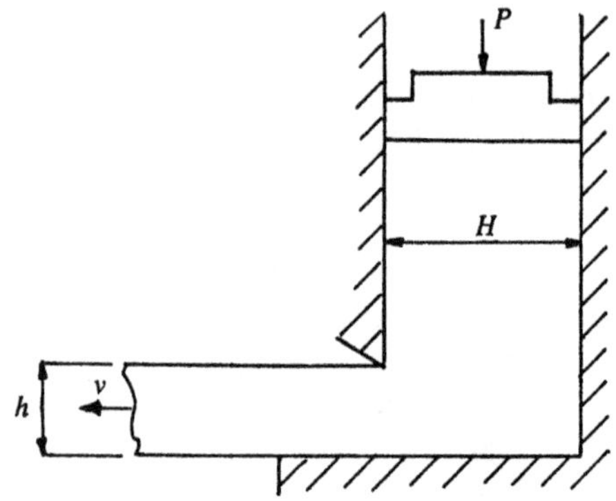

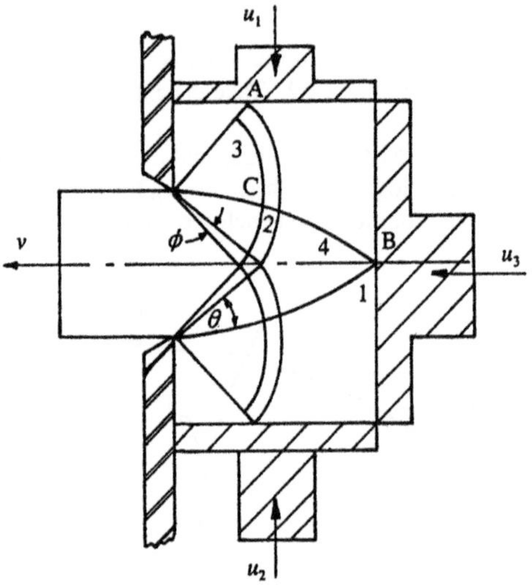

Figure 6.55 **Figure 6.56a**

6.9 Figures 6.56a and 6.56b give the general SLF and hodograph solutions respectively for frictionless extrusion under three rams. The inward speeds u_1 and u_2 of the lateral rams are equal. Determine (i) the correct angle of intersection of the slip lines with the rams, (ii) the α - and β - lines for which θ and ϕ re-appear within the hodograph as shown, (iii) velocity vectors within the hodograph corresponding to the absolute velocities at points 1, 2, 3 and 4 and the pressure exerted by each ram at A and B. How would you modify the SLF (a) with the presence of friction along each lateral ram face and the hodograph in Fig. 6.56b, when $u_1 \neq u_2$ with $u_3 = 0$?

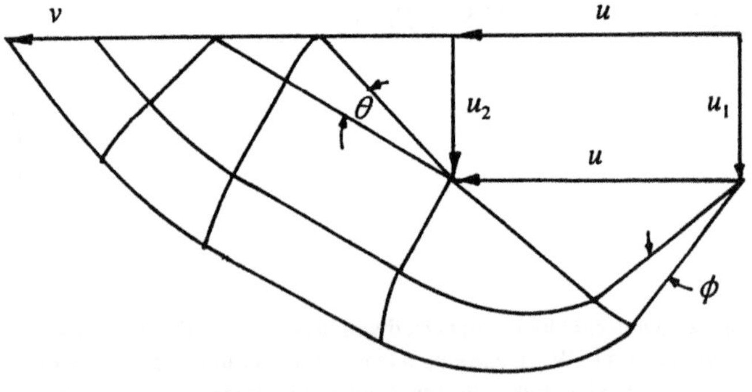

Figure 6.56b

CHAPTER 7

LIMIT ANALYSIS

7.1 Introduction

The plastic behaviour of a structure or a process involving plastic-rigid material may often be represented in lower- and upper-bounds. Two principles are employed to obtain these bounds: (i) that a statically admissible stress field is in equilibrium, satisfying the forces applied to the boundary and (ii) that a kinematically admissible velocity field is compatible with internal strains, satisfying imposed displacements at the boundary. Principle (i) provides a lower-bound solution, in which the admissible stress field will nowhere exceed the yield stress. Principle (ii) is associated with an upper-bound solution, which normally provides the loading associated with the admissible displacement field. Since the work dissipated by the upper bound loading is usually more than that done by the actual forces, the true solution will lie between the two bounds. The latter impose limits to the plastic behaviour but in certain structures, as for the beams that follow, no difference exists between the bounds.

7.2 Collapse of Beams

In consideration of the limit loading for a simply-supported beam in Fig. 7.1a, the lower bound load W_L is supplied from the moment diagram (Fig. 7.1b). In the beam cross-section beneath W there exists an admissible stress field. The collapse moment is given by [1]

$$M_{ult} = \frac{abW}{a + b}$$

from which a lower bound load will be fractionally less than

$$W_L = \frac{M_{ult}(a + b)}{ab} \qquad (7.1a)$$

The beam displacement diagram, in Fig. 7.1c, shows that the work done by an upper-bound load is greater than the energy dissipated with the rotation θ of the hinge. This gives

$$W_U\,\delta = M_{ult}\,\theta$$

Substituting $\theta = \delta/a + \delta/b$ leads to

$$W_U = \frac{M_{ult}(a + b)}{ab} \qquad (7.1b)$$

In general, the true loading W lies in the range: $W_L < W < W_U$, but eqs(7.1a, b) show no distinction between the bounds in this case. If the ends of the beam in Fig. 7.1a were both fixed, collapse would occur from the formation of three plastic hinges: one at each end and one beneath W. The condition for collapse is that equal ultimate moments M_{ult} are attained simultaneously within each hinge.

(a)

(b)

(c)

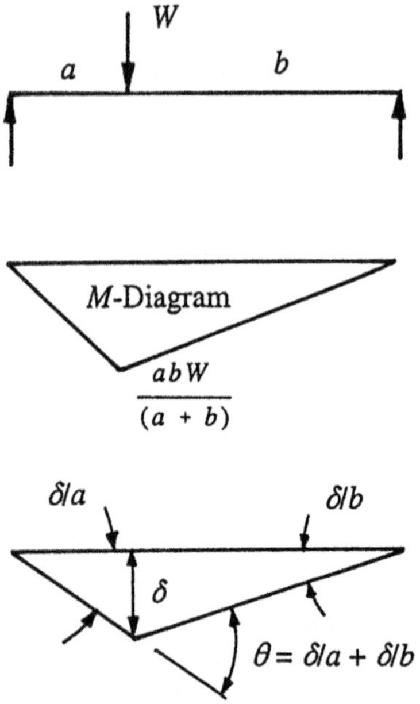

Figure 7.1 Collapse of a simply supported beam

Their magnitudes are each 100% greater than M_{ult} for a simply supported beam. This gives, from eq(7.1a),

$$W_L = \frac{2M_{ult}(a + b)}{a\,b} \tag{7.2a}$$

The deflected shape of the encastre beam is similar to Fig. 7.1c. The work of collapse is given by

$$W_U\,\delta = M_{ult}\,\theta + \left(M_{ult} \times \frac{\delta}{a} \right) + \left(M_{ult} \times \frac{\delta}{b} \right)$$

$$W_U = \frac{2\,M_{ult}(a + b)}{a\,b} \tag{7.2b}$$

Again, there is no distinction between the bounds in eqs(7.2a,b). Note that for both beam fixings, the shape of the cross-section is not specified when deriving the lower-bound collapse load W_L. This is generally true for a beam with a uniform cross-section.

To take an example of a non-uniform section, consider a plane, vee-notched beam subjected to a uniform moment M/unit width (see Fig. 7.2a). A lower bound moment M_L is found from the fully-plastic admissible stress field in the weakest section. This gives

$$M_L = F_L d = Y\left(1 \times \frac{a}{2} \right) \frac{a}{2} = 0.25\,Y a^2 \tag{7.3a}$$

With an upper bound collapse under a single plastic hinge (Fig. 7.2b), the energy dissipated is $2M_U\,\delta\theta$. Now M_U follows from the shear yield stress $k = Y/2$, acting around the plastic zone radius r.

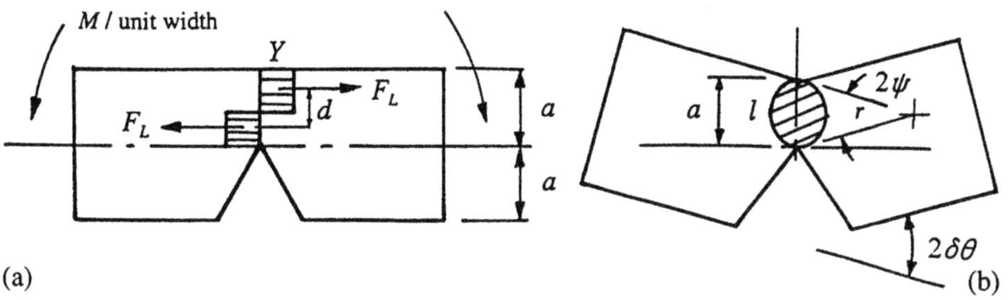

Figure 7.2 Plane vee-notched beam

This gives

$$M_U = klr = \frac{Ylr}{2} \tag{7.3b}$$

where $l = 2\psi r$ is the arc length. From eq(7.3b), collapse under a single plastic hinge is assumed to occur when the hinge product lr is at its minimum:

$$lr = 2\psi r^2 = 2\psi\left(\frac{a^2}{4\sin^2\psi}\right) = \frac{a^2\psi}{2\sin^2\psi}$$

$$\frac{\mathrm{d}}{\mathrm{d}\psi}(lr) = \frac{a^2}{2\sin^2\psi} - \frac{a^2\psi\cos\psi}{\sin^3\psi} = 0$$

$$\tan\psi = 2\psi$$

That is, $\psi = 1.166$ rad and $lr = 0.69a^2$. Substituting into eq(7.3b) gives

$$M_U = 0.345Ya^2 \tag{7.3c}$$

The true collapse moment M lies within the range bounded by eqs(7.3a,c):

$$0.25Ya^2 < M < 0.345Ya^2$$

A slip-line field solution gives $0.315Ya^2$. It is generally true that SLF solutions will be closer to upper-bound estimates.

7.3 Collapse of Structures

7.3.1 Principal of Virtual Work

A simpler method to obtain collapse loading of beams and plane frames is to treat collapse as a rigid mechanism under an equilibrium system of co-planar, non-concurrent forces F_i where $i = 1, 2 \ldots N$. The equibrium conditions are

$$\sum_{i=1}^{N} F_i = 0 \quad \text{and} \quad \sum_{i=1}^{N} M_i = 0 \tag{7.4a,b}$$

where the moments M_i must sum to zero at any two arbitary points. Since the displacements and rotations that occur in a collapsing structure do not alter the equilibrium eqs(7.4a,b), we say that they are *virtual*, written as Δ^v and θ^v, respectively. It follows that the equilibrium system of *real* forces and moments forces will do zero *virtual* work. That is

$$\sum_{i=1}^{N} F_i \Delta_i^v = 0 \quad \text{and} \quad \sum_{i=1}^{N} M_i \theta_i^v = 0 \tag{7.5a,b}$$

Combining eqs(7.5a,b) leads to a useful form

$$\sum_{i=1}^{N} F_i \Delta_i^v = \sum_{i=1}^{N} M_i \theta_i^v = 0 \tag{7.6}$$

In eq(7.6), θ_i^v refers to the rotation at each hinge under a collapse moment and Δ_i^v to the deflections beneath the corresponding forces. To apply eq(7.6), a collapse geometry must be assumed. Where a number of collapse mechanisms are possible, the true collapse load minimises eq(7.6). We now identify M_{ult} with a section-dependent, collapse moment M_P.

7.3.2 Beams

(a) *Propped Cantilever*

To determine the uniformly distributed collapse load w_P for a cantilever with an end prop (see Fig. 7.3a), the mechanism of collapse, given in Fig. 7.3b, is employed. Firstly, the position z_o of the second hinge is required.

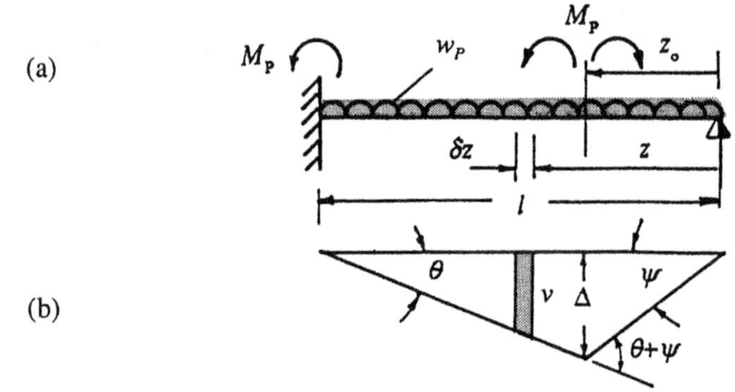

Figure 7.3 Collapse mechanism in a propped cantilever

Applying eq(7.6) and omittting the superscript v for simplicity:

$$M_P\,\theta + M_P(\theta + \psi) = w_P \int_0^l v\,\mathrm{d}z \tag{7.7}$$

where v is the deflection at position z and $w\mathrm{d}z$ is the elemental force at this position. The integral in eq(7.7) is clearly the enclosed area

$$\int_0^l v\,\mathrm{d}z = \frac{l\Delta}{2} \tag{7.8a}$$

The rotation ψ is found from

$$\Delta = (l - z_o)\theta = z_o\,\psi \tag{7.8b}$$

which gives

$$\psi = \frac{(l - z_o)\theta}{z_o} \tag{7.8c}$$

Substituting eqs(7.8a,b,c) into eq(7.7):

$$w_P = \frac{2M_P(l + z_o)}{l z_o(l - z_o)} \tag{7.9}$$

To find the z_o value which minimises eq(7.9) we set $dw_P/dz_o = 0$. This gives

$$z_o^2 + 2z_o l - l^2 = 0$$

for which $z_o = (\sqrt{2} - 1)l = 0.4141l$. Then, from eq(7.9), $w_P = 11.66M_P/l^2$.

(b) *Continuous Beam*

Figure 7.4a shows a stepped, cantilever beam resting upon two supports at C and E. The plastic collapse moment for the reduced section within length AC is one half that for the section within length CE. All the possible modes of failure must be examined (as shown in Figs 7.4b-d) in the search for the least value of the collapse load W.

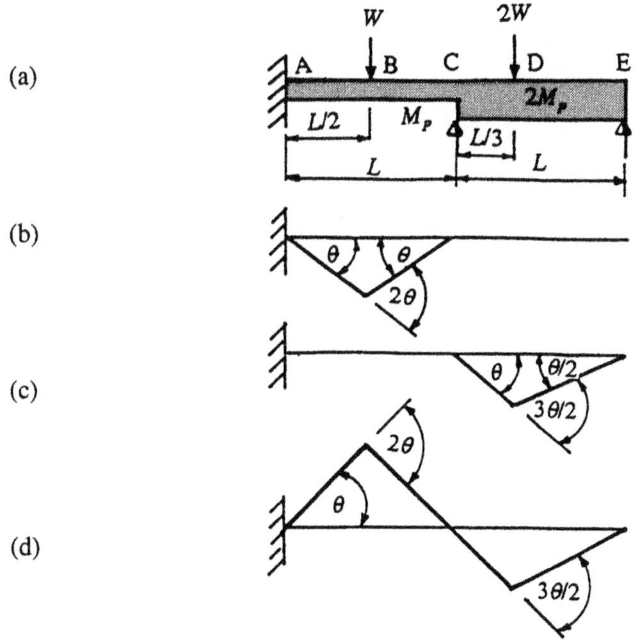

Figure 7.4 Collapse modes in a continuous beam

Now apply eq(7.6) to each mode in turn. In Fig. 7.4b, M_P is reached at A, B and C:

$$M_P\theta + M_P(2\theta) + M_P\theta = W(L/2)\theta \;\Rightarrow\; W = 8M_P/L \tag{7.10a}$$

In Fig. 7.4c, M_P is reached in a hinge to the left of C and $2M_P$ must be reached for collapse at D. The rotation at E is not accompanied by a moment (free-end). This gives

$$M_P\theta + 2M_P(3\theta/2) = 2W(L/3)\theta \;\Rightarrow\; W = 6M_P/L \qquad (7.10b)$$

In Fig. 7.4d, collapse moments of M_P, M_P and $2M_P$ are reached at A, B and D respectively:

$$M_P\theta + M_P(2\theta) + 2M_P(3\theta/2) = -W(L/2)\theta + 2W(L/3)\theta \;\Rightarrow\; W = 36M_P/L \qquad (7.10c)$$

The least load from eqs(7.10a - c) is the collapse condition: $W = 6M_P/L$.

7.3.3 Portal Frames

Application of eq(7.6) to describe collapse in a portal frame requires all the possible mechanisms to be considered [2]. For example, the frame ABCDE in Fig. 7.5 is subjected to both vertical and horizontal loadings with 'hinged' supports at A and E. Since hinges can carry no moment, collapse may occur: by collapse in the horizontal beam BCD (Fig. 7.5b), by sway in the stanchions AB and DE (Fig. 7.5c) or by a combination of the two (Fig. 7.5d).

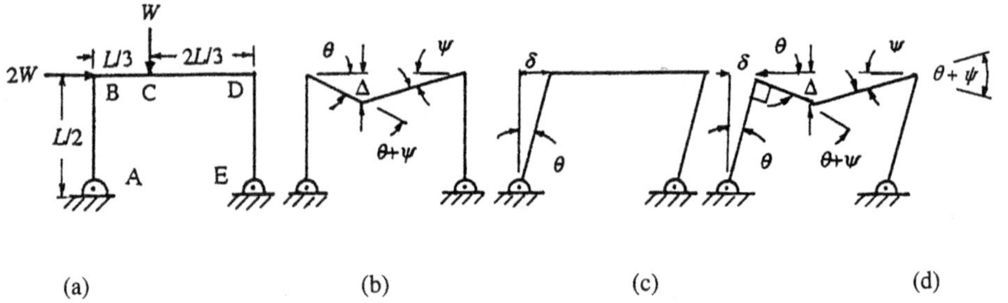

Figure 7.5 Collapse modes in a portal frame

Corner collapse will occur in the limb with the lower collapse moment where different sections exist between the limbs. Assuming that the stanchions have a 50% greater moment capacity than the beam, all the hinges will form on the beam side of the corners under M_P, as shown. Failure from each mechanism occurs with hinges at: (b) B, C and D, (c) B and D and (d) C and D. Applying eq(7.6) to each mode in Figs 7.5 b, c and d, respectively:

$$M_P\theta + M_P(\theta + \psi) + M_P\psi = W\Delta \qquad (7.11a)$$

$$2M_P\theta = 2W\delta \qquad (7.11b)$$

$$M_P(\theta + \psi) + M_P(\theta + \psi) = W\Delta + 2W\delta \qquad (7.11c)$$

Substituting $\psi = \theta/2$, $\delta = L\theta/2$ and $\Delta = L\theta/3$ into eqs(7.11a,b,c) leads to the respective collapse loads: $9M_P/L$, $2M_P/L$ and $9M_P/4L$, showing that a sway collapse would occur. Note, these calculations assume that the collapse moment ($1.5M_P$) for each stanchion is not reached. Therefore, it would be more economical to select a constant section, with collapse moment M_P, for the whole structure.

With encastre fixings, further hinges form at A and E under modes (c) and (d) in the manner of Figs 7.6d and f. Equations (7.11b,c) are modified to

$$2M_P\theta + 2(3M_P/2)\theta = 2W\delta \tag{7.12a}$$

$$M_P(\theta + \psi) + M_P(\theta + \psi) + 2\,(3M_P/2)\theta = W\Delta + 2W\delta \tag{7.12b}$$

The respective collapse loads are $W = 5M_P/L$ and $9M_P/(2L)$, indicating that a combined mode of failure occurs. The structure can be designed more economically when both sway and combined mode failures occur simultaneously. Let M_P' ($< M_P$) be the modified collapse moment for the stanchions in modes of Figs 7.6e,g. Equations (7.12a,b) are modified to

$$2M_P'\theta + 2M_P'\theta = 2W\delta \tag{7.13a}$$

$$2M_P'\theta + M_P(\theta + \psi) + 2M_P'\theta = W\Delta + 2W\delta \tag{7.13b}$$

Equation (7.13a) gives $M_P' = WL/4$. Substituting this into eq(7.13b) gives $M_P = 11WL/36$. Therefore $W = (36M_P)/(11L)$ and $M_P' = 9M_P/11$, which confirms that the given stanchions are over-designed. A graphical interpretation of this optimum design now follows.

7.3.4 Minimum Weight Solution

We may employ virtual work to design a beam or a frame to have minimum weight. To do this, even the most unlikely failure mode has to be accounted for. In the portal frame of Fig.7.6a there are in fact six modes (b-g) as shown. With the deflections δ, rotations θ and collapse moments M_P indicated, the application of virtual work to each mode results in a corresponding system of dimensionless moment equations:

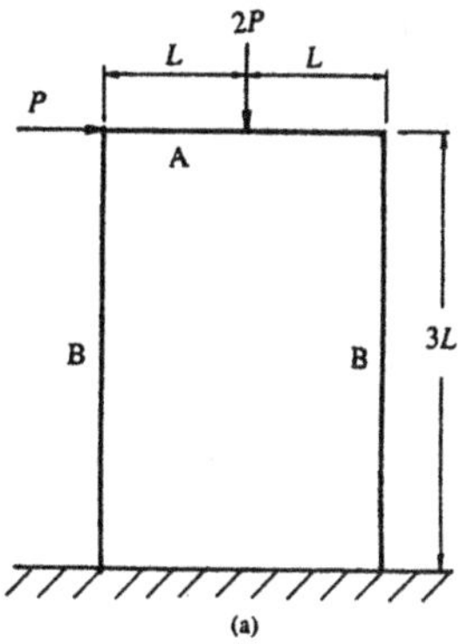

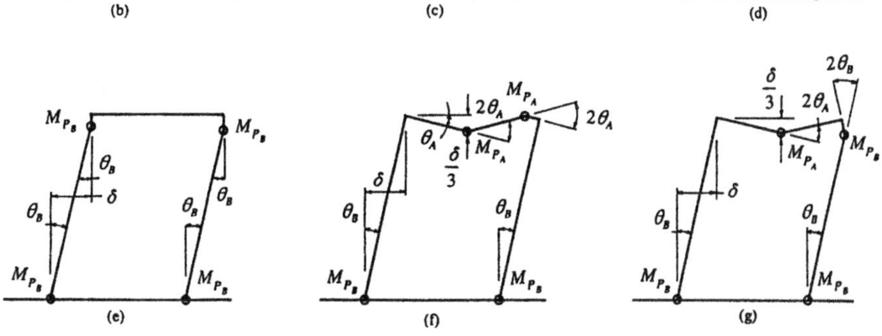

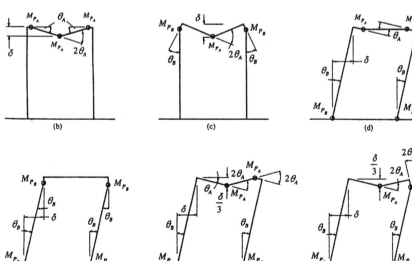

Figure 7.6 All possible failure modes in a simple portal frame

(Fig. 7.6b): $2P\delta = 4M_{P_A}\theta_A \quad \Rightarrow \quad \dfrac{M_{P_A}}{PL} \geq 0.5$ (7.14a)

where $\theta_A = \delta / L$

(Fig. 7.6c): $2P\delta = 2M_{P_A}\theta_A + 2M_{P_B}\theta_B \quad \Rightarrow \quad \dfrac{M_{P_A}}{PL} + \dfrac{M_{P_B}}{PL} \geq 1.0$ (7.14b)

where $\theta_A = \theta_B = \delta / L$

(Fig. 7.6d): $P\delta = 2M_{P_A}\theta_A + 2M_{P_B}\theta_B \quad \Rightarrow \quad \dfrac{M_{P_A}}{PL} + \dfrac{M_{P_B}}{PL} \geq \dfrac{3}{2}$ (7.14c)

where $\theta_A = \theta_B = \delta / (3L)$

(Fig. 7.6e): $P\delta = 4M_{P_B}\theta_B \quad \Rightarrow \quad \dfrac{M_{P_B}}{PL} \geq \dfrac{3}{4}$ (7.14d)

where $\theta_B = \delta / (3L)$

(Fig. 7.6f): $P\delta + 2P\dfrac{\delta}{3} = 4M_{P_A}\theta_A + 2M_{P_B}\theta_B \quad \Rightarrow \quad 2\dfrac{M_{P_A}}{PL} + \dfrac{M_{P_B}}{PL} \geq \dfrac{5}{2}$ (7.14e)

where $\theta_A = \theta_B = \delta / (3L)$

(Fig. 7.6g): $P\delta + 2P\dfrac{\delta}{3} = 2M_{P_A}\theta_A + 4M_{P_B}\theta_B \quad \Rightarrow \quad \dfrac{M_{P_A}}{PL} + 2\dfrac{M_{P_B}}{PL} \geq \dfrac{5}{2}$ (7.14f)

where $\theta_A = \theta_B = \delta / (3L)$.

Each of eqs(7.14a-f) plot as a straight line within the axes given in Fig. 7.7.

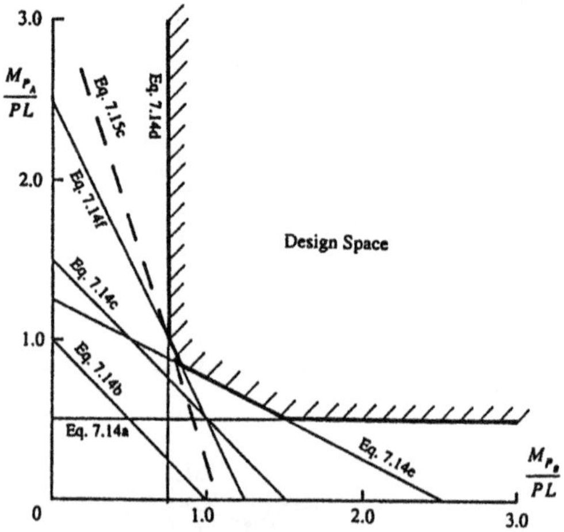

Figure 7.7 Design space for portal frame collapse

The inequalities reveal a working *design space* that is bounded by eqs(7.14a, d, e and f), as shown. If we now superimpose a weight function X upon Fig. 7.7, one point on the design space boundary will provide a safe, minimum weight solution. The weight/unit length depends upon the collapse (ultimate) moment. The latter can be expressed in terms of the yield stress and the dimensions of the section area, as in eq(5.5). From this we may interpret a weight function for a whole structure [3]:

$$X = \sum M_P L \tag{7.15a}$$

and for the portal frame in Fig. 7.6, eq(7.15a) becomes

$$X = (M_{P_A} \times 2L) + 2\,(M_{P_B} \times 3L) \tag{7.15b}$$

The corresponding dimensionless form is

$$\frac{X}{PL^2} = 2\frac{M_{P_A}}{PL} + 6\frac{M_{P_B}}{PL} \tag{7.15c}$$

The right-hand side of eq(7.15c) describes a straight line of gradient −3, as shown. A minimum value of $X/(PL^2)$ has been chosen to lie at the corner where the collapse modes in Figs 7.7e,g occur simultaneously. This requires a simultaneous solution to the corresponding eqs(7.14d and f), which gives the minimum weight conditions:

$$\frac{M_{P_A}}{PL} = 1\,, \quad \frac{M_{P_B}}{PL} = \frac{3}{4} \quad \text{and} \quad \frac{X}{PL^2} = 6.5$$

Commercial sections will not conform exactly to these conditions but the method allows the selection to lie within the design space, close to the minimum weight point.

7.4 Die Indentation

7.4.1 Lower Bounds

In order to estimate a maximum, lower-bound, fields of all possible stress discontinuities are used. Let us apply this approach to flat die indentation of a large plate (see Fig. 6.38). The simplest posssible configuration of stress discontinuities is that given in Fig. 7.8a.

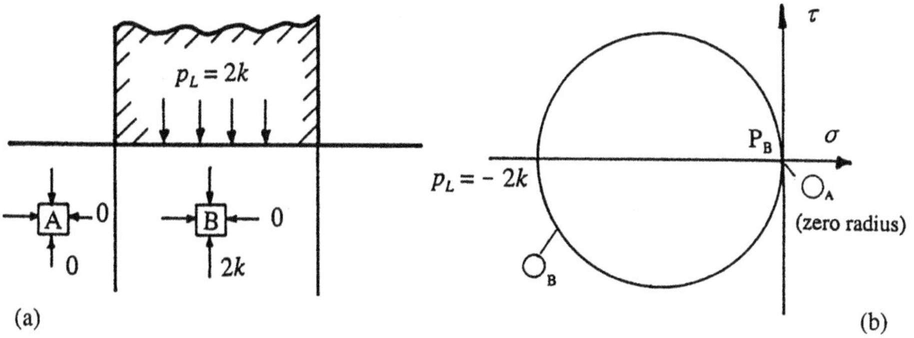

Figure 7.8 Simple stress discontinuity pattern for flat die indentation

Two constant stress regions, A and B, are defined in Fig. 7.8a by the given states: A being unstressed and B satisfying the plane strain yield criterion: $\sigma_1 - \sigma_3 = 2k$. In the stress plane (see Fig. 7.8b), point A defines the origin and B becomes a circle O_B with pole P_B, at the origin. Clearly, the constant die pressure $p_L = 2k$ is far less than that found previously from the SLF solution (6.28): $p_d = 2k(1 + \pi/2) = 5.142k$.

To raise the lower bound p_L, the stress states may be altered to those given in Fig. 7.9a.

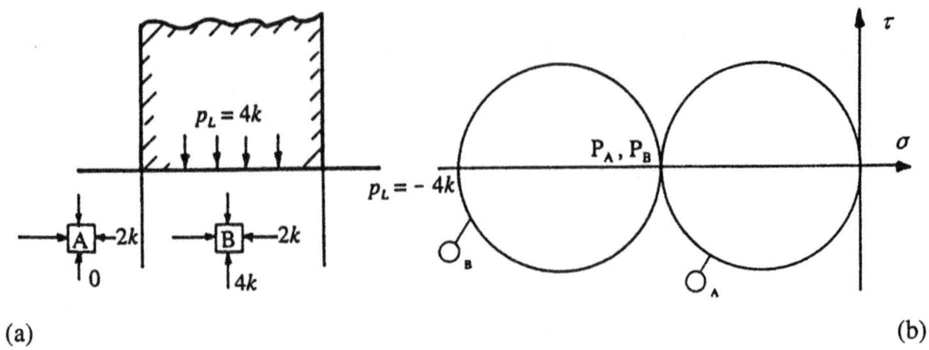

Figure 7.9 Stress discontinuities for flat die indentation

Both regions A and B now satisfy the yield criterion. The stress plane (Fig. 7.9b) consists of two circles $\bigcirc_A$ and $\bigcirc_B$, each of radius k, with common poles, P_A and P_B. Projecting the horizontal plane of the die face through P_B intersects $\bigcirc_B$ to give a die pressure $p_L = 4k$, which is twice that found from Fig. 7.8b.

Let us now examine whether the two alternative patterns of stress discontinuities on either side of the vertical centre line in Fig. 7.10a are admissible.

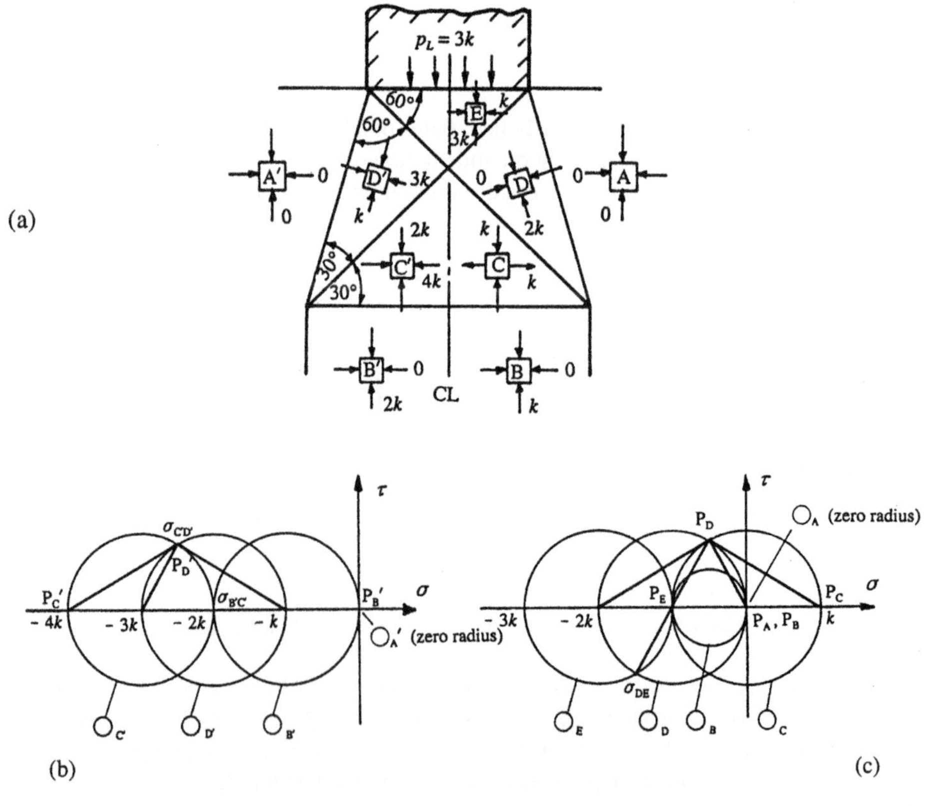

Figure 7.10 Complex stress discontinuities for flat die indentation

We first work in an anticlockwise direction, to the left of the centre-line, starting with a stress-free condition for region A′. Circles $\bigcirc_B'$ and $\bigcirc_C'$, for yielding within regions B′ and C′, are shown in Fig. 7.10b, with their respective pole points P_B' and P_C'. The pole P_D' for $\bigcirc_D'$ is found from projecting plane C′D′ through P_C'. Circle$\bigcirc_D'$ gives the principal stresses: $\sigma_1 = -k$ and $\sigma_3 = -3k$, within region D′, orientated as shown. However, because there is no continuity in the stress normal to plane D′A′ when we return to A′ from region D′, this patten of discontinuities is inadmissible.

Starting again with a stress-free condition for region A in Fig. 7.10a, we now work clockwise in across regions B, C, D and E, to the right of the vertical centre line. This shows elastically stressed material in region B, implying that $\sigma_1 = +k$, $\sigma_3 = -k$ within the yielded region C. Circle $\bigcirc_C$ has its centre at the stress origin in Fig. 7.10c. Projecting plane CD through pole P_C establishes pole P_D for $\bigcirc_D$. Projecting planes parallel and perpendicular to DA through P_D intersects $\bigcirc_D$ to give a principal stress state: $\sigma_1 = 0$, $\sigma_3 = -2k$, within region D. Since state D maintains continuity with zero stress for region A, we can now determine the stress state within region E. Projecting direction DE through P_D gives a normal stress σ_{DE} within $\bigcirc_D$ that must also apply to region E. This establishes the pole P_E for $\bigcirc_E$ and a principal stress state within E: $\sigma_1 = -k$, $\sigma_3 = -3k$, as shown. The latter defines the normal die pressure as $p_L = 3k$. In searching for an even greater lower bound, another discontinuous stress field is considered in Fig. 7.11a.

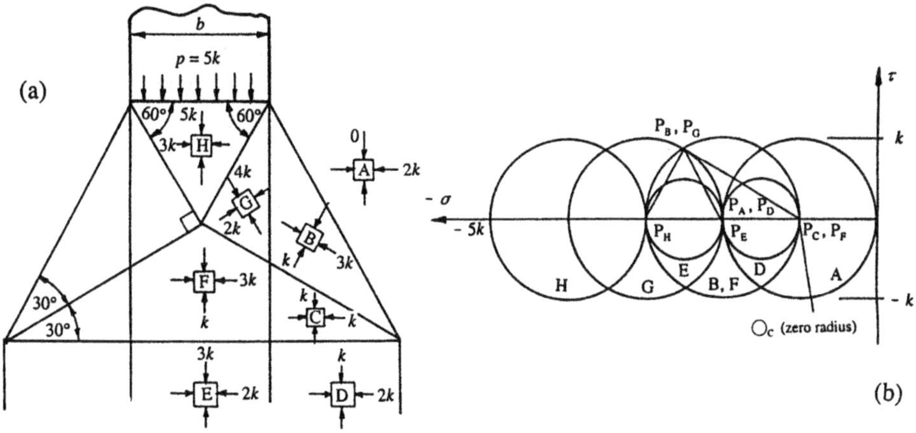

Figure 7.11 Further complex stress discontinuities for flat die indentation

The solution begins with the stress state: $\sigma_1 = 0$ and $\sigma_3 = -2k$, within region A. Following the construction of $\bigcirc_A$ in Fig. 7.11b, the pole point P_B and $\bigcirc_B$ are determined. This gives $\sigma_1 = -k$ and $\sigma_3 = -3k$ for region B. Locating the position of pole P_C reveals that $\sigma_3 = -k$ within region C. It can be seen from route ADC, that poles P_A and P_D are coincident and regions C and D are elastically stressed. That is, the radii of circles $\bigcirc_C$ and $\bigcirc_D$, being zero and $k/2$ respectively, are less than k. Pole P_E is located from the continuous stress state from D to E. Similarly P_F is found from continuity in C to F. The radii of circles $\bigcirc_E$ and $\bigcirc_F$ are $k/2$ and k respectively. Pole P_G follows from P_F and $\bigcirc_G$ gives: $\sigma_1 = -4k$, $\sigma_3 = -2k$. Pole P_H follows from P_G when, finally, $\bigcirc_H$ reveals a die pressure of $p_L = 5k$.

Alternative patterns of stress discontinuities, e.g. that given in Fig. 7.12a, have not served to raise the die pressure above $p_L = 5k$, indicating that this is the true lower-bound solution. Here, the stress plane (Fig. 7.12b) shows elastic material in regions D and G and yielded material elsewhere, when we start with the assumed state in region F.

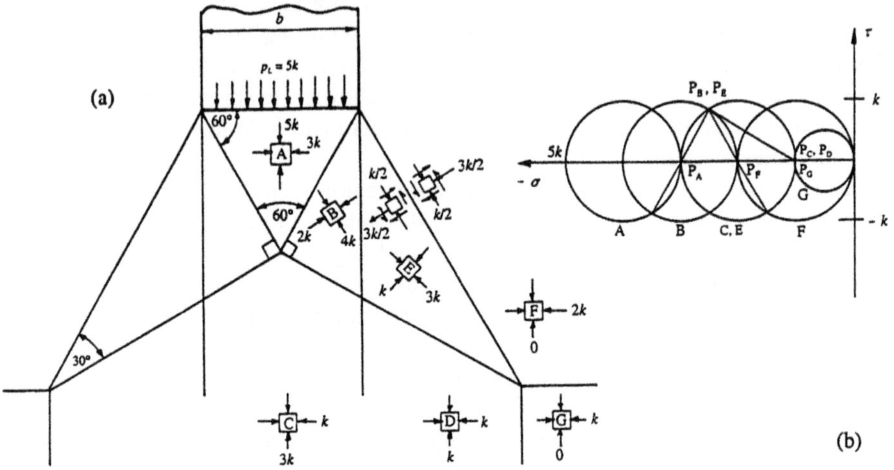

Figure 7.12 Alternative stress discontinuity pattern for flat die indentation

Other principal stress states shown follow from the corresponding circles after locating their pole points in the order of P_F, P_E, P_B P_C, P_D and P_G. A similar pattern of stress discontinuities appears later as a lower bound for extrusion.

7.4.2 Upper Bounds

Upper bound solutions employ patterns of velocity discontinuities. Let the element ABCD in Fig. 7.13 be distorted to A′B′C′D′ after it has crossed the velocity discontinuity YY. The latter could also be a slip line.

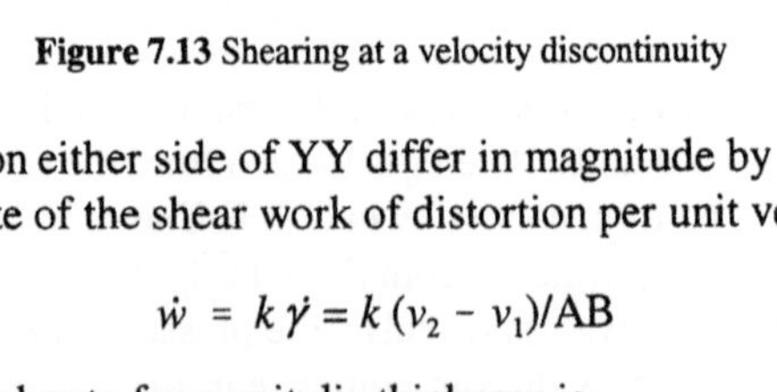

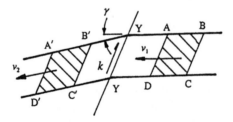

Figure 7.13 Shearing at a velocity discontinuity

The absolute velocities on either side of YY differ in magnitude by $v_2 - v_1$ and in direction by γ, as shown. The rate of the shear work of distortion per unit volume is

$$\dot{w} = k\dot{\gamma} = k(v_2 - v_1)/\text{AB}$$

from which the total work rate for a unit die thickness is

$$\dot{W} = \dot{w}V = k(v_2 - v_1)(\text{AD} \times \text{AB} \times 1) / \text{AB}$$

$$= k(v_2 - v_1)\,\text{AD} = k \times \text{YY} \times v_{YY} \qquad (7.16)$$

where $v_{YY} = v_2 - v_1$ is the magnitude of the velocity discontinuity and AD = YY is the length of the discontinuity line.

Consider two possible upper bound mechanisms for die indentation in Figs 7.14a,b.

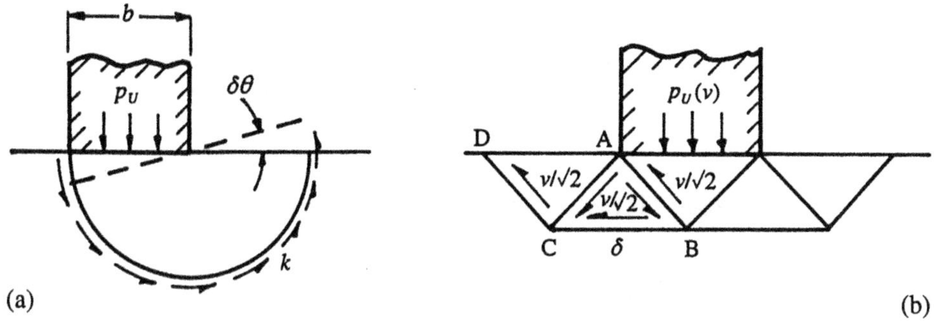

Figure 7.14 Upper bounds for flat die indentation

A simplified rigid rotation $\delta\theta$ of the semi-circular zone of material beneath the die is assumed in Fig. 7.14a. We equate the average external work done by p_U to the work produced by shear yielding on radius b. When rotation $\delta\theta$ occurs in a time δt the tangential velocity discontinuity is $v = b\delta\theta/\delta t = b\,\dot{\theta}$ and eq(7.16) gives, for a die of unit thickness,

$$p_U(b \times 1)(b \times \dot{\theta}/2) = k(\pi b \times 1)(b\,\dot{\theta}) \tag{7.17a}$$

from which $p_U = 2\pi k = 6.28k$, to a first approximation. The true, upper bound will reduce p_U to its minimum value. Consider the rigid motion of five triangular blocks within the SLF in Fig. 7.14b. When the die velocity is v downwards, the resolved velocity discontinuities are as shown. The rate of external work by p_U is equated to the net shear work rate, produced by shear forces acting along the slip lines AB, BC, AC and CD. Equation (7.16) gives, for mirrored slip on both sides of the die, with breadth b and unit thickness:

$$p_U(w \times 1)v = 2k\left\{\left[\frac{w}{\surd 2}\left(\frac{v}{\surd 2} + \frac{v}{\surd 2}\right)\right]_{AB} + (w\,v)_{BC} + \left[\left(\frac{w}{\surd 2}\right)\left(\frac{v}{\surd 2}\right)\right]_{AC} + \left[\left(\frac{w}{\surd 2}\right)\left(\frac{v}{\surd 2}\right)\right]_{CD}\right\} \tag{7.17b}$$

from which $p_U = 6k$. Any further attempt to reduce the upper bound estimate of p_U should aim for the die pressure given by the SLF solution, i.e. $p_d = 5.14k$ from eq(6.28).

7.5 Extrusion

7.5.1 Lower Bounds

A simple, lower-bound estimate for extrusion consists of the two stress discontinuities given in Fig. 7.15a.

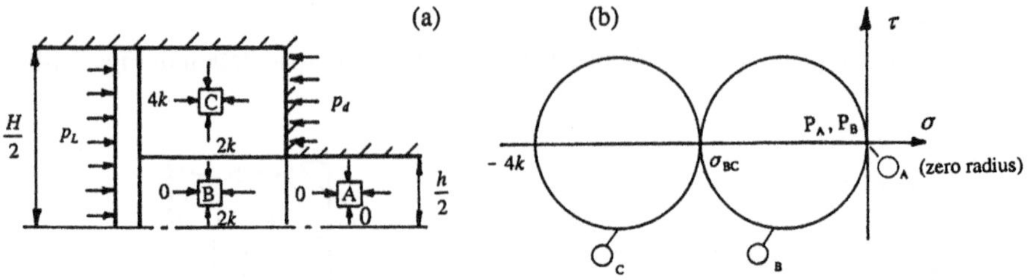

Figure 7.15 Possible stress discontinuity pattern for extrusion

Using the stress plane in Fig. 7.15b, the stress states for the yielded regions B and C are found from the unstressed material in region A, This shows that the die pressure is $p_d = 4k$, from which the lower bound ram pressure p_L becomes

$$4k\,(H - h) \times 1 = (H \times 1)\, p_L$$

$$\frac{p_L}{2k} = 2\left(1 - \frac{1}{R}\right) \tag{7.18}$$

where $R = H/h$. Alexander [4] improved upon eq(7.18), by raising p_L for $R > 3$. The stress discontinuity pattern (Fig. 7.16a) is similar to that given in Fig. 7.12a for flat die indentation. The lettering of uniform regions of stress: A, B, C, D, E, F and G, correspond between these two figures.

(a)

(b)

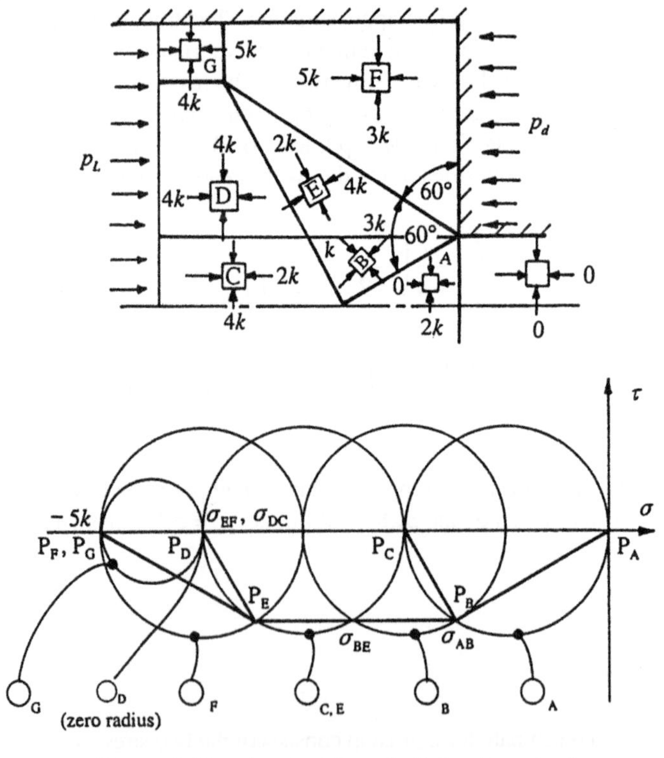

Figure 7.16 Complex stress discontinuity for plane extrusion

However, different stress states apply to each region. Those in Fig.7.16a follow from the stress plane in Fig. 7.16b. It is seen that if the normal stress is to remain continuous between adjacent regions, then $\bigcirc_D$ and $\bigcirc_G$ must have radii of zero and $k/2$ respectively. Within region F, circle $\bigcirc_F$ gives $p_d = 5k$, from which:

$$\frac{p_L}{2k} = 2.5\left(1 - \frac{1}{R}\right) \tag{7.19}$$

7.5.2 Upper Bounds

Upper bound solutions are more useful to manufacturing process design than lower bounds. They are not prone to large overestimations because in practice materials will harden to some degree. A good upper bound solution requires an accurate estimate of the true deformation zone. When slip-line fields are available they may be employed to guide the correct zone shape. Upper-bounding for limit loading is always possible even when SLFs are not available. The procedure then followed is to examine a number of possible velocity discontinuity patterns and select the one which gives the lowest, upper-bound estimate.

Figure 7.17a gives a simple velocity discontinuity pattern for smooth, $R = 2$ extrusion.

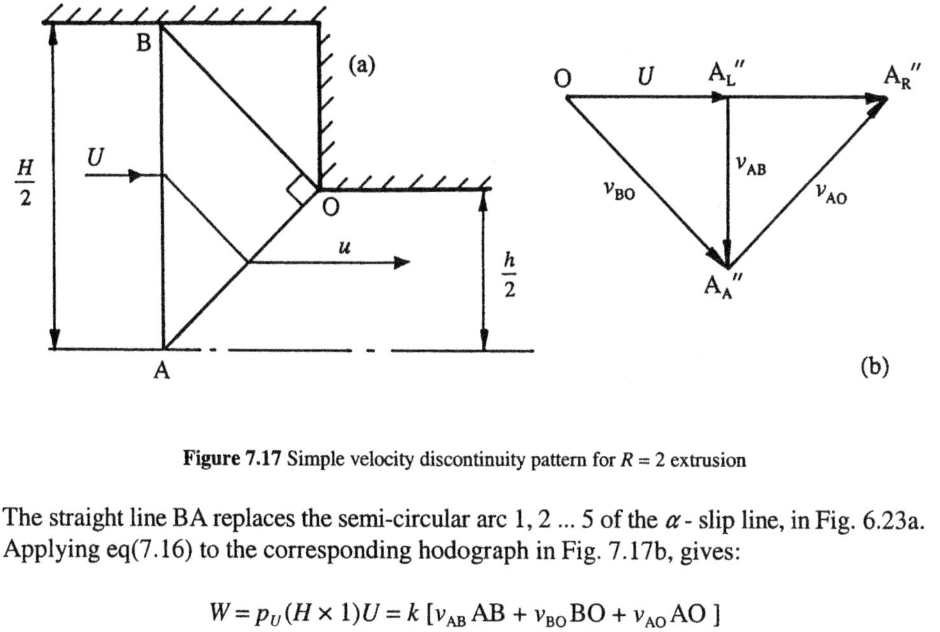

Figure 7.17 Simple velocity discontinuity pattern for $R = 2$ extrusion

The straight line BA replaces the semi-circular arc 1, 2 ... 5 of the α - slip line, in Fig. 6.23a. Applying eq(7.16) to the corresponding hodograph in Fig. 7.17b, gives:

$$W = p_U(H \times 1)U = k\,[v_{AB}\,AB + v_{BO}\,BO + v_{AO}\,AO\,]$$

$$= 2k\left[U\frac{H}{2} + \sqrt{2}U\frac{H}{2\sqrt{2}} + \sqrt{2}U\frac{H}{2\sqrt{2}}\right] = 3k\,UH$$

$$\frac{p_U}{2k} = 1.5$$

which compares with $p_e/(2k) = 1.285$, from the SLF solution (6.17a).

A further pattern of velocity discontinuities for $R = 3$ is given in Fig. 7.18a. Here, the α - slip line's circular arc 12 in Fig. 6.27 has been replaced with a straight line AB. Correspondingly, there results a different hodograph (Fig. 7.18b), which requires the application of eq(7.16) across the full container depth to find the upper bound.

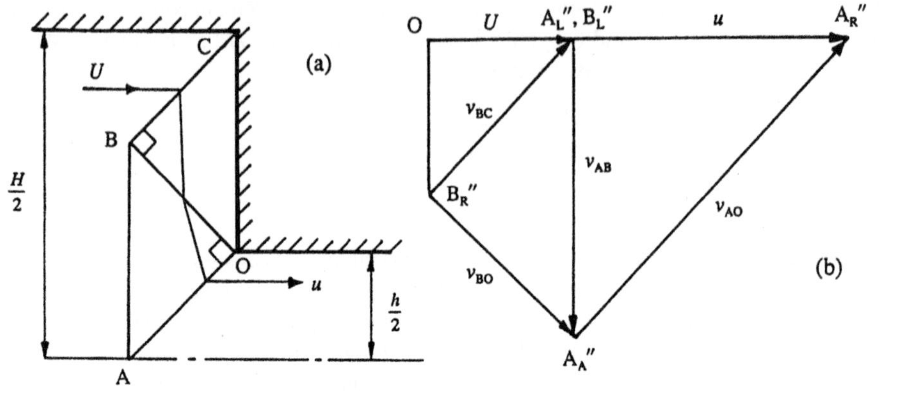

Figure 7.18 Velocity discontinuity pattern for $R = 3$ extrusion

This gives

$$W = p_U(H \times 1)U = k\,[v_{AB}\,AB + v_{BO}\,BO + v_{AO}\,AO + v_{BC}\,BC\,]$$

$$= k\left[\left(2U \times \frac{2H}{3}\right) + \left(\surd 2U \times \frac{\surd 2}{3}H\right) + \left(2\surd 2U \times \frac{\surd 2}{3}H\right) + \left(\surd 2U \times \frac{\surd 2}{3}H\right)\right] = 4kUH$$

$$\therefore \quad \frac{p_U}{2k} = 2$$

which compares with $p_e/2k = 1.71$ from SLF theory in eq(6.22a,b). If there were sticking friction along the inner die wall, an additional velocity discontinuity would arise along OC. Adding $v_{OC}\,OC = U \times 2H/3$ to W, given above, leads to $W = 14kUH/3$ and $p_U/(2k) = 7/3$, which is a 16.7% increase over the frictionless case. With both wall and die friction, we could obtain an upper bound solution from simplifying the corresponding SLF solutions given in Figs 6.34 and 6.35.

Johnson [5] proposed an alternative pattern of velocity discontinuities given in Fig. 7.19a. This provides an upper-bound solution for frictionless extrusion ratios R in a medium range from 3 to 10.

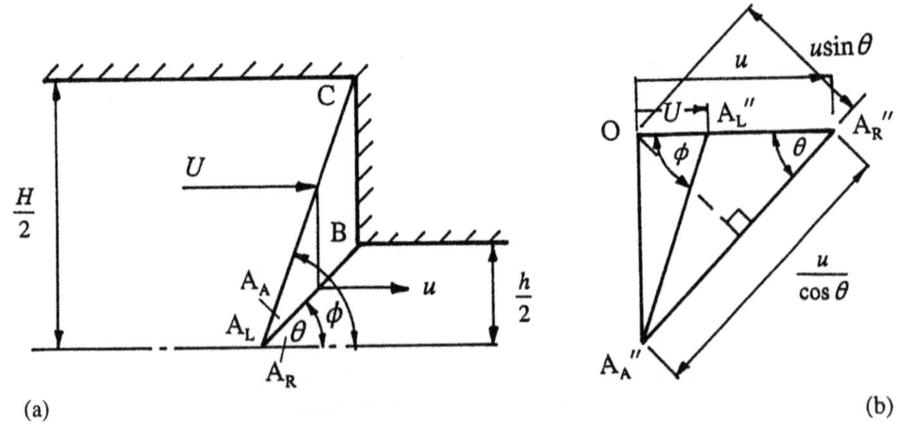

Figure 7.19 Velocity discontinuity pattern for medium R extrusion

The geometry of the physical plane and hodograph in Figs 7.19a,b shows an extrusion ratio:

$$R = \frac{H}{h} = \frac{\tan\phi}{\tan\theta} = \frac{u}{U} \tag{7.20}$$

The rate of external work is $p_U(H \times 1)U$. Working with a unit volume of material extruded in unit time, the specific external work w is simply the extrusion pressure p_U. Let v_N and v_T be the normal and tangential velocities to AB and AC. Equating w to the distortional energy:

$$w = p_U = k\left(\gamma_{AB} + \gamma_{AC}\right)$$

$$p_U = k\left[\left(\frac{v_T}{v_N}\right)_{AB} + \left(\frac{v_T}{v_N}\right)_{AC}\right]$$

$$= k\left(\frac{1}{\cos\theta\,\sin\theta} + \frac{1}{\cos\phi\,\sin\phi}\right)$$

$$= k\left[\frac{(1 + \tan^2\theta)}{\tan\theta} + \frac{(1 + \tan^2\phi)}{\tan\phi}\right] \tag{7.21a}$$

The same result would be found from eq(7.16). Substituting eq(7.20) into eq(7.21a) gives

$$\frac{p_U}{k} = \frac{(1 + R)(1 + R\tan^2\theta)}{R\tan\theta} \tag{7.21b}$$

which has a minimum for $d(p_U/k) = 0$. This gives $R\tan^2\theta = 1$ or $\tan\phi = 1/\sqrt{R}$. Substituting into eq(7.21b), we find

$$\frac{p_U}{2k} = \frac{1 + R}{\sqrt{R}} \tag{7.21c}$$

which is known to overestimate p_U at higher R. Johnson [5] reduced p_U for extrusion ratios exceeding 10 with the alternative velocity discontinuity pattern shown in Fig. 7.20a.

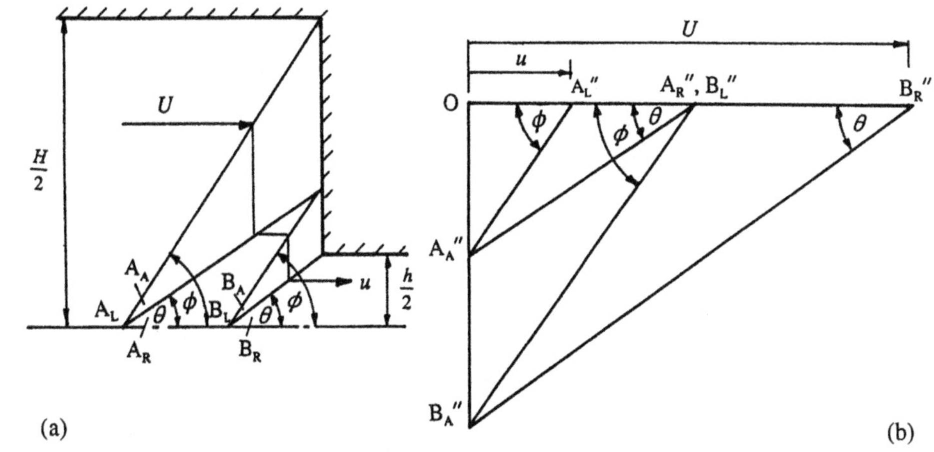

Figure 7.20 Velocity discontinuity pattern for higher R extrusion ratios

For the geometry given in Fig. 7.20a, it follows that

$$R = \frac{H}{h} = \frac{\tan^2\phi}{\tan^2\theta} = \frac{u}{U} \tag{7.22}$$

The hodograph (Fig. 7.20b) shows that the dissipation rate of shear strain energy will be twice that for Fig. 7.19b. That is

$$w = p_U = 2k\left[\frac{(1 + \tan^2\theta)}{\tan\theta} + \frac{(1 + \tan^2\phi)}{\tan\phi}\right] \tag{7.23a}$$

Substituting eq(7.22) into eq(7.23a)

$$\frac{p_U}{2k} = \frac{(1 + \sqrt{R})(1 + \sqrt{R}\tan^2\theta)}{\sqrt{R}\tan\theta} \tag{7.23b}$$

Equation (7.23b) is a minimum for $\sqrt{R}\tan^2\theta = 1$. This gives $\tan^2\theta = 1/\sqrt{R}$, so that

$$\left(\frac{p_U}{2k}\right)^2 = \frac{4(1 + \sqrt{R})^2}{\sqrt{R}} \tag{7.23c}$$

Table 7.1 lists $p/(2k)$ values from the upper and lower-bound estimates given in eqs(7.18), (7.19), (7.21c) and (7.23c).

Table 7.1 Upper and lower bound pressures for given extrusion ratios

R	$p_L/2k$		$p_U/2k$	
	$2(1 - 1/R)$	$2.5(1 - 1/R)$	$(1 + R)/\sqrt{R}$	$[4(1 + \sqrt{R})^2/\sqrt{R}]^{1/2}$
2	1.000	1.250	2.12	4.06
3	1.333	1.666	2.31	4.16
9	1.778	2.222	3.33	4.62
25	1.920	2.400	5.20	5.37
36	1.944	2.430	6.17	5.70

For a given extrusion ratio, the highest lower-bound and the lowest upper-bound are selected from Table 7.1. The extrusion pressure is expected to lie between these two bounds, as exact solutions show. Note that the p_U's in the final column are only appropriate for $R > 30$ and that p_L estimates converge with increasing R.

7.6 Strip Rolling

In the simplest two-high mill in Fig. 7.21a, a plate of original thickness h_1 is rolled to a final strip thickness h_2 by two rolls of radius R. Normally, the thickness at entry is not more than $h_1 = 5$ mm, giving a typical strip width to thickness ratio $w/h_1 \approx 10$, with a roll radius $R \approx 100h_1$. There are a number of approaches that have been used to determine the rolling torque. The elementary plane-strain, slab analysis technique allows for a constant coefficient of friction μ between the deforming material and the rolls. Because plane sections remain plane, the directions of the frictional forces $\mu p(R\delta\phi)$ oppose at positions, A and B, on either side of a neutral plane (broken vertical line in Fig. 7.21a). This plane is neutral since there is no relative movement between roll and strip. In addition, a homogenous compression of the strip thickness is assumed under a constant yield stress. The elastic deformation of the strip is normally ignored, though more exact theories [1, 6] attempt to account for this. Moreover, the roll itself may elastically deform. This becomes important in cold rolling sheet where, under a high radial pressure p, an elastic flattening of the roll increases its radius (see Chapter 14). Here, as with the hot rolling of strip, these effects can be ignored since the radial pressure is far less.

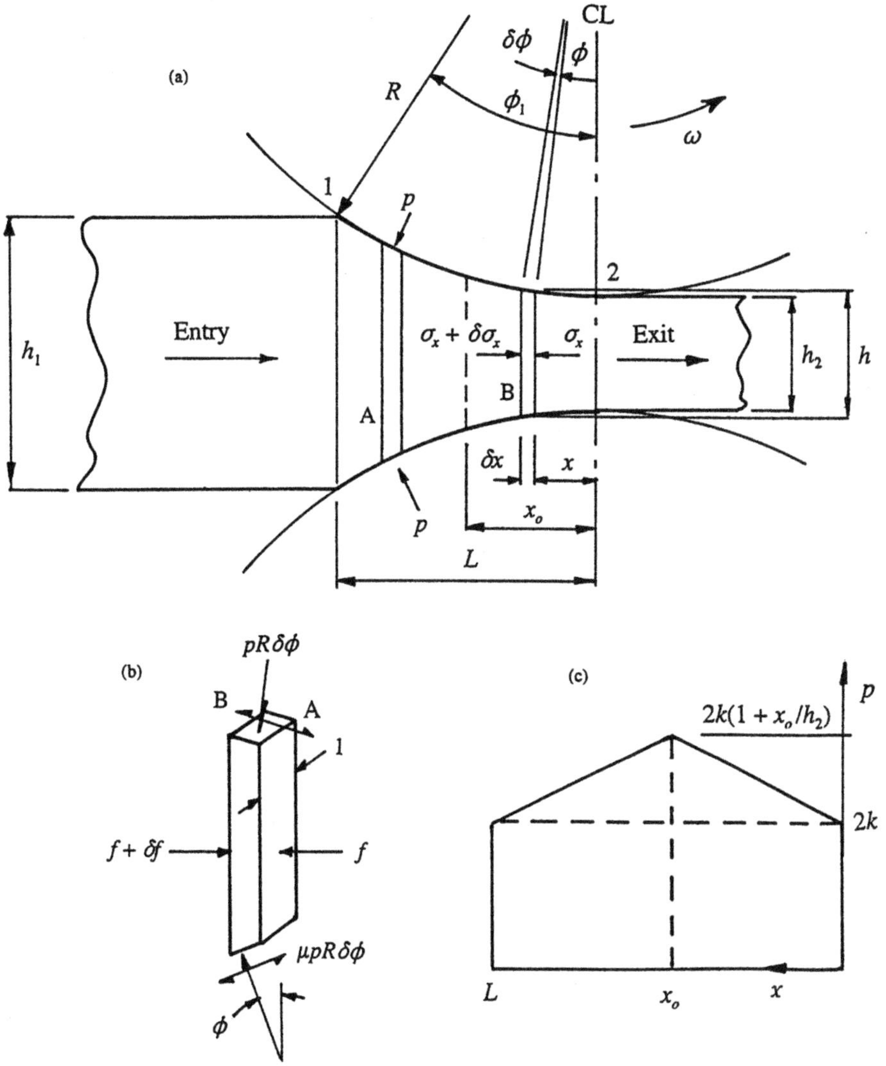

Figure 7.21 Slab analysis for strip rolling of thin, rectangular section

In Fig. 7.21b, the slab forces acting at position B, within the arc of contact, are shown. For the slab at position A the friction forces are reversed as shown. Where σ_x is uniform the horizontal force is $f = \sigma_x h$/unit width. Force equilibrium for each slab's x-direction leads to

$$(f+\delta f)+2\mu pR\delta\phi\cos\phi=f+2pR\delta\phi\sin\phi \quad \text{at A}$$
$$(f+\delta f)=f+2\mu pR\delta\phi\cos\phi+2pR\delta\phi\sin\phi \quad \text{at B}$$

Putting $k=\mu p$ for sticking friction leads to the combined form

$$\delta f=2pR\delta\phi\sin\phi\pm 2kR\delta\phi\cos\phi$$

$$\frac{\mathrm{d}}{\mathrm{d}\phi}(\sigma_x h) = 2(pR\ \delta\phi\sin\phi \pm kR\ \delta\phi\cos\phi) \tag{7.24a}$$

Further equations are required to relate the variables σ_x, p and h before a solution from this elementary theory can be found. The simplest possible relation follows from the assumption that the roll arcs may be replaced with flat plattens. It follows that $h = h_2$ is constant, $\phi = 0$ and $R\delta\phi = \delta x$. Equation (7.24a) becomes

$$h_2\frac{\mathrm{d}\sigma_x}{\mathrm{d}x} = \pm\ 2k \tag{7.24b}$$

where the origin for x lies on the roll's vertical centre line. Integrating eq(7.24b) for the exit side (-) we find

$$\sigma_x = \frac{2kx}{h_2} + C_2 \tag{7.25a}$$

Now, $\sigma_x = 0$ for $x = 0$ and therefore $C_2 = 0$, which reduces eq(7.25a) to

$$\sigma_x = \frac{2kx}{h_2} \quad \text{for} \quad x_o \le x \le 0 \tag{7.25b}$$

It follows that at the neutral plane, where $x = x_o$, eq(7.25b) becomes

$$\sigma_x = \frac{2kx_o}{h_2} \tag{7.25c}$$

Integrating eq(7.24b) for the entry side (+) gives

$$\sigma_x = -\frac{2kx}{h_2} + C_1 \tag{7.26a}$$

where C_1 is found from knowing σ_x on the exit side of the neutral plane. Substituting eq(7.25c) into eq(7.26a) gives

$$\frac{2kx_o}{h_2} = -\frac{2kx_o}{h_2} + C_1 \Rightarrow C_1 = \frac{4kx_o}{h_2}$$

and, hence, eq(7.26a) becomes

$$\sigma_x = \left(\frac{2k}{h_2}\right)(2x_o - x) \quad \text{for} \quad L \le x \le x_o \tag{7.26b}$$

Equations (7.25b) and (7.26b) may now be combined with the plane strain yield criterion, when σ_x and p are taken to be principal stresses. This gives

$$p - \sigma_x = 2k$$

The roll pressures are then

$$p = 2k\left(1 + x/h_2\right) \quad \text{for} \quad 0 \le x \le x_o \tag{7.27a}$$

$$p = 2k\left[1 + (2x_o - x)/h_2\right] \quad \text{for} \quad L \le x \le x_o \tag{7.27b}$$

which are distributed as a 'friction hill'. The area enclosed beneath the hill gives the total roll force/unit width, as

$$P = 2kL\left(1 + \frac{x_o}{2h_2}\right) \tag{7.28}$$

When P is taken to act vertically through the centre of the arc of contact, the torque exerted about the centre line of each roll is simply $PL/2$. The length L of the contact arc is found from the geometry of Fig. 7.21a:

$$\tan\phi_1 = (h_1 - h_2)/L \quad \text{and} \quad \phi_1\,(\text{rad}) \approx L/R$$

The two equations may be combined for small ϕ to give L as

$$L = \sqrt{R\left(h_1 - h_2\right)} \tag{7.29}$$

Thus, from eqs(7.28) and (7.29), the total roll torque (with 2 rolls) becomes

$$T = P\sqrt{R\left(h_1 - h_2\right)}$$

$$= 2k\left[1 + x_o/(2h_2)\right] \times \sqrt{R\left(h_1 - h_2\right)} \tag{7.30a}$$

Equation (7.30a) will provide the drive power $T\omega$, where ω is the roll's angular velocity. Experiment [7] has shown that $x_o \approx L/2$. A factor of 0.8 replaces unity in eq(7.30a) for a similar analysis based upon inclined roll plattens. These modifications give

$$T = 2k\left[0.8 + \frac{\sqrt{R(h_1 - h_2)}}{4h_2}\right] \times \sqrt{R(h_1 - h_2)} \tag{7.30b}$$

Taking, for example, $R = 100$ mm, $h_1 = 10$ mm and $h_2 = 5$ mm in eq(7.30b), gives $T/2k = 42.9$ for T in Nm and k in MPa.

Using their SLF solution as a guide, Johnson and Mellor [8] derived an upper-bound solution to strip rolling based upon the velocity discontinuities given in Fig. 7.22a. It was shown that two discontinuities, composed of the circular arcs AB and BC, were sufficient to ensure that material in contact with the roll rotated with it. Rigid body motion prevails on the entry and exit sides. To construct the discontinuities, an arc BC is drawn tangentially to the roll at exit (where $k = \mu p$) whilst intersecting the strip's principal stress axis (CL) at 45°, Similarly, BC is tangential at A and meets CL at 45°, as shown inset. The chord of arc BC is inclined at 22.5° to the horizontal at point C and its perpendicular bisector locates the centre position $\bigcirc_{BC}$. Similarly, bisecting the chord AB locates the centre, $\bigcirc_{AB}$, of arc AB. The corresponding hodograph is given in Fig. 7.22b.

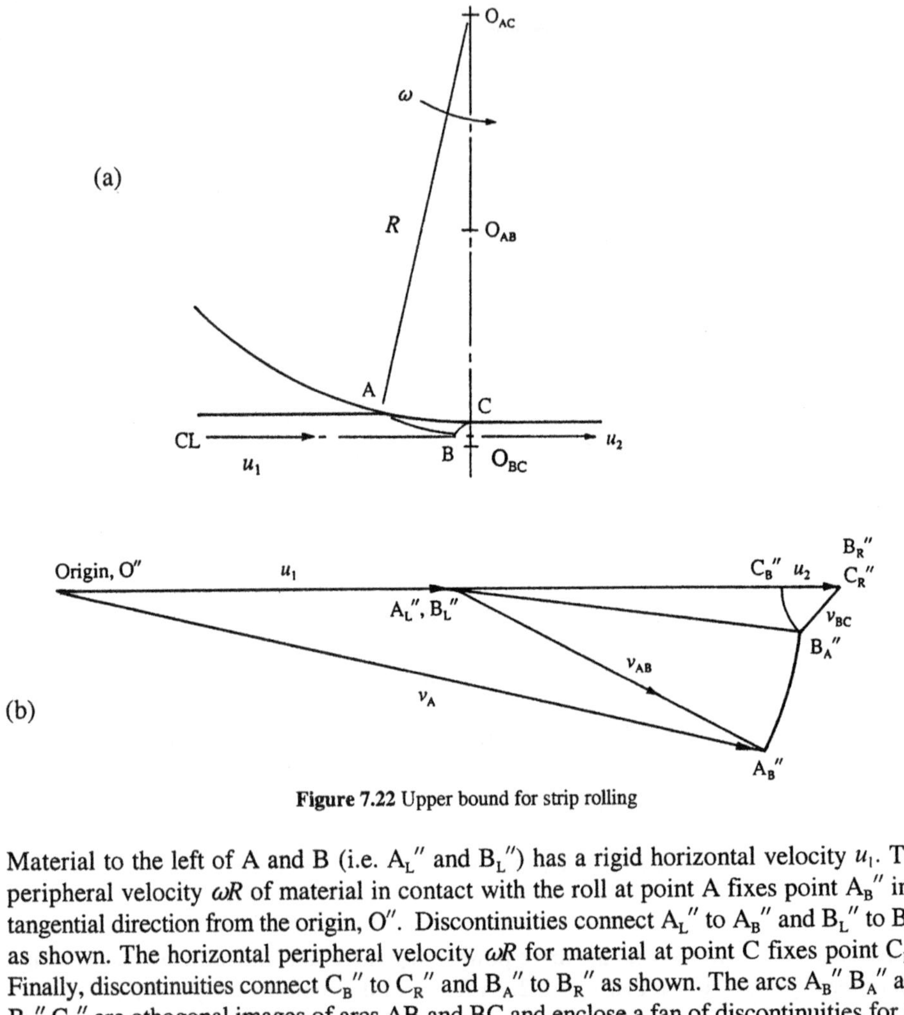

Figure 7.22 Upper bound for strip rolling

Material to the left of A and B (i.e. A_L'' and B_L'') has a rigid horizontal velocity u_1. The peripheral velocity ωR of material in contact with the roll at point A fixes point A_B'' in a tangential direction from the origin, O″. Discontinuities connect A_L'' to A_B'' and B_L'' to B_A'' as shown. The horizontal peripheral velocity ωR for material at point C fixes point C_B''. Finally, discontinuities connect C_B'' to C_R'' and B_A'' to B_R'' as shown. The arcs A_B'' B_A'' and B_A'' C_B'' are othogonal images of arcs AB and BC and enclose a fan of discontinuities for all positions along each arc. The checkpoints B_R'' and C_R'' give $u_2 = (h_1/h_2)u_1$, when the spread in breadth is neglected. To enable a comparison between this upper bound and the solution $T/(2k) = 42.9$ from eq(7.30b), the hodograph is constructed for $R = 100$ mm, $h_1 = 10$ mm and $h_2 = 5$ mm. Applying eq(7.16) gives the total rate of working as

$$W = T\omega = 2k\,[(\text{AB} \times v_{AB}) + (\text{BC} \times v_{BC})]$$

where $\omega = v_A/R$. Substituting for ω and the two discontinuities in terms of u_1, from the scale of the hodograph, leads to a comparable upper bound value $T/(2k) = 48.7$ (same units).

7.7 Transverse Loading of Circular Plates

Up to now, all examples have been plane-strain in nature. We conclude this chapter with a bounding solution to an axially symmetric problem. It is required to find the loading for plastic collapse of circular plates under various lateral loadings. An upper bound to the collapse loading follows from the kinematically admissible velocity field [8].

In Fig. 7.23a, an annular plate of thickness t has an encastre support around its outer radius r_o and carries a distributed force f/unit circumferential length around its inner radius r_i.

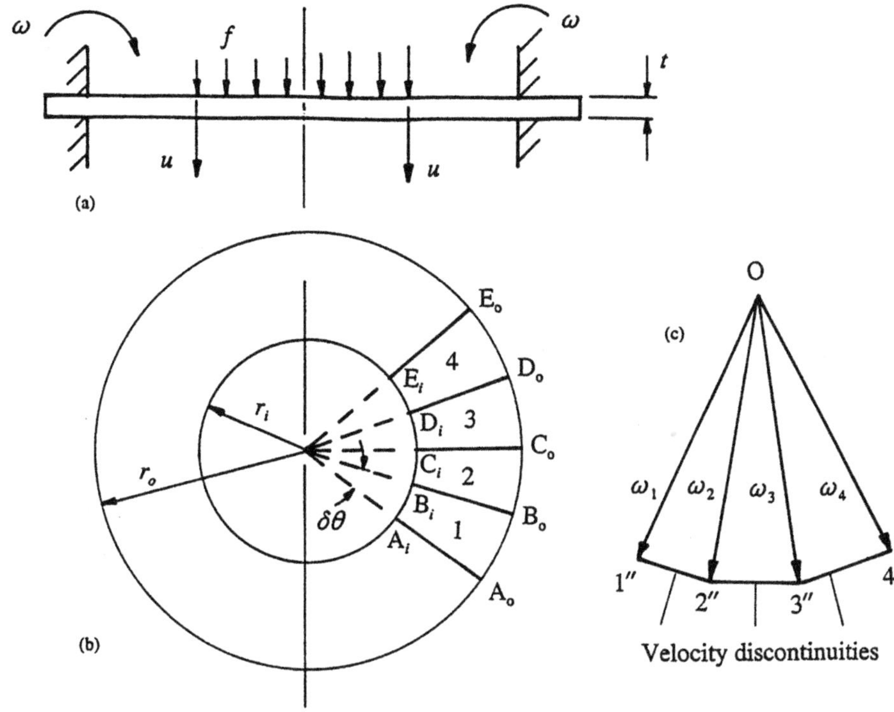

Figure 7.23 Annular, outer encastre plate with inner circumferential loading

One mechanism of plastic collapse involves the formation of radial and circumferential plastic hinges which bound elemental sectors 1, 2, 3, 4 ... etc, as shown in Fig. 7.23b. Plastic hinges A_oB_o and B_iB_o, are formed between sectors 1 and 2. Simultaneously, the hinges B_oC_o and C_iC_o are formed between sectors 2 and 3, etc. As the inner radius r_i descends vertically with velocity u, the sectors rotate rigidly about their respective plastic hinges. That is: sector 1 rotates about A_oB_o, sector 2 rotates about B_oC_o, etc, in an anti-clockwise sense at a constant angular velocity $\omega = u/(r_o - r_i)$. The hodograph in Fig. 7.23c represents these angular velocities as vectors ω_1, ω_2 etc, by applying the right-hand screw rule to each hinge rotation. The hodograph displays velocity discontinuities: 1″2″, 2″3″, 3″4″ etc, across the respective radial hinges: B_iB_o, C_iC_o, D_iD_o, etc. In the limiting case, where $\delta\theta \to 0$, the hodograph becomes a circle of radius ω. At the point of collapse, the rate of external work done $\dot{W}$ is given by

$$\dot{W} = (2\pi r_i) f_U \times u = (2\pi r_i) f_U \times (r_o - r_i)\omega \tag{7.31a}$$

and the rate of internal energy dissipation is

$$\dot{U} = 2\pi r_o \omega M_U + 2\pi (r_o - r_i)\omega M_U \tag{7.31b}$$

In the first term of eq(7.31b), M_U is the upper-bound collapse moment/unit length of the circumferential hinge. In the second term, M_U is expressed per unit length of the radial hinge. The collapse mechanism depends upon equal radial and circumferential moments.

Equating $\dot{W}$ and $\dot{U}$, from eqs(7.31a and b)

$$(2\pi r_i) f_U \times (r_o - r_i)\omega = 2\pi\,\omega M_U (2r_o - r_i)$$

which gives the collapse force f_U/unit of inner circumference, as

$$f_U = \frac{(2r_o - r_i)}{r_i(r_o - r_i)} M_U \qquad (7.32a)$$

The total collapse force becomes

$$F_U = 2\pi r_i f_U = 2\pi \frac{(2r_o - r_i)}{(r_o - r_i)} M_U \qquad (7.32b)$$

With $r_i = 0$ for a solid plate, the central concentrated collapse force is $F_U = 4\pi M_U$. If the outer radius of the annular plate in Fig. 7.23 were to rest upon simple supports, the collapse mechanism is due solely to the formation of radial hinges under the limiting moment M_U. From eq(7.31b), the energy dissipation simplifies to

$$\dot{U} = 2\pi (r_o - r_i)\omega M_U \qquad (7.32c)$$

Equating (7.32c) to eq(7.31a) gives

$$(2\pi r_i) f_U \times (r_o - r_i)\omega = 2\pi (r_o - r_i)\omega M_U$$

from which $f_U = M_U / r_i$ and $F_U = 2\pi r_i f_U = 2\pi M_U$. This total collapse load also applies to an annular plate, simply supported at its inner radius r_i, with a limiting line load $f_U = M_U / r_o$, applied around its outer radius r_o.

Consider next a plate where the inner radius r_i is clamped and the outer radius r_o supports the line load f_U as in Fig. 7.24.

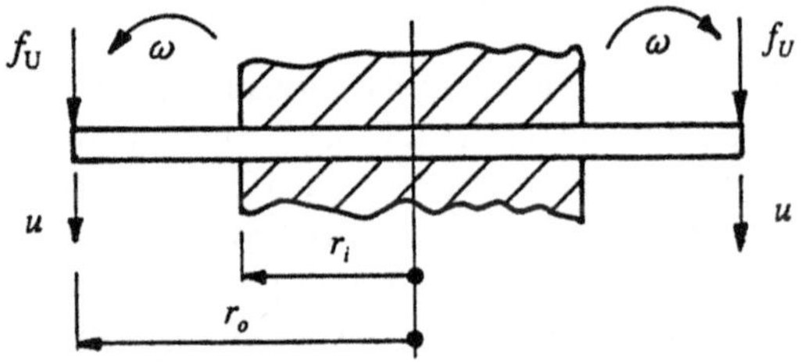

Figure 7.24 Annular plate clamped at inner radius with outer rim circumferential loading

The plate collapses when circumferential and radial plastic hinges form at r_i simultaneously under the limiting moment M_U. The previous boundary conditions are reversed so that we may substitute r_i for r_o in eqs(7.31a,b), to give

$$2\pi r_o f_U u = 2\pi r_i \omega M_U + 2\pi (r_o - r_i)\omega M_U$$
$$2\pi r_o f_U \times (r_o - r_i)\omega = 2\pi\,\omega M_U r_o \qquad (7.33)$$

This leads to the limiting collapse loading:

$$f_U = \frac{M_U}{(r_o - r_i)} \quad \text{and} \quad F_U = 2\pi r_o f_U = \frac{2\pi r_o M_U}{(r_o - r_i)}$$

Where the clamping radius $r_i \to 0$, the total collapse load $F_U \to 2\pi M_u$.

Now, let an annular plate be clamped around its outer edge and carry, instead of line loading, a uniform normal pressure p over its top surface area, as shown in Fig. 7.25.

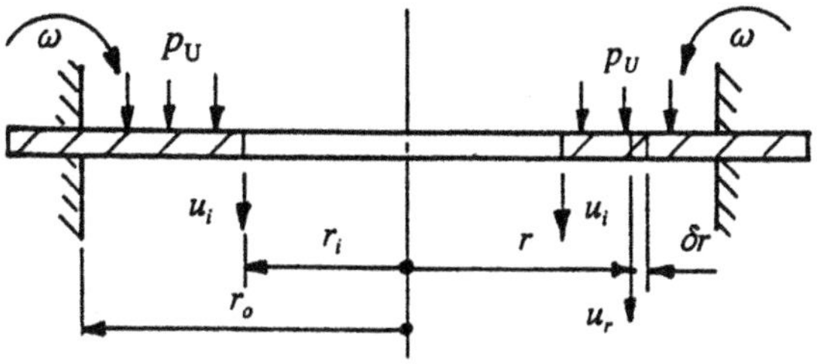

Figure 7.25 Annular plate with outer, encastre fixing and uniform pressure loading

In formulating the external work rate $\dot{W}$, note that the descending velocities u_r are required at all radii: $r_i \le r \le r_o$. Assuming a linear variation in velocity between u_i for $r = r_i$ and zero for $r = r_o$, gives

$$u_r = u_i \frac{(r_o - r)}{(r_o - r_i)} \tag{7.34}$$

At the point of collapse, the external work rate is

$$\dot{W} = \int 2\pi r \, dr \, p_U \times u_r \tag{7.35a}$$

Substituting from eq(7.34) gives

$$\begin{aligned} \dot{W} &= 2\pi p_U u_i \int_{r_i}^{r_o} \frac{(r_o - r)r}{(r_o - r_i)} dr \\ &= \frac{\pi}{3} p_U u_i (r_o - r_i)(r_o + 2r_i) \\ &= \frac{\pi}{3} p_U \omega (r_o - r_i)^2 (r_o + 2r_i) \end{aligned} \tag{7.35b}$$

Equating $\dot{W}$ from eq(7.35b) to $\dot{U}$ from eq(7.31b) gives

$$\tfrac{1}{3}\pi p_U \omega (r_o - r_i)^2 (r_o + 2r_i) = 2\pi \omega M_U (2r_o - r_i)$$

which leads to an upper, limiting normal pressure

$$p_U = \frac{6 M_U (2r_o - r_i)}{(r_o - r_i)^2 (r_o + 2r_i)} \tag{7.36}$$

For a solid plate $r_i = 0$, and eq(7.36) reduces to $p_U = 12M_U / r_o^2$. If the plate in Fig. 7.25 were simply supported around its outer radius, the simplified energy dissipation equation (7.32c) again applies. Equating this to the work rate in eq(7.35b)

$$\tfrac{1}{3}\pi p_U \omega (r_o - r_i)^2 (r_o + 2r_i) = 2\pi (r_o - r_i) \omega M_U$$

gives the upper bound pressure

$$p_U = \frac{6\,M_U}{(r_o - r_i)(r_o + 2r_i)}$$

For a solid, simply supported plate, we take $r_i = 0$ to give $p_U = 6M_U/r_o^2$.

Finally, for the plate in Fig. 7.26, the inner radius r_i is clamped while the remaining surface area is subjected to its uniform collapse pressure p_U, as shown.

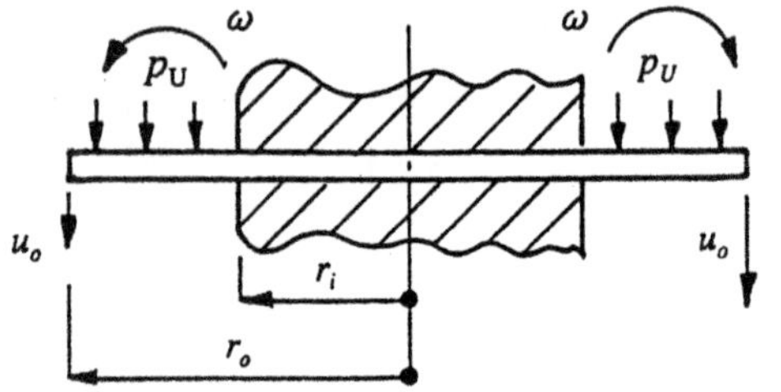

Figure 7.26 Annular plate with inner encastre fixing under uniform pressure loading

This loading configuration reverses the boundary conditions given for Fig. 7.25. Hence, the descent velocity u_r at any radius $r_i \le r \le r_o$, becomes $u_r = u_o(r - r_i)/(r_o - r_i)$. Substituting into eq(7.35a), gives the external work rate

$$\begin{aligned}\dot{W} &= 2\,\pi p_U\,u_o \int_{r_i}^{r_o} \frac{(r - r_i)\,r\,\mathrm{d}r}{(r_o - r_i)}\\ &= \frac{\pi}{3} p_U\,u_o(r_o - r_i)(r_i + 2r_o)\\ &= \frac{\pi}{3} p_U\,\omega\,(r_o - r_i)^2(r_i + 2r_o)\end{aligned} \tag{7.37}$$

The energy dissipation rate is identical to that for the plate in Fig. 7.24. Equation (7.33) gives

$$\tfrac{1}{3}\pi\,p_U\,\omega\,(r_o - r_i)^2(r_i + 2r_o) = 2\pi\,r_i\omega\,M_U + 2\pi\,(r_o - r_i)\,\omega\,M_U$$

$$\tfrac{1}{3}\pi\,p_U\,\omega\,(r_o - r_i)^2(r_i + 2r_o) = 2\pi\,\omega\,M_U\,r_o$$

This supplies the upper limiting pressure as

$$p_U = \frac{6\,M_U\,r_o}{(r_o - r_i)^2\,(r_i + 2r_o)} \tag{7.38}$$

As the clamping radius $r_i \to 0$, eq(7.38) shows that the collapse pressure approaches that for a solid plate, i.e. $p_U \to 3M_U/\,r_o^2$.

7.8 Concluding Remarks

This introduction to limit analysis has demonstrated that an assumption of plastic-rigidity facilitates bounding solutions to the collapse condition in many load-bearing structures. The simplest, upper-bound solution is particularly useful when it is required to obtain, with ease and certainty, the maximum forming forces required. There are many further applications of these techniques, e.g., to forging and machining. These appear in more specialised texts [1, 6, 8, 9, 10], to which the interested reader may refer.

References

1. Ford H. and Alexander J. M. *Advanced Mechanics of Materials*, 1963, Longman.
2. Geirenger H. "Fondements mathematiques de la theorie des corps plastiques isotropes", in *Mem Sci Math*, 1937, 86.
3. Neal B. G. *Plastic Methods of Structural Analysis*, 1964, Chapman and Hall.
4. Alexander J. M. *Quart. Appl.Maths*, 1961, **19**, 32.
5. Johnson W. *Proc. I. Mech. E*, 1959, **173**, 61.
6. Prager W. and Hodge P.G. *The Theory of Perfectly Plastic Solids*, 1951, Wiley, New York.
7. Orowan E. *Proc .I .Mech. E*, 1943, **150**, 140.
8. Johnson W. and Mellor P.B. *Engineering Plasticity*, 1983, Ellis-Horwood.
9. Alexander J. M. and Brewer R. C. *Manufacturing Properties of Materials*, 1963, van Nostrand Co Ltd..
10. Calladine C. R. *Plasticity for Engineers*, 2000, Horwood Pub Ltd.

Exercises

7.1 Obtain upper and lower bounds for a beam with the triangular cut out, base $2a$, height a, shown in Fig. 7.27, when the beam supports a uniform bending moment M. [Ans: $M_L = ka^2$, $M_U = 1.38ka^2$]

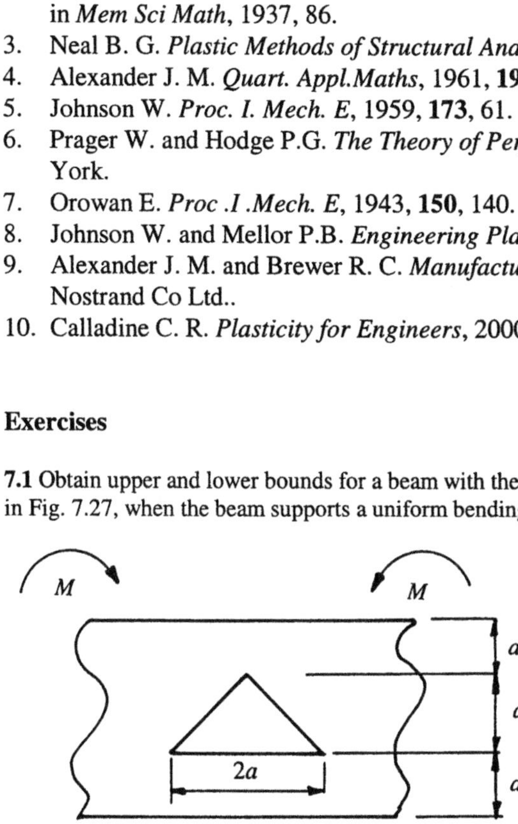

Figure 7.27

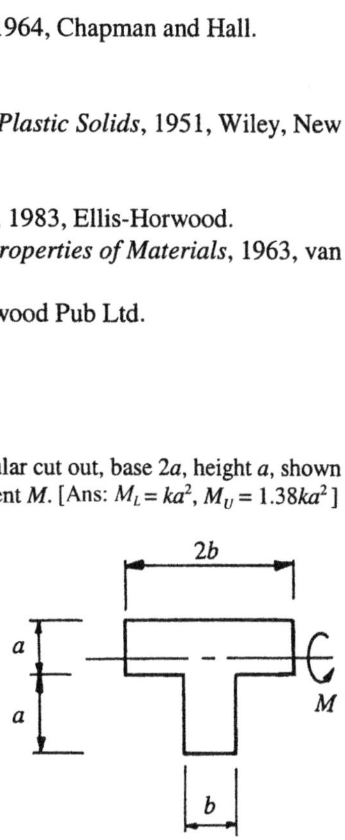

Figure 7.28

7.2 The T-section in Fig. 7.28 comprises the uniform section of a rigid, perfectly-plastic beam with equal tensile and compressive yield stresses, $Y = 2k$. Determine the upper and lower bound conditions for collapse when a bending moment M acts in the sense shown.
[Answer: $M_L = 2.76ka^2b$, $M_U = 2.76ka^2b + 1.38ka^3$]

7.3 Derive, from energy considerations, expressions for the collapse load of the fixed-base portal frame in Fig. 7.29 under: (a) a central vertical load W, (b) a horizontal load P and (c) combined loads W and P, when $P = W/3$.

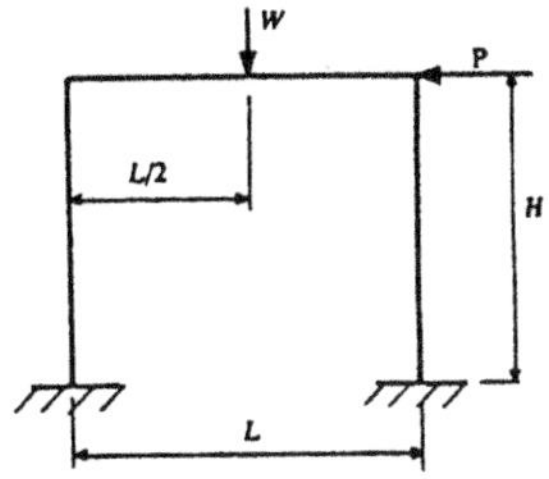

Figure 7.29

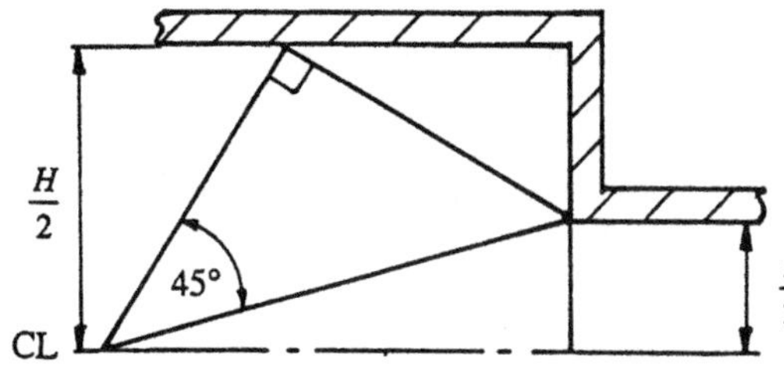

Figure 7.30

7.4 The stress discontinuity pattern in Fig. 7.30 may be used to obtain a lower bound solution to the problem of frictionless plane strain extrusion through a square die. Construct the stress plane and show that the extrusion pressure is given by:

$$\frac{p}{2k} = 2\left[1 - \frac{2R}{\sqrt{(2R - 1)^2 + 1}}\right]$$

where $R = H/h$. Compare the value of $p/2k$ for $R = 3$ withthat from the slip line field (see Fig. 6.27).

7.5 Fig. 7.31 shows ¼ of a testpiece in which the ratio of the notched to un-notched breadths is 2/3. Determine the axial force P required to initiate yield. What is the new position of the notch boundary after the rigid region moves away from the horizontal centre line by 6.5 mm?

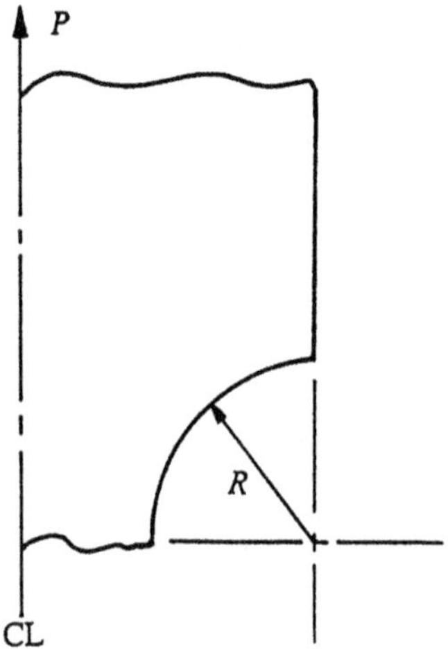

Figure 7.31

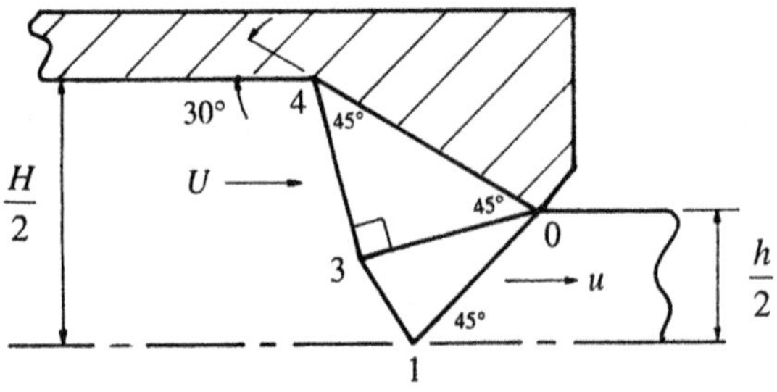

Figure 7.32

7.6 Figure 7.32 shows a velocity discontinuity pattern for frictionless extrusion through an inclined die with a reduction ratio of 2, as derived from an SLF solution (see Fig. 6.31a). Modify the hodograph in Fig. 6.31b to correspond with the straight slip line 13 and from this show that the upper-bound extrusion pressure is $p_U/2k = 0.76$, in agreement with eq(6.24b).

7.7 A square section beam of side a and length L is built in at one end and simply supported at the other end. A vertical load F is applied at a distance x from the built-in end. When the collapse momemt is M, show that F and the position in the length for which F is a minimum become:

$$\frac{F}{M} = \frac{1}{(L - x)} + \frac{2}{x}, \quad \text{where } x = \frac{\sqrt{2}L}{(1 + \sqrt{2})}$$

7.8 A thin, flat triangular plate is fixed in position around its perimeter whilst carrying a normal force F at its centroid. Find an upper bound solution to F in terms of the collapse moment M_U.

7.9 A thin, flat triangular plate is fixed in position along each side whilst carrying a normal pressure p. Find an upper-bound solution to p in terms of the collapse moment M_U.

CHAPTER 8

CRYSTAL PLASTICITY

8.1 Introduction

Much work has been done to increase our understanding of the microstructural mechanisms responsible for plastic flow. It is not the purpose of this chapter to detail all possible dislocation mechanisms by which atomic slip can occur as these have been adequately documented elsewhere [1, 2]. Instead, the discussion is limited to a consideration of those aspects of the micro-mechanics of deformation that lend support to the key elements employed in continuum theories of plasticity. Important among these, as following chapters will show, are the physical interpretation of work hardening, the Bauschinger effect, the shape of the initial and subsequent yield surfaces, the associated flow rule and the manner in which strain within a grain contributes to the observed strain for a polycrystal. The plastic response of single crystals and polycrystals to a tensile stress involves slip with hardening. Anisotropic or directional hardening occurs in a single crystal. A polycrystal with many randomly oriented grains has an aggregated isotropic structure that hardens anisotopically under a directionally imposed plastic strain. However, an isotropic hardening behaviour is often assumed. The two behaviours influence the subsequent yield surface and its manner of representation with the isotropic and kinematic hardening models.

External parameters that influence plastic flow include the environmental conditions and the nature of the loading. Internal, micro-structural parameters involve the orientation of the slip plane, lattice imperfections, alloying and impurity elements as follows:

1. Particular slip systems operating as close-packed planes in close-packed directions, that enable individual crystals to deform plastically.
2. The behaviour of slip mechanisms operating in the presence of existing defects in real lattices. Among these defects are: (i) point defects - interstitials and vacancies, (ii) line defects - dislocations, (iii) plane defects - grain boundaries and (iv) volume defects - cracks and atomic stacking faults.
3. The movement, multiplication and interaction between dislocations and the extent to which these are impeded by other defects acting as obstacles.

The important point to note is that such defects present in metallic crystals facilitate the process of block slip to occur between adjacent planes of atoms at considerably reduced stress levels. It has been observed [1] that the shear stress required for slip is four orders of magnitude lower than a theoretical cohesive strength, calculated from $G/(2\pi)$, in which G is the shear modulus. On a micro-scale, slip is the major contributor to low- temperature plasticity in metals. Other influences such as twinning, void growth and micro-shear banding can contribute to plasticity under certain conditions. For example, void growth occurs in dispersion hardened alloys at higher temperatures and twinning arises in a hexagonal close-packed lattice to enhance the plasticity from slip on its single slip plane. In the main, the contribution to plasticity from these sources will be less than that arising from slip, so these will not be developed further here.

8.2 Resolved Shear Stress and Strain

8.2.1 Simple Tension

In order to understand the nature of hardening in polycrystalline aggregate it is instructive, firstly, to analyse the stress in a single, cylindrical crystal of metal, loaded in tension along it axis (see Fig. 8.1a).

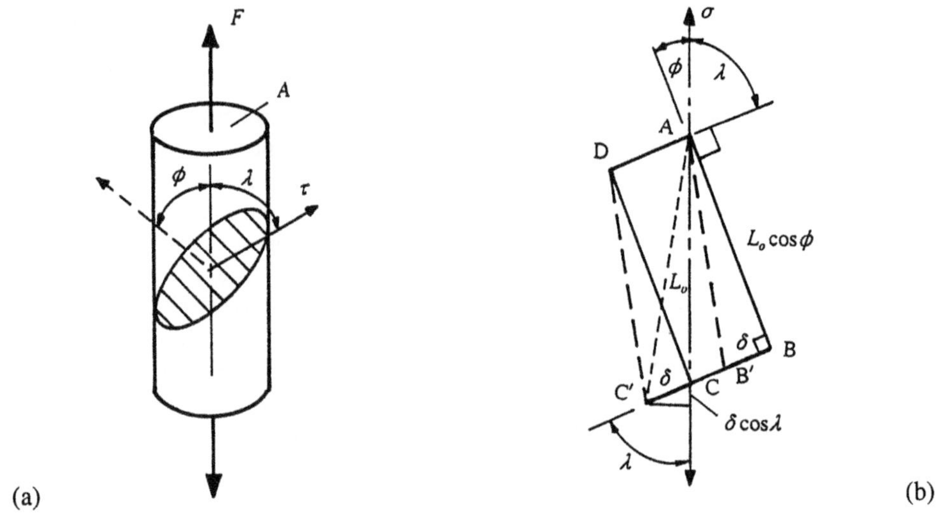

Figure 8.1 Single crystal under tension

Let a tensile force F act normal to section area A of the crystal, to induce a nominal tensile stress: $\sigma = F/A$. The normal to the inclined slip plane and the slip direction make respective angles ϕ and λ with the tensile axis as shown. The resolved shear stress τ is aligned with the slip direction lying in the slip plane. The magnitude of τ is found from dividing that component of F, aligned with the slip direction, by the area of the slip plane. The latter is an ellipse with an area: $A/\cos\phi$. Thus:

$$\tau = \frac{F \cos\lambda}{A / \cos\phi} = \sigma \cos\phi \cos\lambda \tag{8.1}$$

It follows from eq(8.1) that τ is zero for the $\phi = 0°$ and $90°$ principal directions, the former being associated with a cleavage fracture plane. In contrast, τ will reach its maximum value of $\sigma/2$ for a $\phi = 45°$ slip plane. The magnitude of this resolved shear stress that initiates slip upon a 45° plane is called the *critical resolved shear stress* τ_{cr}. Planes with different orientation will slip once τ_{cr} is attained. The corresponding *resolved shear strain*, or glide strain γ, is a more appropriate measure of deformation to use than percentage elongation or nomimal engineering strain. An often-used approximation to γ is applied when the axial strain ε_z is small. Thus, in Fig. 8.1b, ABCD is a rectangular element lying between two parallel slip planes, separated by length L_o along the tensile axis. Let δ equal the amount of slip BB′ and CC′, along the lower slip plane. Within the deformed element AB′C′D, the geometry shows that AC has extended by the amount $\delta\cos\lambda$. Hence, the axial strain is:

$$\varepsilon_z = \frac{\delta}{L_o} \cos\lambda \tag{8.2a}$$

Let AB = $L_o \cos\phi$ be the perpendicular distance between the original slip planes. AB will also rotate towards the tensile axis during deformation. When the rotation is small the glide strain may be defined simply in terms of the angular change to the right angle ∠ DAB. That is, γ becomes the tangent of the angle ∠ BAB′:

$$\gamma = \tan(\text{BAB}') \simeq \frac{\delta}{L_o \cos\phi} \quad (\text{rad}) \tag{8.2b}$$

Combining eqs(8.2a,b) provides the glide strain in terms of the measured axial strain:

$$\gamma \simeq \frac{\varepsilon_z}{\cos\phi \cos\lambda} \tag{8.2c}$$

Honeycombe [2] employed two alternative methods for calculating γ, for larger deformations. These involve the geometry of a normal to the slip plane that rotates towards the tensile axis with increasing strain. Given final orientations ϕ_f and λ_f of a slip plane, the resolved shear strain becomes

$$\gamma = \frac{\cos\lambda_f}{\sin(90 - \phi_f)} - \frac{\cos\lambda}{\sin(90 - \phi)} \tag{8.2d}$$

where ϕ and λ define the initial orientations of this plane (Fig. 8.1a). Alternatively, if the final, nominal axial strain ε_{zf} is measured, γ follows from:

$$\gamma = \frac{1}{\sin(90 - \phi)}\left[\sqrt{(1 + \varepsilon_{zf})^2 - \sin^2\lambda} - \cos\lambda\right] \tag{8.2e}$$

Equations (8.1) and (8.2c-e) define shear stress and shear strain for the slip plane. Their plot will reveal that three distinct regions of hardening accompany slip within a single crystal.

8.2.2 General Stress State

Now, instead of an applied tension, let a general stress tensor σ_{ij} be defined in arbitrary axes x_i (i = 1, 2 and 3). Firstly, σ_{ij} has to be transformed to give the shear stress upon the active slip plane. We may identify this shear stress with component σ_{12}' in eq(1.22b). Let us write this as σ_{12}^s where axes x_1^s and x_2^s lie parallel and normal to an active slip plane s in Fig. 8.2a.

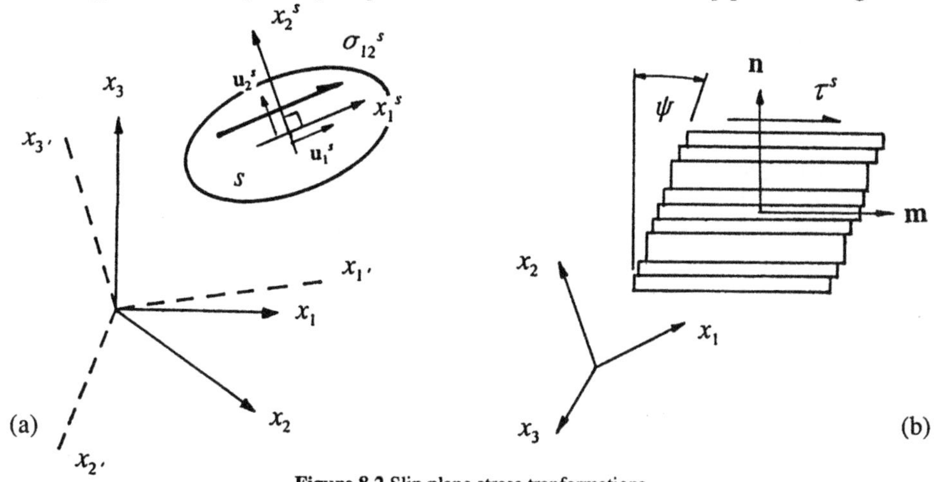

Figure 8.2 Slip plane stress tranformations

Equation (1.22a) gives this shear stress component directly when $i = 1$ and $j = 2$:

$$\sigma_{12}{}^s = l_{1p}{}^s l_{2q}{}^s \sigma_{pq} = (l_{1i}{}^s l_{2j}{}^s)\, \sigma_{ij} = \alpha_{ij}{}^s \sigma_{ij} \tag{8.3a}$$

where the *Schmidt orientation factor* $\alpha_{ij}{}^s = l_{1i}{}^s l_{2j}{}^s$ employs the symmetries of the stress tensor, namely $\sigma_{ij} = \sigma_{ji}$. The Schmidt factor employs a set of direction cosines $l_{1i}{}^s$ for axis $x_1{}^s$ and another set $l_{2j}{}^s$ for axis $x_2{}^s$, with respect to the reference (stress) axes x_i.

It is often more convenient to align reference axes with the cubic axes x_i' ($i = 1, 2$, and 3) of a unit cell, given that the slip plane and slip direction are defined with Miller's indices (see Section 8.3). Referring to Fig. 8.2a, two stress transformations become necessary to find $\sigma_{12}{}^s$: From eq(1.22a), we transform from stress axes x_i to reference (cube) axes x_i':

$$\sigma_{ij}' = l_{ip}\, l_{jq}\, \sigma_{pq}$$

Then, transform from reference axes x_i' to the slip plane:

$$\sigma_{12}{}^s = l_{1p}{}^s\, l_{2q}{}^s\, \sigma_{pq}'$$

These two transformations are combined as follows:

$$\begin{aligned}\sigma_{12}{}^s &= l_{1p}{}^s l_{2q}{}^s \sigma_{pq}' = l_{1p}{}^s l_{2q}{}^s (l_{pr} l_{qs} \sigma_{rs}) \\ &= (l_{1p}{}^s l_{pr}\, l_{2q}{}^s l_{qs})\, \sigma_{rs} = (l_{1p}{}^s l_{pi}\, l_{2q}{}^s l_{qj})\, \sigma_{ij} \\ &= \alpha_{pq}{}^s\, (l_{pi} l_{qj}) \sigma_{ij}\end{aligned} \tag{8.3b}$$

in which the Schmidt factor $\alpha_{pq}{}^s = l_{1p}{}^s l_{2q}{}^s$ again appears. The matrix form of eqs(8.3a) employs column matrices of direction cosines for each unit vector, i.e. $\mathbf{u}_1{}^s = \{l_{11}{}^s\ l_{12}{}^s\ l_{13}{}^s\}^T$ and $\mathbf{u}_2{}^s = \{l_{21}{}^s\ l_{22}{}^s\ l_{23}{}^s\}^T$ (see Fig. 8.2a). Within eq(8.3b), the transformation matrix $\mathbf{L}$ between axes x_i and x_i' must also appear. Equations(8.3a,b) become, respectively:

$$\sigma_{12}{}^s = (\mathbf{u}_1{}^s)^T\, \mathbf{T}\, \mathbf{u}_2{}^s \quad \text{and} \quad \sigma_{12}{}^s = (\mathbf{u}_1{}^s)^T\, (\mathbf{L}\, \mathbf{T}\, \mathbf{L}^T)\mathbf{u}_2{}^s \tag{8.4a,b}$$

where $\mathbf{T} \equiv \sigma_{ij}$ is the Cauchy stress matrix (1.17). For simplicity, we now replace $\sigma_{12}{}^s$ with τ^s and refer unit vectors **n** and **m**, for the normal and slip directions respectively, to cube reference axes x_i, as shown in Fig. 8.2b. Equations (8.3a) and (8.4a) become:

$$\tau^s = m_i n_j \sigma_{ij} = \alpha_{ij}{}^s \sigma_{ij} = \mathbf{m}^T\mathbf{T}\mathbf{n} \tag{8.5a}$$

where, in the matrix notation, m_i and n_j are direction cosine components of column vectors $\mathbf{m} = \{m_1\ m_2\ m_3\}^T$ and $\mathbf{n} = \{n_1\ \ n_2\ \ n_3\}^T$ in cube axes x_i. According to Schmidt's law, it follows from eq(8.5a) that slip occurs when the resolved shear stress attains a critical value:

$$\tau^s = \alpha_{ij}{}^s \sigma_{ij} = \tau_{cr}{}^s \tag{8.5b}$$

In Chapter 1 the identical nature of stress and strain transformations was revealed. Consequently, the shear strain $\gamma^s = \tan\psi$ in Fig. 8.2b may be transformed to the cubic axes x_i by an inverse transformation:

$$\varepsilon_{ij}{}^P = \tfrac{1}{2}\,(n_i m_j + m_i n_j)^s\, \gamma^s = \alpha_{ij}{}^s\, \gamma^s \tag{8.6a}$$

where $\alpha_{ij}^{s} = ½\,(n_i m_j + m_i n_j)^s = ½\,(\mathbf{nm}^T + \mathbf{mn}^T)^s$. The slip plane orientation will be altered by the rotation component of the shear distortion. Figure 8.3a shows plane shear distortions e_{12} and e_{21} from which the shear strain $\gamma = \varepsilon_{12}^{P} + \varepsilon_{21}^{P}$ is derived. These are accompanied by plastic rotation (spin) ω_{12}^{P} and ω_{21}^{P} of the crystal body about the 3-axis (see Fig. 8.3b).

(a) 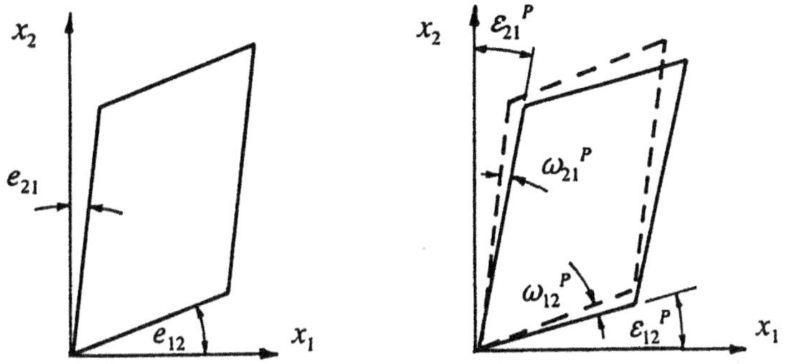 (b)

Figure 8.3 Decomposition of shear distortions

The decomposition shown conforms to eq(2.2b). In general, the 3D rotations are found from dyadic products of the unit vectors **m** and **n** as:

$$\omega_{ij}^{P} = ½\,(n_i m_j - m_i n_j)^s\,\gamma^s = \beta_{ij}^{s}\,\gamma^s \tag{8.6b}$$

where $\beta_{ij}^{s} = ½\,(n_i m_j - m_i n_j)^s = ½\,(\mathbf{nm}^T - \mathbf{mn}^T)^s$.

Where multiple slip systems, $s = 1, 2, 3 \ldots N$ exist within a single crystal they contribute to the overall observed strain in the cubic reference frame. We must sum incremental plastic strain components from eq(8.6a), as follows:

$$\mathrm{d}\varepsilon_{ij}^{P} = \sum_{s=1}^{N} \alpha_{ij}^{s}\,\mathrm{d}\gamma^s = \frac{1}{2}\sum_{s=1}^{N}\left(n_i m_j + m_i n_j\right)^s \mathrm{d}\gamma^s \tag{8.7}$$

For example, with $N = 5$ slip systems, eq(8.7) gives:

$$\begin{aligned}
\mathrm{d}\varepsilon_{11}^{P} &= (n_1 m_1)^1 \mathrm{d}\gamma^1 + (n_1 m_1)^2 \mathrm{d}\gamma^2 + \cdots + (n_1 m_1)^5 \mathrm{d}\gamma^5\\
\mathrm{d}\varepsilon_{22}^{P} &= (n_2 m_2)^1 \mathrm{d}\gamma^1 + (n_2 m_2)^2 \mathrm{d}\gamma^2 + \cdots + (n_2 m_2)^5 \mathrm{d}\gamma^5\\
\mathrm{d}\varepsilon_{33}^{P} &= -\,(\mathrm{d}\varepsilon_{22}^{P} + \mathrm{d}\varepsilon_{22}^{P})\\
\mathrm{d}\varepsilon_{12}^{P} = \mathrm{d}\varepsilon_{21}^{P} &= (n_1 m_2 + m_1 n_2)^1 \mathrm{d}\gamma^1 + (n_1 m_2 + m_1 n_2)^2 \mathrm{d}\gamma^2 + \cdots + (n_1 m_2 + m_1 n_2)^5 \mathrm{d}\gamma^5\\
\mathrm{d}\varepsilon_{13}^{P} = \mathrm{d}\varepsilon_{31}^{P} &= (n_1 m_3 + m_1 n_3)^1 \mathrm{d}\gamma^1 + (n_1 m_3 + m_1 n_3)^2 \mathrm{d}\gamma^2 + \cdots + (n_1 m_3 + s_1 m_3)^5 \mathrm{d}\gamma^5\\
\mathrm{d}\varepsilon_{23}^{P} = \mathrm{d}\varepsilon_{32}^{P} &= (n_2 m_3 + m_2 n_3)^1 \mathrm{d}\gamma^1 + (n_2 m_3 + m_2 n_3)^2 \mathrm{d}\gamma^2 + \cdots + (n_2 m_3 + m_2 n_3)^5 \mathrm{d}\gamma^5
\end{aligned}$$

in which plastic incompressibility $\mathrm{d}\varepsilon_{ii}^{P} = 0$ and strain symmetry $\mathrm{d}\varepsilon_{ij}^{P} = \mathrm{d}\varepsilon_{ji}^{P}$ apply. Mathematically, the 5 independent strain components may be taken to arise from 5 independent slip systems. In a real crystal lattice more or less systems may operate depending upon its atomic arrangement.

Finally, the incremental work done by the stress tensor must equal the work done in shearing the slip plane. To show this we use eqs(8.5a) and (8.6a) to express the specific work done by the applied stress tensor:

$$\sigma_{ij} \times \mathrm{d}\varepsilon_{ij}^{P} = \tau^s (\alpha_{ij}^{s})^{-1} \times \alpha_{ij}^{s}\,\mathrm{d}\gamma^s = \tau^s \times \mathrm{d}\gamma^s \tag{8.8}$$

in which $m_i n_j = ½ (n_i m_j + m_i n_j)$. It is seen that the final term in eq(8.8) is the specific work done upon the slip plane, i.e. the product of its resolved shear stress and shear strain.

8.3 Lattice Slip Systems

Schmidt [3] showed experimentally that τ_{cr} is almost constant for a pure metal. This means that for a crystal of high purity, τ_{cr} is independent of the orientation of the slip plane. Equation (8.1) shows that it is possible to vary σ, ϕ and λ and so maintain τ_{cr} constant. The active slip plane will depend upon the atomic packing within a unit cell of a metallic lattice.

Three *Miller indices* are used to identify a slip plane. The indices are derived from the intercepts between the plane and orthogonal cube axes *x*, *y* and *z*. Reciprocals of the intercepts are then taken and any resulting fractions cleared. Thus, for the shaded plane (110), shown in Fig. 8.4a, the intercepts are 1, 1 and ∞, for which their reciprocals become 1, 1 and 0. Brackets are placed around the Miller indices as follows: ( ) to denote all planes parallel to this and { } for all planes of that type.

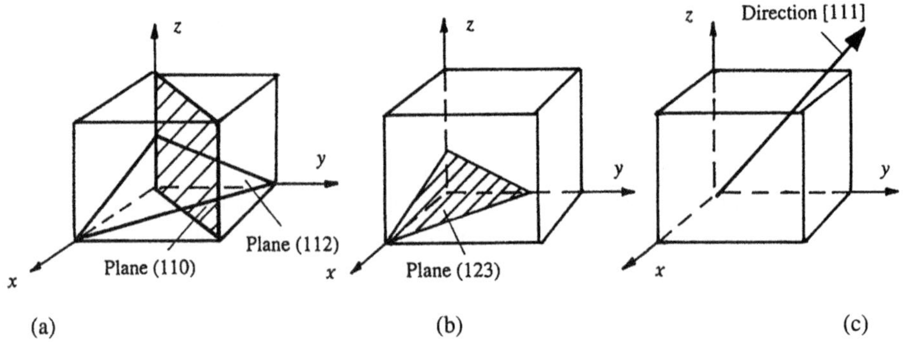

Figure 8.4 Slip planes and directions

Further examples of planes (112) and (123) are shown in Figs 8.4a,b. Three further Miller indices are used to denote a slip direction within a plane. Firstly, the direction vector must be translated to pass through the origin of *x*, *y* and *z* in the sense required. Direction indices become integer co-ordinates for the vector's tip. For example, Fig. 8.4c shows a [111] slip direction. Here, brackets [ ] denotes all parallel directions and ⟨ ⟩ denote all directions of that type. Metals such as Fe, Mo and W, with body-centred cubic (b.c.c.) cells, have no obvious close-packed planes. It is found that plastic flow may activate one or more of twelve slip planes each with slightly different atomic packing density. Of these, three planes in particular: {110}, {112} and {123} (see Fig. 8.4a,b), most commonly activate slip in ⟨111⟩ close-packed directions. With four ⟨111⟩ slip directions and twelve slip planes per b.c.c. cell, there are a maximum of 48 possible slip systems. However, many of these systems are extremely sensitive to temperature and impurity and may not operate. Consequently, a b.c.c. structure can be ductile at high temperature and brittle at low temperature. It follows that the critical resolved shear stress τ_{cr}, required to initiate b.c.c. yielding, is extremely structure sensitive.

Slip in f.c.c.and h.c.p lattices is better defined than in b.c.c. The (111) plane in Fig. 8.5a has three slip directions: $[10\bar{1}]$, $[0\bar{1}1]$ and $[\bar{1}10]$, where a bar denotes a negative intercept.

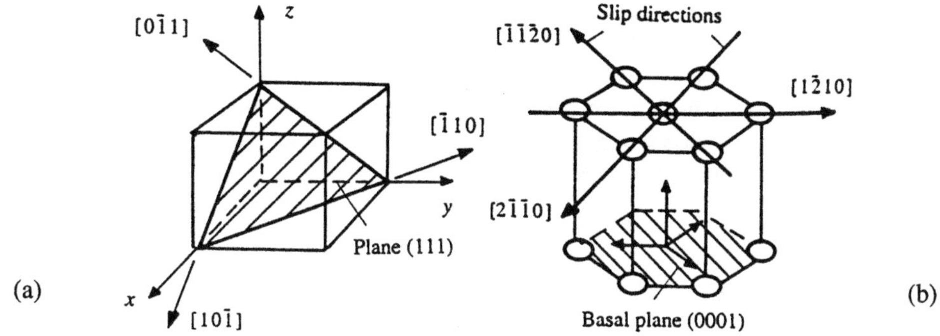

Figure 8.5 Slip planes for f.c.c.and h.c.p. cells

The square brackets shown embrace parallel directions to each of the three shown, for which we could contain them all as $\langle 1\,0\,\bar{1}\rangle$, indicating directions of the $1\,0\,\bar{1}$ type. In fact, Fig. 8.5a applies to a face-centred cubic (f.c.c.) cell, typical of Al, Cu, Ag, Au and Pt. Here, we say that slip occurs on close-packed octahedral $\{111\}$ planes, in close-packed $\langle 1\,0\,\bar{1}\rangle$ directions. Now, as there are four $\{111\}$ planes within a single f.c.c. cell and three slip directions, it follows that there are twelve slip systems by which an f.c.c. metal can deform. These many possibilities for slip account for the ductile nature of an f.c.c. metal.

Planes and directions in a hexagonal close-packed (h.c.p.) structure are specified with four, *Miller-Bravais indices*. A given plane is identified from the reciprocals of its intercepts with four axes originating from the basal plane, shown shaded in Fig. 8.5b. Three axes lie in this plane at 120° and a fourth axis lies perpendicular to it. Directions are again specified with their lowest integer co-ordinates. Slip within a h.c.p. cell can only occur on the close-packed basal plane (0001) in close-packed $\langle 2\,\bar{1}\,\bar{1}\,0\rangle$ directions. The latter includes the three directions shown: $[2\,\bar{1}\,\bar{1}\,0]$, $[\bar{1}\,2\,1\,0]$ and $[\bar{1}\,\bar{1}\,\bar{2}\,0]$. Since there is only one basal plane, containing three slip directions per cell (see Fig. 8.5b), the number of slip systems is restricted to three. As a consequence, h.c.p. metals: Be, Cd, Mg, Ti, Zn and Zr, are less ductile. However, h.c.p metals can also deform by twinning, in which a mirror of homogenous shear occurs across a band of atomic planes, as shown in Fig. 8.6.

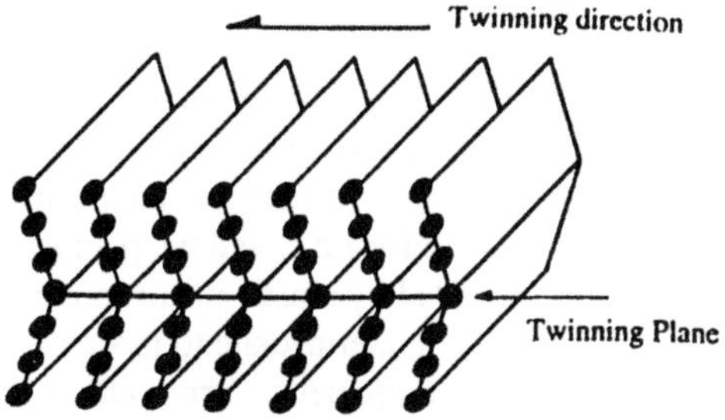

Figure 8.6 Mechanism of twinning

For h.c.p: $\{1\,0\,\bar{1}\,2\}$ defines the twinning planes and $\langle 1\,0\,\bar{1}\,\bar{1}\rangle$ defines the twinning directions. Deformation twinning is less common in a b.c.c. lattice, being restricted to respective planes and directions: $\{112\}$ and $\langle 111\rangle$. Twinning is rarely seen in a deforming f.c.c. lattice, though it has been observed for {111} planes and in ⟨112⟩ directions with low-temperature deformation of copper. The twinning mechanism requires that a critical shear stress displaces adjacent planes by an amount less than the atomic spacing. These sum to yield large shear displacements on layers furthest from the twin plane. The effect of twinning is to rotate the basal plane back following its forward rotation to accommodate slip upon this plane. The reversal produces a geometric softening of a lattice which would otherwise require an incresed shear stress to produce further slip.

8.4 Hardening

8.4.1 Single Crystals

Figures 8.7a and b shows the extent to which the critical resolved shear stress for various cells is dependent upon temperature and purity.

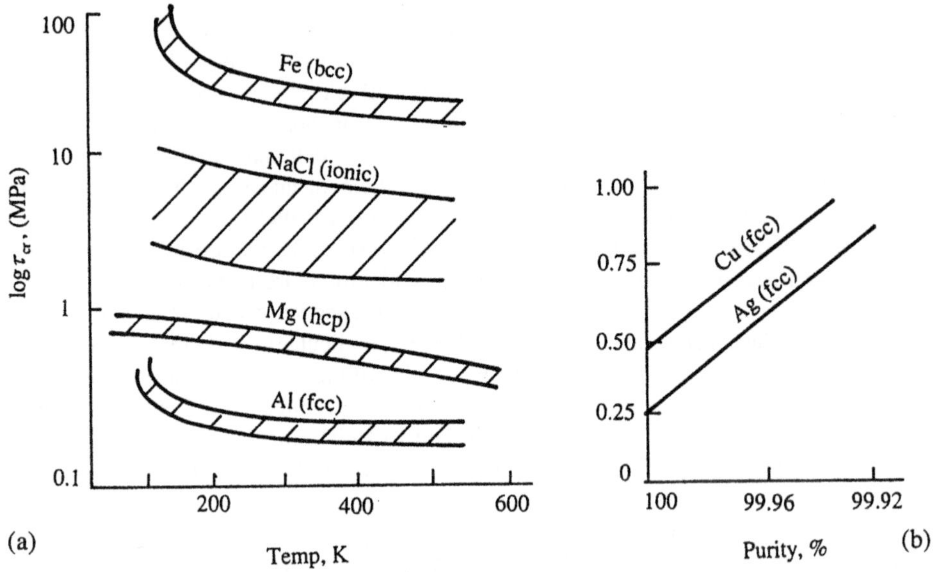

Figure 8.7 Dependence of τ_{cr} upon temperature and purity

Figure 8.7a shows that the resolved shear strength of a b.c.c. metal is far greater than both f.c.c. and h.c.p. metals over a wide temperature range. At room temperature, a typical value for a b.c.c. structure is $\tau_{cr} = 50$ MPa. For f.c.c and h.c.p. structures, $\tau_{cr} \approx 1$ MPa. Despite the even lower values of τ_{cr}, found for f.c.c. and h.c.p. crystals of super purity (Fig. 8.7b), the initial yield point is still preceded by elastic distortion in these structures. The strength of an ionic, NaCl crystal structure is intermediate to those of h.c.p and b.c.c. These results apply only to single crystals with these cells. The scatter band in Fig. 8.7a is due to small variations in purity, the sensitivity of which is revealed for two f.c.c. metals in Fig. 8.7b. It is important to emphasise that τ_{cr} applies to a slip system which is the first to activate under an applied tensile stress. Equation (8.1) identifies this with the slip system that is the first to reach the critical value of the Schmidt factor: $m = \tau_{cr}/\sigma = \cos\phi \cos\lambda$. With the large number of slip systems available in b.c.c. and f.c.c cells, it is possible for two or more systems to operate together as τ_{cr} is attained simultaneously. This is apparent from measurements of the resolved shear stress required to maintain subsequent plastic deformation. For example, Fig. 8.8a shows that two types of behaviour, A and B, are possible for a deforming, f.c.c single crystal.

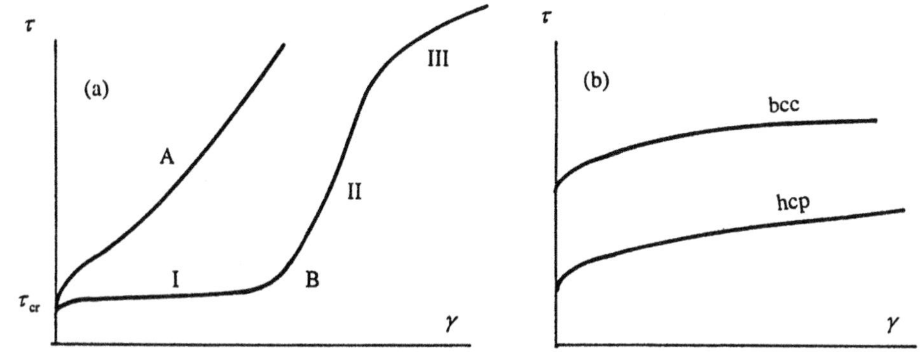

Figure 8.8 Hardening in a single cells: (a) f.c.c., (b) b.c.c and h.c.p.

Both curves commence at the same τ_{cr} value, typically 1-10 MPa, depending upon the metal purity. Subsequent flow behaviour is controlled by the number of active slip systems. For curve A in Fig. 8.8a, several slip systems operate simultaneously, leading to a high rate of work hardening $d\tau/d\gamma$. For curve B, a single slip system operates during stage I at a very low, constant rate of work hardening. The rate and duration of this linear stage depend strongly upon the Schmidt factor m. The occurrence of multiple slip on intersecting slip planes promotes stage II, where the rate of hardening has increased dramatically to attain a constant: $d\tau/d\gamma \approx G/300$. This is followed by stage III in which $d\tau/d\gamma$ falls with increasing resolved shear strain γ. The resolved shear stress τ that initiates stage III depends upon the temperature, purity and the rate of straining, similar to the critical τ_{cr} to commence stage I. Stage III is also associated with thermally activated dynamic recovery. The recovery requires a sufficient density of dislocations to be developed within stage II before it can occur. Recovery processes, climb and cross slip then become active, so reducing the dislocation density and softening the material.

A three-stage curve may not result from continued slip within other cells. For a h.c.p. cell, slip is confined to the basal plane (see Fig. 8.5b) and so it will remain in stage I, as shown in Fig. 8.8b. However, it is possible for dislocations to interact between the basal {0001} and other {1211} planes in certain h.c.p. metals. Dislocation interaction presents sessile barriers to slip and can result in a three-stage curve, similar to that for a f.c.c. cell. Moreover, with twinning present in an h.c.p. cell, there will result a discontinuous (serrated) stress-strain curve. Twinning can also arise in a b.c.c. cell but subsequent slip normally remains in stage I under a low rate of work hardening. The latter is maintained by each of the many slip systems for this cell operating sequentially as they are activated under τ_{cr} by an increasing applied stress. The b.c.c. stage I stress levels are higher that those for h.c.p. (see Fig. 8.8b) and may be high enough to promote a brittle fracture at low temperatures.

8.4.2 Polycrystals

Figure 8.9 shows that the slip directions within individual grains of a polycrystal are randomly oriented to the axis of tensile stress σ.

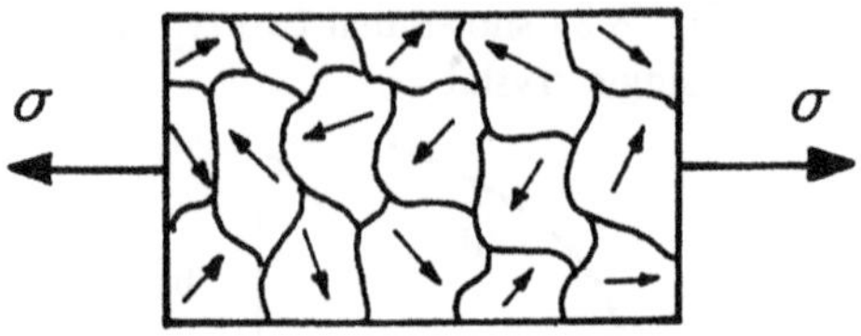

Figure 8.9 Grains in a polycrystalline material

Slip-induced plasticity within a polycrystal is similar to that for a single crystal but, because the grain orientations differ, the Schmidt factor will be high for some grains and low for others. Therefore, before gross yielding by slip can occur, the applied stress must be raised to produce yield in grains with initially low Schmidt factors. In addition, there are constraints imposed by surrounding grains upon the slip within an individual grain. Consequently, the flow stresses in polycrystals are substantially higher than those for single crystals. For example, in annealed copper the respective yield stresses are 3.5 and 0.65 MPa. It is often assumed that polycrystalline flow behaviour is an average of stages I - III (Fig. 8.8a) when they co-exist simultaneously within many differently oriented grains. Increasing the stress beyond the initial yield value activates further slip planes because the increased slip produced in one grain must be relieved by slip in its neighbours. Without the relief necessary to maintain continuity of slip, cracking would occur. The grain boundary presents a barrier to slip which effectively raises both the initial and subsequent flow stresses. The initial yield stress Y increases with decreasing grain size d according to Petch's relation [4]

$$Y = A + B / \sqrt{d} \tag{8.9}$$

where A and B are constants. Also, the rate of macro-hardening $\mathrm{d}\sigma/\mathrm{d}\varepsilon^{P}$ is greater for a crystal with finer grains than one with coarse grains. A varying grain orientation prevents a homogeneity of slip throughout an aggregate. As the slip differential within neighbouring grains is transmitted through the grain boundary the stress and strain within individual grains becomes uneven. The strain is greater at the grain boundary than at its centre but can become more uniform with increasing strain in smaller grains. The linear, stage I hardening of a single crystal is never observed in a fine-grained polycrystal. Consequently, a parabolic flow curve follows its initial yield point. Here this curve becomes an average of simultaneous stage II and stage III hardening as the applied stress can never distinguish between each stage. The result is that when a polycrystalline material is loaded into the plastic region it becomes increasingly harder to deform. The mixed stage II and III hardening within individual grains and the barriers to slip presented by grain boundaries resist plastic flow. The terms *work-hardening or strain-hardening* are used when plasticity occurs under a continuously increasing flow stress.

8.5 Yield Surface

8.5.1 Isotropic Aggregate

We have seen that twelve slip systems are possible in a face-centred cubic lattice. When slip is the only source of plastic strain, each slip system identifies a pair of complementary yield planes $f(\sigma_{ij}) = 0$ that bound an elastic region (Fig. 8.10a). A rule of flow (see Section 8.6) shows that $d\varepsilon_{ij}^{P}$ lies normal to each yield plane for a slip system in which the resolved shear stress is critical. When a number of slip systems operate together in cubic lattices, a yield polyhedron will apply to each grain within a randomly oriented aggregate of grains. Lin [5] and Kocks [6] derived plane sections of this surface for cubic metals. An exampleof an f.c.c. polyhedral yield surface is given in Fig. 8.10b, in which the stress axes are σ_{11}, σ_{12} and σ_{13}.

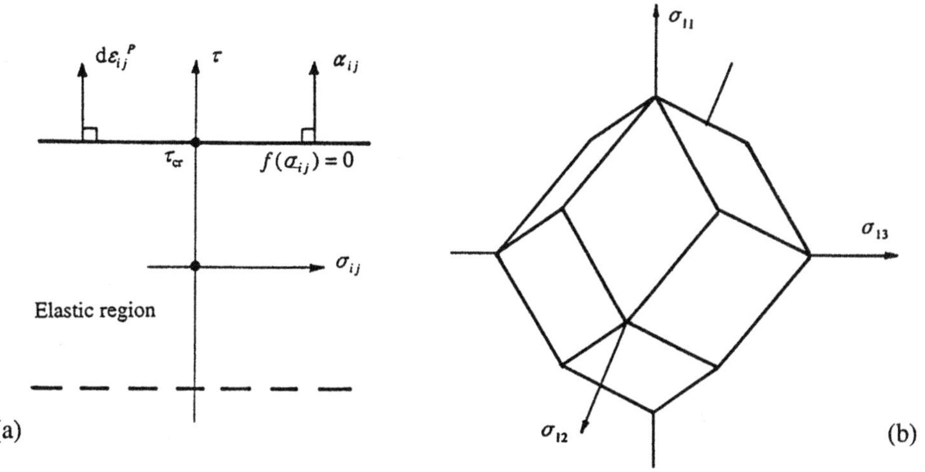

Figure 8.10 Yield planes for slip in an f.c.c. grain

When the stress axes co-incide with the f.c.c. cell axes, the direction cosines for the vectors **n** and **m** in eq(8.5a) equate directly to the Miller indices for the slip plane and slip direction for each of 12 slip systems. The yield planes meet to form edges and corners of the yield polyhedron. Two slip systems operate together along the edges of the yield surface and four or more systems opearate simultaneously at its corners. In an ideal polycrystal, Tresca's maximum shear stress criterion of yielding applies as the critical shear stress is attained on the 24 planes in each grain simultaneously. Thus, Tresca's yield criterion applies to an ideal aggregate of crystals, each with similar elastic moduli and critical resolved shear stresses for the active slip planes. This gives a *lower bound* solution. In a real f.c.c. crystal, where these properties may vary between grains, a homogenous yielding does not occur. We have seen that the constraint exerted upon a grain by its surrounding matrix raises the critical resolved stress. Raising the applied stress enables slip along the most favourably oriented planes to spread into neighbouring grains. The yield stress associated with the initial spread of plasticity will lie on a three-dimensional yield surface framed in the applied stress axes. This smooth convex surface is bounded by the yield planes for slip within all individual grains of an aggregate [7], to give an *upper bound* solution.

Yield loci under plane stress states are again constructed from a generalised Schmidt law (8.5a). A lower-bound yield locus is found when this law is applied to a minimum number of active slip systems within a single grain of a polycrystal. The grain must be taken in isolation under its biaxial stress state. Correspondingly, the sum of the separate shear strains attains a least value [8]. More realistically, the micro-mechanics of yielding in cubic metals should involve a restricted crystallographic slip within an aggregate of grains. The constraints presented to slip in a single grain, from its boundary with differently oriented neighbouring grains, will raise the yield stress. In their account of restricted slip, Bishop and Hill [9] showed that an isotropic, upper-bound yield locus may be established from the *maximum work principal*. This assumes that inhomogenous, simultaneous slip is the only mechanism of spreading plasticity in a polycrystalline solid. To examine the work done in deforming a unit volume we let σ_{ij} be an initially elastic, *macroscopic* stress distribution. This corresponds to point A lying within the yield surface f in Fig. 8.11a. The elastic energy stored is identified with an enclosed triangular area beneath component stress-strain plots, typified by Fig. 8.11b.

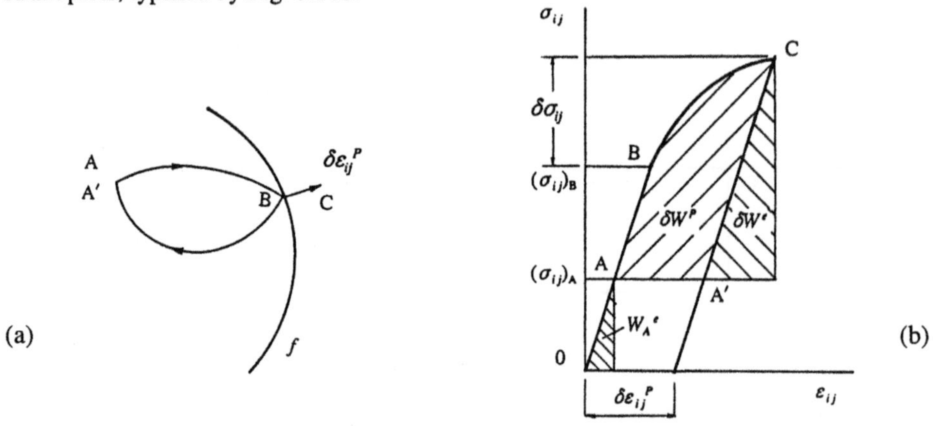

Figure 8.11 Elastic and plastic work for a cycle of loading

The corresponding *microscopic* elastic stress is σ_{ij}^* and here all slip systems s are inactive when $\alpha_{ij}^s\,\sigma_{ij}^* \le \tau_{cr}^s$. Now let the macro-stress be increased to the yield point B and then to point C by a small amount $\delta\sigma_{ij}$, before unloading to the initial stress level at A. Figure 8.11b shows the additional work done δW^P in reaching C. The elastic work has increased by δW^e but is restored to its former level upon unloading to A′. It is seen that the increments in macroscopic plastic strain $\delta\varepsilon_{ij}^P$, produced by $\delta\sigma_{ij}$, are responsible for the permanent change to the stored energy. The incremental plastic work δW^P for loading cycle ABCA′ may be written as:

$$[(\sigma_{ij})_B - (\sigma_{ij})_A]\,\delta\varepsilon_{ij}^P + \delta\sigma_{ij}\,\delta\varepsilon_{ij}^P > 0 \tag{8.10a}$$

When point A lies on the yield surface, then $(\sigma_{ij})_B = (\sigma_{ij})_A$ and eq(8.10a) simplifies to:

$$\delta\sigma_{ij}\,\delta\varepsilon_{ij}^P > 0 \tag{8.10b}$$

Correspondingly, the microscopic stress distribution is altered by $\delta\sigma_{ij}^*$ and an incremental, plastic micro-strain $\delta\varepsilon_{ij}^{P*}$ arises from the slip $\delta\gamma^s$ in systems where $\alpha_{ij}^s\,\sigma_{ij}^* > \tau_{cr}^s$. Here:

$$\delta\sigma_{ij}^*\,\delta\varepsilon_{ij}^{P*} > 0 \tag{8.10c}$$

The *maximum external work principal* combines eqs(8.10b,c) within volume V, as:

$$\delta\sigma_{ij}\,\delta\varepsilon_{ij}^P = \int_V \mathrm{d}\sigma_{ij}^*\,\mathrm{d}\varepsilon_{ij}^{P*}\,\mathrm{d}V \ge 0 \tag{8.11a}$$

No work is done when $\delta\sigma_{ij}$ refers to an unloading or a neutral loading around the yield surface. Within eqs(8.10a,b) the macro-plastic strain increment tensor becomes a volume average of the micro-plastic strain tensor:

$$\delta\varepsilon_{ij}^P = \int_V \mathrm{d}\varepsilon_{ij}^{P*}\,\mathrm{d}V \tag{8.11b}$$

Taylor [8] employed eq(8.8) within a *minimum shear principle*:

$$\sigma_{ij}\, \mathrm{d}\varepsilon_{ij}^{P} \le \tau^{s} \mathrm{d}\gamma^{s} \tag{8.12a}$$

Equation (8.12a) is integrated for a plastic aggregate of volume V

$$\int_V \sigma_{ij}\, \mathrm{d}\varepsilon_{ij}^{P}\, \mathrm{d}V \le \int_V \left(\sum_{s=1}^{N} \tau^{s} \,|\, \mathrm{d}\gamma^{s} \,| \right) \mathrm{d}V \tag{8.12b}$$

where stress and strain distributions may vary continuously whilst satisfying both equilibrium and compatibility. The exact calculation of the yield surface, based upon eq(8.12b), is complex. Certain simplifications [10] will produce upper and lower-bound yield surfaces to satisfy either equilibrium or compatibility, but not both. For example, compatibility is satisfied in Taylor's lower bound with the assumption of uniform strain within the grains equal to the macroscopic strain (elastic strain is neglected). Employing eq(8.8) for N slip systems gives the uniform strain increments as

$$\mathrm{d}\varepsilon_{ij}^{P*} = \mathrm{d}\varepsilon_{ij}^{P} = \sum_{s=1}^{N} \alpha_{ij}^{s}\, \mathrm{d}\gamma^{s} \tag{8.13a}$$

but the condition of equilibrium is violated at grain boundaries. Conversely, when $\sigma_{ij} = \sigma_{ij}^{*}$ is uniform in eq(8.12b), we may write from eq(8.5b):

$$\sigma_{ij}^{*} = \sigma_{ij} = (\alpha_{ij}^{s})^{-1} \tau^{s} \tag{8.13b}$$

Here equilibrium is achieved for a uniform distribution of microscopic stress throughout the grains, but compatibility is violated. Equations (8.13a,b) apply to an isotropic aggregate which hardens uniformly, i.e. it shows no Bauschinger effect. It may be deduced from this that the yield surface lies between the upper and lower bounds. For example, Fig. 8.12a shows these two bounds for an ideal cubic lattice [11]. This restricted the slip under a principal, biaxial stress state to the (111) plane in a $[11\bar{2}]$ direction.

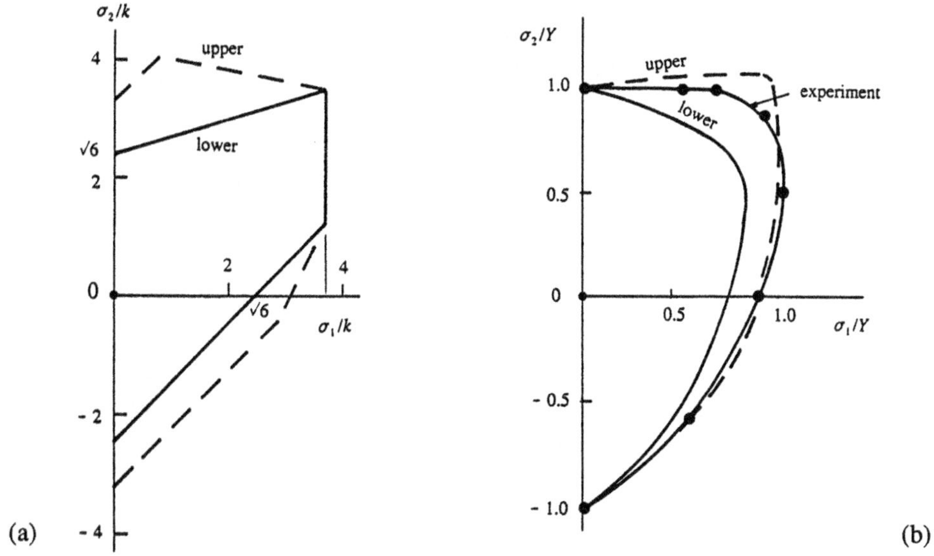

Figure 8.12 Bounding loci for an (a) isotropic (b) anisotropic texture

Further bounding yield loci may be derived for different operating slip systems. When extended from an f.c.c. grain to a polycrystal, the bounding becomes equivalent to taking its yield locus to lie between von Mises and Tresca. We have seen this previously in the continuum theory of yielding for a polycrystalline aggregate (Section 3.3, p. 71), where both criteria identify with functions of the stress deviator invariants. Again, it is implied here that an aggregate of small, randomly orientated grains is macroscopically isotropic, where flow stress is independent of the material direction.

8.5.2 Texture

The plastic flow behaviour of single crystals is anisotropic in that it depends upon the relative orientations of the slip plane and the stress axis. While the yield stress of an aggregate of equiaxed grains is usually direction independent, that for non-equiaxed grains is not. For example, anisotropic yielding is found where the grains of a polycrystal exhibit a preferred orientation, as in a heavily rolled sheet metal. The degree of anisotropy may approach that for a single crystal when the directionality is severe. Anisotropy arising from cold rolling non-ferrous sheets cannot be entirely removed by a recrystallisation anneal. Indeed, the slip and any twinning sites from prior-working become thermally activated at the annealing temperature. The likely influences of crystallographic anisotropy is to distort, translate and rotate the yield surface. Such features are associated with the anisotropic hardening that is known to arise from plastic pre-straining (see p. 265). A wrought, fibrous structure is associated with an alignment of impurities and voids along the forging direction. A hot forged structure is less likely to have been hardened anisotropically than a cold-rolled structure but the former can display anisotropic fracture stress and strain.

Upper and lower-bound estimates of yielding in textured f.c.c aggregates of copper and aluminium were made by Althoff and Wincierz [12]. In their analysis of textures, a bound was taken to be the average of a number of yield loci. Each locus was found from an individual texture component. Their method is similar to placing bounds upon yieldiing within individual grains of an isotropic solid. Figure 8.12b illustrates the result of compounding the residual textures arising from the recrystallisation of a dominant (112), $[11\bar{1}]$ texture in an extruded aluminium tube. The experimental data confirms that the yield locus, defined by 0.1% offset plastic strain, lies between the upper and lower bounds. Figures 8.12a,b, show that the bounds for equiaxed and textured structures do not display: (i) the smooth curved boundary of the measured yield loci and (ii) any difference between yield stresses in tension and compression. These observations imply that strain within a grain is not distributed uniformly. The strain uniformity is assumed when maximising external work for an upper bound. Backofen and Piehler [11, 13] employed an alternative technique for bounding yield loci in rolled sheets with cubic lattices. Here a deformation texture, representative of planar isotropy, was established from rotating an ideal texture about the sheet normal. Figures 8.13a-e illustrate the construction of lower-bound yield loci in normalised stress axes, by applying Schmidt slip to each stage of this rotation. Lower bounds with corners again result from their technique. In Fig. 8.13e, the average lower-bound locus for the full rotation is compared with an upper bound from applying the maximum work principle [13]. It is seen that the region between the bounds, within which the true locus is expected to lie, can accommodate a limited amount of unsymmetrical distortion and a difference between initial yield stresses. Kocks' [7] computer simulations of rotating textures provided upper-bound yield loci for rolled and pre-strained polycrystals. These loci were found to be irregularly shaped with flattenened sides and pointed vertices.

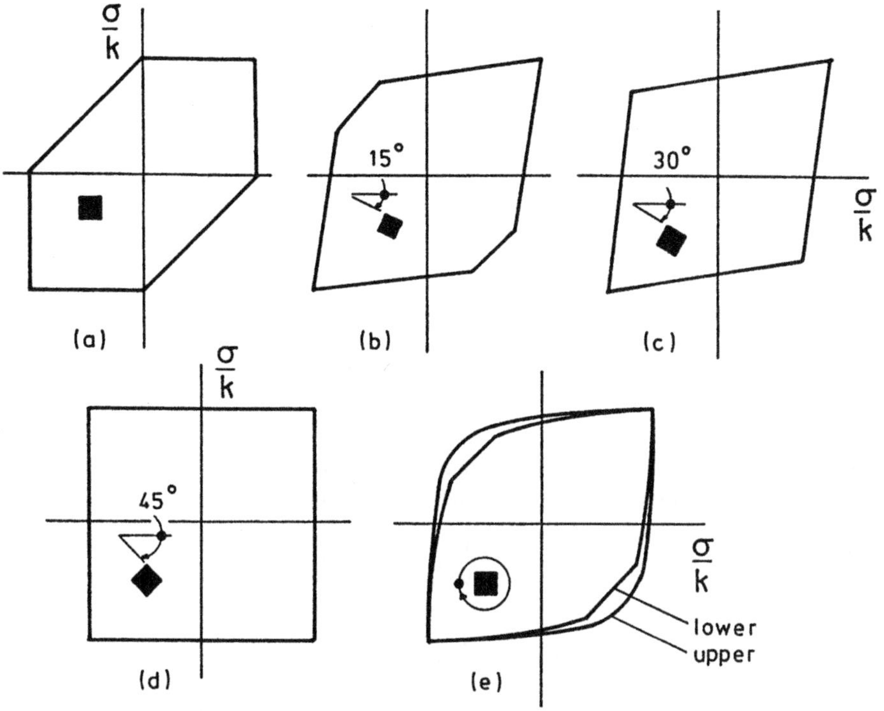

Figure 8.13 Lower bound yield loci for a cubic texture

Good agreement has been found between theoretical and experimental yield loci for textures under biaxial stress states [14-17]. Some typical observations on the initial yield behaviour of rolled sheet metal can be seen in Chapter 11 (p. 365). In the continuum plasticity theory, initial anisotropy appears within a suitable yield function containing the macroscopic stress and plastic strain history. Many yield functions have been employed (see Table 3.3) to represent irregular shaped yield surfaces in polycrystalline aggregates. Here, tensor formulations admit all macro-stress and plastic strain states. An *associated flow rule* identifies the yield function with a plastic potential for which stress derivatives equate to an aggregated stress-strain relation. This classical macroscopic approach to yielding is clearly more convenient than bounding with Schmidt's active micro-slip systems.

8.6 Flow Rule

The decomposition of the total strain into elastic and inelastic components is a common feature in the classical theory of plasticity. The theory assumes that elasticity is due to lattice distortion and that slip on crystallographic planes accounts for all inelastic deformation. The latter is taken to be quasi-static for low rates of straining, so ignoring the effect of a discrete dislocation substructure. Strictly, the dynamics of dislocation motion renders the inelastic strain component as inherently time-dependent or visco-plastic [18]. To admit visco-plastic strain, it becomes necessary to consider a similar decomposition in the total rate of deformation. Verification of the plastic potential, employed in classical theory, follows from a consideration of crystallographic slip and the motion of discrete dislocation lines [19]. When the strains are small, grain rotation can be ignored so that slip alone is responsible for the macroscopic plastic strain ε_{ij}^P. The latter is taken as the surplus $\varepsilon_{ij} - \varepsilon_{ij}^e$, where ε_{ij} defines a homogenous strain field and ε_{ij}^e is the contribution to elastic lattice distortion. When inelastic strain rates $\dot{\varepsilon}_{ij}^P$ result from a general stress state σ_{ij}, the work rate from V_s slip systems within volume V, is expressed from re-writing eq(8.12a) as

$$\sigma_{ij}\,\dot{\varepsilon}_{ij}^P = \frac{1}{V}\int_{V_s} \tau\dot{\gamma}\,\mathrm{d}V \tag{8.14}$$

In eq(8.14), τ is the resolved shear stress when each slip system shears at a rate $\dot{\gamma}$. Let f be the force on a dislocation line L and L_d refer to all such lines within volume V. The work rate may also be expressed in the positive rate of change of slipped area:

$$\sigma_{ij}\,\dot{\varepsilon}_{ij}^P = \frac{1}{V}\int_{L} f\,v\,\mathrm{d}L \tag{8.15}$$

where v is the velocity of a dislocation line segment dL. Let the function $\dot{\varepsilon}_{ij}^P(\sigma_{ij})$ apply to a given slipped state where, correspondingly, $\gamma = \dot{\gamma}(\tau)$ and $v = v(f)$. Following an infinitesimal stress change $\mathrm{d}\sigma_{ij}$, eqs(8.14) and (8.15) become

$$\mathrm{d}\sigma_{ij}\,\dot{\varepsilon}_{ij}^P = \frac{1}{V}\int_{V_s} (\dot{\gamma}\,\mathrm{d}\tau)\,\mathrm{d}V \tag{8.16a}$$

$$\mathrm{d}\sigma_{ij}\,\dot{\varepsilon}_{ij}^P = \frac{1}{V}\int_{L_d} (v\,\mathrm{d}f)\,\mathrm{d}L \tag{8.16b}$$

As both integrands (8.16a,b) must be identical they may be replaced with a potential function $f(\sigma_{ij})$ whose partial derivatives in σ_{ij} supply strain rate components, $\dot{\varepsilon}_{ij}^P$. Introducing a scalar multiplier $\dot{\lambda}$, this leads to a flow rule for strain rates:

$$\dot{\varepsilon}_{ij}^P = \dot{\lambda}\,\frac{\partial f(\sigma_{ij})}{\partial \sigma_{ij}} \quad \text{or} \quad \dot{\mathbf{E}}^P = \dot{\lambda}\,\frac{\partial f(\boldsymbol{T})}{\partial \boldsymbol{T}} \tag{8.17a,b}$$

Equation (8.17a) implies normality of $\dot{\varepsilon}_{ij}^P$ to a convex yield surface. It follows that the yield function $f(\sigma_{ij})$ in eq(8.17a) may be identified with either right-hand side in eqs(8.16a,b):

$$f(\sigma_{ij}) = \frac{1}{V\dot{\lambda}}\int_{V_s}\left[\int_{\tau} \dot{\gamma}\,\mathrm{d}\tau\right]\mathrm{d}V = \frac{1}{V\dot{\lambda}}\int_{L_d}\left[\int_{f} v\,\mathrm{d}f\right]\mathrm{d}L \tag{8.18}$$

Geometrically, the yield function $f(\sigma_{ij})$ in eqs(8.17a) and (8.18) follows from slip considerations alone simply by re-arranging Schmidt's law (8.5b) as

$$f(\sigma_{ij}, \tau_{cr}) = \alpha_{ij}\,\sigma_{ij} - \tau_{cr} = 0 \tag{8.19}$$

for when the resolved shear stress τ_{cr} is allowed to vary within each slip system. Normality is again implied from combining eqs (8.6a) with eq(8.19):

$$\mathrm{d}\varepsilon_{ij}^P = \alpha_{ij}^s\,\mathrm{d}\gamma^s = \mathrm{d}\gamma^s\,\frac{\partial f}{\partial \sigma_{ij}} \tag{8.20}$$

Comparison between eqs(8.17a) and (8.20) identifies $\dot{\lambda}$ with the rate of slip $\mathrm{d}\gamma^s/\mathrm{d}t$. In the continuum theory, the *plastic potential* function f in eq(8.18) is identified with a yield function describing a convex yield surface. Convexity is implied within the maximum work principle through eqs (8.10a,b) (see Drucker's postulate p. 311). Either of the eqs(8.17a,b) is known as the *associated flow rule* of which normality between f and $\dot{\varepsilon}_{ij}^P$ is a consequence.

When quasi-static, eq(8.20) shows there is normality between f and $d\varepsilon_{ij}^P$ for $d\lambda = |d\gamma^s|$. For compressibile materials this association can be too restrictive. Here f and the constitutive relation between stress and plastic strain are disconnected to give a *non-associated flow rule.*

8.7 Micro- to Macro-Plasticity

The macro-plastic strain follows from the volume average of the microsopic strains. The latter may be expressed in the microscopic parameters, either from continuum slip or dislocation motion. They are, respectively:

$$\varepsilon_{ij}^P = \frac{1}{V}\int_{V_s} \tfrac{1}{2}(n_i m_j + m_i n_j)\,\gamma\,dV \quad \text{or} \quad \mathbf{E}^P = \frac{1}{V}\int_{V_s} \tfrac{1}{2}(\mathbf{n}\,\mathbf{m}^T + \mathbf{m}\,\mathbf{n}^T)\,\gamma\,dV \tag{8.21a}$$

$$\varepsilon_{ij}^P = \frac{1}{A}\int_{A_s} \tfrac{1}{2}(n_i b_j + b_i n_j)\,dA \quad \text{or} \quad \mathbf{E}^P = \frac{1}{A}\int_{A_s} \tfrac{1}{2}(\mathbf{n}\,\mathbf{b}^T + \mathbf{b}\,\mathbf{n}^T)\,dA \tag{8.21b}$$

Equation (8.21a) follows from eq(8.6a) in which slip systems are refered to the active volume V_s. In eq(8.21b) A_s is the collective area of active slip planes, **n** is the outward normal vector to a slip plane, **m** defines the slip vector and **b** is Burger's vector, with magnitude $|\mathbf{b}| = \gamma V/A_s$. The shear strain rate in eq(8.16a) is defined as

$$\dot{\gamma} = |\boldsymbol{b}|\,\rho\langle v\rangle = \frac{\gamma V \rho\langle v\rangle}{A_s}$$

where $\rho = L_d/V$ is the mobile dislocation density and $\langle v\rangle$ is the mean dislocation velocity.

A compatibility condition for a polycrystal requires that the deformation in each grain is transmitted to adjoining grains across their grain boundaries. Consequently, the internal stress within an individual grain will depend upon the applied macro-stress and the micro-stress imposed from neighbouring grains. A mathematical averaging technique is used to simplify the evolution of a stress-strain curve from slip within a single grain to the polycrystalline aggregate. Taylor [8] was the first to employ this technique in 1928 for an aluminium polycrystal composed of fine, randomly orientated, equiaxed grains with typical diameters 10^{-4} m. He assumed that at least 5 of the 12 slip systems would activate to accommodate plasticity and preserve volume under tension. Slip systems were activated by the least amount of work when axial strain was uniformly dispersed throughout the grains.

Normally, numerical integration is necessary when evaluating eq(8.21a) to give the strain in a representative volume element. However, a closed-solution arises from integrating slip within a hemispherical volume element composed of ideal grains, e.g. in a single-phase crystalline solid. Grains of similar orientation within the element have parallel slip planes. Euler angles α, β and ϕ determine the orientation of the stress axes x_i relative to the normal **n** and direction **m** for an element of slip plane with solid angle $d\Omega$, as shown in Fig. 8.14a. The number of grains with similar orientation is proportional to the product $(d\Omega \times d\phi)$ and therefore these grains contribute to slip γ by an amount $d\gamma = (d\Omega \times d\phi)\gamma$. The plastic strain ε_{ij}^P in stress axes x_i ($i = 1, 2, 3$) then follows from eq(8.7). This gives the strain arising from all active, parallel slip systems within the unit hemisphere [21]:

$$\varepsilon_{ij}^P = \frac{1}{2}\int_{\Omega} d\Omega \int_{-\pi/2}^{\pi/2} (m_i n_j + n_i m_j)\,\gamma\,d\phi \tag{8.22a}$$

where, in spherical co-ordinates (r, α, β) and with $r = 1$,

$$d\Omega = (r\sin\alpha\,d\beta \times r\,d\alpha) = \sin\alpha\,d\alpha\,d\beta \tag{8.22b}$$

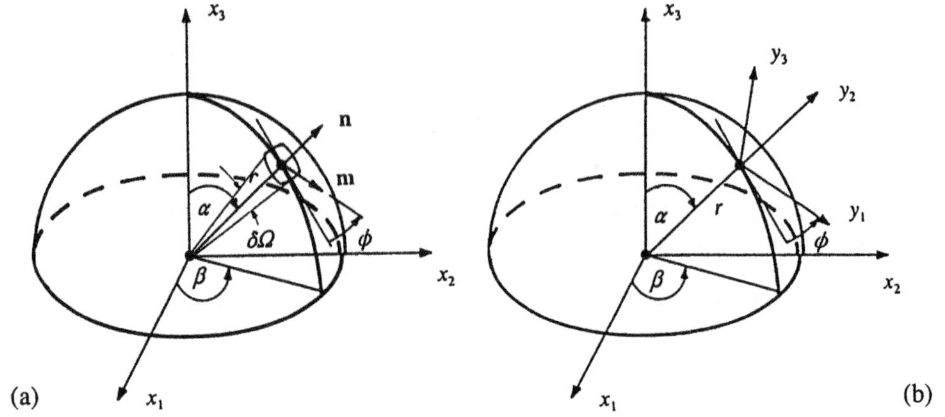

Figure 8.14 Euler angles and loading axes for a hemi-spherical crystal

By setting $\gamma = \gamma(\tau) = \sum c_n (\tau / \tau_{cr} - 1)^n$ as an n-term polynomial in eq(8.22a), it is seen how macroscopic plastic strain accumulates with slip when $\tau \geq \tau_{cr}$. This effect of this simplified approach is to distribute strain uniformly. Alternatively, it may be more realistic to identify the macroscopic strain in a polycrystal with an aggregate of its non-uniform granular strains. To do this, we let y_i ($i = 1, 2, 3$) be the orientation of a point on the slip plane within a single spherical grain (see Fig. 8.14b). Transforming slip from all active planes to axes y_i gives a microscopic strain tensor ε_{ij}^{P*} for that grain. Again, using spherical co-ordinates (r, α, β) and eq(8.22b), we may refer ε_{ij}^{P*} to stress axes x_i. This gives the macroscopic strain tensor for the grain [19]:

$$\varepsilon_{ij}^{P} = \frac{1}{4\pi^2} \int_H \int \sin\alpha \, \mathrm{d}\alpha \, \mathrm{d}\beta \int_{-\pi}^{\pi} \varepsilon_{ij}^{P*} \, \mathrm{d}\phi \tag{8.23a}$$

where eq(8.7) provides the tensor strain components ε_{ij}^{P*} in the crystal axes. Integration applies over the hemisphere H for all grain orientations ϕ. The macroscopic stress becomes a simple volume average of the microscopic stresses in all spherical grains:

$$\sigma_{ij} = \frac{1}{V} \int_V \sigma_{ij}^{*} \, \mathrm{d}V \tag{8.23b}$$

This method shows how the maco stress-strain behaviour evolves from micro-slip. However, many have recognised the need for a micro-macro evolution to account for the restricted, heterogenous deformation within grains. Here, the models of Eshelby [22] and Kroner [23] are useful. The underlying principal is that of amassing all grains of identical orientation into a single spherical or ellipsoidal grain. The 'grain' , with orientation α, lies within an infinite isotropic matrix, as shown in Fig. 8.15a. When the 'composite' structure is stressed under a remote macroscopic stress σ_{ij} , the deformation differs between the spherical inclusion and the surrounding matrix. This results in an internal stress σ_{ij}^{*} within the grain. When matrix and grain are analysed separately under σ_{ij} (Figs 8.15b,c) an internal stress difference $\sigma_{ij} - \sigma_{ij}^{*}$ must exist to maintain compatibility, i.e. the grain must fit the hole. Eshelby [22] examined an elastic interaction between the anisotropic deformation produced from the uniform internal stress σ_{ij}^{*} in the grain (* now refers to the grain) and the homogenous deformation assumed for the surrounding matrix. Kroner [23] extended this approach to include a plastic interaction between grain and matrix. Here the integral (8.23a) supplies the plastic strain in the anisotropic, spherical grain. Equilibrium and compatibility must hold between the stress σ_{ij} and plastic strain ε_{ij}^P within a randomly orientated, isotropic matrix and stress σ_{ij}^* and strain ε_{ij}^{P*} within the anisotropic grain.

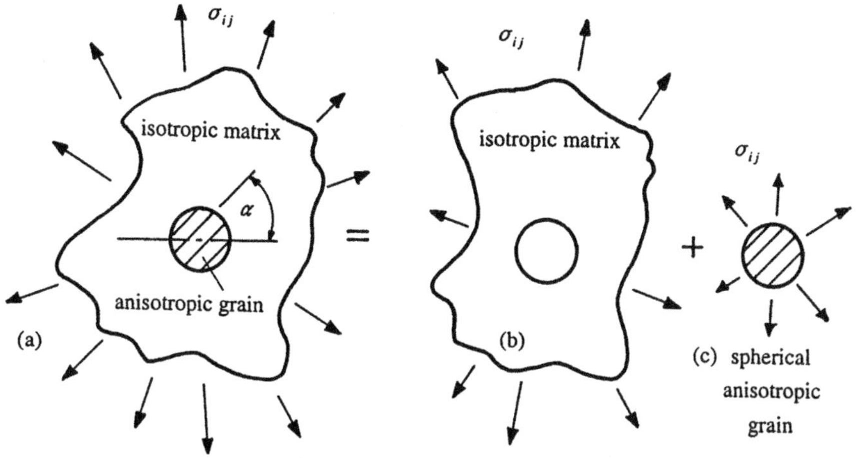

Figure 8.15 The 'self-consistent' Kroner-Eshelby model

The following 'self-consistent' relation achieves the two conditions by connecting the respective stress and plastic strain differences elastically:

$$\sigma_{ij}^* - \sigma_{ij} = -2\mu G\left(\varepsilon_{ij}^{P*} - \varepsilon_{ij}^P\right) \quad \text{or} \quad \mathbf{T}^* - \mathbf{T} = -2\mu G\left(\mathbf{E}^{P*} - \mathbf{E}\right) \tag{8.24}$$

where $\mu = (7 - 5\nu)/[15(1 - \nu)]$, G and ν are the usual elastic constants. The plastic strain difference embodies plastic incompressibility to ensure that the interaction between grain and matrix is independent of hydrostatic stress. Equation (8.24) may be applied in turn to all spherical grains of varying orientation and finally averaged to give the matrix stress and strain: $\varepsilon_{ij}^P = (\varepsilon_{ij}^{P*})_{av}$ and $\sigma_{ij} = (\sigma_{ij}^*)_{av}$. However, eq(8.24) is known to over-estimate the stress difference $\sigma_{ij}^* - \sigma_{ij}$. The coefficient μ was altered by Kroner [23] to account for the non-uniform plastic strain in the matrix but in retaining an elastic interaction eq(8.24) still exaggerated the grain's internal stress, σ_{ij}^*. Hill [24] replaced the coefficient μG in eq(8.24) with a tensor parameter that minimised structural stiffness as the spread of plasticity 'softened' the interaction. Berveiller and Zaoui [25] showed that an approximate method for accommodating a plastic interaction with stress relaxation between grain and matrix, was to replace $2\mu G$ with $2\alpha\mu G$ in eq(8.24) where:

$$\alpha = \left[1 + (3G\bar{\varepsilon}^P)/(2\bar{\sigma})\right]^{-1} \tag{8.25}$$

The use of macroscopic equivalent stress $\bar{\sigma}$ and plastic strain $\bar{\varepsilon}^P$ within eq(8.25) enabled α to be determined under a uniaxial stress. In order to include an instantaneous response from the heterogenous plastic volume V, they [26] added a further 'Hill-term' to eq(8.24):

$$\sigma_{ij}^* = \sigma_{ij} - 2\alpha\mu G\left(\varepsilon_{ij}^{P*} - \varepsilon_{ij}^P\right) + \frac{2\alpha\mu G}{V}\sum N_{ijkl}\left(\varepsilon_{ij}^{P*} - \varepsilon_{ij}^P\right) \tag{8.26}$$

where α is defined in eq(8.25) and N_{ijkl} is a fourth order interaction tensor of tangent moduli.

It is now possible to devise a method by which either of eqs(8.24) and (8.26) are applied to many crystalline slip systems, $s = 1, 2, 3 \ldots N$, within a single grain. Firstly, eq(8.6a) expresses the total strain in the grain as:

$$\varepsilon_{ij}^{P*} = \frac{1}{2}\sum_{s=1}^{N}\left(n_i m_j + n_j m_i\right)^s \gamma^s = \sum_{s=1}^{N} \alpha_{ij}^s \gamma^s \tag{8.27a}$$

or

$$\mathbf{E}^{P*} = \frac{1}{2}\sum_{s=1}^{N}[\mathbf{n}\,\mathbf{m}^{T} + \mathbf{m}\,\mathbf{n}^{T}]^s \gamma^s = \sum_{k=1}^{N} \mathbf{A}^s \gamma^s \tag{8.27b}$$

in which there is no summation over s. Equation (8.6b) gives the corresponding plastic rotation in the grain:

$$\omega_{ij}^{P*} = \frac{1}{2}\sum_{s=1}^{N}\left(n_i m_j - n_j m_i\right)^s \gamma^s = \sum_{s=1}^{N} \beta_{ij}^s \gamma^s \tag{8.28a}$$

or

$$\mathbf{\Omega}^{P*} = \frac{1}{2}\sum_{s=1}^{N}[\mathbf{n}\,\mathbf{m}^{T} - \mathbf{m}\,\mathbf{n}^{T}]^s \gamma^s = \sum_{s=1}^{N} \mathbf{B}^s \gamma^s \tag{8.28b}$$

where $\mathbf{A}^s \equiv \alpha_{ij}^s$ and $\mathbf{B}^s \equiv \beta_{ij}^s$ are the orientation factors for a given slip system s. Let a number of grains, $q = 1, 2, 3 \ldots M$, each with volume V_q, lie within the polycrystalline volume V. The macroscopic plastic strain and rigid rotation of the matrix are found from the volume averages of eqs(8.27a) and (8.28a):

$$\varepsilon_{ij}^{P} = \frac{1}{V}\int_V \varepsilon_{ij}^{P*}\,\mathrm{d}V = \frac{1}{V}\sum_{q=1}^{M}\sum_{s=1}^{N} V_q\, \alpha_{ij}^{qs}\, \gamma^{qs} \tag{8.29a}$$

$$\omega_{ij}^{P} = \frac{1}{V}\int_V \omega_{ij}^{P*}\,\mathrm{d}V = \frac{1}{V}\sum_{q=1}^{M}\sum_{s=1}^{N} V_q\, \beta_{ij}^{qs}\, \gamma^{qs} \tag{8.29b}$$

Now, eq(8.5b) shows that for a given slip system s, within a single grain q, the critical resolved shear stress τ_{cr}^s may be expressed in terms of the uniform stress field σ_{ij}^* within that grain as:

$$\tau_{cr}^s = \sigma_{ij}^* n_i^s m_j^s = \sigma_{ij}^* \alpha_{ij}^s \tag{8.30a}$$

or

$$\tau_{cr}^s = \mathrm{tr}\,[\mathbf{T}^*\,\mathbf{n}^s\,(\mathbf{m}^s)^{T}] = \mathrm{tr}\,(\mathbf{T}^*\mathbf{A}^s) \tag{8.30b}$$

where $\alpha_{ij}^s = n_i^s m_j^s$, or $\mathbf{A}^s = \mathbf{n}^s(\mathbf{m}^s)^{T}$. The macroscopic stress σ_{ij}, for M grains in volume V, derives from the volume average eq(8.23b)

$$\sigma_{ij} = \frac{1}{V}\sum_{q=1}^{M} \sigma_{ij}^* V_q \tag{8.31}$$

Each resolved shear stress in an f.c.c. lattice will depend upon all possible $N = 24$ slip systems:

$$\tau_{cr}^s = h^s(\gamma^1, \gamma^2, \gamma^3 \ldots\ldots \gamma^{24}) \tag{8.32}$$

in which the hardening function h^s depends upon Burger's vector and the dislocation density. The solution to the macro stress-strain behaviour combines a suitable micro-constitutive hardening law (8.31) with the volume averages. Assume that a known σ_{ij} is applied and ε_{ij}^P is to be calculated. Firstly, σ_{ij}^* is estimated from eq(8.31). Then, τ_{cr}^s is found from eq(8.30a) for each slip system within a grain. It follows that the amounts of slip γ^s in eq(8.32), depend upon the chosen hardening law. Taylor [8] employed a linear isotropic hardening law

$$\tau_{cr}^{\,s} = \tau_o + \sum_{k=1}^{N} h^{sk} \gamma^k \tag{8.33}$$

where τ_o is the shear stress necessary to initiate slip. Equation (8.33) allows the constant value of tangent modulus $h^{sk} = \partial \tau_{cr}^{\,s} / \partial \gamma^k$, to differ between slip systems. Other hardening laws have been employed. For example, Brown [27] employed a non-linear relation for each slip system

$$\tau_{cr}^{\,s} = c\,(\gamma^s)^n \tag{8.34}$$

where c and n are empirical constants. The micro- and macro-plastic strains follow from the application of eqs(8.27a) and (8.29a) respectively. Finally, eq(8.24) enables $\sigma_{ij}^{\,*}$ to be re-calculated within an iteration procedure. Berveiller and Zaoui [25] and Hutchinson [28] compared uniaxial, stress-strain curves from this method by making various assumptions (see Fig. 8.16). Firstly, with ideal, elastic-perfect plastic f.c.c. grains, $h^{sk} = 0$ in eq(8.33).

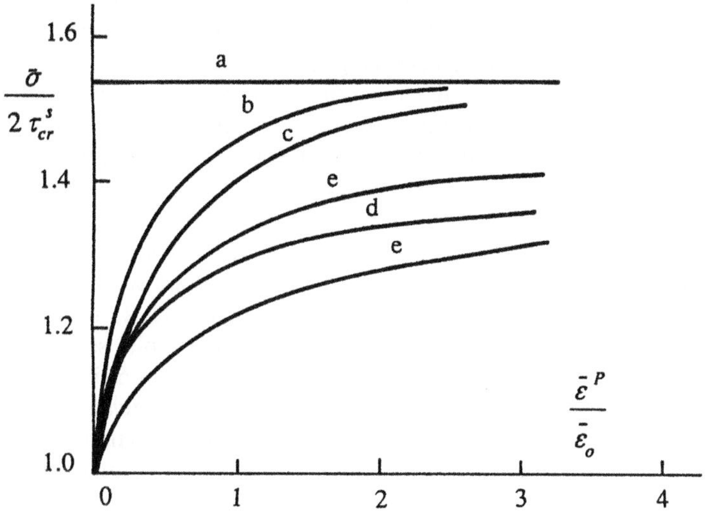

Figure 8.16 Theories of polycrystalline hardening ($\bar{\varepsilon}_o = Y/E$)

Further assumptions were followed in producing the respective curves a-d in Fig. 8.16:

(a) Taylor [8]: Large, uniform strain satisfying compatibility (trivially), elastic strain and work hardening ignored. Equilibrium condition for grain boundary stress is violated.
(b) Lin [5]: Uniform strain satisfying compatibility (trivially) but not restricted to large strain. Includes elastic strain component and plastic hardening. Equilibrium is violated.
(c) Eshelby [22]and Kroner [23]: Both equilibrium and compatibility satisfied. Uniform strain only within grains of identical orientations. Inhomogenous strain distribution elsewhere with an elastic interaction between grains.
(d) Hill [24]: As for (c) but with a rigorous analysis of the inelastic interaction between grains given.
(e) Berveiller and Zaoui [25]: As for (d), but with a less rigorous analysis of the inelastic interaction between grains. Consequently, application of the theory is simplified.

Method (d) is believed to be the most exact though it requires the greatest amount of computation. It was adopted by Weng [29, 30] to account for hardening in a two-phase alloy and distortion in its subsequent yield locus. With α defined from eq(8.25), it is seen that curve e is a reasonable approximation to curve d and can be found with less effort. The account of heterogenous deformation within the surrounding matrix (curve e) has reproduced the observed non-linear, uniaxial stress-strain for a polycrystal to $\varepsilon^P \approx 0.1\%$. Further modifications to the Eshelby-Kroner self-consistent relation (8.24) were made to account for texture development and a Bauschinger effect in the grain [31].

Other methods of treating interactions between grains adapt further the ideas of Taylor et al [8, 32], in preference to following a model of self-consistency. Taylor neglected elastic strains and equated the strain in each grain to an average macroscopic strain. Figure 8.16 shows that Taylor's curve a is realistic only at large strain, where individual grains do not work-harden further. A more realistic prediction, curve b, applies to grains that behave in an elastic-plastic manner, with linear hardening from eq(8.33). Both the Taylor and Lin predictions overestimate the observed response since the condition of equilibrium is violated at grain boundaries. Nonetheless, the simplifications afforded by this method explains why it is often used for predicting polycrystalline deformation from single crystal plasticity. To improve the accuracy a further refinement [34] has incorporated a non-linear hardening law to account for interaction and rotation between grains.

In practice, alloys are employed more often than metals to bear load. This is particularly so in hostile, high-temperature, corrosive environments. Understanding the microstructural processes that constitute the observed strain is far more complex. Alloying raises the yield stress and the rate of hardening. The reason for this is that the stress necessary for a stage III dynamic recovery is increased as alloying atoms resist and impede the motion of dislocations. Additionally, as the Petch eq(8.9) shows, the yield stress is raised if the grain size is reduced by alloying. Theories of solid solution hardening are generally more successful than those of precipitation hardening [1, 2] in explaining these effects. This is largely because the dislocation interactions in solid solutions are more clearly defined. Raising temperature and/or reducing the rate of deformation assists with the processes of climb and cross slip of pinned dislocations by thermal activation. These reduce the yield stress and the hardening rate. Furthermore, when precipitation hardened alloys operate at higher temperatures, it becomes necessary to account for other diffusional processes: grain coarsening, phase changes, micro-voids and crack formation [35].

8.8 Subsequent Yield Surface

Qualitatively, most of the theories discussed in the previous section predict a parabolic stress-strain curve for a polycrystal. A further validation test would lie in their ability to describe the subsequent yield surface for a work-hardened material. For example, the assumption of plastic rigidity, employed by Taylor [8], implies that the initial yield surface applies to all subsequent work-hardened states. This simplification suggests that the initial distribution of grain orientations is not altered by plasticity. Clearly, this model ignores those influences of deformation texture upon the yield surface, as shown in Fig. 8.13.

8.8.1 Isotropic Hardening

Consider Taylor's linear hardening eq(8.33). This gives equal increases in the resolved shear stress for both active and non-active (latent) slip planes. Thus, where the hardening behaviour neither depends upon the sense nor the direction of slip, a polycrystal will display *isotropic hardening*. Such behaviour is most likely to be found in tension and compression of a fully annealed material, i.e. one whose structure is free from the history of previous deformation. With isotropic hardening under other stress states, the initial yield surface is understood to expand uniformly, retaining its shape and orientation. Figure 8.17a illustrates this behaviour from within a principal biaxial stress space of σ_1 versus σ_2.

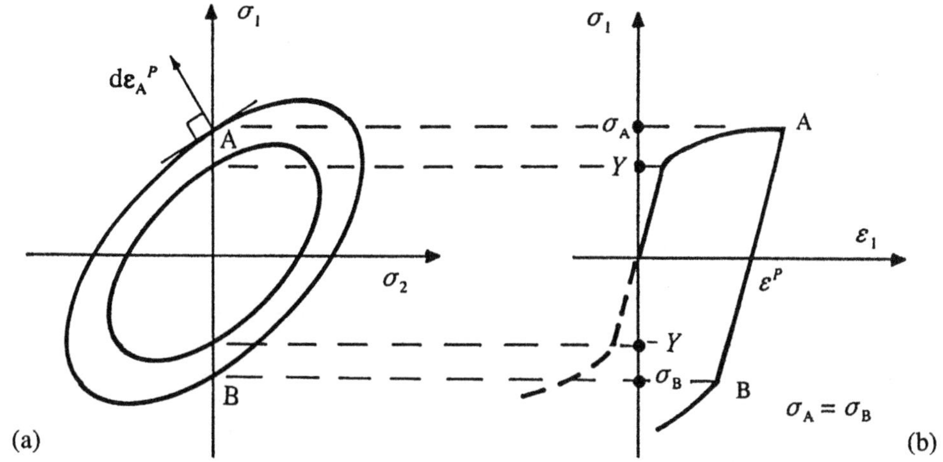

Figure 8.17 The rule of isotropic hardening

Figure 8.17b has projected the expanded yield locus into a stress-strain response to simple tension. The rule shows that when the direction of deformation is reversed upon reaching the work-hardened state A, yielding re-commences when the stress at B equals that at A. The major criticism of isotropic hardening is that it does not show a *Bauschinger effect.* The latter states $\sigma_B < \sigma_A$, which is always observed in stress reversal experiments. Thus, we should only expect the subsequent yield surface be an inflation of the initial yield surface for the backward-extrapoation definition of yield, employed by Taylor and Quinney [32]. Their yield point is detected from the low hardening rate pertaining to large plastic strain. In contrast, alternative definitions of yield employ little or no plastic strain. The identification of the forward and reversed yield stresses σ_A and σ_B with proportional limits (as shown) or, with small offset strains, involves a Bauschinger effect. However, we may employ the isotropic hardening rule for radial outward loading paths, i.e. no stress reversal. Figure 8.17a shows how combining a uniform expansion with the *normality rule* connects σ_A to an incremental plastic strain vector $d\varepsilon_A{}^P$ at the load point. Chapter 10 shows how the rule of isotropic hardening can be combined with the flow rule to match certain experimental data.

8.8.2 The Minimum Surface

We have seen that the slip theory of Batdorf et al [20, 21, 36] employs active slip systems within individual crystals comprising the polycrystalline aggregate. With only one independent slip system per crystal, this theory predicts a subsequent yield surface to be the minimum surface enclosing the prestress point and the initial yield surface. When the latter is defined by von Mises, it projects as a circle on the deviatoric plane (see Fig. 9.1, p. 270). The circle is the cross-section of a cylinder with an axis of hydrostatic stress equally inclined to the axes of principal, deviatoric stress σ_1', σ_2' and σ_3'.

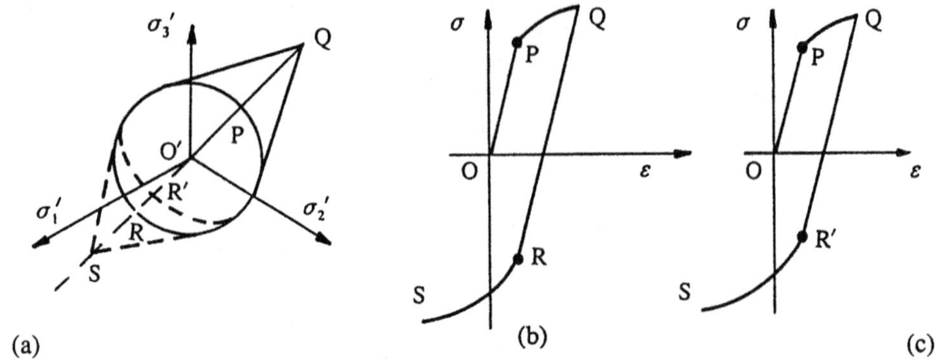

Figure 8.18 Minimum surface showing Bauschinger effect

Once the stress path pierces the initial surface at P, a subsequent surface is formed from projecting tangents from the initial surface to pass through the current stress point Q. This construction leaves the rear portion of the original surface unchanged. The minimum surface reveals that an isotropic material becomes anisotropic. It is seen from the associated stress-plastic strain response (Fig. 8.18b) how a minimum surface displays a Bauschinger effect: $\sigma_R < \sigma_Q$. Figure 8.18c shows a stronger Bauschinger effect, where reversed yielding occurs at R′. Here, the rear of the minimum surface has been correspondingly modified [37, 38], as illustrated with broken lines in Fig. 8.18a.

Much experimental work was conducted in the 1950s and 1960s to confirm whether the subsequent yield surface displayed the corner or pointed vertex at the load point, predicted by slip theory. When the findings from this work were reviewed objectively [39], they appeared to be inconclusive. The majority of investigations confirmed a local distortion of the subsequent yield surface. A blunt nose, not a corner, appears in the region where the surface was originally pierced by the stress vector. It can then be said that a unique normal exists for stress points on all subsequent surfaces, e.g. at point 1 on the subsequent surface f_1, in Fig. 8.19a.

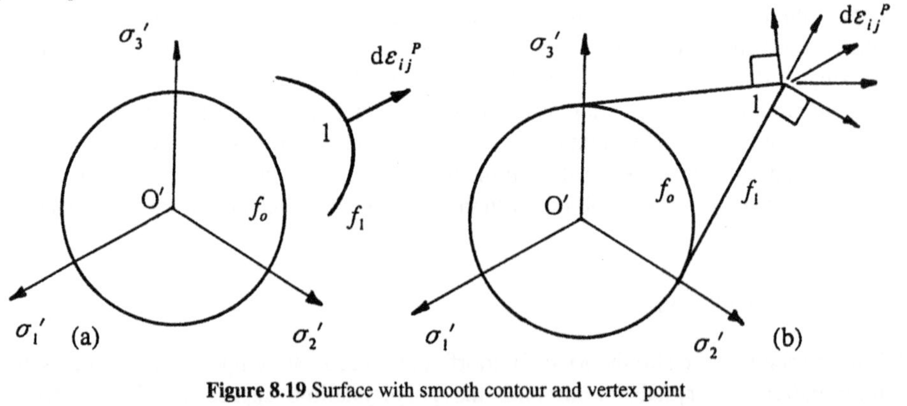

Figure 8.19 Surface with smooth contour and vertex point

The slip theory was validated indirectly from experiments [40, 41] that demonstrated a fan of normals at stress point 1 in Fig. 8.19b. The fan implies an undefined direction of the plastic strain increment vector at the load point, which would be expected in the presence of a corner in the yield surface. Corners may be indeed be present for yielding in Tresca-type single crystals [7]. In contrast, when the many slip systems are activated within the grains of a polycrystal, the aggregate yielding behaviour does provide the unique normal at the load point, as given by macro-plasticity theory.

8.8.3 Kinematic Hardening

When the applied stress is removed from a plastically deformed polycrystal an instantaneous elastic recovery occurs but there remains a self-equilibrating, residual stress distribution. Microscopically, an internal stress builds as the increasing dislocation density of forward deformation becomes less mobile. Dislocations entangle and are impeded by obstacles lying in active slip planes. A pile-up of dislocations promotes a back-stress which resists further slip under the resolved stress in these slip planes. Many of the crystals remain permanently slipped following unloading of the aggregate from the plastic region. The remaining heterogenous distribution of residual stress carries a sense in opposition to the applied stress.

Consider a reversal in the direction of uniaxial deformation, say at point A in Fig. 8.20a. The stress required to initiate reversed yielding at B is lowered because of the additive contribution from an internal stress of the same sense. Thus, the Bauschinger effect, $\sigma_B < \sigma_A$, can be explained from the existence of a residual micro-stress [42, 43].

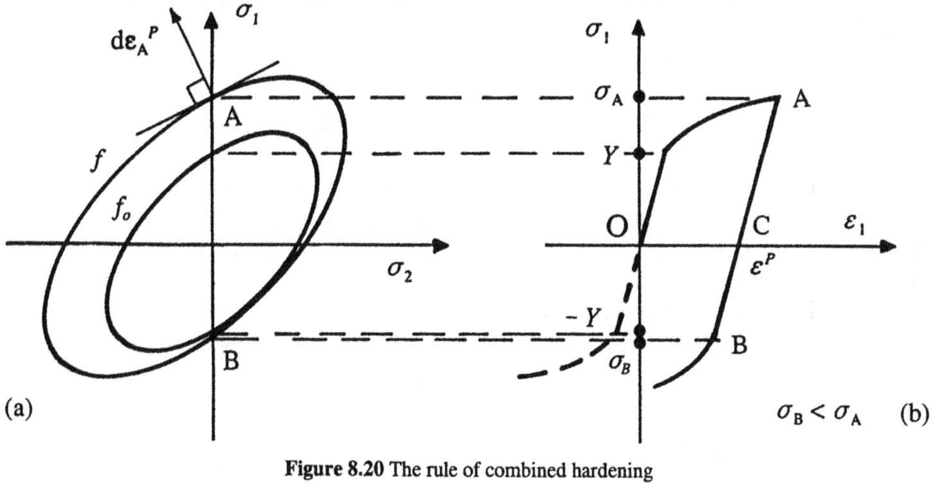

Figure 8.20 The rule of combined hardening

Next, consider uniaxial loading OA, unloading AC and re-loading CA in Fig. 8.20b. The sense of the internal stress, arising from a plastic loading-unloading cycle, opposes the applied stress in re-loading. This explains, qualitatively, why a forward yielding will not occur until the stress level again reaches that at point A. The increase in the forward yield stress (i.e. $\sigma_A > Y$) is associated with the plastic prestrain, ε^P. Clearly, if new yield points A and B are to lie on the subsequent yield surface f, then the initial yield surface f_o must combine an expansion with a translation from its origin, as shown in Fig. 8.20a.

The true *kinematic hardening rule* refers to a simplified, rigid translation of the yield surface in the absence of an expansion (see Fig. 8.21a). Figure 8.21b shows that linear hardening is implied in which the initial elastic range $2Y$ remains unaltered by plastic strain, e.g. $\sigma_A - \sigma_B = 2Y$, where Y is the initial yield stress. The amount of hardening is measured by an increase in the flow stress: $\sigma_A - Y$. The amount of softening is measured by a decrease in flow stress: $Y - \sigma_B$. Each change to Y will have the same magnitude for a given ε^P.

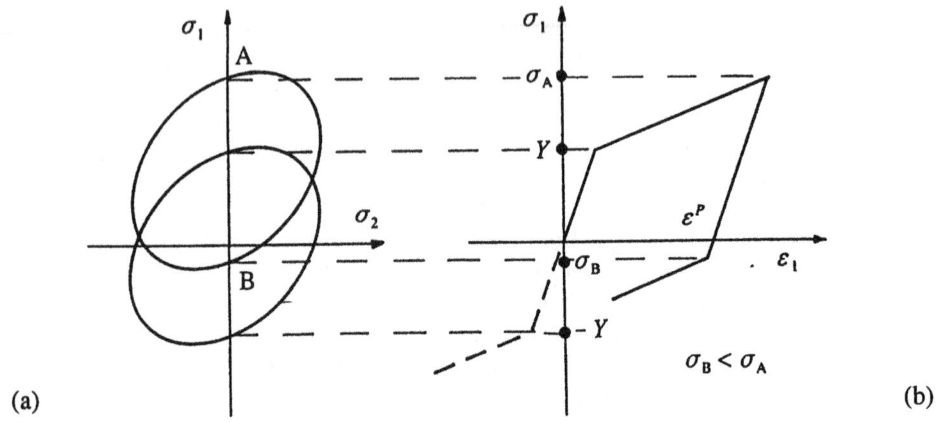

Figure 8.21 The rule of kinematic hardening

These observations reveal an equivalence between the internal stress and the centre of a translated yield surface. Centre co-ordinates a_{ij} define a strain history dependent translation In the Kroner-Eshelby model a_{ij} becomes the difference between the micro and macro stress tensors. Re-arranging eq(8.24) gives an incremental translation corresponding to respective stress changes:

$$\mathrm{d}a_{ij} = \mathrm{d}\sigma_{ij}^{*} - \mathrm{d}\sigma_{ij} = 2\mu G\,(\mathrm{d}\varepsilon_{ij}^{P} - \mathrm{d}\varepsilon_{ij}^{P*})$$

or

$$\mathrm{d}\mathbf{A} = \mathrm{d}\mathbf{T}^{*} - \mathrm{d}\mathbf{T} = 2\mu G\,(\mathrm{d}\mathbf{E}^{P} - \mathrm{d}\mathbf{E}^{P*})$$

The theoretical predictions c - e in Fig. 8.16 employ this model. By reversing the direction of deformation they also predict softening and can therefore be expected to give a realistic Bauschinger effect. Here a dislocation pile-up is the major contributor to the development of back (or internal) stress.

Often, it is more appropriate to model the behaviour in Fig. 8.20 by combining the rules of isotropic and kinematic hardening. For example, Brown [44] described the translation for prestrained aluminium with the non-linear hardening eq(8.34). However, subsequent flow potentials for increasing times under stress (i.e. creep), were concentric Mises ellipsoids with a common centre a_{ij}. Experimental investigations have shown that both initial and subsequent yield surface for a polycrystal are sensitive to the chosen definition of yield (see p. 71). Where a linear region of elasticity exists in pre-strained material this is evidence [45] for a translation in the subsequent yield surface (see Fig. 8.21). This behaviour suggests that kinematic hardening is the preferred model of plasticity with load reversal.

8.9 Summary

This chapter has presented a micro-mechanical description of yield, flow and hardening behaviour. These are the key elements employed in the development of continuum theories of plasticity. Much of micro-mechanics involves the study of slip in single crystals and polycrystals for metals. It appears from the literature that additional sources of plasticity arise in metal alloys. This is not surprising since the essential purpose of alloying is to inhibit slip. Consequently, in order to accommodate a loading strain internally, the identification of those slip planes which remain active would have to be considered along with crack and void formation. Despite these additional processes, the elements of continuum plasticity theory remains largely unaltered. In its use of macro-stress and strain across a poycrystalline aggregate, the continuum theory does not differentiate between metals and alloys specifically. By employing an appropriate yield function and a universal flow rule it supplies the macroscopic plastic strain within the bulk volume of any polycrystal. It will be seen that this is achieved with the empirical representation of hardening behaviour, derived from uniaxial stress-strain curves. The latter depend upon the service conditions, among which the strain rate, test temperature and heat treatment are important.

References

1. Cottrell A. *Theory of Crystal Dislocations*, 1964, Blackie & Son, London.
2. Honeycombe R. W. K. *The Plastic Deformation of Metals*, 1985, Arnold.
3. Schmidt E. and Boas W. *Plasticity of Crystals*, 1950, Hughes, London.
4. Petch N. J. *Jl Iron and Steel Inst*, 1953, **173**, 25.
5. Lin T. F. *Advances in Applied Mechanics*, (ed Chia-Shun,Y.) 1971, 255, Academic Press, New York.
6. Kocks U. F. *Metallurgical Trans*, 1970, **1**, 1121.
7. Kocks U. F. *Unified Constitutive Equations for Creep and Plasticity*, (ed Miller A. K.) 1987, **1**, Elsevier.
8. Taylor G. I. *Jl Inst Metals*, 1938, **62**, 307.
9. Bishop J. F. W. and Hill R. *Phil. Mag*, 1951, **42**, 418.
10. Bishop J. F. W. and Hill R. *Phil. Mag*, 1951, **42**, 1298.
11. Backofen W. A. *Deformation Processing*, 1972, Addison-Wesley.
12. Althoff J. and Wincierz P. *Z Metallkde*, 1972, **63**, 623.
13. Piehler H. R. and Backofen W. A. *Textures in Research and Fracture* (eds Grewen J. and Wassermann G.) 1969, 436, Springer-Verlag.
14. Grzesik D. *Z. Metallkde*, 1972, **63**, 618.
15. Grzesik D. *Mechanische Anisotropie*, (ed Stuwe H. P.) 1974, 231, Springer-Verlag.
16. Matucha K. H. and Wincierz P. ibid, 201.
17. Avery D. H., Hosford W. F. and Backofen W. A. *Trans AIME*, 1965, **233**, 71.
18. Johnson W. G. and Gilman J. J. *Jl Appl Phy*, 1959, **30**, 129.
19. Rice J. R. *Trans ASME, Jl Appl Mech*, 1970, **37**, 728.
20. Batdorf B. and Budianski B. "A mathematical theory of plasticity based upon the concept of slip", *NACA Tech Note* 1871, April 1949.
21. Batdorf S. B. and Budiansky B. *Jl Appl Mech*, 1954, **21**, 323.
22. Eshelby J. D. *Progress in Solid Mechanics*, 1961, **2**, 87, North Holland.
23. Kroner E. *Acta Metallurgica*, 1961, **9**, 155.
24. Hill R. *Jl Mech Phys Solids*, 1965, **13**, 89.
25. Berveiller M. and Zaoui A. *Jl Mech Phys Solids*, 1979, **26**, 325.
26. Berveiller M. and Zaoui A. *Trans ASME, Jl Eng Mat and Tech*, 1984, **106**, 295.
27. Brown G. M. *Jl Mech Phys Solids*, 1970, **18**, 367.
28. Hutchinson J. W. *Proc Roy Soc*, 1970, **A319**, 247.
29. Weng G. J. *Int Jl Plasticity*, 1985, **3**, 275.
30. Weng G. J. *Int Jl Plasticity*, 1987, **3**, 315.
31. Tokuda M., Kratochvil J. and Ohashi Y. *Phys Stat Sol (a)*, 1981, **68**, 629.
32. Taylor G. I. and Quinney H. *Phil Trans Roy Soc*, 1931, **A203**, 323.
33. Lin T. H. *Jl Mech Phys Solids*, 1957, **5**, 143.

34. Havner K. S. *Int Jl Plasticity*, 1985, **1**, 111.
35. Argon A. S. *Constitutive Equations in Plasticity*, 1975, MIT Press.
36. Budiansky B. and Wu T.T. Proc: *4th U.S. Nat Cong Applied Mechanics*, ASME, New York, 1963, **2**, 1175, Pergamom.
37. Como M. and D'Agostino S. *Meccanica*, 1969, **4**, 146.
38. Como M. and D'Agostino S. *Archives of Mechanics*, 1973, **25**, 685.
39. Michno M. J. and Findlay W. N. *Int Jl Non-Linear Mech*, 1976, **11**, 59.
40. Naghdi P. M., Rowley J. C. and Beadle C. W. *Jl. Appl. Mech*, 1955, **22**, 416.
41. Phillips A. and Gray G. A. *Trans ASME, Jl Basic Eng*, 1961, **83**, 275.
42. Como M. and D'Agostino S. *Meccanica*, 1969, **4**, 146.
43. Kafka V. *ZAMM*, 1968, **48**, 265.
44. Brown G. M. *Jl Mech Phys Solids*, 1970, **18**, 383.
45. Ikegami K. Translation *BISI* 14420, 1976, The Metals Society.

Exercises

8.1 Apply eq(8.1) to show when a tensile stress σ is applied to a single crystal, the critical resolved shear stress τ_{cr} is at a maximum for $\phi = \lambda = 45°$. Plot the variation in τ_{cr}/σ when the slip plane is incined at various angles to the stress axis

8.2 Sketch isometric views of: (a) a cube with the portion above the (111) plane removed and (b) a hexagonal prism with the portion above plane (0112) removed, indicating the [1120] direction.

8.3 Show that eqs(8.11a,b) appears in matrix notation as

$$\mathrm{tr}\,(\delta \mathbf{T}\,\delta \mathbf{E}^{\mathrm{P}}) = \int_V \mathrm{tr}\,(\mathrm{d}\mathbf{T}^{*}\,\mathrm{d}\mathbf{E}^{\mathrm{P}*})\,\mathrm{d}V$$

where

$$\delta \mathbf{E}^{\mathrm{P}} = \int_V \mathrm{d}\mathbf{E}^{\mathrm{P}*}\,\mathrm{d}V$$

in which **T** is Cauchy stress matrix (1.46) and $\delta\mathbf{E}^{\mathrm{P}}$ is the incremental plastic strain matrix (2.5b)

8.4 Show that observed values $\tau_{cr} = 6$, 4 and 20 ($\times\, 10^6$ dynes/cm^2) for silver, aluminium and iron respectively, are approximately four orders of magnitude less than their theoretical values given from $\tau_{cr} = G/(2\pi)$. Respective elastic constants (Young's modulus and Poisson's ratio) for these metals are: $E = 70$ GPa, $\nu = 0.37$; $E = 71$ GPa, $\nu = 0.34$ and $E = 206$ GPa, $\nu = 0.29$.

8.5 List the particular point, line, plane and volume defects most responsile for the large discrepancy referred to in Exercise. 8.4

8.6 Examine the application of eq(8.23a) to a spherical f.c.c. grain when the distribution of plastic strain ε_{ij}^{P*} is according to eq(8.8).

8.7 Examine what modifications to eq(8.23a) would be required to accommodate an ellipsoidal grain.

CHAPTER 9

THE FLOW CURVE

9.1 Introduction

In the earlier chapters ideal materials were assumed to flow under a constant yield stress. Some metals do approximate to this behaviour but most will require an increase in stress to progress plastic flow. The hardening rate $d\sigma/d\varepsilon^P$ refers to this increase in stress with plastic strain. This rate is at its greatest immediately beyond the yield point, where there is a rapid increase in dislocation density. Thus, it is in the lower strain region of plasticity where an account of hardening is made, usually under a monotonic loading. If slip processes are to continue, the stress must be raised further to overcome the barriers presented by a dislocated structure. The rate of hardening will, however, diminish to zero at large strain, where the energy of the structure can activate additional climb and cross slip recovery mechanisms.

Theories of plasticity account for hardening by employing various empirical descriptions of uniaxial or torsional flow behaviour. These are found from standard laboratory tests, the choice depending upon the range of plastic strain required. Testing is normally performed under an increasing loading so that a derived description applies to forward deformation. Tension and compression tests are most reliable when limited to the smaller strain region, say, for a theory of an elastic-plastic structure. The appearance of necking in the tension test and barrelling and buckling under compression, prevents each test from attaining large uniaxial strain values. To overcome these experimental difficulties, a modified uniaxial compression test, plain strain and bulge testing have been employed. These allow flow behaviour to be represented at much larger strains, typical of those reached in metal forming processes.

In this chapter, these standard tests are described and their data is employed to demonstrate empirical representations to flow curves. The latter appear in axes of equivalent stress versus equivalent plastic strain. A definition of equivalence is required to correlate flow curves under different stress states in a material with a known initial condition. Particular attention is paid to annealed, isotropic materials obeying the von Mises and Drucker yield conditions. The distinction between the hypotheses of strain and work hardening in these materials is also examined.

9.2 Equivalence in Plasticity

Consider an initially isotropic material whose yield surface is described by a function of the stress deviator invariants: $f(J_2', J_3')$. The von Mises function [1]: $f = J_2'$, is by far the most common form since it relies on the second invariant alone. However, the determination of one constant, say in the Drucker function [2]: $f = J'^{\,2}_2 - cJ'^{\,2}_3$, accommodates a wider range of isotropic behaviour. Here it will be shown that equivalence expressions for stress and plastic strain rest with the choice between a work and strain hardening hypothesis.

9.2.1 *Equivalent Stress*

(a) *von Mises*

When the isotropic yield function f is equated to the second invariant of deviatoric stress this implies that hydrostatic stress does not contribute to plastic deformation:

$$f = J_2' = \frac{1}{2}\,\sigma_{ij}'\,\sigma_{ij}' \tag{9.1a}$$

Yielding begins under a given stress state when J_2' attains a critical value. The latter is determined from uniaxial yield stress Y as $J_2' = Y^2/3$. We then have

$$J_2' = \frac{1}{3}\,Y^2 = \frac{1}{2}\,\sigma_{ij}'\,\sigma_{ij}' \tag{9.1b}$$

from which the yield stress is

$$Y = \sqrt{\frac{3}{2}\,\sigma_{ij}'\,\sigma_{ij}'} \tag{9.2a}$$

Thus, Y may be regarded as an effective, or equivalent, yield stress for a general stress state. When this yield stress is exceeded, an equivalent flow stress $\bar{\sigma}$ ($\bar{\sigma} > Y$) for the plasticity that ensues is correspondingly defined as

$$\bar{\sigma} = \sqrt{\frac{3}{2}\,\sigma_{ij}'\,\sigma_{ij}'} \tag{9.2b}$$

Written in its fully expanded form, eq(9.2b) becomes:

$$\bar{\sigma} = \frac{1}{\sqrt{2}}\sqrt{(\sigma_{11} - \sigma_{22})^2 + (\sigma_{22} - \sigma_{33})^2 + (\sigma_{11} - \sigma_{33})^2 + 6\,(\sigma_{12}^2 + \sigma_{23}^2 + \sigma_{13}^2)} \tag{9.2c}$$

For a principal stress system, the shear stress components σ_{12}, σ_{13} and σ_{23}, are absent so that eq(9.2c) reduces to

$$\bar{\sigma} = \frac{1}{\sqrt{2}}\sqrt{(\sigma_1 - \sigma_2)^2 + (\sigma_2 - \sigma_3)^2 + (\sigma_1 - \sigma_3)^2} \tag{9.2d}$$

Equations (9.2c,d) are the most commonly used equivalent stresses. Equation (9.2b) provides an alternative $\bar{\sigma}$ expression for principal deviatoric stress space (see Fig. 9.1).

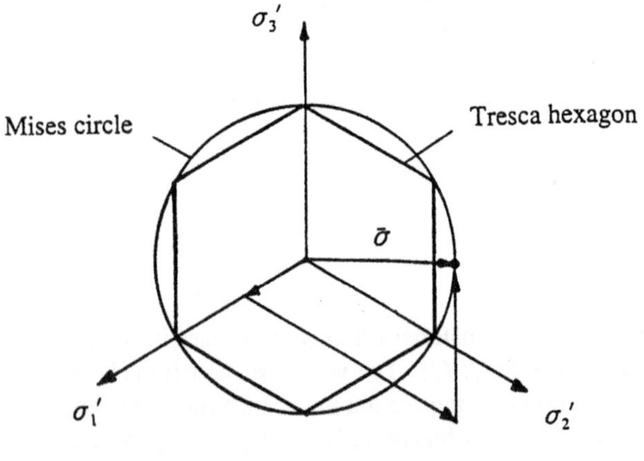

Figure 9.1 The deviatoric plane

Geometrically, eq(9.2b) shows that $\bar{\sigma}$ describes the radius of the circle that projects from the yield surface upon the deviatoric plane, as shown in Fig. 9.1. That is

$$\bar{\sigma} = \sqrt{\frac{3}{2}\left(\sigma_1'^2 + \sigma_2'^2 + \sigma_3'^2\right)} \tag{9.2e}$$

where $\sigma_1' = \sigma_1 - \frac{1}{3}(\sigma_1 + \sigma_2 + \sigma_3)$ etc and $\sigma_1' + \sigma_2' + \sigma_3' = 0$. This circle is the cross-section of the uniform, von Mises cylinder whose axis is equally inclined to orthogonal axes of deviatoric stress. Tresca's cylinder, found from eq(3.14c), appears as a hexagonal section inscribed within the Mises circle upon the deviatoric plane. Since the axes of both cylinders are coincident with the axis of hydrostatic stress, the latter does not affect the radius of the cylinder cross-section, i.e. the equivalent flow stress. However, as $\bar{\sigma}$ increases with plastic hardening, the initial yield surface is taken to expand concentrically by the isotropic hardening rule to contain the current equivalent stress point.

(b) *Drucker* [3]
Here the yield function is of the form:

$$f(J_2', J_3') = J_2'^3 - J_3'^2 \tag{9.3a}$$

where for a uniaxial equivalent stress, the second and third stress deviator invariants become

$$J_2' = \frac{1}{2}\sigma_{ij}'\sigma_{ij}' = \frac{1}{3}\bar{\sigma}^2 \quad \text{and} \quad J_3' = \frac{1}{3}\sigma_{ij}'\sigma_{jk}'\sigma_{ki}' = \frac{2}{27}\bar{\sigma}^3 \tag{9.3b,c}$$

Combining eqs(9.3a-c), Drucker's function gives

$$f = J_2'^3 - cJ_3'^2 = \left(\frac{\bar{\sigma}^2}{3}\right)^3 - c\left(\frac{2\bar{\sigma}^3}{27}\right)^2$$

from which the equivalent stress is

$$\bar{\sigma} = \left[\frac{J_2'^3 - cJ_3'^2}{\left(\frac{1}{3}\right)^3 - c\left(\frac{2}{27}\right)^2}\right]^{\frac{1}{6}} \tag{9.4a}$$

$$= \left[\frac{\frac{27}{8}\left(\sigma_{ij}'\sigma_{ij}'\right)^3 - 3c\left(\sigma_{ij}'\sigma_{jk}'\sigma_{ki}'\right)^2}{1 - \frac{4c}{27}}\right]^{\frac{1}{6}} \tag{9.4b}$$

The general, expanded forms of eq(9.4a,b) are cumbersome though they can be reduced to give $\bar{\sigma}$ for any stress subspace without difficulty. For yielding under torsion, the numerator in eq(9.4b) is simply the shear yield stress $(k^6)^{1/6} = k$. This enables the constant c to be found from the initial tensile yield stress $\bar{\sigma} = Y$, as

$$Y = \frac{k}{\left[\left(\frac{1}{3}\right)^3 - c\left(\frac{2}{27}\right)^2\right]^{\frac{1}{6}}} \tag{9.4c}$$

It will be shown later in this chapter that by including J_3' in the yield function f can account for a greater range of experimental observations.

9.2.2 *Equivalent Plastic Strain*

Equivalent plastic strain expressions may be deduced from the corresponding equivalent stresses. A non-rigorous approach [3] exploits the fact that the symmetrical, second order tensors of deviatoric stress and plastic strain increment have identical properties:

$$\mathbf{T}' = \begin{bmatrix} \sigma_{11}' & \sigma_{12}' & \sigma_{13}' \\ \sigma_{21}' & \sigma_{22}' & \sigma_{23}' \\ \sigma_{31}' & \sigma_{32}' & \sigma_{33}' \end{bmatrix}, \quad \mathrm{d}\mathbf{E}^P = \begin{bmatrix} \mathrm{d}\varepsilon_{11}^P & \mathrm{d}\varepsilon_{12}^P & \mathrm{d}\varepsilon_{13}^P \\ \mathrm{d}\varepsilon_{21}^P & \mathrm{d}\varepsilon_{22}^P & \mathrm{d}\varepsilon_{23}^P \\ \mathrm{d}\varepsilon_{31}^P & \mathrm{d}\varepsilon_{32}^P & \mathrm{d}\varepsilon_{33}^P \end{bmatrix}$$

Correspondence within the matrix notation is: $\sigma_{ii}' = 0$ (tr $\mathbf{T}' = 0$) and $\mathrm{d}\varepsilon_{ii}^P = 0$ (tr $\mathrm{d}\mathbf{E}^P = 0$). Provided the tensor shear strains are used: $\mathrm{d}\varepsilon_{12}^P = \mathrm{d}\varepsilon_{21}^P = ½\,\mathrm{d}\gamma_{12}^P$, $\mathrm{d}\varepsilon_{13}^P = \mathrm{d}\varepsilon_{31}^P = ½\,\mathrm{d}\gamma_{13}^P$ and $\mathrm{d}\varepsilon_{23}^P = \mathrm{d}\varepsilon_{32}^P = ½\,\mathrm{d}\gamma_{23}^P$, then the three invariants of deviatoric stress (J_1', J_2', J_3') are identical to those of plastic strain (I_1, I_2, I_3). These are:

$$J_1 = \sigma_{ii}' = \sigma_{11}' + \sigma_{22}' + \sigma_{33}' = 0, \qquad I_1 = \mathrm{d}\varepsilon_{ii}^P = \mathrm{d}\varepsilon_{11}^P + \mathrm{d}\varepsilon_{22}^P + \mathrm{d}\varepsilon_{33}^P = 0$$

$$J_2' = \frac{1}{2}\,\sigma_{ij}'\,\sigma_{ij}', \qquad I_2 = \frac{1}{2}\,\mathrm{d}\varepsilon_{ij}^P\mathrm{d}\varepsilon_{ij}^P \qquad (9.5a,b)$$

$$J_3' = \frac{1}{3}\,\sigma_{ij}'\,\sigma_{jk}'\,\sigma_{ki}', \qquad I_3 = \frac{1}{3}\,\mathrm{d}\varepsilon_{ij}^P\mathrm{d}\varepsilon_{jk}^P\mathrm{d}\varepsilon_{ki}^P$$

By employing the similarity between eqs(9.5a,b) the equivalent strain can be deduced from the equivalent stress for a given yield function.

(a) *von Mises*

Let $\mathrm{d}\varepsilon_{11}^P$, $\mathrm{d}\varepsilon_{22}^P$, $\mathrm{d}\varepsilon_{33}^P$, $\mathrm{d}\varepsilon_{12}^P$, $\mathrm{d}\varepsilon_{13}^P$ and $\mathrm{d}\varepsilon_{23}^P$, be the incremental plastic strains arising from a change in a general deviatoric stress state from σ_{ij}' to $\sigma_{ij}' + \mathrm{d}\sigma_{ij}'$. The equivalent plastic strain increment expression is identified with the corresponding change to I_2. Let this change be equivalent to the change in I_2 for uniaxial tension within its three, principal strain increments $\mathrm{d}\varepsilon_1^P$, $-½\mathrm{d}\varepsilon_1^P$ and $-½\mathrm{d}\varepsilon_1^P$, for an isotropic solid. Substituting the three strain components into eq(9.5b), leads to $I_2 = ¾(\mathrm{d}\varepsilon_1^P)^2 = ¾(\mathrm{d}\bar{\varepsilon}^P)^2$, from which

$$\mathrm{d}\bar{\varepsilon}^P = \sqrt{\frac{4I_2}{3}} = \sqrt{\frac{2}{3}\left(\mathrm{d}\varepsilon_{ij}^P\,\mathrm{d}\varepsilon_{ij}^P\right)} = \sqrt{\frac{2}{3}\,\mathrm{tr}\left(\mathrm{d}\mathbf{E}^P\right)^2} \qquad (9.6a)$$

Equation (9.6a) expands into its general component form:

$$\mathrm{d}\bar{\varepsilon}^P = \frac{\sqrt{2}}{3}\Big\{(\mathrm{d}\varepsilon_{11}^P - \mathrm{d}\varepsilon_{22}^P)^2 + (\mathrm{d}\varepsilon_{22}^P - \mathrm{d}\varepsilon_{33}^P)^2 + (\mathrm{d}\varepsilon_{11}^P - \mathrm{d}\varepsilon_{33}^P)^2$$

$$+ 6\left[\left(½\,\mathrm{d}\gamma_{12}^P\right)^2 + \left(½\,\mathrm{d}\gamma_{23}^P\right)^2 + \left(½\,\mathrm{d}\gamma_{13}^P\right)^2\right]\Big\}^{\frac{1}{2}} \qquad (9.6b)$$

where $d\varepsilon_{12}{}^P = \tfrac{1}{2} d\gamma_{12}{}^P$ etc. The most widely used $d\bar{\varepsilon}^P$ expression employs principal plastic strain increments. Putting $d\gamma_{12}{}^P = d\gamma_{13}{}^P = d\gamma_{23}{}^P = 0$ in eq(9.6b) gives this as

$$d\bar{\varepsilon}^P = \frac{\sqrt{2}}{3}\sqrt{\left(d\varepsilon_1^P - d\varepsilon_2^P\right)^2 + \left(d\varepsilon_2^P - d\varepsilon_3^P\right)^2 + \left(d\varepsilon_1^P - d\varepsilon_3^P\right)^2} \qquad (9.6c)$$

An alternative, principal strain form follows directly from eq(9.6a):

$$d\bar{\varepsilon}^P = \sqrt{\frac{2}{3}\left[\left(d\varepsilon_1^P\right)^2 + \left(d\varepsilon_2^P\right)^2 + \left(d\varepsilon_3^P\right)^2\right]} = \sqrt{\frac{2}{3}\,\mathrm{tr}\,(d\mathbf{E}^P)^2} \qquad (9.6d)$$

Note that the incompressibility condition $d\varepsilon_{ii}{}^P = 0$ connects eq(9.6c) to eq(9.6d).

A difference in the numerical factor arises between the $\bar{\sigma}$ and $d\bar{\varepsilon}^P$ expressions because a uniaxial stress does not produce uniaxial plastic strain. Lateral strains, $d\varepsilon_2{}^P$ and $d\varepsilon_3{}^P$ accompany the axial strain $d\varepsilon_1{}^P$, where $d\varepsilon_2{}^P = d\varepsilon_3{}^P = -\tfrac{1}{2} d\varepsilon_1{}^P$. Substituting this relationship into eq(9.6c or d) it may be checked that these lead correctly to $d\bar{\varepsilon}^P = d\varepsilon_1{}^P$, i.e. they ensure that the axial strain is the equivalent strain under a uniaxial stress.

The octahedral shear stress and strain increment have also been used as equivalent measures for multi-axial plasticity [4]. These are (see eq(1.34b)) defined as:

$$\tau_o = \frac{1}{3}\sqrt{(\sigma_1 - \sigma_2)^2 + (\sigma_2 - \sigma_3)^2 + (\sigma_1 - \sigma_3)^2}$$

$$d\gamma_o^P = \frac{2}{3}\sqrt{\left(d\varepsilon_1^P - d\varepsilon_2^P\right)^2 + \left(d\varepsilon_2^P - d\varepsilon_3^P\right)^2 + \left(d\varepsilon_1^P - d\varepsilon_3^P\right)^2}$$

If we now reduce these to a uniaxial deformation with the respective substitutions (σ_1, 0, 0) and ($d\varepsilon_1{}^P$, $-\tfrac{1}{2} d\varepsilon_1{}^P$, $-\tfrac{1}{2} d\varepsilon_1{}^P$) we find $\tau_o = \sqrt{2}\sigma_1/3$ and $d\gamma_o{}^P = \sqrt{2} d\varepsilon_1{}^P$, which leads to a more general connection between the equivalent and octahedral stress and strain measures:

$$\bar{\sigma} = 3\tau_o/\sqrt{2} \quad \text{and} \quad d\bar{\varepsilon}^P = d\gamma_o{}^P/\sqrt{2}$$

(b) *Drucker*

To correspond with the use J_3' in defining $\bar{\sigma}$ so I_3 is used to define $d\bar{\varepsilon}^P$. Thus, a strain equivalence is sought between changes to the invariant expression: $I_2{}^3 - cI_3{}^2$, arising from the general and uniaxial systems. For the latter, we have the two invariants $I_2 = \tfrac{3}{4}(d\varepsilon_1{}^P)^2$ and $I_3 = \tfrac{1}{4}(d\varepsilon_1{}^P)^3$, giving

$$I_2^3 - cI_3^2 = [\tfrac{3}{4}(d\varepsilon_1{}^P)^2]^3 - c\,[\tfrac{1}{4}(d\varepsilon_1{}^P)^3]^2$$

Since the axial strain is the equivalent strain:

$$(d\varepsilon_1{}^P)^6 = (d\bar{\varepsilon}^P)^6 = (I_2^3 - cI_3^2)/\,[\,(\tfrac{3}{4})^3 - c(\tfrac{1}{4})^2]$$

Substituting for I_2 and I_3 from eq(9.5b), in the general case

$$d\bar{\varepsilon}^P = \left[\frac{2(d\varepsilon_{ij}^P d\varepsilon_{ij}^P)^3 - \frac{16c}{9}(d\varepsilon_{ij}^P d\varepsilon_{jk}^P d\varepsilon_{ki}^P)^2}{\frac{27}{4} - c}\right]^{\frac{1}{6}} \qquad (9.7a)$$

The principal strain form follows from the expansion to eq(9.7a):

$$d\bar{\varepsilon}^P = \left\{ \frac{2\left[(d\varepsilon_1^P)^2 + (d\varepsilon_2^P)^2 + (d\varepsilon_3^P)^2\right]^3 - \frac{16c}{9}\left[(d\varepsilon_1^P)^3 + (d\varepsilon_2^P)^3 + (d\varepsilon_3^P)^3\right]^2}{\frac{27}{4} - c} \right\}^{\frac{1}{6}} \quad (9.7b)$$

or

$$d\bar{\varepsilon}^P = \left\{ \frac{2\left[\mathrm{tr}\,(d\mathbf{E}^P)^2\right]^3 - \frac{16c}{9}\left[\mathrm{tr}\,(d\mathbf{E}^P)^3\right]^2}{\frac{27}{4} - c} \right\}^{\frac{1}{6}} \quad (9.7c)$$

Equivalent stress expressions corresponding to alternative isotropic yield functions $f(J_2', J_3')$, may be found in a similar manner. Literature reviews [5, 6] show an almost exclusive use of the von Mises $d\bar{\varepsilon}^P$ expressions (9.6a-d).

It has been observed [3] that the constant c appearing in eqs(9.4a-c), may be sufficient to correlate flow stress data without it re-appearing in the $d\bar{\varepsilon}^P$ expressions (9.7a-c). However, equivalence is required in plastic strain as well as in stress. For example, the determination of c from eq(9.4c) will only ensure that the initial yield points in tension and torsion assume the same value of equivalent stress. As the material subsequently hardens there is no guarantee that their equivalent stresses would agree for similar amounts of Mises equivalent plastic strain, where in torsion $\bar{\varepsilon}^P = \gamma^P/\sqrt{3}$, from eq(9.6b). Logically, we might seek better agreement in the flow behaviour between the two tests with $\bar{\varepsilon}^P$ defined from eq(9.7b). That is, by putting $d\varepsilon_1^P = ½ d\gamma^P$, $d\varepsilon_2^P = -½ d\gamma^P$ and $d\varepsilon_3^P = 0$ in eq(9.7b) and integrating, we get $\bar{\varepsilon}^P = \gamma^P/(27 - 4c)^{1/6}$. Experiment will readily show which approach is to be preferred but, firstly, it is necessary to examine in more detail the experimental errors that can arise between equivalent flow data. These normally apply to tension, torsion and compression tests. However, alternative tests may be used to extend the range of plastic strain, e.g. plane strain compression and balanced biaxial tension (bulge test).

9.3 Uniaxial Tests

When metallic materials harden with increasing plastic strain the flow stress increases. A suitable empirical account of this behaviour is normally derived from stress-strain curves under uniaxial tension or compression.

9.3.1 Simple Tension

The engineering stress and strain expressions employ the original testpiece area and length. Errors will arise in stress and strain calculations when large changes to testpiece dimensions occur within the plastic range. For example, the ultimate tensile strength of a metal will not be a true fracture stress if no account is made of the reduction in testpiece area that occurs during necking. Moreover, the stress in a material will actually continue to rise, even though the load falls, as the neck develops before final fracture. It is possible to correct the nominal tensile stress-strain curve to reflect this behaviour. The true stress σ and natural strain increment $\delta\varepsilon$ are defined as

$$\sigma = \frac{W}{A} \quad \text{and} \quad \delta\varepsilon = \frac{\delta l}{l} \tag{9.8a,b}$$

where A and l are the current section area and gauge length respectively.

(a) *True Stress and Strain*

Unlike elasticity, the plastic deformation that is produced by stressing beyond the elastic limit occurs without volume change. Thus, if A_o and l_o are the original area and length respectively, the nominal (engineering) stress and strain are

$$\sigma_o = \frac{W}{A_o} \quad \text{and} \quad \varepsilon_o = \frac{l - l_o}{l_o} \tag{9.9a,b}$$

The incompressibility (constant volume) condition gives

$$A\,l = A_o l_o \tag{9.10a}$$

$$A = \frac{A_o l_o}{l} = \frac{A_o}{1 + \varepsilon_o} \tag{9.10b}$$

Substituting eq(9.10b) into eqs(9.8a) leads to the true stress:

$$\sigma = \frac{W}{A_o}(1 + \varepsilon_o) = \sigma_o(1 + \varepsilon_o) \tag{9.11}$$

Integrating eq(9.8b) provides a measure of the true, or logarithmic, total strain between the original and current limits of length as

$$\varepsilon = \ln\left(\frac{l}{l_o}\right) = \ln(1 + \varepsilon_o) \tag{9.12a}$$

Removing the elastic strain component $\varepsilon^E = \sigma/E$ from eq(9.12a) leaves the plastic component of strain ε^P:

$$\varepsilon^P = \varepsilon - \varepsilon^E = \ln(1 + \varepsilon_o) - \frac{\sigma}{E} \tag{9.12b}$$

Equation (9.11) modifies stress in a nominal curve $\sigma_o = \sigma_o(\varepsilon_o)$, in the manner of Fig. 9.2a. In Figs 9.2b,c, eqs(9.12a,b) have also modified the strain axes to natural (true) strain and natural plastic strain respectively.

(a)

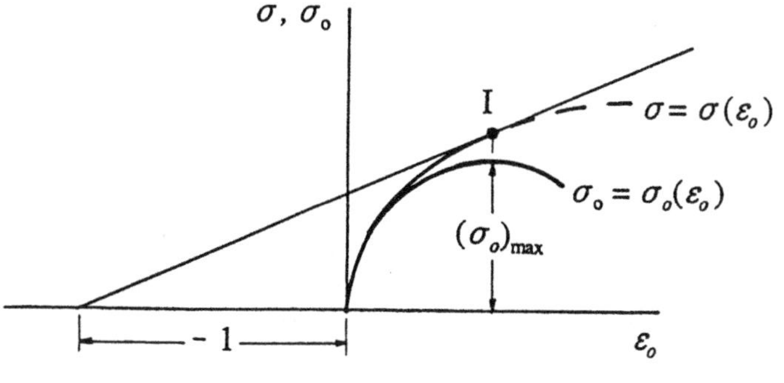

Figure 9.2 Hardening in nominal and true stress-strain axes

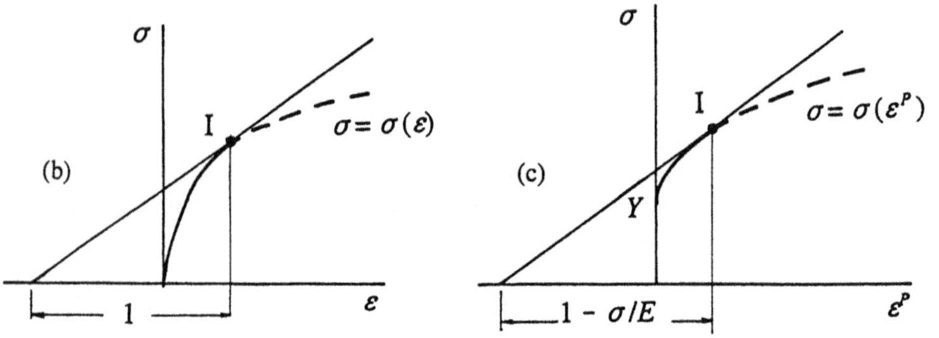

Figure 9.2 Hardening in true stress-strain axes (continued)

It has been shown previously that the axial plastic strain ε^P in tension (Fig. 9.2c) is the equivalent plastic strain.

(b) *Necking*

One limitation of a tension test is that the range of uniform strain is limited by the formation of a neck. In Considere's construction [7] the true stress at the point of instability corresponds to the maximum ordinate for σ_o in Fig. 9.2a. This condition is expressed in

$$\frac{d\sigma_o}{d\varepsilon_o} = 0 \tag{9.13a}$$

Substituting from eq(9.11) and differentiating the quotient leads to

$$\frac{d\sigma_o}{d\varepsilon_o} = (1 + \varepsilon_o)\frac{d\sigma}{d\varepsilon_o} - \sigma = 0$$
$$\frac{d\sigma}{d\varepsilon_o} = \frac{\sigma}{(1 + \varepsilon_o)} \tag{9.13b}$$

Equation (9.13b) shows that the tangent to a curve with an ordinate of true stress intersects the engineering strain axis at $\varepsilon_o = -1$ (see Fig. 9.2a). Now, from eq(9.12a),

$$\frac{d\varepsilon}{d\varepsilon_o} = \frac{1}{(1 + \varepsilon_o)} \tag{9.13c}$$

We may convert the instability condition (9.13b) to true strain using

$$\frac{d\sigma}{d\varepsilon} = \frac{d\varepsilon_o}{d\varepsilon} \times \frac{d\sigma}{d\varepsilon_o} \tag{9.14a}$$

Substituting eqs(9.13b,c) into eq(9.14a) gives

$$\frac{d\sigma}{d\varepsilon} = \sigma \tag{9.14b}$$

Equation (9.14b) shows that a subtangent value of unity appears on a true, total strain axis in Fig. 9.2b. Note that if eq(9.12b) had been used for this derivation instead of eq(9.12a), the instability condition in axes of true stress versus true plastic strain becomes

$$\frac{d\sigma}{d\varepsilon^P} = \frac{\sigma}{(1 - \sigma/E)} \tag{9.14c}$$

Here, the flow curve: $\sigma = \sigma(\varepsilon^P)$ commences at the yield stress Y, not at the origin (see Fig. 9.2c). Since there is no appreciable difference between eqs(9.14b,c), a unit subtangent is again used to locate the point of instability in these axes.

True stress and true strain are adequate for describing the uniform state of plastic deformation in a tensile testpiece up to the start of necking. However, the critical, uniaxial strain is often less than 20%, even in ductile materials. Continued straining is complicated by the non-uniform, triaxial stress state within the neck. Bridgeman [8] was the first to analyse the triaxial state of stress in the neck. He examined the dependence of radial σ_r and tangential σ_θ stresses upon the applied axial stress σ_z. Taking $\sigma_r = \sigma_\theta$, enables an equivalent stress to be derived for the triaxial stress state shown in Fig. 9.3.

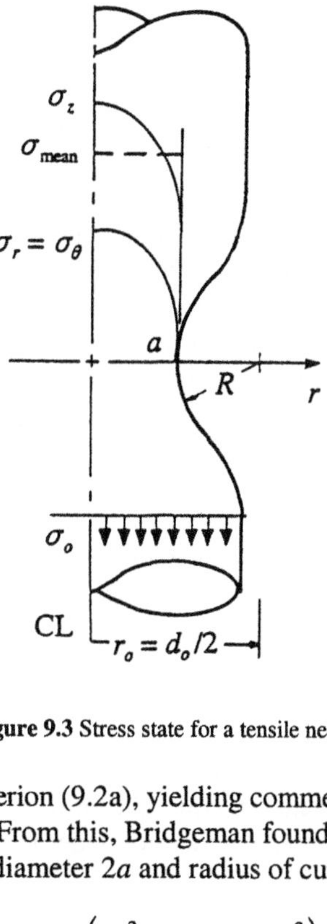

Figure 9.3 Stress state for a tensile neck

According to the von Mises criterion (9.2a), yielding commences in proportion to the root-mean-square of these stresses. From this, Bridgeman found the radial and axial stresses at the narrowest section, of neck diameter $2a$ and radius of curvature is R (see Fig. 9.3):

$$\sigma_r = \sigma_\theta = \tilde{\sigma} \ln \left(\frac{a^2 + 2aR - r^2}{2aR} \right) \tag{9.15a}$$

$$\sigma_z = \tilde{\sigma} \left[1 + \ln \left(\frac{a^2 + 2aR - r^2}{2aR} \right) \right] \tag{9.15b}$$

where $0 \le r \le a$ is an intermediate radiius within this section and $\tilde{\sigma}$ is the true equivalent stress in the neck. The distribution of σ_r and σ_z is shown in Fig. 9.3 . The mean axial stress σ_m in the neck must equilibrate the axial load W:

$$W = \pi a^2 \sigma_m = 2\pi \int_0^a r \sigma_z \, \mathrm{d}r$$

Substituting from eq(9.15b) gives

$$\pi a^2 \sigma_m = 2\pi\bar{\sigma}\int_0^a \left[r + r\ln\left(\frac{a^2 + 2aR - r^2}{2ar}\right)\right] dr$$

$$= 2\pi\bar{\sigma}\left\{\left(aR + \frac{a^2}{2}\right)\left[\ln(a^2 + 2aR) - \ln(2aR)\right]\right\}$$

$$= \pi\bar{\sigma}\left[(a^2 + 2aR)\ln\left(1 + \frac{a}{2R}\right)\right]$$

$$\frac{\bar{\sigma}}{\sigma_m} = \frac{1}{\left(1 + \frac{2R}{a}\right)\ln\left(1 + \frac{a}{2R}\right)} = \frac{1}{\left(1 + \frac{1}{X}\right)\ln(1 + X)} \qquad (9.16)$$

where $X = a/(2R)$. Bridgeman derived correction curves from eq(9.16) in which a/r_o was plotted against $\bar{\sigma}/\sigma_m$ for a given material (e.g. see Fig. 9.22, p. 307). Hence, if the smallest neck diameter $d = 2a$ is measured during an incremental loading to W, beyond the point of instability, the true equivalent stress for each load increment may be calculated.

The true strain can also be estimated from the neck diameter d using eq(9.12a) and the constant volume condition $\pi d_o^2 l_o/4 = \pi d^2 l/4$:

$$\varepsilon = \ln\left(\frac{l}{l_o}\right) = 2\ln\left(\frac{d_o}{d}\right) \qquad (9.17)$$

The broken lines in Figs 9.2b,c show how Bridgeman's necking-phase corrections extend the true stress-strain curves.

9.3.2 Simple Compression

Larger strains can be achieved from compression tests on short cylinders (see Fig. 9.4a) but unless the ends are well lubricated, a non-uniform deformation arises. Barrelling at the testpiece centre (Fig. 9.4b) is caused by cones of undeformed material penetrating inwards from each end, as shown.

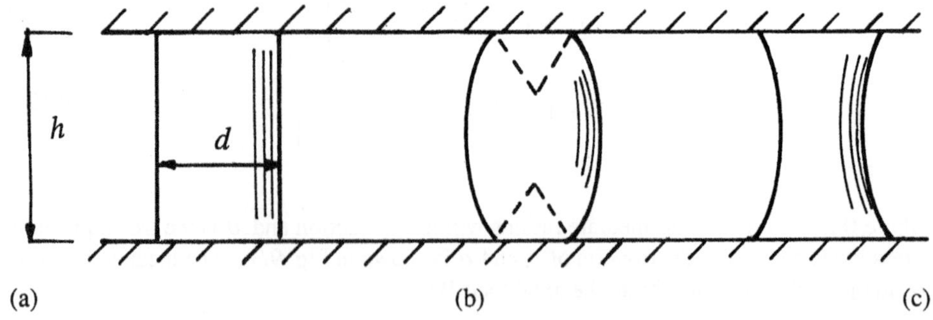

Figure 9.4 Uniform deformation, barrelling and bollarding under compression

The end diameter, in contact with the platens, forms the base of each cone. It follows that any transverse cross-section, cutting the cone, has an inner, elastic-core surrounded by an elastic-plastic annulus. When the apex of each cone has not penetrated to half the depth, there exists a fully-plastic central region at the barrel's maximum diameter. Within the central region, the stress state is triaxial, being composed of equal radial and circumferential stresses, in addition to the axial compession. It has been shown [9] that an effective stress $\bar{\sigma}$ may be related to the mean axial pressure p in the following form:

$$p = \left(1 + \frac{\mu d_o}{3h_o}\right)\bar{\sigma} \tag{9.18}$$

where d_o and h_o are the cylinder's original diameter and height respectively. Equation (9.18) shows that a uniaxial stress state $p = \sigma_e$ is achieved from compression only when the friction coefficient $\mu = 0$.

(a) *Lubrication*

Two methods ensure good lubrication of a short cylinder during compression: (i) to entrap molybdenum disulphide (MoS_2) paste within concentric grooves machined into its ends and (ii) to insert thin sheets of polytetraflouroethylene (ptfe) over the contact areas. The latter is particularly suitable for eliminating barrelling. In fact, with a ptfe thickness of ≈ 0.15 mm, the reverse behaviour of central waisting or 'bollarding' (see Fig. 9.4c) can occur [10]. This behaviour may be minimised with the use of an optimum ptfe thickness, between 0.05 and 0.15 mm, depending upon testpiece size. To attain large strains in the central region, the method is best employed with fresh sheets applied to the ends in a repeated loading-unloading test.

(b) *True stress and strain*

For a homogoneous compression (Fig. 9.4a), the true strain may be calculated from either expression in eq(9.17), depending upon whether the current height h or diameter d of the cylinder is measured. A minus sign is normally avoided by inverting the height ratio to give the natural compressive strain ε, as

$$\varepsilon = \ln\left(\frac{h_o}{h}\right) \tag{9.19a}$$

where $h_o > h$. Similarly, the engineering strain is defined positive:

$$\varepsilon_o = \frac{(h_o - h)}{h_o} \tag{9.19b}$$

Percentage deformation is another measure of compressive strain. It is derived from eq(9.19b) as $\%D = 100\varepsilon_o$. Combining eqs(9.19a and b) leads to

$$\varepsilon = \ln\left(\frac{1}{1 - \varepsilon_o}\right) \tag{9.20}$$

Correspondingly, the true compressive stress is positive when eq(9.11) is written as

$$\sigma = \frac{|W|}{A_o}(1 - \varepsilon_o) \tag{9.21}$$

where $|W|$ is the magnitude of an increasing compressive force. From eqs(9.20) and (9.21), the plastic component of strain is positive when:

$$\varepsilon^P = \varepsilon - \varepsilon^E$$
$$= \ln\left(\frac{1}{1 - \varepsilon_o}\right) - \frac{\sigma}{E}$$
$$= \ln\left(\frac{1}{1 - \varepsilon_o}\right) - \frac{|W|(1 - \varepsilon_o)}{A_o E} \tag{9.22a}$$

Equations (9.21) and (9.22a) allow a true stress versus natural plastic strain curve to be derived from a monotonic loading in compression, when h and W are the measured quantities. Given A_o and h_o are constants, Young's modulus E is derived from the elastic region as: $E = \sigma_o/\varepsilon_o = |W|\, h_o / [A_o(\,h_o - h)]$. An alternative to eq(9.22a) applies when the elastic strain is allowed to recover with an unloading from compression. The natural plastic strain is simply

$$\varepsilon^P = \ln\left(\frac{h_o}{h_p}\right) \tag{9.22b}$$

where h_p is the permanently deformed height.

9.4 Torsion Tests

9.4.1 Thin-Walled Tubes

Torsion tests are conducted on thin-walled cylindrical tubes when it is required to minimise the wall shear stress gradient. The theory of torsion [11] gives the mean shear stress as

$$\tau = \frac{T r_m}{J} \tag{9.23a}$$

where $r_m = \frac{1}{4}(d_i + d_o)$ is the mean wall radius, shown in Fig. 9.5, and J is the polar second moment of area:

$$J = \pi\,(d_o^{\,4} - d_i^{\,4})/32$$

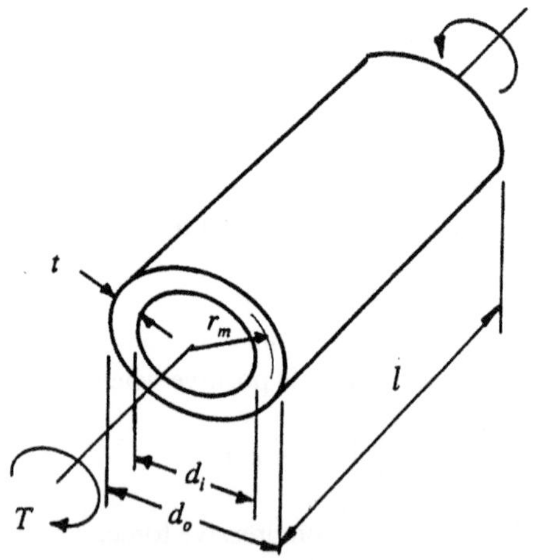

Figure 9.5 Thin-walled tube under torsion

Alternatively, where the wall thickness $t = \frac{1}{2}(d_o - d_i)$ is small compared to the mean diameter $d_m = 2r_m$, the Bredt-Batho theory of torsion [11] gives the mean shear stress as

$$\tau = \frac{T}{2At} \tag{9.23b}$$

where $A = \pi r_m^2$ is the area enclosed by the mean wall circle. The engineering shear strain is also referred to the mean radius as

$$\gamma = \frac{r_m \theta}{l} \tag{9.24a}$$

where θ is the angular twist (in radians) over a fixed length l. The plastic component of shear strain is

$$\gamma^P = \gamma - \gamma^E = \gamma - \frac{\tau}{G} \tag{9.24b}$$

Unlike a tension, the torsion of isotropic tube does not alter its length or cross-section and so a conversion to true stress and strain does not arise. However, large angular twist requires that shear strain be calculated from eq(1.4): $\gamma = \tan\phi$, where $\phi = r_m\theta/l$, since the small-strain approximation: $\gamma \approx \phi$ becomes invalid. If this simple shearing occurs in the 12 plane we may identify non-zero components, $\tau \equiv \sigma_{12}$ and $\gamma^P \equiv \varepsilon_{12}^P + \varepsilon_{21}^P = 2\varepsilon_{12}^P$, within eqs(9.2c) and (9.6c). This leads to an equivalent stress and equivalent plastic strain for torsion

$$\bar{\sigma} = \sqrt{3}\tau \quad \text{and} \quad \bar{\varepsilon}^P = \frac{\gamma^P}{\sqrt{3}} \tag{9.25a,b}$$

9.4.2 Solid Bar

Thin-walled tubes are prone to torsional buckling well before fracture, so limiting the range of shear strain. It is possible to achieve far greater strain from a torsion test on a solid cylindrical bar of hardening material. Nadai's account [12] of the shear stress gradient enables the stress-strain curve for the outer diameter (i.e. τ_o versus γ_o) to be derived from the torque-unit twist curve (T versus θ/l) in Fig. 9.6.

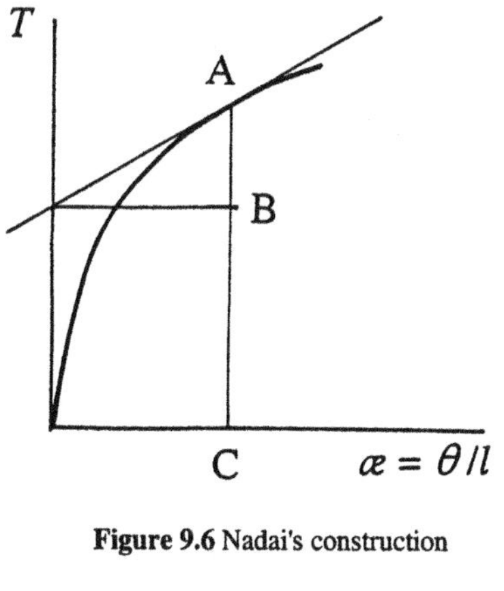

Figure 9.6 Nadai's construction

A solid bar, with outer radius r_o, is in equilibrium with the applied torque T, when

$$T = 2\pi \int_0^{r_o} \tau r^2 \, dr \tag{9.26a}$$

Substituting $\gamma = r\alpha$ for $\alpha = \theta / l$ = constant, into (9.26a) gives,

$$T = \frac{2\pi}{\alpha^3}\int_0^{\gamma_o} \tau(\gamma)\gamma^2 \mathrm{d}\gamma \tag{9.26b}$$

where the function $\tau = \tau(\gamma)$ applies at radius r. Differentiating eq(9.26b) with respect to γ_o,

$$\frac{1}{2\pi}\frac{\mathrm{d}}{\mathrm{d}\gamma_o}(T\alpha^3) = \tau_o(\gamma_o)\gamma_o^2$$

Since $\gamma_o = r_o \alpha$, it follows that

$$\frac{1}{2\pi}\frac{\mathrm{d}}{\mathrm{d}\alpha}(T\alpha^3) = \tau_o r_o^3 \alpha^2$$

giving

$$\tau_o = \frac{1}{2\pi r_o^3 \alpha^2} \times \frac{\mathrm{d}}{\mathrm{d}\alpha}(T\alpha^3) = \frac{1}{2\pi r_o^3}\left(3T + \alpha\frac{\mathrm{d}T}{\mathrm{d}\alpha}\right) \tag{9.27a}$$

The geometrical interpretation of eq(9.27a), is given by the Nadai construction in Fig. 9.6, in which

$$\tau_o = \frac{1}{2\pi r_o^3}(3\mathrm{AC} + \mathrm{AB}) \tag{9.27b}$$

Nadai's method will establish hardening behaviour at the outer diameter. This takes the form of an equivalent stress versus equivalent plastic strain plot when we convert τ_o and γ_o to $\bar{\sigma}$ and $\bar{\varepsilon}^P$ from eqs(9.25a,b). For example, assume that the torque-unit twist diagram for a solid bar can be expressed in the form $T = B\alpha + A\alpha^n$. The linear term represents the elastic line. With $n < 1$, the parabolic term dominates in the plastic range.

The slope to the T versus α curve is: $\mathrm{d}T/\mathrm{d}\alpha = B + nA\alpha^{n-1}$. Substituting into eq(9.27a) gives

$$\tau_o = \frac{1}{2\pi r_o^3}\left[3(B\alpha + A\alpha^n) + \alpha(B + nA\alpha^{n-1})\right]$$

$$= \frac{1}{2\pi r_o^3}\left[A\alpha^n(3 + n) + 4B\alpha\right]$$

The corresponding shear stress-strain relationship for the outer radius r_o is found from setting $\alpha = \gamma_o / r_o$:

$$\tau_o = \frac{2B}{\pi r_o^4}\left[\gamma_o + (3 + n)\left(\frac{A r_o}{4B}\right)\left(\frac{\gamma_o}{r_o}\right)^n\right]$$

$$= \frac{B}{J}\left[\gamma_o + \frac{A(3 + n) r_o^{1-n}\gamma_o^n}{4B}\right]$$

where $J = \pi r_o^4/2$. Finally, the equivalent stress versus equivalent plastic strain relationship is found from setting $\bar{\sigma} = \sqrt{3}\,\tau_o$ and $\bar{\varepsilon}^P = \gamma_o - \tau_o / (\sqrt{3}G)$.

9.5 Uniaxial and Torsional Equivalence

The equivalent stress and plastic strain, defined in Sections 9.3 and 9.4, are employed for correlating the hardening behaviour of an isotropic material under various stress states. For example, with perfect lubrication of an isotropic material, the equivalent stress-strain curve ($\bar{\sigma}$ versus $\int d\bar{\varepsilon}^P$) for compression would expect to coincide with those for tension and torsion. Later, we shall examine explicit forms of the strain and work hardening functions:

$$\bar{\sigma} = H\left(\int d\bar{\varepsilon}^P\right) = F\left(\int \bar{\sigma}\, d\bar{\varepsilon}^P\right)$$

9.5.1 Tension and Compression of Carbon Steel

Consider, firstly, second invariant (von Mises) correlations between tension and compression test data. Figure 9.7 applies to solid circular cylinders of 0.17% carbon steel with a common diameter of 9.75 mm.

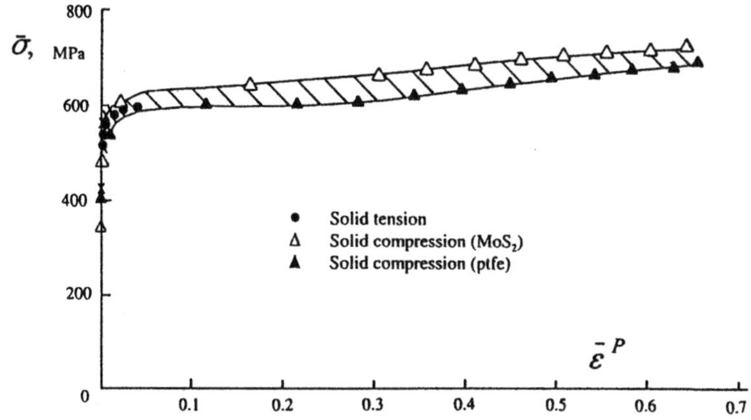

Figure 9.7 Steel flow curves from tension and compression tests on solid bars

Following the application of eqs(9.11), (9.12b), (9.21) and (9.22b) to convert nominal to true stress-strain data, two obvious points arise from Fig. 9.7. Firstly, the tension test upon a solid testpiece is limited to a maximum logarithmic strain, $\varepsilon^P \approx 5\%$, by the formation of a neck. Secondly, large compressive strains appear to show an irregular dependence upon stress. Normally a smooth, convex curve is observed for a hardening material. MoS_2 was used for a lubricant in producing the upper curve. For the lower curve, the testpiece was compressed between ptfe sheets to improve lubrication. Both compression tests were repeatedly interrupted by unloading to allow permanent heights to be measured and re-lubrication of the platens. Despite the precations taken, the compressive data for solid cylinders is unlikely to be purely uniaxial.

9.5.2 Tension and Torsion

(a) *Mild steel*
Figure 9.8 shows that an equivalence correlation can be found between torsion and tension up to the point of necking. The data applies to identical, tubular mild steel (En3B) testpieces: 16.5 mm outer diameter and 12.5 mm inner diameter.

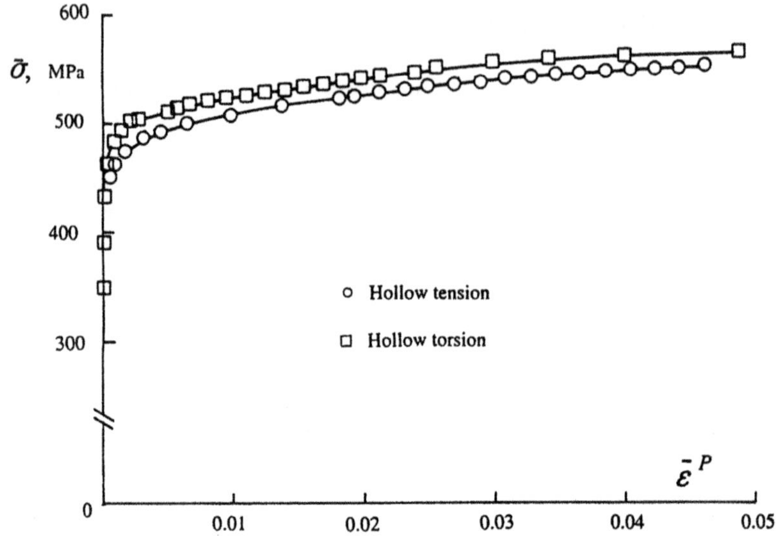

Figure 9.8 Equivalence between tension and torsion for En3B steel

The mean wall formulae (9.23a) and (9.24a,b) were applied to the torsion testpiece for r_m = 7.25 mm, where the angular twist θ was measured on a l = 50 mm gauge length. A similar gauge length was employed for the attatchment of the tensile extensometer. The elastic moduli in tension and torsion were E = 207 GPa and G = 81.6 GPa respectively. The curves are geometrically similar, showing a diminishing rate of hardening $\mathrm{d}\,\bar{\sigma}/\mathrm{d}\,\bar{\varepsilon}^P$ with increasing strain. The equivalent strain of 23% was achieved from torsion before the testpiece succumbed to buckling, as shown in Fig. 9.9a.

(a)

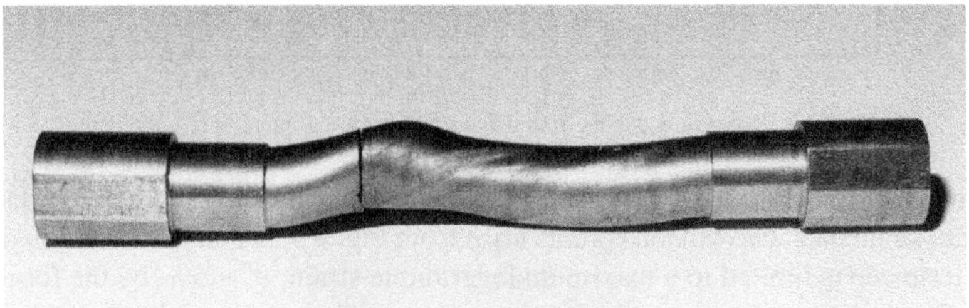

(b)

Figure 9.9 Fracture of En3B under torsion and tension

The critical buckling strain greately exceeds the scale of Fig. 9.8, this being chosen to accommodate the 5% tensile instability strain. Tensile necking is apparent in Fig. 9.9b, though this would be more pronounced in a solid testpiece. Triaxiality of stress exists in the tube wall but not at its two free surfaces. Consequently, the stress gradients are far less severe than in a solid section.

(b) *Black Mild Steel*

A further equivalence correlation between tension and torsion of black mild steel is given in Fig. 9.10. The range of strain has been extended by the respective Bridgeman and Nadai methods.

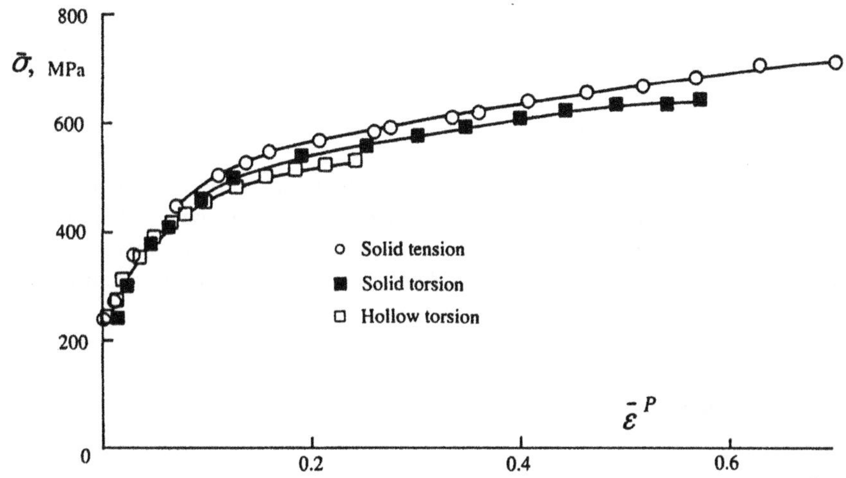

Figure 9.10 Equivalence between tension and torsion for black mild steel

Equations (9.16) and (9.27a) were applied to solid testpiece dimensions: $d_o = 12.7$ mm for tension and $r_o = 12$ mm, $l = 76.2$ mm for torsion. It is seen that Bridgeman's method results in a considerable extension to the tensile strain. In fact, Bridgeman's equivalent strain accommodates tensile instability and is comparable with that found from Nadai's torsion, i.e. lying in a range of $\bar{\varepsilon}^P$ from 60 to 70%. An additional torsion test was conducted on a thin-walled cylinder with $r_m = 10.8$ mm and $t = 2.56$ mm. Having applied eqs(9.25a,b), the plot of $\bar{\sigma}$ versus $\mathrm{d}\bar{\varepsilon}^P$ reveals that the maximum $\bar{\varepsilon}^P$ is curtailed to 25% by the onset of torsional instability. This is consistent with the maximum strain found for the En3B tube in Fig. 9.9. There is reasonable correlation between the three curves though the tensile curve lies consistently above those of torsion over the full range of $\bar{\varepsilon}^P$.

(c) *Pure Aluminium*

For a von Mises material, it is expected that second invariant equivalence definitions will correlate hardening behaviour under all stress states, i.e they result in a single curve of $\bar{\sigma}$ versus $\int \mathrm{d}\bar{\varepsilon}^P$. In practice, data points from different tests may lie within a narrow scatter band in $\bar{\sigma}$ versus $\bar{\varepsilon}^P$ axes. This suggests that the equivalence may also depend upon the third invariants of stress and plastic strain. For example, Fig. 9.11a shows that eqs(9.25a,b) do not provide an equivalence between tension and torsion for annealed, E1A aluminium cylinders of 99.8% commercial purity. Figure 9.11b shows an improved correlation where stress equivalence is based upon eq(9.4c). Here $c = 2\frac{1}{4}$ lies at the lower limit of convexity ($2\frac{1}{4} \le c \le 3\frac{3}{8}$), for a Drucker yield surface. Since this material does not show a well-defined yield point c is calculated from applying eq(9.4c) to tensile and torsional yield stresses (Y and k) at equivalent von Mises offset strains.

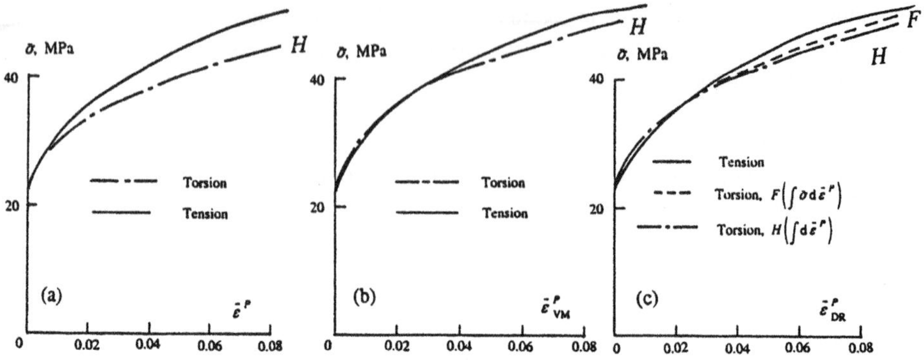

Figure 9.11 Equivalence between tension and torsion for E1A aluminium

In Fig. 9.11c the strain equivalence is altered corresponding to eq(9.7c). The chain line shows no better affinity with tension and demonstrates that it may not be necessary to change both definitions $\bar{\sigma}$ and $\bar{\varepsilon}^P$. It is possible that the curves can be made to coincide for another c value, found by trial. In the absence of a sharp yield point, it is far more convenient to employ a Mises equivalent offset strain to give the $\bar{\sigma}$ required.

9.6 Modified Compression Tests

9.6.1 Cook and Larke

Longer cylinders under uniaxial compression are more likely to buckle than to barrel. In order to avoid bucking, Cooke and Larke [13] proposed a method of extrapolating to the true stress in a cylinder of infinite length. Their method required the compression of cylinders with equal initial diameters d_o but with initial heights h_o varying within the chosen ratios $d_o/h_o = ½$, 1, 2 and 3.

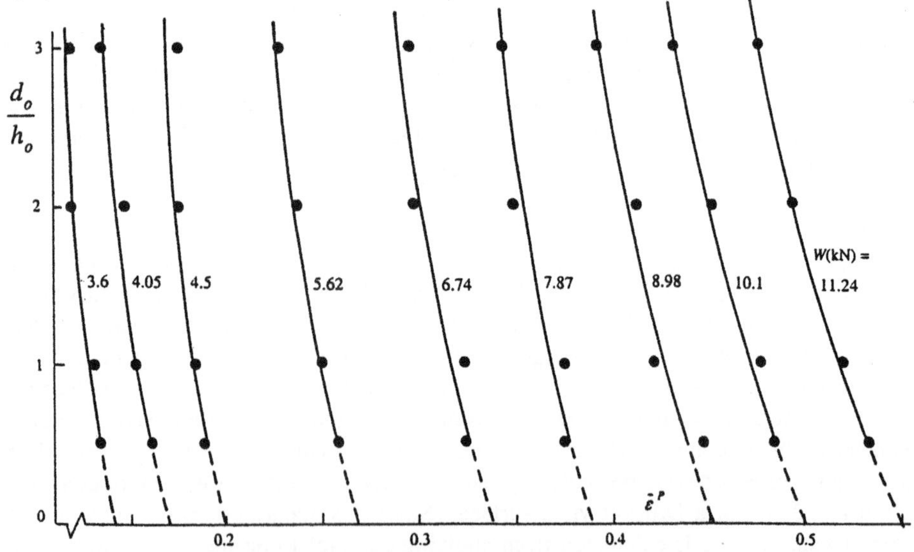

Figure 9.12 Extrapolation to infinite length in compression

The stress $\bar{\sigma}$ was extrapolated to $d_o/h_o = 0$ for equal percentage reductions to each cylinder. Watts and Ford [14] modified this test with an extrapolation that gave $\bar{\varepsilon}^P$ for $d_o/h_o = 0$ from applying a series of equal loads to cylinders with these ratios (see Fig. 9.12). This requires the determination of plastic strain in permanent height reductions, following unloading from increasing load levels to each cylinder. The true stress and strain are calculated from eqs(9.21) and (9.22b). The plastic strain that would arise from applying these loads to an infinitely long cylinder, i.e. $d_o/h_o = 0$, follows from the extrapolation.

Figure 9.12 applies this method to 99.9% pure, annealed copper (B.S. 2870, C101). Testpieces were compressed in a sub-press between the platens of a 500 kN test machine. Watts and Ford's convenient experimental modification enables height reductions to be measured precisely and allows re-lubrication between loads. The results from this test also reveal the effect of *d*/*h* on the flow curve, as shown in Fig. 9.13.

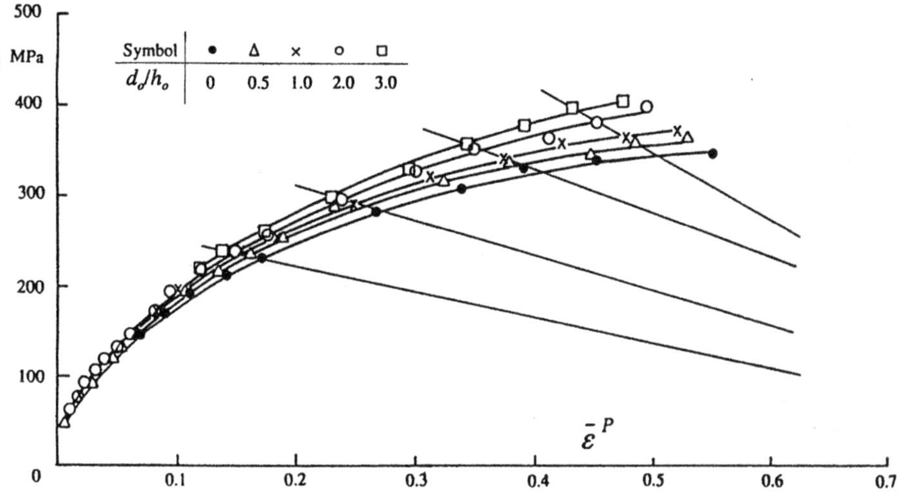

Figure 9.13 Effect of length on compressive flow

The true, lower limiting, flow curve applies to a cylinder of infinite height. As the cylinder height is reduced, so the flow behaviour diverges from the true compressive behaviour. For this reason, compressive curves in Fig. 9.8 would be expected to lie above those from tension. In Fig. 9.13, points ($\bar{\sigma}$, $\bar{\varepsilon}^P$) under constant load, lie on the straight lines shown. It follows from eq(9.21) that the equation of each line is given by

$$\bar{\sigma} = \frac{W}{A_o}\left(1 - \bar{\varepsilon}^P\right)$$

where A_o is the section's original area. Since W is constant, each line connects a common point of unity on the plastic strain axis to the ordinate $\bar{\sigma} = W/A_o$. In the following section the extrapolated curve for $d_o/h_o = 0$ (see Fig. 9.13) will be compared to the flow curve obtained from a plane-strain compression test on this material.

9.6.2 Plane-Strain Compression

In the plane-strain compresion test (see Fig. 6.38), strain in the direction of the width *w* is prevented by the unstressed material to either side of the die. When the die faces are polished and lubricated, Fig. 6.50 shows that the following conditions avoid a widely fluctuating die pressure to give a repeatable flow behaviour [15]:

$$2 \le \frac{b}{t} \le 4 \quad \text{and} \quad 5 \le \frac{w}{b} \le 12 \tag{9.28a,b}$$

The force F may be applied in increments to produce thickness reductions between 1 and 2% though they may be as large as 10 - 15%. A die change to smaller breadth b will avoid violating eqs(9.28a,b), following large amounts of deformation. The variables to be measured are simply the current thickness t, corresponding to an unloading from a given F.

(a) *Theory*

Equivalent stress and strain expressions follow from the Levy-Mises flow rule, i.e. by setting $f = J_2'$ in eq(3.22). For a point beneath the die, this rule gives the principal, plastic strain increments for the thickness, length and width directions respectively:

$$d\varepsilon_1^P = \tfrac{2}{3}\, d\lambda\, [\sigma_1 - \tfrac{1}{2}(\sigma_2 + \sigma_3)] \tag{9.29a}$$

$$d\varepsilon_2^P = \tfrac{2}{3}\, d\lambda\, [\sigma_2 - \tfrac{1}{2}(\sigma_1 + \sigma_3)] \tag{9.29b}$$

$$d\varepsilon_3^P = \tfrac{2}{3}\, d\lambda\, [\sigma_3 - \tfrac{1}{2}(\sigma_1 + \sigma_2)] \tag{9.29c}$$

The co-ordinates are shown in Fig. 6.38. Now, $d\varepsilon_3^P = 0$ for plane strain and, with no longitudinal constraint, $\sigma_2 = 0$. Equation (9.29c) then gives $\sigma_3 = \tfrac{1}{2}\sigma_1$, where the compressive stress is $\sigma_1 = -F/(bw)$. Substituting into eq(9.2d) gives the equivalent stress:

$$\bar{\sigma} = \frac{1}{\sqrt{2}}\sqrt{\left(\sigma_1 - 0\right)^2 + \left(0 - \frac{\sigma_1}{2}\right)^2 + \left(\sigma_1 - \frac{\sigma_1}{2}\right)^2}$$

$$= \frac{\sqrt{3}}{2}\sigma_1 = -\frac{\sqrt{3}}{2}\left(\frac{F}{bw}\right) \tag{9.30b}$$

Similar substitutions into eqs(9.29a,b) result in principal, plastic strain increments:

$$d\varepsilon_1^P = \tfrac{1}{2}\, d\lambda\sigma_1 \quad \text{and} \quad d\varepsilon_2^P = -\tfrac{1}{2}\, d\lambda\sigma_1 \tag{9.31a,b}$$

Substituting eqs(9.31a,b) into eq(9.6e) gives the equivalent plastic strain increment:

$$d\bar{\varepsilon}^P = \sqrt{\frac{2}{3}\left[\left(d\varepsilon_1^P\right)^2 + \left(-d\varepsilon_1^P\right)^2 + (0)^2\right]}$$

$$= \frac{2}{\sqrt{3}}\, d\varepsilon_1^P = \frac{2}{\sqrt{3}}\frac{dt}{t}$$

and integrating for the total plastic strain gives

$$\bar{\varepsilon}^P = \frac{2}{\sqrt{3}}\ln\left(\frac{t}{t_o}\right) = -\frac{2}{\sqrt{3}}\ln\left(\frac{t_o}{t}\right) \tag{9.32b}$$

The minus signs in eqs(9.30b) and (9.32b) are often ignored in the plot of $\bar{\sigma}$ versus $\bar{\varepsilon}^P$. This is consistent with taking a negative square root in the $\bar{\sigma}$ and $\bar{\varepsilon}^P$ expressions.

(b) *Test Results*

A plane-strain compression test was conducted on annealed, pure copper plate of initial dimensions $w = 50$ mm and $t_o = 3.65$ mm. Die breadths $b = 6.37, 4.12$ and 3.19 mm were selected to ensure $b/t < 4$, throughout deformation to a final, 84% reduction in thickness. Loads were applied with MoS_2 lubricant applied to each die face. To prevent misalignment, the dies were guided vertically in a jig. The deformed thickness t was measured following unloading from each of 26 rising loads betwen 10 kN and 99.75 kN. The equivalent flow curve, as derived from eqs(9.30b) and (9.32b), is shown in Fig. 9.14.

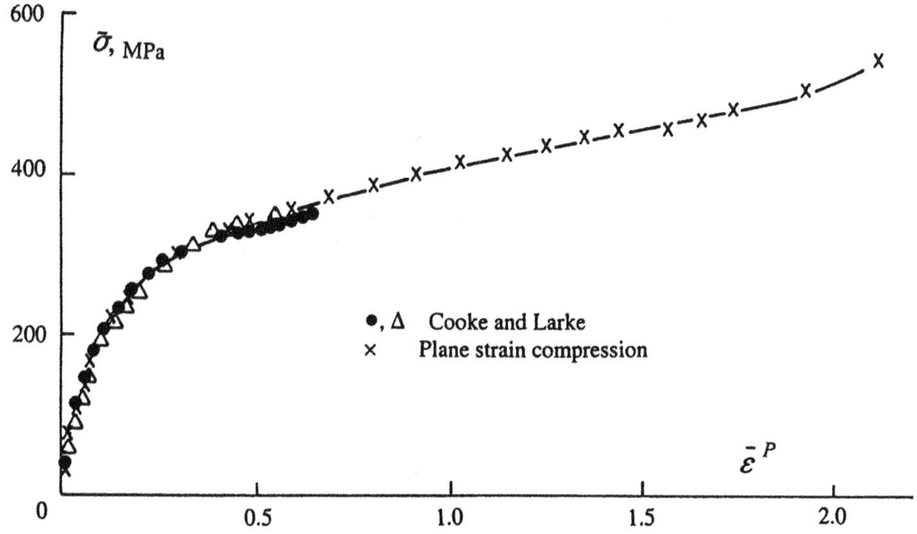

Figure 9.14 Compressive flow under plane strain and from extrapolation

9.6.3 Equivalence in Compression Tests

It is seen that very large equivalent plastic strains, approaching 200%, can be achieved from the plane-strain compression test. Figure 9.14 shows that this magnitude of strain is far more than can be achieved from the maximum extrapolated strain ≃ 60%, obtained from the Cooke and Larke method. It is seen that within a comparable range of strain, the flow behaviour for annealed copper, derived by these two methods, is almost coincident.

Results from the two Cooke and Larke experiments in Fig. 9.14 were obtained from different test machines and, expectedly, this has not altered the material's flow behaviour. The flow behaviour of copper is, however, very sensitive to its initial condition. In the absence of annealing an as-received copper, Fig. 9.15 shows that the flow stresses are higher while the rate of hardening is initially lower. Beyond 30% strain, all flow curves begin to converge as each structure attains a comparable amount of slip and twinning. Figure 9.15 further shows that there is reasonable consistency in plane strain and direct compression when the as-received copper is lubricated with MoS_2 and ptfe. Plane strain provides a smooth curve over a wider strain range.

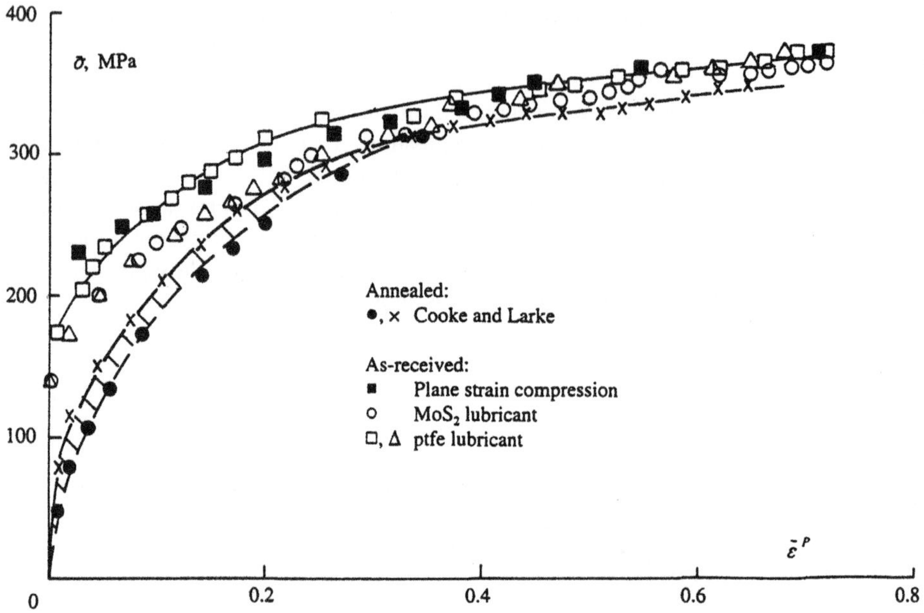

Figure 9.15 Superimposed compressive flow curves for copper

9.7 Bulge Test

Large plastic strains can be achieved from a balanced biaxial tension. In practice, this condition prevails in a bulge test, where pressure is applied to the underside of a thin circular plate that is clamped around its periphery. The flow behaviour is calculated from the current pole thickness t and deflection h at the centre of the bulge, where the pole radius R, is as shown in Fig. 9.16.

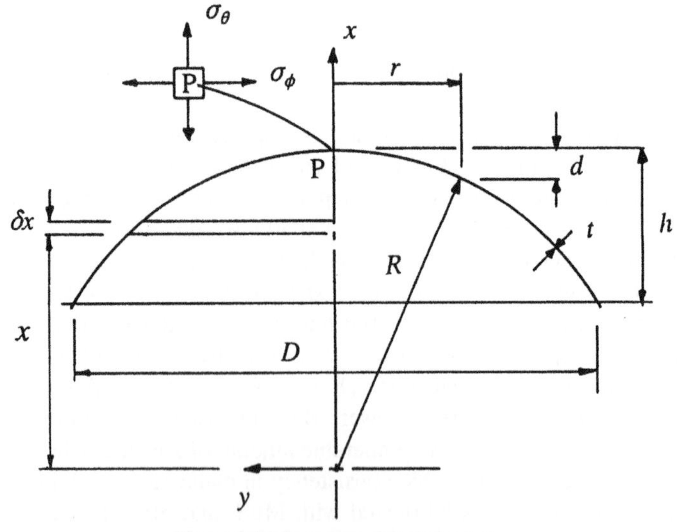

Figure 9.16 Geometry of the bulge test

9.7.1 Theory

Provided the plate is thin, compared to its diameter D, giving $D/t > 100$, the stresses due to bending and shear force may be ignored. Neglecting also the radial stress σ_r, this leaves only the membrane pressure stresses σ_θ and σ_ϕ. At the pole P, the meridional stress σ_ϕ is found from a consideration of force equilibrium vertical to the y-plane. This gives

$$2\pi R\, t\, \sigma_\phi = p\pi R^2 \;\Rightarrow\; \sigma_\phi = pR/2t \tag{9.33a}$$

Similarly, for a second section, perpendicular to this $\sigma_\theta = pR/2t$, showing that an equi-biaxial stress state $\sigma_\theta = \sigma_\phi$ exists everywhere. Substituting eq(9.33a) into eq(9.2d), with $\sigma_r = 0$, leads to an equivalent stress

$$\bar{\sigma}^2 = \sigma_\theta^2 - \sigma_\theta\sigma_\phi + \sigma_\phi^2$$

$$\bar{\sigma} = \sigma_\phi = \frac{pR}{2t} \tag{9.33b}$$

where R and t are current pole values. The Levy-Mises flow rule $d\varepsilon_{ij}^P = d\lambda\,\sigma_{ij}'$, gives the three plastic strain increments as

$$d\varepsilon_\theta^P = ⅔\, d\lambda\,[\sigma_\theta - ½(\sigma_\phi + \sigma_r)] = ⅓\, d\lambda\,\sigma_\theta \tag{9.34a}$$

$$d\varepsilon_\phi^P = ⅔\, d\lambda\,[\sigma_\phi - ½(\sigma_\theta + \sigma_r)] = ⅓\, d\lambda\,\sigma_\theta \tag{9.34b}$$

$$d\varepsilon_r^P = ⅔\, d\lambda\,[\sigma_r - ½(\sigma_\theta + \sigma_\phi)] = -⅔\, d\lambda\,\sigma_\theta \tag{9.34c}$$

and from eqs(9.34a-c):

$$d\varepsilon_\theta^P = d\varepsilon_\phi^P \;\text{ and }\; d\varepsilon_r^P = -2\,d\varepsilon_\theta^P \tag{9.35a,b}$$

which is readily confirmed from the constant volume condition $d\varepsilon_{kk}^P = 0$. Substituting eqs(9.35a,b) into eq(9.6e) gives the equivalent plastic strain increment

$$d\bar{\varepsilon}^P = \sqrt{\frac{2}{3}\left[(d\varepsilon_\theta^P)^2 + (d\varepsilon_\phi^P)^2 + (d\varepsilon_r^P)^2\right]}$$

$$= \sqrt{\frac{2}{3}\left[(-½\,d\varepsilon_r^P)^2 + (-½\,d\varepsilon_r^P)^2 + (d\varepsilon_r^P)^2\right]} = d\varepsilon_r^P \tag{9.36a}$$

By definition, the radial strain is

$$\varepsilon_r^P = \int_{t_o}^{t} \frac{dt}{t} = \ln\left(\frac{t}{t_o}\right) = -\ln\left(\frac{t_o}{t}\right) \tag{9.36b}$$

where t_o is the initial sheet thickness. From eqs(9.36a,b),

$$\bar{\varepsilon}^P = -\ln\left(\frac{t_o}{t}\right) \tag{9.36c}$$

The negative sign may be omitted when the negative root of eq(9.36a) is taken. It follows from eqs(9.33b) and (9.36c) that to construct the plot between $\bar{\sigma}$ and $\mathrm{d}\bar{\varepsilon}^P$, the instantaneous values of the variables p, R and t are required. The radius R is calculated from the geometry in Fig. 9.16:

$$R^2 = r^2 + (R - d)^2 \Rightarrow R = \frac{(r^2 + d^2)}{2d} \tag{9.37a}$$

In early bulge tests [16, 17], r and d in eq(9.37a) were determined from dimensional changes to concentric circles scribed on the undeformed plate. The pole radius R was found by graphical extrapolation to $r \simeq 0$. Equation (9.35b) shows that $\mathrm{d}t/t = -2\mathrm{d}r/r$ so that thickness could also be calculated from this extrapolation:

$$\int_{r_o}^{r} \frac{\mathrm{d}r}{r} = -\frac{1}{2}\int_{t_o}^{t} \frac{\mathrm{d}t}{t}$$

$$\ln\left(\frac{r}{r_o}\right) = \frac{1}{2}\ln\left(\frac{t_o}{t}\right) \tag{9.37b}$$

The left-hand side of eq(9.37b) becomes the hoop strain when extrapolated to $r = 0$.

9.7.2 *Instrumented Test*

In an instrumented bulge test [18], tranducers are incorporated within a spherometer to measure t at the pole and h above a horizontal plane of diameter D (Fig. 9.16). In addition to a pressure transducer, a flowmeter, placed within the pressure line, provides the volume of pressurising fluid beneath the dome. This volume can be expressed in terms of the spherical dome radius R. Revolving the elemental strip in Fig. 9.16 about the x - axis gives $\delta V = \pi y^2 \delta x$. Substituting $y^2 = R^2 - x^2$ and integrating between limits of x gives

$$V = \pi\int_{R-h}^{R} (R^2 - x^2)\,\mathrm{d}x = \pi\left|R^2 x - \frac{x^3}{3}\right|_{R-h}^{R}$$

$$= \pi h^2\left(R - \frac{h}{3}\right) \Rightarrow R = \frac{V}{\pi h^2} + \frac{h}{3} \tag{9.38a}$$

Equating (9.38a) to (9.37a), with $r = D/2$ and $d = h$, leads to

$$\frac{V}{\pi h^2} + \frac{h}{3} = \frac{[(D/2)^2 + h^2]}{2h} \tag{9.38b}$$

Equation (9.38b) gives h as one root to the cubic equation:

$$h^3 + \frac{3}{4}Dh - \frac{6V}{\pi} = 0 \tag{9.38c}$$

Equation (9.38c) shows that it is possible theoretically to dispense with h measurement but this is not advisable when deviations from a spherical dome occur, particularly in the region of the clamped rim. Such deviation leads to greater error in the calculation of h than of R.

This is particularly true for bulging orthotropic sheet metals when eqs(9.33b) and (9.36a) appear with a multiplication factor:

$$\bar{\sigma} = K\sigma_\phi \quad \text{and} \quad \mathrm{d}\bar{\varepsilon}^P = \mathrm{d}\varepsilon_r^P / K$$

where K depends upon the anisotropy coefficients r_1 and r_2 for the sheet (refer to p. 342):

$$K = \{3(r_1 + r_2) / [2\,(r_1 + r_1 r_2 + r_2)\,]\}^{1/2}$$

9.7.3 Equivalence with Torsion

In Fig. 9.17 a comparison is made between equivalent flow curves for a bulge test and a tubular torsion test for annealed brass of B.S. 2874 composition (40% Zn, 57% Cu, 3% Pb).

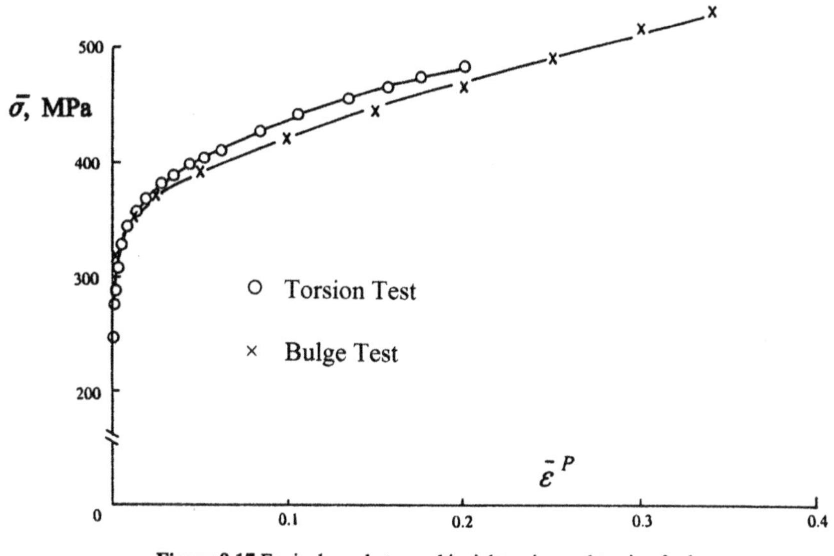

Figure 9.17 Equivalence between biaxial tension and torsion for brass

The figure shows the two curves to be in reasonable agreement over a similar range of strain. The material hardens rapidly initially before falling to attain a steady rate at higher strain.

The dimensions chosen for the instrumented bulge test were $t_o = 0.39$ mm, $D = 100$ mm and $d = 25.4$ mm. The tubular torsion test was conducted for similar conditions to those in Fig. 9.8 (i.e. with $d_i = 12.5$ mm, $d_o = 16.5$mm and $l = 50$mm). Within the elastic region a shear modulus $G = 39$ GPa, was calculated from the torque and twist measurements made. The elasticity modulus, $E = 102$ GPa, was calculated from $E = 2G\,(1 + \nu)$, assuming Poisson's ratio $\nu = 1/3$. Under monotonic loading, E and G are used to subtract the elastic components of strain from the total strain. In torsion, eq(9.25b) is applied to give the equivalent plastic strain. The strain at the point of torsional failure was $\bar{\varepsilon}^P \approx 20\%$, corresponding to 140° twist. Note that this strain is consistent with a tensile elongation of 20 - 30% for this material. The equivalent plastic strain in a continuous bulge test modifies eq(9.36c) to

$$\bar{\varepsilon}^P = -\left[\ln\left(\frac{t_o}{t}\right) - \frac{\bar{\sigma}}{E}\right]$$

where $\bar{\sigma}$ is defined in eq(9.33b). Equation (9.36c) applies when t is measured following a pressure release from the plastic range. Here, however, t is measured under an increasing pressure, and so elastic strain is subtracted. The application of eq(9.36c) to find $\bar{\varepsilon}^P$ would involve negligible error due to elastic strain. The advantage of the bulge test is that a greater fracture strain is reached compared to torsion. In Fig. 9.17, the bulge would extend the strain to 64% at the pole before bursting. The state of deformation just prior to failure is shown in Fig. 9.18.

Figure 9.18 Deformation under equi-biaxial tension

9.8 Equations to the Flow Curve

Most empirical representations [9, 19] of the uniaxial flow curve obey the strain hardening function: $\sigma = H(\int d\varepsilon^P)$. The corresponding non-dimensional forms of the more commonly used functions H are shown in Figs 9.19a-f. Here σ_o is the yield stress and ε_o is a constant.

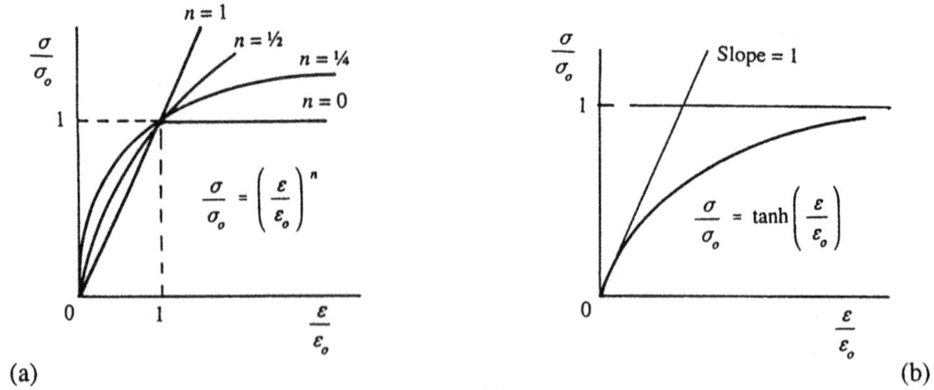

Figure 9.19 Empirical representations of flow curves

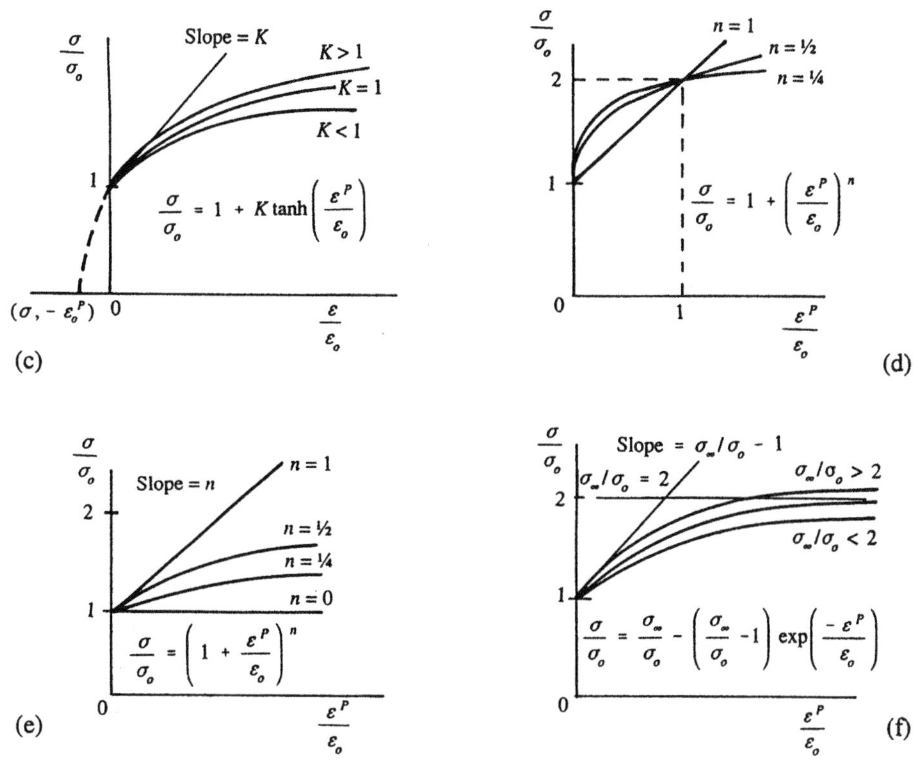

Figure 9.19 Empirical representations of flow curves (continued)

9.8.1 Hollomon [20]

The simplest and most popular function is the Hollomon power law (see Fig. 9.19a):

$$\sigma = A\,\varepsilon^n \quad \text{or} \quad \frac{\sigma}{\sigma_o} = \left(\frac{\varepsilon}{\varepsilon_o}\right)^n \tag{9.39a,b}$$

where A ($= \sigma_o/\varepsilon_o^{\,n}$) is a strength coefficient and n is a strain hardening exponent, which lies in the range $0.1 \le n \le 0.55$ for metals. Because eqs(9.39a,b) pass through the origin, they strictly should not be applied to a curve with an intercept on the stress axis. When the abscissa is plastic strain ε^P, this intercept becomes the yield stress. Consequently, when eqs(9.39a,b) are to represent a flow curve $\sigma = H\,(\int \mathrm{d}\varepsilon^P)$, the greatest deviation will arise for low total strains $\varepsilon^t < 0.2\%$. For $\varepsilon > 5\%$, the error becomes negligible and many have then associated ε in eqs(9.39a,b) with ε^P.

9.8.2 Ramberg-Osgood [21]

This well-known equation describes total strain from the addition of elastic strain to eq(9.39a). It is usually written in the following form:

$$\varepsilon^t = \frac{\sigma}{E} + \left(\frac{\sigma}{E_o}\right)^m \tag{9.40}$$

where E is Young's modulus and E_o and m are constants. As with all continuous functions H, eq(9.40) is unable to represent discontinuous flow curves, e.g. the upper and lower yield points in low carbon steel.

9.8.3 Prager [22]

Prager's hyperbolic flow curve is shown in Fig. 9.19b. The original expressions

$$\sigma = Y \tanh\left(E\,\varepsilon / Y\right) \quad \text{or} \quad \sigma/\sigma_o = \tanh\left(\varepsilon/\varepsilon_o\right) \qquad (9.41a,b)$$

describe a curve originating from the origin, with slope E, that rapidly approaches an asymptote $\sigma = \sigma_o$. As with eqs(9.39a,b), no yield point is identified. Where a hyperbolic expression is required to give $\sigma = \sigma_o$ for $\varepsilon^P = 0$, we may modify eqs(9.41a,b) to become

$$\sigma = Y + K \tanh\left(E\varepsilon / Y\right) \quad \text{or} \quad \sigma/\sigma_o = 1 + K \tanh\left(\varepsilon/\varepsilon_o\right) \qquad (9.42a,b)$$

where, as shown in Fig. 9.19c, K may be varied to give the asymptote desired.

9.8.4 Ludwik [23]

The simplest modification to eqs(9.39a,b), which give $\sigma = \sigma_o$ (or $\sigma = Y$) for $\varepsilon^P = 0$, lead to Ludwik's power law (see Fig. 9.19d):

$$\sigma = Y + A\left(\varepsilon^P\right)^n \quad \text{or} \quad \frac{\sigma}{\sigma_o} = 1 + \left(\frac{\varepsilon^P}{\varepsilon_o}\right)^n \qquad (9.43a,b)$$

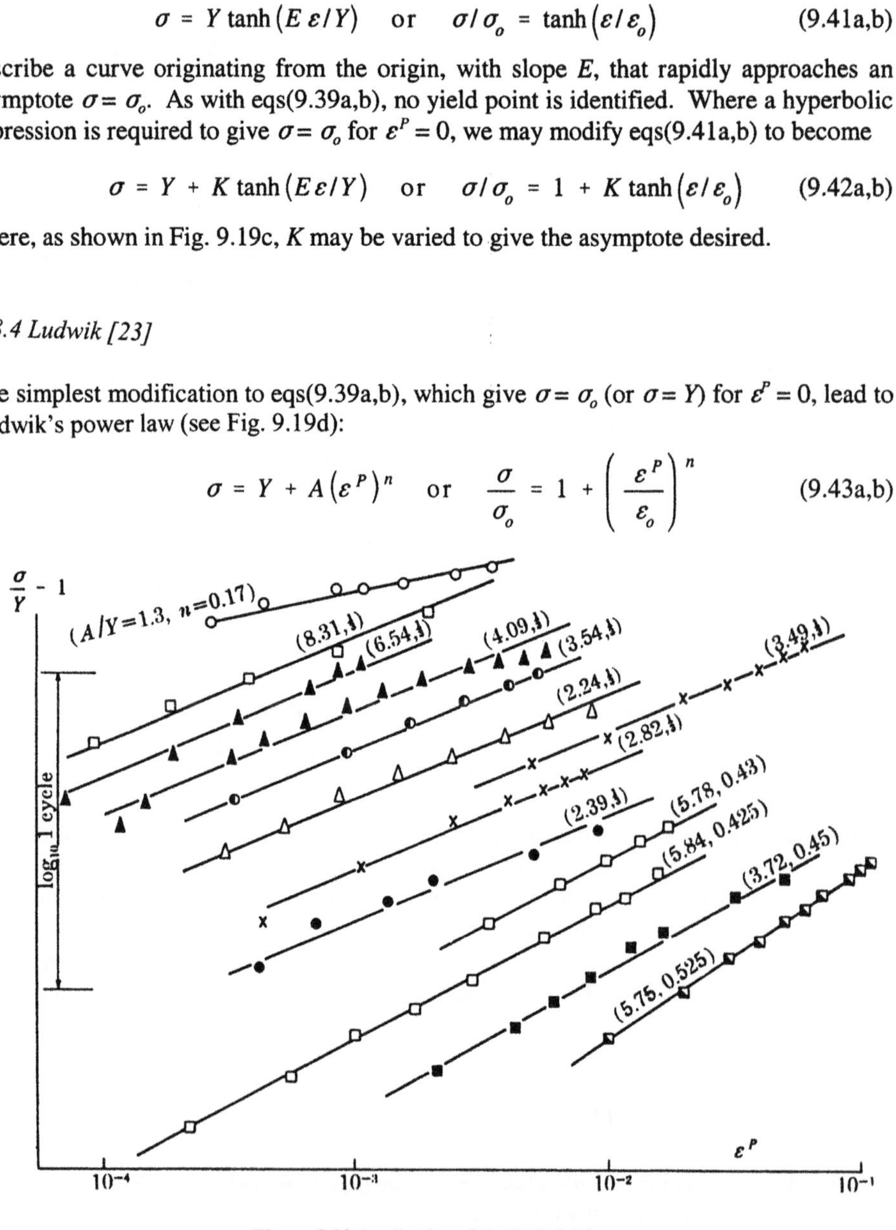

Figure 9.20 Application of the Ludwick law

Key: ○, Δ steels; □,○, Δ, ● Al -alloys; × brass; ■ Ti-alloy, □ Mg-alloy (Points separated for clarity)

Equations (9.43a,b) display parabolic strain hardening in which $\sigma > Y$ for $n < 1$. Here A and n are empirical constants, similar to those in Hollomon's law. It will be shown in Chapter 10 that eq(9.43a) is particularly useful to the theory of plasticity, allowing incremental strain expressions to be integrated for the majority of loading paths. In particular, when $n = ⅓$, closed solutions apply. Ludwick's logarithmic plot (see Fig. 9.20) identifies n and A in eq(9.43a) with the slope and intercept respectively of the resulting straight line. Figure 9.20 shows that the flow behaviour of many engineering alloys in tension and torsion conforms to eq(9.43a) for plastic strains up to 10% [24]. Some initial deviation is found with the indicated constants because of a large increase in the rate of hardening $d\sigma/d\varepsilon^P$ as $\varepsilon^P \to 0$. When $\varepsilon^P = 0$, eq(9.43a) gives $d\sigma/d\varepsilon^P = \infty$, in contrast to finite rates observed. Revised constants A and n can improve accuracy for their application to a small strain range.

9.8.5 Swift [25]

Swift's power function, illustrated in Fig. 9.19e, represents hardening within three constants:

$$\sigma = K\left(A + \varepsilon^P\right)^n \quad \text{or} \quad \frac{\sigma}{\sigma_o} = \left(1 + \frac{\varepsilon^P}{\varepsilon_o}\right)^n \qquad (9.44a,b)$$

In eq(9.44a), constants A, K and n bear a relationship to the yield stress: $Y = KA^n$. The finite hardening rate $d\sigma/d\varepsilon^P = nKA^{n-1}$ for $\varepsilon^P = 0$ is preferable to the infinite rate given by eq(9.43a). However, to separate the constants, one further point is required either, within a mid-range of strain (co-ordinates σ, ε^P), or at the negative strain intercept $(0, -\varepsilon_o)$. Alternatively, the normalised eq(9.44b) gives, $\sigma/\sigma_o = 1$ and $\varepsilon^P = 0$ with $d(\sigma/\sigma_o)/d(\varepsilon^P/\varepsilon_o) = n$, this allowing n to be separated from ε_o. For example, half-hard aluminium has been represented [9] with the following form of eq(9.44b):

$$\sigma / 105 = (1 + \varepsilon^P / 0.222)^{0.25}$$

with σ in MPa over a plastic strain range $0 \le \varepsilon^P \le 0.4$. If we choose the point of instability in a tension test and apply eq(9.14c) to eq(9.44b), the true ultimate stress and plastic strain σ_u and ε_u^P are related to n and ε_o in the following ways:

$$\frac{\sigma_u}{\sigma_o} = \left(\frac{n}{\varepsilon_o}\right)^n \left(1 - \frac{\sigma_u}{E}\right)^n \approx \left(\frac{n}{\varepsilon_o}\right)^n \qquad (9.44c)$$

$$\frac{\varepsilon_u^P}{\varepsilon_o} = \frac{n}{\varepsilon_o}\left(1 - \frac{\sigma_u}{E}\right) - 1 \approx \frac{n}{\varepsilon_o} - 1 \qquad (9.44d)$$

The binomial expansion approximations to eq(9.44b) are sometimes useful. These apply to $\varepsilon^P / \varepsilon_o < 1$:

$$\frac{\sigma}{\sigma_o} = 1 + n\left(\frac{\varepsilon}{\varepsilon_o}\right) + \frac{n}{2}(n - 1)\left(\frac{\varepsilon}{\varepsilon_o}\right)^2 + \frac{n}{6}(n - 1)(n - 2)\left(\frac{\varepsilon}{\varepsilon_o}\right)^3 + \ldots$$

and to $\varepsilon^P / \varepsilon_o > 1$:

$$\frac{\sigma}{\sigma_o} = \left(\frac{\varepsilon}{\varepsilon_o}\right)^n \left[1 + n\left(\frac{\varepsilon_o}{\varepsilon}\right) + \frac{n}{2}(n - 1)\left(\frac{\varepsilon_o}{\varepsilon}\right)^2 + \frac{n}{6}(n - 1)(n - 2)\left(\frac{\varepsilon_o}{\varepsilon}\right)^3 + \ldots\right]$$

9.8.6 Vocé [26]

Finally, Vocé's exponential law (see Fig. 9.19f) can be adapted as a strain hardening function

$$\sigma = A + (Y - A)\,\mathrm{e}^{-B\varepsilon^P} \quad \text{or} \quad \frac{\sigma}{\sigma_o} = \frac{\sigma_\infty}{\sigma_o} - \left(\frac{\sigma_\infty}{\sigma_o} - 1\right)\mathrm{e}^{-\frac{\varepsilon^P}{\varepsilon_o}} \tag{9.45a,b}$$

Equation (9.45a) is consistent with $\sigma = Y$ for $\varepsilon^P = 0$, where there is a finite hardening rate $\mathrm{d}\sigma/\mathrm{d}\varepsilon^P = B(Y - A)$. Constant A is determined from matching eq(9.45a) to a stress chosen at large strain. In eq(9.45b), the initial hardening rate is

$$\frac{\mathrm{d}(\sigma/\sigma_o)}{\mathrm{d}(\varepsilon^P/\varepsilon_o)} = \frac{\sigma_\infty}{\sigma_o} - 1$$

Should we match the curve to the ultimate strength point (σ_u, ε_u^P), eqs(9.14c) and (9.45b) may be combined. This provides the following relationships between σ_∞ and ε_o at the point of tensile instability:

$$\frac{\sigma_u}{\sigma_o} = \frac{\sigma_\infty}{\sigma_o}\left(\frac{1 - \sigma_u/E}{1 - \sigma_u/E + \varepsilon_o}\right) \approx \frac{\sigma_\infty}{\sigma_o(1 + \varepsilon_o)} \tag{9.45c}$$

$$\mathrm{e}^{\frac{\varepsilon_u^P}{\varepsilon_o}} = \frac{(\sigma_\infty - \sigma_o)(1 - \sigma_u/E + \varepsilon_o)}{\varepsilon_o\,\sigma_\infty} \approx \frac{(\sigma_\infty - \sigma_o)(1 + \varepsilon_o)}{\varepsilon_o\,\sigma_\infty} \tag{9.45d}$$

The non-dimensional eqs(9.39b - 9.45b) provide a more rational mathematical basis of hardening function. In these, σ_o is the yield stress, n, ε_o, K and σ_∞ are material constants. The graphically summary in Fig. 9.19a - f gives the initial rate of hardening and shows trends appropriate to the choice of n, K and σ_∞/σ_o. Of the various functions mentioned here, the Hollomon, Ludwik and Swift forms are most often used in plasticity theory.

9.9 Strain and Work Hardening Hypotheses

In the rule of isotropic hardening, the flow potential f appears as a scalar function of the stress deviator invariants:

$$f(J_2', J_3') = \chi(\textstyle\int \mathrm{d}\psi) \tag{9.46}$$

where χ is an increasing function of a suitable scalar measure ψ of hardening. We have shown previously how to derive $\bar{\sigma}$ for a Drucker function $f(J_2', J_3')$ in eq(9.4b). The corresponding definition of $\mathrm{d}\bar{\varepsilon}^P$ will depend upon the hardening hypothesis used, e.g. strain hardening is given in eqs(9.7a-c). Using the work hypothesis ensures that the flow rule and equivalence relationships are associated with the same yield function but it is more difficult to derive an equivalent plastic strain [3]. However, it is simplified given the yield function in a reduced stress space. Take, for example, the Drucker yield function $f = J_2'^3 - cJ_3'^2$, for torsion. The equivalent flow stress follows from eq(9.4c) when $\bar{\sigma}$ replaces Y:

$$\bar{\sigma} = \frac{\tau}{\left[(1/3)^3 - 4c/9^3\right]^{1/6}}$$

Applying the work hypothesis:

$$\bar{\sigma}\, \mathrm{d}\bar{\varepsilon}^P = \sigma_{ij}\, \mathrm{d}\varepsilon_{ij}^P = \tau\, \mathrm{d}\gamma^P$$

$$\mathrm{d}\bar{\varepsilon}^P = \frac{\tau\, \mathrm{d}\gamma^P}{\bar{\sigma}} = \left(\frac{1}{3^3} - \frac{4c}{9^3}\right)^{\frac{1}{6}} \mathrm{d}\gamma^P$$

which differs from $\mathrm{d}\bar{\varepsilon}^P = \mathrm{d}\gamma^P/(27 - 4c)^{1/6}$ by strain hardening (see p. 274). Consequently, the flow curves correlate differently between the two hypotheses, as illustrated for tension and torsion in Fig. 9.11c. Recall that the chain line H is found from the strain hardening hypothesis with $\bar{\sigma}$ and $\bar{\varepsilon}^P$ derived from similar functions in the stress and strain invariants. The broken line F applies to the new position of the equivalent torsional flow curve when $\bar{\varepsilon}^P$ is defined by the work hypothesis. The latter provides a better correlation with tension for the same constant c in the yield function. We may regard the deviation between F and H being due to the influence of J_3' within the chosen hardening hypothesis. Only when J_3' is absent in the yield function are the two hardening hypotheses indistinguishable.

9.9.1 Strain Hardening

Let f in eq(9.46), be equated to the second invariant of deviatoric stress

$$f = J_2' = \tfrac{1}{2}\, \sigma_{ij}'\, \sigma_{ij}' = \tfrac{1}{2}\, \mathrm{tr}\, \mathbf{T}'^2$$

and $\mathrm{d}\psi$, be the increment in the second invariant of plastic strain

$$\mathrm{d}\psi = \mathrm{d}I_2 = \tfrac{1}{2}\, \mathrm{d}\varepsilon_{ij}^P\, \mathrm{d}\varepsilon_{ij}^P = \tfrac{1}{2}\, \mathrm{tr}\, (\mathrm{d}\mathbf{E}^P)^2$$

These give one form of eq(9.46) as

$$J_2' = \chi\left(\textstyle\int \mathrm{d}I_2\right) \tag{9.47a}$$

Moreover, with $J_2' = \tfrac{1}{3}(\bar{\sigma})^2$ and $\mathrm{d}I_2 = \tfrac{3}{4}(\mathrm{d}\bar{\varepsilon}^P)^2$, eq(9.47a) becomes

$$\frac{\bar{\sigma}^2}{3} = \frac{3}{4}\chi\left[\int (\mathrm{d}\bar{\varepsilon}^P)^2\right] \tag{9.47b}$$

Equation (9.47b) shows that the current equivalent stress depends upon equivalent plastic strain. For convenience, strain hardening is expressed directly as $\psi = \int \mathrm{d}\bar{\varepsilon}^P$. Taken again with $f = J_2'$, eq(9.46) becomes

$$\frac{\bar{\sigma}^2}{3} = \chi\left(\int \mathrm{d}\bar{\varepsilon}^P\right)$$

from which a hypothesis of strain hardening [27] is normally written:

$$\bar{\sigma} = H\left(\int \mathrm{d}\bar{\varepsilon}^{P}\right) \tag{9.48}$$

where, from eqs(9.2) and (9.6),

$$\bar{\sigma} = \sqrt{\frac{3}{2}\sigma_{ij}'\sigma_{ij}'} = \sqrt{\frac{3}{2}\left[\mathrm{tr}\,(\mathbf{T}')^{2}\right]} \tag{9.49a}$$

$$\mathrm{d}\bar{\varepsilon}^{P} = \sqrt{\frac{2}{3}\mathrm{d}\varepsilon_{ij}^{P}\,\mathrm{d}\varepsilon_{ij}^{P}} = \sqrt{\frac{2}{3}\left[\mathrm{tr}\,(\mathrm{d}\mathbf{E}^{P})^{2}\right]} \tag{9.49b}$$

The numerical factors in eqs(9.49a,b) enable the correct reduction to a uniaxial stress state. That is, $\bar{\varepsilon}^{P}$ and $\bar{\sigma}$ become the axial plastic strain and stress in a tension test. This test will establish which of the empirical formulae, discussed in the previous section, is the most suitable hardening function H in eq(9.48). The function χ which connects alternative $\bar{\sigma}$ and $\mathrm{d}\bar{\varepsilon}^{P}$ definitions, corresponding to $f(J_2', J_3')$ in eq(9.46), may be also be appraised in this way. These alternative definitions become necessary when $f(J_2', J_3')$ provides the required equivalence between tests. For an isotropic material, $\bar{\sigma}$ and $\mathrm{d}\bar{\varepsilon}^{P}$ are derived from the invariants of the tensors of deviatoric stress and incremental plastic strain. For example, in the Drucker function eqs(9.4) and (9.7) apply to eq(9.48). Here, the current equivalent stress becomes a function of the equivalent plastic strain. There in no further connection between $\bar{\sigma}$ and $\mathrm{d}\bar{\varepsilon}^{P}$ other than the similarity between the two invariants from which they derive.

9.9.2 Work Hardening

Equating ψ to the specific work of plastic deformation, W^P/unit volume, gives

$$\psi = \int \mathrm{d}W^{P} = \int \sigma_{ij}\,\mathrm{d}\varepsilon_{ij}^{P} = \int \mathrm{tr}\left(\mathbf{T}\,\mathrm{d}\mathbf{E}^{P}\right) \tag{9.50}$$

Taking eq(9.50) with a Mises potential $f = J_2' = \bar{\sigma}^2/3$, the hardening function in eq(9.46) becomes

$$\frac{\bar{\sigma}^{2}}{3} = \chi\left(\int \mathrm{d}W^{P}\right)$$

This hypothesis of work hardening [26] is normally given as

$$\bar{\sigma} = F\left(\int \mathrm{d}W^{P}\right) \tag{9.51a}$$

Equation (9.51a) shows that the current equivalent stress is a function of the specific work of plastic deformation. The hardening function F may be established from a uniaxial test where the argument $\int \mathrm{d}W^P$ is the area beneath the stress versus plastic strain curve. To show this, we can apply eq(9.50) to a uniaxial stress:

$$W^{P} = \int \sigma_{1}\,\mathrm{d}\varepsilon_{1}^{P} = \int \bar{\sigma}\,\mathrm{d}\bar{\varepsilon}^{P} \tag{9.51b}$$

In general, the increment of plastic work $\mathrm{d}W^P$ in eq(9.51a) may be expressed in absolute or deviatoric stress forms. For an absolute stress σ_{ij}:

$$\delta W^P = \sigma_{ij}\,\delta\varepsilon_{ij}^{\,P} = \mathrm{tr}\,(\mathbf{T}\,\delta\mathbf{E}^P) \tag{9.52a}$$
$$= \sigma_{11}\,\delta\varepsilon_{11}^{\,P} + \sigma_{22}\,\delta\varepsilon_{22}^{\,P} + \sigma_{33}\,\delta\varepsilon_{33}^{\,P} + 2\,(\sigma_{12}\,\delta\varepsilon_{12}^{\,P} + \sigma_{13}\,\delta\varepsilon_{13}^{\,P} + \sigma_{23}\delta\varepsilon_{23}^{\,P})$$

Substituting eq(3.9a) into eq(9.52a) leads to the deviatoric form

$$\delta W^P = (\sigma'_{ij} + \tfrac{1}{3}\,\delta_{ij}\sigma_{kk})\delta\varepsilon_{ij}^{\,P} = \sigma_{ij}'\,\delta\varepsilon_{ij}^{\,P} + \tfrac{1}{3}\,\sigma_{kk}\delta\varepsilon_{ii}^{\,P}$$
$$= \sigma_{ij}'\,\delta\varepsilon_{ij}^{\,P} = \mathrm{tr}\,(\mathbf{T}'\delta\mathbf{E}^P) \tag{9.52b}$$
$$= \sigma_{11}'\delta\varepsilon_{11}^{\,P} + \sigma_{22}'\delta\varepsilon_{22}^{\,P} + \sigma_{33}'\delta\varepsilon_{33}^{\,P} + 2(\sigma_{12}\,\delta\varepsilon_{12}^{\,P} + \sigma_{13}\delta\varepsilon_{13}^{\,P} + \sigma_{23}\delta\varepsilon_{23}^{\,P})$$
$$= \sigma_{11}'\delta\varepsilon_{11}^{\,P} + \sigma_{22}'\delta\varepsilon_{22}^{\,P} + (\sigma_{11}' + \sigma_{22}')(\delta\varepsilon_{11}^{\,P} + \delta\varepsilon_{22}^{\,P}) + 2(\sigma_{12}\delta\varepsilon_{12}^{\,P} + \sigma_{13}\delta\varepsilon_{13}^{\,P} + \sigma_{23}\delta\varepsilon_{23}^{\,P})$$

in which we have employed $\delta\varepsilon_{ii}^{\,P} = \mathrm{tr}\,(\delta\,\mathbf{E}^P) = 0$ and $\sigma_{ii}' = \mathrm{tr}\,\mathbf{T}' = 0$. We may identify equivalent stress $\bar{\sigma}$ and plastic strain increment $\mathrm{d}\bar{\varepsilon}^{\,P}$ from eq(9.51b), to give

$$\delta W^P = \bar{\sigma}\,\delta\bar{\varepsilon}^{\,P} = \sigma_{ij}\,\delta\varepsilon_{ij}^{\,P} = \sigma_{ij}'\,\delta\varepsilon_{ij}^{\,P} \tag{9.52c}$$

From eqs(9.49a,b), equivalent matrix forms of eq(9.52c) are

$$\delta W^P = \sqrt{\left[\mathrm{tr}\,(\mathbf{T}')^2\right]\left[\mathrm{tr}\left(\delta\mathbf{E}^P\right)^2\right]}$$
$$= \mathrm{tr}\left(\mathbf{T}\delta\mathbf{E}^P\right) = \mathrm{tr}\left(\mathbf{T}'\delta\mathbf{E}^P\right) \tag{9.52d}$$

In the absence of a flow rule, the selection of $\delta\bar{\varepsilon}^{\,P}$ in eq(9.52c) can be arbitrary. While $\bar{\sigma}$ depends upon the chosen yield function, $\delta\bar{\varepsilon}^{\,P}$ derives from a similar function in the plastic strain invariants, as was shown for the strain hardening hypothesis. When the flow rule is associated with the yield function, then $\bar{\sigma}$ and $\delta\bar{\varepsilon}^{\,P}$ in eq(9.52c) are connected. For example, with a von Mises yield function, $\bar{\sigma}$ is defined from eq(9.49a). The associated flow rule is $\mathrm{d}\varepsilon_{ij}^{\,P} = \mathrm{d}\lambda\sigma_{ij}'$. Substituting these into eq(9.52c) gives

$$\bar{\sigma}\,\delta\bar{\varepsilon}^{\,P} = \sqrt{(\sigma_{ij}'\,\delta\varepsilon_{ij}^{\,P})(\sigma_{ij}'\,\delta\varepsilon_{ij}^{\,P})} = \sqrt{(\sigma_{ij}'\,\delta\lambda\,\sigma_{ij}')\,(\delta\varepsilon_{ij}^{\,P}\,\delta\varepsilon_{ij}^{\,P})/\,\delta\lambda}$$
$$= \sqrt{(\sigma_{ij}'\,\sigma_{ij}')(\delta\varepsilon_{ij}^{\,P}\,\delta\varepsilon_{ij}^{\,P})} = \sqrt{\tfrac{2}{3}\,\bar{\sigma}^2\,(\delta\varepsilon_{ij}^{\,P}\,\delta\varepsilon_{ij}^{\,P})}$$
$$\delta\bar{\varepsilon}^{\,P} = \sqrt{\tfrac{2}{3}\,(\delta\varepsilon_{ij}^{\,P}\,\delta\varepsilon_{ij}^{\,P})} = \sqrt{\tfrac{2}{3}\left[\mathrm{tr}\,(\delta\mathbf{E}^P)^2\right]}$$

The fact that $\mathrm{d}\bar{\varepsilon}^{\,P}$ is identical to eq(9.49b) confirms that a Mises material is independent of the hardening hypothesis. Thus, a plot of $\bar{\sigma}$ versus $\int\bar{\sigma}\,\mathrm{d}\bar{\varepsilon}^{\,P}$ will not correlate flow curves any better than a plot of $\bar{\sigma}$ versus $\int\mathrm{d}\bar{\varepsilon}^{\,P}$. However, equivalence plots for other materials obeying $f(J_2', J_3')$ will depend upon the hardening hypothesis. Consequently, this yield function provides a choice between an abscissa of equivalent plastic strain or specific plastic work.

9.9.3 Comparisons Between H and F

The respective strain and work hardening functions H and F, appearing in eqs(9.48) and (9.51a), are found by changing the absissae within the following standard tests (H to the left, F to the right).

(a) *Tension*

$$\bar{\sigma} \text{ versus } \int (\mathrm{d}l/l - \mathrm{d}\bar{\sigma}/E) \qquad \bar{\sigma} \text{ versus } \int \bar{\sigma}(\mathrm{d}l/l - \mathrm{d}\bar{\sigma}/E)$$

$$\bar{\sigma} \text{ versus } \int \mathrm{d}l/l - \bar{\sigma}/E \qquad \bar{\sigma} \text{ versus } \int \bar{\sigma}\,\mathrm{d}l/l - \bar{\sigma}^2/(2E) \quad (9.53a,b)$$

(b) *Uniaxial Compression*

A similar expression applies to simple compression with lubrication, say with ptfe sheet, under continuous loading.

$$\bar{\sigma} \text{ versus } \int \mathrm{d}h/h - \bar{\sigma}/E \qquad \bar{\sigma} \text{ versus } \int \bar{\sigma}\,\mathrm{d}h/h - \bar{\sigma}^2/(2E) \quad (9.54a,b)$$

(c) *Torsion*

$$\sqrt{3}\tau \text{ versus } \int \mathrm{d}\gamma^P/\sqrt{3} \qquad \sqrt{3}\tau \text{ versus } \int (\sqrt{3}\tau)(\mathrm{d}\gamma^P/\sqrt{3})$$

$$\sqrt{3}\tau \text{ versus } \int (r\,\mathrm{d}\theta/l - \mathrm{d}\tau/G) \qquad \sqrt{3}\tau \text{ versus } \int \tau(r\,\mathrm{d}\theta/l - \mathrm{d}\tau/G)$$

$$\sqrt{3}\tau \text{ versus } \int r\,\mathrm{d}\theta/l - \tau/G \qquad \sqrt{3}\tau \text{ versus } \int \tau(r\,\mathrm{d}\theta/l) - \tau^2/(2G) \quad (9.55a,b)$$

(d) *Bulge Test (Equi-Biaxial Tension)*

$$\bar{\sigma} \text{ versus } (\mathrm{d}\varepsilon_r^t - \mathrm{d}\varepsilon_r^e) \qquad \bar{\sigma} \text{ versus } \int \bar{\sigma}(\mathrm{d}\varepsilon_r^t - \mathrm{d}\varepsilon_r^e) \quad (9.56a,b)$$

where, in eqs(9.56a,b), the total and elastic radial strain increments are

$$\mathrm{d}\varepsilon_r^t = \mathrm{d}t/t \text{ and } \mathrm{d}\varepsilon_r^e = -(\nu/E)(\mathrm{d}\sigma_\theta + \mathrm{d}\sigma_\phi) = -2(\nu/E)\,\mathrm{d}\bar{\sigma} \quad (9.56c,d)$$

Substituting eqs(9.56c,d) into eqs(9.56a,b) leads to

$$\bar{\sigma} \text{ versus } \int (\mathrm{d}t/t + 2\nu\,\mathrm{d}\bar{\sigma}/E) \qquad \bar{\sigma} \text{ versus } \int \bar{\sigma}(\mathrm{d}t/t + 2\nu\,\mathrm{d}\bar{\sigma}/E)$$

$$\bar{\sigma} \text{ versus } \int \mathrm{d}t/t + 2\nu\bar{\sigma}/E \qquad \bar{\sigma} \text{ versus } \int \bar{\sigma}\,\mathrm{d}t/t + \nu\bar{\sigma}^2/E \quad (9.56e,f)$$

Note, that $\mathrm{d}l$, $\mathrm{d}h$, $\mathrm{d}\theta$ and $\mathrm{d}t$ in eqs(9.53) to (9.56), denote the total dimensional changes arising from elastic plus plastic strain. In interrupted tests, plastic lengths l, h and t, are measured under no-load conditions. The first integral term then defines plastic strain and plastic work directly, a technique next followed in modifying compression tests.

(e) *Cook and Larke*

Under no-load measurements of the cylinder height, the equivalence plots are simplified to:

$$\bar{\sigma} \text{ versus } \int \mathrm{d}h/h \qquad \bar{\sigma} \text{ versus } \int \bar{\sigma}\,\mathrm{d}h/h \quad (9.57a,b)$$

(f) *Plane Strain Compression*

For no-load, permanent thickness measurements, the equivalence relation is derived as

$$\bar{\sigma} \text{ versus } (2/\sqrt{3}) \int \mathrm{d}t / t \qquad \bar{\sigma} \text{ versus } (2/\sqrt{3}) \int \bar{\sigma}\, \mathrm{d}t / t \tag{9.58a,b}$$

The calculation of the equivalent stress $\bar{\sigma}$ within eqs (9.57) and (9.58) has been given with the separate description of each test in Section 9.6.

9.9.4 Relationships between F and H

The functions H and F are found from the above plots empirically. Alternatively, F can be derived from a particular strain hardening function H.

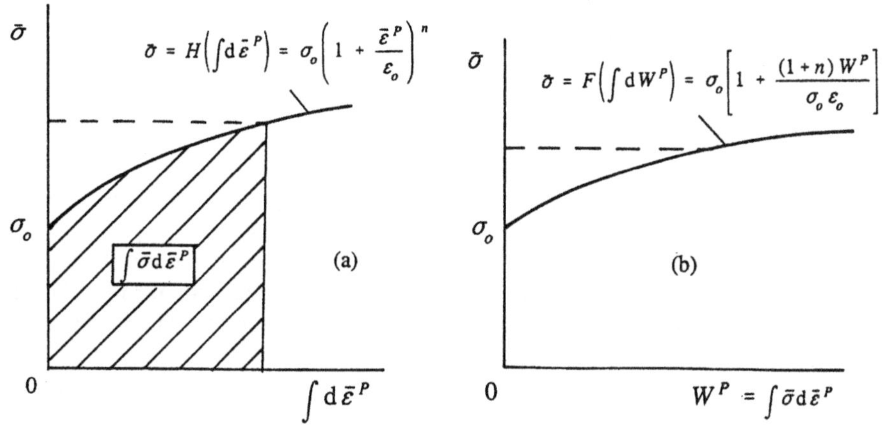

Figure 9.21 Arguments for the hypotheses of strain and work hardening

For example, in Fig. 9.21a H is identifed with the Swift eq(9.44b). The argument of plastic work in eq(9.53b) becomes

$$W^P = \int \bar{\sigma}\, d\bar{\varepsilon}^P = \int_0^{\bar{\varepsilon}^P} \sigma_o \left(1 + \frac{\bar{\varepsilon}^P}{\varepsilon_o}\right)^n d\bar{\varepsilon}^P$$

$$= \frac{\sigma_o \varepsilon_o}{n+1} \left| \left(1 + \frac{\bar{\varepsilon}^P}{\varepsilon_o}\right)^{n+1} \right|_0^{\bar{\varepsilon}^P} = \frac{\sigma_o \varepsilon_o}{n+1} \left[\left(\frac{\bar{\sigma}}{\sigma_o}\right)^{\frac{n+1}{n}} - 1 \right]$$

Transposing for $\bar{\sigma}$, determines the function F as

$$\bar{\sigma} = F(W^P) = \sigma_o \left[1 + \frac{(1+n)\, W^P}{\sigma_o \varepsilon_o}\right]^{\frac{n}{1+n}}$$

The explicit relationship between F and H is illustrated graphically in Fig. 9.21b.

On the other hand, with a function H, defined from Ludwik's eq(9.43b), the argument of plastic work becomes

$$W^P = \int_0^{\bar{\varepsilon}^P} \sigma_o \left[1 + \left(\frac{\bar{\varepsilon}^P}{\varepsilon_o} \right)^n \right] \mathrm{d}\bar{\varepsilon}^P$$

$$= \sigma_o \bar{\varepsilon}^P \left[1 + \frac{1}{n+1} \left(\frac{\bar{\varepsilon}^P}{\varepsilon_o} \right)^n \right] = \sigma_o \varepsilon_o \left(\frac{\bar{\sigma}}{\sigma_o} - 1 \right)^{\frac{1}{n}} \left[1 + \frac{1}{n+1} \left(\frac{\bar{\sigma}}{\sigma_o} - 1 \right) \right]$$

We then see that H does not supply a closed solution for F. In Chapter 10 (see p. 314) F and H will be used to define an isotropic hardening rule.

9.10 Concluding Remarks

Experiments with a von Mises material (i.e. $f = J_2'$) cannot distinguish between the work and strain hardening hypotheses. Here, the strain hardening hypothesis is preferred, for simplicity, since it identifies directly with the common hardening expressions of Hollomon, Swift, Ludwik etc. The choice between these expressions depends upon the material and the range of strain. A number of standard tests are available to establish an appropriate expression with the minimum of experimental error. For isotropic materials, obeying a more general yield function $f(J_2', J_3')$, we have seen how it is possible to distinguish between the two hardening hypotheses, resulting in different equivalent plastic strain expressions. While $\mathrm{d}\bar{\varepsilon}^P$ from strain hardening may be defined arbitrarily, $\mathrm{d}\bar{\varepsilon}^P$ from work hardening is more exacting as it must satisfy the incremental plastic work expression: $\bar{\sigma}\,\mathrm{d}\bar{\varepsilon}^P = \sigma_{ij}\,\mathrm{d}\varepsilon_{ij}^P$.

References

1. von Mises R. *Natchr. Ger. Wiss.* Gottingen, 1913, 582.
2. Drucker D. *Quart. Appl Math*, 1950, **7**, 411.
3. White G. N. and Drucker D. C. *Jl Appl Phys*, 1950, **21**, 1013.
4. Johnson A. E. *Complex Stress, Creep, Relaxation and Fracture of Metallic Alloys*, 1962, HMSO, Edinburgh.
5. Ikegami K. *Brit. Ind & Sci Int. Trans Ser* (The Metals Society), **1442**, 1976.
6. Michno M. J. and Findley W. N. *Acta Mech*, 1973, **18**, 163.
7. Considere A. *Ann. Ponts et Chaussees*, 1885, **9**, 574.
8. Bridgeman P. W. *Studies in Large Plastic Flow and Fracture*, 1952, McGraw-Hill, New York.
9. Johnson W. and Mellor P. B. *Engineering Plasticity*, 1983, Ellis-Horwood.
10. Hsu T. C. *ASME*, 1967, W.A. / Met.11
11. Bredt R. *Zeitschrift des Vereines Deutscher Ingenieure*, 1896, **40**, 785, 813.
12. Nadai A. *Theory of Flow and Fracture of Solids*, 1950, McGraw-Hill.
13. Cooke M, and Larke E. C. *Jl Inst. Metals*, 1945, **71**, 371.
14. Watts A. B. and Ford H. *Proc. I. Mech. E*, 1952, **1B**, 449.
15. Watts A. B. and Ford H. *Proc. I. Mech. E*, 1955, **169**, 1141.
16. Brown W. F. and Sachs G. *Trans ASME*, 1948, **70**, 241.
17. Mellor P. *Jl Mech. Phys. Solids*, 1956, **5**, 41.

18. Easterbrook K. and Grieve R. *Jl Mech. Work. Tech*, 1985, **11**, 229.
19. Zyczkowski M. *Combined Loadings in the Theory of Plasticity*, 1981, PWN, Warsaw.
20. Hollomon J. H. *Trans Amer Inst Mech Eng*, 1945, **162**, 269.
21 Ramberg W. and Osgood W. *Nat. Adv. Comm for Aero*, 1943, TN902.
22. Prager W. *Proc 5th ICAM Conference*, 1938, 234.
23. Ludwik P. *Elements der Technologischen Mechanik*, 1909, Springer, Berlin.
24. Rees D.W.A. *Proc Roy Soc*, 1987, **A410**, 443.
25. Swift H. W. *Jl Mech. Phys. Solids*, 1952, **1**, 1.
26. Vocé E. *Jl Inst. Metals*, 1948, **74**, 537.
27. Bland D. R. *Proc. 9th Int Conf Appl Mech*, 1957, **8**, 45.

Exercises

9.1 Confirm the stress and strain predictions, defining the point of tensile instability, as given by the Swift eqs(9.44c,d).

9.2 Confirm the stress and strain predictions, defining the point of tensile instability, as given by the Vocé eqs(9.45c,d).

9.3 Examine whether it is possible to derive the function F in the work hardening hypothesis for the flow curves described by Hollomon, modified Prager and Vocé in Figs 9.19a, c and f respectively.

9.4 Is the strain hardening hypothesis, which defines $\bar{\varepsilon}^P$ in eqs(9.7c), consistent with a meaningful work increment expression? Examine this from a consideration of deformation under pure shear.

9.5 The following data applies to the hydrostatic bulging of 0.8 mm thick steel sheet through a circular die 180 mm diameter:

p, bar	1.45	3.67	12.09	19.62	24.9	32.46	39.11	
	41.77	46.64	52.39	59.04	63.47	68.79	74.55	76.0
h, mm	0.15	3.00	8.85	12.9	16.35	20.55	24.6	
	25.95	28.95	32.55	36.9	40.65	46.05	53.7	58.35
t, mm	0.8	0.794	0.769	0.744	0.72	0.689	0.664	
	0.651	0.629	0.601	0.564	0.492	0.486	0.402	0.372

Construct the equivalent stress versus equivalent strain curve based upon the current pole heights and thickness values given.

9.6 The following data applies to a plane strain compression test upon an annealed aluminium strip with initial dimensions: breadth 50 mm and thickness 6 mm and a die of width 6 mm:

W, kN	4.98	7.48	9.97	12.46	14.95	17.44	19.94	22.43	24.92
	27.41	29.90	34.89	39.87	44.86	49.84	54.82	59.81	64.79
Δt, mm	0.30	0.32	0.36	0.40	0.58	0.75	0.90	1.11	1.39
	1.78	2.29	3.15	3.75	4.25	4.50	4.75	5.35	5.75

The thickness change Δt applies to an unloaded condition following the removal of each applied load level W. Plot the equivalent stress versus equivalent plastic strain curve and determine whether a Hollomon law can be applied to describe the curve.

9.7 The following torque-twist data applies to a torsion test upon a solid bar of mild steel: diameter 20.27 mm and gauge length 203.2 mm. Use the Nadai analysis to obtain a true, equivalent stress versus strain plot for the material.

T, Nm	28.25	45.21	75.73	96.08	118.7	144.7	168.0	191.0	198.9
	237.4	260.0	274.7	282.6	276.9	282.6	282.6	282.6	288.2
	288.2	291.1	295.0	301.8	310.8	327.8	340.22	350.4	361.7
	373.0	382.0	401.3	418.2	435.2	439.1	452 1	491.7	514.3
	542.5	553.9	582.1	587.8	594.5	605.8	616.0	626.2	640.9
	649.9	678.2	689.5	700.8					
$\theta°$	0.2	0.4	0.6	0.8	1.0	1.2	1.4	1.6	1.8
	2.0	2.2	2.4	2.6	2.8	3.0	3.2	3.4	5.0
	17.5	20	22	25	30	35	40	45	50
	55	60	70	80	90	100	110	140	160
	200	220	250	270	300	330	360	420	480
	540	600	660	720					

9.8 The following force-extension data applies to tensile loading upon a cylindrical bar of mild steel, diameter 11.43 mm and gauge length 57.15 mm. Construct both the nominal and true stress-strain curves for the material, superimposed upon the same axes.

W, kN	9.47	13.71	17.54	21.93	25.42	29.61	33.19	36.08	38.68
	40.87	44.16	45.85	47.05	47.75	48.35	49.44	50.34	50.84
	51.43	52.33	53.33	53.93	54.63	55.32	56.02	57.02	58.01
	58.81	59.41	60.01	60.51	61.30	62.0	62.15	62.3	62.5
	62.7	62.8	62.8	62.8	62.5	62.2	61.3	44.36	
x, mm	0.0254	0.0508	0.0762	0.102	0.127	0.152	0.178	0.203	0.229
	0.254	0.305	0.356	0.406	0.457	0.508	0.635	0.762	0.889
	1.016	1.27	1.524	1.78	2.03	2.29	2.54	3.05	3.56
	4.06	4.57	5.08	5.59	6.604	7.62	8.128	8.636	9.144
	9.652	10.16	10.67	11.176	11.684	11.938	12.192	12.7	

9.9 The following force-extension data applies to loading followed by unloading in a tension test upon a cylindrical bar of mild steel: diameter 11.43 mm and gauge length 50.8 mm.

Loading:

W, kN	2.59	5.78	7.88	9.77	11.86	13.95	16.05	18.04	20.23
	22.53	26.62	30.60	33.89	36.78	39.08	40.77	42.26	43.46
	44.46	45.45	46.15	47.45	49.64	50.04	50.64	51.04	51.44
	51.73	52.03	52.23	52.53					
x, mm	0.0051	0.0102	0.0152	0.0203	0.0254	0.0305	0.0356	0.0406	0.0457
	0.051	0.061	0.071	0.081	0.091	0.102	0.112	0.122	0.132
	0.142	0.152	0.178	0.203	0.254	0.279	0.305	0.330	0.356
	0.381	0.406	0.432	0.457					

Unloading:

W, kN	41.07	32.09	22.73	13.46	5.085	0
x, mm	0.432	0.406	0.381	0.356	0.330	0.315

Determine the initial modulus of elasticity for the material and the proof stress for offset engineering strains of 0.1% and 0.5%. What does the modulus become when unloading from the plastic range?

9.10 The following data applies to a tensile test upon a 12.7 mm diameter mild steel testpiece in which the diameter of the neck was recorded through to fracture. Calculate and plot the true stress versus the natural strain in the neck when based upon the Bridgeman correction curve given in Fig. 9.22.

W, kN	34.27	37.83	44.5	46.73	51.18	53.4	56.29	57.85	58.43
	59.36	59.54	59.72	59.72	59.81	59.63	59.63	59.45	59.27
	58.74	58.3	57.94	57.85	56.96	56.52	55.63	55.18	54.29
	53.85	53.13	52.51	51.18	49.84	49.75	48.45	46.73	46.28

d, mm	12.61	12.57	12.496	12.446	12.357	12.268	12.116	12.014	11.938
	11.836	11.735	11.659	11.532	11.465	11.354	11.227	11.151	11.049
	10.846	10.719	10.592	10.363	10.236	10.058	9.931	9.804	9.728
	9.55	9.347	9.246	8.992	8.915	8.814	8.636	8.484	8.433

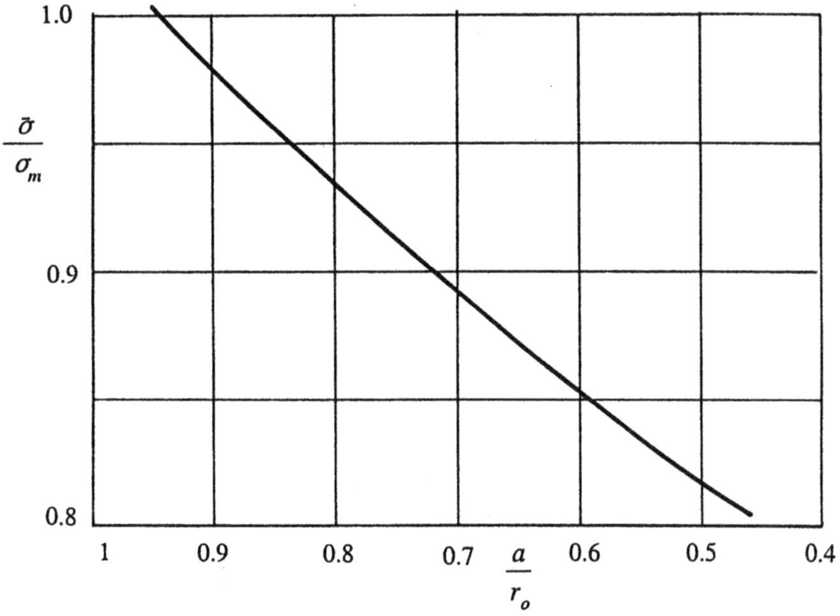

Figure 9.22

9.11 The following height reduction data applies to a Cooke and Larke compression test upon three solid copper cylinders, each with a common diameter d_o = 18 mm and initial heights h_o of 9, 18 and 36 mm respectively.

Δh, mm	0.03	0.14	0.30	0.50	0.73	1.00	1.26	1.53	1.83
	2.13	2.45	2.73	3.0	3.24	3.46	3.75	3.97	4.17
	4.37	4.53	4.81	5.0	5.22	(for h_o = 9 mm)			

Δh, mm	0.08	0.34	0.72	1.15	1.65	2.20	2.82	3.46	4.11
	4.77	5.37	5.95	6.54	7.03	7.53	8.01	8.43	8.86
	9.22	9.70	10.0	10.31	10.77	(for h_o = 18 mm)			

Δh, mm	0.05	0.44	1.15	2.02	3.04	4.14	5.44	6.72	8.12
	9.47	10.74	12.04	13.29	14.49	15.57	16.59	17.57	18.39
	19.14	20 29	20.82	21.34	22.29	(for h_o = 36 mm)			

Given that the following loads apply to the Δh values in each cylinder , construct the true stress versus natural strain curve for the material when based upon an extrapolation to a cylinder with $d_o/h_o = 0$.

W (kN)	10	20	30	40	50	60	70	80	90
	100	110	120	130	140	150	160	170	180
	190	200	220	230	250	(for all h_o values)			

9.12 The following load versus extension data applies to a tensile test upon a ductile, thin sheet steel. Testpiece dimensions were, width 9.75 mm, thickness 0.2 mm and gauge length 50 mm. Derive the true stress and natural plastic strain values and show that these conform to a Hollomon law $\sigma = A\varepsilon^n$ in which A = 540 and n = 0.17. Take E = 180 GPa.

x, mm	0.2	0.4	0.6	0.8	1.0	1.2	1.4	1.6	1.8
W, kN	0.499	0.527	0.545	0.564	0.581	0.598	0.606	0.618	0.625

x, mm	2.0	2.4	2.8	3.2	3.6	4.0	4.4	4.8	5.2
W, kN	0.636	0.648	0.660	0.676	0.687	0.696	0.701	0.715	0.713

CHAPTER 10

PLASTICITY WITH HARDENING

10.1 Introduction

When the initial yield stress of a hardening material is exceeded, a subsequent yield surface is required to express the current flow stress for the strained material. This shows that it becomes necessary to account for the effect of plastic strain on the yield surface. In this chapter we restrict our attention to the account provided by the rules of isotropic and kinematic hardening. Thus the initial yield surface may expand uniformly or translate rigidly to contain the current stress point by each respective rule. It appears that subsequent yield functions are mostly based upon the initial von Mises and Tresca conditions for mathematical simplicity. For example, when the Mises function $f = J_2'$, is employed within the simpler rule of isotropic hardening it leads to the flow rule of Levy-Mises. Any number of flow rules arise from applying each rule to alternative, initial yield functions. Important among these are the influences of J_3' and initial anisotropy on the subsequent yield surface. It will be shown how alternative flow rules can be based upon: (i) an initial function $f(J_2', J_3')$, (ii) the orthotropic function, discussed in the following chapter, and (iii) Hencky's deformation theory, extended to account for hardening.

It is also appropriate to examine the elements of the two rules with experimental data. This is done to enable limitations of their usefulness to be stated. It will appear from this that a more versatile model is one that combines the two models of hardening. This leads to a combined hardening concept embracing sources of hardening from within lattice slip.

10.2 Conditions Associated with the Yield Surface

The fundamental laws of isothermal plasticity apply to its irreversibile behaviour. These laws govern the shape of both initial and subsequent boundaries between the elastic and plastic regions and provide the magnitude and path of plastic strain. Classical theories express plasticity in the following manner and couple this with Hookean elasticity.

10.2.1 Loading Function

The subsequent yield surface is expressed as a *loading function* of stress σ_{ij}, plastic strain ε_{ij}^P and a hardening parameter χ:

$$f(\sigma_{ij}, \varepsilon_{ij}^P, \chi) = 0 \tag{10.1}$$

where it has been seen previously that χ is either a function of plastic strain or work (see eqs(9.46)). The total differential of f in eq(10.1) is

$$\mathrm{d}f = \left(\frac{\partial f}{\partial \sigma_{ij}}\right)\mathrm{d}\sigma_{ij} + \left(\frac{\partial f}{\partial \varepsilon_{ij}^{P}}\right)\mathrm{d}\varepsilon_{ij}^{P} + \left(\frac{\partial f}{\partial \chi}\right)\mathrm{d}\chi = 0 \tag{10.2a}$$

Where all changes occur in time dt, the rate form of eq(10.2a) becomes

$$\frac{\mathrm{d}f}{\mathrm{d}t} = \left(\frac{\partial f}{\partial \sigma_{ij}}\right)\left(\frac{\mathrm{d}\sigma_{ij}}{\mathrm{d}t}\right) + \left(\frac{\partial f}{\partial \varepsilon_{ij}^{P}}\right)\left(\frac{\mathrm{d}\varepsilon_{ij}^{P}}{\mathrm{d}t}\right) + \left(\frac{\partial f}{\partial \chi}\right)\left(\frac{\mathrm{d}\chi}{\mathrm{d}t}\right) = 0 \tag{10.2b}$$

where $\mathrm{d}\varepsilon_{ij}^{P}$ and $\mathrm{d}\chi$ depend upon the direction of the incremental stress vector $\mathrm{d}\sigma_{ij}$ from a point P on the surface $f = 0$. Three *loading conditions* arise, as shown in Fig. 10.1.

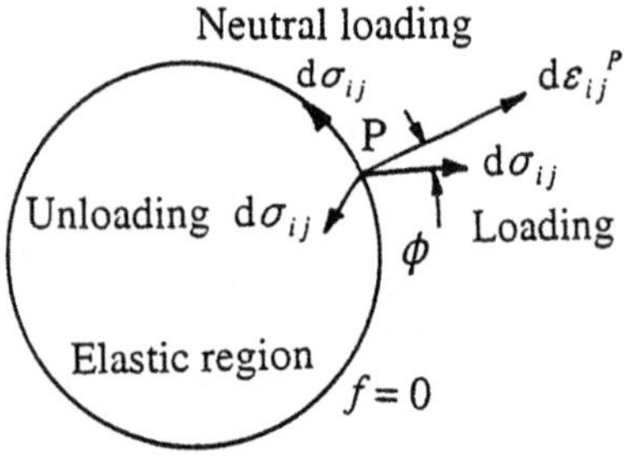

Figure 10.1 The three conditions of loading

They are loading, unloading and neutral loading. Each condition is associated with the change, or rate of change, to f produced by the corresponding stress change. Writing the change to f as $\mathrm{d}f_\sigma$, it follows from eqs(10.2a,b) that

$$\mathrm{d}f_\sigma = \frac{\partial f}{\partial \sigma_{ij}}\mathrm{d}\sigma_{ij} \quad \text{and} \quad \frac{\mathrm{d}f_\sigma}{\mathrm{d}t} = \frac{\partial f}{\partial \sigma_{ij}} \times \frac{\mathrm{d}\sigma_{ij}}{\mathrm{d}t} \tag{10.3a,b}$$

For an elastic-plastic loading we have $\mathrm{d}f_\sigma > 0$ so that $\mathrm{d}\varepsilon_{ij}^{P} > 0$ and $\mathrm{d}\chi > 0$. With an elastic unloading from the plastic region $\mathrm{d}f_\sigma < 0$ and $\mathrm{d}\varepsilon_{ij}^{P} = \mathrm{d}\chi = 0$. Neutral loading refers to a change of stress state along the yield surface. This gives $\mathrm{d}f_\sigma = 0$ and $\mathrm{d}\varepsilon_{ij}^{P} = \mathrm{d}\chi = 0$. Note that a change to the elastic strain will result from each loading.

10.2.2 Flow Rule

Hill [1] and Drucker [2] were among the first to employ a plastic potential function for a plastic loading condition. The potential describes a closed, convex surface in strain space whose outward normal defines the direction of the plastic strain increment vector [3]. It is convenient to let this surface lie in stress space, where it differs from the plastic potential in elastic dimension only. It may then be taken that the incremental vector of plastic loading strain lies normal to the stress surface $g(\sigma_{ij})$. This normality condition is expressed in

$$\mathrm{d}\varepsilon_{ij}^{P} = \mathrm{d}\lambda \frac{\partial g(\sigma_{ij})}{\partial \sigma_{ij}} \tag{10.4a}$$

The elastic conversion between the strain and stress spaces identifies $g(\sigma_{ij})$ with the yield function f. The flow rule is then 'associated' with the yield function and eq(10.4a) becomes

$$d\varepsilon_{ij}^{P} = d\lambda \frac{\partial f}{\partial \sigma_{ij}} \tag{10.4b}$$

The positive scalar $d\lambda$ is defined from combining eqs(10.2a) and (10.4b) as

$$d\lambda = \frac{-\left(\partial f/\partial \sigma_{ij}\right) d\sigma_{ij} - \left(\partial f/\partial \chi\right) d\chi}{\left(\partial f/\partial \varepsilon_{mn}\right)\left(\partial f/\partial \sigma_{mn}\right)} \tag{10.5}$$

Explicit forms of $d\lambda$ follow from eq(10.5) when f is based upon the Mises yield function and an isotropic hardening rule:

$$f = \tfrac{1}{2}\, \sigma_{ij}'\, \sigma_{ij}' - \chi(\psi) = 0 \tag{10.6}$$

where $\psi = \int d\bar{\varepsilon}^{P}$ or $\psi = \int dW^{P}$. It is then straightforward to derive $d\lambda$ from the reduction of eqs (10.5) and (10.6) to simple tension. It will be seen that this allows χ to be identified with functions H and F from the strain and work hypotheses (see Section 10.6).

10.2.3 Drucker's Postulate

Consider a unit volume of material in elastic equilibrium under a stress state σ_{ij}^{*} at point A in Fig. 10.2a. Drucker [2] considered the net change to the elastic energy stored when some external action, force moment, torque etc, applied a loading cycle. The sequence of loading produces a new stress state σ_{ij} at a point B on the yield surface, then pierces the surface with an infinitesimal stress change $d\sigma_{ij}$ before restoring it to σ_{ij}^{*}. We may relate this loading sequence to a cyclic response ABCA′ within each stress-strain component plot (Fig. 10.2b).

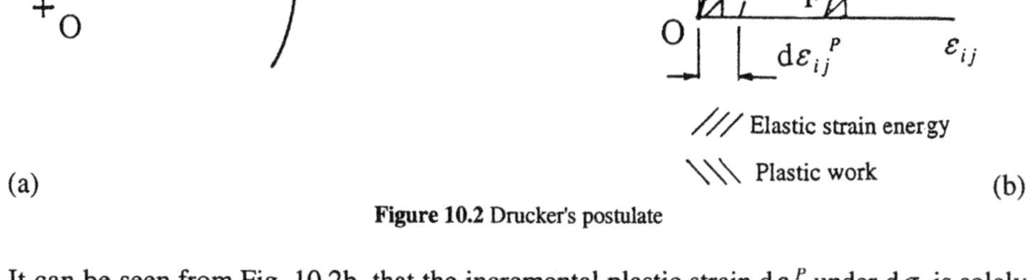

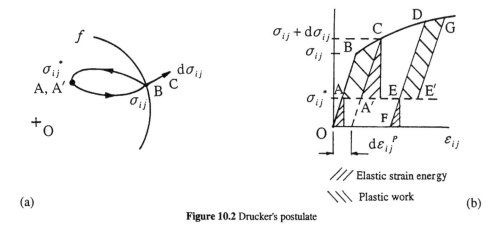

Figure 10.2 Drucker's postulate

It can be seen from Fig. 10.2b, that the incremental plastic strain $d\varepsilon_{ij}^{P}$ under $d\sigma_{ij}$ is solely responsible for a permanent change to the stored energy. It was shown previously for a similar cycle (Fig. 8.11) that elastic energy recovers, leaving the plastic work done as

$$dW^{P} = (\sigma_{ij} - \sigma_{ij}^{*})d\varepsilon_{ij}^{P} + d\sigma_{ij}\, d\varepsilon_{ij}^{P} > 0 \tag{10.7}$$

At its extremes σ_{ij}^* may either define an unstressed state or $\sigma_{ij}^* = \sigma_{ij}$. The inequality (10.7) applies to the loading condition in Fig. 10.1. When $d\sigma_{ij}$ corresponds to unloading or neutral loading, then $dW^P = 0$. Splitting eq(10.7), it follows that two conditions apply to loading:

$$\left(\sigma_{ij} - \sigma_{ij}^*\right) d\varepsilon_{ij}^P > 0 \quad \text{or} \quad \left(\bar{\sigma} - \bar{\sigma}^*\right) d\bar{\varepsilon}^P \geq 0 \tag{10.8a,b}$$

$$d\sigma_{ij}\, d\varepsilon_{ij}^P > 0 \quad \text{or} \quad d\bar{\sigma}\, d\bar{\varepsilon}^P \geq 0 \tag{10.8c,d}$$

Drucker's postulate also applies to any subsequent yield surface where σ_{ij} identifies a new yield condition, e.g. point D instead of B in Fig. 10.2b. Thus when $d\sigma_{ij}$ loads to point G, the postulate will apply to the cycle EDGE′. The difference is that the initial store of recoverable elastic energy at point E (small triangle) is accompanied by the irrecoverable work of previous inelastic deformation (i.e. area OABCDEF).

It is convenient to visualise each loading condition (10.8b,d) with scalar products of equivalent stress and equivalent plastic strain increment, $\bar{\sigma}$ and $d\bar{\varepsilon}^P$. With the von Mises definition of $\bar{\sigma}$, the initial yield surface f appears as a circle of radius O′B on the deviatoric plane in Fig. 10.3a.

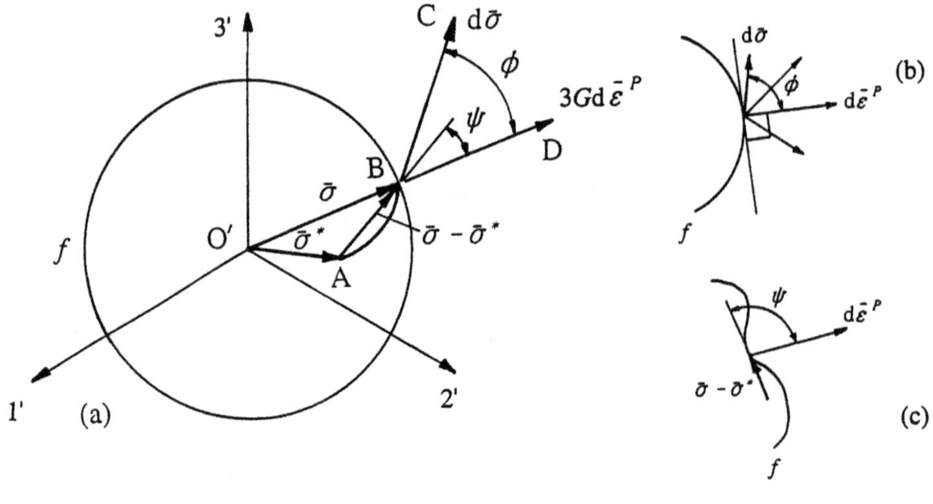

Figure 10.3 Drucker's postulate on deviatoric plane

Within eq(10.8b), $\bar{\sigma}^*$ and $\left(\bar{\sigma} - \bar{\sigma}^*\right)$ appear as respective vectors **O′A** and **AB** on this plane. Vector **BC** is the change to the initial yield stress $d\bar{\sigma}$, this being accompanied by an equivalent plastic strain increment $d\bar{\varepsilon}^P$. The latter may be scaled to stress vector **BD** on the deviatoric plane. To do this the principal, incremental, deviatoric plastic strains are converted to dimensions of deviatoric stress with the constant multiplier $2G$ [4]. Thus, the magnitude $|\mathbf{BD}|$ becomes the sum of ' stress components ' $2Gd\varepsilon_1'^P$, $2Gd\varepsilon_2'^P$ and $2Gd\varepsilon_3'^P$:

$$|\mathbf{BD}| = 2G\sqrt{3/2}\,\sqrt{(d\varepsilon_1'^P)^2 + (d\varepsilon_2'^P)^2 + (d\varepsilon_3'^P)^2} = 2G \times \frac{3}{2}\, d\bar{\varepsilon}^P \tag{10.9a}$$

Equation (9.6e) and $d\varepsilon_{ij}^P = d\varepsilon_{ij}'^P$ have been used with a projection factor √(3/2) from orthogonal axes 1', 2' and 3'. Equation (10.9a) shows that |**BD**| is proportional to $d\bar{\varepsilon}^P$ and extends the $\bar{\sigma}$ vector **O′B**. The incremental stress and strain vectors form the scalar product:

$$\mathbf{BC} \bullet \mathbf{BD} = |\mathbf{BC}|\,|\mathbf{BD}| \cos\phi = 3G\,(\bar{\sigma}\, d\bar{\varepsilon}^P) \cos\phi \tag{10.9b}$$

It follows from eq(10.8d) and (10.9b) that the included angle ϕ, between $\mathrm{d}\bar{\sigma}$ and $\mathrm{d}\bar{\varepsilon}^P$, must be less than 90°. Hence, for loading, it may be deduced [2, 5, 6] that $\mathrm{d}\bar{\varepsilon}^P$ remains in the position of an exterior normal, independently of the direction of $\mathrm{d}\bar{\sigma}$ in Fig. 10.3b. Furthermore, eq(10.8b) shows an included angle $\psi < 90°$, between vectors **AB** and **BD** in Fig. 10.3a. The implication is that the yield surface f must always be convex. Figure 10.3c shows how it is possible for $\psi > 90°$ when a surface is concave.

10.3 Isotropic Hardening

In the rule of isotropic hardening the initial yield surface expands uniformly in stress space as a material plastically deforms under an outward loading. Subsequent yield surfaces thus retain the shape and orientation of the initial surface. Figure 10.4 illustrates this rule within two common plane stress states: (a) tension-torsion and (b) principal biaxial stress.

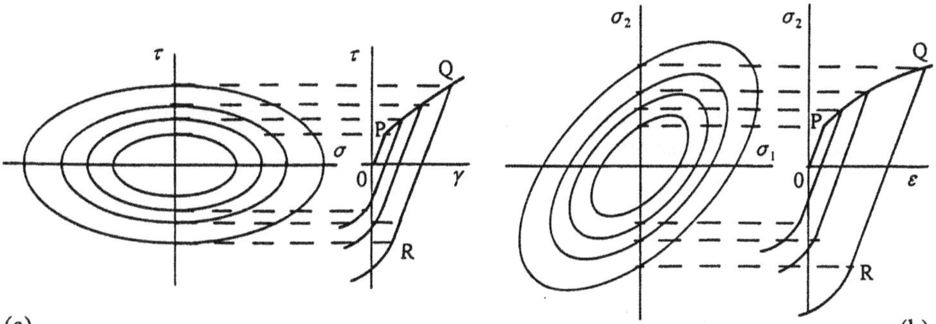

(a) (b)

Figure 10.4 Isotropic hardening under plane stress

Figures 10.4a,b connect the rule to simple shear and uniaxial hardening curves OPQ, as yield loci expand to contains flow stresses between P and Q. Clearly, the expanding elastic interior must be bounded by forward and reversed yield stresses of the same magnitude, i.e. $\tau_R = \tau_Q$ and $\sigma_{2R} = \sigma_{2Q}$ in Figs 10.4a,b. Consequently, a Bauschinger effect (in which $\tau_Q > \tau_R$ and $\sigma_{2Q} > \sigma_{2R}$) is absent in this rule. The mathematical condition for isotropic hardening is that stress σ_{ij} in eq(10.1) may be separated from ε_{ij}^P and χ in the form of eq(9.46). Hence the yield (or loading) function depends solely upon σ_{ij} while its size will depend upon χ. This is expressed as

$$f(\sigma_{ij}) = \chi(\psi) \tag{10.10a}$$

where χ is a monotonically increasing hardening function of a suitable scalar quantity ψ of plastic strain or plastic work. The change to f in eq(10.10a), corresponding to a change in stress $\mathrm{d}\sigma_{ij}$, is given by eq(10.3a):

$$\mathrm{d}f = \frac{\partial f}{\partial \sigma_{ij}}\,\mathrm{d}\sigma_{ij} \tag{10.10b}$$

This change in f will now depend upon the three aforementioned conditions of loading:

(i) $\mathrm{d}f > 0$, $\mathrm{d}f = \mathrm{d}\chi$ and $\mathrm{d}\varepsilon_{ij}^P > 0$ for elastic-plastic *loading*

(ii) $\mathrm{d}f < 0$, $\mathrm{d}\chi = 0$ and $\mathrm{d}\varepsilon_{ij}^P = 0$ for elastic *unloading*

(iii) $\mathrm{d}f = \mathrm{d}\chi = 0$ and $\mathrm{d}\varepsilon_{ij}^P = 0$ for *neutral loading*

No further hardening can occur as a material is either unloaded or is subjected to a neutral stress change. Note that in the case of a rigid plastic material, or one that has reached saturation hardening, it is possible for $\mathrm{d}\varepsilon_{ij}^{P} > 0$ under the condition $\mathrm{d}f = \mathrm{d}\chi$ in (iii).

The specific forms of eq(10.10a) that follow depend upon the function f and the argument ψ. The initial condition of the material governs the choice of yield function f, while ψ will depend upon the hypothesis of hardening.

10.3.1 Levy-Mises

The isotropic function f in eq(10.10a) is most often identified with the second invariant of deviatoric stress: $J_2' = \frac{1}{2}\sigma_{ij}'\sigma_{ij}'$, so implying that hydrostatic stress does not contribute to plastic deformation. Equation (10.10a) becomes

$$J_2' = \chi(\psi) \tag{10.11a}$$

$$\tfrac{1}{2}\sigma_{ij}'\sigma_{ij}' = \chi(\psi) \tag{10.11b}$$

We know from the previous chapter that the scalar measure of hardening ψ is normally taken to be the equivalent plastic strain $\bar{\varepsilon}^{P}$ or, the specific work of plastic deformation W^{P}/unit volume [1]. These give, respectively,

$$\psi = \bar{\varepsilon}^{P} = \frac{2}{3}\int\left(\mathrm{d}\varepsilon_{ij}^{P}\,\mathrm{d}\varepsilon_{ij}^{P}\right)^{\frac{1}{2}} \tag{10.12a}$$

$$\psi = W^{P} = \int\sigma_{ij}\,\mathrm{d}\varepsilon_{ij}^{P} = \int\sigma_{ij}'\,\mathrm{d}\varepsilon_{ij}^{P} \tag{10.12b}$$

The left-hand side of eq(10.11b) may be identified with an equivalent stress $\bar{\sigma}$:

$$\bar{\sigma} = \sqrt{3J_2'} = \sqrt{\frac{3}{2}\sigma_{ij}'\sigma_{ij}'} \tag{10.12c}$$

The numerical factors in eq(10.12a) and (10.12c) enable the correct reduction to a uniaxial stress state, i.e. $\bar{\varepsilon}^{P}$ and $\bar{\sigma}$ become the axial plastic strain and stress in a tension test. Substituting eqs(10.12a) and (10.12c) into eq(10.11b), the strain hardening hypothesis appears in its simplest equivalent form (see eq(9.48)):

$$\bar{\sigma} = H\left(\int\mathrm{d}\bar{\varepsilon}^{P}\right) \tag{10.13a}$$

It follows that the hardening function H in eq(10.13a) may be established from simple tension. The most useful explicit forms of eq(10.13a) are those given by Hollomon [7], Ludwik [8] and Swift [9] (e.g. see pp. 295-297). Alternatively, substituting from eqs(10.12b,c), the hardening law (10.11b) appears as a work hypothesis (see eq(9.51a)):

$$\bar{\sigma} = F\left(\int\mathrm{d}W^{P}\right) \tag{10.13b}$$

Equation (10.13b) is less convenient because function F must be established from the area beneath a curve of stress versus plastic strain. Equations (10.11b) and (10.12c) provide the relationship between the hardening functions in eqs(10.13a,b) as

$$\chi(\psi) = \frac{\bar{\sigma}^2}{3} \quad \text{where} \quad \bar{\sigma} = H\left(\int \mathrm{d}\bar{\varepsilon}^P\right) = F\left(\mathrm{d}W^P\right) \tag{10.13c}$$

Taking both stress deviator invariants $J_2' = \tfrac{1}{2}\sigma_{ij}'\sigma_{ji}'$ and $J_3' = \tfrac{1}{3}\sigma_{ij}'\sigma_{jk}'\sigma_{ki}'$ into function $f(\sigma_{ij})$ in eq(10.10a) will account for most experimental observations on isotropic yielding [10]. Note that the previous chapter showed that including J_3' in f required a re-definition of $\bar{\sigma}$ and $\bar{\varepsilon}^P$ through the chosen yield function and hardening hypothesis.

The single most attractive feature of isotopic hardening over alternative hardening rules is its mathematical simplicity. The latter imposes limitations, as we shall see, but for most outward loading paths, isotropic hardening can predict plastic strain paths acceptably. The incremental theory of plasticity is completed when isotropic hardening is combined with the flow rule (10.4b). Levy [11] and von Mises [12] did this simply by associating the plastic potential function $g(\sigma_{ij})$, in eq(10.4a), with a von Mises flow potential $f(\sigma_{ij}) = J_2' = \tfrac{1}{2}\sigma_{ij}'\sigma_{ij}'$. This leads to a linear constitutive relation between the incremental plastic strain tensor $\mathrm{d}\varepsilon_{ij}^P$ and the current deviatoric stress tensor σ_{ij}':

$$\mathrm{d}\varepsilon_{ij}^P = \mathrm{d}\lambda\, \sigma_{ij}' \tag{10.14}$$

where $\sigma_{ij}' = \sigma_{ij} - \tfrac{1}{3}\delta_{ij}\sigma_{kk}$. Kronecker's delta, in which $\delta_{ij} = 1$ for $i = j$ and $\delta_{ij} = 0$ for $i \neq j$, ensures that normal stress components are reduced by their mean (or hydrostatic) stress value. The incremental scalar $\mathrm{d}\lambda$ is a factor of proportionality, linking the current equivalent stress to an increment of equivalent plastic strain, i.e. the co-ordinates ($\bar{\sigma}$, $\mathrm{d}\bar{\varepsilon}^P$) of a uniaxial flow curve. Here mean stress $\sigma_m = \tfrac{1}{3}\bar{\sigma}$ and we have $\sigma_1' = \bar{\sigma} - \tfrac{1}{3}\bar{\sigma} = \tfrac{2}{3}\bar{\sigma}$. Also, the tensile strain $\mathrm{d}\varepsilon_1^P = \mathrm{d}\bar{\varepsilon}^P$. Equation (10.14) gives $\mathrm{d}\varepsilon_1^P = \mathrm{d}\lambda\, \sigma_1'$, from which

$$\mathrm{d}\lambda = \frac{3\,\mathrm{d}\bar{\varepsilon}^P}{2\,\bar{\sigma}} \tag{10.15}$$

Equation (10.15) shows that $\mathrm{d}\lambda$ is inversely proportional to the incremental plastic modulus, i.e. proportional to the compliance, $\mathrm{d}\bar{\varepsilon}^P / \bar{\sigma}$. For non-linear hardening, it follows from the strain and work hardening hypotheses, eqs(10.13a,b), that

$$H' = \frac{\mathrm{d}\bar{\sigma}}{\mathrm{d}\bar{\varepsilon}^P} \tag{10.16a}$$

$$F' = \frac{\mathrm{d}\bar{\sigma}}{\mathrm{d}W^P} = \frac{\mathrm{d}\bar{\sigma}/\mathrm{d}\bar{\varepsilon}^P}{\mathrm{d}W^P/\mathrm{d}\bar{\varepsilon}^P} = \frac{\mathrm{d}\bar{\sigma}/\mathrm{d}\bar{\varepsilon}^P}{\bar{\sigma}} \tag{10.16b}$$

Differentiating the Ludwik eq(9.43b), for example, we find H' in eq(10.16a):

$$H' = \frac{\mathrm{d}\bar{\sigma}}{\mathrm{d}\bar{\varepsilon}^P} = n\left(\frac{\sigma_o}{\varepsilon_o}\right)\left(\frac{\bar{\sigma}}{\sigma_o} - 1\right)^{\frac{n-1}{n}} \tag{10.16c}$$

The compliance may be expressed, from eqs(10.16a,b), in one of two forms:

$$\frac{\mathrm{d}\bar{\varepsilon}^P}{\bar{\sigma}} = \frac{\mathrm{d}\bar{\sigma}}{\bar{\sigma}H'} \quad \text{or} \quad \frac{\mathrm{d}\bar{\varepsilon}^P}{\bar{\sigma}} = \frac{\mathrm{d}\bar{\sigma}}{F'\bar{\sigma}^2} \tag{10.17a,b}$$

Combining eqs(10.15) and (10.17a,b) reveals how $d\lambda$ changes monotonically with an increasing $\bar{\sigma}$ during loading:

$$d\lambda = \frac{3d\bar{\sigma}}{2H'\bar{\sigma}} \quad \text{or} \quad d\lambda = \frac{3d\bar{\sigma}}{2F'\bar{\sigma}^2} \tag{10.18a,b}$$

Note, that eqs(10.18a,b) also derive from eq(10.5) provided f is expressed in an isotropic hardening form of eq(10.6). For a principal triaxial stress system: σ_1, σ_2 and σ_3, eq(10.14) expands to give principal, plastic strain increments:

$$d\varepsilon_1^P = \frac{d\bar{\varepsilon}^P}{\bar{\sigma}}\left[\sigma_1 - \frac{1}{2}(\sigma_2 + \sigma_3)\right] \tag{10.19a}$$

$$d\varepsilon_2^P = \frac{d\bar{\varepsilon}^P}{\bar{\sigma}}\left[\sigma_2 - \frac{1}{2}(\sigma_1 + \sigma_3)\right] \tag{10.19b}$$

$$d\varepsilon_3^P = \frac{d\bar{\varepsilon}^P}{\bar{\sigma}}\left[\sigma_3 - \frac{1}{2}(\sigma_1 + \sigma_2)\right] \tag{10.19c}$$

where $d\bar{\varepsilon}^P/\bar{\sigma}$ is given by either of eqs(10.17a,b). Numerical integration of eqs(10.19a,b,c) is usually required except in the case of uniform biaxial and triaxial stressings, where components increase proportionately or follow a single step during loading. Equations (10.19a-c) are often compared with triaxial elastic relations. In the latter, ν replaces ½ for compressible elasticity and $d\bar{\varepsilon}^P/\bar{\sigma}$ is replaced by an elastic compliance $1/E$. More important than the similar structure between the linear elastic and non-linear plastic equations is a recognition of the entirely different material responses they are to represent.

10.3.2 Drucker

Employing the Drucker loading function [13] as a flow potential, eq(10.10a) becomes

$$f(\sigma_{ij}) = J_2'^3 - cJ_3'^2 = \chi(\psi) \tag{10.20}$$

Now, for any isotropic function $f(J_2', J_3')$, the associated flow rule becomes

$$\begin{aligned} d\varepsilon_{ij}^P = d\lambda\frac{\partial f}{\partial \sigma_{ij}'} &= d\lambda\left[\left(\frac{\partial f}{\partial J_2'}\right)\left(\frac{\partial J_2'}{\partial \sigma_{ij}'}\right) + \left(\frac{\partial f}{\partial J_3'}\right)\left(\frac{\partial J_3'}{\partial \sigma_{ij}'}\right)\right] \\ &= d\lambda\left[\left(\frac{\partial f}{\partial J_2'}\right)\sigma_{ij}' + \left(\frac{\partial f}{\partial J_3'}\right)\left(\sigma_{ik}'\sigma_{kj}' - \frac{2}{3}J_2'\delta_{ij}\right)\right] \\ &= d\lambda\left[\left(\frac{\partial f}{\partial J_2'}\right)\sigma_{ij}' + \left(\frac{\partial f}{\partial J_3'}\right)t_{ij}'\right] \end{aligned} \tag{10.21a}$$

where $t_{ij}' = \partial J_3'/\partial \sigma_{ij}' = \sigma_{ik}'\sigma_{kj}' - ⅔ J_2'\delta_{ij}$. Substituting $f(\sigma_{ij})$ from eq(10.20) into eq(10.21a) gives Drucker's flow rule:

$$d\varepsilon_{ij}^{P} = d\lambda\,(3\,J'^{2}_{2}\,\sigma'_{ij} - 2cJ'_{3}\,t'_{ij}) \qquad (10.21b)$$

Principal plastic strain increments $d\varepsilon_1^P$, $d\varepsilon_2^P$ and $d\varepsilon_3^P$ follow from eq(10.21b) with the following substitutions:

$$J_2' = \tfrac{1}{2}(\sigma_1'^2 + \sigma_2'^2 + \sigma_3'^2), \quad J_3' = \tfrac{1}{3}(\sigma_1'^3 + \sigma_2'^3 + \sigma_3'^3)$$
$$t_1' = \sigma_1'^2 - \tfrac{2}{3}J_2' = \sigma_1'^2 - \tfrac{1}{3}(\sigma_1'^2 + \sigma_2'^2 + \sigma_3'^2)$$
$$= \tfrac{1}{3}(2\sigma_1'^2 - \sigma_2'^2 - \sigma_3'^2)$$

and similarly

$$t_2' = \tfrac{1}{3}(2\sigma_2'^2 - \sigma_1'^2 - \sigma_3'^2)$$
$$t_3' = \tfrac{1}{3}(2\sigma_3'^2 - \sigma_1'^2 - \sigma_2'^2)$$

Converting to absolute stress components

$$\sigma_1' = \sigma_1 - \tfrac{1}{3}(\sigma_1 + \sigma_2 + \sigma_3)$$
$$\sigma_2' = \sigma_2 - \tfrac{1}{3}(\sigma_1 + \sigma_2 + \sigma_3)$$
$$\sigma_3' = \sigma_3 - \tfrac{1}{3}(\sigma_1 + \sigma_2 + \sigma_3)$$

we then have

$$J_2' = [(\sigma_1 - \sigma_2)^2 + (\sigma_1 - \sigma_3)^2 + (\sigma_2 - \sigma_3)^2]\,/\,6$$
$$J_3' = [(2\sigma_1 - \sigma_2 - \sigma_3)^3 + (2\sigma_2 - \sigma_1 - \sigma_3)^3 + (2\sigma_3 - \sigma_1 - \sigma_2)^3]\,/\,81$$
$$t_1' = [(\sigma_1 - \sigma_2)^2 + (\sigma_1 - \sigma_3)^2 - 2(\sigma_2 - \sigma_3)^2]\,/\,9$$
$$t_2' = [(\sigma_1 - \sigma_2)^2 + (\sigma_2 - \sigma_3)^2 - 2(\sigma_1 - \sigma_3)^2]\,/\,9$$
$$t_3' = [(\sigma_1 - \sigma_3)^2 + (\sigma_2 - \sigma_3)^2 - 2(\sigma_1 - \sigma_2)^2]\,/\,9$$

The scalar $d\lambda$ is again determined from the given yield function by setting $\sigma_2 = \sigma_3 = 0$ for a uniaxial stress. Equation (10.21b) leads to

$$d\varepsilon_1^P = \frac{2}{9}\,d\lambda\,\sigma_1^5\left(1 - \frac{4c}{27}\right) \qquad (10.22a)$$

so that

$$d\lambda = \frac{9\,d\bar{\varepsilon}^P}{2\bar{\sigma}^5(1 - 4c/27)} \qquad (10.22b)$$

where the equivalent stress is given by eq(9.4a). The equivalent strain is either defined by the strain hypothesis (eq 9.7c) or, from the work hardening hypothesis (eq 9.52c). According to each hypothesis, $d\lambda$ follows from eqs(10.16a,b) and (10.22b) respectively:

$$d\lambda = \frac{9\,d\bar{\sigma}}{2\bar{\sigma}^5 H'(1 - 4c/27)} \quad \text{and} \quad d\lambda = \frac{9\,d\bar{\sigma}}{2\bar{\sigma}^6 F'(1 - 4c/27)} \qquad (10.23a,b)$$

The derivatives $F' = dF/d\bar{\varepsilon}^P$ and $H' = dH/d\bar{\varepsilon}^P$ eliminate strain in their arguments so allowing eqs (10.23a,b) to express the monotonic dependence of $d\lambda$ upon $\bar{\sigma}$. Note that eq(10.20) provides a uniaxial relationship between the hardening functions H and F:

$$\bar{\sigma}^6\left[\frac{1}{27} - \frac{4c}{9^3}\right] = \chi(\psi) \quad \text{where} \quad \psi = \int d\bar{\varepsilon}^P \quad \text{or} \quad \psi = \int dW^P \qquad (10.23c)$$

10.4 Validation of Levy-Mises and Drucker Flow Rules

Many workers have conducted radial outward loadings upon thin-walled cylinders by combining tension, torque and internal pressure. The linear plastic strain paths observed in annealed material have confirmed eq(10.14). That is, for a given uniform stress state within the wall, flow theory predicts a constant ratio between the plastic strain increment components under a proportional loading (e.g. see Fig. 10.5).

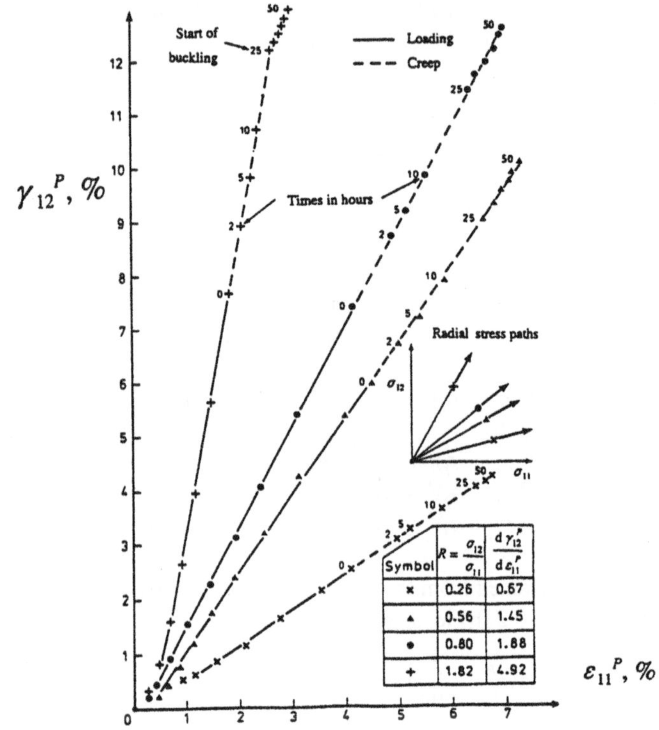

Figure 10.5 Linear plastic strain paths under radial tension-torsion

10.4.1 Combined Tension-Torsion

The linear plastic strain paths shown in Fig. 10.5 apply to radial loading under combined tension-torsion of a 99.8% commercially pure, annealed aluminium (E1A). Four constant stress ratios were employed: $R = \sigma_{12}/\sigma_{11} = 0.26, 0.56, 0.80$ and 1.82 [14]. In the absence of torsional buckling, Fig. 10.5 also shows that when constant, combined loads are maintained for a period of 50 hours, the strain paths continue to extend in time without changing their gradients. The occurrence of room-temperature creep conforms to earlier observations for a similar purity 1100 aluminium [15, 16]. The creep behaviour suggests that a strain rate form of eqs(10.19a,b,c) applies when the essential slip mechanisms of low-temperature deformation are common to plasticity and creep. This may not apply to higher temperatures, where creep deformation involves thermally activated climb and cross-slip of dislocations, together with grain coarsening and cavitation. Linearity has often been seen in many plastic strain paths arising from tension-torsion loading [15 - 20]. Their gradients, including those from Fig. 10.5, have been used to construct Fig. 10.6.

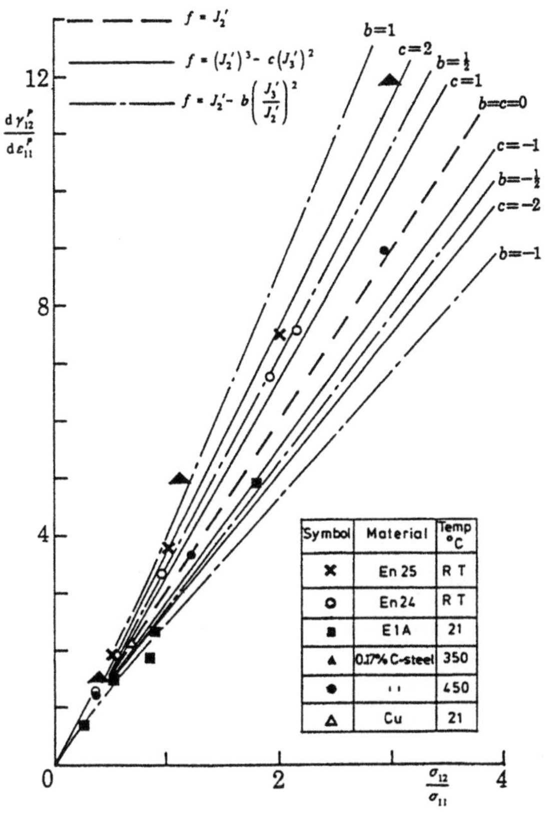

Figure. 10.6 Stress-dependent plastic strain increment ratio

These data apply to radial loadings in En steel [17, 18], low C-steels at elevated temperature [18, 19] and commercially pure copper [20]. Predictions from eq(10.14) are simply

$$d\gamma_{12}^P / d\varepsilon_{11}^P = 3\sigma_{12} / \sigma_{11} = 3R \quad (10.24a)$$

$$d\varepsilon_{22}^P / d\varepsilon_{11}^P = -1/2 \quad (10.24b)$$

Equation (10.24a) gives the broken line in Fig. 10.6, this being representative of J_2' materials only. Improved predictions are found [10, 21] for three homogenous flow potentials involving both invariants $f(J_2', J_3')$. For example, eq(10.21b) gives plastic strain increment ratios associated with Drucker's flow potential:

$$\frac{d\gamma_{12}^P}{d\varepsilon_{11}^P} = \frac{3R\left(R^2 + \frac{1}{3}\right)^2 - \frac{2Rc}{9}\left(R^2 + \frac{2}{9}\right)}{\left(R^2 + \frac{1}{3}\right)^2 - \frac{c(2 + 3R^2)}{27}\left(R^2 + \frac{2}{9}\right)} \quad (10.25a)$$

$$\frac{d\varepsilon_{22}^P}{d\varepsilon_{11}^P} = \frac{-\left(R^2 + \frac{1}{3}\right)^2 - \frac{2c}{9}\left(R^2 + \frac{2}{9}\right)\left(R^2 - \frac{1}{3}\right)}{2\left(R^2 + \frac{1}{3}\right)^2 - \frac{2c(2 + 3R^2)}{27}\left(R^2 + \frac{2}{9}\right)} \quad (10.25b)$$

It appears from Fig. 10.6 that eqs(10.25a,b) represent the data for certain steels better than $f = J_2'$ can. Heat treatment rendered these alloys with a uniform, isotropic grain size which permits the selection of a suitable alternative isotropic function $f(J_2', J_3')$. Figure 10.7 presents a more convincing demonstration of the J_3' influence. This is provided by experiments where both diametral and axial strains were found in combined tension-torsion experiments upon: En 25 steel [22], copper, aluminium and two mild steels [23]. These show that the strain ratio $d\varepsilon_{22}^P / d\varepsilon_{11}^P$, differs from the $-\tfrac{1}{2}$ prediction from eq(10.24b).

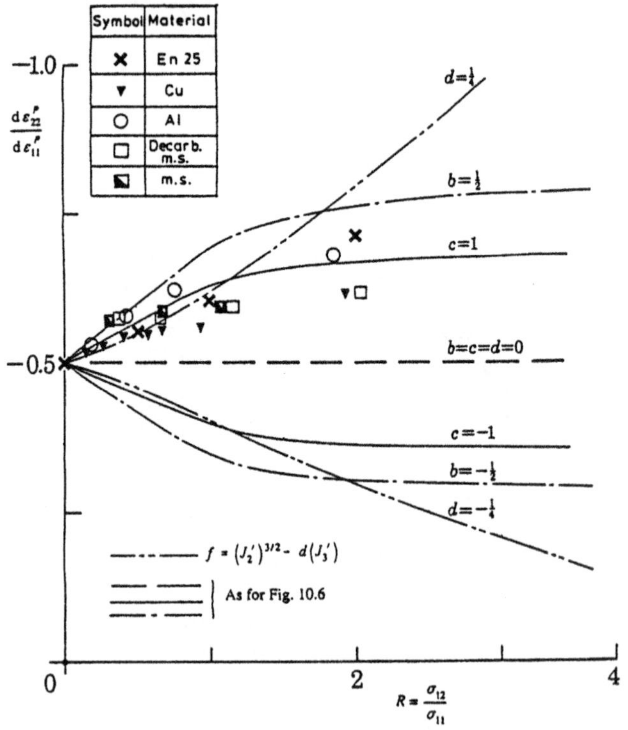

Figure 10.7 Departure from $f = J_2'$ theory

It is seen that the effect that stress ratio R has on decreasing the strain ratio can be predicted from eq(10.25b) and two further homogenous yield functions in J_2' and J_3' as indicated.

An alternative check on eq(10.14) is provided by Lode's [24] respective stress and strain parameters μ and ν:

$$\nu = \frac{2d\varepsilon_3^P - d\varepsilon_1^P - d\varepsilon_2^P}{d\varepsilon_1^P - d\varepsilon_2^P} = \frac{2d\lambda\,\sigma_3' - d\lambda\,\sigma_1' - d\lambda\,\sigma_2'}{d\lambda\,\sigma_1' - d\lambda\,\sigma_2'} = \frac{2\sigma_3 - \sigma_1 - \sigma_2}{\sigma_1 - \sigma_2} = \mu \qquad (10.26)$$

An observed equality between μ and ν for a given radial path will validate eq(10.14). Substituting the principal plastic strains and stresses (p. 21) under tension-torsion (σ_{11}, σ_{12}) into eq(10.26) gives Lode's strain parameter:

$$\nu = -\frac{d\varepsilon_{11}^P}{\sqrt{\left(d\varepsilon_{11}^P\right)^2 + \frac{4}{9}\left(d\gamma_{12}^P\right)^2}} = -\frac{1}{\sqrt{1 + \frac{4}{9}\left(\frac{d\gamma_{12}^P}{d\varepsilon_{11}^P}\right)^2}} \qquad (10.27a)$$

Lode's stress parameter becomes

$$\mu = -\frac{\sigma_{11}}{\sqrt{\sigma_{11}^2 + 4\sigma_{12}^2}} = -\frac{1}{\sqrt{1 + 4(\sigma_{12}/\sigma_{11})^2}} \qquad (10.27b)$$

Employing the observed plastic strain gradient $d\gamma_{12}^P / d\varepsilon_{11}^P$ for each stress ratio σ_{12}/σ_{11} within eqs(10.27a,b), has enabled the construction of Fig. 10.8.

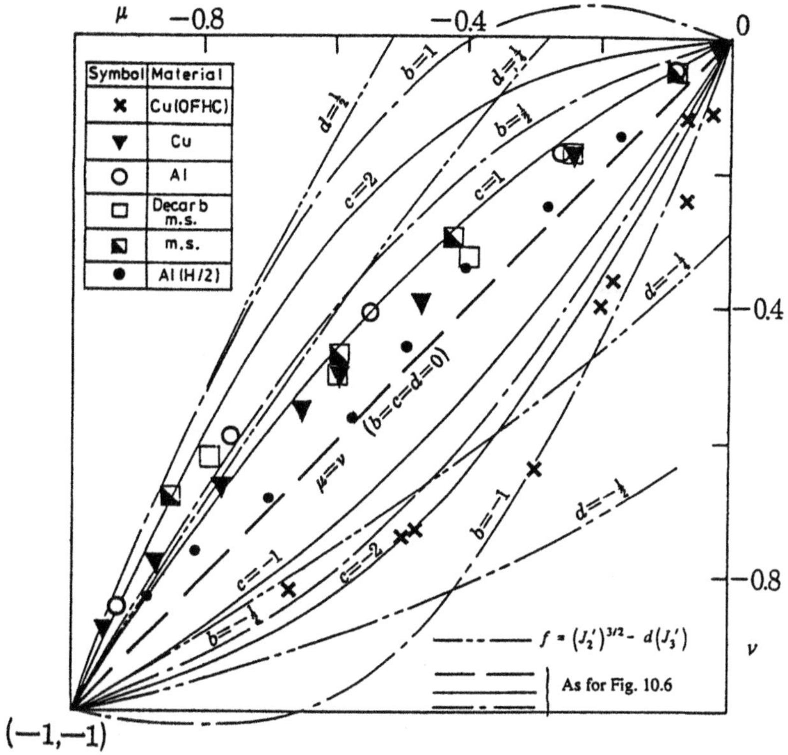

Figure 10.8 Lode parameter plot

This Lode plot applies to steels [22, 23], aluminium [23, 25] and copper [23, 26]. It also reproduces the μ, ν data from pioneering experiments of Taylor and Quinney in 1931 [23]. They were the first to reveal a systematic deviation from $\mu = \nu$ in each material. The precise reason for the deviation remains uncertain. It may be due to a residual anisotropy in these materials because tests were not conducted to establish beyond doubt a condition of initial isotropy. However, each material in Fig. 10.8 was annealed so, when isotropy is assumed, it implies that any deviation from $\mu = \nu$ is due to the influence of the third invariant J_3'. For example, using Drucker's yield function in eq(10.20) it can be shown [10] that the non-linear relationship between μ and ν is given by

$$\nu = \mu \left[\frac{27(\mu^2 + 3)^2 - 4c(\mu^2 - 3)(\mu^2 - 9)}{27(\mu^2 + 3)^2 + 8c\mu^2(\mu^2 - 9)} \right] \qquad (10.28)$$

Figure 10.8 shows that eq(10.28) and a further homogenous function $f(J_2', J_3')$, can account for the trends observed through the choice of constants b and c. An exception is found in the function containing constant d because J_3' changes sign with stress. Consequently, this function leads to a discontinuity at $\mu = 0$.

10.4.2 Principal Biaxial Stress

Consider proportional stressing under combined internal pressure-axial loading of a thin-walled cylinder. Principal stresses σ_1 and σ_2 in the axial and circumferential directions respectively, increase proportionately in a ratio $R = \sigma_1 / \sigma_2$.

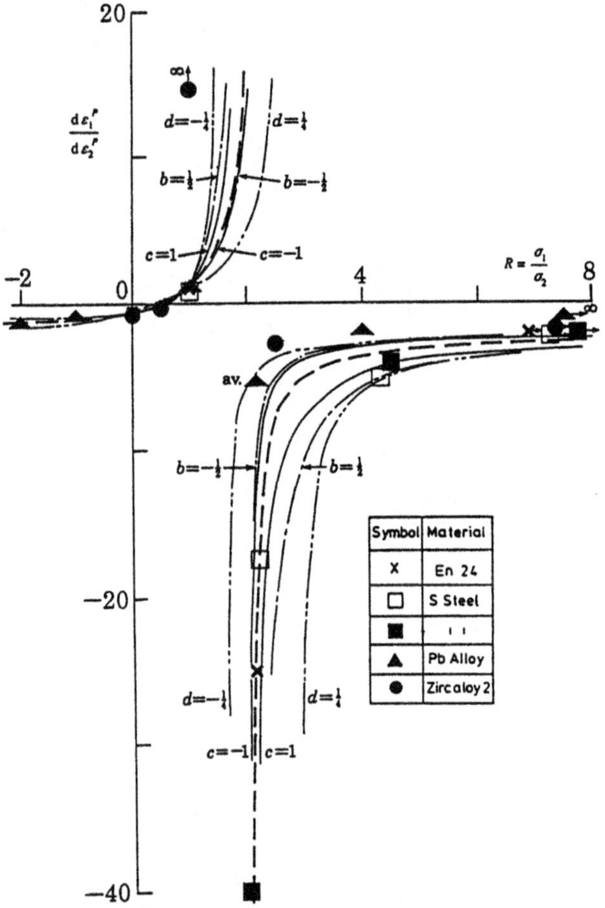

Figure 10.9 Dependence of strain increment ratio upon principal stress

The Levy-Mises prediction sets the radial stress $\sigma_3 = 0$ in eq(10.19a,b) to give a strain ratio:

$$\frac{d\varepsilon_1^P}{d\varepsilon_2^P} = \frac{2R - 1}{2 - R} \tag{10.29}$$

Figure 10.9 compares the prediction from eq(10.29) with experimental data for En 24 [17], stainless steel [27] and alloys of lead [28] and zirconium [29]. A similar conclusion can be drawn in relation to the observed deviations from eq(10.29). That is, predictions from the alternative homogenous functions $f(J_2', J_3')$, given in Fig. 10.9, usually match the observed behaviour in a given alloy. For example, we find Drucker's prediction to the strain ratio from putting $\sigma_3 = 0$ in eq(10.21b):

$$\frac{d\varepsilon_1^P}{d\varepsilon_2^P} = \frac{81(1 + R^2 - R)^2(2R - 1) - 2c(2R^2 - 1 - 2R)[(2 - R)^3 + (2R - 1)^3 - (1 + R)^3]}{81(1 + R^2 - R)^2(2 - R) - 2c(2 - R^2 - 2R)[(2 - R)^3 + (2R - 1)^3 - (1 + R)^3]} \tag{10.30}$$

Here, however, the influence of initial anisotropy in lead and zirconium alloys is more likely for the discrepancy found with isotropic predictions. The reported initial condition of each alloy [28, 29] suggests that a strain ratio, based upon an orthotropic yield function, would be a more approriate choice than eq(10.30) when explaining these results (see Chapter 11).

10.4.3 Flow Equivalence

Consider again proportional loading of aluminium under combined tension-torsion. Figure 10.5 shows linear plastic strain paths, between shear and axial components, resulting from increasing the corresponding stress components under constant ratios. The measurement of hoop strain under simple tension confirmed eq(10.24b), indicating that an isotropic yield function applies. A slight deviation from the same gradient, $d\varepsilon_{22}^P / d\varepsilon_{11}^P = -½$, occurred with tension and torsion combined. Moreover, the observed slopes are less than the Mises prediction $d\gamma_{12}^P / d\varepsilon_{11}^P = 3R$, from eq(10.24a).

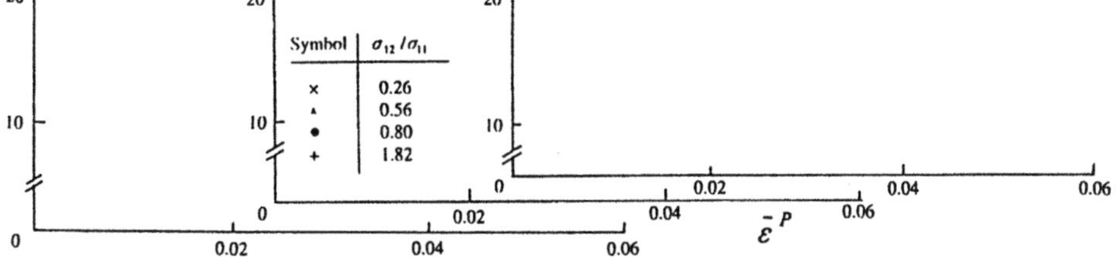

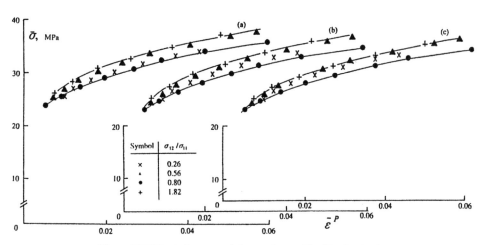

Figure 10.10 Equivalence correlations under combined tension-torsion

With co-ordinates 1, 2 and 3, aligned with the axial, circumferential and radial directions respectively, the von Mises equivalent stress and strain are found from eqs(9.2c) and (9.6c):

$$\bar{\sigma} = \sqrt{\sigma_{11}^2 + 3\sigma_{12}^2} \tag{10.31a}$$

$$d\bar{\varepsilon}^P = \sqrt{(d\varepsilon_{11}^P)^2 + \frac{1}{3}(d\gamma_{12}^P)^2} = d\varepsilon_{11}^P \sqrt{1 + \frac{1}{3}\left(\frac{d\gamma_{12}^P}{d\varepsilon_{11}^P}\right)^2} \tag{10.31b}$$

Equations (10.31a,b) provide a von Mises correlation in Fig. 10.10a, this being independent of the hardening hypotheses employed for the four radial tests.

Using Drucker's function, the stress invariants J_2' and J_3' in eq(9.5a), reduce to

$$J_2' = \tfrac{1}{2}\,\sigma_{ij}'\,\sigma_{ij}' = \sigma_{12}^{\,2} + \sigma_{11}^{\,2}/3, \tag{10.32a}$$

$$J_3' = \tfrac{1}{3}\,\sigma_{ij}'\,\sigma_{jk}'\,\sigma_{ki}' = 2\sigma_{11}^{\,3}/27 + \sigma_{11}\,\sigma_{12}^{\,2}/3 \tag{10.32b}$$

Substituting eqs(10.32a,b) into eq(9.4a) gives the equivalent stress expression

$$\bar{\sigma} = \left[\frac{(\sigma_{12}^2 + \sigma_{11}^2/3)^3 - c\left(2\sigma_{11}^3/27 + \sigma_{11}\sigma_{12}^2/3\right)^2}{1/27 - 4c/9^3}\right]^{\frac{1}{6}} \tag{10.33}$$

By including J_3' in the yield function, the expression for $\mathrm{d}\bar{\varepsilon}^P$ will depend upon the choice between the work and strain-hardening hypothesis.

(a) *Strain Hypothesis*

When the strain-hardening hypothesis defines $\mathrm{d}\bar{\varepsilon}^P$, the invariants in eq(9.5b) become

$$I_2 = \frac{1}{2}\,\mathrm{d}\varepsilon_{ij}^P\mathrm{d}\varepsilon_{ij}^P = \frac{1}{2}\,[(\mathrm{d}\varepsilon_{11}^P)^2 + (\mathrm{d}\varepsilon_{12}^P)^2 + (\mathrm{d}\varepsilon_{21}^P)^2 + (\mathrm{d}\varepsilon_{22}^P)^2 + (\mathrm{d}\varepsilon_{33}^P)^2]$$

$$I_3 = \frac{1}{3}\,\mathrm{d}\varepsilon_{ij}^P\mathrm{d}\varepsilon_{jk}^P\mathrm{d}\varepsilon_{ki}^P = \frac{1}{3}\,[(\mathrm{d}\varepsilon_{11}^P)^3 + \mathrm{d}\varepsilon_{11}^P(\mathrm{d}\varepsilon_{12}^P)^2 + \mathrm{d}\varepsilon_{11}^P(\mathrm{d}\varepsilon_{21}^P)^2$$
$$+ \mathrm{d}\varepsilon_{22}^P(\mathrm{d}\varepsilon_{21}^P)^2 + \mathrm{d}\varepsilon_{11}^P(\mathrm{d}\varepsilon_{12}^P)^2 + \mathrm{d}\varepsilon_{22}^P(\mathrm{d}\varepsilon_{21}^P)^2 + \mathrm{d}\varepsilon_{22}^P(\mathrm{d}\varepsilon_{12}^P)^2 + (\mathrm{d}\varepsilon_{22}^P)^3 + (\mathrm{d}\varepsilon_{33}^P)^3]$$

Now the substitution of $\mathrm{d}\varepsilon_{11}^P = \mathrm{d}\bar{\varepsilon}^P$ with $\mathrm{d}\varepsilon_{22}^P = \mathrm{d}\varepsilon_{33}^P = -\tfrac{1}{2}\mathrm{d}\bar{\varepsilon}^P$ and $\mathrm{d}\varepsilon_{12}^P = \mathrm{d}\varepsilon_{21}^P = \tfrac{1}{2}\mathrm{d}\gamma_{12}^P$, simplifies these invariant expression to

$$I_2 = \frac{1}{4}\,[(3\mathrm{d}\varepsilon_{11}^P)^2 + (\mathrm{d}\gamma_{12}^P)^2] \tag{10.34a}$$

$$I_3 = \frac{1}{4}\,\mathrm{d}\varepsilon_{11}^P[(\mathrm{d}\varepsilon_{11}^P)^2 + \frac{1}{2}\,(\mathrm{d}\gamma_{12}^P)^2] \tag{10.34b}$$

Substituting eqs(10.34a,b) into eq(9.7a) gives

$$\mathrm{d}\bar{\varepsilon}^P = \left\{\frac{\frac{1}{4}\left[(3\mathrm{d}\varepsilon_{11}^P)^2 + (\mathrm{d}\gamma_{12}^P)^2\right]^3 - c\left[(\mathrm{d}\varepsilon_{11}^P)^3 + \frac{1}{2}\mathrm{d}\varepsilon_{11}^P(\mathrm{d}\gamma_{12}^P)^2\right]^2}{\frac{27}{4} - c}\right\}^{\frac{1}{6}} \tag{10.35}$$

The constant c in eq(10.35) is found from matching observed $\mathrm{d}\gamma_{12}^P/\mathrm{d}\varepsilon_{11}^P$ ratios in Fig. 10.5 with the corresponding prediction. Derivatives of Drucker's function: $f(J_2', J_3') = J'^{\,3}_2 - cJ'^{\,2}_3$ are $\partial f/\partial J_2' = 3J'^{\,2}_2$ and $\partial f/\partial J_3' = -2cJ_3'$ so that eq(10.21b) gives the axial, circumferential and shear strain increments respectively:

$$\mathrm{d}\varepsilon_{11}^P = \mathrm{d}\lambda\,[3(\sigma_{11}^2/3 + \sigma_{12}^2)^2(2\sigma_{11}/3) - 2c(2\sigma_{11}^3/27 + \sigma_{11}\,\sigma_{12}^2/3)\,t_{11}'] \tag{10.36a}$$

$$\mathrm{d}\varepsilon_{22}^P = \mathrm{d}\lambda\,[3(\sigma_{11}^2/3 + \sigma_{12}^2)^2(-\sigma_{11}/3) - 2c(2\sigma_{11}^3/27 + \sigma_{11}\sigma_{12}^2/3)\,t_{22}'] \tag{10.36b}$$

$$\mathrm{d}\varepsilon_{12}^P = \mathrm{d}\gamma_{12}^P/2 = \mathrm{d}\lambda\,[3(\sigma_{11}^2/3 + \sigma_{12}^2)^2\sigma_{12} - 2c\,(2\sigma_{11}^3/27 + \sigma_{11}\,\sigma_{12}^2/3)\,t_{12}'] \tag{10.36c}$$

where $t_{11}' = (2\sigma_{11}^2 + 3\sigma_{12}^2)/9$, $t_{22}' = (-\sigma_{11}^2 + 3\sigma_{12}^2)/9$ and $t_{12}' = \sigma_{11}\,\sigma_{12}/3$. Dividing eqs(10.36b,c) by eq(10.36a) leads to the theoretical plastic strain increment ratios in eqs(10.25a,b). Identifying eqs(10.25a,b) with the gradients in Fig. 10.5, $c = -2$ is chosen within its limiting range, $-27/8 \le c \le 9/4$, for a convex yield function. Equations (10.33), (10.35) and (10.25a) provide theoretical equivalence relationships from strain hardening but the corresponding plot in Fig. 10.10b appears to be no better than that found from the simplified Mises J_2' correlation (see Fig. 10.10b).

(b) *Work hypothesis*
Here $\bar{\sigma}$ remains as defined in eq(10.33) but $\mathrm{d}\bar{\varepsilon}^P$ is re-defined from the work hypothesis (see eq 9.52c):

$$\bar{\sigma}\,\mathrm{d}\bar{\varepsilon}^P = \sigma_{ij}\,\mathrm{d}\varepsilon_{ij}^P = \sigma_{11}\,\mathrm{d}\varepsilon_{11}^P + \sigma_{12}\,\mathrm{d}\gamma_{12}^P$$

$$\mathrm{d}\bar{\varepsilon}^P = (\sigma_{11}/\bar{\sigma})\left(\mathrm{d}\varepsilon_{11}^P + R\,\mathrm{d}\gamma_{12}^P\right) \tag{10.37a}$$

Substituting eq(10.33) into eq(10.37a) gives:

$$\mathrm{d}\bar{\varepsilon}^P = \left[\frac{1/27 - 4c/9^3}{(R^2 + 1/3)^3 - c(R^2/3 + 2/27)^2}\right]^{\frac{1}{6}}\left(\mathrm{d}\varepsilon_{11}^P + R\,\mathrm{d}\gamma_{12}^P\right) \tag{10.37b}$$

where eq(10.25a) expresses the dependence of R upon the ratio $\mathrm{d}\gamma_{12}^P/\mathrm{d}\varepsilon_{11}^P$. Using the observed ratios for R and $\mathrm{d}\gamma_{12}^P/\mathrm{d}\varepsilon_{11}^P$ in eqs(10.25a) and (10.37b), Fig. 10.10c shows that the work hypothesis provides a slightly improved correlation compared to Figs 10.10a,b. The point to note here is that the associated flow rule will match the strain ratios independently of the work hardening hypothesis.

10.5 Non-Associated Flow Rules

Up to now only isotropic yield functions $f(J_2', J_3')$, have been considered. This implies that the chosen expressions for $\bar{\sigma}$ and $\mathrm{d}\bar{\varepsilon}^P$ (i.e. by Mises and Drucker) are restricted to material whose flow behaviour is independent of direction. Here, a non-coincidence of flow curves, when based upon the conventional Mises definitions of equivalent stress and strain, has been attributed to the influence of the third invariant. When a material is initially anisotropic to a measurable degree, isotropic invariants should not be used to define its equivalence. For this we must use a potential that characterises the nature of anisotropy, e.g. a normal or orthotropic form typical of, say, rolled sheet (see Chapter 11).

The literature reveals instances where the flow rule has not been associated with the yield criterion. This allows the function employed within the flow rule (the flow potential) to differ from that used to define the yield criterion. This flexible approach has improved predictions to observed behaviour as compared to making the classical association between the yield criterion and flow potential. Using both the Tresca and von Mises functions will serve to illustrate a theory with unrelated equivalence and flow. For a Tresca yield function, the equivalent stress is simply:

$$\bar{\sigma} = \sigma_1 - \sigma_3 \tag{10.38a}$$

where σ_1 and σ_3 are the greatest and least principal stresses, i.e. the roots to the characteristic eq(1.24a). A Tresca's equivalent strain may be found from eq(10.38a), corresponding to a uniaxial work expression (9.51b):

$$\mathrm{d}\bar{\varepsilon}^P = \frac{2}{3}\left(\mathrm{d}\varepsilon_1^P - \mathrm{d}\varepsilon_3^P\right) \tag{10.38b}$$

While eqs(10.38a,b) reduce correctly to $\bar{\sigma} = \sigma_1$ and $\mathrm{d}\bar{\varepsilon}^P = \mathrm{d}\varepsilon_1^P$ for uniaxial deformation, they will otherwise differ from a von Mises equivalence. Consider pure shear, where the stresses are $\sigma_1 = \tau$, $\sigma_2 = 0$ and $\sigma_3 = -\tau$ and the strain increments are $\mathrm{d}\varepsilon_1^P = \mathrm{d}\gamma^P/2$, $\mathrm{d}\varepsilon_2^P = 0$ and $\mathrm{d}\varepsilon_3^P = -\mathrm{d}\gamma^P/2$. Substituting these in eqs(9.2d) and (9.6d), von Mises gives $\bar{\sigma} = \sqrt{3}\,\tau$ and $\mathrm{d}\bar{\varepsilon}^P = \mathrm{d}\gamma^P/\sqrt{3}$. With similar substitutions into eqs(10.38a,b), Tresca gives $\bar{\sigma} = 2\tau$ and $\mathrm{d}\bar{\varepsilon}^P = 2\mathrm{d}\gamma^P/3$. The incremental work of pure shear is:

$$\mathrm{d}W^P = \bar{\sigma}\mathrm{d}\bar{\varepsilon}^P = \sigma_1\mathrm{d}\varepsilon_1^P + \sigma_3\mathrm{d}\varepsilon_3^P$$

which is consistent only with von Mises equivalence definitions. Similar plastic work violations would be found from $\mathrm{d}\bar{\varepsilon}^P$ expressions derived earlier from $f(J_2', J_3')$ and the strain hypothesis (see Chapter 9). However, this has not precluded the use of a Tresca equivalent stress in the non-associated approach. For example, Marin [30] and Soderberg [31] employed a Mises potential in a flow rule for rate dependent plasticity (creep). Equation (8.17a) gave this as:

$$\dot{\varepsilon}_{ij}^P = \dot{\lambda}\,\partial(J_2')/\partial\sigma_{ij} = \dot{\lambda}\,\sigma_{ij}' \tag{10.39a}$$

where $\dot{\lambda}$ is readily defined in terms of $\bar{\sigma}$ and $\dot{\bar{\varepsilon}}^P$ for simple tension:

$$\dot{\lambda} = \frac{3\dot{\bar{\varepsilon}}^P}{2\bar{\sigma}} \tag{10.39b}$$

For their equivalence, the authors defined $\bar{\sigma}$ and $\dot{\bar{\varepsilon}}^P$ in eq(10.39b) by both von Mises (associated) and Tresca (non associated). For the latter $\bar{\sigma}$ is given by eq(10.38a) and $\dot{\bar{\varepsilon}}^P$ is the time derivative of eq(10.38b):

$$\dot{\bar{\varepsilon}}^P = \frac{\mathrm{d}\bar{\varepsilon}^P}{\mathrm{d}t} = \frac{2}{3}\left(\dot{\varepsilon}_1^P - \dot{\varepsilon}_3^P\right) \tag{10.39c}$$

Their experiments revealed less spread within the Tresca equivalent stress-strain rate plot compared to von Mises. Consequently, greater accuracy was found from non-associated predictions to creep rates in rotating discs, and tubes under internal pressure combined with axial tension [32 - 34]. Combining von Mises equivalence with a Tresca flow rule is less realistic practically but has been used for its mathematical simplicity.

10.6 Prandtl-Reuss Flow Theory

In deforming beyond the elastic limit an elastic component of strain continues to accompany the plastic strain. When the latter is large and elastic strain is small enough to be ignored, the Levy-Mises theory estimates the total deformation in an elastic-plastic solid. Such predictions will be unacceptable where elastic and plastic strain components are comparable in their magnitudes. Here, both elastic and plastic incremental strains arising from a loading condition $\mathrm{d}\sigma_{ij} > 0$ are required. The incremental elastic component of strain $\mathrm{d}\varepsilon_{ij}^e$, follows from Hooke's law and the plastic component $\mathrm{d}\varepsilon_{ij}^P$ from the Levy-Mises eq(10.14).

10.6.1 Tensor Subscript Notation

The Prandtl-Reuss, total incremental strain theory [35, 36] sums the elastic and plastic incremental strains. The theory was originally presented in the tensor subscript notation. Since $\sigma_{kk}' = 0$, five independent deviatoric stress components may act simultaneously to give six total strain increments:

$$\mathrm{d}\varepsilon_{ij}^{t} = \mathrm{d}\varepsilon_{ij}^{P} + \mathrm{d}\varepsilon_{ij}^{e}$$

$$\mathrm{d}\varepsilon_{ij}^{t} = \mathrm{d}\lambda\,\sigma_{ij}' + \frac{\mathrm{d}\sigma_{ij}'}{2G} + \frac{(1-2\nu)}{3E}\delta_{ij}\,\mathrm{d}\sigma_{kk} \tag{10.40}$$

The expanded form of eq(10.40) represents three direct and three shear strain increments:

$$\mathrm{d}\varepsilon_{11}^{P} = \frac{2\mathrm{d}\lambda}{3}\left[\sigma_{11} - \frac{1}{2}(\sigma_{22} + \sigma_{33})\right] + \frac{1}{E}\left[\mathrm{d}\sigma_{11} - \nu(\mathrm{d}\sigma_{22} + \mathrm{d}\sigma_{33})\right], \quad \text{etc}$$

$$\mathrm{d}\varepsilon_{12}^{P} = \mathrm{d}\gamma_{12}^{P}/2 = \mathrm{d}\lambda\sigma_{12} + \mathrm{d}\sigma_{12}/(2G) \quad \text{etc}$$

Integration of eq(10.40) shows that the permanent, plastic component of total strain depends upon the history of stress, while the recoverable elastic component depends upon the current stress. This applies to the low-temperature, small-strain regime of many polycrystalline materials where they remain insensitive to the rate of straining.

10.6.2 Experimental Validation of Prandtl-Reuss Theory

The Prandtl-Reuss theory has matched experimental observations on metal plasticity with both linear and parabolic hardening under a variety of non-radial loading paths [37]. Using Ludwik's hardening eq(9.43a), the constants A and $n < 1$ may be found from reported data for tension or torsion tests. Alternatively, they may be found from within a loading path that begins with either of these modes. Values of A/Y and n, so found are given in Fig. 9.20 for many engineering alloys, some of which will be converted to eq(9.43b) and used here. An exponent $n = 1$ applies to ideal linear hardening material. Near uniform stress states are achievable from loading thin-walled tubular testpieces with combinations of tension, compression, torsion and internal pressure.

(a) *Stepped Path Under Tension-Torsion*

Equations (10.18a) and (10.40) provide the total, incremental shear and axial strains under uniform axial and shear stress components σ_{11} and σ_{12}, as follows:

$$\mathrm{d}\gamma_{12}^{t} = \frac{3\,\mathrm{d}\bar{\sigma}}{H'\bar{\sigma}}\sigma_{12} + \frac{\mathrm{d}\sigma_{12}}{G} \tag{10.41a}$$

$$\mathrm{d}\varepsilon_{11}^{t} = \frac{\mathrm{d}\bar{\sigma}}{H'\bar{\sigma}}\sigma_{11} + \frac{\mathrm{d}\sigma_{11}}{E} \tag{10.41b}$$

$$\bar{\sigma}^{2} = \sigma_{11}^{2} + 3\sigma_{12}^{2} \tag{10.41c}$$

Let the tensile stress σ_{11} be first applied to exceed the initial yield stress σ_o then held constant ($\sigma_{11} = \sigma_p$) while σ_{12} increases within the plastic range. Equations (9.43b), (10.16c) and (10.41c) are employed as follows:

$$H' = \mathrm{d}\bar{\sigma}/\mathrm{d}\bar{\varepsilon}^P = (n\sigma_o/\varepsilon_o)(\bar{\varepsilon}^P/\varepsilon_o)^{n-1} = (n\sigma_o/\varepsilon_o)(\bar{\sigma}/\sigma_o - 1)^{\frac{n-1}{n}} \tag{10.42a}$$

$$\bar{\sigma}^2 = \sigma_p^2 + 3\sigma_{12}^2 \quad \Rightarrow \quad \bar{\sigma}\,\mathrm{d}\bar{\sigma} = 3\sigma_{12}\,\mathrm{d}\sigma_{12} \tag{10.42b,c}$$

The total incremental strains under an increasing σ_{12} are found from eqs(10.41) and (10.42):

$$\mathrm{d}\gamma_{12}^t = \frac{9\varepsilon_o}{n}\frac{\left[(S_p^2 + 3T^2)^{\frac{1}{2}} - 1\right]^{\frac{1-n}{n}}}{\left[3 + (S_p/T)^2\right]}\mathrm{d}T + \frac{\sigma_o}{G}\mathrm{d}T \tag{10.43a}$$

$$\mathrm{d}\varepsilon_{11}^t = \frac{3\varepsilon_o}{n}\frac{\left[(S_p^2 + 3T^2)^{\frac{1}{2}} - 1\right]^{\frac{1-n}{n}}}{(S_p/T + 3T/S_p)}\mathrm{d}T \tag{10.43b}$$

where the dimensionless stresses are $S_p = \sigma_p/\sigma_o$ and $T = \sigma_{12}/\sigma_o$. Normally, numerical integration of eqs(10.43a,b) is required unless $n = ⅓$, when closed solutions are found. For example, Ivey [38] conducted an experiment of this kind on a 24S-T4 aluminium alloy where, from Fig. 9.20, $n \approx ⅓$ and $\varepsilon_o = (Y/A)^3 = 2.24^{-3} = 0.089$. Here $S_p = 1.46$ exceeded the tensile yield stress and therefore the lower limit of integration for eqs(10.43a,b) is $T = 0$ (see the inset Fig. 10.11a). Under the shear stress branch, eqs(10.43a,b) become

$$\gamma_{12}^P = 27\varepsilon_o\int_0^T\left[T^2 - \frac{2T^2}{(S_p^2 + 3T^2)^{1/2}} + \frac{T^2}{(S_p^2 + 3T^2)}\right]\mathrm{d}T + \frac{\sigma_o}{G}\int_0^T \mathrm{d}T \tag{10.43c}$$

$$\varepsilon_{11}^t = \varepsilon_{11}^P = 9S_p\varepsilon_o\int_0^T\left[T - \frac{2T}{(S_p^2 + 3T^2)^{1/2}} + \frac{T}{(S_p^2 + 3T^2)}\right]\mathrm{d}T \tag{10.43d}$$

Equations (10.43c,d) are composed of standard integrals which lead to total strains:

$$\gamma_{12}^t = 9\varepsilon_o\left[T^3 - T(S_p^2 + 3T^2)^{1/2} + \frac{S_p^2}{\sqrt{3}}\ln\left\{\left[1 + 3\left(\frac{T}{S_p}\right)^2\right]^{\frac{1}{2}} + \frac{\sqrt{3}T}{S_p}\right\}\right.$$
$$\left. + T - \frac{S_p}{\sqrt{3}}\tan^{-1}\left(\frac{\sqrt{3}T_p}{S_p}\right)\right] + \frac{\sigma_o T}{G} \tag{10.44a}$$

$$\varepsilon_{11}^t = \varepsilon_{11}^P = 9S_p\varepsilon_o\left\{\frac{T^2}{2} - \frac{2}{3}(S_p^2 + 3T^2)^{1/2} + \frac{1}{6}\ln\left[1 + 3\left(\frac{T}{S_p}\right)^2\right] + \frac{2S_p}{3}\right\} \tag{10.44b}$$

This integration shows that under the increasing shear stress: (i) the final terms in eqs(10.43a) and (10.44a) is the elastic shear strain, and (ii) plastic axial strain is present despite the absence of elastic axial strain. Figures 10.11a,b show that the predictions from eqs(10.44a,b) match the trends observed in plastic strain paths and flow stresses.

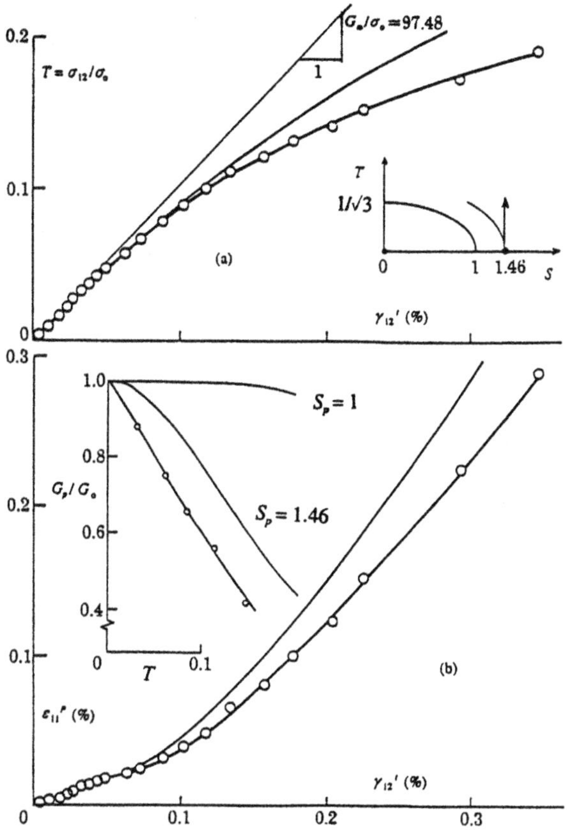

Figure 10.11 Stepped tension-torsion loading of 24S-T4 aluminium alloy

The variation in shear modulus under the shear stress path is shown inset in Fig. 10.13b. It is known that the minimum surface model (see Fig. 8.18) and Hencky's total strain theory both predict an initial shear modulus $G_p = \mathrm{d}\sigma_{12} / \mathrm{d}\gamma_{12}'$, of plastically prestrained material different from the elastic shear modulus G_o. Taking an isotropic expansion of the yield surface to the prestress allows a differentiation of eq(10.44a) to supply a modulus ratio:

$$\frac{G_p}{G_o} = \left\{ 1 + \frac{27\,\varepsilon_o T^2 G_o}{\sigma_o} \left[1 - \frac{2}{(S_p^2 + 3T^2)^{1/2}} + \frac{1}{(S_p^2 + 3T^2)} \right] \right\}^{-1} \tag{10.45}$$

Equation (10.45) shows that $G_p = G_o$ when $T = 0$, irrespective of S_p. Further confirmation of this was found for similar experiments conducted on steel [39], aluminium [40] and various aluminium alloys [39 - 43]. As T increases, the fall in G_p/G_o, observed in the inset diagram, is replicated with eq(10.45). Note that when $S_p = 1$, the fall in G_p/G_o is so gradual under increasing σ_{12} that one could be deceived into expecting a region of elasticity from Prandtl-Reuss though, strictly, the condition for this is $S_p < 1$.

(b) *Stepped Path Under Tension-Internal Pressure*

One method of achieving stepped stressed paths in principal biaxial stress space is to subject thin-walled cylinders to combined internal pressure and axial load. Radial stress (σ_3) is ignored and the respective axial and circumferential stress components σ_1 and σ_2 may be assumed to be uniform. Equations(10.12c) and (10.40) supply

$$\mathrm{d}\varepsilon_1^t = \frac{\mathrm{d}\bar{\sigma}}{H'\bar{\sigma}}\left(\sigma_1 - \frac{\sigma_2}{2}\right) + \frac{1}{E}(\mathrm{d}\sigma_1 - v\,\mathrm{d}\sigma_2) \tag{10.46a}$$

$$\mathrm{d}\varepsilon_2^t = \frac{\mathrm{d}\bar{\sigma}}{H'\bar{\sigma}}\left(\sigma_2 - \frac{\sigma_1}{2}\right) + \frac{1}{E}(\mathrm{d}\sigma_2 - v\,\mathrm{d}\sigma_1) \tag{10.46b}$$

$$\bar{\sigma}^2 = \sigma_1^2 - \sigma_1\sigma_2 + \sigma_2^2 \quad \Rightarrow \quad 2\bar{\sigma}\,\mathrm{d}\bar{\sigma} = (2\sigma_1 - \sigma_{2p})\,\mathrm{d}\sigma_1 \tag{10.46c,d}$$

Equation (10.46d) applies where σ_1 increases in the presence of a constant σ_2 (i.e. σ_{2p}). Substituting eqs(10.42a) and (10.46d) into eqs(10.46a,b) and setting $S_1 = \sigma_1/\sigma_o$, $S_{2p} = \sigma_{2p}/\sigma_o$ and $\bar{S} = \bar{\sigma}/\sigma_o = (S_1^{\,2} - S_1 S_{2p} + S_{2p}^{\,2})^{1/2}$, leads to total, principal strains under increasing σ_1:

$$\varepsilon_1^t = \frac{\varepsilon_o}{4n}\int_{S_{1p}}^{S_1} \frac{(2S_1 - S_{2p})^2(\bar{S} - 1)^{\frac{1-n}{n}}}{\bar{S}^2}\,\mathrm{d}S_1 + \frac{\sigma_o}{E}\int_0^{S_1}\mathrm{d}S_1 \tag{10.47a}$$

$$\varepsilon_2^t = \frac{\varepsilon_o}{4n}\int_{S_{1p}}^{S_1} \frac{(2S_1 - S_{1p})(2S_{2p} - S_1)(\bar{S} - 1)^{\frac{1-n}{n}}}{\bar{S}^2}\,\mathrm{d}S_1 - \frac{v\sigma_o}{E}\int_0^{S_1}\mathrm{d}S_1 \tag{10.47b}$$

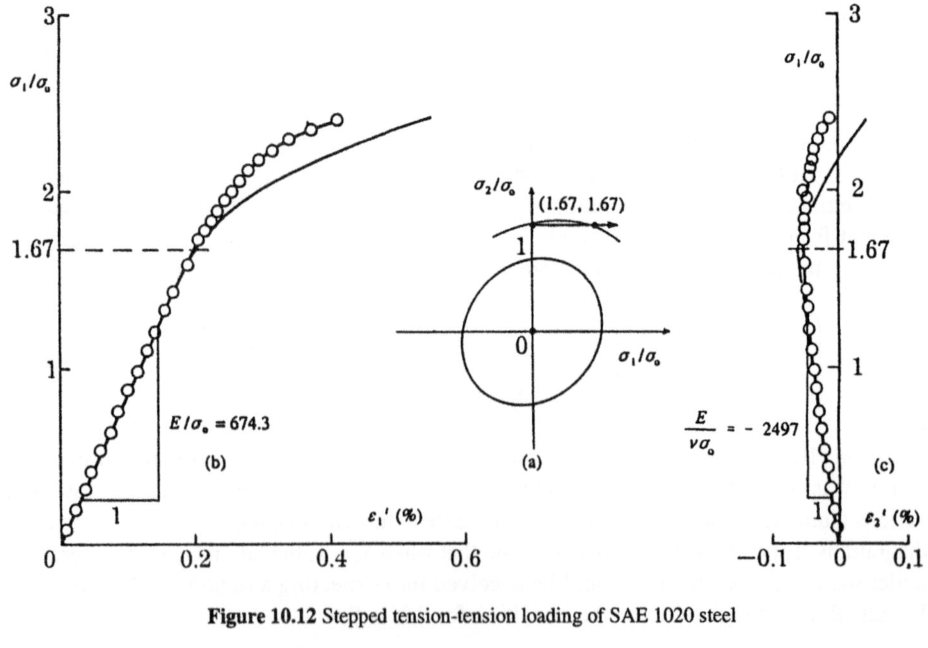

Figure 10.12 Stepped tension-tension loading of SAE 1020 steel

In the experiment of Marin and Hu [44], a mild steel testpiece was prestressed to $S_{2p} = 1.67$. Thereafter, S_1 was increased, as shown in Fig. 10.12a. Figures 10.12b,c show that elastic stress-strain responses prevail until $S_{1p} = 1.67$, this being the lower limit for the first integral in eqs(10.47a,b). Taking from Fig. 9.20, $n \approx ⅓$, $\varepsilon_o = 6.54^{-3} = 3.575 \times 10^{-3}$ for this material, eqs(10.47a,b) become

$$\varepsilon_1^t = \frac{3\varepsilon_o}{4}\int_{S_{1p}}^{S_1}(2S_1 - S_{2p})^2\left[1 - \frac{2}{(S_1^2 - S_{2p}S_1 + S_{2p}^2)^{1/2}} + \frac{1}{(S_1^2 - S_{2p}S_1 + S_{2p}^2)}\right]\mathrm{d}S_1 + \frac{\sigma_o}{E}\int_0^{S_1}\mathrm{d}S_1 \qquad (10.47c)$$

$$\varepsilon_2^t = \frac{3\varepsilon_o}{4}\int_{S_{1p}}^{S_1}(2S_1 - S_{1p})(2S_{2p} - S_1)\left[1 - \frac{2}{(S_1^2 - S_{2p}S_1 + S_{2p}^2)^{1/2}} + \frac{1}{(S_1^2 - S_{2p}S_1 + S_{2p}^2)}\right]\mathrm{d}S_1 - \frac{\nu\sigma_o}{E}\int_0^{S_1}\mathrm{d}S_1 \qquad (10.47d)$$

where $\sigma_o/E = (674.3)^{-1}$ and $\nu = 0.27$ from within Figs 10.12b,c. Closed solutions can be found from the integration of eqs(10.47c,d). However, for the predictions shown, it was simpler to apply Simpson's rule repeatedly, with a step length $\Delta S_1 = 0.2$. These reproduce the trends observed and suggest that the deviation found can be attributed to plastic anisotropy in the cold-drawn tube material. The elastic response observed is predictable from isotropic hardening in its range and magnitude. In this test, where the elastic and plastic strain magnitudes are similar, an omission of elastic strain would lead to unacceptable error. We see that the Prandtl-Reuss theory is essential to give the full strain response to S_1.

10.7 Kinematic Hardening

It has been seen that the rule of isotropic hardening is attractive from a mathematical standpoint when applied to plasticity arising from outward loading paths. Though intended for initially isotropic material a similar hardening rule can be applied to initially anisotropic material with an appropriate change to the yield function (see chapter 11). Moreover, the same rule can provide for the rate dependence of stress by allowing a strain rate vector to lie normal to an expanding visco-plastic, or creep, potential. The latter may again be associated with the yield function but in the flow rule, the incremental plastic strains are replaced by viscous strain rates. The isotropic hardening rule is only useful for outward loading paths and works best for radial, or stepped loadings in which plastic strain history does not exert a strong influence. In Chapter 8 (pp. 262 - 266) we saw that the isotropic hardening rule could not account for the Bauschinger effect. Moreover, it will not account for plastic strain arising from a ' neutral ' loading condition [45], where a stress path follows the boundary of an isotropic expansion of the initial yield locus. Clearly, the subsequent yield condition needs to be re-defined under more complex paths.

10.7.1 Bauschinger Effect

The laboratory testing referred to in Chapter 9 was conducted under monotonically increasing loading. Correspondingly, the descriptions of flow curves given apply to forward deformation. One of the critical validation tests for any theory is whether it can represent, with sufficient accuracy, the stress required to reverse the direction of plastic strain. Bauschinger [46] first observed a common feature of polycrystalline materials in which a reduction in flow stress accompanied a reversal to the plastic strain (see Fig. 8.20b). There is also a marked increase in the reversed hardening rate under the lower flow stress. Physically, a lesser stress is required to activate slip processes in a reversed direction than that required to continue with forward slip on active slip planes. To explain this Kadashevitch and Novozhilov [47] identified a field of residual micro-stresses within a given slipped state. The Bauschinger effect was explained from identifying the sense of this micro-stress distribution to oppose forward flow but assist reversed flow. A lower applied stress is therefore required to promote reversed yielding.

10.7.2 Translation Rules

Prager [48] modelled the Bauschinger effect with a rigid translation of the initial yield surface away from its origin. He assumed that the direction of translation followed the incremental plastic strain vector $\mathrm{d}\varepsilon_{ij}^{P}$, as shown in Fig. 10.13a.

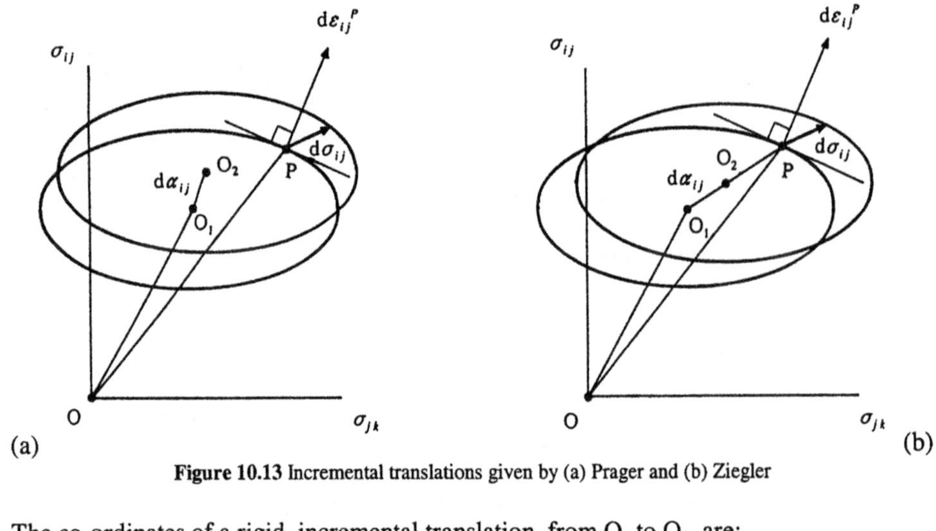

Figure 10.13 Incremental translations given by (a) Prager and (b) Ziegler

The co-ordinates of a rigid, incremental translation, from O_1 to O_2, are:

$$\mathrm{d}\alpha_{ij} = c\,\mathrm{d}\varepsilon_{ij}^{P} \tag{10.48}$$

When eq(10.48) is used to describe the translation of an initial Mises surface, the subseqent function is referred to the co-ordinates for O_2 as:

$$f = \tfrac{1}{2}\,(\sigma_{ij}' - c\int \mathrm{d}\varepsilon_{ij}^{P})^2 = k^2 \tag{10.49}$$

where k is the shear yield stress. Applying eq(10.49) to simple tension:

$$\tfrac{1}{2}[(\sigma_{11}' - c\varepsilon_{11}^P)^2 + (\sigma_{22}' - c\varepsilon_{22}^P)^2 + (\sigma_{33}' - c\varepsilon_{33}^P)^2] = k^2$$

Substituting $\sigma_{11}' = \tfrac{2}{3}\sigma_{11}$, $\sigma_{22}' = -\tfrac{1}{3}\sigma_{11}$ and $\sigma_{33}' = -\tfrac{1}{3}\sigma_{11}$, with $\varepsilon_{22}^P = \varepsilon_{22}^P = -\tfrac{1}{2}\varepsilon_{11}^P$, leads to the uniaxial hardening law:

$$\sigma_{11} = \sigma_o + (3c/2)\,\varepsilon_{11}^P \qquad (10.50)$$

where $\sigma_o = \sqrt{3}k$ is the tensile yield stress. Equation (10.50) shows that Prager's translation describes linear hardeming at a rate proportional to c. A useful consequence of this is that the path of the centre translation is proportional to the plastic strain path.

The reader should note the difference between the translations shown in Figs 10.13a,b. The latter corresponds to an alternative rigid translation rule given by Ziegler [49]:

$$\mathrm{d}\alpha_{ij} = \mathrm{d}\mu\,(\sigma_{ij} - \alpha_{ij}) \qquad (10.51)$$

where scalar $\mathrm{d}\mu = \mathrm{d}\bar{\sigma}/\sigma_o = (c/\sigma_o)\mathrm{d}\bar{\varepsilon}^P$ is defined from the linear hardening rate c in simple tension. Equation (10.51) ensures a translation along the radial path in the manner of Fig. 3.7 and this may be identified with a linear tensile hardening material, as shown in Fig. 8.21. Of the two rules, Prager's has received most attention, particularly with its modification to non-linear hardening [50]:

$$\mathrm{d}\alpha_{ij} = c_1\,\mathrm{d}\varepsilon_{ij}^P - (c_2/\sigma_o)\,\alpha_{ij}\,\mathrm{d}\bar{\varepsilon}^P \qquad (10.52a)$$

where c_1 and c_2 are functions of the plastic strain invariants I_2 and I_3 (see eqs(9.5a,b)). A similar, non-linear modification to Ziegler's law [51] gives

$$\mathrm{d}\alpha_{ij} = (c_1/\sigma_o)\,\mathrm{d}\bar{\varepsilon}^P(\sigma_{ij} - \alpha_{ij}) - (c_2/\sigma_o)\,\alpha_{ij}\,\mathrm{d}\bar{\varepsilon}^P \qquad (10.52b)$$

10.7.3 Reversed Yield Stress

The usual way to determine the parameters c_1 and c_2 in eqs(10.52a,b) is to combine either equation with a standard non-linear hardening law, e.g. Ludwick, Swift, Vocé etc and then to fit the pair of equations to a flow curve with a reversal (see Fig. 8.20b). In this case both expansion and translation of the yield surface are permitted (Fig. 8.20a). For example, integrating eq(10.52a) gives the translation $\alpha_f = \alpha(\varepsilon^P)$ for OA under uni-axial deformation

$$\frac{\alpha_f}{\sigma_o} = \frac{c_1}{c_2}\left(1 - \mathrm{e}^{-c_2\varepsilon^P/\sigma_o}\right) \qquad (10.53)$$

The forward flow stress becomes the sum of two parts $\sigma_f = \alpha_f + \mathfrak{R}(\varepsilon^P)$ where $\mathfrak{R}(\varepsilon^P)$ is a 'radius' of the expanded yield surface (see Fig. 10.14) If we let σ_f be given by Ludwick's eq(9.43b), with α from eq(10.53), the radius becomes

$$\mathfrak{R}(\varepsilon^P) = \sigma_f - \alpha_f = \sigma_o\left[1 + (\varepsilon^P/\varepsilon_o)^n\right] - \frac{\sigma_o c_1}{c_2}\left(1 - \mathrm{e}^{-c_2\varepsilon^P/\sigma_o}\right) \qquad (10.54a)$$

The reversed yield stress σ_P at point P, immediately follows as

$$\sigma_P = \alpha_f - \mathfrak{R}(\varepsilon^P) \qquad (10.54b)$$

$$\frac{\sigma_P}{\sigma_o} = \frac{2c_1}{c_2}\left(1 - \mathrm{e}^{-c_2\varepsilon^P/\sigma_o}\right) - \left[1 + \left(\varepsilon^P/\varepsilon_o\right)^n\right] \qquad (10.54c)$$

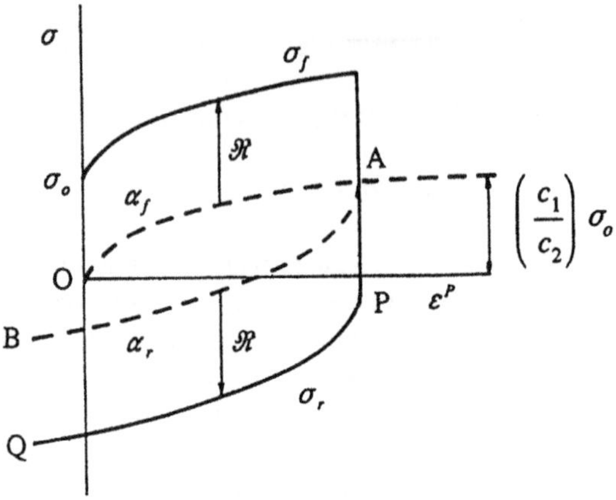

Figure 10.14 Uniaxial plastic flow curve with reversal

10.7.4 Reversed Flow Curve

Equations (10.53) and (10.54c) give normalised stresses for the centre translation OA and the reversed yield point P respectively in Fig. 10.14. To describe the reversed flow curve PQ it is necessary to find the corresponding centre translation AB. Strain reversal requires a change in the sign of the second term in eq(10.52a) [52]. Applying this with lower limits of integration for point A (ε_A^P, α_A):

$$\int_{\alpha_A}^{\alpha} \frac{d\alpha}{c_1 + c_2\alpha/\sigma_o} = \int_{\varepsilon_A^P}^{\varepsilon^P} d\varepsilon^P$$

$$\frac{\alpha}{\sigma_o} = \left(\frac{c_1}{c_2} + \frac{\alpha_A}{\sigma_o}\right) e^{c_2(\varepsilon^P - \varepsilon_A^P)/\sigma_o} - \frac{c_1}{c_2} \tag{10.55a}$$

Substituting α_A/σ_o from eq(10.53), with $\varepsilon^P = \varepsilon_A^P$, into eq(10.55a) leads to α_r for AB:

$$\frac{\alpha_r}{\sigma_o} = \frac{c_1}{c_2}\left(2e^{c_2(\varepsilon^P - \varepsilon_A^P)/\sigma_o} - e^{c_2(\varepsilon^P - 2\varepsilon_A^P)/\sigma_o} - 1\right) \tag{10.55b}$$

Now the reversed flow curve PQ is $\sigma_r = \alpha_r - \Re$ where $\Re = \sigma_f - \alpha_f$ is based upon the forward flow curve. That is, $\Re(\varepsilon^P)$ in eq(10.54a) defines both σ_f and σ_r curves at a given ε^P, as shown in Fig. 10.14:

$$\frac{\sigma_r}{\sigma_o} = \frac{c_1}{c_2}\left(2e^{c_2(\varepsilon^P - \varepsilon_A^P)/\sigma_o} - e^{c_2(\varepsilon^P - 2\varepsilon_A^P)/\sigma_o} - e^{-c_2\varepsilon^P/\sigma_o}\right) - \frac{\sigma_f}{\sigma_o} \tag{10.56}$$

where from eq(10.53) c_1/c_2 is an asymptote for α/σ_o at large strain The above analysis is based upon a reversal between tension and compression but this can be difficult to achieve in practice. Consequently, torsion of thin-walled tubes and bending of thin strips may be used, in which unwanted friction is removed and stress gradients have been reduced. For torsion, the translation OA is equivalent to writing eq(10.53) as:

$$\frac{\beta}{k} = \frac{c_1}{c_2'}\left(1 - e^{-c_2'\gamma^P/k}\right) \tag{10.57}$$

and eq(10.56) is re-written for the reversed flow curve PQ as

$$\frac{\tau_r}{k} = \frac{c_1}{c_2'}\left(2e^{c_2'(\gamma^P - \gamma_A^P)/k} - e^{c_2'(\gamma^P - 2\gamma_A^P)/k} - e^{-c_2'\gamma^P/k}\right) - \frac{\tau_f}{k} \tag{10.58}$$

where $c_2' = c_2/3$ arises from the use of engineering shear strain. Figure 10.15 shows reversed torsional flow curves for annealed En3B tubes with inner diameter 12.7 mm, outer diameter 16.1 mm and a gauge length of 80 mm.

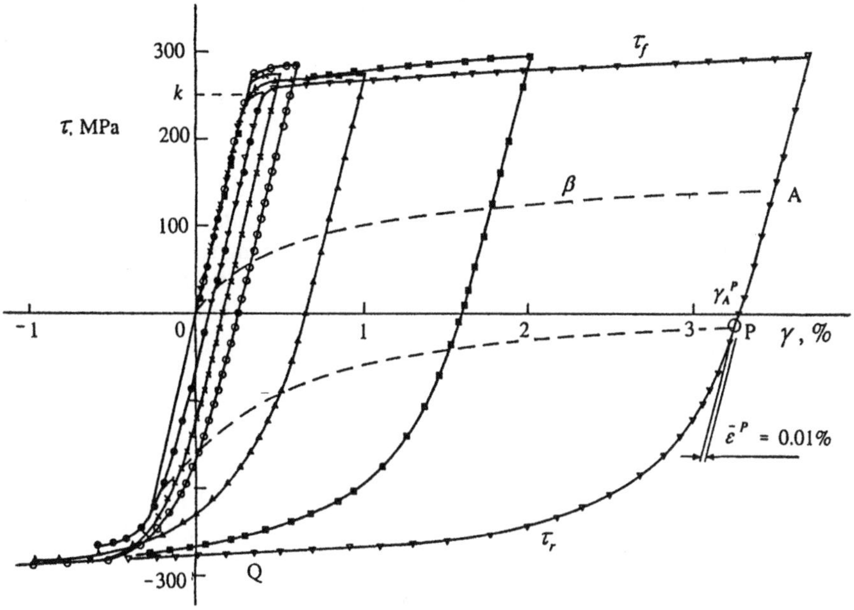

Figure 10.15 Reversed flow curves under torsion

At each γ_A^P value, the reversed yield point P is determined for 0.01% Mises proof strain. The mid-point translation locus OA reveals an asymptote in the region of $c_1/c_2' = 0.74$. If we now fit eq(10.58) to the reversed flow curve PQ, having subtracted elastic strain appropriate to each stress level, then c_1/k will vary with γ^P in the manner of Fig. 10.16. This identifies the function $c_1(I_2)$ with a complex polynomial in which $I_2 = (\gamma^P)^2/4$. Note $c_2'/k = (3 \times 0.74)\, c_1/k$.

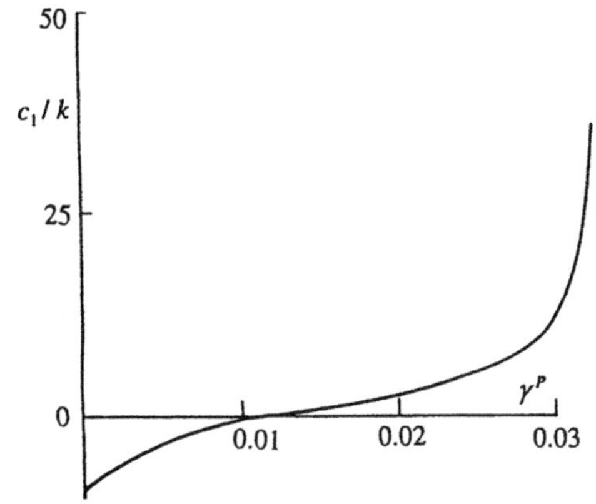

Figure 10.16 Dependence of c_1/k upon γ^P

For multiaxial stress states a subsequent combined hardening yield function is required. We write this to contain both the translation and expansion of an initial von Mises yield surface as

$$\frac{3}{2}\left(\sigma_{ij}' - \alpha_{i,j}'\right)\left(\sigma_{ij}' - \alpha_{i,j}'\right) = \left[\mathscr{R}\left(\bar{\varepsilon}^{P}\right)\right]^{2}$$

in which $\alpha_{ij}' = \alpha_{ij} - ⅓\delta_{ij}\,\alpha_{kk}$ and eq(10.54a) will again define the right-hand side when an equivalent Mises plastic strain $\bar{\varepsilon}^{P}$ identifies with the tensile strain ε^{P}.

10.8 Concluding Remarks

The rule of isotropic hardening is suitable for radial loading paths. The size may be expressed from a hardening hypothesis though it appears difficult to distinguish between hypotheses of work and strain hardening as applied to an $f(J_2', J_3')$ isotropic material. The strain hardening hypothesis may be preferred, for simplicity, since the chosen equivalent strain expression is directly identified with an empirical hardening function (e.g. Hollomon, Swift etc). However, the work hypothesis is more exacting since it will ensure non-violation of the plastic work. Stress reversal involves a Bauschinger effect that is best modelled with a shift in the origin of the yield locus coupled to its isotropic expansion.

References

1. Hill R. *The Mathematical Theory of Plasticity*, 1989, Oxford University Press.
2. Drucker D. C. *Proc. 1st U.S. Nat. Congr of Applied Mechanics*, 1952, 487, ASME.
3. Prager W. *Jl Appl Phy*, 1949, **20**, 235.
4. Ford H. and Alexander J. M. *Advanced Mechanics of Materials*, 1963, Longmans.
5. Zyczkowski M. *Combined Loadings in the Theory of Plasticity*, 1981, Polish Sci Pub.
6. Naghdi P. *Plasticity, Proc 2nd Symp on Naval Struct Mech*, (eds Lee E. H. and Symonds P. S.) 1960, 121, Pergamon Press.
7. Hollomon J. H. *Am Inst Mech Eng*., 1945, **162**, 268.
8. Ludwick P. *Elemente der Technologischen Mechanik*, 1909 Springer, Berlin.
9. Swift H. W. *Jl Mech Phys Solids*, 1952, **1**, 1
10 Rees D. W. A. *Proc Roy Soc*., 1982, **A383**, 333.
11. Levy M. *Jl Math Pures Appl*. 1871, **16**, 369.
12. von Mises R. *Ges Wiss Gottingen*, 1913, 582.
13. Drucker D. C. *Trans ASME, Jl Appl Mech*., 1949, **16**, 349.
14. Rees D. W. A. and Mathur S. B. *Proc: Non Linear Problems in Stress Analysis*, (ed Stanley. P.) 1978, 185, Applied Science.
15. Phillips A and Ricciuti M. *Int. Jl Solids and Struct*., 1976, **12**, 159.
16. Phillips A. and Lee, C-W. *Int Jl Solids and Struct*., 1979, **15**, 715.
17. Shahabi S. N. and Shelton A. *Jl Mech Eng Sci*., 1975, **17**, 93.
18. Johnson A. E., Frost N. E. and Henderson J. *The Engineer*, 1955, **199**, 366, 402, 457.
19. Johnson A. E. and Frost N. E. *The Engineer*, London, 1952, **194**, 713.
20. Khan A. S. and Parikh Y. *Eng Fract Mech*., 1985, **21**, 697.
21. Freudental A. M. and Gou P. F. *Acta Mech*., 1969, **8**, 34.
22. Rogan J. Ph.D.Thesis, University of London, 1966.
23. Taylor G. I. and Quinney H. *Phil Trans Roy Soc*., London, 1931, **A230**, 323.
24. Lode W. *Z. Phys* 1926, **36**, 913.

25. Baraya G. L. and Parker J. *Int Jl Mech Sci.*, 1963, **5**, 353.
26. Ellington P. *Jl Mech Phys Solids*, 1958, **6**, 276.
27. Dillamore I., Hazel R. J., Watson T. W. and Hadden P. *Int Jl Mech Sci*, 1971, **13**, 1049.
28. Fessler H. and Hyde T. Proc: *Non Linear Problems in Stress Analysis*, (ed Stanley P.) 1978, 233, Applied Science.
29. Mehan R. L. *Trans ASME, Jl Basic Engng*, 1961, **83**, 499.
30. Marin J. *Jl. Appl. Mech.*, 1937, **4**, 55.
31. Soderberg C. R. *Trans ASME*, 1936, **58**, 733.
32. Drucker D. C. and Stockton F. D. *Proc Soc Exp Stress Anal.*, 1953, **10**, 127.
33. Mair W. M. *Jl Strain Anal.*, 1967, **2**, 188.
34. Paul B., Chen W. and Lee L. *Proc. 4th US Nat Congr Appl Mech ASME*, 1962, 1013.
35. Prandtl L. *ZAMM*, 1928, **8**, 85.
36. Reuss A. *ZAMM*, 1930, **10**, 266.
37. Rees D. W. A. *Proc R. Soc*, 1987, **A410**, 443.
38. Ivey H. J. *Jl Mech Engng Sci.*, 1961, **3**, 15.
39. Morrison J. L. M. and Shepherd W. M. *Proc I Mech. E*, 1950, **163**, 1.
40. Daneshi G. H. and Hawkyard J. B. *Int Jl Mech Sci*, 1976, **18**, 195.
41. Peters R. W., Dow N. F. and Batdorf S. B. *Proc Soc Exp Stress Anal.*, 1950, **7**, 127.
42. Feigen M. *Proc: 2nd US Nat Cong Appl Mech*, 1955, 469, Pergamon.
43. Naghdi P. M., Essenberg F. and Koff W. *Jl Appl Mech.*, 1958, **25**, 201.
44. Marin J. and Hu L. W. *Trans ASME*, 1956, **78**, 499.
45. Rees D. W. A. *Int Jl Plasticity*, 1988, **4**, 91.
46. Bauschinger J. *Civilingenieur*, 1881, 289.
47. Kadashevitch and Novozhilov V. V. *Prikl Mat Mekh*, 1958, **22**, 78.
48. Prager W. *Jl Appl. Mech.*, 1956, **23**, 483.
49. Shield R. T. and Ziegler H. *ZAMP*, 1958, **9a**, 260.
50. Armstrong P. J. and Frederick, C. O. 1966, *CEGB Rpt.* RD/B/N, 731, 1966.
51. Brunet M., Morestin F. and Walter H. *Int Jl Forming Proc*, 2002, **5**, 225.
52. Chaboche J. L. *Int Jl Plasticity*, 1986, **2**, 149.

Exercises

10.1 Examine whether it is possible to derive the function F in the work-hardening hypothesis for the flow curves described by Hollomon, modified Prager and Vocé (see Figs 9.19a, c and f respectively).

10.2 Show that the Levy-Mises flow rule may also be written, using the Mises equivalent stress and equivalent plastic strain increment definitions, as

$$\mathrm{d}\varepsilon_{ij}^{P} = \left(\frac{\mathrm{d}\bar{\varepsilon}^{P}}{\bar{\sigma}}\right)\frac{\partial \bar{\sigma}}{\partial \sigma_{ij}}$$

10.4 Is the strain hardening hypothesis, which defines $\bar{\varepsilon}^{P}$ in eq(10.35), consistent with a meaningful work expression? Examine from a consideration of deformation under simple shear.

10.5 Show that $\delta\lambda$ in eqs(10.18a,b) may be derived from eq(10.5) and the Mises potential eq(10.6).

10.6 Show that $\delta\lambda$ in eqs(10.23a,b) may be derived from eq(10.5) and the Drucker potential (10.20).

10.7 Derive, for a thin-walled cylinder, the Prandtl-Reuss predictions to total strain components ε_{11}' and γ_{12}', that arise from increasing torque under a constant, initially elastic tension. Assume that in the wall the uniform shear stress increases in the presence of a constant, initially elastic axial stress.

10.8 A thin-walled cylinder with closed ends is constructed from a von Mises hardening material. It is subjected to a constant, principal stress ratio $\sigma_2/\sigma_1 = 4/3$ where σ_1 and σ_2 are axial and circumferential stresses due to independent axial load and internal pressure. Show that the corresponding total strains in the axial, circumferential and radial directions are, respectively:

$$\varepsilon_1^t = \ln\left(\frac{\sigma_2}{p}\right)^{\frac{1}{2}} + \frac{\sigma_2}{4K}, \quad \varepsilon_2^t = \ln\left(\frac{\sigma_2}{p}\right)^{\frac{1}{4}} + \frac{\sigma_2}{4K} \quad \text{and} \quad \varepsilon_3^t = \ln\left(\frac{p}{\sigma_2}\right)^{\frac{3}{4}} + \frac{\sigma_2}{4K}$$

where p is the internal pressure and K is the bulk modulus of the cylinder material. Determine the angle between the vectors of equivalent stress and equivalent plastic strain increment on the principal stress plane | [Answer: 15.07°]

10.9 A thin-walled closed tube with an initial mean diameter d_o and thickness t_o is subjected to an internal pressure p. Neglecting elasticity and assuming a Mises material with Hollomon hardening $\bar{\sigma}/\sigma_o = (\bar{\varepsilon}^P/\varepsilon_o)^n$, show that the circumferential plastic strain is:

$$\frac{\varepsilon_2^P}{\varepsilon_o} = 2\left(\frac{\sqrt{3}}{4}\right)^{\frac{n+1}{n}}\left(\frac{pd}{\sigma_o t}\right)^{\frac{1}{n}}$$

Confirm that the relationship between pressure and current geometry which preserves volume is:

$$\left(\frac{pd^2}{d_o t_o}\right)\left(\frac{\varepsilon_o^n}{\sigma_o}\right) = 2\left(\frac{2}{\sqrt{3}}\right)^{1+n} \ln\left(\frac{d}{d_o}\right)^n$$

10.10 A thin-walled cylinder with closed ends is constructed from a Mises hardening material. It is subjected to constant stress ratios $\sigma_{22}/\sigma_{12} = \sigma_{22}/\sigma_{11} = 2$ where σ_{11}, σ_{12} and σ_{22} are axial, shear and circumferential stresses due to applied torsion and internal pressure p. Show that the total strain components are

$$\varepsilon_{11}^t = \frac{\sigma_{22}}{6K}, \quad \varepsilon_{22}^t = \ln\left(\frac{\sigma_{22}}{p}\right)^{\frac{1}{2}} + \frac{\sigma_{22}}{6K},$$

$$\varepsilon_{33}^t = \ln\left(\frac{p}{\sigma_{22}}\right)^{\frac{1}{2}} + \frac{\sigma_{22}}{6K}, \quad \varepsilon_{12}^t = \ln\left(\frac{\sigma_{22}}{p}\right)^{\frac{1}{2}}$$

where subscripts 1, 2 and 3 denote the axial, circumferential and radial directions respectively and K is the bulk modulus of the cylinder material. Show that the stress vector and the plastic strain increment vector are co-incident in the σ_{12} versus σ_{22} plane.

10.11 In a torsion test a thin-walled tube is twisted into the plastic range to a plastic shear strain γ^P. The torque is then reversed to twist the tube to its reversed shear yield point. Show that the normalised, shear yield stress for this reversal is expressed as

$$\frac{\tau_r}{k} = \frac{3c_1}{c_2}\left(e^{\frac{c_2\gamma^P}{3k}} - 1\right) - \left(1 + \frac{\gamma^P}{\gamma_o}\right)^n$$

Assume a Mises material ($\bar{\sigma} = \sqrt{3}\tau$, $\bar{\varepsilon}^P = \gamma^P/\sqrt{3}$), a Prager translation (10.52a) and a Swift forward hardening law: $\tau_r/k = (1 + \gamma^P/\gamma_o)^n$, where k is the initial shear yield stress, γ_o and n are constants.

CHAPTER 11

ORTHOTROPIC PLASTICITY

11.1 Introduction

Often, when isotropic definitions of $\bar{\sigma}$ and $\bar{\varepsilon}^P$ are employed to describe equivalence, it is found that flow data from different tests become scattered when plotted on $\bar{\sigma}$ versus $\bar{\varepsilon}^P$ axes. This is because many materials are initially anisotropic to some degree. To quantify this the following analysis assumes that orthogonal axes of orthotropy lie within both bar and sheet metal stock. In this way, the anisotropy is defined within three orthogonal planes as it arises from different processes, including the production of rolled and drawn sheet, extruded and forged billets. The residual stress resulting from each process may be eliminated by post heat treatment but there may still remain an orthotropic anisotropy. Klinger and Sachs [1] were among the first to show this for a heat treated 24ST aluminium alloy plate. They found for tensile testpieces, aligned with orthogonal directions, that the incremental plastic strain ratios did not conform to the isotropic value $d\varepsilon_{22}^P / d\varepsilon_{11}^P = -\frac{1}{2}$. This was attributed to the partial retention of the original preferred grain orientation. These authors so identified a crystallographic anisotropy, originating from within the highly anisotropic single crystals that comprise the polycrystalline matrix. It was also shown that laminar inclusions, cavities and arrangements of alloy phases can produce similar deviations from isotropic plasticity.

11.2 Orthotropic Flow Potential

Hill's quadratic yield function (3.27) accounts for the directionally dependent yield stresses in orthotropic, metallic materials. To describe the plastic flow behaviour a similar function is taken for the plastic potential in the flow rule (3.22). This is simplified when the Cartesian stress co-ordinates coincide with the material's orthogonal axes. The isotropic hardening rule is taken to define the subsequent yield surface in a generalised quadratic form (3.29). Combining this with eq(10.10a) describes isotropic hardening in initially anisotropic, incompressible material:

$$\tfrac{1}{2} H_{ijkl}\, \sigma_{ij}'\, \sigma_{kl}' = \chi(\psi) \tag{11.1}$$

For an orthotropic condition and with plastic incompressibility imposed by the stress deviators, it has been shown previously (see p. 84) that the fourth order tensor H_{ijkl} is equivalent to the six anisotropy parameters F, G, H, L, M and N in eq(3.27). Equation (11.1) ensures that an initial orthotropy remains unchanged for constant ratios F/G, G/H etc, during hardening. For this, it is sufficient that the argument ψ be identified with an appropriate anisotropic form of equivalent plastic strain. We shall see how $d\bar{\varepsilon}^P$ may be derived from either the work or strain hardening hypothesis so that the choice between them remains within the argument $\psi = \bar{\varepsilon}^P$ in eq(11.1). Furthemore, because we shall identify the left-hand side of eq(11.1) with the square of an orthotropic equivalent stress expression, the harening rule is simplified to $\bar{\sigma} = \bar{\sigma}(\bar{\varepsilon}^P)$. In fact, most equivalent stress-plastic strain correlations for anisotropic materials [2-4] have been derived from this hardening function.

Othotropic, incremental constitutive relations are supplied by the flow rule (3.22). These relations must be used with appropriate definitions of $\bar{\sigma}$ and $d\bar{\varepsilon}^P$ to establish the scalar multiplier $d\lambda$. The following theories of othotropic plasticity are similar to those proposed by Jackson, Smith and Lankford [5], Dorn [6], Prager [7] and Fisher [8]. Certain modifications to quadratic potential theories were proposed by Hu [9] and Jones and Gillis [10] to provide for a wider range of anisotropic behaviour.

11.2.1 Equivalent Stress

Firstly, it is necessary to derive expressions the equivalent stress and plastic strain increment for an orthotropic material in terms of its anisotropy parameters H_{ijkl} within eq(11.1). Hill [11] employed the reduced form with six coefficients F, G, H, L, M and N (see eq(3.27)). When the six stress components σ_{ij} align with principal material (orthotropic) axes 1, 2 and 3, the plastic potential becomes

$$2f(\sigma_{ij}) = F(\sigma_{22} - \sigma_{33})^2 + G(\sigma_{33} - \sigma_{11})^2 + H(\sigma_{11} - \sigma_{22})^2 + 2L\sigma_{23}^{\;2} + 2M\sigma_{13}^{\;2} + 2N\sigma_{12}^{\;2} = 1 \quad (11.2)$$

The unity value on the right hand side is preserved during hardening. It follows that F, G, H etc, (with units stress^{-2}) must diminish as the stress components σ_{11}, σ_{22} etc, increase. For uniaxial flow along each axis, eq(11.2) shows

$$(G + H)\,\sigma_1^{\;2} = 1 \quad (11.3a)$$

$$(F + H)\,\sigma_2^{\;2} = 1 \quad (11.3b)$$

$$(F + G)\,\sigma_3^{\;2} = 1 \quad (11.3c)$$

When the flow stresses σ_1, σ_2 and σ_3 are constrained to increase proportionately with hardening along 1, 2 and 3, Hill connected an equivalent stress to function $f(\sigma_{ij})$ in eq(11.2):

$$(F + G + H)\,\bar{\sigma}^2 = 3f(\sigma_{ij}) \quad (11.4a)$$

Substituting eq(11.2) into eq(11.4a), gives an equivalent stress expression

$$\bar{\sigma} = \sqrt{\frac{3}{2}\left[\frac{F(\sigma_{22} - \sigma_{33})^2 + G(\sigma_{33} - \sigma_{11})^2 + H(\sigma_{11} - \sigma_{22})^2 + 2L\sigma_{23}^2 + 2M\sigma_{13}^2 + 2N\sigma_{12}^2}{(F + G + H)}\right]} \quad (11.4b)$$

Setting $F = G = H = 1$ and $L = M = N = 3F$ in eq(11.4b) restores the Mises definition (9.2c).

Finnie and Heller [12] defined an alternative form of equivalent stress for creep of orthotropic material, simply by taking F, G, H etc as constants:

$$2\bar{\sigma}^2 = F(\sigma_{22} - \sigma_{33})^2 + G(\sigma_{33} - \sigma_{11})^2 + H(\sigma_{11} - \sigma_{22})^2 + 6L\sigma_{23}^{\;2} + 6M\sigma_{13}^{\;2} + 6N\sigma_{12}^{\;2} \quad (11.5a)$$

The reduction in eq(11.5a) to a Mises isotropic form is easily achieved from setting all constants to unity to correspond with $f(\sigma_{ij}) = J_2' = \bar{\sigma}^2/3$. Consequently, we may identify $\bar{\sigma}$ in eq(11.5a) with its orthotropic potential as:

$$6f(\sigma_{ij}) = 2\bar{\sigma}^2 \quad (11.5b)$$

11.2.2 Equivalent Flow Rule

The flow rule provides plastic strain increments when the plastic potential is either that given in eq(11.4a) or (11.5b). Alternatively, the flow rule can be based upon a Hill or Finnie $\bar{\sigma}$ definition from eq(10.4b):

$$d\varepsilon_{ij}^{P} = d\lambda \frac{\partial f}{\partial \sigma_{ij}} = d\lambda \frac{\partial f}{\partial \bar{\sigma}} \frac{\partial \bar{\sigma}}{\partial \sigma_{ij}} \tag{11.6}$$

Clearly, the relationship between $f(\sigma_{ij})$ and $\bar{\sigma}$ will differ between eqs(11.4a) and (11.5b). From Hill, we substitute eq(11.4a) into eq(11.6):

$$d\varepsilon_{ij}^{P} = \frac{2}{3} d\lambda (F + G + H) \bar{\sigma} \left(\frac{\partial \bar{\sigma}}{\partial \sigma_{ij}} \right) = \left(\frac{d\lambda}{\bar{\sigma}} \right) \left(\frac{\partial \bar{\sigma}}{\partial \sigma_{ij}} \right) \tag{11.7a}$$

and from Finnie, substitute eq(11.5b) into eq(11.6):

$$d\varepsilon_{ij}^{P} = d\lambda \left(\frac{2\bar{\sigma}}{3} \right) \left(\frac{\partial \bar{\sigma}}{\partial \sigma_{ij}} \right) \tag{11.7b}$$

Since the function $f(\sigma_{ij})$ is homogenous, of degree $n = 2$ in stress, Euler's theorem states

$$nf = 2f = \sigma_{ij} \left(\frac{\partial f}{\partial \sigma_{ij}} \right) \tag{11.8a}$$

Combining eq(11.8a) with the work hypothesis and the flow rule gives

$$dW^{P} = \bar{\sigma}\, d\bar{\varepsilon}^{P} = \sigma_{ij}\, d\varepsilon_{ij}^{P} = d\lambda\, \sigma_{ij} \left(\frac{\partial f}{\partial \sigma_{ij}} \right) = 2f\, d\lambda \tag{11.8b}$$

when, from eq(11.8b), it follows

$$d\lambda = \frac{\bar{\sigma}\, d\bar{\varepsilon}^{P}}{2f} \tag{11.8c}$$

Correspondingly, we substitute into eq(11.8c): (i) $2f = 1$ from eq(11.2) and (ii) $2f = 2\,\bar{\sigma}^2/3$, from eq(11.5b). Thus, $d\lambda$ is defined by Hill and Finnie respectively:

$$d\lambda = \bar{\sigma}\, d\bar{\varepsilon}^{P} \quad \text{and} \quad d\lambda = \frac{3\, d\bar{\varepsilon}^{P}}{2\bar{\sigma}} \tag{11.9a,b}$$

These scalars ensure that making the substitutions from eq(11.9a,b) into eqs(11.7a,b) will lead to an identical *equivalent rule of flow*:

$$d\varepsilon_{ij}^{P} = d\bar{\varepsilon}^{P} \left(\frac{\partial \bar{\sigma}}{\partial \sigma_{ij}} \right) \tag{11.10}$$

11.2.3 Incremental Constitutive Relations

Substituting eq(11.2) into eq(11.6) gives the following relations between the six independent components of stress and incremental plastic strains:

$$\mathrm{d}\varepsilon_{11}{}^P = \mathrm{d}\lambda\,[H(\sigma_{11} - \sigma_{22}) - G(\sigma_{33} - \sigma_{11})] \quad (11.11a)$$

$$\mathrm{d}\varepsilon_{22}{}^P = \mathrm{d}\lambda\,[F(\sigma_{22} - \sigma_{33}) - H(\sigma_{11} - \sigma_{22})] \quad (11.11b)$$

$$\mathrm{d}\varepsilon_{33}{}^P = \mathrm{d}\lambda\,[G(\sigma_{33} - \sigma_{11}) - F(\sigma_{22} - \sigma_{33})] \quad (11.11c)$$

$$\mathrm{d}\gamma_{12}{}^P = 2\mathrm{d}\lambda\,N\,\sigma_{12},\ \mathrm{d}\gamma_{23}{}^P = 2\mathrm{d}\lambda\,L\,\sigma_{23} \text{ and } \mathrm{d}\gamma_{13}{}^P = 2\mathrm{d}\lambda\,M\,\sigma_{13} \quad (11.11d,e,f)$$

Summing eqs(11.11a-c) confirms incompressible plasticity (i.e. $\mathrm{d}\varepsilon_{ii}{}^P = 0$). Finnie's potential (11.5b) gives identical engineering shear strains (11.11d-f) but the coefficient $\mathrm{d}\lambda$ in eqs(11.11a-c) is replaced by $\mathrm{d}\lambda/3$. Consequently, we must take $\mathrm{d}\lambda$ appropriately from eqs(11.9a,b) to ensure that the strain components agree.

11.2.4 Equivalent Plastic Strain

To determine Hill's equivalent plastic strain increment $\mathrm{d}\bar{\varepsilon},^P$ we write from eqs(11.9a) and (11.11a-c):

$$F\,\mathrm{d}\varepsilon_{11}{}^P - G\,\mathrm{d}\varepsilon_{22}{}^P = (FG + GH + HF)(\sigma_{11} - \sigma_{22})(\bar{\sigma}\,\mathrm{d}\bar{\varepsilon}^P) \quad (11.12a)$$

$$G\,\mathrm{d}\varepsilon_{22}{}^P - H\,\mathrm{d}\varepsilon_{33}{}^P = (FG + GH + HF)(\sigma_{22} - \sigma_{33})(\bar{\sigma}\,\mathrm{d}\bar{\varepsilon}^P) \quad (11.12b)$$

$$H\,\mathrm{d}\varepsilon_{33}{}^P - F\,\mathrm{d}\varepsilon_{11}{}^P = (FG + GH + HF)(\sigma_{33} - \sigma_{11})(\bar{\sigma}\,\mathrm{d}\bar{\varepsilon}^P) \quad (11.12c)$$

Combining eqs(11.4b) with eqs(11.12a-c) gives, eventually,

$$\mathrm{d}\bar{\varepsilon}^P = \sqrt{\frac{2(F+G+H)}{3}}\left[F\left(\frac{G\mathrm{d}\varepsilon_{22}^P - H\mathrm{d}\varepsilon_{33}^P}{FG+GH+HF}\right)^2 + G\left(\frac{H\mathrm{d}\varepsilon_{33}^P - F\mathrm{d}\varepsilon_{11}^P}{FG+GH+HF}\right)^2\right.$$
$$\left. + H\left(\frac{F\mathrm{d}\varepsilon_{11}^P - G\mathrm{d}\varepsilon_{22}^P}{FG+GH+HF}\right)^2 + \frac{(\mathrm{d}\gamma_{23}^P)^2}{2L} + \frac{(\mathrm{d}\gamma_{13}^P)^2}{2M} + \frac{(\mathrm{d}\gamma_{12}^P)^2}{2N}\right]^{\frac{1}{2}} \quad (11.13)$$

Alternatively, from Finnie's eqs(11.5a) and (11.10),

$$\mathrm{d}\bar{\varepsilon}^P = \sqrt{2}\left[F\left(\frac{G\mathrm{d}\varepsilon_{22}^P - H\mathrm{d}\varepsilon_{33}^P}{FG+GH+HF}\right)^2 + G\left(\frac{H\mathrm{d}\varepsilon_{33}^P - F\mathrm{d}\varepsilon_{11}^P}{FG+GH+HF}\right)^2\right.$$
$$\left. + H\left(\frac{F\mathrm{d}\varepsilon_{11}^P - G\mathrm{d}\varepsilon_{22}^P}{FG+GH+HF}\right)^2 + \frac{(\mathrm{d}\gamma_{23}^P)^2}{6L} + \frac{(\mathrm{d}\gamma_{13}^P)^2}{6M} + \frac{(\mathrm{d}\gamma_{12}^P)^2}{6N}\right]^{\frac{1}{2}} \quad (11.14)$$

Equation (11.14) reduces to an isotropic definition (9.6b) for *constants* of unity. To achieve a similar reduction from eq(11.13) the *coefficients* obey $F = G = H$ and $L = M = N = 3F$.

11.2.5 Anisotropy Parameters

The parameters F, G, H, L, M and N can be expressed either in terms of the yield stresses or the plastic strain increment ratios obtained from tension tests. Identifying σ_{1y}, σ_{2y} and σ_{3y} with the uniaxial yield stresses in principal material directions and σ_{12y}, σ_{23y} and σ_{13y}, with the shear yield stresses in each plane, Hill's equivalent stress (11.4b) gives coefficients F, G, H etc (units of $(\mathrm{MPa})^{-2}$):

$$F = \tfrac{1}{2}(1/\sigma_{3y}^2 + 1/\sigma_{2y}^2 - 1/\sigma_{1y}^2)$$
$$G = \tfrac{1}{2}(1/\sigma_{1y}^2 + 1/\sigma_{3y}^2 - 1/\sigma_{2y}^2) \qquad (11.16a\text{-}f)$$
$$H = \tfrac{1}{2}(1/\sigma_{2y}^2 + 1/\sigma_{1y}^2 - 1/\sigma_{3y}^2)$$
$$N = 1/(2\sigma_{12y}^2),\ L = 1/(2\sigma_{23y}^2) \text{ and } M = 1/(2\sigma_{13y}^2)$$

Finnie's equivalent stress (11.5a) gives F, G, H etc as dimensionless constants:

$$F = (1/\sigma_{3y}^2 + 1/\sigma_{2y}^2 - 1/\sigma_{1y}^2)\,\bar{\sigma}^2$$
$$G = (1/\sigma_{1y}^2 + 1/\sigma_{3y}^2 - 1/\sigma_{2y}^2)\,\bar{\sigma}^2 \qquad (11.17a\text{-}f)$$
$$H = (1/\sigma_{2y}^2 + 1/\sigma_{1y}^2 - 1/\sigma_{3y}^2)\,\bar{\sigma}^2$$
$$N = \bar{\sigma}^2/(3\sigma_{12y}^2),\ L = \bar{\sigma}^2/(3\sigma_{23y}^2) \text{ and } M = \bar{\sigma}^2/(3\sigma_{13y}^2)$$

where $\bar{\sigma}^2 = (\sigma_{1y}^2 + \sigma_{2y}^2 + \sigma_{3y}^2)$. Since absolute values of F, G, H etc, are rarely necessary, it is more common to identify the ratios between them with plastic strain increment ratios. For example, let $r = \mathrm{d}\varepsilon_w^P/\mathrm{d}\varepsilon_t^P$ be the gradient to a plot between width and thickness plastic strains from tensile testing sheet material along its orthotropic axes. The tensile stress lies in the plane of the sheet so that in one test σ_1 is aligned with 1 and in another test σ_2 is aligned with 2. A through-thickness compressive stress σ_3 aligns with 3. Eqs (11.11a-c) give corresponding r values:

$$r_1 = \mathrm{d}\varepsilon_2^P/\mathrm{d}\varepsilon_3^P = H/G \qquad (11.18a)$$
$$r_2 = \mathrm{d}\varepsilon_1^P/\mathrm{d}\varepsilon_3^P = H/F \qquad (11.18b)$$
$$r_3 = \mathrm{d}\varepsilon_2^P/\mathrm{d}\varepsilon_1^P = F/G \qquad (11.18c)$$

in which $r_3 = r_1/r_2$. The departure from unity in each r-value provides: (i) a sensitive measure of degree of orthotropy and (ii) a useful measure of formability in sheet metals.

11.3 Orthotropic Flow Curves

11.3.1 Work Hypothesis

Hill's theory, which is based upon the work hardening hypothesis, is employed for the following analysis. When in-plane stresses σ_{11}, σ_{22} and σ_{12} are applied to a rolled sheet material, eq(11.4b) reduces to

$$\bar{\sigma} = \sqrt{\frac{3}{2}}\left[\frac{F\sigma_{22}^2 + G\sigma_{11}^2 + H(\sigma_{11} - \sigma_{22})^2 + 2N\sigma_{12}^2}{F + G + H}\right]^{\frac{1}{2}} \qquad (11.19)$$

in which direction 1 is aligned with the rolling direction. Reducing the stress components in eq(11.19) and substituting from eqs(11.18a,b) gives equivalent, uniaxial flow stresses for the rolling and transverse directions respectively:

$$\bar{\sigma} = \sqrt{\frac{3r_2(1 + r_1)}{2(r_1 + r_2 + r_1 r_2)}}\ \sigma_1 \tag{11.20a}$$

$$\bar{\sigma} = \sqrt{\frac{3r_1(1 + r_2)}{2(r_1 + r_2 + r_1 r_2)}}\ \sigma_2 \tag{11.20b}$$

The equivalent stresses are the same when, from eqs(11.20a,b),

$$\sigma_2 = \sqrt{\frac{r_2(1 + r_1)}{r_1(1 + r_2)}}\ \sigma_1 \tag{11.21}$$

If $d\varepsilon_1^P$ is the plastic strain increment under σ_1, the two lateral strain increments become

$$d\varepsilon_3^P = -\frac{d\varepsilon_1^P}{1 + r_1} \quad \text{and} \quad d\varepsilon_2^P = -\frac{r_1\, d\varepsilon_1^P}{1 + r_1} \tag{11.22a,b}$$

in which eq(11.18a) has been combined with eqs(11.11a-c). Similarly, for a strain increment $d\varepsilon_2^P$ arising from the tensile stress σ_2, the lateral strains become

$$d\varepsilon_1^P = -\frac{r_2\, d\varepsilon_2^P}{1 + r_2} \quad \text{and} \quad d\varepsilon_3^P = -\frac{d\varepsilon_2^P}{1 + r_2} \tag{11.23a,b}$$

Here $d\varepsilon_1^P + d\varepsilon_2^P + d\varepsilon_3^P = 0$ applies to each test. Substituting eqs(11.22a,b) and (11.23a,b) into eq(11.13) and omitting shear strain terms provides equivalent plastic strain increments for each direction:

$$d\bar{\varepsilon}^P = \sqrt{\frac{2(r_1 + r_2 + r_1 r_2)}{3r_2(1 + r_1)}}\ d\varepsilon_1^P \tag{11.24a}$$

$$d\bar{\varepsilon}^P = \sqrt{\frac{2(r_1 + r_2 + r_1 r_2)}{3r_1(1 + r_2)}}\ d\varepsilon_2^P \tag{11.24b}$$

The equivalent plastic strain increments are the same, when from eqs (11.24a,b):

$$d\varepsilon_2^P = \sqrt{\frac{r_1(1 + r_2)}{r_2(1 + r_1)}}\ d\varepsilon_1^P \tag{11.25}$$

Equations (11.21) and (11.25) enable a transverse flow curve (2 - direction) to be predicted from the axial flow curve (1 - direction). The validity of the theory may be checked from the experimental measurement of the two flow curves.

It is also possible to predict compressive flow in the thickness (3 - direction) for a thin sheet given its tensile flow curve for the 1 - direction. Equations (11.4c) and (11.20a) provide the following stress relationships:

$$\bar{\sigma} = \sqrt{\frac{3(r_1 + r_2)}{2(r_1 + r_1 r_2 + r_2)}}\,\sigma_3 \tag{11.26}$$

$$\sigma_3 = \sqrt{\frac{r_2(1 + r_1)}{(r_1 + r_2)}}\,\sigma_1 \tag{11.27}$$

Taking the thickness strain $d\varepsilon_3^P$ to be positive, the lateral strains are found from the incompressibility condition:

$$d\varepsilon_1^P = -\frac{r_3\, d\varepsilon_3^P}{1 + r_3} = -\frac{r_2\, d\varepsilon_3^P}{r_1 + r_2} \tag{11.28a}$$

$$d\varepsilon_2^P = -\frac{d\varepsilon_3^P}{1 + r_3} = -\frac{r_1\, d\varepsilon_3^P}{r_1 + r_2} \tag{11.28b}$$

where eq(11.18c) defines r_3. Substituting eqs(11.28a,b) into eq(11.13) gives the equivalent thickness strain:

$$d\bar{\varepsilon}^P = \sqrt{\frac{2(r_1 + r_2 + r_1 r_2)}{3(r_1 + r_2)}}\, d\varepsilon_3^P \tag{11.29}$$

Equating (11.24a) and (11.29) relates incremental strains for the 1- and 3-directions:

$$d\varepsilon_3^P = \sqrt{\frac{(r_1 + r_2)}{r_2(1 + r_1)}}\, d\varepsilon_1^P \tag{11.30}$$

Where it is not practical to compress a thin sheet, the hydraulic bulge test (see p. 290) will provide the required equivalence between in-plane and thickness strains. To show this, put $\sigma_1 = \sigma_2$ and $\sigma_3 = 0$ in eq(11.4b). This gives a similar equivalent stress to eq(11.26) when σ_1 replaces σ_3. From eq(11.11a-c), the principal strain increments become

$$d\varepsilon_1^P = d\lambda G\sigma_1,\ d\varepsilon_2^P = d\lambda F\sigma_1 \ \text{ and } \ d\varepsilon_3^P = -(G + F)\,d\lambda\sigma_1$$

Expressing $d\varepsilon_1^P$ and $d\varepsilon_2^P$ in terms of r_1, r_2 and $d\varepsilon_3^P$ confirms strain increments (11.28a,b) and therefore eq(11.29).

11.3.2 Strain Hardening Hypothesis

Chakrabarty [13] defined an in-plane equivalent stress as

$$\bar{\sigma} = \sqrt{\frac{(G + H)\sigma_{11}^2 - 2H\sigma_{11}\sigma_{22} + (H + F)\sigma_{22}^2 + 2N\sigma_{12}^2}{G + H}} \tag{11.31}$$

Equation (11.31) gives $\bar{\sigma} = \sigma_{11}$ for uniaxial flow in the 1-direction, which differs from eq(11.19). In taking $f = \bar{\sigma}$, the flow rule (11.10) gave the equivalent strain increment as

$$d\bar{\varepsilon}^P = (G + H)\sqrt{\frac{(d\varepsilon_{11}^P)^2 + (d\varepsilon_{22}^P)^2 + (d\varepsilon_{33}^P)^2 + 2(d\varepsilon_{12}^P)^2}{2(G^2 + GH + H^2)}} \quad (11.32)$$

The following principal stress and strain relationships, between uniaxial flow in the 1- and 2- directions, are found from eqs(11.31) and (11.32):

$$\sigma_2 = \sqrt{\frac{r_2(1 + r_1)}{r_1(1 + r_2)}}\,\sigma_1 \quad (11.33)$$

$$d\varepsilon_2^P = \frac{(1 + r_2)}{(1 + r_1)}\sqrt{\frac{1 + r_1 + r_1^2}{1 + r_2 + r_2^2}}\,d\varepsilon_1^P \quad (11.34)$$

Equations (11.21) and (11.33) agree, but eqs(11.25) and (11.34) show that the work and strain hypotheses connect the two in-plane strains differently.

Thickness compression is assumed identical to in-plane, equi-biaxial tension. Hence the 1- and 3- flow relationships follow from substituting $\sigma_{11} = \sigma_{22} = \sigma_3$ and $\sigma_{12} = 0$ in eq(11.31):

$$\bar{\sigma} = \sigma_1 = \sqrt{\frac{(r_1 + r_2)}{r_2(1 + r_1)}}\,\sigma_3 \quad (11.35a)$$

$$\sigma_3 = \sqrt{\frac{r_2(1 + r_1)}{(r_1 + r_2)}}\,\sigma_1 \quad (11.35b)$$

Equation (11.35b) agrees with eq(11.27). The reduction to $d\bar{\varepsilon}^P$ is found from substituting eqs(11.28a,b) into eq(11.32), with $d\bar{\varepsilon}^P = d\varepsilon_1^P$:

$$d\bar{\varepsilon}^P = \frac{(1 + r_1)}{(r_1 + r_2)}\sqrt{\frac{r_1^2 + r_2^2 + r_1 r_2}{1 + r_1 + r_1^2}}\,d\varepsilon_3^P \quad (11.36a)$$

$$d\varepsilon_3^P = \frac{(r_1 + r_2)}{(1 + r_1)}\sqrt{\frac{1 + r_1 + r_1^2}{r_1^2 + r_2^2 + r_1 r_2}}\,d\varepsilon_1^P \quad (11.36b)$$

11.3.3 Comparisons With Experiment

In Fig. 11.1, a comparison is made between the longitudinal and transverse flow in as-rolled, 6.35 mm (¼″) thick, copper sheet. The specification C101 (see B.S. 2870) refers to an electrolytic, tough-pitch copper of 99.9% purity.

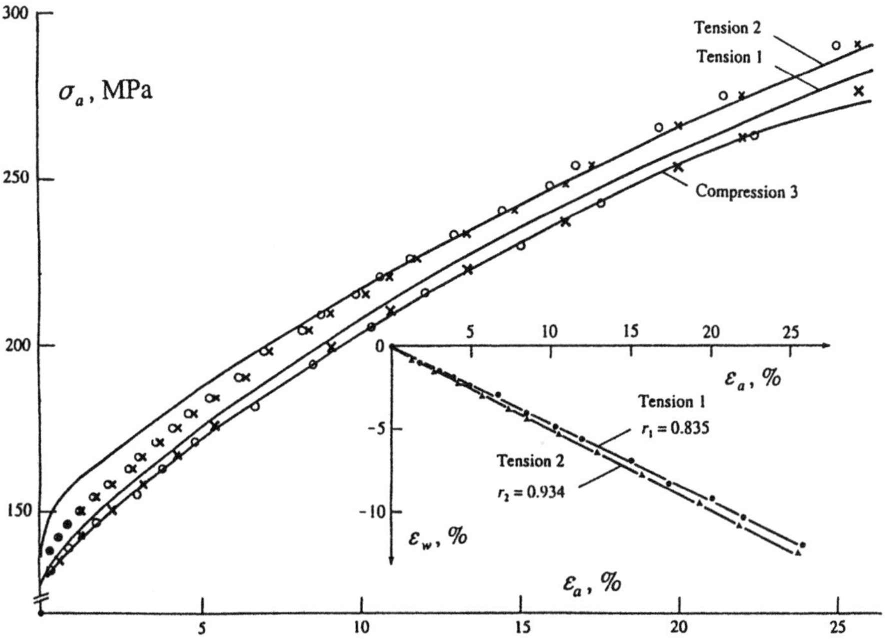

Figure 11.1 Longitudinal, transverse and thickness flow in copper sheet

(Key: 1, 2 and 3 are experimental flow curves, o work hardening predictions to curves 2 and 3 as calculated from curve 1; × strain hardening predictions to curves 2 and 3 calculated from curve 1. Inset figure shows axial versus width plastic strain paths for 1- and 2- directions, from which r_1 and r_2 are calculated from their gradients)

Waisted testpieces were machined from the 1- and 2- directions with their parallel gauge dimensions 55 mm long by 12.5 mm wide, according to B S 18. Incremental tension tests were performed in a tensile test machine. Permanent length and width changes were made with the load removed. The continuous lines 1 and 2 in Fig. 11.1 are the experimental true stress versus natural axial plastic strain curves for the 1- and 2- directions. Shown inset are plots between the width strain and axial plastic strains. Using $d\varepsilon_{ii}^{P} = 0$, enables the calculation of $r_1 = 0.835$ and $r_2 = 0.934$ from the slope to each path. Linear strain paths show that ratios between F, G and H remain constant in plastic flow. This fact enables the curve for the 2 - direction to be derived from the 1- direction curve in the manner outlined above. Work hardening eqs(11.21) and (11.25) give $\sigma_2 = 1.03\sigma_1$ and $\varepsilon_2^{P} = 0.971\varepsilon_1^{P}$. Strain hardening eqs(11.33) and (11.34) give $\sigma_2 = 1.03\sigma_1$ and $\varepsilon_2^{P} = 1.0012\varepsilon_1^{P}$. Applying these relations to curve 1 provides the calculated points from work and strain hardening. Their coincidence with curve 2 is good beyond 10% strain, where strain hardening appears slightly superior. The thickness flow curve 3 in Fig. 11.1 applies to lubricated uniaxial compression. To predict compressive flow (i.e. σ_3 versus ε_3^{P}) from tensile flow, substitute $r_1 = 0.835$ and $r_2 = 0.934$, in eqs(11.27) and (11.30). This gives $\sigma_3 = 0.984\sigma_1$ and $\varepsilon_3^{P} = 1.016\varepsilon_1^{P}$, for work hardening. Making similar substitutions into eqs(11.35b) and (11.36b) gives equally good predictions, $\sigma_3 = 0.984\sigma_1$ and $\varepsilon_3^{P} = 1.001\varepsilon_1^{P}$, from strain hardening.

Figure 11.2 confirms that each hypothesis provides an equivalent account of the spread in flow behaviour between the rolling and thickness directions. Points predicted from work and strain hardening are indistinguishable, these all being consistent with the slightly lower flow stresses observed under compression. With r_1 and r_2 values marginally less than unity, the sheet anisotropy was not severe.

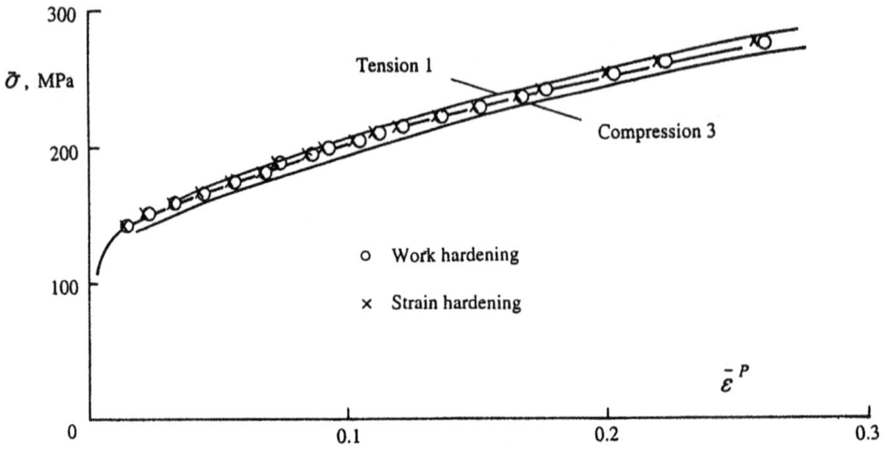

Figure 11.2 Equivalent tensile and compressive flow in copper sheet

11.4 Planar Isotropy

When shear stress is absent along material directions 1 and 2, eq(11.19) reduces to a principal, plane stress form:

$$\tfrac{2}{3}(F+G+H)\,\bar{\sigma}^2 = (F+H)\sigma_2^2 + (G+H)\sigma_1^2 - 2H\sigma_1\sigma_2 = 1 \qquad (11.37a)$$

Dividing throughout by F, eq(11.37a) beomes

$$r_1(1+r_2)\sigma_2^2 + r_2(1+r_1)\sigma_1^2 - 2r_1 r_2\sigma_1\sigma_2 = \text{constant} \qquad (11.37b)$$

Putting $\sigma_2 = 0$ and $\sigma_1 = \sigma_{1y}$ in eq(11.37b) gives $r_2(1+r_1)\sigma_{1y}^2 = \text{constant}$, where σ_{1y} is the yield strength for the 1- direction. Equation (11.37b) becomes

$$r_1(1+r_2)\sigma_2^2 + r_2(1+r_1)\sigma_1^2 - 2r_1 r_2\sigma_1\sigma_2 = r_2(1+r_1)\sigma_{1y}^2 \qquad (11.37c)$$

where from eqs(11.18a,b), $r_1 = H/G$, $r_2 = H/F$. An r_3 value is defined from a thickness compression test as either $r_3 = G/F$ or $r_3 = F/G$, depending upon how the width and thickness is arranged (see Fig. 11.3). It follows that one of the corresponding relationships holds:

$$r_1 r_3 = r_2 \quad \text{or} \quad r_2 r_3 = r_1 \qquad (11.38a,b)$$

Let us consider any metal stock production processes, i.e. extrusion, cogging, hot and cold rolling with interstage anneals. Assume that a given process results in a grain structure in one of three directional forms shown in Figs 11.3a-c. Each schematic block arrangement of grains indicates a preferred orientation. These impart a planar isotropic character to the material when the preferred direction aligns with each orthogonal axes 1, 2 and 3, as shown.

(a) *Flow behaviour independent of orientation in the 1 - 2 plane (Fig. 11.3a)*

When grains have been compressed uniformly in the 1-2 plane, $r_1 = r_2 = r_{12}$ and $\sigma_{1y} = \sigma_{2y}$. For uniaxial flow in the 3- direction, e.g. from a through-thickness compression test, both eqs(11.38a,b) show $r_3 = 1$. The yield function (11.37c) becomes

$$\sigma_1^2 + \sigma_2^2 - \frac{2r_{12}}{1 + r_{12}}\sigma_1\sigma_2 = \sigma_{1y}^2 \tag{11.39a}$$

$$x^2 + y^2 - \frac{2r_{12}}{1 + r_{12}}xy = 1 \tag{11.39b}$$

where $x = \sigma_1/\sigma_{1y}$ and $y = \sigma_2/\sigma_{2y}$. Equation (11.39b) describes a yield criterion for sheet metal with *normal anisotropy*. Here the thickness flow stress differs from the in-plane flow stress, the latter being invariant to its orientation (see also Section 11.8). We shall see that in practice an r variation often occurs in the 1-2 plane of rolled sheet metal. If the variation is slight, a single, weighted $\bar{r}$ value replaces r_{12} in eq(11.39b), i.e. $\bar{r} = \frac{1}{4}(r_0 + 2r_{45} + r_{90})$ in which 0°, 45° and 90° are orientations to the roll.

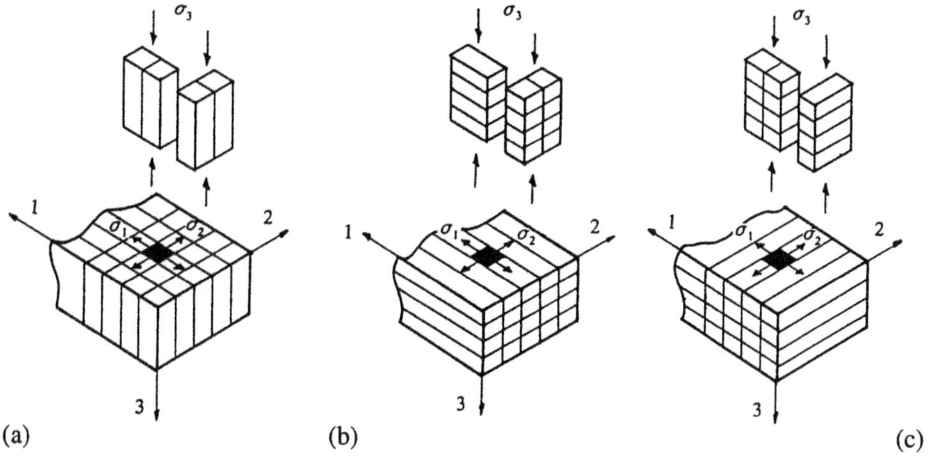

Figure 11.3 Planar isotropy in three forms

(b) *Flow behaviour independent of orientation in the 2 - 3 plane (Fig. 11.3b)*

Grains with a preferential orientation aligned with 1 are taken to be transversely isotropic in their 2-3 plane. This gives $r_2 = r_3 = r_{23}$ and $\sigma_{2y} = \sigma_{3y}$. Equation (11.38a) gives $r_1 = 1$ so that eq(11.37c) becomes

$$\sigma_1^2 + \frac{1 + r_{23}}{2r_{23}}\sigma_2^2 - \sigma_1\sigma_2 = \sigma_{1y}^2 \tag{11.40a}$$

Normalising in the following manner we find

$$\left(\frac{\sigma_1}{\sigma_{1y}}\right)^2 + \frac{(1 + r_{23})}{2r_{23}}\left(\frac{\sigma_2}{\sigma_{2y}}\right)^2\left(\frac{\sigma_{2y}}{\sigma_{1y}}\right)^2 - \left(\frac{\sigma_1}{\sigma_{1y}}\right)\left(\frac{\sigma_2}{\sigma_{2y}}\right)\left(\frac{\sigma_{2y}}{\sigma_{1y}}\right) = 1 \tag{11.40b}$$

where, from eq(11.37b),

$$\left(\frac{\sigma_{2y}}{\sigma_{1y}}\right)^2 = \frac{r_2(1 + r_1)}{r_1(1 + r_2)} = \frac{2r_{23}}{1 + r_{23}} \tag{11.40c}$$

Substituting eq(11.40c) into eq(11.40b) gives

$$x^2 + y^2 - \left(\frac{2r_{23}}{1 + r_{23}}\right)^{\frac{1}{2}} xy = 1 \tag{11.41a}$$

Equation (11.38b) also gives $r_1 = r_{23}^2$, which leads to an alternative yield criterion:

$$x^2 + y^2 - \left(\frac{4r_{23}^3}{(1 + r_{23})(1 + r_{23}^2)}\right)^{\frac{1}{2}} xy = 1 \tag{11.41b}$$

The two yield criteria (11.41a,b) together define an inner elastic bound of the yield locus.

(c) *Flow behaviour independent of orientation in the 1 - 3 plane*

Grains aligned with 2 are transversely isotropic in the 1-3 plane. This gives $\sigma_{1y} = \sigma_{3y}$ with $r_1 = r_3 = r_{13}$. Equations (11.38a,b) show that $r_2 = r_{13}^2$ or $r_2 = 1$. Two corresponding forms of normalised yield criteria follow from eqs(11.37c):

$$x^2 + y^2 - \left(\frac{4r_{13}^3}{(1 + r_{13})(1 + r_{13}^2)}\right)^{\frac{1}{2}} xy = 1 \tag{11.42a}$$

and

$$x^2 + y^2 - \left(\frac{2r_{13}}{1 + r_{13}}\right)^{\frac{1}{2}} xy = 1 \tag{11.42b}$$

For a given r_{13} value, eqs(11.42a,b) identify an inner elastic bound for the yield locus. In Fig. 11.4, a comparison is made between eqs(11.39b), (11.41a) and (11.42b).

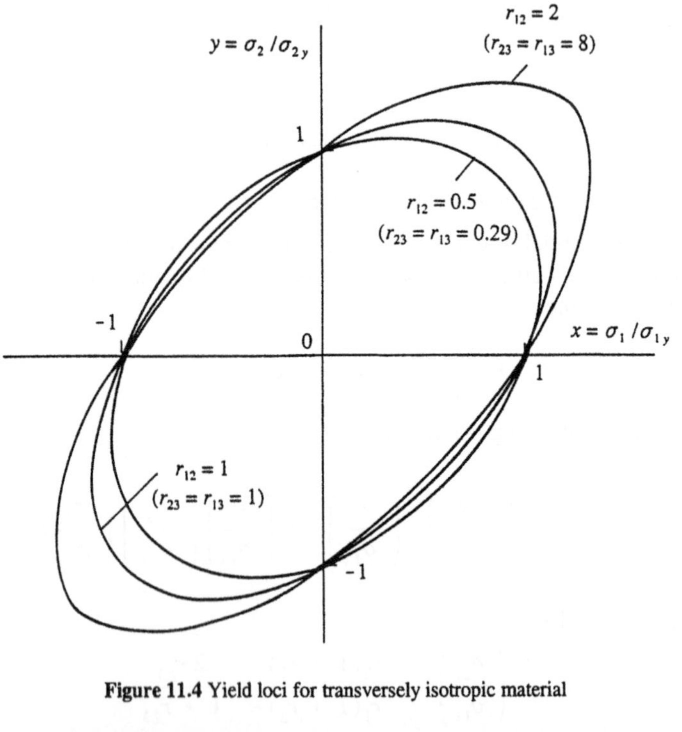

Figure 11.4 Yield loci for transversely isotropic material

Clearly, the loci from eqs(11.41a) and (11.42b) will coincide for similar r_{23} and r_{13} values. However, these loci can only coincide with the locus defined by eq(11.39b) for r values of unity in a fully isotropic material. Otherwise, it is seen that the r_{12}, r_{23} and r_{23} values in the plane of isotropy must differ if the three loci are to coincide. Within the first quadrant of Fig. 11.4, it is seen that the resistance to yielding increases as r exceeds unity. The yield stresses are more sensitive to changes in r_{12} than to changes in r_{13} or r_{23}.

11.5 Rolled Sheet Metals

Heavily rolled sheet metals are orthotropic in character, having their principal directions aligned with the rolling and transverse directions. Assume that the strength and, therefore, the flow behaviour (i.e. the stress-strain curve) varies with orientation in the plane of the sheet but that it shows transverse isotropy as in Fig. 11.3b. The r values for orthotropic plasticity are again used to characterise the dependence of flow upon in-plane orientation.

11.5.1 In-Plane r Variation

The off-axis tensile test will quantify in-plane sheet anisotropy, when the precise processing history is not known. Conducting a number of tensile tests upon testpieces machined at different orientations θ to the roll direction (see Fig. 11.5) reveals the variation in r with θ. Here the respective axes 1 and 2 refer to the roll and transverse directions in the sheet while axes 1′ and 2′ refer to the testpiece axial and width directions.

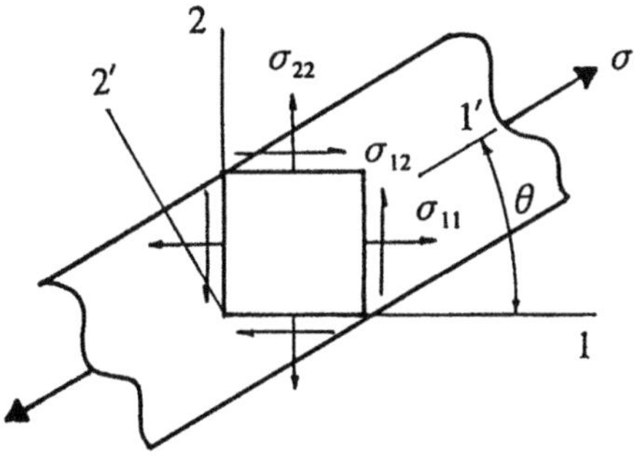

Figure 11.5 Off-axis tensile testpiece in the 1-2 plane

The plane strain transformation eqs(2.9a-c) give off-axis strain increments

$$d\varepsilon'_{11}{}^{P} = d\varepsilon_{11}{}^{P}\cos^2\theta + d\varepsilon_{22}{}^{P}\sin^2\theta + d\varepsilon_{12}{}^{P}\sin 2\theta \qquad (11.43a)$$

$$d\varepsilon'_{22}{}^{P} = d\varepsilon_{11}{}^{P}\sin^2\theta + d\varepsilon_{22}{}^{P}\cos^2\theta - d\varepsilon_{12}{}^{P}\sin 2\theta \qquad (11.43b)$$

$$d\varepsilon'_{12}{}^{P} = -\tfrac{1}{2}\,(d\varepsilon_{11}{}^{P} - d\varepsilon_{22}{}^{P})\sin 2\theta + d\varepsilon_{12}{}^{P}\cos 2\theta \qquad (11.43c)$$

The in-plane stress state σ_{11}, σ_{22} and σ_{12} is aligned with material directions, 1 and 2, as shown. The material strain increment eqs(11.11a-d) are reduced to

$$d\varepsilon_{11}{}^{P} = d\lambda\,[(H+G)\sigma_{11} - H\,\sigma_{22}] \qquad (11.44a)$$

$$d\varepsilon_{22}{}^{P} = d\lambda\,[(F+H)\sigma_{22} - H\,\sigma_{11}] \qquad (11.44b)$$

$$d\varepsilon_{33}{}^{P} = -\,d\lambda\,(G\sigma_{11} + F\sigma_{22}) \qquad (11.44c)$$

$$d\varepsilon_{12}{}^{P} = d\lambda\,N\,\sigma_{12} \qquad (11.44d)$$

A tensile stress σ, aligned with the testpiece 1′- axis, transforms to give stress components along the material axis as

$$\sigma_{11} = \sigma \cos^2\theta \qquad (11.45a)$$

$$\sigma_{22} = \sigma \sin^2\theta \qquad (11.45b)$$

$$\sigma_{12} = \tfrac{1}{2}\sigma \sin 2\theta \qquad (11.45c)$$

Substituting eqs(11.45a-c) and (11.44a,b,d) into eqs(11.43a-c), leads to

$$d\varepsilon'_{11}{}^{P} = d\lambda\sigma[(H+G)\cos^4\theta - 2H\sin^2\theta\cos^2\theta + (F+H)\sin^4\theta + (N/2)\sin^2 2\theta] \qquad (11.46a)$$

$$d\varepsilon'_{22}{}^{P} = d\lambda\sigma[(2H+F+G)\sin^2\theta\cos^2\theta - H(\sin^4\theta + \cos^4\theta) - (N/2)\sin^2 2\theta] \qquad (11.46b)$$

$$d\varepsilon'_{12}{}^{P} = d\lambda\sigma\{\tfrac{1}{2}\sin 2\theta[(F+H)\sin^2\theta - (H+G)\cos^2\theta] + (H/2)(\sin^2\theta\sin 2\theta - \cos^2\theta\sin 2\theta) + (N/2)\sin 2\theta\cos 2\theta\} \qquad (11.46c)$$

$$d\varepsilon_{33}{}^{P} = d\varepsilon_3{}^{P} = -\,d\lambda\sigma(G\cos^2\theta + F\sin^2\theta) \qquad (11.46d)$$

where $\varepsilon_3{}^{P}$ is the principal thickness strain. Dividing eqs(11.46b,d) gives the r - θ variation

$$r_\theta = \frac{d\varepsilon'^{P}_{22}}{d\varepsilon^{P}_3} = -\frac{(2H+F+G)\sin^2\theta\cos^2\theta - H(\sin^4\theta + \cos^4\theta) - (N/2)\sin^2 2\theta}{G\cos^2\theta + F\sin^2\theta} \qquad (11.47a)$$

The coefficients F, G, H and N in eq(11.47a) may be found from measured r_θ values for testpieces with orientations: $\theta = 0°$, 45° and 90°. Equation(11.47a) shows

$$r_1 = r_{0°} = H/G,\ \ r_{45°} = (2N - G - F)/[2(G+F)] \text{ and } r_2 = r_{90°} = H/F \qquad (11.47b)$$

which confirm eqs(11.18a,b). Equation (11.47a) is now re-written in these three r_θ-values:

$$r_\theta = \frac{2r_0 r_{90}(\sin^4\theta + \cos^4\theta) - [r_0 r_{90} - r_{45}(r_0 + r_{90})]\sin^2 2\theta}{2(r_0\sin^2\theta + r_{90}\cos^2\theta)} \qquad (11.47c)$$

11.5.2 Comparison with Experiment

Consider the off-axis tensile plasticity of four rolled sheet materials: C101 annealed copper, C638 copper-alloy (3% Al, 2% Si, 0.46% Co), each 6 mm thick, and two plated steels 1 mm and ¼ mm, used in the automotive and canning industries. Tensile testpieces were machined with parallel gauge dimensions: 12 mm wide × 62 mm long. End widths were enlarged to 25 mm for mounting in wedge grips. Axial, width and thickness changes were all measured following unloading in steps from the plastic range. An axial extensometer, with 50 mm gauge length, was mounted upon each testpiece. The greater ductilities of the copper and steel facilitated micrometer measurement of current width and thickness. A thickness extensometer and post-yield strain gauges were used to determine the smaller thickness changes and plastic width strain in copper-alloy.

(a) *Copper*
The linear, natural plastic strain paths found for annealed C101 copper sheet (Fig. 11.6) confirm that the anisotropy parameters remain constant during deformation. The linearity applies to 20% plastic axial strain, i.e. $\varepsilon'_{11}{}^{P} = \ln(l/l_o) = 0.2$.

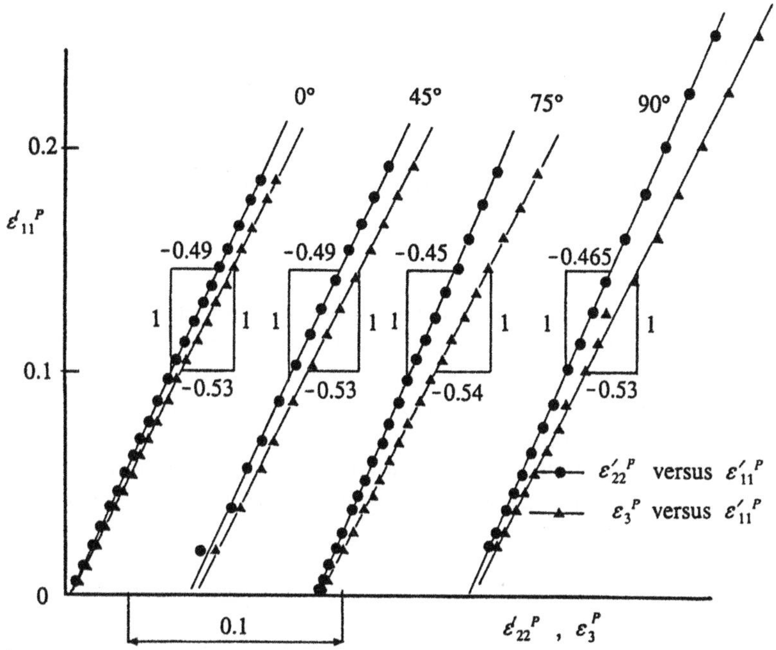

Figure 11.6 Plastic strain paths for off-axis tension in copper sheet

Note that engineering strain axes would not produce a linear plot. Natural width and thickness plastic strains are similarly defined: $\varepsilon'^{P}_{22} = \ln(w/w_o)$ and $\varepsilon'^{P}_{3} = \ln(t/t_o)$. Each gradient in Fig. 11.6 conforms to the incompressibility condition

$$\frac{d\varepsilon'^{P}_{22}}{d\varepsilon'^{P}_{11}} + \frac{d\varepsilon^{P}_{3}}{d\varepsilon'^{P}_{11}} = -1 \tag{11.48}$$

Applying eq(11.48) to the 0°, 45° and 90° strain gradients gives $r_0 = 0.924$, $r_{90} = 0.877$ and $r_{45} = 0.924$. These enable eq(11.47c) to predict the observed r variation for all intermediate orientations, as shown in upper curve of Fig. 11.7.

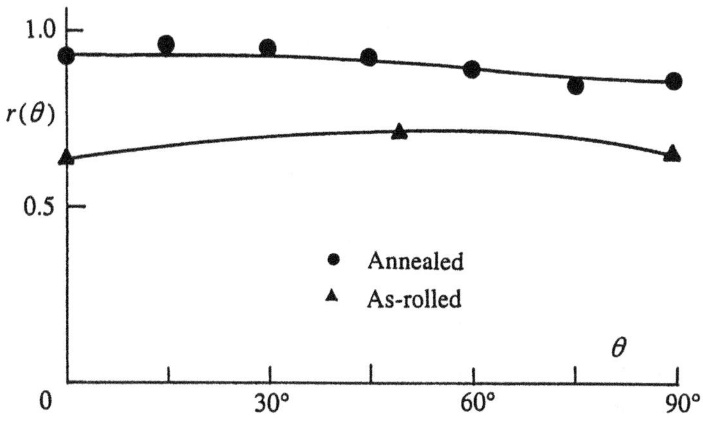

Figure 11.7 Variation in strain ratios with orientation for C101 copper

The agreement between theory and experiment is acceptable but this sheet does not depart greatly from an isotropic condition ($r_o = r_{45} = r_{90} = 1$). A second, as-rolled C101 copper sheet gave lower r values, $r_0 = 0.631$, $r_2 = 0.658$ and $r_{45} = 0.705$, which have again been used with eq(11.47c) to predict an r_θ variation, shown in Fig. 11.7. It would be appropriate to describe the near planar isotropic condition of each material with a single $\bar{r}$-value for a measure of their formability. For example, the $\bar{r}$ value is often correlated with the largest diameter blank that can be drawn through a die without tearing [17].

(b) *Copper-Alloy*

The strain paths for a copper-alloy sheet (C638) revealed a far greater degree of in-plane anisotropy with $r_0 = 0.6$, $r_{45} = 1.15$ and $r_{90} = 2.03$. The r value for the 90° orientation is more than 100% greater than the isotropic, unity value within a range where axial plastic strains did not exceed 1%. The orthotropy and limited ductility is a consequence of a 52% thickness reduction by cold-rolling in three passes. Figure 11.8 again shows acceptable predictions to the the r variation from eq(11.47c). Within r_θ there is a more sensitive measure of in-plane anisotropy than would be given by a width to axial strain ratio.

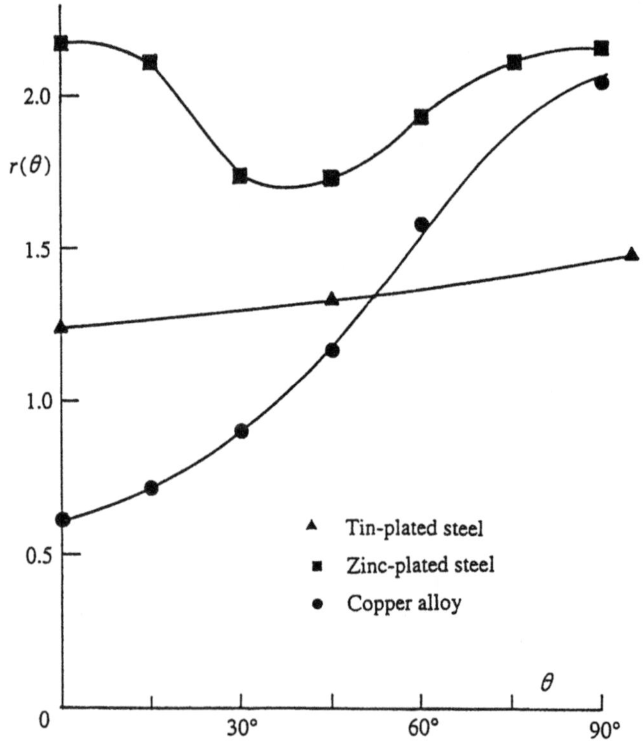

Figure 11.8 Variation in r value with orientation for copper alloy and coated steel sheets

(c) *Plated Steels*

In practice, consistently high r values are most desirable if a sheet is to resist thinning when being formed at moderately high strain, particularly under in-plane, biaxial tension [14-16]. Figure 11.8 shows that $r > 1$ for all θ in a 0.85 mm zinc-plated, automotive steel (zintec) and a 0.3 mm can body steel (tin plate). We have seen from eq(11.47c) that a homogenous, quadratic yield function expresses the full r_θ variation in terms of three known r values.

Consequently, this prediction may not fit intermediate r_θ values precisely. An improved fit is provided by a higher order yield function. For example, a homogenous, cubic yield function: $f(\sigma_{ij}) = C_{ijklmn}\,\sigma_{ij}\,\sigma_{kl}\,\sigma_{mn}$, gives r_θ as

$$r_\theta = \frac{d\varepsilon'^{P}_{22}}{d\varepsilon^{P}_{3}} = \frac{C_1\cos^6\theta + C_2\sin^6\theta + C_3\cos^4\theta\sin^2\theta + C_4\cos^2\theta\sin^4\theta}{C_5\cos^4\theta + C_6\sin^4\theta + \cos^2\sin^4\theta} \tag{11.49}$$

The six parameters, C_1, C_2 .... C_6, in eq(11.49) allow a good fit to most observed variations in r_θ [18]. Moreover, a cubic function can describe earing in deep drawn cups [19]. There appears to be no single yield function available to match all manifestations of plastic anisotropy across a range of engineering alloys. The choice of function (Table 3.3, p. 84) will reflect a match to more important influences with minimum mathematical complexity.

11.5.3 Equivalence Correlation

To establish an equivalence between the individual off-axis flow curves, the parameters *F*, *G*, *H* and *N* in eqs (11.4b) and (11.19) are required. For example, with the *r* values quoted above for the second C101 sheet, eqs(11.18) and (11.47b) give $G = 1.583H$, $F = 1.52H$ and $N = 3.74H$. Substituting into eq(11.13), leads to an equivalent plastic strain increment:

$$(d\bar{\varepsilon}^{P})^2 = 1.25(d\varepsilon_{11}^{P})^2 + 0.993\,(d\varepsilon_{11}^{P})(d\varepsilon_{22}^{P}) + 1.282\,(d\varepsilon_{22}^{P})^2 + 0.366\,(d\varepsilon_{12}^{P})^2 \tag{11.50}$$

where $d\varepsilon_{11}^{P}$, $d\varepsilon_{22}^{P}$ and $d\varepsilon_{12}^{P}$, refer to increments of plastic strain aligned with the axes of anisotropy (Fig. 11.5). In the off-axis test, the strain components, $d\varepsilon'^{P}_{11}$ and $d\varepsilon'^{P}_{22}$ were measured. An inverse strain transformation $\mathbf{E} = \mathbf{M}^T\mathbf{E}'\mathbf{M}$ (from eq 2.5a) gives:

$$d\varepsilon_{11}^{P} = d\varepsilon'^{P}_{11}\cos^2\theta + d\varepsilon'^{P}_{22}\sin^2\theta - 2d\varepsilon'^{P}_{12}\sin\theta\cos\theta \tag{11.51a}$$

$$d\varepsilon_{22}^{P} = d\varepsilon'^{P}_{11}\sin^2\theta + d\varepsilon'^{P}_{22}\cos^2\theta + 2d\varepsilon'^{P}_{12}\sin\theta\cos\theta \tag{11.51b}$$

$$d\varepsilon_{12}^{P} = d\varepsilon_{21}^{P} = \tfrac{1}{2}\,(d\varepsilon'^{P}_{11} - d\varepsilon'^{P}_{22})\sin 2\theta + d\varepsilon'^{P}_{12}(\cos^2\theta - \sin^2\theta) \tag{11.51c}$$

$$d\gamma_{12}^{P} = (d\varepsilon_{12}^{P} + d\varepsilon_{21}^{P}) = (d\varepsilon'^{P}_{11} - d\varepsilon'^{P}_{22})\sin 2\theta + d\gamma'^{P}_{12}\cos 2\theta \tag{11.51d}$$

Substituting eqs(11.51a,b,d) into eq(11.50) provides the equivalent plastic strain increment in terms of incremental plastic strain ratios:

$$(d\bar{\varepsilon}^{P})^2 = 1.25\left\{\left[\cos^2\theta + \left(\frac{d\varepsilon'^{P}_{22}}{d\varepsilon'^{P}_{11}}\right)\sin^2\theta - \frac{1}{2}\left(\frac{d\gamma'^{P}_{12}}{d\varepsilon'^{P}_{11}}\right)\sin 2\theta\right]^2\right.$$

$$+\ 0.993\left[\cos^2\theta + \left(\frac{d\varepsilon'^{P}_{22}}{d\varepsilon'^{P}_{11}}\right)\sin^2\theta - \frac{1}{2}\left(\frac{d\gamma'^{P}_{12}}{d\varepsilon'^{P}_{11}}\right)\sin 2\theta\right]$$

$$\times\left[\sin^2\theta + \left(\frac{d\varepsilon'^{P}_{22}}{d\varepsilon'^{P}_{11}}\right)\cos^2\theta + \frac{1}{2}\left(\frac{d\gamma'^{P}_{12}}{d\varepsilon'^{P}_{11}}\right)\sin 2\theta\right]$$

$$+\ 1.282\left[\sin^2\theta + \left(\frac{d\varepsilon'^{P}_{22}}{d\varepsilon'^{P}_{11}}\right)\cos^2\theta + \frac{1}{2}\left(\frac{d\gamma'^{P}_{12}}{d\varepsilon'^{P}_{11}}\right)\sin 2\theta\right]^2$$

$$+\ 0.366\left[1 - \left(\frac{d\varepsilon'^{P}_{22}}{d\varepsilon'^{P}_{11}}\right)\sin 2\theta + \left(\frac{d\gamma'^{P}_{12}}{d\varepsilon'^{P}_{11}}\right)\cos 2\theta\right]^2\Bigg\}(d\varepsilon'^{P}_{11})^2 \qquad (11.52)$$

We can substitute into eq(11.52), either the measured ratios $d\varepsilon'^{P}_{22}/d\varepsilon'^{P}_{11}$ and $d\varepsilon'^{P}_{12}/d\varepsilon'^{P}_{11}$, for each orientation, or, theoretical ratios derived from eqs(11.46a-d). Since no attempt was made to measure the small order of shear strain $d\gamma^{P}_{12}$, in the off-axis tests, it is necessary to employ the theoretical expression (11.46c) with eq(11.52). For example, with $\theta = 45°$ we find a strain ratio $d\gamma'^{P}_{12}/d\varepsilon'^{P}_{11} = -\ 0.0237$. Substituting this and $d\varepsilon'^{P}_{22}/d\varepsilon'^{P}_{11} = -\ 0.414$ in eq(11.52), gives $d\bar{\varepsilon}^{P} = 1.0232\, d\varepsilon'^{P}_{11}$. Hazlett, Robinson and Dorn [20] demonstrated the appearance of shear strain in off-axis tension of an anisotropic magnesium-aluminium alloy (AZ31X). Transverse lines, marked on the surface for all but the 0° and 90° testpieces, were rotated by applying perpendicular tension. Note that engineering shear strain $d\gamma'^{P}_{12}$ is the sum of the rotations of transverse and longitudinal fibres, but gripping prevents the latter from rotating. The equivalent stress corresponding to eq(11.52) is found from eq(11.19):

$$\bar{\sigma}^2 = 0.944\,\sigma_{11}^2 - 0.731\,\sigma_{11}\,\sigma_{22} + 0.922\,\sigma_{22}^2 + 2.735\,\sigma_{12}^2 \qquad (11.53a)$$

Substituting eqs(11.45a-c) into eq(11.53a) gives

$$\bar{\sigma} = \sqrt{[0.944\cos^4\theta - 0.731\cos^2\theta\sin^2\theta + 0.922\sin^4\theta + 0.684\sin^2 2\theta]}\,\sigma \qquad (11.53b)$$

where σ is the stress applied parallel to the orientation θ. For example, with $\theta = 45°$, eq(11.53b) gives $\bar{\sigma} = 0.982\sigma$. Figure 11.9 gives the plot between eqs(11.52) and (11.53b).

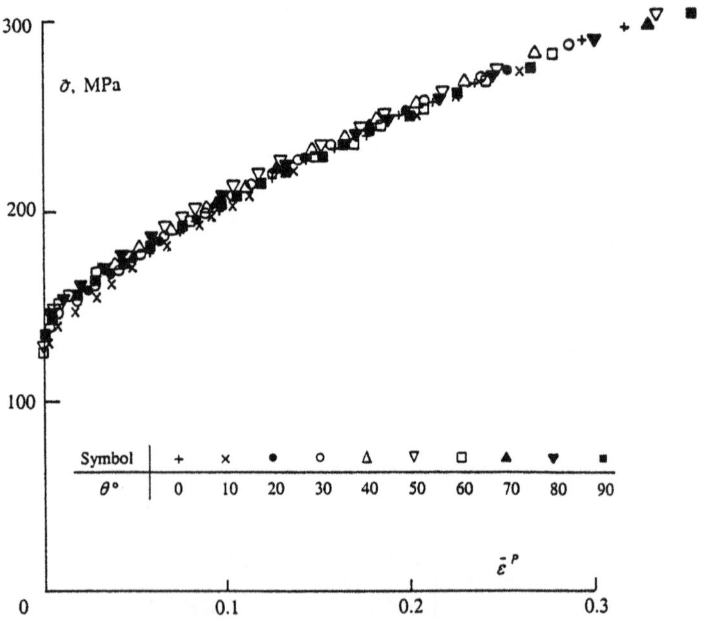

Figure 11.9 Equivalence correlation for off-axis tension in as-rolled copper

The excellent correlation confirms a description of anisotropy in the sheet's plane with three ratios: F/H, G/H and N/H. The ratios identify conveniently with the three r values supplied by tension tests in longitudinal, transverse and 45° directions.

11.6 Extruded Tubes

Let direction 1 be taken as the the tube's extruded direction so that the principal material directions 1, 2 and 3 become aligned with the tube's axial, circumferential and radial directions respectively. Consider a thin-walled tube subjected to various combined loadings that produce plane stress states aligned with these material directions.

11.6.1 Biaxial Tension

The combination of axial force and internal pressure gives axial and hoop (i.e. principal) stresses, σ_1 and σ_2 respectively, in a ratio $R = \sigma_1/\sigma_2$. Provided the wall thickness is less than one tenth of the mean diameter, radial stress σ_3 in the wall is small enough to be ignored.

$$\frac{d\varepsilon_1^p}{d\varepsilon_2^p} = \frac{2\sigma_1 - \sigma_2}{2\sigma_2 - \sigma_1}$$

von Mises

Symbol	Material	ref.
×	En 24	21
□	s. steel	22
■	s. steel	22
▲	Pb alloy	23
●	Zircaloy	24

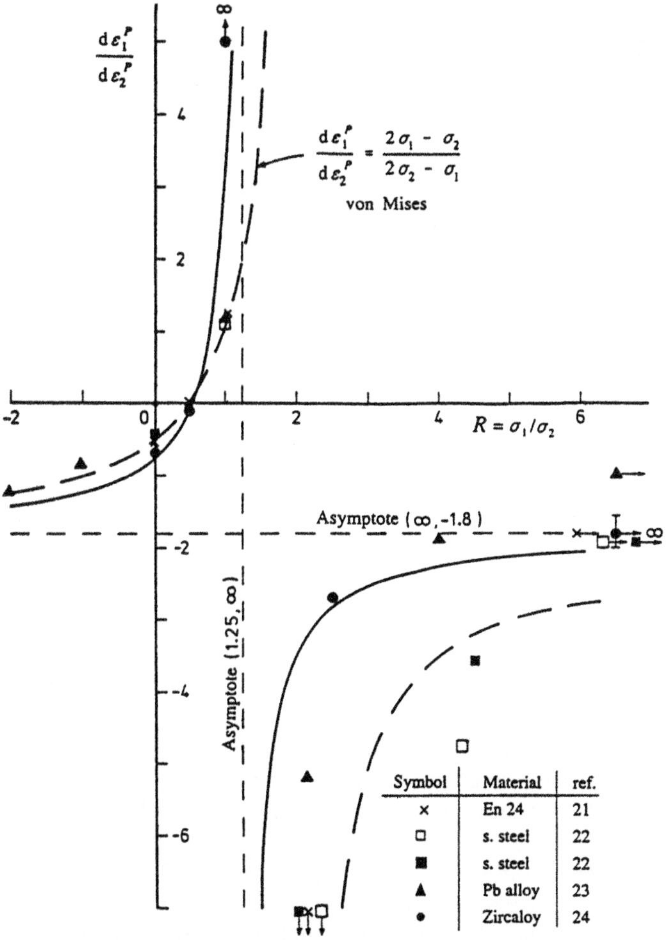

Figure 11.10 Dependence of strain increment ratio upon biaxial stress ratio

Making appropriate reductions to eqs(11.11a,b) gives the ratio between the axial and circumferential (hoop) plastic strain increments:

$$\frac{\mathrm{d}\varepsilon_1^P}{\mathrm{d}\varepsilon_2^P} = \frac{H(R-1)+GR}{F-H(R-1)} = \frac{(H/G)(R-1)+R}{(F/H)(H/G)-(H/G)(R-1)}$$

$$= \frac{r_1(R-1)+R}{(r_1/r_2)-r_1(R-1)} = \frac{r_1 r_2(R-1)+Rr_2}{r_1-r_1 r_2(R-1)} \quad (11.54)$$

where r_1 and r_2 in eqs(11.18a,b) are found from separate axial and circumferential tension tests. Figure 11.10 shows that eq(11.54) accounts for the manner in which the strain increment ratio depends upon the stress ratio for five materials [21 - 24]. With the exception of En24, which conforms to a von Mises condition, it can be seen how materials with r values different from unity in eq(11.54) depart from this condition. For example, the asymptotes given for a zirconium alloy (Zircaloy) correspond approximately to $r_1 = 1.3$ and $r_2 = 0.7$.

11.6.2 Tension-Torsion

(a) *Strain ratios*

Combining an axial force with torsion produces two independent stress components: σ_{11} and σ_{12} respectively. Dividing eqs(11.11a,b,d) gives the following ratios between increments of shear, axial, diametral and thickness plastic strains, $\mathrm{d}\gamma_{12}{}^P$, $\mathrm{d}\varepsilon_{11}{}^P$, $\mathrm{d}\varepsilon_{22}{}^P$ and $\mathrm{d}\varepsilon_{33}{}^P$, respectively:

$$\frac{\mathrm{d}\gamma_{12}^P}{\mathrm{d}\varepsilon_{11}^P} = \frac{2\mathrm{d}\varepsilon_{12}^P}{\mathrm{d}\varepsilon_{11}^P} = \frac{2N\sigma_{12}}{(H+G)\sigma_{11}} = \frac{(r_1+r_2)(1+2r_{45})\sigma_{12}}{r_2(1+r_1)\sigma_{11}} \quad (11.55a)$$

$$\frac{\mathrm{d}\varepsilon_{22}^P}{\mathrm{d}\varepsilon_{11}^P} = -\frac{H}{(H+G)} = -\frac{r_1}{(1+r_1)} \quad (11.55b)$$

$$\frac{\mathrm{d}\varepsilon_{33}^P}{\mathrm{d}\varepsilon_{11}^P} = -\frac{G}{(H+G)} = -\frac{1}{(1+r_1)} \quad (11.55c)$$

in which eq(11.47b) has been applied. Either eqs(11.55b,c) provides r_1 when diametral or thickness changes are measured with twist and extension. A circumferential tension provides r_2 while r_{45} is calculated from eq(11.55a) for a given stress ratio $R = \sigma_{12}/\sigma_{11}$. Alternatively, r values can be found from miniature tensile testpieces machined from the tube wall. Equation (11.55a) predicts a strain increment ratio linearly dependent upon R. Figure 11.11 presents evidence of this behaviour for 5 materials [3, 21, 25, 26], previously considered with isotropic potentials (see Fig. 10.6). When anisotropy is present, eq(11.55a) shows how r values explain a gradient different from 3 (by Levy-Mises). Test temperatures 300°C and 400°C apply to the 0.17% carbon steels (m.s.). On steel conforms to Mises condition at the higher temperature. For the other materials at ambient temperature, it is seen how having r values different from unity explains their departure from a von Mises condition.

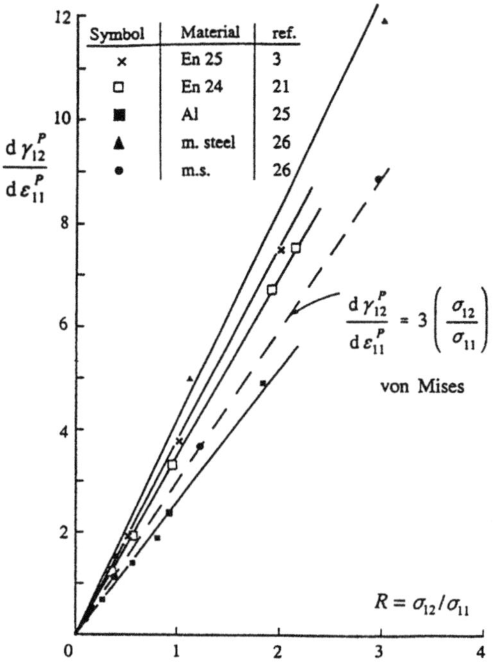

Figure 11.11 Dependence of strain increment ratio upon combined tension-torsion stress ratio

(b) *Equivalence Correlation*

The work hypothesis gives an equivalent stress in eq(11.19) as

$$\frac{\bar{\sigma}}{\sigma_{11}} = \frac{3}{2}\sqrt{\frac{r_2(1 + r_1) + (r_1 + r_2)(1 + 2r_{45})R^2}{r_1 + r_1 r_2 + r_2}} \tag{11.56a}$$

Setting $g = d\gamma_{12}^P / d\varepsilon_{11}^P$ in eq(11.13), the equivalent plastic strain appears as

$$\frac{d\bar{\varepsilon}^P}{d\varepsilon_{11}^P} = \frac{2}{3}\sqrt{\frac{(r_1 + r_1 r_2 + r_2)[1 + r_2(1 + r_1)g^2]}{r_2(1 + r_1)(1 + 2r_{45})(r_1 + r_2)}} \tag{11.56b}$$

In the hypothesis of strain hardening, eqs(11.31) and (11.32) correspondingly reduce to

$$\frac{\bar{\sigma}}{\sigma_{11}} = \sqrt{1 + \frac{(r_1 + r_2)(1 + 2r_{45})g^2}{r_2(1 + r_1)}} \tag{11.57a}$$

$$\frac{d\bar{\varepsilon}^P}{d\varepsilon_{11}^P} = \sqrt{1 + \frac{(1 + r_1)^2 g^2}{4(1 + r_1 + r_1^2)}} \tag{11.57b}$$

where eqs(11.55b,c) have been used. To compare correlations of $\bar{\sigma}$ versus $\int d\bar{\varepsilon}^P$ from each hypothesis, four radial loading tests with $R = \sigma_{12}/\sigma_{11} = 0.39$, 0.92, 1.77 and 7.1, were conducted on tubes machined from as-received, 99.8% pure aluminium bar. The testpiece diameters were 25.4 mm i.d. and 28.4 mm o.d. Of the 75 mm parallel length, 50 mm defined a gauge length for attachment of an extensometer. The latter housed displacement transducers for the measurement of extension and angular twist simultaneously. Reductions to the outer diameter were measured with a micrometer. Force and torque were applied to the testpiece through threads and a square register machined into each of its enlarged ends [26]. Elastic strains were removed from total strains using elastic constants determined from initial elastic loading, i.e. E = 68 GPa, G = 25.5 GPa and $\nu = E/(2G) - 1 = 0.33$. The resulting inelastic strains enabled the strain paths to be plotted for plasticity under incremental loading and, thereafter, creep under a constant load, as shown in Fig. 11.12.

(a)

(b)

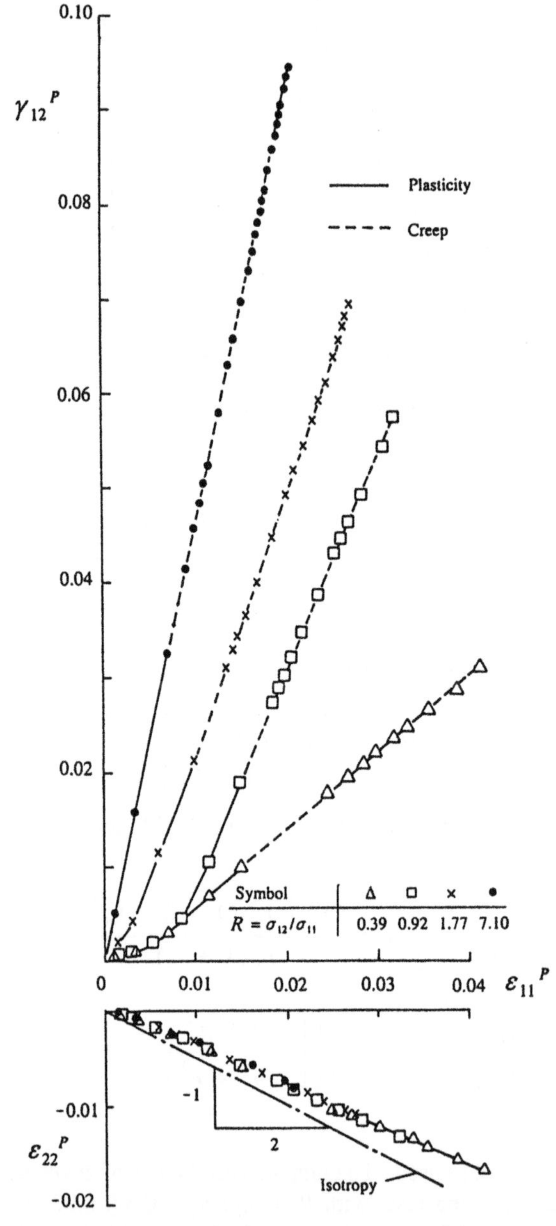

Figure 11.12 Anisotropic plastic strain paths under combined tension-torsion

The axes in each figure correspond to the inelastic strains

$$\gamma_{12}^{P} = \gamma_{12} - \frac{\sigma_{12}}{G}$$

$$\varepsilon_{11}^{P} = \varepsilon_{11} - \frac{\sigma_{11}}{E}$$

$$\varepsilon_{22}^{P} = \varepsilon_{22} - \frac{\nu\sigma_{11}}{E}$$

Note that the paths soon become linear with increasing plastic strain. An account of the initial rotations, which are characteristic of a material with pre-strain history, will be given in the following section. The ratios between the incremental strains in eqs (11.55a,b) are identified with the slopes of linear paths in Figs 11.12a,b. The r values were also found from miniature tensile testpieces, machined from the bar in longitudinal and transverse directions. Equations (11.56a,b) give the equivalent stress and equivalent plastic strain from the work hardening hypothesis:

$$\bar{\sigma} = 0.907\,(1 + 2.05R^2)^{1/2}\,\sigma_{11} \tag{11.58a}$$

$$\bar{\varepsilon}^{P} = 1.103\,[\,1 + 0.488\,g^2\,]^{1/2}\,\varepsilon_{11}^{P} \tag{11.58b}$$

in which $R = \sigma_{12}/\sigma_{11}$ is the stress ratio and $g = \mathrm{d}\gamma_{12}^{P}/\mathrm{d}\varepsilon_{11}^{P}$ is the plastic strain increment ratio, i.e. the constant slope to a linear plastic strain path. In the strain-hardening hypothesis, equivalence is expressed through eqs(11.57a,b):

$$\bar{\sigma} = (1 + 2.05R^2)^{1/2}\,\sigma_{11} \tag{11.59a}$$

$$\bar{\varepsilon}^{P} = [\,1 + 0.33\,g^2\,]^{1/2}\,\varepsilon_{11}^{P} \tag{11.59b}$$

Substituting the observed values of g for each R enables equivalence plots in Figs. 11.13a,b to be constructed for each hypothesis.

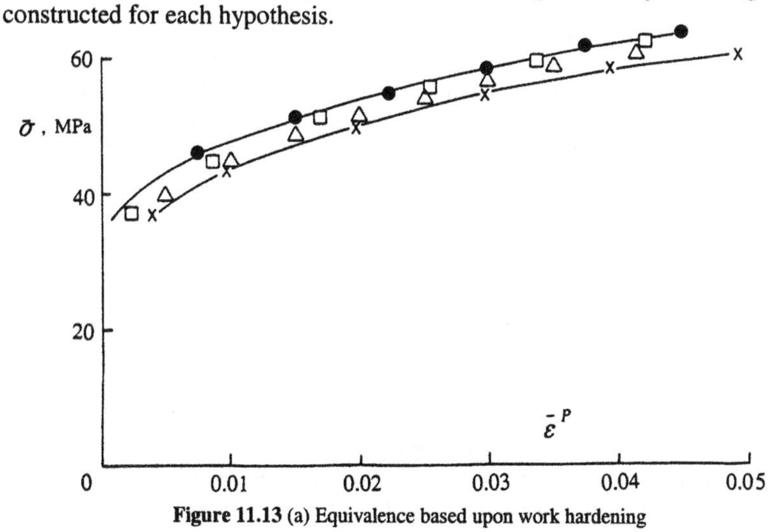

Figure 11.13 (a) Equivalence based upon work hardening

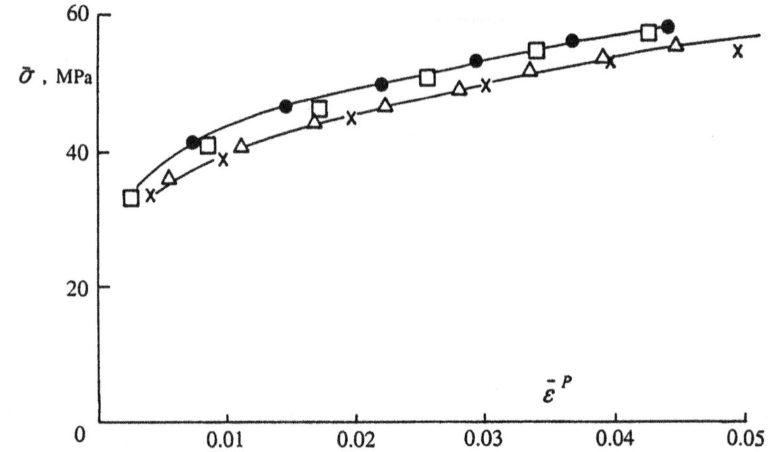

Figure 11.13 (b) Equivalence based upon strain hardening

Clearly, these experiments are unable to reveal which is the better hypothesis. Narrow scatter bands result from the observed strain paths and confirm that orthotropic plasticity theory may be employed equally well with either hypothesis.

11.7 Non-Linear Strain Paths

The foregoing equivalence correlations require linear plastic strain paths. Consequently, the constant ratios between the anisotropy parameters *F*, *G*, *H*, *L*, *M* and *N* describe an unchanging anisotropy. For this condition, Hill [27] showed that the parameters decrease as yield stresses, in eq(11.2), increase in strict proportion to the amount of cold work. In Finnie's alternative formulation (11.5a) *F*, *G*, *H*, etc, are dimensionless constants. The literature reports linear plastic strain paths for orthotropic sheets of aluminium [28], aluminium alloy [1, 20, 25] and mild steel [20, 28]. In contrast, curved plastic strain paths were found for alloys of magnesium [20, 29], titanium [30 - 32] and aluminium 3S [33]. Here, the gradient, i.e. the incremental plastic strain ratio, changes with increasing plastic strain. An example of non-linear behaviour of aluminium alloy strip (HE 30TF) under tension is given in Figs 11.14a,b. The stress-strain plots in Fig. 11.14a apply to a testpiece with gauge dimensions: 12.7 mm wide × 50.8 mm long, machined from 9.5 mm thick, rolled sheet. Testpiece axes were aligned with the roll direction, 1 and the transverse direction, 2. Total strains were measured with bonded, post-yield, strain gauges which also provided elastic constants $E = 78.3$ GPa and $\nu = 0.30$ for the 1- direction, $E = 77.4$ GPa and $\nu = 0.336$ for the 2- direction. Corresponding E and ν were used in Fig. 11.14b to remove elastic strains from the total axial and lateral strains:

$$\varepsilon'^{P}_{11} = \varepsilon'_{11} - \sigma / E$$
$$\varepsilon'^{P}_{22} = \varepsilon'_{22} - \nu\sigma / E$$

where $1'$ and $2'$ denote the testpiece length and width co-ordinates, i.e. in Fig. 11.5, $1'$coincides with 1 and $1'$ coincides with 2 in each testpiece. The quadratic theory supplies an incremental strain ratio $\mathrm{d}\varepsilon'^{P}_{22} / \mathrm{d}\varepsilon'^{P}_{11}$ from eqs(11.46a,b) with $\theta = 0°$ and $90°$. Using the normality rule and preservation of volume, it follows that the quadratic theory can match a non-linear plastic strain path when *F*, *G*, *H*, *L*, *M* and *N*, become functions of plastic strain.

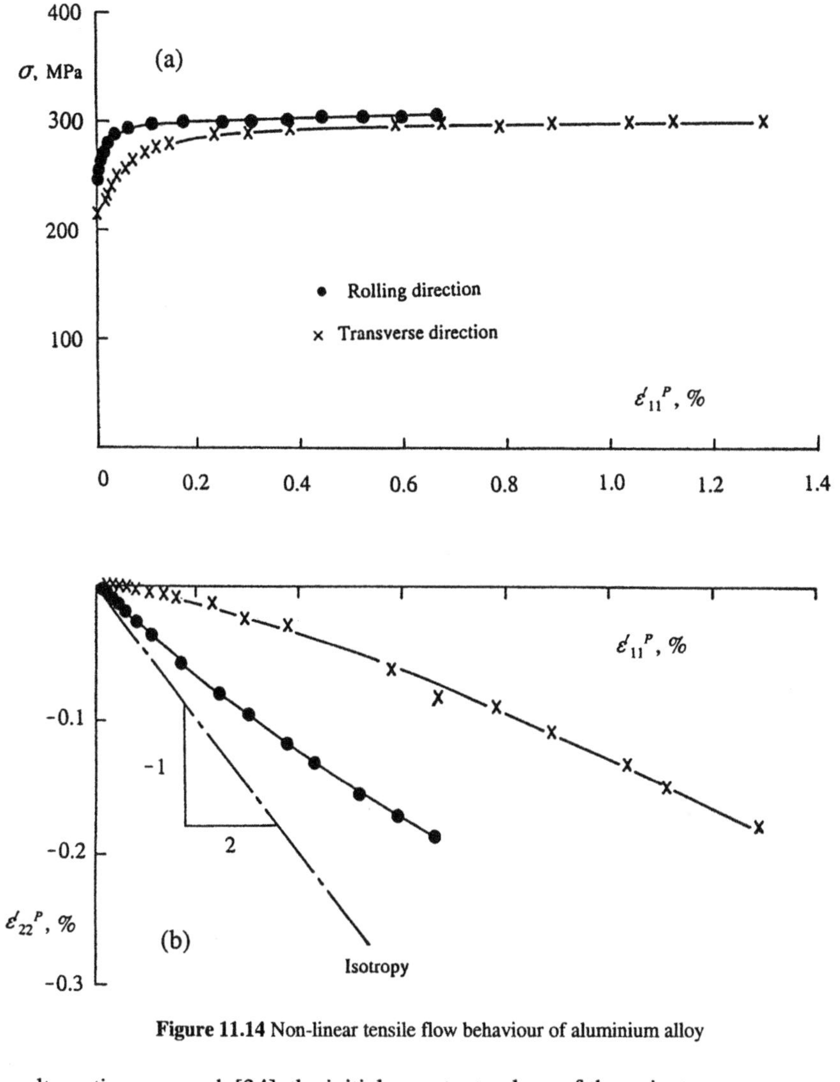

Figure 11.14 Non-linear tensile flow behaviour of aluminium alloy

In an alternative approach [34], the initial, constant values of the anisotropy parameters are retained within a combined linear-quadratic theory (3.30). A reduction to plane stress gave

$$f(\sigma_{ij}') = H_{ij}\,\sigma_{ij}' + H_{ijkl}\,\sigma_{ij}'\sigma_{kl}' \tag{11.60a}$$

$$= L_1\sigma_{11} + L_2\sigma_{22} + L_3\sigma_{12} + Q_1\sigma_{11}^{\,2} + Q_2\sigma_{22}^{\,2} + Q_3\sigma_{11}\,\sigma_{22} + Q_4\sigma_{12}^{\,2} + Q_5\sigma_{11}\,\sigma_{12} + Q_6\sigma_{22}\sigma_{12} \tag{11.60b}$$

when, from eq(10.4b), the incremental stress-strain relations become

$$\mathrm{d}\varepsilon_{11}^{\,P} = \mathrm{d}\lambda\,(\partial f/\partial\sigma_{11}) = \mathrm{d}\lambda\,(L_1 + 2Q_1\,\sigma_{11} + Q_3\,\sigma_{22} + Q_5\,\sigma_{12}\,) \tag{11.61a}$$

$$\mathrm{d}\varepsilon_{22}^{\,P} = \mathrm{d}\lambda\,(\partial f/\partial\sigma_{22}) = \mathrm{d}\lambda\,(L_2 + 2Q_2\,\sigma_{22} + Q_3\,\sigma_{11} + Q_6\,\sigma_{12}\,) \tag{11.61b}$$

$$\mathrm{d}\varepsilon_{12}^{\,P} = \mathrm{d}\lambda\,(\partial f/\partial\sigma_{12}) = \mathrm{d}\lambda\,(L_3 + 2Q_4\,\sigma_{12} + Q_5\,\sigma_{11} + Q_6\,\sigma_{22}\,) \tag{11.61c}$$

$$\mathrm{d}\varepsilon_3^{\,P} = -\,(\mathrm{d}\varepsilon_{11}^{\,P} + \mathrm{d}\varepsilon_{22}^{\,P})$$

$$= -\,\mathrm{d}\lambda[(L_1 + L_2) + (2Q_1 + Q_3)\sigma_{11} + (Q_3 + 2Q_2)\sigma_{22} + (Q_5 + Q_6)\sigma_{12} \tag{11.61d}$$

Applying the stress and strain transformations (11.45a-c) and (11.43a-c) to an off-axis testpiece, the axial and transverse strains become

$$\begin{aligned} d\varepsilon'^{P}_{11} &= d\lambda\{[L_1 + 2Q_1(\sigma\cos^2\theta) + Q_3(\sigma\sin^2\theta) + Q_5(½\sigma\sin 2\theta)]\cos^2\theta \\ &+ [L_2 + 2Q_2(\sigma\sin^2\theta) + Q_3(\sigma\cos^2\theta) + Q_6(½\sin 2\theta)]\sin^2\theta \\ &+ [L_3 + 2Q_4(½\sigma\sin 2\theta) + Q_5(\sigma\cos^2\theta) + Q_6(\sigma\sin^2\theta)]\sin 2\theta\} \end{aligned} \quad (11.62a)$$

$$\begin{aligned} d\varepsilon'^{P}_{22} &= d\lambda\{[L_1 + 2Q_1(\sigma\cos^2\theta) + Q_3(\sigma\sin^2\theta) + Q_5(½\sigma\sin 2\theta)]\sin^2\theta \\ &+ [L_2 + 2Q_2(\sigma\sin^2\theta) + Q_3(\sigma\cos^2\theta) + Q_6(½\sin 2\theta)]\cos^2\theta \\ &- [L_3 + 2Q_4(½\,\sigma\sin 2\theta) + Q_5(\sigma\cos^2\theta) + Q_6(\sigma\sin^2\theta)]\sin 2\theta\} \end{aligned} \quad (11.62b)$$

$$\begin{aligned} d\varepsilon'^{P}_{12} &= d\lambda\{-[L_1 + 2Q_1(\sigma\cos^2\theta) + Q_3(\sigma\sin^2\theta) + Q_5(½\sigma\sin 2\theta)]½\sin 2\theta \\ &+ [L_2 + 2Q_2(\sigma\sin^2\theta) + Q_3(\sigma\cos^2\theta) + Q_6(½\sin 2\theta)]½\sin 2\theta \\ &+ [L_3 + 2Q_4(½\sigma\sin 2\theta) + Q_5(\sigma\cos^2\theta) + Q_6(\sigma\sin^2\theta)]\cos 2\theta\} \end{aligned} \quad (11.62c)$$

$$d\varepsilon_3^P = -d\lambda[(L_1 + L_2) + (2Q_1 + Q_3)\sigma\cos^2\theta + (Q_3 + 2Q_2)\sigma\sin^2\theta + (Q_5 + Q_6)½\sin 2\theta] \quad (11.62d)$$

Taking the ratios between the strain increments shows that the applied stress σ is retained within the resulting expression. For example, putting $\theta = 0°$ and $90°$ in eqs(11.62a-d), gives two plastic strain increment ratios for each orientation:

$$r_{0°} = \frac{d\varepsilon'^{P}_{22}}{d\varepsilon_3^P} = \frac{-(L_2 + Q_3\sigma)}{(L_1 + L_2) + (2Q_1 + Q_3)\sigma} \quad (11.63a)$$

$$\frac{d\varepsilon'^{P}_{22}}{d\varepsilon'^{P}_{11}} = \frac{L_2 + Q_3\sigma}{L_1 + 2Q_1\sigma} \quad (11.63b)$$

$$r_{90°} = \frac{d\varepsilon'^{P}_{22}}{d\varepsilon_3^P} = \frac{-(L_1 + Q_3\sigma)}{(L_1 + L_2) + (2Q_2 + Q_3)\sigma} \quad (11.63c)$$

$$\frac{d\varepsilon'^{P}_{22}}{d\varepsilon'^{P}_{11}} = \frac{L_1 + Q_3\sigma}{L_2 + 2Q_2\sigma} \quad (11.63d)$$

It follows that when eqs(11.63b,d) are applied to Fig. 11.14b, with L_1, L_2, Q_1, Q_2 and Q_3 assumed constant, the applied stress σ governs the ratio. Thus, by taking $\sigma = \sigma(\varepsilon'^{P}_{11})$ from Fig. 11.14a, enables each strain path in Fig. 11.14b to be found by numerical integration.

The yield function (11.60b), assumes different tensile and compressive yield stresses, as seen in Fig. 3.12. Note that in using an associated flow rule, normality is implied between the plastic strain increment vector and the yield surface. The stress-dependent, plastic strain increment ratios, given in eqs(11.63a-d), suggest that the curvature of the yield surface is continuously changed in the vicinity of an advancing stress. Clearly, this contravenes the assumption that the yield surface expands uniformly within an isotropic hardening rule.

11.8 Alternative Yield Criteria

There have been inconsistencies reported between the predicted and observed equi-biaxial yield stress in materials with r-values less than unity [35]. To overcome this, further generalisations have been made to Hill's criterion, including modifications made by that author [36] (see also Table 3.3). The reduction of these criteria to a plane, principal stress state is the most common form and these will now be examined.

If we identify Hill's parameters F, G and H with the r values in the usual way and omit shear stress terms, the principal stress form of Hill's yield function (11.2) becomes

$$r_1(\sigma_2 - \sigma_3)^2 + r_2(\sigma_1 - \sigma_3)^2 + r_1 r_2(\sigma_1 - \sigma_2)^2 = r_2(1 + r_1)\sigma_{1y}^2 \tag{11.64}$$

where σ_{1y} is the yield stress for the material 1- direction. Here the principal stresses are aligned with the material axes. In the modification proposed by Hosford [35], eq(11.64) was written as

$$r_1 |\sigma_2 - \sigma_3|^a + r_2 |\sigma_1 - \sigma_3|^a + r_1 r_2 |\sigma_1 - \sigma_2|^a = r_2(1 + r_1)\sigma_{1y}^a \tag{11.65}$$

where $a > 2$ is an integer and each modulus $|\ |$ requires a positive stress difference. Values of a were suggested lying in a range from 6 to 8 for pencil glide and from 8 to 10 for bcc materials in general.

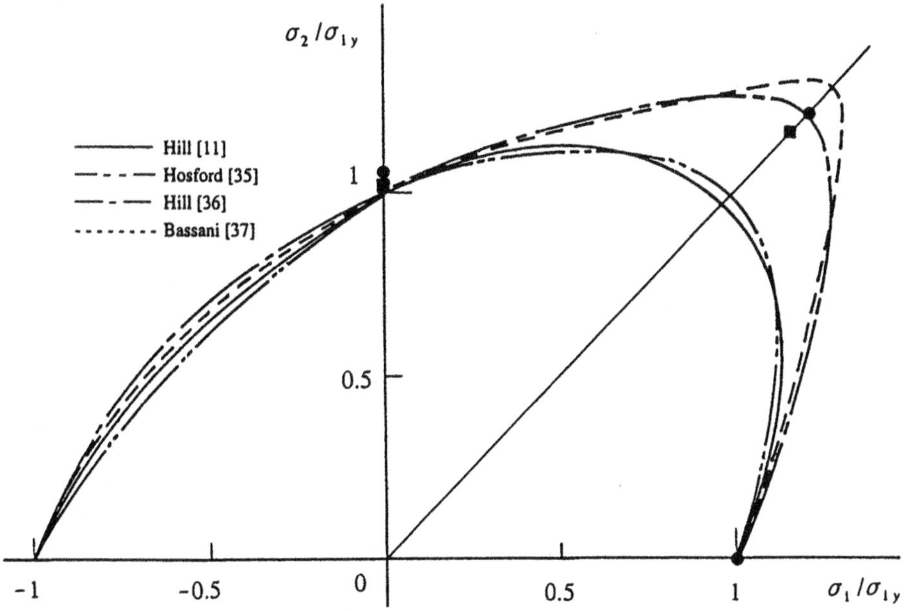

Figure 11.15 Comparison between planar-isotropic yield loci

For a planar isotropic yield criterion we set $\sigma_3 = 0$ and $r = r_1 = r_2$ in eq(11.65), to give

$$\sigma_1^a + \sigma_2^a + r|\sigma_1 - \sigma_2|^a = (1 + r)\sigma_{1y}^a \tag{11.66a}$$

If $a = 2$ in eq(11.66a), the corresponding quadratic form becomes

$$(\sigma_1 + \sigma_2)^2 + (1 + 2r)(\sigma_1 - \sigma_2)^2 = 2(1 + r)\sigma_{1y}^2 \tag{11.66b}$$

which can be written as

$$\sigma_1^2 + \sigma_2^2 - \frac{2r}{1+r}\sigma_1\sigma_2 = \sigma_{1y}^2 \qquad (11.66c)$$

This is the planar isotropic form previously identified in eq(11.39a). In Hill's 1979 modification to eq(11.66b), a non-quadratic potential was proposed [36]:

$$|\sigma_1 + \sigma_2|^m + (1 + 2r)\,|\sigma_1 - \sigma_2|^m = 2(1 + r)\sigma_{1y}{}^m \qquad (11.66c)$$

where $2 \geq m \geq 1$. Bassani [37] employed different exponents in a non-quadratic yield criterion:

$$|\sigma_1 + \sigma_2|^q + (q/p)(1 + 2r)\sigma_{1y}{}^{q-p}\,|\sigma_1 - \sigma_2|^p = [1 + (q/p)(1 + 2r)]\,\sigma_{1y}{}^q \qquad (11.66d)$$

where $p \geq 1$, $q \geq 1$ are integer exponents and $r \equiv \bar{r}$. A comparison between eqs(11.66a-d) and flow data for two ¼ mm can-end steels is given in Fig. 11.15. Taking $m = 1.5$, $a = 6$, $p = 1$ and $q = 2$ in the respective criteria demonstrates the divergence between them for a material with $\bar{r} = 0.874$. Here, the equi-biaxial yield stresses, found from bulge forming, exceed the tensile yield stress in each $\bar{r} < 1$ material [38]. The loci are sensitive to their stress exponents, which control the curvature (and hence the strain path) in a region of biaxial tension. It is seen that Hill's modification (11.66c) best describes this. Note, however, that the small difference observed between tensile yield stresses in the rolling and transverse directions, i.e. $\sigma_{2y}/\sigma_{1y} \neq 1$, do not appear within these symmetrical loci. Hill [39] later corrected for this by combining cubic and quadratic dimensionless stress terms in a criterion:

$$\frac{\sigma_1^2}{\sigma_{1y}^2} - \frac{c\,\sigma_1\sigma_2}{\sigma_{1y}\sigma_{2y}} + \frac{\sigma_2^2}{\sigma_{2y}^2} + \left[(p + q) - \frac{(p\sigma_1 + q\sigma_2)}{\sigma_{by}}\right]\frac{\sigma_1\sigma_2}{\sigma_{1y}\sigma_{2y}} = 1$$

where σ_{by} is the biaxial yield stress, p and q are constants and c combines the three yield points in Fig. 11.15:

$$\frac{c}{\sigma_{1y}\sigma_{2y}} = \frac{1}{\sigma_{1y}^2} + \frac{1}{\sigma_{2y}^2} + \frac{1}{\sigma_{by}^2}$$

11.9 Concluding Remarks

The choice of yield function (i.e. the plastic potential) depends upon the initial condition of a material. Generally, an isotropic potential will predict the plasticity in heat treated material. Otherwise, an account needs to be made of crystallographic anisotropy resulting from processing, particularly in rolled sheet and extruded bars. Hill's 1948 homogenous, quadratic yield function, with its associated flow rule, provides a good account of plasticity under radial loading of an orthotropic solid, e.g. a rolled sheet metal. However, certain anomalies have arisen with a quadratic potential. Among these are the absence of a Bauschinger effect and a positive difference between equi-biaxial and uni-axial yield stresses. This has led to modifications with the addition of linear and/or cubic stress terms and from using non-quadratic stress exponents. Increasing flexibility by these means allows the anisotropy parameters to match observed behaviour more closely. As with isotropic material there appears no particular advantage in using the work or strain hypotheses of hardening for an orthotropic solid. The equivalent stress-plastic strain correlations from each hypothesis would appear to be equallly suitable within a normality rule of flow.

References

1. Klinger L. J. and Sachs G. *Jl. Aero. Sci.* 1948, **15**, 599.
2. Cunninghan D. M., Thompsen E. G. and Dorn J. E. *Proc ASTM,* 1947, **47**, 546.
3. Rogan J. and Shelton A. *Jl Strain Analysis*, 1969, **4**, 127.
4. Shahabi S. N. and Shelton A. *Jl Mech Eng Sci*, 1975, **17**, 93.
5. Jackson L. R., Smith K. F. and Lankford W. T. *Jl. Metals* 1949, **323**, 1.
6. Dorn J. E. *Jl Appl Phys*, 1949, **20**, 15.
7. Prager W. *Jl Appl Phys*, 1949, **20**, 234.
8. Fisher J. C. *Trans ASME*, 1949, **71**, 349.
9. Hu L. W. *Trans ASME, Jl Appl Mech*, 1956, **23**, 444.
10. Jones S. E. and Gillis P. P. *Met Trans*, 1984, **15**A, 129.
11. Hill R. *Proc Roy Soc A*, 1948, **A193**, 281.
12. Finnie I. and Heller W. R. *Creep of Engineering Materials*, 1959, McGraw-Hill, N.Y.
13. Chakrabarty J. *Int. Jl Mech. Sci*, 1970, **12**, 169.
14. Whiteley R. L. *Trans ASM*, 1960, **52**, 154.
15. Bramley A. N. and Mellor P. B. *Int. Jl Mech. Sci*, 1966, **8**, 101.
16. Bramley A. N. and Mellor P. B. *Int. Jl Mech. Sci*, 1968, **10**, 211.
17. Atkinson M. *Sht Met. Ind.*, 1967, **44**, 167.
18. Li K. Y. *A Study of Certain Classes of Constitutive Equations for Anisotropic Solid Materials*, Ph.D. thesis, Trinity College, Dublin, 1985.
19. Bourne L. and Hill R. *Phil Mag.*, 1950, **41**, 671.
20. Hazlett T. H., Robinson A. T. and Dorn J. E. *Trans ASM*, 1950, **42**, 1326.
21. Shahabi S. N. and Shelton A. *Jl Mech Eng Sci*, 1975, **17**, 82.
22 Dillamore I. L, Hazel R. J, Watson T.W and Hadden P. *Int Jl Mech Sci*, 1971, **13**, 1049.
23. Fessler H. and Hyde T. H. *Proc: Non Linear Problems in Stress Analysis*, p. 233, (Ed. Stanley P.) Inst. of Physics, 1977.
24. Mehan R. L. *Trans ASME, J Basic Engng*, 1961, **83**, 499.
25. Rees D. W. A. and Mathur S. B. *Proc: Non Linear Problems in Stress Analysis*, p. 185, (Ed. Stanley P.) Inst of Physics, 1977.
26. Johnson A. E., Frost N. E. and Henderson J. *The Engineer*, 1955, 199, 366, 402, 457.
27. Hill R. *The Mathematical Theory of Plasticity*, Clarendon Press, Oxford 1989.
28. Rees D. W. A. *Proc. Roy. Soc.*, 1982, **A383**, 333.
29. Avery D. H., Hosford W. F and Backofen W. A. *Trans Met. Soc., AIME*, 1965, **223**, 71.
30. Lee D. and Backofen W. A. *Trans Met. Soc. AIME*, 1966, **236**, 1077.
31. Mellor P. B. *Mechanics of Solids*, (eds Hopkins H. G. and Sewell M. J.) 1982, 383, Pergamon Press.
32. Larson F. R. *Trans ASM* 1964, **27**, 321.
33. McEvily A. J. and Hughes P. J. *NACA Tech. Note* 3248, 1954.
34. Rees D. W. A. *Acta Mech*, 1982, **43**, 223.
35. Hosford W.F. *Proc: 7th Nth Amer Metal Working Res Conf*, SME, Dearbon Michigan, p. 191, 1979.
36. Hill R. *Math. Proc Camb Phil Soc*, 1979, **85**, 179.
37. Bassani, J. L. *Int Jl Mech Sci*, 1977, **19**, 651
38. Rees, D.W.A. *Jl de Phy IV*, 2003, **105**, 19, see also Procedings: *Non-Linear Mechanics of Anisotropic Materials*, EMMC 6, p. 33, (Eds Cescotto S. et al) Euromech-Mecamat, Univesity of Liége, 2002.
39. Hill, R. *Int Jl Mech Sci.*, 1993, **35**, 19.

Exercises

11.1 Derive the relationships between the components of H_{ijkl} in eq(11.1) and the six anisotropy coefficients employed with Hill's orthotropic yield function in eq(11.14).

11.2 Figure 11.16 shows an off-axis, combined tension-torsion test cylinder. Axial and shear stress components (σ_{11}', σ_{12}') are produced in the wall of a thin-cylinder by the combined action of tension and torque (W, T). The material axes of orthotropy 1, 2 are inclined to tube axis, while axis 3 remains radial, as shown.

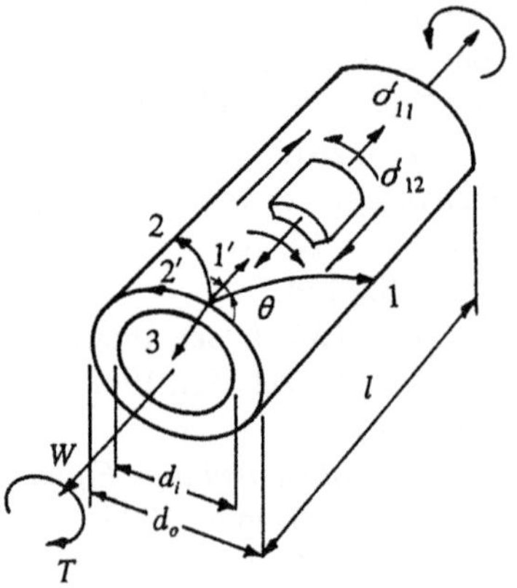

Figure 11.16

Show, from eq(11.4b), that the equivalent stress can be expressed as

$$2\bar{\sigma}^2 = \beta_1(\sigma'_{11})^2 + 2\beta_2\sigma'_{11}\sigma'_{12} + \beta_3(\sigma'_{12})^2$$

where the β - coefficients are given by

$$\beta_1 = (H+G)\cos^4\theta + (H+F)\sin^4\theta - 2H\sin^2\theta\cos^2\theta + (3N/2)\sin^2 2\theta$$
$$\beta_2 = (H+G)\cos 2\theta\sin 2\theta - (2H+F)\sin^2\theta\sin 2\theta + H\cos^2\theta\sin 2\theta - 3N\cos 2\theta\sin 2\theta$$
$$\beta_3 = (4H+G+F)\sin^2 2\theta + 6N\cos^2 2\theta$$

11.3 Show that the plastic strain increment ratios for the cylinder in Fig. 11.16 are given by

$$\frac{\mathrm{d}\gamma'^{P}_{12}}{\mathrm{d}\varepsilon'^{P}_{11}} = \frac{\beta_2\sigma'_{11} + \beta_3\sigma'_{12}}{\beta_1\sigma'_{11} + \beta_2\sigma'_{12}} \quad \text{and} \quad \frac{\mathrm{d}\varepsilon'^{P}_{22}}{\mathrm{d}\varepsilon'^{P}_{11}} = \frac{\beta_4\sigma'_{11} + \beta_5\sigma'_{12}}{\beta_1\sigma'_{11} + \beta_2\sigma'_{12}}$$

where coefficients β_1, β_2 and β_3 are defined in Exercise 11.2 and

$$\beta_4 = (2H+G-6N)\sin^2\theta\cos^2\theta - H(\sin^4\theta + \cos^4\theta)$$
$$\beta_5 = (2H+G)\sin 2\theta\sin^2\theta - (2H+F)\sin 2\theta\cos^2\theta + 3N\sin 2\theta\cos 2\theta$$

11.4 If the stresses for the tube in Fig. 11.16 are applied in a constant ratio $R = \sigma'_{12}/\sigma'_{11}$, show that the principal axes of stress coincide with the principal material axes when $\theta = ½\tan^{-1}(2R)$. What then is the incremental plastic strain ratio $\mathrm{d}\gamma_{12}^{P}/\mathrm{d}\varepsilon_{11}^{P}$? Employ eqs(11.11a-c) to show that the principal plastic strain increments for this condition are

$$\mathrm{d}\varepsilon_1^P = \frac{\delta\lambda\,\sigma}{3}\left[\frac{G}{2} + \frac{H + G/2}{1 + 4R^2}\right]$$

$$\mathrm{d}\varepsilon_2^P = \frac{\delta\lambda\,\sigma}{3}\left[\frac{F}{2} - \frac{H + F/2}{1 + 4R^2}\right]$$

$$\mathrm{d}\varepsilon_3^P = -\frac{\delta\lambda\,\sigma}{6}\left[F + G + \frac{G - F}{1 + 4R^2}\right]$$

Hint: The principal stress directions are found from by $\tan 2\theta = 2\sigma'_{12}/\sigma'_{11}$

11.5 If the off-axis tube in Fig. 11.16 is subjected to a separate tensile force W and a torque T, determine the corresponding $\mathrm{d}\gamma'^P_{12}/\mathrm{d}\varepsilon'^P_{11}$ ratios. Show, (i) in the absence of torsion, that the axial plastic strain increment becomes

$$\mathrm{d}\varepsilon'^P_{11} = (\mathrm{d}\bar{\varepsilon}^P / 2\bar{\sigma})\,\beta_2\,\sigma'_{11}$$

and, (ii) in the absence of tension, the plastic shear strain increment becomes

$$\mathrm{d}\gamma^P_{12} = (\mathrm{d}\bar{\varepsilon}^P / 2\bar{\sigma})\,\beta_3\,\sigma'_{12}.$$

For what orientations θ, do $\mathrm{d}\varepsilon'^P_{11}$ and $\mathrm{d}\gamma'^P_{12}$ attain their maxima?

11.6 Show that when the thin-walled tubular specimen in Fig. 11.16 is subjected to torsion only, the equivalent stress reduces to

$$\bar{\sigma} = (\sigma_{12y}/\sqrt{2})\,[\,(4H + G + F)\sin^2 2\theta + 6N\cos^2 2\theta\,]^{1/2}$$

where σ_{12} is the shear flow stress for the given orientation θ.

11.7 Confirm that the $\bar{\sigma}$ expression in Exercise 11.6 also applies when, under torsion, θ is identified with the relative rotation of the material's 1 - axis to the testpiece axis 1′. Take the two axes to be aligned initially.

11.8 Show that the ratio between the shear yield stress σ_{12y} and the tensile yield stress σ_{1y} is given by eq(11.2) as

$$\sigma_{12y}/\sigma_{1y} = [(H + G)/(6N)]^{1/2}$$

when the stress components are aligned with the principal axes of orthotropy. What are the corresponding ratios between yield stresses: (i) σ_{12y} and σ_{2y} and (ii) σ_{12y} and σ_{3y} ?

[Answer: $\sigma_{12y}/\sigma_{2y} = [(H + F)/(6N)]^{1/2}$, $\sigma_{12y}/\sigma_{3y} = [(F + G)/(6N)]^{1/2}$]

11.9 Establish further relationships between the following shear and tensile yield stresses for an orthotropic material: (i) σ_{23y} and σ_{1y} (ii) σ_{23y} and σ_{2y}, (iii) σ_{23y} and σ_{3y}, (iv) σ_{31y} and σ_{1y}, (v) σ_{31y} and σ_{2y} and (vi) σ_{31y} and σ_{3y}.

[Answer: $\sigma_{23y}/\sigma_{1y} = [(H + G)/(6L)]^{1/2}$, $\sigma_{23y}/\sigma_{2y} = [(H + F)/(6L)]^{1/2}$, $\sigma_{23y}/\sigma_{3y} = [(F + G)/(6L)]^{1/2}$,
$\sigma_{31y}/\sigma_{1y} = [(H + G)/(6M)]^{1/2}$, $\sigma_{31y}/\sigma_{2y} = [(H + F)/(6M)]^{1/2}$, $\sigma_{31y}/\sigma_{3y} = [(F + G)/(6M)]^{1/2}$]]

11.10 A sheet is transversely isotropic within the 1 - 2 plane ($r = r_{0°} = r_{90°}$). Show that plane strain compression and simple tension are related through

$$p = (1 + r)\sigma_1/(1 + 2r)^{1/2} \quad \text{and} \quad \mathrm{d}t/t = (1 + 2r)^{1/2}\,\varepsilon_1^P/(1 + r)$$

where σ_1, ε_1^P are the stress and plastic strain for a tension test. In plane compression p is the pressure, dt/t is an incremental through-thickness strain where t is the curent thickness (see Fig. 6.38).

11.11 Show that the variation in flow stress for tensile testpieces lying in the 1 - 2 plane with their axes inclined at θ to the 1- direction is given by

$$\bar{\sigma} = \sqrt{2}\sigma[\,(H + G)\cos^4\theta + (H + F)\sin^4\theta + \tfrac{1}{2}(N - H)\sin^2 2\theta]^{-1/2}$$

when $\bar{\sigma}$ is defined from eq(11.4b). Hence show that the condition for a maximum or minimum yield stress is given by

$$\tan^2\theta = (2H + G - 3N)\,/\,(2H + F - 3N)$$

11.12 Repeat Exercise 11.11 for testpieces lying in the 1 - 3 and 2 - 3 planes. That is, show that the yield stress variations are respectively

$$\bar{\sigma} = \sqrt{2}\sigma[(H + G)\cos^4\theta + (G + F)\sin^4\theta + \tfrac{1}{2}(M - G)\sin^2 2\theta]^{-1/2}$$
$$\bar{\sigma} = \sqrt{2}\sigma[\,(H + F)\cos^4\theta + (G + F)\sin^4\theta + \tfrac{1}{2}(L - F)\sin^2 2\theta]^{-1/2}$$

for which the maximum or minimum correspond to

$$\tan 2\theta = \frac{2G + H - 3M}{2G + F - 3M}\,, \qquad \tan 2\theta = \frac{2F + H - 3L}{2F + G - 3L}$$

11.13 What is the condition of instability when the off-axis tensile equivalence correlation for copper, shown in Fig. 11.9, obeys the Swift law (9.44b)? Find the particular instability condition for a testpiece orientation of $\theta = 45°$.

11.14 A planar isotropic rolled material resembles that given in Fig. 11.3a. Following a further plane strain process, which allows deformation to occur in the 2 and 3- directions, but prevents further deformation in the original rolling direction 1, the structure is converted to that shown in Fig. 11.3b. Assuming that the r value for each isotropic plane is similar, i.e. $r_{12} = r_{13}$, compare the elastic boundary resulting from each process for plane stress states within plane 1 - 2.

CHAPTER 12

PLASTIC INSTABILITY

12.1 Introduction

There are many examples where the influence of plasticity upon the load-carrying capacity of a structure must be considered. A purely elastic analysis will supply the critical loading at which plasticity first appears but the analysis will not extend into the plastic regime without modifications. These can take the form of a change to the modulus of a material to reflect the loss in stiffness of a strut, the introduction of empirically-based plasticity reduction factors for plate buckling and the derivation of an appropriate sub-tangent under multiaxial stressing. We may treat such problems with the plasticity theory appropriate to the initial condition of the material. The Levy-Mises theory is adequate for an isotropic condition but in orthotropic sheets such problems should employ anisotropy parameters, e.g. the r values as discussed in the previous chapter. An example of the latter arises in the determination of the limiting strains in sheet metal forming. The theory may be coupled to either a diffuse or local instability criterion for the sheet material to determine the combination of limiting, in-plane plastic strains. These strains are used in the construction of a forming limit diagram.

12.2 Inelastic Buckling of Struts

12.2.1 Tangent Modulus

Engesser's modification [1] to the Euler's buckling theory accounts for inelastic buckling simply by replacing the elastic modulus with a plastic tangent modulus. With this reduction in stiffness, the section stress is given by

$$\sigma = \frac{P}{A} = \frac{\pi^2 E_t}{\left(L_e/k\right)^2} \tag{12.1a}$$

where L_e is an effective length that accounts for the particular rotational restraint exerted by the end fixing. Within the range of slenderness ratios $50 \leq L_e/k \leq 100$, the effect of end constraint on the plastic buckling load P is less than that of an elastic strut with similar end fixings. For example, we should not assume that the plastic buckling load of a pinned-end strut, lying in the same range, will be doubled by fixing its ends. However, for $L_e/k \leq 50$ an elastic constraint is assumed and an equivalent elastic length $L_e = cL$ may be employed, particularly when a safety factor is used. This gives: $L_e = L$ for pinned ends, $L_e = L/2$ for fixed ends, $L_e = L/\sqrt{2}$ for pinned-encastre end fixings and $L_e = 2L$ for fixed-free ends fixings. The tangent modulus $E_t = \mathrm{d}\sigma/\mathrm{d}\varepsilon$ in eq(12.1a) is the gradient of the tangent to the uniaxial compressive stress-strain curve within its plastic range (see Fig. 12.1). It follows from this definition of E_t that the buckling stress σ in eq(12.1a) must satisfy the following condition:

$$\sigma = \frac{\pi^2 (\mathrm{d}\sigma/\mathrm{d}\varepsilon)}{\left(L_e/k\right)^2} \tag{12.1b}$$

Since σ appears on both sides of eq(12.1b) it may be solved by trial. That is, a plastic buckling stress σ, is selected to be greater than the yield stress σ_o. Assuming that a stress-strain curve is available, $E_t = \mathrm{d}\sigma/\mathrm{d}\varepsilon$ is found and eq(12.1b) is solved for L_e/k. The solution is correct only when L_e/k matches that for the given strut. The procedure is aided when all such solutions to eq(12.1b) appear as points on a plot of σ versus L_e/k. Alternatively, a suitable empirical representation to the σ versus ε curve may be employed, the simplest being the Hollomon law (9.39b), which is represented graphically in Fig. 12.1.

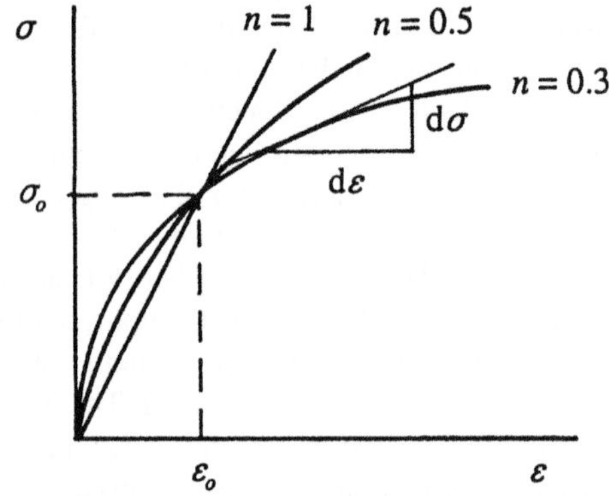

Figure 12.1 Hollomon curve with tangent modulus

Given that the stress-strain curve is described by

$$\frac{\sigma}{\sigma_o} = \left(\frac{\varepsilon}{\varepsilon_o}\right)^n \tag{12.2a}$$

the gradient of its tangent is

$$E_t = \frac{\mathrm{d}\sigma}{\mathrm{d}\varepsilon} = n\left(\frac{\sigma_o}{\varepsilon_o}\right)\left(\frac{\sigma}{\sigma_o}\right)\left(\frac{\sigma_o}{\sigma}\right)^{1/n} \tag{12.2b}$$

where ε_o and n are material constants. Substituting eq(12.2b) into eq(12.1b) results in an equation that is soluble in σ:

$$\frac{\sigma}{\sigma_o} = \left(\frac{n}{\varepsilon_o}\right)^n \left(\frac{\pi}{L_e/k}\right)^{2n} \tag{12.3}$$

Equation (12.3) defines the Engesser's curve 1 in Fig. 12.2, this curve being valid for a net section stress $\sigma > \sigma_o$. In contrast, Euler's elastic curve 2 in this figure applies to section stresses $\sigma < \sigma_o$ and is expressed as:

$$\sigma = \frac{P}{A} = \frac{\pi^2 E}{\left(L_e/k\right)^2} \tag{12.4}$$

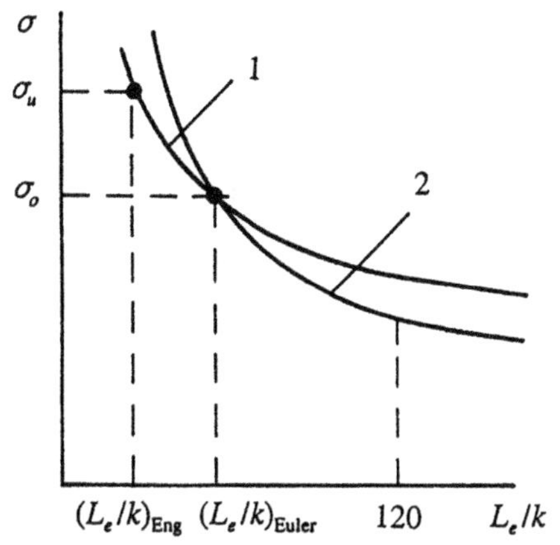

Figure 12.2 Buckling curves (Key: curve 1 - Engesser, curve 2 - Euler)

To ensure an intersection between the two curves at $\sigma = \sigma_o$, we first find Euler's critical slenderness ratio from eq(12.4):

$$\left(\frac{L_e}{k}\right)_{\text{Euler}} = \sqrt{\left(\frac{\pi^2 E}{\sigma_o}\right)} \tag{12.5}$$

Since $(L_e/k)_{\text{Euler}}$ defines an intersection co-ordinate, substitution of eq(12.5) into eq(12.3) leads to the condition

$$n\sigma_o = E\varepsilon_o \tag{12.6}$$

We then see that Euler becomes a special case of eq(12.3) when $n = 1$. Thus, a curve of any $0 < n < 1$ value will pass through this common point $[(L_e/k)_{\text{Euler}}, \sigma_o]$. Correspondingly, the intersection occurs at the common yield point $(\varepsilon_o, \sigma_o)$ for Hollomon curves with different n values (see Fig. 12.1).

Contrary to the Hollomon law, the stress in a plastic strut cannot be increased indefinitely. The cut-off ordinate in Fig. 12.2 occurs at the ultimate compressive strength σ_u as shown. Thus, the critical value of L_e/k for an Engesser strut follows from eqs(12.3) and (12.6) as

$$\left(\frac{L_e}{k}\right)_{\text{Eng}} = \sqrt{\frac{\pi^2 E}{\sigma_o\left(\sigma_u/\sigma_o\right)^{1/n}}} \tag{12.7}$$

12.2.2 Empirical Formulae

Empirical plastic strut formulae account for a range of short struts with L_e/k initially less than the critical Engesser value (from eq 12.7) but they extend L_e/k partly into the elastic range. Figures 12.3a,b show the range of interest in which section stresses may vary from the ultimate value σ_u to less than the yield stress σ_o. In each diagram the Engesser and Euler curves, 1 and 2 respectively, intersect at point A. The empirical approach replaces the two segments of each curve with a continuous curve that plots to the safe side and terminates at σ_u on the σ - axis. The following strut formulae are available for this, the choice between them depending upon the strut material.

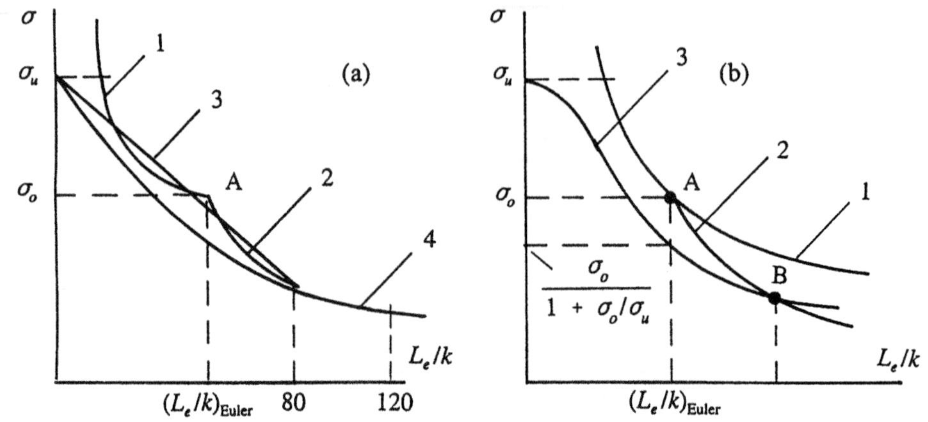

Figure 12.3 Empirical approximations (3) and (4) compared to Euler (1) and Engesser (2)

(a) *Straight Line*
The straight line 3 in Fig. 12.3a is is simply written as

$$\sigma = \sigma_u [\, 1 - q\,(L_e / k)\,] \tag{12.8}$$

where q is a material constant. For aluminium alloy struts, eq(12.8) is arranged to intersect the Euler curve at $L_e/k = 80$ while for other non-ferrous metals an intersection at $L_e/k = 120$ may be preferred. The Euler stress ordinate at each intersection is found from eq(12.4). Substituting this into eq(12.8) q is found. Typically, $E = 75$ GPa, $\sigma_u = 320$ MPa and $q = 8 \times 10^{-3}$ for an aluminium alloy.

(b) *Parabola*
The parabola 4 in Fig. 12.3a provides a safer prediction for steel struts with effective slenderness ratios lying in the range $L_e/k \le 120$. This is written as

$$\sigma = \sigma_u [\, 1 - b\,(L_e / k)^2\,] \qquad (12.9]$$

where b is a material constant, found from known co-ordinates (σ, L_e/k) at the intersection with the Euler curve. Values of b for steels lie in the range (40 - 50) $\times 10^{-6}$ for ultimate strengths between 400 and 500 MPa and with a modulus $E = 210$ GPa.

(c) *Rankine-Gordon*
Their critical plastic buckling load P_{RG} is found from the condition

$$1/P_{RG} = 1/P_{Eng} + 1/P_u \tag{12.10a}$$

where $P_u = A\sigma_u$ is the ultimate load and $P_{Eng} = A\sigma$. Re-arranging eq(12.10a) for P_{RG} gives

$$P_{RG} = \frac{P_u}{1 + P_u / P_{Eng}} \tag{12.10b}$$

Substituting P_{Eng} separately from eqs(12.1a) and (12.3) with $P_o = \sigma_o A$ gives two alternative forms for P_{RG}:

$$P_{RG} = \frac{A\sigma_u}{1 + \dfrac{\sigma_u}{\pi^2 E_t}\left(\dfrac{L_e}{k}\right)^2} = \frac{A\sigma_u}{1 + \dfrac{\sigma_u}{\sigma_o}\left(\dfrac{\sigma_o}{\pi^2 E}\right)^n \left(\dfrac{L_e}{k}\right)^{2n}} \quad (12.11a,b)$$

Equation (12.11b) is a more useful for combining with eq(12.6). This reveals material constants σ_o, σ_u, E and n within the section stress:

$$\sigma = \frac{\sigma_u}{1 + \dfrac{\sigma_u}{\sigma_o}\left(\dfrac{\sigma_o}{\pi^2 E}\right)^n \left(\dfrac{L_e}{k}\right)^{2n}} \quad (12.12)$$

Equation(12.12) appears as curve 3 in Fig. 12.3b for which the stress is $\sigma = \sigma_o/(1 + \sigma_o/\sigma_u)$ at $(L_e/k)_{\text{Euler}}$ and lies on the safe side of the Euler curve 2. The slenderness ratio at an intersection point B with curve 2 is found from equating eqs(12.4) and (12.12).

Example 12.1 An 800 mm long steel strut has a thin-walled, elliptical cross-section shown in Fig. 12.4a. The mean lengths of the major and minor axes are 80 and 30 mm respectively and the wall thickness is 3 mm. At its end fixings, Fig. 12.4a shows that the strut is free to rotate about a pin aligned with its y-axis but it is prevented from rotating about its x-axis by the rigid walls shown. Compare the allowable compressive plastic loads according to the Engesser, parabolic and Rankine-Gordon formulae, using a safety factor of 1.5. For steel take $\sigma_o = 300$ MPa, $\sigma_u = 450$ MPa, $E = 210$ GPa and $n = ⅓$.

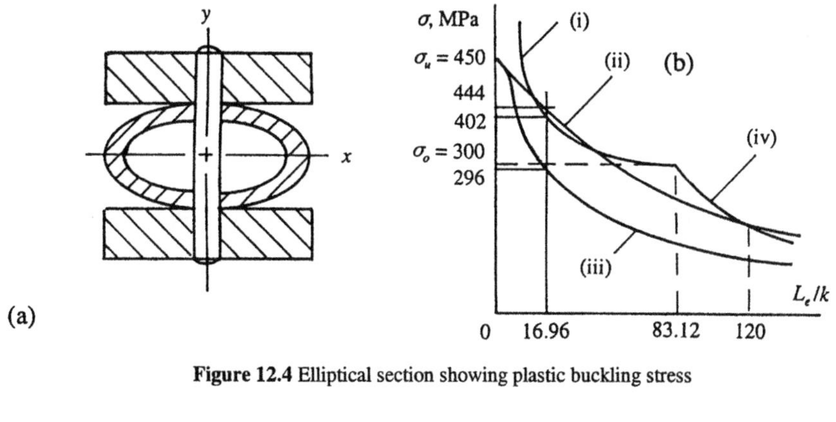

Figure 12.4 Elliptical section showing plastic buckling stress

The required properties of the cross-section are:

$A = \pi ab = \pi[(83 \times 35) - (77 \times 27)] = 2073.45 \text{ mm}^2$

$I_x = \pi ab^3/4 = (\pi/4)[83(33)^3 - 77(27)^3] = 1.1523 \times 10^6 \text{ mm}^4$

$k_x = \sqrt{(I_x/A)} = 23.583 \text{ mm}$

$I_y = \pi ba^3/4 = (\pi/4)[33(83)^3 - 27(77)^3] = 5.1385 \times 10^6 \text{ mm}^4$

$k_y = \sqrt{(I_y/A)} = 49.782 \text{ mm}$

The strut will buckle about the axis for which L_e/k has the greater value (this gives the lower load). For the x-axis the ends are fixed, so $L_e = 400$ mm and

$$L_e / k_x = 400 / 23.583 = 16.96$$

For the y-axis the ends are pinned so that $L_e = 800$ mm giving

$$L_e / k_y = 800 / 49.782 = 16.07$$

Having established that buckling is more likely to occur about the x-axis the following safe loads can be found.

(i) *Engesser*
Applying eqs(12.3) and (12.6) Engesser's plastic buckling stress is

$$\sigma / \sigma_o = (E/\sigma_o)^n[\pi / (L_e / k)]^{2n}$$
$$= (210 \times 10^3/300)^{1/3}(\pi/16.96)^{2/3} = 1.339$$
$$\sigma = 1.339 \times 300 = 401.76 \text{ MPa}$$

Hence the safe stress is 401.76 /1.5 = 267.84 MPa, from which the safe load is 267.84 × 2073.45 = 555.35 kN.

(ii) *Parabola*
Taking the intersection with the Euler curve to occur at $L_e/k = 120$, the corresponding stress in eq(12.4) is

$$\sigma = \pi^2 E / (L_e / k)^2 = 143.93 \text{ MPa}$$

and b is found from eq(12.9) as

$$b = (1 - 143.93 / 450) / (120)^2 = 47.23 \times 10^{-6}$$

Hence the plastic bucking stress is

$$\sigma = 450[1 - (47.23 \times 10^{-6})(16.96)^2] = 443.89 \text{ MPa}$$

Thus, the safe stress and load are 295.92 MPa and 613.58 kN respectively.

(iii) *Rankine-Gordon*
The buckling stress follows from eq(12.12), in units of N and mm, as:

$$\sigma = 450 \Bigg/ \left[1 + \frac{450}{300}\left(\frac{300}{\pi^2 \times 210000}\right)^{\frac{1}{3}} (16.96)^{\frac{2}{3}}\right] = 296.1 \text{ MPa}$$

Thus, the safe stress and load are 197.38 MPa and 409.27 kN respectively. Figure 12.4b compares the three buckling stress predictions graphically. It is seen that Engesser gives a valid stress ($\sigma < \sigma_u$) and that Rankine-Gordon ($\sigma < \sigma_o$) is rather conservative. The intersection between curve (i) and Euler's curve (iv) is found from eq(12.5) as $L_e/k = 83.12$.

12.2.3 Wide Strut

Consider the compression of a wide, thin plate, thickness t, with similar length and width dimensions, a and b respectively (see Fig. 12.5a). With unsupported sides, the strut deflects laterally with uniform curvature while the cross-section $b \times t$ remains rectangular during bending. Consequently, biaxial, in-plane stresses, σ_x and σ_y, exist in the body of the plate.

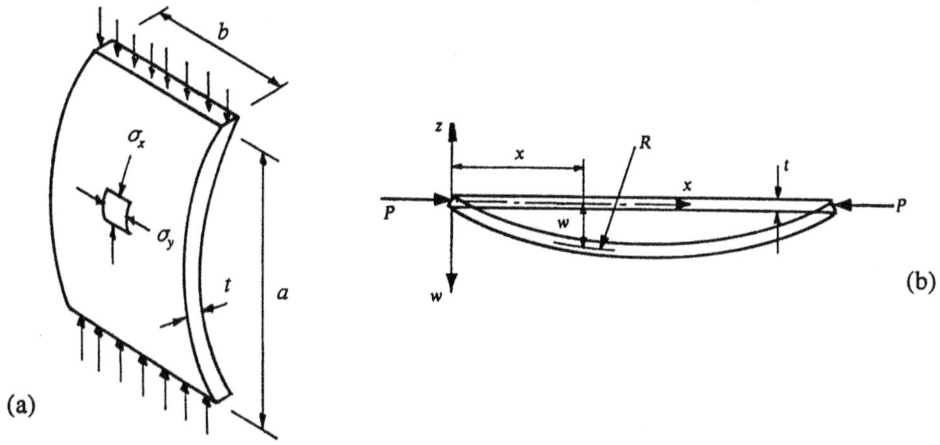

Figure 12.5 Buckling of a wide plate strut

With no dimensional change in the y - direction:

$$\varepsilon_y = 0 = (1/E)(\sigma_y - \nu\,\sigma_x)$$

$$\varepsilon_x = (1/E)(\sigma_x - \nu\,\sigma_y) = (1 - \nu^2)\sigma_x/E \tag{12.13}$$

When b is small compared to a, the absence of ε_y results in *anticlastic curvature*, i.e. the cross-section does not remain rectangular. The distortion arises from opposing lateral strains induced within the tensile and compressive surfaces (the Poisson effect). Here $\sigma_y = 0$ and the strains along the length and width are simply: $\varepsilon_x = \sigma_x/E$ and $\varepsilon_y = -\nu\,\sigma_x/E$. Comparing with eq(12.13) reveals a difference in ε_x between narrow and wide struts, i.e. it is necessary to modify the flexure equation for a wide strut by the factor $(1 - \nu^2)$. From Fig. 12.5b,

$$\mathrm{d}^2w / \mathrm{d}x^2 = 1/R = \varepsilon_x / z$$

Substituting from eq(12.13), with $M/I = \sigma_x/z$, it follows that

$$\mathrm{d}^2w/\mathrm{d}x^2 = (1 - \nu^2)\sigma_x/(Ez) = (1 - \nu^2)M / (EI)$$

With bending moment $M = -Pw$, the solution to the critical buckling load for a pinned-end plate is

$$P_{cr} = \frac{\pi^2 EI}{(1 - \nu^2)a^2}$$

Substituting $P_{cr} = \sigma_{cr} bt$ and $I = Ak^2$ where $k^2 = t^2/12$ leads to the section stress:

$$\sigma_{cr} = \frac{\pi^2 E}{(1 - \nu^2)(a/k)^2} \tag{12.14a}$$

Setting $L_e = a(1 - \nu^2)^{½}$ for simply supported ends, other edge fixings may be accounted for by re-writing eq(12.14a) as

$$\sigma_{cr} = \frac{\pi^2 E}{(L_e / k)^2} \tag{12.14b}$$

in which

$L_e = (a/2)(1 - \nu^2)^{½}$ for fixed ends

$L_e = (a/\sqrt{2})(1 - \nu^2)^{½}$ for pinned-fixed ends

$L_e = 2a(1 - \nu^2)$ for fixed-free ends

Engesser's inelastic theory applies to these equivalent lengths when E in eq(12.14b) is replaced by E_t. For example, in the simply supported case $(L_e/k)_{Eng}$ in Fig. 12.2 is identified with the plate term: $(1 - \nu^2)^{½}(a/k)$. All other relationships, derived previously for plastic buckling of struts, including the empirical approaches, can then be applied to wide struts.

12.3 Buckling of Plates

12.3.1 Uni-Directional Compressive Loading

(a) *Simple Supports*

Under a uni-directional compression, the critical buckling stress σ_x, for a plate with all four sides simply supported, is given as [2]:

$$(\sigma_x)_{cr} = (D\pi^2 / t)\,[(m/a)^2 + (n/b)^2]^2 (a/m)^2$$

$$(\sigma_x)_{cr} = (D\pi^2 / t\, b^2)\,[mb/a + n^2 a/mb]^2 \tag{12.15}$$

where $D = Et^3 / [12(1 - \nu^2)]$ is the flexural stiffness, t, a and b are the plate thickness, length and breadth respectively. The number of half-waves in the x- and y- directions are denoted by m and n respectively. For simple supports, the interior buckles with a half-wave ($n = 1$) in the y - direction, as shown in Fig. 12.6.

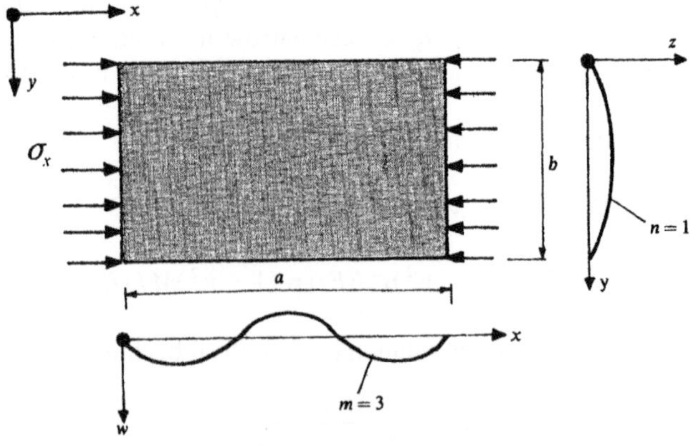

Figure 12.6 Buckling of a thin, simply-supported plate under uniaxial compression

From eq(12.15), the buckling stress becomes

$$(\sigma_x)_{cr} = (D\pi^2 / t\, b^2)[m/r + r/m]^2 \tag{12.16}$$

where $r = a/b$. Differentiating for m-values that minimise eq(12.16) with integral r values:

$$d(\sigma_x)_{cr}/dm = 2(m/r + r/m)(1/r - r/m^2) = 0$$

$$m/r^2 + 1/m - 1/m - r^2/m^3 = 0$$

$$m^4 - r^4 = 0$$

$$(m - r)(m + r)(m^2 + r^2) = 0$$

The condition $m = r$ implies that the plate interior will buckle into an integral number of square cells $a \times a$ each under the same stress. That is, from eq(12.16)

$$(\sigma_x)_{cr} = \frac{4D\pi^2}{tb^2} = \frac{\pi^2 E}{3(1 - \nu^2)}\left(\frac{t}{b}\right)^2 \qquad (12.17)$$

The graph in Fig. 12.7 provides the buckling stress when non-integer values of r are taken with particular values of m in eq(12.16).

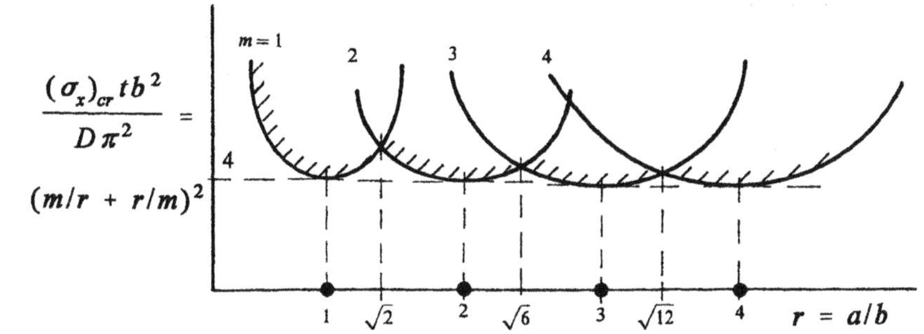

Figure 12.7 Effect of r and m on uniaxial buckling stress

The trough in each curve corresponds to eq(12.17) and at the intersections of these curves where $r = \sqrt{2}, \sqrt{6}, \sqrt{12}$ etc, m has been increased by one. This occurs when, from eq(12.16),

$$[\, m/r + r/m \,] = [\, (m + 1)/r + r/(m + 1) \,]$$

$$\therefore \quad r = \sqrt{m(m + 1)}$$

from which: $r = \sqrt{2}$ for $m = 1$, $r = \sqrt{6}$ for $m = 2$, $r = \sqrt{12}$ for $m = 3$ etc. Equation (12.17) should not be confused with the buckling stress for a thin, wide plate acting as a strut with its longer, parallel sides are unsupported (see Section 12.2.3).

(b) *Other Edge Fixings*

A number of approaches have been proposed for constrained edges. The simplest of these employs the expression

$$(\sigma_x)_{cr} = (K_r D\pi^2) / (t\, b^2) \qquad (12.18a)$$

where the restraint coefficient K_r applies to any edge fixings when

$$K_r = (m/r)^2 + p + q\,(r/m)^2 \qquad (12.18b)$$

Clearly, with restraint factors $p = 2$ and $q = 1$, eqs (12.18a,b) contain eq(12.16) in the special case of simple supports. The dependence of K_r upon the two rotational edge restraint factors

(p and q), plate dimensions ($r = a/b$) and buckling mode m, has been established experimentally in certain cases. Table 12.1 applies to the case of fixed sides.

Table 12.1 Restraint coefficients for a plate with fixed edges

$r = a/b$	0.75	1.0	1.5	2.0	2.5	3.0
K_r	11.69	10.07	8.33	7.88	7.57	7.37

As r increases, the effect of edge restraint lessens and K_r approaches the minimum value of 4 as found from eq(12.17) for a plate with simply supported edges. An alternative graphical approach [3] employs design curves derived from the restraint coefficients. The latter supply the ratio between the critical elastic buckling stress $(\sigma_{cr})_e$ for a plate with a given edge fixing to that of a simply supported plate (eq 12.17) with a similar geometry $r = a/b$. For example, Fig. 12.8 shows how the stress ratio varies with r for clamped and various mixed-edge fixings. A similar graphical presentation has extended this to biaxially loaded plates [3].

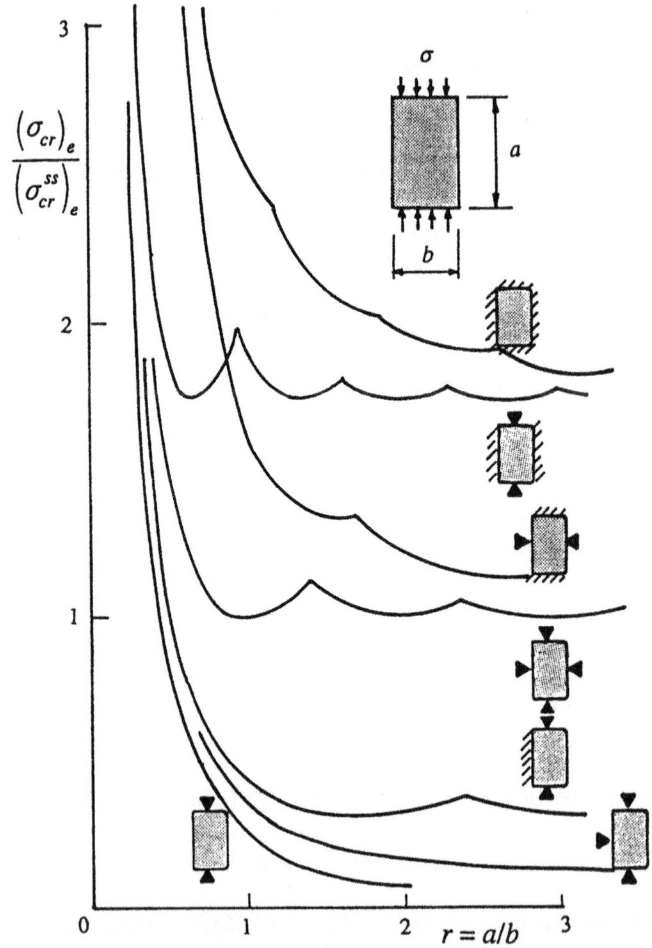

Figure 12.8 Plate buckling under compression

(c) *Inelastic Buckling*
In thicker plates the critical elastic stress $(\sigma_{cr})_e$, as calculated from eq(12.17), can exceed the yield stress Y of the plate material. The solution will be invalid because, with the use of E and ν in eq(12.17), linear elasticity is assumed at the critical stress level. Figure 12.9a shows that where plasticity has occurred it reduces the buckling stress to a lower level $(\sigma_{cr})_p$. Figure 12.9b shows how tangent and secant moduli, E_t and E_s respectively, are used to account for a plastic stress level.

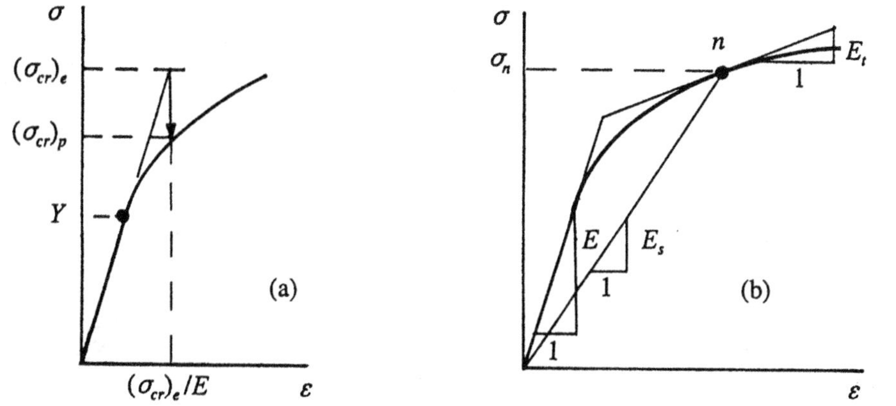

Figure 12.9 Tangent and secant moduli

The secant modulus gives the total strain at a reference point n (see Fig. 12.9b) as

$$\varepsilon_n = \sigma_n / E_s \tag{12.19}$$

The Ramberg-Osgood description [4] to a stress-strain curve gives the total strain under a plastic stress:

$$\varepsilon = \sigma / E + \alpha(\sigma / E)^m \tag{12.20}$$

Combining eqs(12.19) and (12.20) for the reference stress σ_n gives

$$E / E_s - 1 = \alpha(\sigma_n / E)^{m-1} \tag{12.21a}$$

Differentiating eq(12.20) gives the gradient $E_t = \mathrm{d}\sigma/\mathrm{d}\varepsilon$ at point n:

$$(1/m)(E / E_t - 1) = \alpha(\sigma / E)^{m-1} \tag{12.21b}$$

Let σ_n be the stress level at which $E_t = E/2$. Equation (12.21b) gives

$$\alpha = (1/m)(\sigma_n / E)^{1-m} \tag{12.21c}$$

Substituting eq(12.21c) into eq(12.20) and multiplying through by E/σ_n lead to a normalised stress-total strain relationship

$$\varepsilon = \frac{\sigma_n}{E}\left[\frac{\sigma}{\sigma_n} + \frac{1}{m}\left(\frac{\sigma}{\sigma_n}\right)^m\right] \tag{12.22}$$

where σ_n and m are material properties found from fitting eq(12.22) to a stress-strain curve.

Typical m values lie the range 16-29 for aluminium alloys and in the range 5-17 for steels. The m value depends upon whether the material is in sheet or bar form. Figure 12.9a shows that the elastic strain under $(\sigma_{cr})_e$ is $\varepsilon = (\sigma_{cr})_e/E$. Substituting this value for ε in eq(12.22) allows $\sigma = (\sigma_{cr})_p$ to be found. Design data [5] employ a *plasticity reduction factor* $\mu < 1$ with a graphical solution for $(\sigma_{cr})_p$:

$$\mu = (\sigma_{cr})_p / (\sigma_{cr})_e \tag{12.23}$$

Both m and σ_n influence μ in the manner shown in Fig. 12.10.

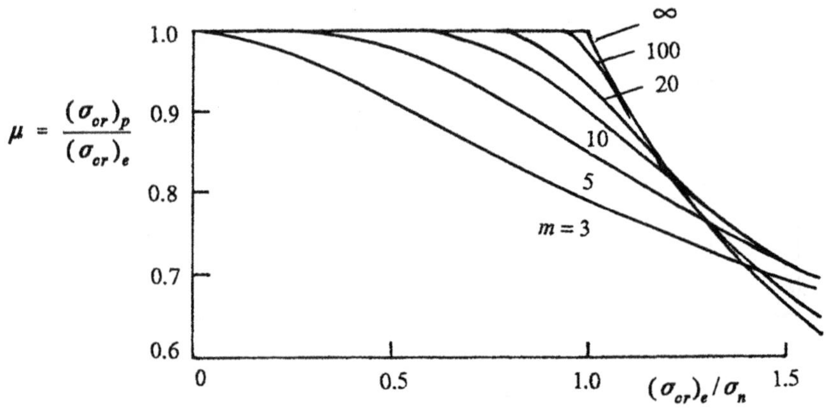

Figure 12.10 Plasticity reduction factor

Firstly, we find from eq(12.17) a critical elastic stress. This determines the ratio $(\sigma_{cr})_e/\sigma_n$ from which a μ value is found from Fig. 12.10. Equation (12.23) is then employed to find $(\sigma_{cr})_p$ as in the following example.

Example 12.2 A 320 mm square steel plate is 7 mm thick. It is is simply supported along all sides and carries a uni-axial compressive stress. Determine the critical elastic and plastic buckling stresses. What is the influence of clamping the unloaded sides upon the plastic buckling stress? Take: $E = 210$ GPa, $Y = 310$ MPa, $\sigma_n = 450$ MPa, $\nu = 0.27$ and $m = 5$.

The plate aspect ratio is an integral number $r = 1$. Equation (12.17) provides the critical elastic buckling stress for simply-supported sides:

$$\left(\sigma_{cr}\right)_e = \frac{\pi^2 \times 210 \times 10^3}{3(1 - 0.27^2)}\left(\frac{7}{320}\right)^2 = 356.59 \text{ MPa}$$

Since this exceeds the compressive yield stress $Y = 310$ MPa, a correction for plasticity becomes necessary. Thus, $(\sigma_{cr})_e/\sigma_n = 356.59/450 = 0.792$, when Fig. 12.10 gives $\mu = 0.916$ for $m = 5$. Hence we find from eq(12.23) a critical plastic buckling stress:

$$(\sigma_{cr})_p = 0.916 \times 356.59 = 326.64 \text{ MPa}$$

With the loaded sides clamped, Fig. 12.8 gives $(\sigma_{cr})_e = 1.7 \times 356.59 = 606.2$ MPa. Correcting for plasticity as before: $(\sigma_{cr})_e/\sigma_n = 606.2/450 = 1.347$. Figure 12.10 gives a reduction factor $\mu = 0.753$ when, from eq(12.23), the plastic buckling stress has been increased to $(\sigma_{cr})_p = 0.753 \times 606.2 = 456.47$ MPa.

12.3.2 Bi-axial Compression

Consider a simply-supported, thin rectangular plate $a \times b$ with thickness t. Let uniform compressive stresses σ_x act normal to $b \times t$ and σ_y act normal to $a \times t$, as shown in Fig. 12.11.

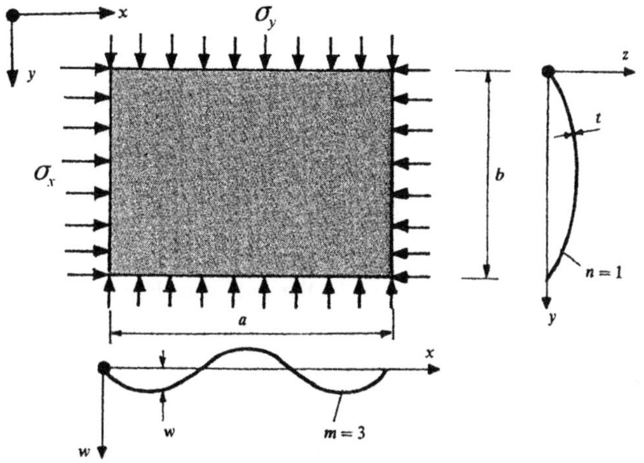

Figure 12.11 Buckling of a thin plate under biaxial stress

When the stresses increase proportionately in the ratio $\beta = \sigma_y/\sigma_x$ the actual number of half waves of buckling are those which minimise σ_x. In general, this stress is expressed as [2]

$$(\sigma_x)_{cr} = \frac{D\pi^2[(m/a)^2 + (n/b)^2]^2}{t[(m/a)^2 + \beta(n/b)^2]} \tag{12.24}$$

where D is the flexural stiffness and m and n are the respective number of half waves m and n for buckling in the x and y directions. In Fig. 12.11, for example, $m = 3$ and $n = 1$ is shown. For a square plate, where $a = b$, buckling occurs with one half wave in each direction and $m = n = 1$. Equation (12.24) reduces to

$$(\sigma_x)_{cr} = \frac{\pi^2 E}{3(1 - \nu^2)(1 + \beta)}\left(\frac{t}{a}\right)^2 \tag{12.25a}$$

and in the case of equi-biaxial compression $\beta = 1$, when eq(12.25a) gives

$$(\sigma_x)_{cr} = \frac{\pi^2 E}{6(1 - \nu^2)}\left(\frac{t}{a}\right)^2 \tag{12.25b}$$

Another useful buckling stress $(\sigma_x)_{cr}$ applies to $b << a$. Firstly, we must find the number of half wavelengths of the buckled shape for the x - direction. Taking $n = 1$ for the much smaller b dimension, eq(12.24) becomes

$$(\sigma_x)_{cr} = \frac{D\pi^2(m^2 + r^2)^2}{a^2 t(m^2 + \beta r^2)} = \frac{D\pi^2(m/r + r/m)^2}{b^2 t[1 + \beta(r/m)^2]} \tag{12.26}$$

and m is found from the condition that $(\sigma_x)_{cr}$ is a minimum. This is

$$d(\sigma_x)_{cr}/dm = [1 + \beta(r/m)^2]\, 2(m/r + r/m)(1/r - r/m^2) - (m/r + r/m)^2 \times 2\beta(r/m)(-\, r/m^2) = 0$$

$$[1 - (r/m)^2][1 + \beta(r/m)^2] + \beta(r/m)^2[1 + (r/m)^2] = 0$$

$$\therefore (r/m)^2(1 - 2\beta) = 1$$

This gives

$$m = r\,(1 - 2\beta)^{1/2} \tag{12.27}$$

with the corresponding half wavelength $a/m \approx r/m = (1 - 2\beta)^{-1/2}$. Substituting eq(12.27) into eq(12.26) gives

$$(\sigma_x)_{cr} = \frac{D\pi^2[(1 - 2\beta) + 2 + 1/(1 - 2\beta)]}{bt^2[1 + \beta/(1 - 2\beta)]} = \frac{4D\pi^2}{b^2 t}(1 - \beta)$$

It is apparent from eq(12.28a) that as $\beta \rightarrow$ ½ the half wavelength in the x-direction approaches infinity as the buckling stress becomes

$$(\sigma_x)_{cr} \rightarrow (2D\pi^2)\,/\,(b^2 t) \tag{12.28b}$$

The critical lateral stress follows from each of these solutions as $(\sigma_y)_{cr} = \beta\,(\sigma_x)_{cr}$. These stresses remain elastic provided they satisfy a von Mises yield criterion:

$$\sigma_x^2 - \sigma_x\sigma_y + \sigma_y^2 \leq Y^2$$

which gives

$$(\sigma_x)_{cr} \leq Y\,/\,(1 - \beta + \beta^2)^{1/2}$$

If $(\sigma_x)_{cr}$ exceeds this limiting elastic value a correction for plasticity may be made with a reduction factor as before. When applying Fig. 12.10 to one component of a biaxial stress it becomes necessary to re-define the abscissa as $(\sigma_x)_{cr}/\sigma_n'$ where $\sigma_n' = \sigma_n / (1 - \beta - \beta^2)^{1/2}$.

12.3.3 Local Buckling of Plate Sections

The straight, thin walls of an open section may distort without translation or rotation. Localised stress concentrations at corners may exceed the yield stress and cause either a local crippling failure or a reduction in the resisitance to buckling by other modes. Strut sections, consisting of straight sides such as I, ⊤,] ,⌊ and ⊔, as well as thin-walled closed tubes, are prone to local buckling at sharp corners. When the strut length is at least four times the section depth h, each limb may be treated as a plate with simple side support from one or both its neighbouring limbs. A supported side will restrain an unsupported side. The solution to the local elastic compressive buckling stress takes the common form

$$\sigma_{be} = KE\,(t\,/\,h)^2 \tag{12.29}$$

where the thickness t of the sides $t < d/5$ where d is the semi-flange length. The buckling coefficient K depends upon the d/h ratio and the neighbouring restraints within the four sections as shown in Fig. 12.12. When $\nu \neq 0.3$, K should be corrected by a multiplying factor $(1 - 0.3^2)/(1 - \nu^2)$. If σ_{be} from eq(12.29) is found to exceed the yield stress of the strut material it again becomes necessary to employ the plasticity reduction factor $\sigma_b = \mu\,\sigma_{be}$ with a suitable description to the stress-strain curve, as in the previous example.

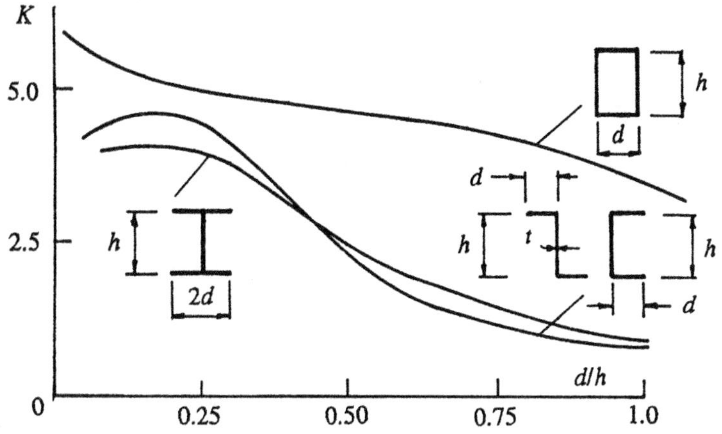

Figure 12.12 Local buckling coefficient for uniform thin sections

In medium length struts, local instability results in a loss of stiffness but without complete failure. However, the influence of local instability upon global buckling arising from flexure and torsion is important. If the strut is very long the global buckling stresses are attained well before the onset of local buckling. On the other hand, a short strut may carry more compression beyond that producing local buckling. Final failure is estimated to occur under a crippling stress σ_c, found from

$$\sigma_c = (\sigma_b\, \sigma_{0.1})^{1/2} \tag{12.30}$$

where $\sigma_{0.1}$ refers to the 0.1% compressive proof stres for the strut material. Test data shows that eq(12.30) is accurate to within 10%.

Example 12.3 Find the buckling and crippling stresses for a short strut of ⅂- section, given its dimensions: h = 125 mm, d = 25 mm and t = 4 mm (see Fig. 12.12). Take E = 74 GPa, $\nu = 0.3$, m = 10, $\sigma_{0.1}$ = 325 MPa, and σ_n = 280 MPa.

From Fig. 12.12, we find K = 4 for d/h = 0.2. Equation (12.29) gives the local elastic buckling stress

$$\sigma_{be} = 4 \times 74 \times 10^3 (4\,/\,125)^2 = 303.1 \text{ MPa}$$

This exceeds the yield stress because σ_n = 280 MPa corresponds to $E_t = 2E$ as in Fig. 12.9b. The ratio σ_{be}/σ_n = 303.1/280 = 1.083 is used with Fig. 12.10 to give a plasticity reduction factor μ = 0.877. Equation (12.23) provides the local buckling stress

$$\sigma_b = \mu\ \sigma_{be} = 0.877 \times 303.1 = 265.82 \text{ MPa}$$

and using eq(12.30), a crippling stress is estimated:

$$\sigma_c = (265.82 \times 325)^{1/2} = 293.92 \text{ MPa}$$

12.3.4 Post-Buckling of Flat Plates

When a plate buckles the load may be increased further as the axial compressive stress increases in the material along the side supports (see Fig. 12.13).

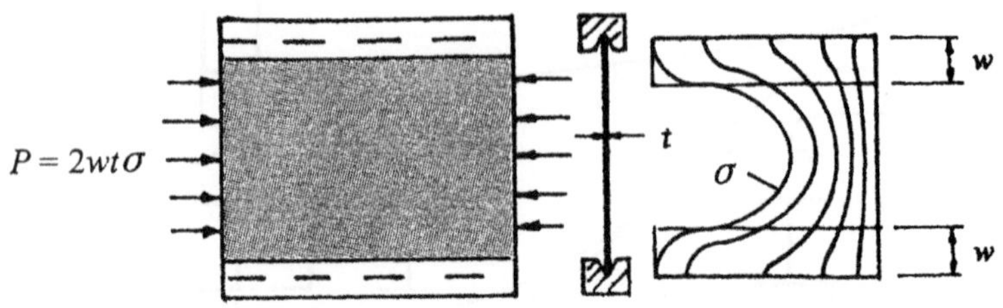

Figure 12.13 Stress distribution in a buckled plate

Only a slight increase in axial stress occurs in the central buckled material. Consequently, if we assume that the whole load P is carried by two edge strips of effective width $2w$, over which σ is assumed constant, the load supported becomes

$$P = 2\,w\,t\,\sigma$$

With the edges of our equivalent elastic plate all simply supported, the critical buckling stress is [2]:

$$\left(\sigma_x\right)_{cr} = \frac{\pi^2 E}{3(1 - \nu^2)}\left(\frac{t}{2w}\right)^2 \tag{12.31a}$$

from which w may be found once $(\sigma_x)_{cr}$ attains the yield stress Y:

$$w = \frac{\pi t}{2}\sqrt{\frac{E}{3Y(1 - \nu^2)}} \tag{12.31b}$$

Taking $\nu = 0.3$ in eq(12.31b) gives

$$w = 0.95\,t\sqrt{E/Y}$$

Experiment shows that the coefficient is nearer 0.85. With other edge fixings along the long sides, the asymptotic values of the critical elastic buckling stress ratio applies. These appear in Fig. 12.8 in which eq(12.31a) defines the denominator. For example, when one side is simply supported and the other is free, Fig. 12.8 gives $\sigma_{cr}/\sigma_b = 0.106$ and this modifies the semi-effective width:

$$w = 0.31\,t\sqrt{E/Y}$$

12.3.5 Buckling of Plates in Shear

When the sides of a thin plate $a \times b \times t$ are subjected to shear stress τ, the principal stress state (σ_1, σ_2) within the plate becomes one of diagonal tension and compression. The principal stresses are of equal magnitude: $\sigma_1 = \tau$ and $\sigma_2 = -\tau$ (see Fig. 12.14a). Hence, we have a principal, biaxial stress ratio $\beta = -1$. Shear buckling occurs with parallel wrinkles lying perpendicular to the compressive stress σ_2 (see Fig. 12.14b).

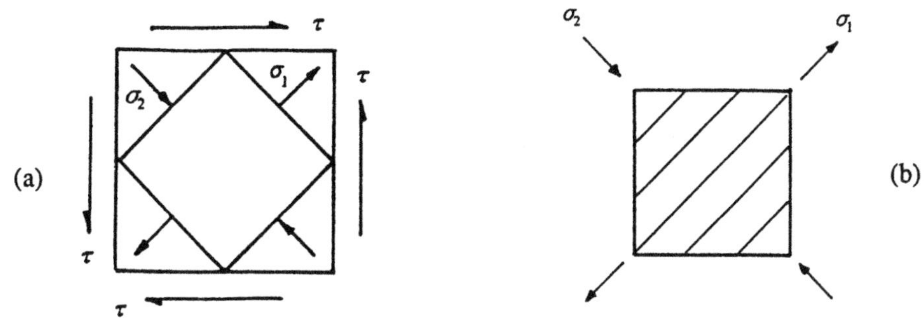

Figure 12.14 Plate in shear showing principal stress directions

Under this condition the plate cannot sustain a further increase in diagonal compression though an increase in diagonal tension is possible. Thus, a compressive buckling of flat plates arises under shear. The critical elastic shear stress is given by [6]

$$(\tau_{cr})_e = K_s E\,(t\,/\,b)^2 \tag{12.32}$$

where b is the lesser side length. The shear buckling coefficient K_s depends upon the edge fixing in the manner of Fig. 12.15 when $\nu = 0.3$.

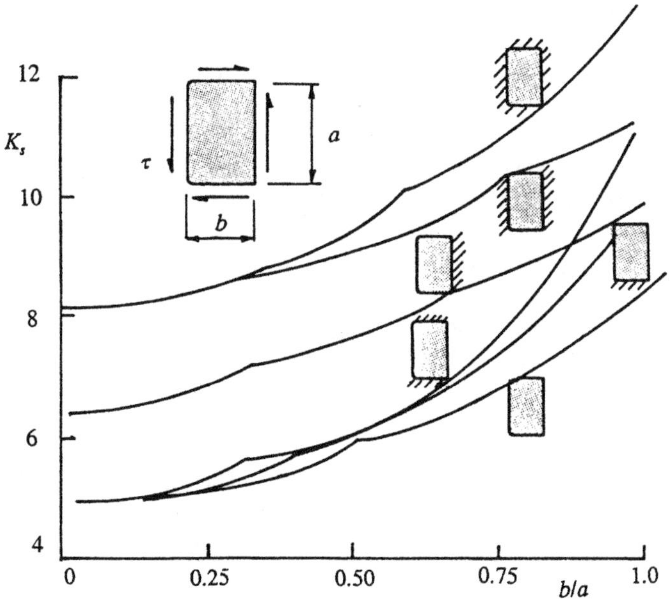

Figure 12.15 Dependence of shear buckling coefficient upon geometry and edge-fixing

If $\nu \neq 0.3$ the K_s value is factored by $0.91/(1 - \nu^2)$. Equation (12.32) supplies $(\tau_{cr})_e$ but this critical elastic stress will need further correction if it exceeds the shear yield stress k of the plate material [7]. The plasticity reduction factor μ is obtained from Fig. 12.16 knowing σ_n and m for the material. This gives:

$$(\tau_{cr})_p = \mu\,(\tau_{cr})_e \tag{12.33}$$

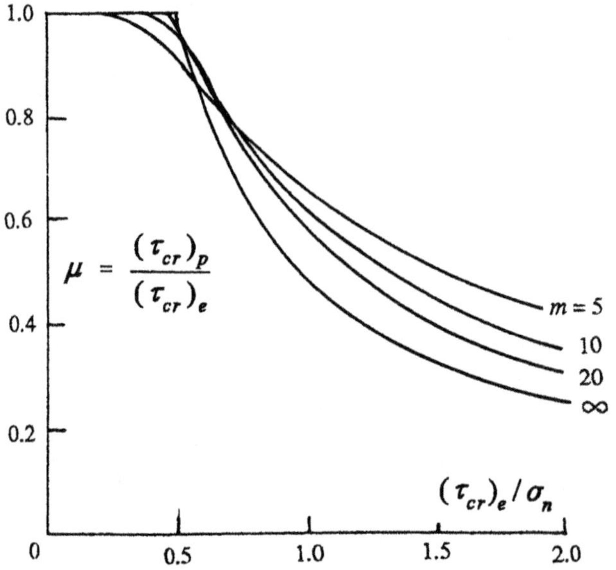

Figure 12.16 Plasticity reduction factor in shear

Example 12.4 Find the critical buckling stress in shear loading an aluminium plate 2.5 mm thick, 100 mm wide and 200 mm long. The longer sides are clamped and the shortes sides are simply supported. Plate material properties are: $k = 200$ MPa, $\sigma_n = 350$ MPa, $m = 15$, $E = 70$ GPa and $\nu = 0.33$.

For $b/a = 0.5$, the appropriate curve in Fig. 12.15 gives $K_s = 9.0$. Correcting for Poisson's ratio gives: $K_s = 9.0 \times 0.91/(1 - 0.33^2) = 9.19$. The elastic buckling stress is therefore, from eq(12.32),

$$(\tau_{cr})_e = 9.19 \times 70 \times 10^3 (2.5 / 100)^2 = 402.1 \text{ MPa}$$

Since this far exceeds the shear yield stress $k = 200$ MPa, a plasticity reduction factor is required. The abscissa in Fig. 12.16 now has a value:

$$(\tau_{cr})_e / \sigma_n = 402.1/ 350 = 1.149$$

Interpolating for $m = 15$ gives an ordinate of $\mu = 0.52$. Therefore from eq(12.33):

$$(\tau_{cr})_p = 0.52 \times 402.1 = 209.1 \text{ MPa}$$

12.4 Tensile Instability

12.4.1 Uniaxial Tension

In Chapter 9 it was shown that the maximum load condition (9.13a) provided the relation between nominal stress and strain at the inception of necking in a tension test. The various Considére expressions (9.14a-c) employ nominal or true stress, engineering or natural (logarithmic) strain according to the manner of presenting a stress-strain diagram. The empirical equations used to describe flow curves (see Figs 9.19a-f) employ true stress and natural strain so these are to be used with eq(9.14b) as the following example shows.

Example 12.5 Determine the true stress and true strain at the point of tensile instability for the Hollomon and Ludwik functions.

From eq(9.39b),

$$d\sigma / d\varepsilon = n\,(\sigma_o / \varepsilon_o)(\varepsilon / \varepsilon_o)^{n-1}$$

and from eq(9.14b), at the point of instability $d\sigma/d\varepsilon = \sigma$:

$$n(\sigma_o / \varepsilon_o)(\varepsilon / \varepsilon_o)^{n-1} = \sigma_o (\varepsilon / \varepsilon_o)^n$$

which shows that:

$$\varepsilon = n \quad \text{and} \quad \sigma = \sigma_o (n / \varepsilon_o)^n$$

That is, the strain at which necking begins equals the work hardening exponent for a given material. The Ludwick law (9.43a) gives

$$d\sigma / d\varepsilon^P = n\,A\,(\varepsilon^P)^{n-1}$$

Combining this with eqs (9.14c) and (9.43a) we find the instability strain

$$\varepsilon^P \simeq n\,(1 - Y / \sigma)$$

giving a quadratic in the instability stress

$$\sigma^2 - (Y + An^n)\,\sigma + A\,n^{n+1}\,Y = 0$$

12.4.2 Orthotropic Sheet

The previous example assumes an isotropic condition, i.e. the analysis is independent of the testpiece orientation. In orthotropic sheet both the condition of the material and the orientation of the testpiece need to be considered. Recall that equivalence expresses a current stress and strain state aligned with the material's orthogonal directions. Using eqs(11.2) and (11.4a), a plane stress yield function f appears in terms of an anisotropy parameters F, G, H and the equivalent stress $\bar{\sigma}$ as

$$2f = (2/3)(F + G + H)\,\bar{\sigma}^2 = (G + H)\sigma_1^2 - 2H\sigma_1\sigma_2 + (F + H)\sigma_2^2 = 1 \qquad (12.34)$$

Re-writing eq(12.34) in terms of the r values: $r_1 = H/G$ and $r_2 = H/F$, leads to an equivalent plane stress expression:

$$\bar{\sigma}^2 = \frac{r_2(1 + r_1)\sigma_1^2 - 2r_1 r_2 \sigma_1 \sigma_2 + r_1(1 + r_2)\sigma_2^2}{\frac{2}{3}(r_1 + r_1 r_2 + r_2)} \qquad (12.35)$$

The equivalent plastic strain increment is found from the plastic work expression (11.8b):

$$\bar{\sigma}\,d\bar{\varepsilon}^P = \sigma_{ij}\,d\varepsilon_{ij}^P$$

From eq(12.35),

$$\left(d\bar{\varepsilon}^{P}\right)^2 = \frac{2(r_1 + r_1 r_2 + r_2)\left[(1/r_2)(d\varepsilon_1^P)^2 + (1/r_1)(d\varepsilon_2^P)^2 + (d\varepsilon_3^P)^2\right]}{3(1 + r_1 + r_2)} \tag{12.36}$$

When a tensile testpiece is aligned with the sheet's rolling direction the equivalent stress follows from eqs(12.35) as

$$\bar{\sigma}^2 = \frac{3r_2(1 + r_1)\sigma_1^2}{2(r_1 + r_1 r_2 + r_2)} \tag{12.37}$$

where σ_1 is the tensile stress. The lateral strain increments are

$$d\varepsilon_2^P = -\frac{r_1 d\varepsilon_1^P}{1 + r_1} \quad \text{and} \quad d\varepsilon_3^P = -\frac{d\varepsilon_1^P}{1 + r_1} \tag{12.38a,b}$$

Substituting eqs(12.38a,b) into eq(12.36) gives the total equivalent plastic strain

$$\left(\bar{\varepsilon}^{P}\right)^2 = \frac{2(r_1 + r_1 r_2 + r_2)}{3r_2(1 + r_1)}\left(\varepsilon_1^P\right)^2 \tag{12.39}$$

If we write eq(12.37) as $\bar{\sigma} = K\sigma_1$ then eq(12.39) becomes $\bar{\varepsilon}^P = (\varepsilon_1^P)/K$ where K is the square root of the coefficient containing r_1 and r_2. It follows that the instability condition (9.14b) applies to the true stress-strain curve for the 1- direction as

$$d\sigma_1/d\varepsilon_1^P = \sigma_1 \tag{12.40a}$$

Equation (12.40a) may be written as

$$\frac{d\sigma_1}{d\bar{\sigma}} \times \frac{d\bar{\varepsilon}^P}{d\varepsilon_1} \times \frac{d\bar{\sigma}}{d\bar{\varepsilon}^P} = \frac{\bar{\sigma}}{K}$$

giving

$$\frac{d\bar{\sigma}}{d\bar{\varepsilon}^P} = K\bar{\sigma} \tag{12.40b}$$

so that the reciprocal of K becomes the strain intercept (i.e. z in Fig. 12.17) made by the tangent to the $\bar{\sigma}$ versus $\bar{\varepsilon}^P$ curve at the point of instability.

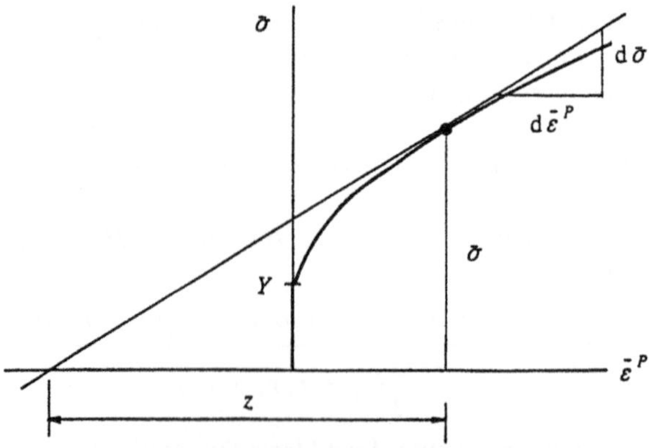

Figure 12.17 Sub-tangent to an equivalent flow curve

If we approximate this curve with the Hollomon law (9.39b) this gives $\mathrm{d}\bar{\sigma}/\mathrm{d}\bar{\varepsilon}^P = n\bar{\sigma}/\bar{\varepsilon}^P$ from which $K = n/\bar{\varepsilon}^P$. The equivalent strain at the point of instability is therefore $\bar{\varepsilon}^P = n/K$. It is also possible to predict this strain for off-axis testpieces using a single flow curve equation and the strain transformation equations in a similar manner [8].

12.4.3 *Subtangent*

We may extend the previous instability analysis to problems involving more than one stress with the use of equivalent stress and plastic strain. Equation (12.40b) shows that this will modify Considére's condition to

$$\mathrm{d}\bar{\sigma}/\mathrm{d}\bar{\varepsilon}^P = \bar{\sigma}/z \qquad (12.40c)$$

where z is the subtangent that defines the particular problem. Figure 12.17 gives a geometrical interpretation to eq(12.40c) in which z is the intercept along the strain axis made with the gradient $\mathrm{d}\bar{\sigma}/\mathrm{d}\bar{\varepsilon}^P$ of the tangent and the ordinate $\bar{\sigma}$ as shown. Clearly, for a tension test upon isotropic material $z = 1$ since $\bar{\sigma}$ and $\bar{\varepsilon}^P$ are, by definition, the axial stress and axial plastic strain in that test. For tension upon an orthotropic sheet, discussed above,

$$z = \frac{1}{K} = \sqrt{\frac{2(r_1 + r_2 + r_1 r_2)}{3r_2(1 + r_1)}}$$

Other problems require z to be found usually from combining a maximum pressure instability criterion with constitutive relations defining the material condition. The conversions will enable $\bar{\sigma}$ and $\bar{\varepsilon}^P$ to be written in terms of a single dependent variable allowing the application of eq(12.40c), as the following examples will show.

(a) *Spherical Pressure Vessel*

Let the pressure vessel's isotropic steel have an equivalent stress-strain curve represented by a Swift law:

$$\bar{\sigma} = \sigma_o\left(1 + \bar{\varepsilon}^P/\varepsilon_o\right)^n$$

It is required to find expressions for (i) the internal pressure at which a bi-axial tensile instability occurs, (ii) the equivalent strain and (iii) the maximum hoop stress and strain reached at the point of instability. The hoop and meridional stresses are $\sigma_\theta = \sigma_\phi = pr/2t$, from which

$$\ln \sigma_\theta = \ln (p/2) + \ln r - \ln t$$

$$\mathrm{d}\sigma_\theta/\sigma_\theta = \mathrm{d}p/p + \mathrm{d}r/r - \mathrm{d}t/t$$

Since $\mathrm{d}p/p = 0$ at the point of instability and $\mathrm{d}\varepsilon_\theta^P = \mathrm{d}r/r$, $\mathrm{d}\varepsilon_r^P = \mathrm{d}t/t$ then

$$\mathrm{d}\sigma_\theta/\sigma_\theta = \mathrm{d}\varepsilon_\theta^P - \mathrm{d}\varepsilon_r^P \qquad (12.41)$$

For the Mises equivalent stress (9.2d) we replace subscripts 1, 2 and 3 by θ, ϕ and r to identify the stress components: $\sigma_1 = \sigma_\theta$, $\sigma_2 = \sigma_\phi = \sigma_\theta$ and $\sigma_3 = \sigma_r = 0$. Thus:

$$\bar{\sigma} = \frac{1}{\sqrt{2}}\sqrt{0 + \sigma_\phi^2 + \sigma_\theta^2} = \sigma_\theta \qquad (12.42)$$

The Levy-Mises equation (9.29a-c) gives

$$d\varepsilon_\theta{}^P = (2\,d\lambda/3)\sigma_\theta/2$$

$$d\varepsilon_\phi{}^P = (2\,d\lambda/3)\sigma_\theta/2 = d\varepsilon_\theta{}^P$$

$$d\varepsilon_r{}^P = -\,(d\varepsilon_\theta{}^P + d\varepsilon_\phi{}^P) = -\,2\,d\varepsilon_\theta{}^P$$

$$\therefore\ d\varepsilon_\theta{}^P = d\varepsilon_\phi{}^P = -\,d\varepsilon_r{}^P/2 \qquad (12.43)$$

when from eq(9.6d) the equivalent plastic strain increment is

$$d\bar{\varepsilon}^P = \surd(2/3)\,\surd[(d\varepsilon_\theta{}^P)^2 + (d\varepsilon_\theta{}^P)^2 + (-\,2d\varepsilon_\theta{}^P)^2] = 2\,d\varepsilon_\theta{}^P \qquad (12.44)$$

From eqs(12.42) and (12.44):

$$d\bar{\sigma}/d\bar{\varepsilon}^P = \tfrac{1}{2}\,d\sigma_\theta/d\varepsilon_\theta^P$$

where, from eqs(12.41) and (12.43)

$$d\sigma_\theta/d\varepsilon_\theta^P = \sigma_\theta(1 -\ d\varepsilon_r{}^P/\,d\varepsilon_\theta^P) = 3\sigma_\theta$$

$$\therefore\ d\bar{\sigma}/d\bar{\varepsilon}^P = 3\,\bar{\sigma}/2$$

Thus, in the equivalent plot of Fig. 12.17, the subtangent value is $z = 2/3$. The equivalent plastic instability strain is found from the Swift law:

$$d\bar{\sigma}/d\bar{\varepsilon}^P = n\,\bar{\sigma}/[\varepsilon_o(1 + \bar{\varepsilon}^P/\varepsilon_o)] = 3\,\bar{\sigma}/2$$

$$\therefore\ \bar{\varepsilon}^P = 2n/3 - \varepsilon_o$$

from which the maximum hoop strain is $\varepsilon_\theta^P = \tfrac{1}{2}\,\bar{\varepsilon}^P$ and the equivalent stress at instability is $\bar{\sigma} = \sigma_o[2n/(3\,\varepsilon_o)]^n$. Since $\bar{\sigma} = \sigma_\theta = pr/2t$, this gives the critical pressure as

$$p = \frac{2t\sigma_o}{r}\left(\frac{2n}{3\,\varepsilon_o}\right)^n \qquad (12.45)$$

(b) *Cylindrical Pressure Vessel*

For a thin-walled, closed cylinder the stresses are $\sigma_r = 0$ and $\sigma_\theta = pr/t = 2\sigma_z$. Hence eq(12.41) again applies to the point of instability. The equivalent stress eq(9.2d) gives

$$\bar{\sigma} = (1/\surd2)\,\surd[(\sigma_\theta - \sigma_z)^2 + \sigma_\theta{}^2 + \sigma_z{}^2]$$

$$= (1/\surd2)\,\surd[\tfrac{1}{4}(\sigma_\theta)^2 + \sigma_\theta{}^2 + \tfrac{1}{4}(\sigma_\theta)^2] = \surd3\,\sigma_\theta/2 \qquad (12.46)$$

and from eq(9.29a-c), it follows that $d\varepsilon_z{}^P = 0$ and:

$$d\varepsilon_\theta{}^P = (2\,d\lambda/3)[\sigma_\theta - \tfrac{1}{2}(0 + \tfrac{1}{2}\sigma_\theta)] = (d\lambda/2)\sigma_\theta$$

$$d\varepsilon_r{}^P = (2\,d\lambda/3)[0 - \tfrac{1}{2}(\sigma_\theta + \tfrac{1}{2}\sigma_\theta)] = -\,(d\lambda/2)\sigma_\theta$$

$$\therefore\ d\varepsilon_r{}^P = -\,d\varepsilon_\theta{}^P \qquad (12.47)$$

The equivalent plastic strain is, from eq(9.6e),

$$d\bar{\varepsilon}^P = (\sqrt{2/3})\sqrt{[(d\varepsilon_\theta^P)^2 + (-\,d\varepsilon_\theta^P)^2 + (0)^2]} = (2/\sqrt{3})d\varepsilon_\theta^P \qquad (12.48)$$

Combining eqs(12.46) and (12.48) gives

$$d\bar{\sigma}/d\bar{\varepsilon}^P = (3d\sigma_\theta)/(4d\varepsilon_\theta^P)$$

where, from eqs(12.41) and (12.46),

$$d\bar{\sigma}/d\bar{\varepsilon}^P = \tfrac{3}{4}\sigma_\theta(d\varepsilon_\theta^P - d\varepsilon_r^P)/(d\varepsilon_\theta^P) = 3\sigma_\theta/2$$

$$\therefore\quad d\bar{\sigma}/d\bar{\varepsilon}^P = \sqrt{3}\,\bar{\sigma}$$

i.e. a subtangent of $1/\sqrt{3}$ in this case. Hence from the given law

$$d\bar{\sigma}/d\bar{\varepsilon}^P = n\bar{\sigma}/(\varepsilon_o + \bar{\varepsilon}^P) = \sqrt{3}\,\bar{\sigma}$$

$$\therefore\quad \bar{\varepsilon}^P = n/\sqrt{3} - \varepsilon_o$$

The corresponding equivalent stress is $\bar{\sigma} = \sigma_o(n/\sqrt{3}\,\varepsilon_o)^n$, giving a critical pressure as

$$p = \frac{2t\sigma_o}{\sqrt{3}r}\left(\frac{n}{\sqrt{3}\,\varepsilon_o}\right)^n$$

12.5 Circular Bulge Instability

The attainment of maximum pressure in a bulge test is similar to that for a spherical vessel. However, eq(12.45) is inappropriate because the bulge continuously forms out of the plane of the sheet. Consider the section of the bulge shown in Fig. 9.16 in which an instability arises from bulging isotropic material in a circular die. The radius of curvature at an instant where the current height is h, is given by:

$$R = (a^2 + h^2)/(2h) \qquad (12.49)$$

where $a = D/2$ is the radius of the circular die and h is the bulge height. The in-plane, equi-biaxial stress at the pole consists of hoop and meridional components (see eqs(9.33a,b)), as with a spherical vessel:

$$\sigma_\theta = \sigma_\phi = pR/2t \qquad (12.50)$$

Differentiating eq(12.50) gives

$$\delta\sigma_\theta/\sigma_\theta = \delta p/p + \delta R/R - \delta t/t \qquad (12.51a)$$

where $\delta\varepsilon_r^P = \delta t/t$ defines the pole thickness strain. Setting $\delta p/p = 0$ in eq(12.51a) gives an instability condition

$$(1/\sigma_\theta)(\delta\sigma_\theta/\delta\varepsilon_r^P) = -\,1 + (1/R)(\delta R/\delta\varepsilon_r^P) \qquad (12.51b)$$

The hoop and meridional pole strain increments are each equal to $\delta h/R$ so it follows from the incompressibility condition that $\delta\varepsilon_r^P = -2\delta h/R$ giving $\delta h/\delta\varepsilon_r^P = -\tfrac{1}{2}R$. Also, from differentiating eq(12.49), $\delta R/\delta h = 1 - R/h$ and, therefore

$$(1/R)(\delta R/\delta\varepsilon_r^P) = (1/R)(\delta R/\delta h)(\delta h/\delta\varepsilon_r^P) = -\tfrac{1}{2}(1 - R/h) \qquad (12.52)$$

Substituting eq(12.52) into eq(12.51b) gives

$$\frac{1}{\sigma_\theta}\frac{\delta\sigma_\theta}{\delta\varepsilon_r^P} = \frac{1}{\bar{\sigma}}\frac{\delta\bar{\sigma}}{\delta\bar{\varepsilon}^P} = -\frac{3}{2} + \frac{R}{2h} \qquad (12.53)$$

We have readily converted the left-hand side of eq(12.53) into equivalent stress and strain since $\bar{\sigma} = \sigma_\phi = \sigma_\theta$ and $d\bar{\varepsilon}^P = d\varepsilon_r^P$ at the pole of a bulge (see eqs(9.33b) and (9.36a)). The term $R/2h$ within the right-hand side of eq(12.53) appears from an integration involving eq(12.49) and a series expansion as follows:

$$\varepsilon_r^P = -2\int dh/R = -2\int_0^h 2h\,dh/(a^2 + h^2) = -2\ln[2Rh/(2Rh - h^2)]$$

$$1 - h/2R = \exp(\varepsilon_r^P/2) = 1 + \varepsilon_r^P/2 + (\varepsilon_r^P)^2/8 + \ldots$$

$$h/2R = -\tfrac{1}{2} \times \varepsilon_r^P(1 + \tfrac{1}{4}\varepsilon_r^P)$$

$$R/2h = -[1/(2\varepsilon_r^P)](1 + \varepsilon_r^P/4)^{-1}$$

$$= -[1/(2\varepsilon_r^P)][1 - \varepsilon_r^P/4 + (\varepsilon_r^P)^2/16 \ldots]$$

$$\simeq -1/(2\varepsilon_r^P) + 1/8 \qquad (12.54)$$

in which squared strain is neglected. Note that ε_r^P is negative since the pole material thins but, by taking $\bar{\varepsilon}^P = \varepsilon_r^P$ as a positive quantity, it follows from eqs(12.53) and (12.54) that

$$\frac{d\bar{\sigma}}{d\bar{\varepsilon}^P} = \left(\frac{11}{8} - \frac{1}{2\bar{\varepsilon}^P}\right)\bar{\sigma} \qquad (12.55)$$

Comparing eq(12.55) with eq(12.40b), the sub-tangent is written as [9]

$$\frac{1}{z} = \frac{11}{8} - \frac{1}{2\bar{\varepsilon}_I^P} \qquad (12.56)$$

where $\bar{\varepsilon}_I^P$ is the equivalent instability strain. Combined with Swift law (9.44b), this subtangent becomes

$$\frac{1}{z} = \frac{n}{\bar{\varepsilon}_I^P + \varepsilon_o} \qquad (12.57)$$

When eqs(12.56) and (12.57) are combined $\bar{\varepsilon}_I^P$ becomes one root to the quadratic:

$$(\bar{\varepsilon}_I^P)^2 + [\varepsilon_o - (4/11)(2n + 1)]\bar{\varepsilon}_I^P - 4\varepsilon_o/11 = 0 \qquad (12.58a)$$

In the case of an annealed material, where $\varepsilon_o \simeq 0$, eq(12.58a) provides

$$\bar{\varepsilon}_I^P = (4/11)(2n + 1) \qquad (12.58b)$$

Equation (12.58b) will also apply for a Hollomon hardening law (9.39b). Alternatively, using his own hardening law, Swift [10] identified the critical pole strain as a root to the cubic equation in $\bar{\varepsilon}_I^{\,P}$. If we use a Hollomon's hardening law instead, Swift's analysis results in the following quadratic equation:

$$(\bar{\varepsilon}_I^{\,P})^2 - (n + 33/10)\,\bar{\varepsilon}_I^{\,P} + (9/10)[1 + \sqrt{(363/50)}] = 0 \qquad (12.59)$$

For example, with a hardening exponent $n = 0.6$ for annealed stainless steel, we find $\bar{\varepsilon}_I^{\,P}$ as 0.8 and 1.26 from eqs(12.58b) and (12.59). These compare with $\bar{\varepsilon}_I^{\,P} = 0.77$ from eq(12.58a) on taking $\varepsilon_o = 0.048$ for this material.

12.6 Ellipsoidal Bulging of Orthotropic Sheet

We may extend the circular bulge analysis to a more general problem of ellipsoidal bulging of orthotropic sheet metal. The problem is simplified when the rolling direction of the sheet is aligned with the minor axis of the die aperture (i.e. the major principal strain direction), as shown in Fig. 12.18.

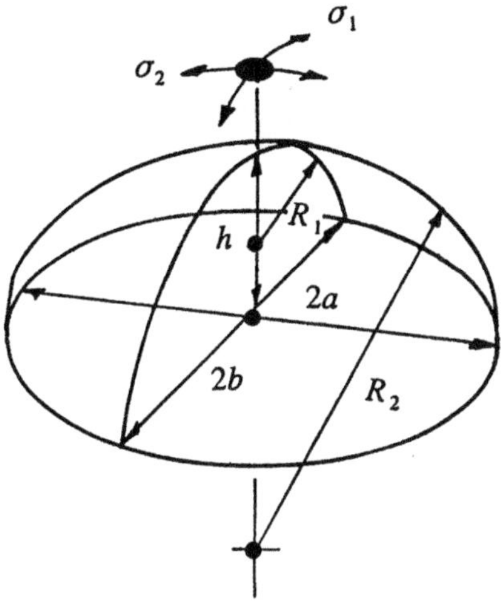

Figure 12.18 Ellipsoidal bulge geometry

The principal radii of curvature are

$$R_1 = (h^2 + b^2)\,/\,2h \quad \text{and} \quad R_2 = (h^2 + a^2)\,/\,2h \qquad (12.60a,b)$$

where h is the bulge height and a, b are the semi-lengths of the major and minor axes of the die aperture respectively. Figure 12.18 shows the principal stress state at the pole, from which the corresponding strain increments are

$$d\varepsilon_1^{\,P} = dh/R_1 = d\lambda\,[H(\sigma_1 - \sigma_2) + G\sigma_1] \qquad (12.61a)$$

$$d\varepsilon_2^{\,P} = dh/R_2 = d\lambda\,[F\sigma_2 - H(\sigma_1 - \sigma_2)] \qquad (12.61b)$$

Dividing eqs(12.61a,b) allows the stress ratio to be found:

$$Q = \sigma_2/\sigma_1 = [R_2 + R_1(1 + 1/r_1)]\,/\,[R_1 + R_2(1 + 1/r_2)] \qquad (12.62)$$

Given a lateral pressure p and current thickness t, the membrane equilibrium equation for the pole's in plane principal stresses is [9]

$$\sigma_1 / R_1 + \sigma_2 / R_2 = p / t \tag{12.63a}$$

which may be written as

$$\sigma_1 = pR / t \tag{12.63b}$$

where

$$R = R_1 R_2 / (R_2 + QR_1) \tag{12.63c}$$

12.6.1 Instability Strains

Differentiating eq(12.63b) and setting $\delta p/p = 0$ gives the condition of instability as

$$\delta\sigma_1/\sigma_1 = \delta R/R - \delta\varepsilon_3{}^P \tag{12.64}$$

where

$$\delta R/R = (\partial R/\partial R_1)(\delta R_1/R) + (\partial R/\partial R_2)(\delta R_2/R) \tag{12.65a}$$

Now from eqs(12.60a,b) and (12.63c), eq(12.65a) becomes

$$\delta R/R = \Phi\, \delta R_1/R_1 + \Omega\, \delta R_2/R_2 \tag{12.65b}$$

where

$$\Phi = \frac{2 + (1 + 1/r_2)[2q - (1 + 1/r_1) + q^2(1 + 1/r_2)]}{[2 + q(1 + 1/r_2) + (1 + 1/r_1)/q][1 + (1 + 1/r_2)q]} \tag{12.66a}$$

$$\Omega = \frac{(1 + 1/r_2) + 2(1 + 1/r_1)(1 + 1/r_2)/q + (1 + 1/r_1)/q^2]}{[2 + q(1 + 1/r_2) + (1 + 1/r_1)/q][1/q + (1 + 1/r_2)]} \tag{12.66b}$$

in which $q = R_2/R_1$. Dividing eq(12.64) by $\delta\varepsilon_3{}^P$ and introducing eqs(12.65b) leads to

$$-(1/\sigma_1)(\delta\sigma_1/\delta\varepsilon_3{}^P) = -1 + (\Phi/R_1)(\delta R_1/\delta\varepsilon_3{}^P) + (\Omega/R_2)(\delta R_2/\delta\varepsilon_3{}^P) \tag{12.67a}$$

Introducing, from eqs(12.60a,b), the following relationships:

$$\delta R_1/\delta h = 1 - R_1/h,\ \delta R_2/\delta h = 1 - R_2/h \text{ and } \delta h/\delta\varepsilon_3{}^P = -R_1R_2/(R_1 + R_2)$$

where $\delta\varepsilon_3 P = -(\delta\varepsilon_1{}^P + \delta\varepsilon_2{}^P)$, eq(12.67a) becomes

$$-(1/\sigma_1)(\delta\sigma_1/\delta\varepsilon_3{}^P) = -1 - \Phi(1 - R_1/h)/(1 + R_1/R_2) - \Omega(1 - R_2/h)/(1 + R_2/R_1) \tag{12.67b}$$

Equation (12.67b) will supply the critical subtangent at instability when σ_1 and $\delta\varepsilon_3{}^P$ are converted to equivalent stress and strain. To do this we use reduce eqs(11.4b) and (11.13) to one dependent variable as follows:

$$\bar{\sigma} = X\sigma_1,\quad d\bar{\varepsilon}^P = Z\,d\varepsilon_1^P \tag{12.68a,b}$$

where

$$X = \sqrt{\frac{3\left[r_2(1 + r_1) - 2Qr_1r_2 + Q^2r_1(1 + r_2)\right]}{2(r_1 + r_2 + r_1r_2)}} \qquad (12.69a)$$

$$Z = \sqrt{a + b\left(\frac{d\varepsilon_2^P}{d\varepsilon_1^P}\right)^2 + c\left(\frac{d\varepsilon_2^P}{d\varepsilon_1^P}\right)} = \sqrt{a + b\left(\frac{R_1}{R_2}\right)^2 + c\left(\frac{R_1}{R_2}\right)} \qquad (12.69b)$$

The coefficients a, b and c in eq(12.69b) are

$$a = [2\,(r_1 + r_2 + r_1r_2)(1 + r_2)]/\,[3\,r_2(1 + r_1 + r_2)]$$
$$b = [2\,(r_1 + r_2 + r_1r_2)(1 + r_1)]/\,[3\,r_1(1 + r_1 + r_2)]$$
$$c = [4\,(r_1 + r_2 + r_1r_2)]/\,[3\,(1 + r_1 + r_2)]$$

The thickness strain follows from the incompressibility condition as

$$\delta\varepsilon_3^P = -\,(\delta\varepsilon_1^P + \delta\varepsilon_2^P) = -\,(1 + R_1/R_2)\delta\varepsilon_1^P \qquad (12.69c)$$

Combining eqs(12.67b), (12.68a,b) and (12.69c) and applying eq(12.40b) leads to the sub-tangent expression

$$\frac{1}{z} = \frac{(R_1 + R_2)}{ZR_2}\left\{1 + \frac{\Phi[(1 + 1/r_1) - Q]}{(1/r_1 + Q/r_2)}\left(\frac{3}{4} - \frac{Z}{2\bar{\varepsilon}^P}\right)\right.$$
$$\left. + \frac{\Omega[Q(1 + 1/r_2) - 1]}{(1/r_1 + Q/r_2)}\left(\frac{3}{4} - \frac{ZR_2}{2R_1\bar{\varepsilon}^P}\right)\right\} \qquad (12.70)$$

Taking $1/z = n/\bar{\varepsilon}^P$ from the Hollomon law enables eq(12.70) to be solved for the equivalent strain at instability. The pole strain follows from eqs(12.69a,b,c) as

$$\varepsilon_1^P = \frac{\bar{\varepsilon}^P}{Z}, \quad \varepsilon_2^P = \frac{R_1}{R_2}\varepsilon_1^P \quad \text{and} \quad \varepsilon_3^P = -(\varepsilon_1^P + \varepsilon_2^P)$$

For example, with a circular bulge setting $R_1 = R_2$, eq(12.70) gives

$$\bar{\varepsilon}^P = \frac{Z\left\{n(1/r_1 + Q/r_2) + \Phi[(1 + 1/r_1) - Q] + \Omega[Q(1 + 1/r_2) - 1]\right\}}{2\left\{(1/r_1 + Q/r_2) + \tfrac{3}{4}\left\{\Phi[(1 + 1/r_1) - Q] + \Omega[Q(1 + 1/r_2) - 1]\right\}\right\}}$$

Also, by setting $r_1 = r_2 = 1$ for an isotropic sheet gives $\Phi = 1/3$, $\Omega = 2/3$, $Q = 1$ and $Z = 2$, so recovering eq(12.56) from eq(12.70).

12.6.2 Pressure Versus Height

So far we have dealt with the pole instability condition only. One ready measure of prior pole deformation is the pressure versus height curve, which may be predicted from eq(12.63b) and (12.68a) as

$$p = \bar{\sigma} t / R X \tag{12.71}$$

The current thickness t follows from the thickness strain ε_3^P as

$$\varepsilon_3^P = \ln(t / t_o) \Rightarrow t = t_o \exp(\varepsilon_3^P) \tag{12.72a,b}$$

Now from eqs(12.60a,b)

$$1 + \frac{R_1}{R_2} = \frac{2h^2 + a^2 + b^2}{h^2 + a^2} \tag{12.73}$$

Substituting eq(12.73) into eq(12.69c) and integrating:

$$\varepsilon_3^P = -\frac{2h^2 + a^2 + b^2}{h^2 + a^2}\,\varepsilon_1^P \tag{12.74}$$

where

$$\varepsilon_1^P = \int \frac{dh}{R_1} = 2\int_0^h \frac{h\,dh}{h^2 + b^2} = \ln\left(1 + \frac{h^2}{b^2}\right) \tag{12.75a}$$

$$\therefore \quad \bar{\varepsilon}^P = Z\varepsilon_1^P = Z\ln\left(1 + \frac{h^2}{b^2}\right) \tag{12.75b}$$

Combining eqs(12.74) and (12.75a) and substituting into eq(12.72b) gives the current pole thickness:

$$t = t_o \Bigg/ \left(1 + \frac{h^2}{b^2}\right)^{\frac{2h^2 + a^2 + b^2}{h^2 + a^2}} \tag{12.76}$$

Thus for any given h value we may find $\bar{\varepsilon}^P$ from eq(12.75b), when $\bar{\sigma}$ follows from the flow law chosen for the material.

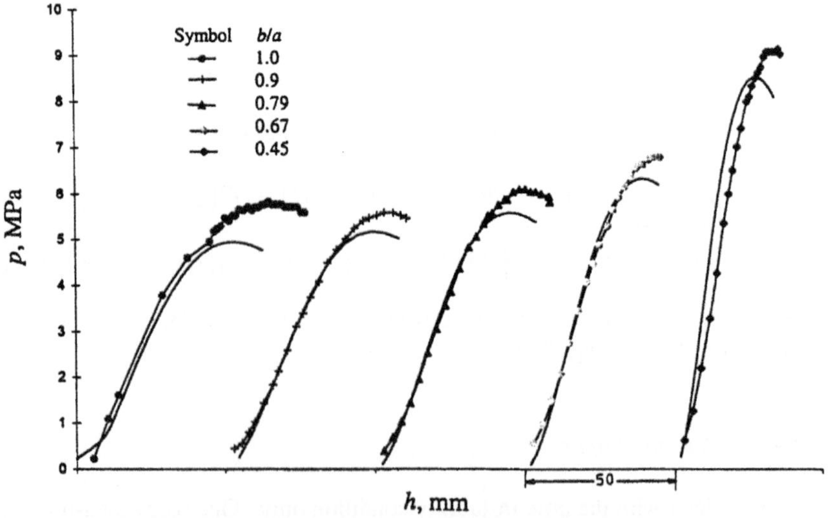

Figure 12.19 Pressure versus height for ellipsoidal bulging through 5 dies

The predicted pressure follows from eqs(12.71) and (12.76). For example, this procedure has been used [11] with Hollomon's law to predict the p, h curves for elliptical bulging of an automotive steel (where $A = 570$, $n = 0.385$). In each of 5 dies oil is pumped to the underside of the disc so forming a circular or elliptical bulge depending upon the die aperture. The major axis of the latter was constant at 180 mm while the minor axis varied with the die aspect ratios: $b/a = 1, 0.90, 0.79, 0.67$, and 0.45. Sealing was achieved with a rubber O-ring in the lower ring and by a circumferential bead in the blank which formed when the top and bottom rings were bolted together. A pressure transducer was connected to the oil feed. A dial gauge measured the pole height h above the die surface. In all the present tests the rolling direction was aligned with the minor axis. Though there is not complete agreement with the experimental curves, Fig. 12.19 shows that the S-shaped curve and a pressure maximum at instability are reproduced satisfactorily.

12.7 Plate Stretching

There are a number of theoretical predictions to the instability arising from in-plane, biaxial stressing of thin sheets. Those reviewed in [12-14] all appear as particular sub-tangents z, within the Considére expression (12.40b). In this section we shall examine the possibility of both diffuse and local instability conditions promoting necking in a thin sheet. Since either condition places a theoretical limit on the strain available for forming, the two predictions may be used to construct a material's forming limit diagram (FLD).

12.7.1 Diffuse Instability

In the bulge test, where both in-plane strains are positive, a visible neck does not form instantly with the attainment of maximum pressure. Instead, thinning spreads around the pole at this pressure showing that there still remains useful strain for forming before a local neck finally forms into a crack. A similar diffuse instability condition arises in thin flat sheets under in-plane biaxial tension. Swift [10] and Moore and Wallace [15] identified this condition with the simultaneous attainment of maxima in applied perpendicular forces. Figure 12.20 shows a sheet with rolling direction (1) aligned to the major principal stress.

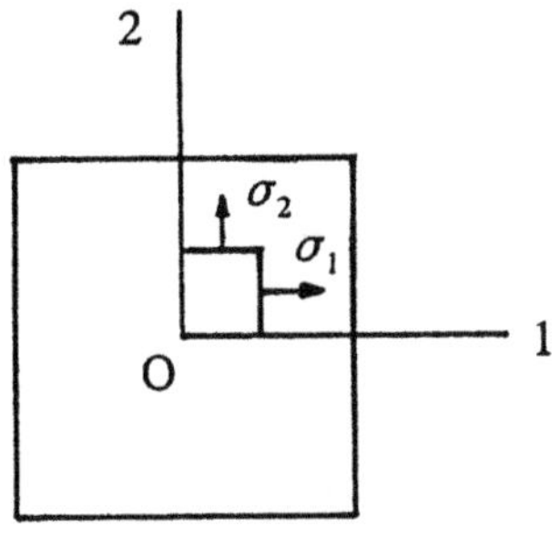

Figure 12.20 Sheet orientation and co-ordinates

The following plastic strain increment ratios are found from eqs(12.61a,b):

$$w = \frac{\delta\varepsilon_2^P}{\delta\varepsilon_1^P} = \frac{r_1[Q - r_2(1 - Q)]}{r_2[r_1(1 - Q) + 1]} \tag{12.77a}$$

$$r = \frac{\delta\varepsilon_2^P}{\delta\varepsilon_3^P} = \frac{r_1[r_2(1 - Q) - Q]}{Qr_1 + r_2} \tag{12.77b}$$

where $-1 < (Q = \sigma_2/\sigma_1) < 1$. The equivalent stress is again given by eqs(12.68a), i.e. $\bar{\sigma} = X\sigma_1$, where X has been defined in eq(12.69a). The equivalent strain is written from eq(12.68b) as $\bar{\varepsilon}^P = Z\varepsilon_1^P$ where Z now follows from eqs(12.36) and (12.77a) as

$$Z = (a + bw^2 + cw)^{1/2} \tag{12.78}$$

in which a, b and c appear with eq(12.69b). Now, at any instant in the deformation,

$$\delta\bar{\sigma} = \frac{\partial\bar{\sigma}}{\partial\sigma_1}\delta\sigma_1 + \frac{\partial\bar{\sigma}}{\partial\sigma_2}\delta\sigma_2 = \left[\frac{\partial\bar{\sigma}}{\partial\sigma_1} + \frac{\partial\bar{\sigma}}{\partial\sigma_2}\frac{\delta\sigma_2}{\delta\sigma_1}\right]\delta\sigma_1 \tag{12.79}$$

where from eq(12.35):

$$\frac{\partial\bar{\sigma}}{\partial\sigma_1} = \frac{3\sigma_1\left[r_2(1 + r_1) - Qr_1r_2\right]}{2\bar{\sigma}\left(r_1 + r_2 + r_1r_2\right)} \tag{12.80a}$$

$$\frac{\partial\bar{\sigma}}{\partial\sigma_2} = \frac{3\sigma_1\left[Qr_1(1 + r_2) - r_1r_2\right]}{2\bar{\sigma}\left(r_1 + r_2 + r_1r_2\right)} \tag{12.80b}$$

Assuming homogenous deformation and that a neck band forms perpendicular to the major principal stress direction, the simultaneous instability conditions apply:

$$\frac{\delta\sigma_1}{\delta\varepsilon_1^P} = \sigma_1 \quad \text{and} \quad \frac{\delta\sigma_2}{\delta\varepsilon_2^P} = \sigma_2 \tag{12.81a,b}$$

Equations (12.81a,b) and (12.77a) lead to

$$\frac{\delta\sigma_2}{\delta\sigma_1} = \frac{\sigma_2\,\delta\varepsilon_2^P}{\sigma_1\,\delta\varepsilon_1^P} = Qw \tag{12.82}$$

Substituting eqs(12.80a,b) and (12.82) into eq(12.79):

$$\bar{\sigma}\,\delta\bar{\sigma} = K\sigma_1\,\delta\sigma_1 \tag{12.83a}$$

where K depends upon r_1, r_2 and Q as follows:

$$K = \frac{3\left\{r_2[r_2(1 + r_1) - Qr_1r_2][r_1(1 - Q) + 1] + Qr_1[Qr_1(1 + r_2) - r_1r_2][Q - r_2(1 - Q)]\right\}}{2r_2(r_1 + r_2 + r_1r_2)[r_1(1 - Q) + 1]} \tag{12.83b}$$

The sub-tangent z is identified from combining eqs(12.68a,b) and (12.83a):

$$\frac{\delta\bar{\sigma}}{\delta\bar{\varepsilon}^P} = \frac{K\bar{\sigma}}{X^2 Z} \tag{12.84}$$

Comparing eq(12.84) with eq(12.40c) shows $z = X^2 Z / K$ and we may substitute for X, Z and K from eqs(12.69a), (12.78) and (12.83b) to give

$$z = \frac{\sqrt{\left[\frac{2}{3}(1 + r_1 + r_2)\right]\left[(1 + 1/r_1) - 2Q + Q^2(1 + 1/r_2)\right]^3}}{(1 + 1/r_1)^2 - Q(1 + 2/r_1) - Q^2(1 + 2/r_2) + Q^3(1 + 1/r_2)^2} \tag{12.85a}$$

Knowing z allows $\bar{\varepsilon}^P$ to be found from which the component strains are $\varepsilon_1^P = \bar{\varepsilon}^P / Z$ and $\varepsilon_2^P = w\varepsilon_1^P$. An alternative expression [10] leading to the diffuse instability subtangent z_D in eq(12.85a) employs the flow rule with the yield criterion f in eq(12.34):

$$\frac{1}{z_D} = \frac{\sigma_1(\partial f/\partial\sigma_1)^2 + \sigma_2(\partial f/\partial\sigma_2)^2}{[\sigma_1(\partial f/\partial\sigma_1) + \sigma_2(\partial f/\partial\sigma_2)]\,\mathrm{d}f/\mathrm{d}\bar{\sigma}} \tag{12.85b}$$

12.7.2 Local Instability

For simple tension of a thin strip, the transverse strain is negative and a local neck can occur obliquely to the applied stress. Similarly, under in-plane, biaxial plate stressing a negative transverse strain is crucial to local neck band formation. As with the onset of necking in a tensile test, it has been suggested [16] that local neck formation in a plate requires plane strain deformation, irrespective of the original straining path. This means that for a groove to form strain is prohibited along its length as widening and thinning continues. The condition for the initiation of local necking is that the maximum force per unit width of plate reaches a maximum. Thus, in Fig. 12.20 when $Q = \sigma_2/\sigma_1$ is constant, a condition of local instability occurs when force $T_1 = \sigma_1 t$ aligned with the sheet's rolling direction reaches its maximum. This gives

$$\delta T_1/T_1 = \delta\sigma_1/\sigma_1 + \delta t/t = 0 \tag{12.86}$$

Now from eqs(12.68a,b) and (12.77a),

$$\delta\varepsilon_3^P = \delta t/t = -(1 + w)\,\delta\bar{\varepsilon}^P/Z \quad \text{and} \quad \bar{\sigma} = X\sigma_1 \tag{12.87a,b}$$

Substituting eqs(12.87a,b) into eq(12.86) gives

$$\frac{\delta\bar{\sigma}}{\delta\bar{\varepsilon}^P} = \frac{(1 + w)\bar{\sigma}}{Z} \tag{12.88}$$

from which the subtangent of local instability is

$$z = Z/(1 + w) \tag{12.89}$$

Substituting for Z, w, a, b and c from eqs(12.69b,c) and (12.77a),

$$1 + w = \frac{r_2 + Q r_1}{r_2[r_1(1 - Q) + 1]} \tag{12.90a}$$

$$= \sqrt{a + bw^2 + cw} \tag{12.90b}$$

$$= \frac{\sqrt{\tfrac{2}{3}(r_1 + r_2 + r_1 r_2)(1 + r_1 + r_2)\,[Q^2 r_1 + r_2 + r_1 r_2(1 - Q)^2}}{r_2[r_1(1 - Q) + 1]}$$

From eq(12.89), we divide eq(12.90b) by eq(12.90a) to give z:

$$z = \frac{\sqrt{\tfrac{2}{3}(1 + 1/r_1 + 1/r_2)[(1 + 1/r_1) - 2Q + Q^2(1 + 1/r_2)]}}{1/r_1 + Q/r_2} \tag{12.91a}$$

An alternative expression leading to the subtangent (z_L) for local necking uses the flow rule and yield function f in eq(12.34) as follows [17]:

$$\frac{1}{z_L} = \frac{\partial f/\partial \sigma_1 + \partial f/\partial \sigma_2}{\mathrm{d}f/\mathrm{d}\bar{\sigma}} \tag{12.91b}$$

in which $\bar{\sigma}$ appears in eq(12.35). The equivalent instability strain follows from Hollomon as $\bar{\varepsilon}_I^{\,P} = nz$ or, from Swift (12.57), as $\bar{\varepsilon}_I^{\,P} = nz - \varepsilon_o$. The components of $\bar{\varepsilon}_I^{\,P}$ are $\varepsilon_1^{\,P} = \bar{\varepsilon}_I^{\,P}/Z$ and $\varepsilon_2^{\,P} = w\varepsilon_1^{\,P}$. Equation (12.91a) is important when $\varepsilon_2^{\,P}$ is negative for $-1 \le Q \le 0.5$ since a local groove will form at a fixed orientation to the stress axes 1 and 2 (see Figs 12.21a,b).

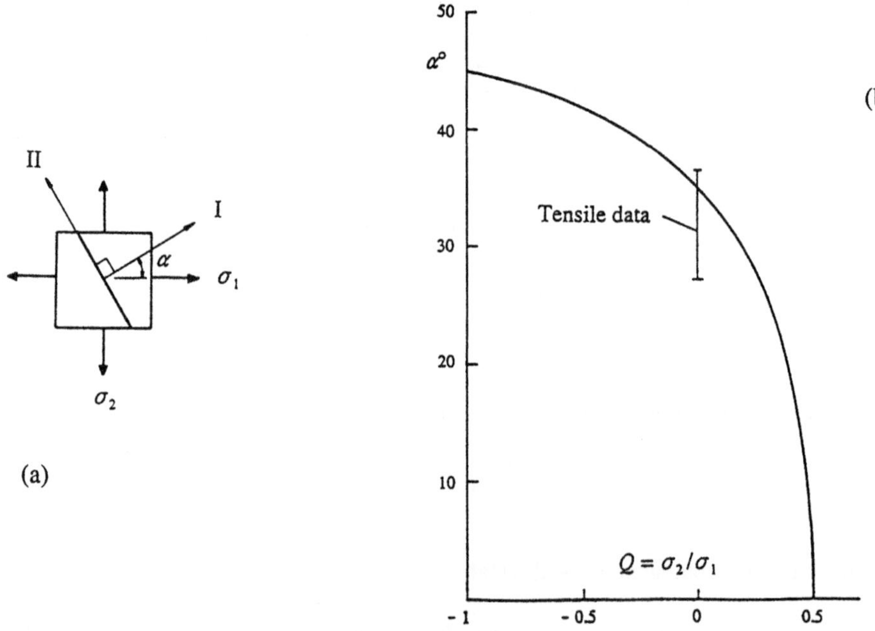

Figure 12.21 Groove formation under local instability

Here the axes of stress coincide with the material axes (i.e. with 1 parallel to the roll and 2 transverse to the roll). Let an axis I lie normal to the groove at an orientation α to 1. Axis II lies parallel to the groove as shown. Transforming the principal strain increments $d\varepsilon_1{}^P$ and $d\varepsilon_2{}^P$ along the axis II gives the normal strain component aligned with the groove as

$$d\varepsilon_{II}{}^P = d\varepsilon_1{}^P \sin^2\alpha + d\varepsilon_2{}^P \cos^2\alpha$$

Setting $d\varepsilon_{II}{}^P = 0$ leads to

$$\cos 2\alpha = (1 + w)/(1 - w) \qquad (12.92a)$$

where $w = d\varepsilon_2{}^P/d\varepsilon_1{}^P$ depends upon the stress ratio $Q = \sigma_2/\sigma_1$, as in eq(12.77a). Substituting eq(12.77a) into eq(12.92a) gives the groove orientation in terms of Q and the two r values:

$$\cos 2\alpha = \frac{r_2 + Qr_1}{r_2(1 + 2r_1) - Qr_1(1 + 2r_2)} \qquad (12.92b)$$

In the case of a simple tension test under an axial stress σ_1 with $\sigma_2 = 0$ (see Fig. 12.22a) we set $Q = 0$ in eq(12.92b) to give $\cos 2\alpha = 1/(1 + 2r_1)$. If the material is isotropic then $r_1 = 1$ and we find $\alpha = ½ \cos^{-1}(⅓) = 35.26°$ (i.e. the groove lies at 90° to this). The Mohr's strain circle (see Fig. 12.22b) provides a geometrical interpretation of this solution.

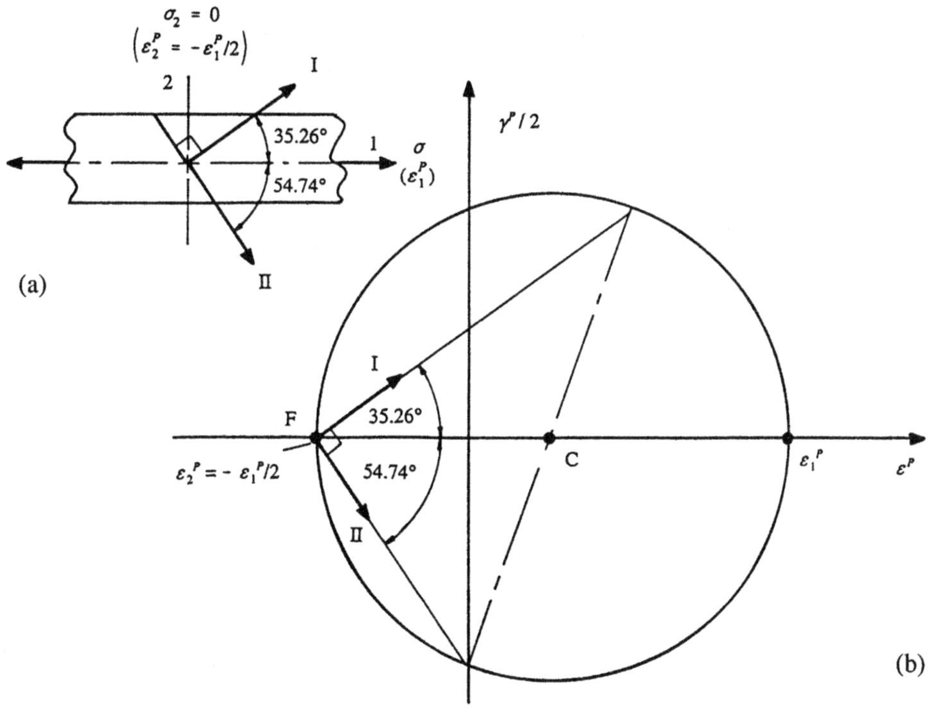

Figure 12.22 Mohr's circle showing groove orientation

Setting $\varepsilon_1{}^P$ and $\varepsilon_2{}^P = -½\varepsilon_1{}^P$ along the plastic strain ε^P- axis, the centre C bisects these points and the circle passes through them. The groove direction II passes though the focus F and a point of zero ε^P as shown. For other stress ratios, applied to isotropic material, the groove orientation follows from eq(12.92b) as

$$\cos 2\alpha = (1 + Q)/[3(1 - Q)] \qquad (12.92c)$$

As Q increases from zero the centre C moves to the right along its ε^P axis. With $Q = ½$, under plane strain, the circle becomes tangential to the vertical, semi-shear strain axis $\gamma^P/2$. Correspondingly, $\alpha = 0°$ and the groove becomes vertically aligned. For $Q > ½$, it follows that the circle will lie to the right of the $\gamma^P/2$ axis, when no further inclined grooves can form. Figure 12.21b shows the dependence of groove inclination α upon stress ratio Q in the range $-1 \le Q \le ½$. Superimposed tensile data at $Q = 0$ reveal that rolled sheets of aluminium alloy and tin-plate form grooves across a range of inclinations deviating from the isotropic prediction. For stress ratios $Q > ½$, failure may be diffuse or a groove may form perpendicular to σ_1 by another mechanism, as outlined in the following section.

12.7.3 The Forming Limit Diagram

The local and diffuse theories of instability given above provide a combination of critical principal strains (ε_1^P, ε_2^P) depending upon the stress ratio. When the full range of stress ratios $-1 \le Q \le 1$ are considered across crucial instability regimes, the critical strains plot to define a forming limit diagram (FLD) for the material. Different instability criteria result in different FLD's in co-ordinates of ε_1^P and ε_2^P, i.e. the criteria control both the shape and position of the FLD within a given quadrant of strain. The question as to which prediction supplies the most appropriate limiting strains for forming operations may best be answered from a comparison with experimental data. Figure 12.23 compares results from laboratory tests upon an automotive steel with predictions using sub-tangents from: (i) maximum pressure, (ii) diffuse and (iii) local instability [18]. The sub-tangents appear in eqs(12.70), (12.85a) and (12.91a) respectively.

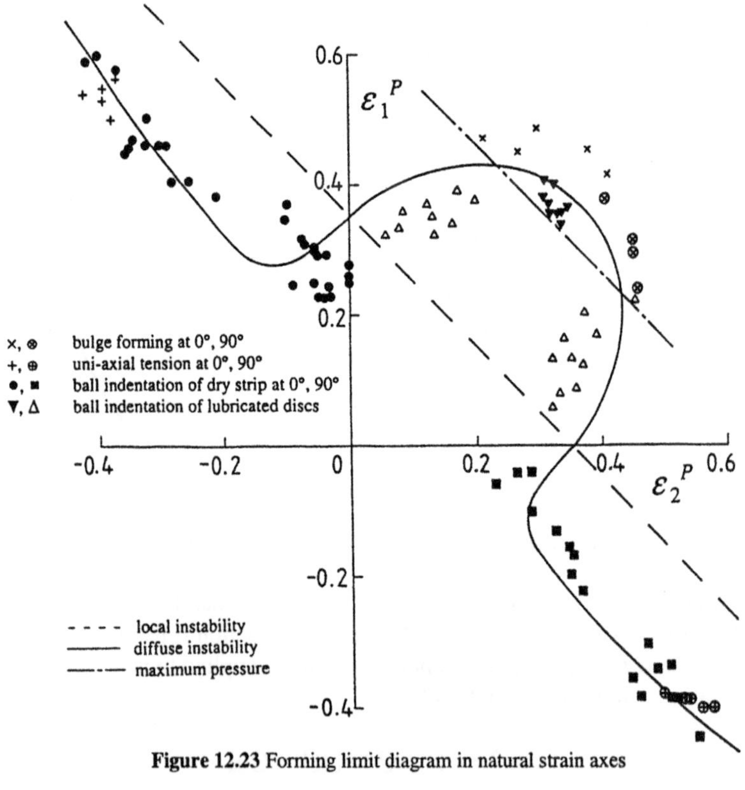

Figure 12.23 Forming limit diagram in natural strain axes

The subtangents were calculated with material properties $r_1 = 2.31$ and $r_2 = 2.54$ for $n = 0.37$, $\varepsilon_o = 1/25$ and $\sigma_o = 200$ MPa in Swift's eq(9.44b). Stress ratios Q cover three strain quadrants. For example, $Q = 0$ for tension in the 1-direction, $Q = \infty$ for tension in the 2-direction and $1 < Q < 2$ for elliptical bulging, depending upon a/b for the die and $r_{1,2}$ for the material. The conversion to the equivalent instability strain uses $\bar{\varepsilon}_I^{\,P} = nz - \varepsilon_o$. The component strains $\varepsilon_1^{\,P}$ and $\varepsilon_2^{\,P}$ are derived fron $\bar{\varepsilon}_I^{\,P}$ in the manner outlined above. Natural strains apply to these calculations and to the plot in Fig. 12.23 though they may be converted to engineering strains for a convenient working FLD using the relationship $e = \exp\varepsilon - 1$. In Fig. 12.23 the sheet's othotropic axes are taken as the co-ordinates. Thus, data in quadrants 2 and 4 apply to 0° and 90° orientations respectively for tension testpieces and for punch indentation of strips with different widths. The experimental data lying in the first quadrant applies to bulge forming, with 0° and 90° orientations to the roll and to hemispherical punch indentation of discs. The symmetry of the latter test reflects all its data about a 45° line in this quadrant. The strains in all tests were measured from the deformation to a circular grid pattern in the vicinity of a failure. Of the three predictions shown, that based upon diffuse instability appears the most representative of the limit strains measured. Veerman and Neve [19] postulated that a true FLD applies in the absence of strain gradients at the instant localised necking begins. They were able to detect this from the plot of strains across adjacent circles within the necked region of spherically punched testpieces. Hence we should expect local instability to provide the true fracture strains in quadrants 2 and 4. The two predictons agree only at points of intersection with the axes, i.e. the plane strain limit. The local instability prediction is not appropriate to quadrant 1. The maximum pressure line is adequate for providing limiting strains under near bi-axial tension in bulge forming.

Figure 12.23 shows that instability predictions are likely to provide a conservative estimate to the forming limits in a low carbon, ductile sheet steel. This is because the forming limit strains are sensitive to the practical definition of that limit. Strains found from the distortion to grid circles or from thickness measurements in the fracture zone, can exceed the predictions by the extent of diffuse straining. This is most likely in the positive strain quadrant of an FLD where we should predict the onset of diffuse straining as a lower forming limit. Consequently, we should not expect precise agreement between predicted and measured fracture strains. Alternative approaches to FLD construction: (i) rely solely upon experimental data to give a shape resembling a combination of local and diffuse instability predictions between quadrants 1 and 2 [20, 21] and (ii) assume that a groove will form from a pre-existing defect [22], as outlined in the following section.

12.7.4 FLD from Groove Formation

A modification to the local instability prediction is clearly required for quadrant 1. In the original Marciniak hypothesis [22] a pre-existing defect provides a local reduction in thickness between the two zones A and B in Fig. 12.24a With original thicknesses t_{Bo} and t_{Ao} in each respective zone, the defect size factor is defined:

$$f_o = t_{Bo} / t_{Ao} \tag{12.93a}$$

Under in-plane, biaxial tension, the current thickness reduction factor is modified as follows:

$$f = \frac{t_B}{t_A} = \frac{t_{Bo}}{t_{Ao}} \times \frac{t_B}{t_{Bo}} \times \frac{t_{Ao}}{t_A} \tag{12.93b}$$

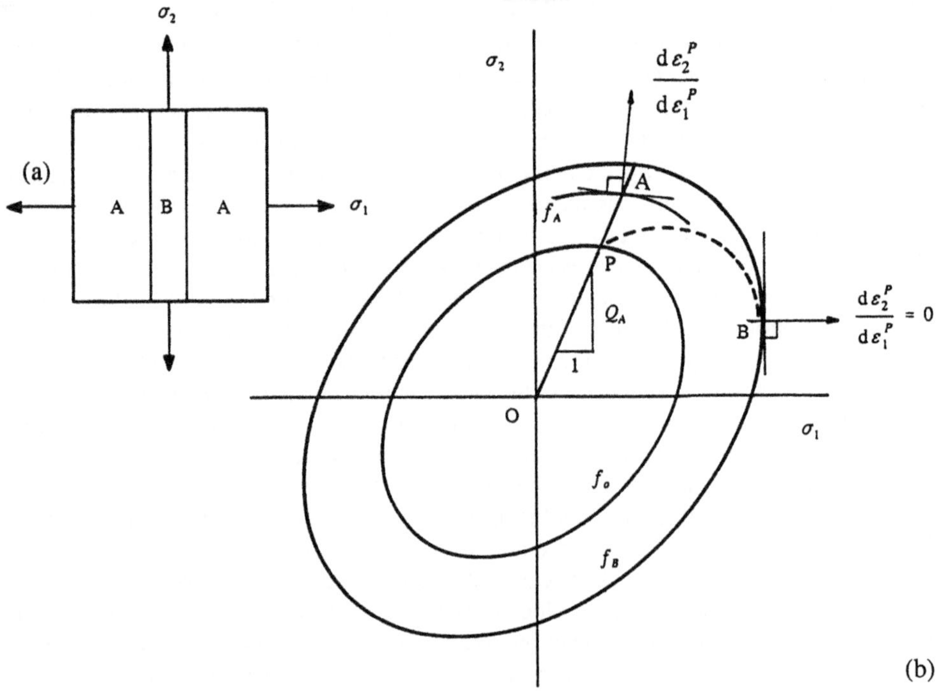

Figure 12.24 Local groove necking within a stretched plate

Using the following thickness strain definitions

$$\varepsilon_{3A} = \ln (t_A / t_{Ao}) \text{ and } \varepsilon_{3B} = \ln (t_B / t_{Bo}) \quad (12.94a,b)$$

it follows from eqs(12.93b) and (12.94a,b) that

$$f = f_o \exp(\varepsilon_{3B} - \varepsilon_{3A}) \quad (12.95)$$

Figure 12.24b shows that the principal, in-plane plastic strain ratio $d\varepsilon_2^P/d\varepsilon_1^P$ in region A remains constant. That is, the direction of the outward normal to locus f_A is constant as f_A expands isotropically from the initial yield point P. Point A contains the current stress state $(\sigma_{2A}, \sigma_{1A})$ as it increases in a constant ratio $Q_A = \sigma_{2A}/\sigma_{1A}$ In contrast, the corresponding stress and strain ratios within groove B continuously change to approach the plane strain condition $d\varepsilon_2^P/d\varepsilon_1^P = 0$, as shown. At this point ε_{1B}^P increases rapidly to failure when the strain state in the adjacent region A is taken to constitute the forming limit. To find the limiting strain in A first note that the major principal stress σ_{1B} in the groove leads that within A to satisfy force equilibrium:

$$\sigma_{1A} t_A = \sigma_{1B} t_B \quad (12.96a)$$

from which

$$\sigma_{1B} = \sigma_{1A} / f \quad (12.96b)$$

where $f < 1$. Also, a minor principal strain compatibility condition applies:

$$d\varepsilon_{2A}^P = d\varepsilon_{2B}^P \quad (12.97a)$$

In an orthotropic sheet the principal axes of stress and plastic strain coincide when the stress axes align with material axes. This permits a substitution from eq(11.11b) into eq(12.97a):

$$[F(\sigma_2 - \sigma_3) - H(\sigma_1 - \sigma_2)]_A = [F(\sigma_2 - \sigma_3) - H(\sigma_1 - \sigma_2)]_B \qquad (12.97b)$$

Setting $\sigma_3 = 0$ and $r_2 = H/F$ leads to the minor principal stress within the groove:

$$\sigma_{2B} = \sigma_{2A} - (\sigma_{1A} - \sigma_{1B})\, r_1 r_2 / (1 + r_2) \qquad (12.98)$$

Dividing eqs (11.11a,b) connects the stress and strain ratios within each region as

$$\frac{d\varepsilon_2^P}{d\varepsilon_1^P} = \frac{r_1[(Q - 1)r_2 + Q]}{r_2[(1 - Q)r_1 + 1]} \qquad (12.99)$$

where $Q = \sigma_2/\sigma_1$. Also required are the equivalent stress and plastic strain in region A and an appropriate relation between them, e.g. Hollomon, Swift etc. We may write these as $\bar{\sigma} = X\sigma_1$ and $d\bar{\varepsilon}^P = Z d\varepsilon_1^P$ where X and Z follow from eqs(12.34) and (12.36) as

$$X = \sqrt{\frac{r_2(1 + r_1) - 2Qr_1r_2 + Q^2 r_1(1 + r_2)}{2(r_1 + r_2 + r_1r_2)/3}} \qquad (12.100a)$$

in which $r_1 = H/G$ and

$$Z = \sqrt{a + b\left(\frac{d\varepsilon_2^P}{d\varepsilon_1^P}\right)^2 + c\left(\frac{d\varepsilon_2^P}{d\varepsilon_1^P}\right)} \qquad (12.100b)$$

where a, b and c depend upon the two r-values (see eq 12.69b):

$$a = \frac{2(r_1 + r_2 + r_1r_2)(1 + r_2)}{3r_2(1 + r_1 + r_2)}, \quad b = \frac{2(r_1 + r_2 + r_1r_2)(1 + r_1)}{3r_1(1 + r_1 + r_2)}, \quad c = \frac{4(r_1 + r_2 + r_1r_2)}{3(1 + r_1 + r_2)}$$

Table 12.2 Iteration for a groove plane strain failure

---

1) Assume, or find from measurement, an initial f_o value, normally: $0.95 < f_o < 1$
2) Apply an increment of equivalent plastic strain to region A, i.e. $d\bar{\varepsilon}_A^P$
3) Find the three principal plastic strain increments:
   $d\varepsilon_{1A}^P = d\bar{\varepsilon}_A^P / Z$, $d\varepsilon_{2A}^P = w_A\, d\varepsilon_{1A}^P$ and $d\varepsilon_{3A}^P = -(d\varepsilon_{1A}^P + d\varepsilon_{2A}^P)$
4) Form sums: $\sum d\varepsilon_{1A}^P$ and $\sum d\varepsilon_{1A}^P$ over successive iterations
5) Find $\bar{\sigma}_A$ from $\sum d\bar{\varepsilon}_A^P$ using Swift's law: $\bar{\sigma} = \sigma_o\left(1 + \bar{\varepsilon}^P/\varepsilon_o\right)^n$
6) Find the principal stresses in region A:
   $\sigma_{1A} = \bar{\sigma}_A / X$ and $\sigma_{2A} = Q_A\, \sigma_{1A}$
7) Determine the principal stresses in region B:
   $\sigma_{1B} = \sigma_{1A}/f_o$ (initially) or $\sigma_{1B} = \sigma_{1A}/f$ (subsequently) with σ_{2B} from eq(12.98)
8) For $Q_B = \sigma_{2B}/\sigma_{1B}$, find w_B from eq(12.99). Stop when $w_B \to 0$.
9) Find the principal plastic strain increments within the groove:
   $d\varepsilon_{1B}^P = d\varepsilon_{2A}^P/w_B$ and $d\varepsilon_{3B}^P = -(d\varepsilon_{1B}^P + d\varepsilon_{2B}^P)$ where $d\varepsilon_{2A}^P = d\varepsilon_{2B}^P$
10) Sum the thickness strains increments $\varepsilon_{3A}^P = \sum d\varepsilon_{3A}^P$ and $\varepsilon_{3B}^P = \sum d\varepsilon_{3B}^P$ within successive iterations and re-define f from eq(12.95).
11) Return to step 2.

Given the r values and Swift's flow properties quoted earlier, an iteration (Table 12.2) is applied to a constant stress ratio Q_A within the system of eqs(12.93) - (12.100). Let the radial stress path, defined by Q_A, extend beyond the yield point P in Fig. 12.24b due to an increment $d\bar{\varepsilon}^P$. A repeated calculation of the strain increment ratio $w = d\varepsilon_2^P/d\varepsilon_1^P$ is made from following the steps given. The iteration stops as $w \to 0$ within stage 8 so that the strain sums formed from stage 4 constitute the forming limits. These limiting strains can be found for any stress ratio but more often the analysis is confined to quadrant 1. In quadrant 2, the limit strains would coincide with those found from the sub-tangent for an inclined local neck, considered earlier. Figure 12.25 shows that the quadrant 1 limit line from Table 12.2 lies above the diffuse instability limit, thereby extending the bi-axial stretch strain limits for an automotive steel at the chosen defect ratio f_o value.

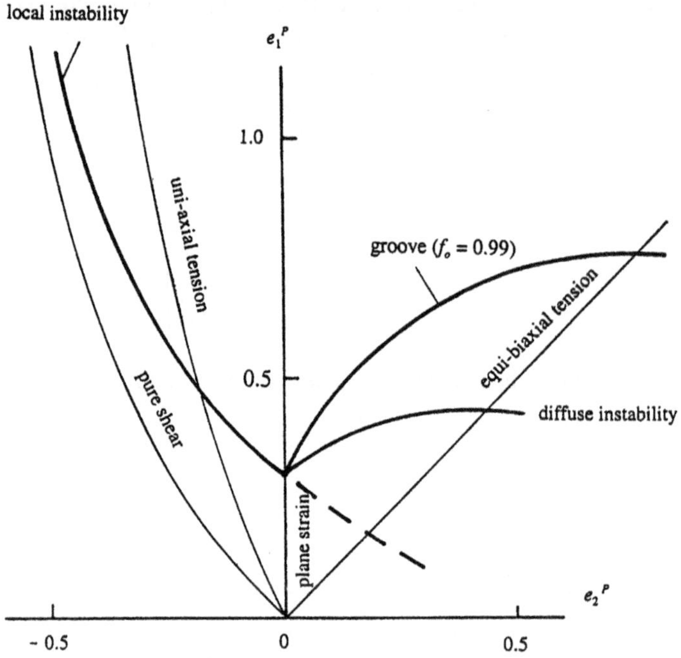

Figure 12.25 Comparison of FLDs from diffuse, local and groove formation

Either prediction given in quadrant 1, when combined with the local instability line in quadrant 2, provides an FLD of in-plane limit strains under any stress ratio Q. For example, in axes of engineering strain, Fig. 12.25 shows that the strain path is linear for equi-biaxial tension ($Q = 1$) but is non-linear for uni-axial tension ($Q = 0$) and pure shear ($Q = -1$). Limit strains are provided by their intersections with the chosen FLD. Note that plane strain constitutes the lowest strain limit and is responsible for most product failures in the region of bends and folds.

12.8 Concluding Remarks

Plastic instability appears in many forms: from the buckling of long struts and thin plates at loads beyond the elastic limit to the material instabilities that control fracture. We have seen that among the analytical methods used are extensions to elastic analyses using suitable plastic moduli and the realisation of pressure and load maxima coincident with the onset of failure, e.g. as with the Considére condition in simple tension. The extension to biaxial stress states may involve local or diffuse instability depending upon the applied stress ratio. In practice, where biaxial stress states occur in pressure vessels and in sheet metal forming, the initial orthotropy of the rolled sheet may become an important factor in these analyses. In the most common plane strain failure mode, anisotropy is almost always admitted along with an initial imperfection, when constructing the forming limit diagram based upon failure from a narrow, inclined local neck.

References

1. Shanley F. R. *Jl Aero Sci.*, 1947, **14**, 261.
2. Vinson J. R *Structural Mechanics*, 1974, Wiley.
3. ESDU 72019 *Buckling of Flat Isotropic Plates Under Uniaxial and Biaxial Loading*, August 1972.
4. Ramberg W. and Osgood W. R. *Nat. Adv. Comm for Aero.*, 1943, TN902.
5. ESDU 83044 *Plasticity Correction Factors for Plate Buckling*, December 1983.
6. ESDU 71005 *Buckling of Flat Plates in Shear*, February 1971.
7. Gerard G. *Jl Appl Mech.*, 1948, **15**(1), 7.
8. Rees D. W. A. *Jl Strain Anal.*, 1995, **30**, 305.
9. Hill R. *Phil Mag.*, 1950, **41**, 1133.
10. Swift H. W. *Jl Mech Phys Solids*, 1952, **1**, 1.
11. Rees D. W. A. *Jl Mats Proc Tech.*, 1999, **92-93**, 508.
12. Lee S. H. and Kobayashi S. *Proc: 3rd North. Amer Metal Working Res. Conf*, Pittsburgh, Carnegie Press, p. 277, 1975.
13. Hasek V.V. *Proc: ICF4 Fracture*, Waterloo, Canada, Pergamon Press, **2**, p. 475, 1977.
14. Korhonen A. S. *Trans ASME, Jl Eng Mat and Tech.*, 1978, **100**, 303.
15. Moore G. G. and Wallace J. F. *Jl Inst Met.*, 1964-65, **93**, 33.
16. Marciniak Z. and Kuczynski K. *Int Jl Mech Sci.*, 1967, **9**, 609.
17. Hill R. *Jl Mech Phys Solids*, 1952, **1**, 19.
18. Rees D. W. A. *Jl Mats Proc Tech.*, 1995, **55**, 146.
19. Veerman C and Neve P. V. *Sht Met Ind.*, 1972, **49**, 421.
20. Keeler S. *Sht Met Ind.*, 1965, **42**, 683.
21. Pearce R. *Sht Met Ind.*, 1971, **48**, 943.
22. Marciniak Z. and Duncan J. *Mechanics of Sheet Metal Forming*, 1972, E. Arnold.

Exercises

12.1 It is required to find the maximum compressive load that a brass strut with an elliptical thin-walled section can support without plastic buckling. If the strut length is 825 mm, the major and minor axes are 80 mm and 30 mm respectively and the wall thickness is 3 mm, estimate the buckling load for the minor axis where the ends may be assumed pinned. Take the Hollomon stress-strain law for brass as: $\sigma/\sigma_c = 7.83\varepsilon^{1/3}$ where the compressive yield stress $\sigma_c = 144$ MPa.

12.2 A pinned-end strut 1.5 m long has the cross-section in Fig. 12.26. Find Engesser's plastic buckling load given that Hollomon's law: $\sigma/\sigma_c = 5\varepsilon^{1/3}$ describes the compressive stress-strain curve in the region beyond a yield stress of $\sigma_c = 300$ MPa. Compare Engesser's load with predictions from the straight line and parabolic formulae using mid-range constants for steel on p. 374: $E = 207$ GPa. [Answer: 817.2 kN]

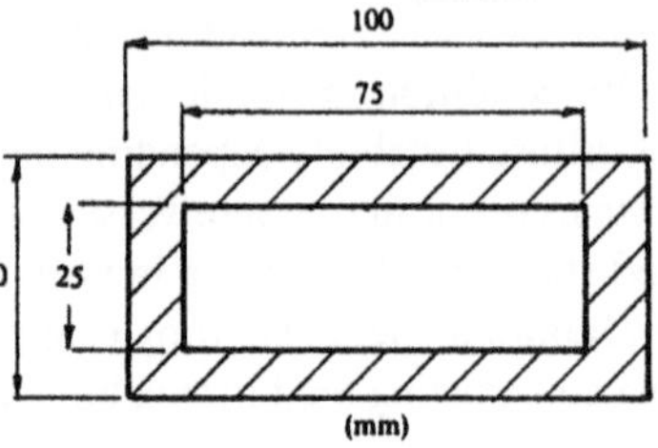

Figure 12.26

12.3 Determine the elastic compressive stress which, when applied to the shorter sides of a thin plate 390 × 200 × 7.5 mm, will cause it to buckle in the presence of a constant stress of 80 MPa which acts: (i) in compression and (ii) in tension along its longer sides. Take $E = 72.4$ GPa and $\nu = 0.33$ with all sides simply supported. Correct the stresses for plasticity effects using Fig. 12.10. Take $\sigma_n = 405$ MPa and $m = 16$. [Answer: 265 MPa, 371 MPa, 455 MPa]

12.4 A plate $a \times b \times t$ is subjected to a normal tensile stress $\sigma_y = \sigma$ on its $a \times t$ sides and a compressive stress $\sigma_x = -\sigma$ on its $b \times t$ sides. If all edges are simply supported show that the minimum value of critical buckling stress is given by $(\sigma_x)_{cr} = (8\pi^2 D)/(tb^2)$ for one halfwave in the y-direction. To what values of $r = a/b$ does this minimum apply? [Answer: $1/\sqrt{3}$, $2/\sqrt{3}$ etc]

12.5 Find the shear stress for which a rectangular plate 200 mm wide × 260 mm long × 4.5 mm thick buckles when the short sides are clamped and the long sides are simply supported. Using Figs 12.15 and 12.16, correct for plasticity given $\sigma_n = 340$ MPa, $m = 16$, $E = 73$ GPa and $\nu = 0.3$. [Ans: 198 MPa]

12.6 Determine the equivalent stress and strain at the point of instability for a thin-walled, open-ended pressurised cylinder of mean diameter d and thickness t, if the material hardens according to the law: $\bar{\sigma} = \sigma_o(1 + B\bar{\varepsilon}^P)^n$. What is the pressure that causes this instability?
[Answer: $\bar{\sigma} = \sigma_o(2nB/3)^n$, $\bar{\varepsilon}_u{}^P = (1/B)(2nB/3 - 1)$, $p = (4t\sigma_o/d)(2nB/3)^n$]

12.7 Determine the true fracture strain of a ductile material in terms of: (i) the % elongation at fracture and (ii) the % reduction of area at fracture in a tensile test.

12.8 The following table of results applies to a tension test on a cylindrical testpiece 13 mm diameter and 50 mm long:

Force, kN	10	20	30	40	52.29	61.5	67.2	72	74.8
	76.6	77	78	78.3	78.53	78.72	78.79	78.73	78.6
	78.37	78.16	77.82	77.50	77.12	76.71			
Extension, mm	0.02	0.04	0.05	0.07	0.50	1.20	2.03	3.45	4.3
	5.0	6.2	7.37	8.21	9.01	9.90	11.23	12.20	13.0
	14.2	15.0	16.1	16.9	17.5	18.1			

Determine graphically, from the Considére condition, the true stress and natural strain at the point of instability. Determine the strength coefficient, hardening exponent and instability strain from a Hollomon description of the given data. Compare with Considére's instability strain.

12.9 Construct an FLD for conditions of simple local and diffuse instability within quadrants 1 and 3 respectively. The sheet metal has a yield stress 150 MPa with a Hollomon exponent $n = 0.25$. The in-plane anisotropy is characterised with $r_1 = 1.7$ and $r_2 = 2.0$. Hint: Determine the sub-tangents z form eqs(12.85) and (12.91) for a range of stress ratios. Convert each z to an equivalent instability strain which decomposes into the two principal in-plane engineering strains.

CHAPTER 13

STRESS WAVES IN BARS

13.1 Introduction

This introduction to elastic and plastic wave generation within round bars applies to slow heavy strikers and lighter struck members, as in a forging operation and faster impacts from projectiles upon heavier bodies. By equating external work to the strain energy stored, the dynamic stresses and deflections may be found, usually from assuming that the static and dynamic deflection curves are the same. This approach usually gives a good approximate solution though it provides little information on the mechanics of loading. Where a hard, fast moving striker is of similar weight to the struck member, as from an explosive charge, the stress waves generated apply to relatively short times after impact. The full solution for longer times after impact would require an examination of normal vibration modes using the exact equations of elasticity. However, the approximate dynamic solutions are more realistic, usually, than those as found from a static stress analysis and thereby can predict certain observed effects, including spalling and scabbing. The analysis begins with a geometrical interpretation of the elastic impact wave equations from which bar stresses and particle velocities are found. This is followed with an analysis of plastic waves using a simple bi-linear approximation to the stress-strain curve. For a more expansive discussion the reader is referred to dedicated works [1-10] upon this topic.

13.2 The Wave Equation

Let the longitudinal wave propogation in a bar be c_o m/s and the particle velocity in the bar be v m/s. The latter is intimately related to the stress in the bar and is readily converted to an absolute velocity from knowing the bar velocity after impact. A given cross-section in the bar, distance x from the left end origin, displaces by an amount u where the compressive stress is σ. At a further distance δx away from this cross section, the displacement and stress increase by the respective amounts $(\partial u/\partial x)\delta x$ and $(\partial \sigma/\partial x)\delta x$ respectively (see Fig. 13.1).

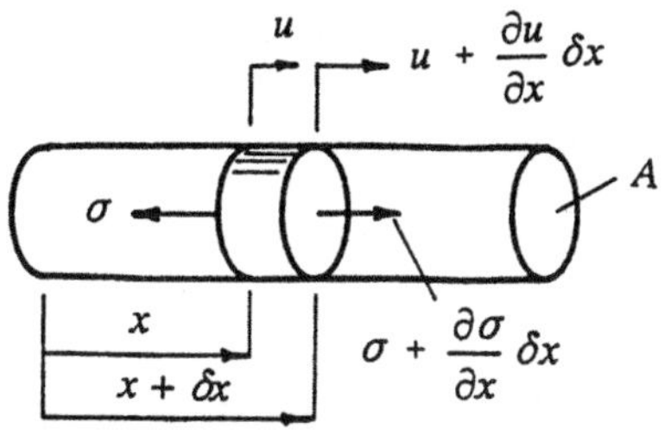

Figure 13.1 Elastic stress wave in a bar

Hooke's law gives the elastic stress as

$$\sigma = E\varepsilon = E\,\partial u/\partial x$$

so that its rate of change with respect to x becomes

$$\partial\sigma/\partial x = E(\partial^2 u/\partial x^2) \tag{13.1}$$

where E is the elastic modulus of the bar. Newton's second law of motion connects the axial force due to the change in stress with the force required to accelerate the plane through δx in time δt. Writing the mass $m = \rho A\,\delta x$, where ρ is the density and A is the bar's section area, this becomes

$$F = \rho A\,\delta x \times \frac{\partial^2 u}{\partial t^2} = A\frac{\partial \sigma}{\partial x}\,\delta x$$

and substituting from eq(13.1)

$$\rho A\frac{\partial^2 u}{\partial t^2}\,\delta x = A E\frac{\partial^2 u}{\partial x^2}\,\delta x$$

This results in a one-dimensional wave equation:

$$\frac{\partial^2 u}{\partial t^2} = c_o^2\frac{\partial^2 u}{\partial x^2} \tag{13.2}$$

where $c_o = \surd(E/\rho)$. The general solution [1] to eq(13.2) is

$$u = f_1\,(c_o t - x) + f_2\,(c_o t + x) \tag{13.3a}$$

in which the two functions f_1 and f_2 may be interpreted from a simple harmonic motion of the displacement: $u = a\sin\omega t$. With a period T, the wave frequency is $\omega = 2\pi/T$ and its wavelength is $\lambda = c_o T$. Thus

$$u = a\sin(2\pi/\lambda)c_o t \tag{13.3b}$$

To match eq(13.3b) with the first function f_1 in eq(13.3a), the former equation must be modified with a time delay x/c_o for position x as follows:

$$u = a\sin(2\pi/\lambda)(c_o t - x) \tag{13.3c}$$

and this reveals a wave propagating in direction x-positive. Similarly, within the second function f_2 in eq(13.3a), a wave propagates in direction x-negative. Thus, the solution given in eq(13.3c) reveals two waves in opposing directions both travelling at the propagation velocity $c_o = \surd(E/\rho)$ but in opposite directions.

13.3 Particle Velocity

Consider the bar in Fig. 13.2 with a compressive stress maintained at its left-hand end. The velocity v of particles, due to the effect of the stress wave produced, may be found from equating impulse to the change of linear momentum.

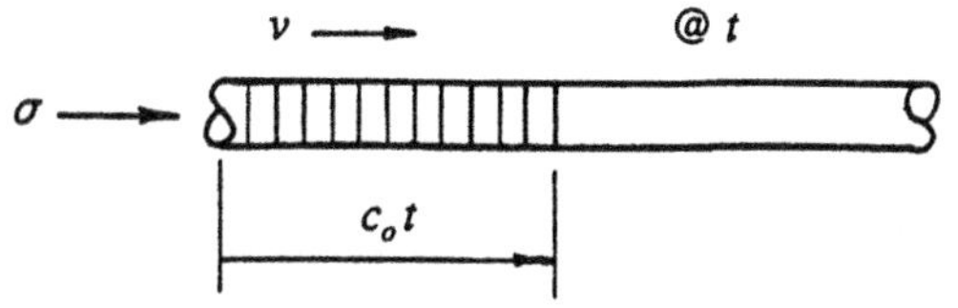

Figure 13.2 Impulse and particle velocity due to a stress application

That is, at time t in Fig. 13.2:

$$A\sigma t = A\rho\,(c_o t)v$$

from which

$$\sigma = \rho c_o v \quad \text{or} \quad v = \sigma/(\rho c_o) \tag{13.4a,b}$$

in which ρc_o is the acoustic impedance. Equations (13.4a,b) show that both σ and v are independent of the bar's dimensions. The two waves identified above may now be associated with the *sense* of the particle velocity as follows:

- A *compressive wave* has both c_o and v in direction x-positive (same sense).
- A *tensile wave* has c_o in direction x-positive and v in direction x-negative (opposite sense).

13.3.1 Reflection at a Boundary

A compressive wave will be reflected at the free end of a rod to become a tensile wave, i.e. compression returns as tension and vice versa (see Figs 13.3a,b).

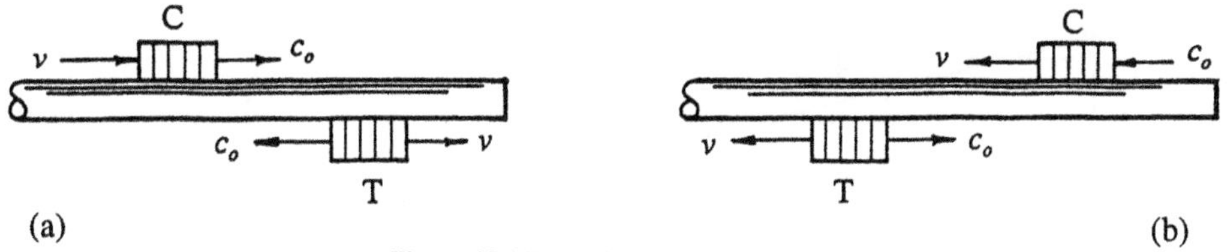

Figure 13.3 Reflection at a free boundary

The stress reversal must produce a condition of zero stress at the free end if an identical wave of opposite particle sign is reflectd by the boundary. As a consequence both the velocity and displacement are doubled at the free end. In contrast, at a fixed boundary the end displacement must remain zero so that the reflected wave remains in the same sense as the incident wave, i.e. it remains either in tension or compression (see Figs 13.4a,b).

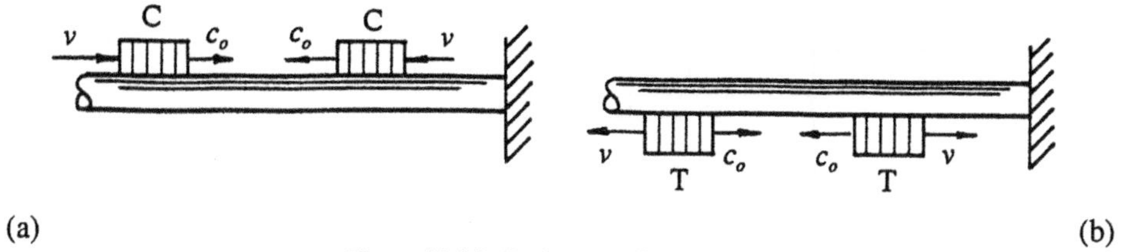

Figure 13.4 Reflections at a fixed boundary

It follows that upon reflection at a fixed boundary the stress is doubled but the velocity and hence the displacement must cancel to zero.

13.3.2 Displacement-Time Diagram

The position of both compressive and tensile elastic wave fronts at their propagation and particle velocities $c_o = \sqrt{(E/\rho)}$ and $v = \sigma / (\rho\, c_o)$ can appear with similar stress levels but opposing directions on an x-t diagram. In any region of this diagram the force at all points must be in equilibrium if additional stress waves due to discontinuities in section area are to be avoided. Moreover, a compatibility condition is ensured when all points within this region have the same absolute velocity.

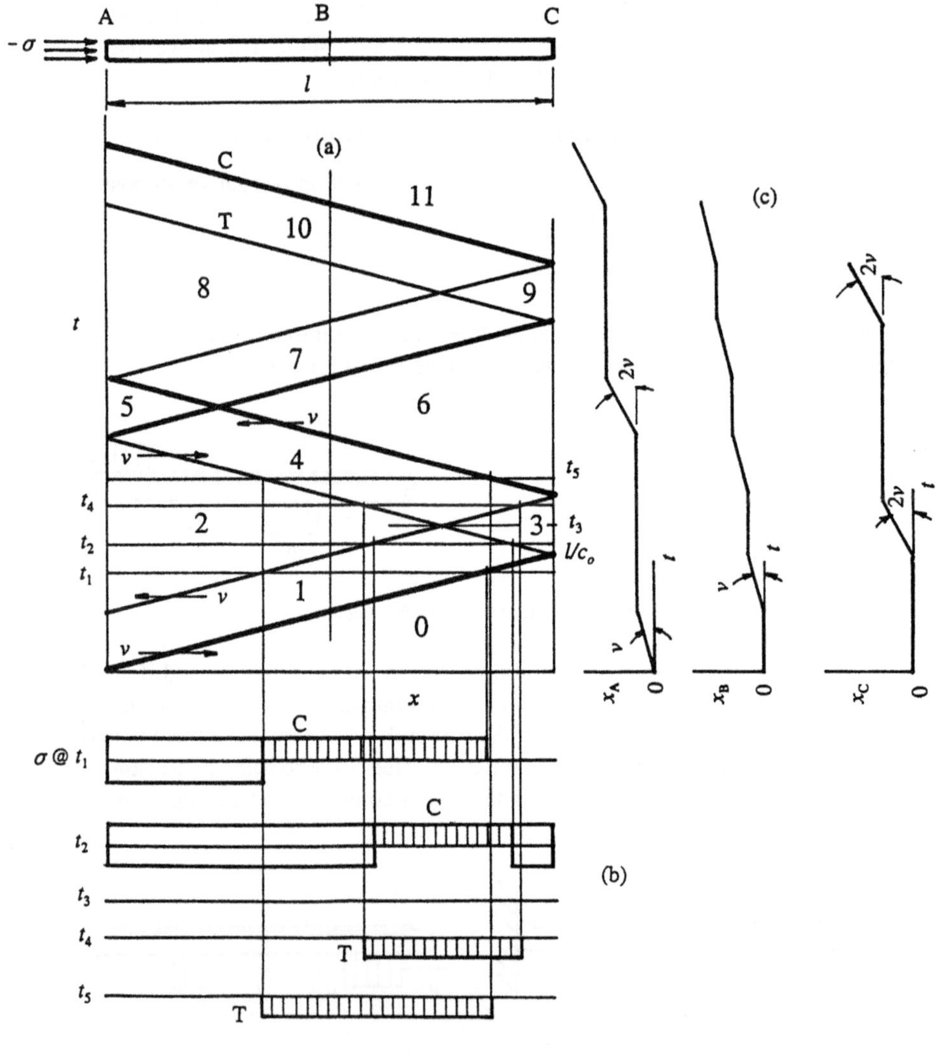

Figure 13.5 The x - t diagram for a compressive stress pulse

Finally, to conserve momentum at a given time, the momentum change must equal the impulse applied up to that time. Example 13.1 embodies these principles within the construction the x - t diagram in Fig. 13.5 when a stress pulse is applied to one end of a bar.

Example 13.1 A uniform circular bar of length l is subjected to a constant compressive stress at its left-hand end for a pulse time $t = l/(2c_o)$. Construct the x-t diagram showing the tensile and compressive waves. Use this diagram to show how the displacements of the bar's ends and centre vary with time.

The x-t diagram consists of a repeated pattern of wave reflections with alternating sense as shown in Fig. 13.5a. The given pulse is equivalent to applying a compressive wave followed by a tensile wave separated by the pulse time (see Fig. 13.5b). The time for this wave to reach the right-hand end is l/c_o. Hence, the compressive wave reaches the bar centre before the tensile wave begins. Within the regions 0, 1, 2 etc in Fig.13.5a, the net stress and absolute velocities within Table 13.1 apply.

Table 13.1 Stress and velocity within given regions of the x - t diagram

region	0	1	2	3	4	5	6	7	8	9	10	11
stress	0	$-\sigma$	0	0	σ	0	0	$-\sigma$	0	0	σ	0
velocity	0	v	0	$2v$	v	$2v$	0	v	0	$2v$	v	0

The time-dependent displacements of the bar ends and centre, A, C and B respectively, may be constructed from the wave intersections at these positions, as shown in Fig. 13.5c. The gradients dx/dt of the sloping parts of these diagrams define the particle velocities v and $2v$.

13.4 Longitudinal Impact of Bars

13.4.1 Equal Section Areas

Impact between circular bars with similar section areas but different lengths must satisfy a dynamic equilibrium and a compatibility condition. Let the velocity of bar 1 be v_1 and that of bar 2 be v_2, where $v_1 > v_2$ (see Fig. 13.6a). At the instant of longitudinal impact each bar experiences a compressive wave of identical magnitude σ at a particle velocity v, found from

$$v_1 - v = v_2 + v \;\Rightarrow\; v = \tfrac{1}{2}(v_1 - v_2) \qquad (13.5\text{a,b})$$

Substituting eq(13.5b) into either side of eq(13.5a) shows that the absolute velocity of each bar becomes $\frac{1}{2}(v_1 + v_2)$. The x-t diagram in Fig. 13.6b applies to this condition. From this, the time at separation point S is found as $t_s = 6l/c_o$. The conditions of stress and velocity across each region is given in Table 13.2.

Table 13.2 Stress and velocity regions

Region	0	1	2	3	4	5	6
stress	0	0	$-\sigma$	0	0	σ	0
abs vel	v_2	v_1	$\frac{1}{2}(v_1 + v_2)$	v_2	v_1	$\frac{1}{2}(v_1+v_2)$	v_1

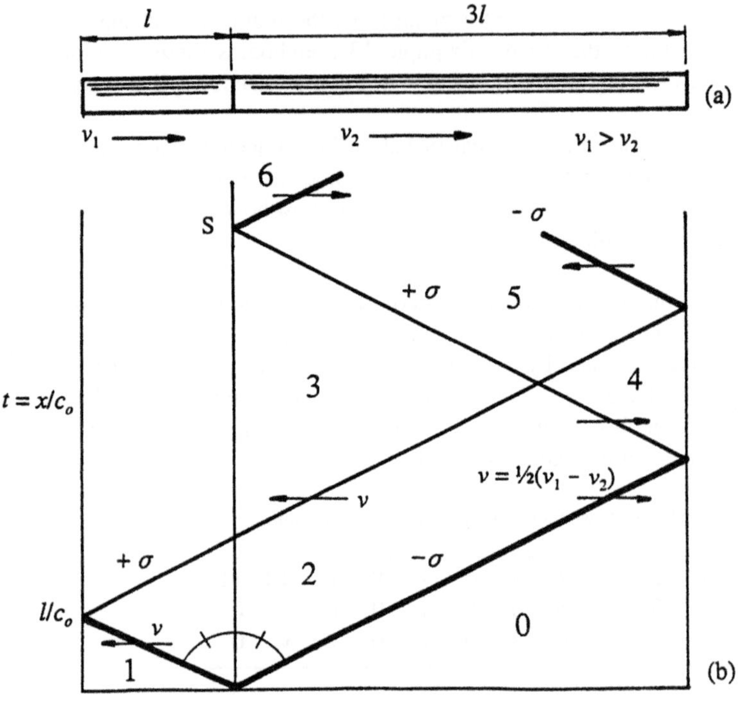

Figure 13.6 Longitudinal impact of two bars

13.4.2 Bar or Striker with Stepped Change in Section Areas

When a striker impacts a larger longitudinal bar, a wave reflection occurs at the change in section (see Fig. 13.7a). Firstly, let the incident wave *a* exist in the smaller section. Waves *b* and *c* reflect and transmit respectively at the section change. Their stress levels σ_a, σ_b and σ_c, where $\sigma_a > \sigma_b$, are denoted positive when compressive, as shown in Fig. 13.7b.

(a)

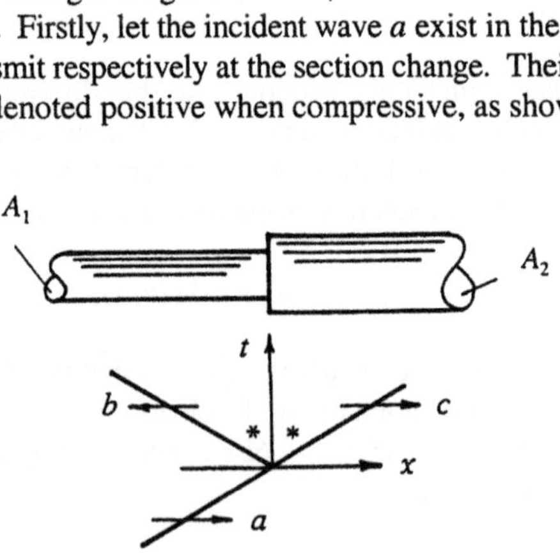

(b)

Figure 13.7 Reflection with a step-up in area

By assuming compression within the three waves *a*, *b* and *c*, equilibrium and compatibility conditions provide simultaneous equations:

$$(\sigma_a + \sigma_b)A_1 = \sigma_c A_2 \qquad (13.6a)$$

$$(\sigma_a - \sigma_b) / (\rho c_o) = \sigma_c / (\rho c_o) \qquad (13.6b)$$

Given $A_2 = 2A_1$, for example, the solution to eqs(13.6a,b) shows $\sigma_b = \sigma_a/3$ and $\sigma_c = 2\sigma_a/3$. However, if A_2 is very large the effect is similar to incidence upon a fixed end. Now let the incident wave exist in the larger section (see Fig. 13.8a).

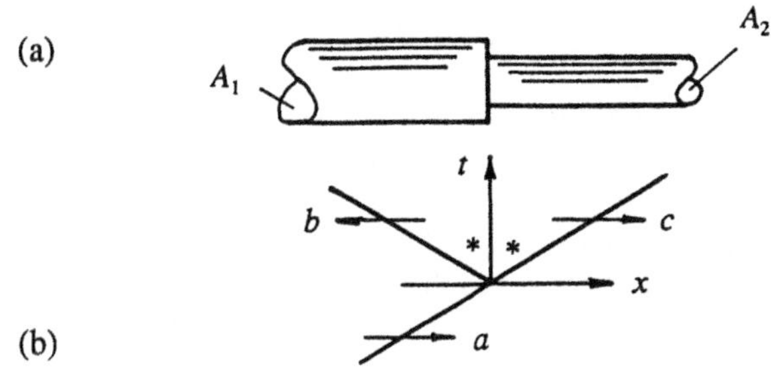

Figure 13.8 Reflection with a step-down in area

The two conditions become

$$(\sigma_a + \sigma_b)\,A_1 = \sigma_c\,A_2 \qquad (13.7a)$$

$$\sigma_a - \sigma_b = \sigma_c \qquad (13.7b)$$

and, if $A_1 = 2A_2$, we find from eqs(13.7a,b) stress levels $\sigma_b = -\,\sigma_a/3$ and $\sigma_c = 4\sigma_a/3$, showing that the reflected wave b changes sign, i.e. it becomes tensile.

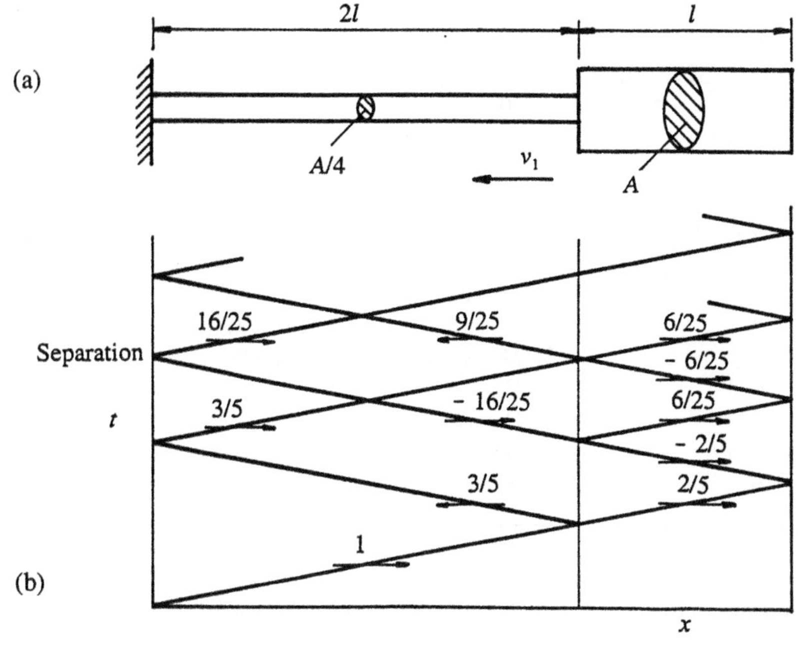

Figure 13.9 x - t diagram for a pneumatic striker showing magnitude, direction and sense of stress waves

Example 13.2 Figure 13.9a shows the stepped steel striker of a pneumatic tool which impacts upon a fixed target at its left end. Plot the x - t diagram and find the highest impact velocity if the dynamic elastic stress limit is 675 MPa. The initial stress may be taken as unity by setting $\sigma_o = \rho\, c_o\, v = 1$. Take $E = 210$ GPa and $\rho = 7890$ kg/m³.

With zero net velocity at the target end: $v_1 - v = 0$. From eqs(13.4b), at the stress limit:

$$v_1 = \sigma / (\rho c_o) = \sigma / \surd(E\rho)$$
$$= (675 \times 10^6) / \surd(210 \times 10^9 \times 7890) = 16.58 \text{ m/s}$$

Applying eqs(13.6a,b) across the stepped increase in the cross-section, where $A_2 = 4A_1$, gives $\sigma_b = 3\sigma_a/5$ and $\sigma_c = 2\sigma_a/5$. Applying eqs(13.7a,b) to the step-down section change, where $A_2 = A_1/4$, gives $\sigma_b = -\,3\sigma_a/5$ and $\sigma_c = 8\sigma_a/5$ and. The x - t diagram (see Fig. 13.9b) is constructed from the repeated application of these stress multiplication factors to transmission and reflection of incident waves as they meet the section change from either side. Recall that at the fixed, target end, the reflected wave retains the sense of the stress within the incident wave but the particle velocity is reversed.

Example 13.3 A striker travels lengthwise with a velocity $3v_o/2$ to impact the left end of a bar that is rigidly fixed at its right end (see Fig. 13.10a). Assuming elastic impact conditions, plot the x - t diagram, showing the separation point on this diagram and from it produce stress-time and velocity-time plots for locations X, Y and Z. Assume that the striker and bar material is the same.

Table 13.3 Regional wave stresses and velocities

Region	Stress	Absolute velocity
0	0	0
1	$-\,2\sigma$	$v = 3v_o/4$
2	$(-\,2\sigma + 4\sigma/3) = -\,2\sigma/3$	$-\,v/3 = -\,v_o/4$
3	$(-\,2\sigma - 2\sigma) = -\,4\sigma$	0
4	$(-\,4\sigma + 4\sigma/3) = -\,8\sigma/3$	$-\,v/3 - v = -\,4v/3 = -\,v_o$
5	$(-\,8\sigma/3 + 2\sigma) = -\,2\sigma/3$	$-\,4v/3 - v = -\,7v/3 = -7v_o/4$
6	$(-\,8\sigma/3 + 4\sigma/3) = -\,4\sigma/3$	$-\,4v/3 + 4v/3 = 0$
7	$(-\,2\sigma/3 + 4\sigma/3) = 2\sigma/3$	$-\,v = -3v_o/4$
8	$(+\,2\sigma/3 + 2\sigma) = 8\sigma/3$	0
L	0	$2v = 3v_o/2$
M	$-\,\sigma$	$3v_o/2 - v = (3v_o/2 - 3v_o/4) = 3v_o/4$
N	0	$3v_o/4 - v = (3v_o/4 - 3v_o/4) = 0$
O	$-\,\sigma/3$	$-\,v/3 = -\,v_o/4$
P	0	$-\,v/3 - v/3 = -\,v_o/2$
Q	$-\,\sigma/3$	$-\,2v/3 - v/3 = -\,v = -\,3v_o/4$
R	0	$-\,v - v/3 = -\,4v/3 = -\,v_o$

Force balance gives $\sigma_o A = \sigma_1 A\,/\,2$, which shows that the bar stress is twice that of the striker. The absolute velocities of particles at the contact surfaces must be the same:

$$3v_o/2 - v = v + v_1$$

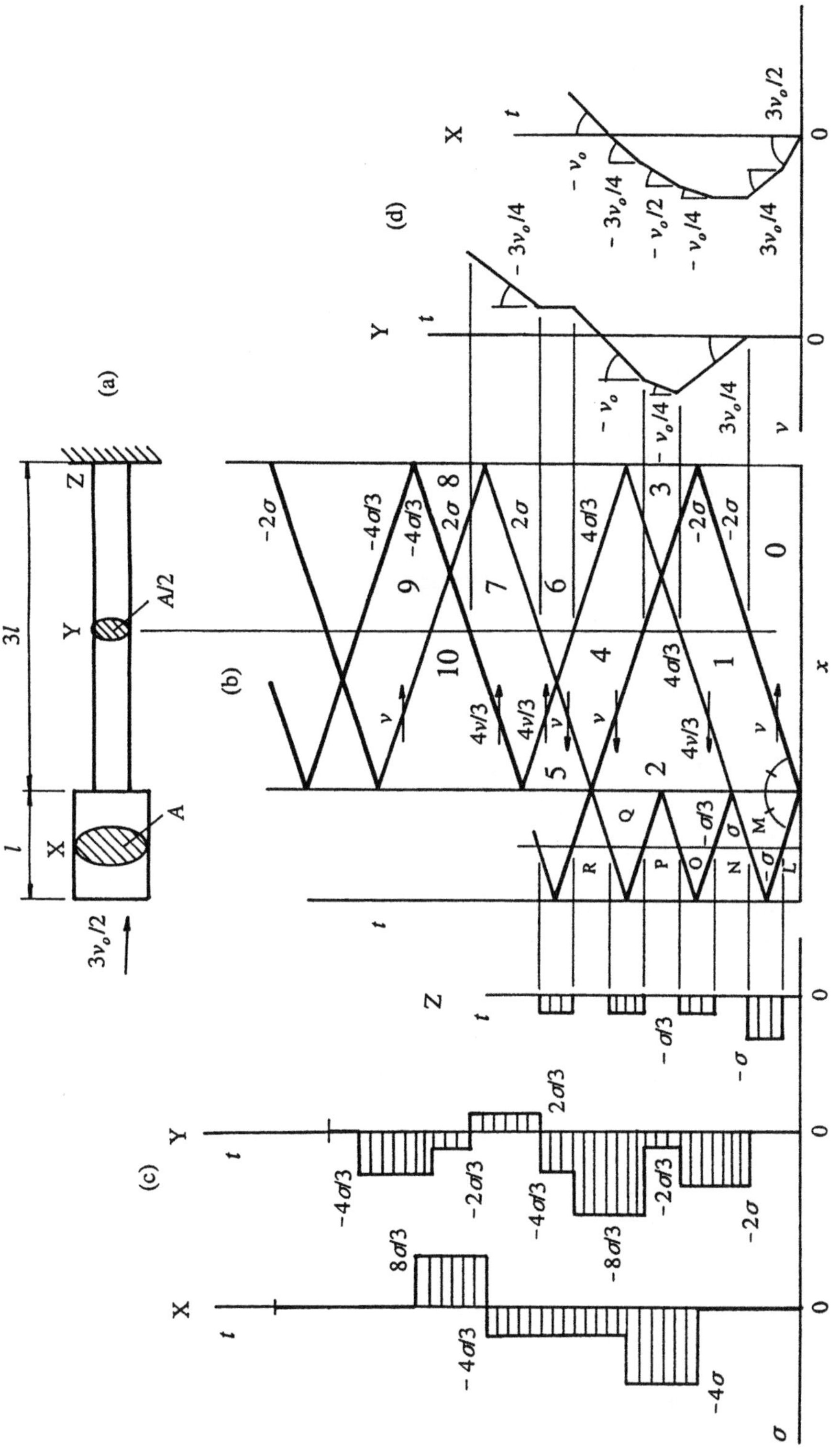

Figure 13.10 (a) Striker and bar showing (b) the x - t, (c) σ - t and (d) v - t diagrams; Key —— tension, ━━ compression

and since $v_1 = 0$, this shows that $v = 3v_o/4$ is the particle velocity of two identical compressive waves directed away from the contacting surfaces. Because the area of the bar section is halved, the multiplication factors in eqs(13.6) and (13.7) apply. These provide a sufficient portion of the x - t diagram (Fig. 13.10b) to reveal that a bar separation occurs at point 2. Within the enclosed regions 0, 1, 2... and L, M, N... , the stress and absolute velocities are calculated within Table 13.3. The stress column in Table 13.3 enables the accompanying stress plots for positions X, Y and Z to be constructed to the left side of the x - t diagram, as shown in Fig. 13.10c. The absolute velocity column in Table 13.3 allows the displacement plots for points X and Y to be constructed to the right of the x - t plot (see Fig. 13.10d), there being no displacement at Z.

13.4.3 Change of Material

Where a change of material occurs in the bar length, a transmission and reflection occurs at the interface, similar to the effect of a section change. Let E, ρ and c describe the modulus, density and propagation velocity in each bar material 1 and 2, as shown in Fig. 13.11a.

(a)

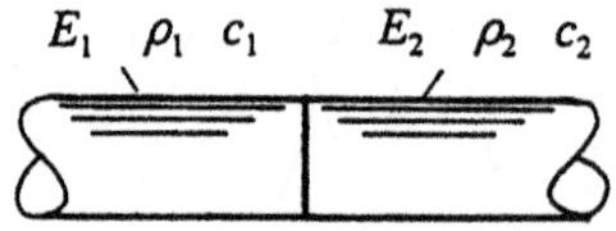

(b)

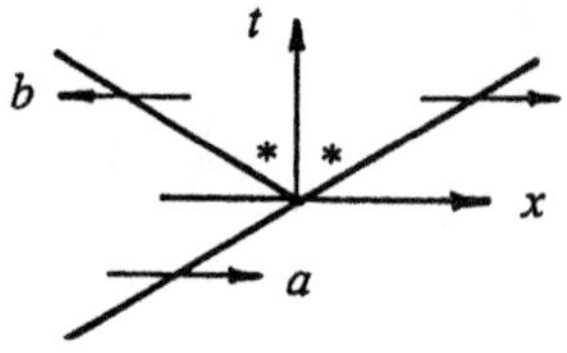

Figure 13.11 Reflection at the interface of a composite bar

For an incident wave a the force and velocity balance equations yield

$$\sigma_a + \sigma_b = \sigma_c \tag{13.8a}$$

$$(\sigma_a - \sigma_b) / (\rho_1 c_1) = \sigma_c / \rho_2 c_2 \tag{13.8b}$$

Solving eqs (13.8a,b) leads to the stress within the reflected and transmitted waves b and c (see Fig. 13.11b):

$$\sigma_b = \left[\frac{\sqrt{E_2\rho_2} - \sqrt{E_1\rho_1}}{\sqrt{E_2\rho_2} + \sqrt{E_1\rho_1}}\right]\sigma_a, \qquad \sigma_c = \left[\frac{2\sqrt{E_2\rho_2}}{\sqrt{E_2\rho_2} + \sqrt{E_1\rho_1}}\right]\sigma_a \tag{13.9a,b}$$

Example 13.4 Find the proportion of stress for an incident wave within epoxy resin that is transmitted and reflected at an interface with glass. Repeat for the same arrangement but with an incident wave within the glass. For epoxy: $E = 3.24$ GPa and $\rho = 1245$ kg/m^3 and for glass: $E = 62$ GPa and $\rho = 2215$ kg/m^3.

For a transmission from epoxy (subscript 1) to glass (subscript 2):

$$\sqrt{(E_1\rho_1)} = \sqrt{(3.24 \times 10^9 \times 1245)} = 2.008 \times 10^6\ \text{Nm/s}^2$$
$$\sqrt{(E_2\rho_2)} = \sqrt{(62 \times 10^9 \times 2215)} = 11.718 \times 10^6\ \text{Nm/s}^2$$

Substituting into eq(13.9a,b) gives

$$\sigma_b = (11.718 - 2.008)\sigma_a / (11.718 + 2.008) = 0.707\sigma_a$$
$$\sigma_c = (2 \times 11.718)\sigma_a / (11.718 + 2.008) = 1.707\sigma_a$$

Interchanging subscripts 1 and 2 for glass to epoxy, eq(13.9a,b) gives

$$\sigma_b = (2.008 - 11.718)/(2.008 + 11.718) = -\ 0.707\sigma_a$$
$$\sigma_c = (2 \times 2.008) / (2.008 + 11.718) = 0.293\sigma_a$$

Note that for air $\sqrt{(E\rho)}$ is small enough to be neglected compared to glass (say). So, for a glass to air interface eqs(13.9a,b) show $\sigma_b \approx -\ \sigma_a$ and $\sigma_c \approx 0$. That is, the wave is fully reflected with a change in sign, as was shown earlier for a free-end condition.

13.5 Plastic Waves

We have seen when the end of a long rod is struck in such a way that a constant elastic stress is maintained at its end, then a wave of compression with constant amplitude will travel along the rod at a speed of $c_o = \sqrt{(E/\rho)}$. Where the stress level in a longitudinal wave exceeds the yield stress of the bar material, i.e. $\sigma > \sigma_o$, two wave fronts arise: (i) an elastic wave of amplitude σ_o travelling at c_o and (ii) a plastic wave of amplitude σ which follows the elastic wave with a lower velocity c (see Fig. 13.12a). The distances x travelled by each wave at time t are respectively $c_o t$ and ct. Since $c < c_o$, at the plastic wave front, there will be a step up in the stress from σ_o to σ, where σ remains constant between the plastic wave front and the left struck end of the bar.

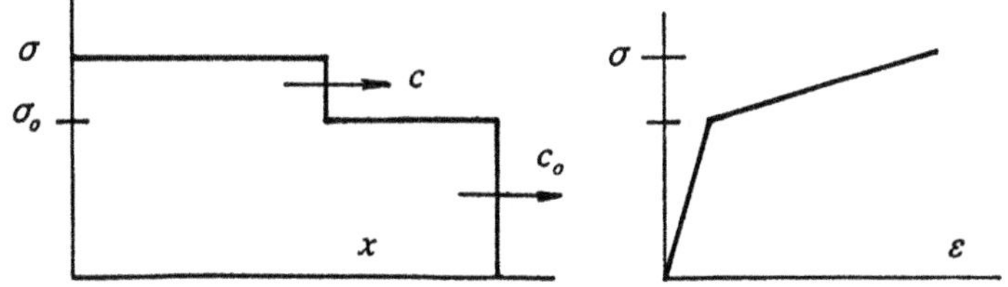

(a) (b)

Figure 13.12 Elastic and plastic wave fronts in a bar at time t

Where the material flow curve can be represented in a bi-linear form (see Fig. 13.12b), the corresponding strains in the bar are simply proportional to the stresses within Fig. 13.12a. Otherwise, an account of the plastic particle velocity c is required. Plastic wave theory applies to a static stress-strain curve [11, 12] in which the hardening rate $S = \mathrm{d}\sigma/\mathrm{d}\varepsilon$ decreases with strain. Let a particle displacement occur at position x within a circular bar of area A, due to a change in stress over the infinitesimal length δx, as shown previously in Fig. 13.1. We now replace E with S in eq(13.2) to give

$$\frac{\partial^2 u}{\partial t^2} = \left(\frac{S}{\rho}\right)\frac{\partial^2 u}{\partial x^2} \tag{13.10}$$

where $S = S(\varepsilon)$ is a variable. Kolsky [1] outlines von Karman's rigorous solution to eq(13.10). Taylor's simplified solution [11]: $x/t = \surd(S/\rho)$ employs the following boundary conditions:

(i) for time t at $x = 0$: $u\,(0, t) = -\,v_o t$

(ii) for time t at $x = \infty$, $u\,(\infty, t) = 0$

Since the strain is found from $\varepsilon = \partial u/\partial x$, the displacement u follows as

$$u = \int_\infty^x \varepsilon \, \mathrm{d}x \tag{13.11a}$$

Set $\beta = x/t$ so that the strain $\varepsilon = f(\beta)$ for a given time $t = \mathrm{d}x/\mathrm{d}\beta$. Equation (13.11a) becomes

$$u = t\int_\infty^\beta f(\beta)\,\mathrm{d}\beta \tag{13.11b}$$

Using conditions (ii) above, the substitution of the lower limit must give zero displacement. Equation eq(13.11b) integrates to

$$u = t\,F(\beta) \tag{13.12}$$

Now if t varies, successive differentiation of eq(13.12) reveals

$$\frac{\partial u}{\partial t} = F(\beta) + t\,F'(\beta)\frac{\partial \beta}{\partial t}$$

$$\frac{\partial^2 u}{\partial t^2} = 2\,F'(\beta)\frac{\partial \beta}{\partial t} + t\,F''(\beta)\left(\frac{\partial \beta}{\partial t}\right)^2 + t\,F'(\beta)\frac{\partial^2 \beta}{\partial t^2} \tag{13.13}$$

where, $F'(\beta) = f(\beta)$ and $F''(\beta) = f'(\beta)$. With $\beta = x/t$, the following partial derivatives apply to the right hand side of eq(13.13)

$$\partial\beta/\partial t = -\,x/t^2,\;\; (\partial\beta/\partial t)^2 = x^2/t^4,\;\; \partial^2\beta/\partial t^2 = 2x/t^3 \tag{13.14a,b,c}$$

The left side of eq(13.13) is identified with the right side of eq(13.10), so that

$$\frac{\partial^2 u}{\partial x^2} = \frac{\partial \varepsilon}{\partial x} = \frac{\partial f(\beta)}{\partial x} = \frac{1}{t}\,f'(\beta) \tag{13.15}$$

Substituting eqs(13.14a,b,c) and (13.15) into (13.13)

$$\frac{S}{\rho t}\,f'(\beta) = -\,\frac{2x}{t^2}\,f(\beta) + \frac{x^2}{t^3}\,f'(\beta) + \frac{2x}{t^2}\,f(\beta)$$

$$\frac{1}{t}\,f'(\beta)\left[\left(\frac{x^2}{t^2}\right) - \frac{S}{\rho}\right] = 0$$

$$\frac{1}{t}\,f'(\beta)\left(\beta^2 - \frac{S}{\rho}\right) = 0 \tag{13.16}$$

Equation (13.16) is satisfied either by:

$$f'(\beta) = \frac{\partial f(\varepsilon)}{\partial \beta} = 0 \quad \Rightarrow \quad \varepsilon = \text{constant w.r.t. } \beta \tag{13.17a}$$

or

$$\beta = \sqrt{\frac{S}{\rho}} \tag{13.17b}$$

Figure 13.13 provides a geometrical interpretation to the solutions (13.17a,b) within a region of work hardening between the yield strain ε_o and a plastic strain ε_1 (see Fig. 13.13a). The derived curve in Fig. 13.13b shows that only the second solution (13.17b) is valid.

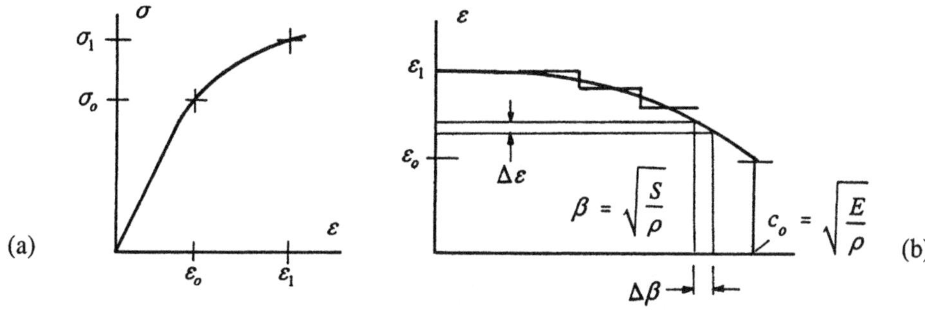

Figure 13.13 Variation in β within the plastic range

Using an incremental solution enables curve (b) to be approximated with a series of small steps in which β is taken to be constant for an increment of inelastic strain. This is equivalent to approximating the plastic region of the stress-strain curve with linear segments.

13.5.1 Plastic Particle Velocity

Combining condition (i) above with eq(13.11b) gives

$$-\,v_o = \int_{\infty}^{0} f(\beta)\,\mathrm{d}\beta = -\int_{0}^{\infty} \varepsilon\,\mathrm{d}\beta \tag{13.18a}$$

The integrand in eq(13.18a) may be replaced by its equivalent strip area $\varepsilon\,\mathrm{d}\beta = \beta\,\mathrm{d}\varepsilon$, as shown in Fig. 13.13b.

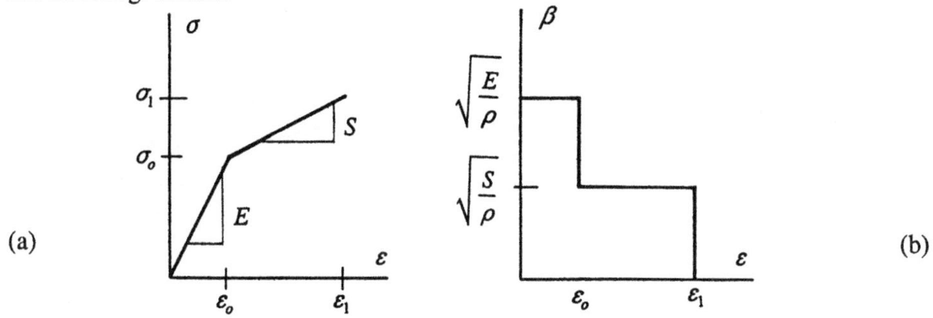

Figure 13.14 Bi-linear stress-strain curve

This transforms eq(13.18a) to

$$v_o = \int_0^{\varepsilon_1} \beta \, d\varepsilon \tag{13.18b}$$

With the simplest bi-linear approximation to a stress-strain curve (see Fig. 13.14a), the moduli within the elastic and plastic regions remain constant as E and S respectively. Hence eq(13.18b) may be integrated as follows:

$$v_o = \int_0^{\varepsilon_o} \sqrt{E/\rho} \, d\varepsilon + \int_{\varepsilon_o}^{\varepsilon_1} \sqrt{S/\rho} \, d\varepsilon$$

$$= \sqrt{\frac{E}{\rho}} \varepsilon_o + \sqrt{\frac{S}{\rho}} (\varepsilon_1 - \varepsilon_o) \tag{13.19a}$$

Substituting $\varepsilon_o = \sigma_o / E$, $\varepsilon_1 - \varepsilon_o = (\sigma_1 - \sigma_o)/S$, $\sqrt{(E\rho)} = \rho c_o$ and $\sqrt{(S\rho)} = \rho c$ into eq(13.19a):

$$v_o = \frac{\sigma_o}{\rho c_o} + \frac{(\sigma_1 - \sigma_o)}{\rho c_o} \left(\frac{c_o}{c} \right) \tag{13.19b}$$

in which the first term is the particle velocity arising from an elastic impact at the yield stress. The second term gives the plastic particle velocity when the stress level exceeds the yield stress. Here the identity $c = \beta = \sqrt{(S/\rho)}$ defines the plastic wave propagation velocity, which appears in an inverse ratio with the elastic propagation velocity c_o:

$$\frac{c_o}{c} = \frac{\sqrt{E/\rho}}{\sqrt{S/\rho}} = \sqrt{\frac{E}{S}} \tag{13.20}$$

Figure 13.14b suggests a simple extension for a multi-linear plastic stress-strain curve in which further plastic terms may be added to eq(13.19b) for each plastic stress increment.

13.5.2 Reflection at a Fixed Boundary

Recall that the stress within an elastic wave retains its sense and magnitude while the particle velocity changes direction upon reflection at a fixed boundary. Consequently, the stress in the material must double at the fixed end of the bar in Fig. 13.15.

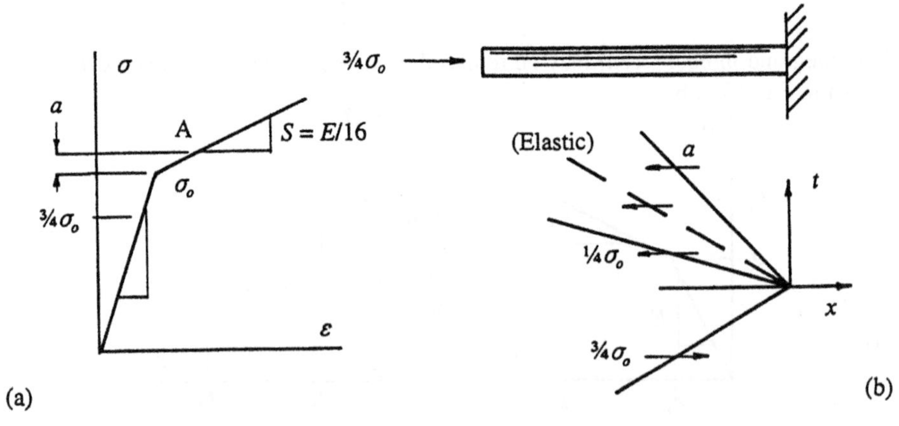

Figure 13.15 Elastic and plastic reflections at a fixed boundary

It follows that plasticity effects must be considered for fixed boundary reflections of incident elastic waves with stress magnitudes greater than $\frac{1}{2}\sigma_o$. Under these conditions both the elastic and plastic waves, as expressed by respective terms within eq(13.19b), will reflect to ensure zero displacement. To show this, let the stress within an incident elastic wave be $\frac{3}{4}\sigma_o$ in the material model of Fig. 13.15a. Firstly, we assume that an elastic stress reflection $\frac{1}{4}\sigma_o$ (Fig. 13.15b) gives a net stress with a magnitude of the yield stress σ_o. Let a stress increment a, beyond the yield stress σ_o, also exist within a reflected plastic wave in Fig. 13.15b. From eq(13.20), $c_o/c = 4$ and setting $v = 0$ after reflection, gives

$$\tfrac{3}{4}\,\sigma_o - (\tfrac{1}{4}\,\sigma_o + 4a) = 0$$

from which $a = \sigma_o/8$. Note that if the incident stress had equalled σ_o then only a plastic wave can reflect. As an alternative to a bi-linear model in Fig. 13.15a, let the plastic gradients in Fig. 13.16a give c_o/c = 2, 4, 8 and 16 etc, these being separated as far as possible by stress increments $\sigma_o/16$.

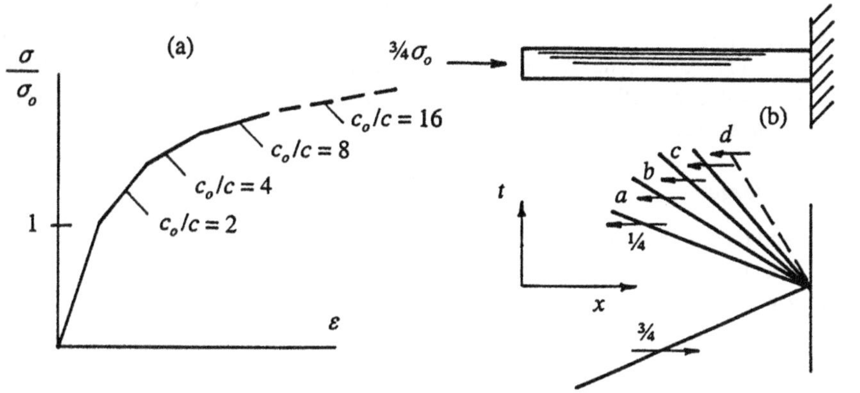

Figure 13.16 Multi-linear approximation to plastic region

The slopes dx/dt of the reflected waves in Fig. 13.16b under plastic stress increments a, b and c are found by factoring the initial elastic slope by the respective c_o/c values. The zero velocity condition shows that stress increments beyond c cannot exist and therefore c must conform to a lesser increment within:

$$\frac{3\sigma_o}{4} - \frac{\sigma_o}{4} - 2 \times \frac{\sigma_o}{16} - 4 \times \frac{\sigma_o}{16} - 8c = 0$$

from which $c = \sigma_o/64$

13.5.3 Unloading Waves

Consider loading a bar at time $t = 0$ to a plastic stress level σ_1 at its left-hand end. At a time $t = t_1$ this stress is released With a bi-linear approximation to the bar's stress-strain curve, loading is represented by two waves: (i) elastic under the yield stress σ_o and (ii) plastic under the stress increment $\sigma_1 - \sigma_o$ (Figs 13.17a,b). The elastic unloading wave is parallel to that for loading under σ_o but offset by time t_1. Consequently, given its faster particle velocity, the unloading wave will intersect the plastic increment loading wave at point A in Fig. 13.17b.

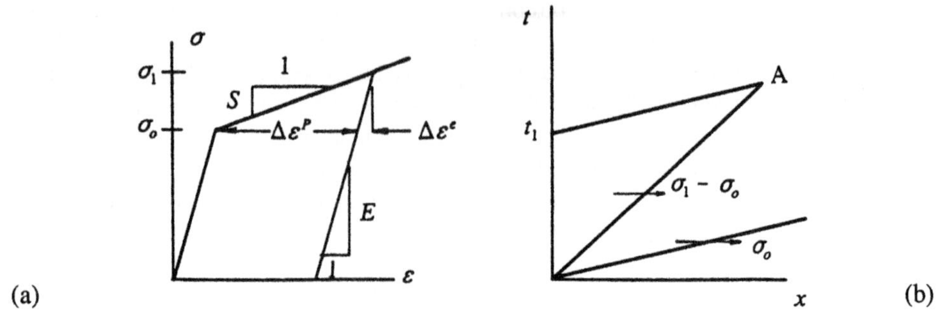

Figure 13.17 Loading and unloading waves

What happens at A will depend upon the magnitude of the stress increment $\sigma_1 - \sigma_o$. While the stress within the unloading wave will always exceed that in the plastic wave, the respective particle velocities of the two waves will not necessarily be in the proportion expressed by eq(13.19b). The plastic wave velocity will depend upon whether both stress and velocity can be overcome by the interaction. Depending upon the stress increment the following may occur, further plastic straining, a general elastic unloading into the elastic region and elastic reloading.

(I) $(\sigma_1 - \sigma_o)$ *large*

The σ - ε and the x - t diagrams in Figs 13.18a,b are now interlinked with state points B, C D etc. The particle velocity within the plastic increment will exceed that within the unloading wave when $(\sigma_1 - \sigma_o)c_o/c_1 > \sigma_1$. Consequently, a reduced plastic wave under stress increment $b = \sigma_2 - \sigma_o$ will continue to propagate, to give state point E in Fig. 13.18b.

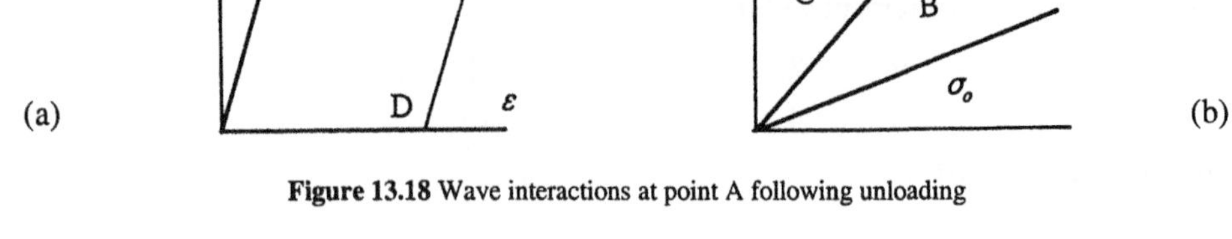

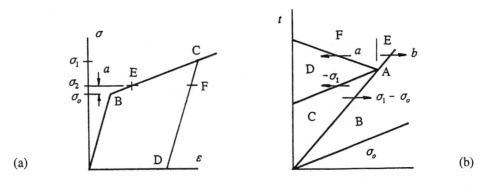

Figure 13.18 Wave interactions at point A following unloading

This figure also shows that when the left-hand end is unloaded to D, following the wave interaction at A, a further elastic reloading wave under stress a travels back from A giving state point F. The stress levels a and b must satisfy both the equilibrium and compatibility conditions. These conditions are applied to the left and right of the vertical section through A, using the signs and directions shown:

$$(\sigma_1 - \sigma_o) - \sigma_1 + a = b \tag{13.21a}$$

$$(\sigma_1 - \sigma_o)c_o/c - \sigma_1 - a = b \times c_o/c \tag{13.21b}$$

(II) $\sigma_1 - \sigma_o$ *small*

The particle velocity within a plastic wave of small stress increment may be overcome by the velocity of the elastic unloading wave. The latter will then have sufficient energy to continue unloading beyond the interaction point A (i.e. within region E) under a reduced stress level $-b\ (= \sigma_o - \sigma_2)$, as shown in Figs 13.19a,b.

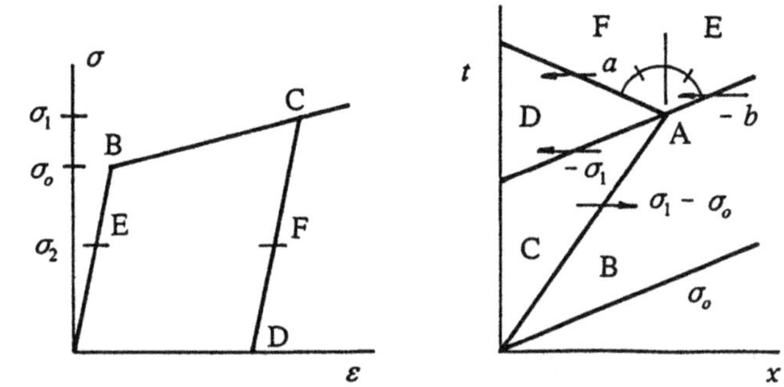

(a) (b)

Figure 13.19 Interaction at A for a small plastic stress increment

Following the interaction, a re-loading wave under stress a must also travel back to the left hand end (state F). Both stress levels a and b must preserve equilibrium and compatibility:

$$(\sigma_1 - \sigma_o) - \sigma_1 + a = -b \tag{13.22a}$$

$$(\sigma_1 - \sigma_o)\, c_o / c - \sigma_1 - a = -b \tag{13.22b}$$

(III) *Special case*

At a critical stress difference $\sigma_1 - \sigma_o$, lying between cases I and II above, no further plasticity can occur and nor can a continued unloading. To find the corresponding stress level σ_1, refer to Figs 13.20a,b.

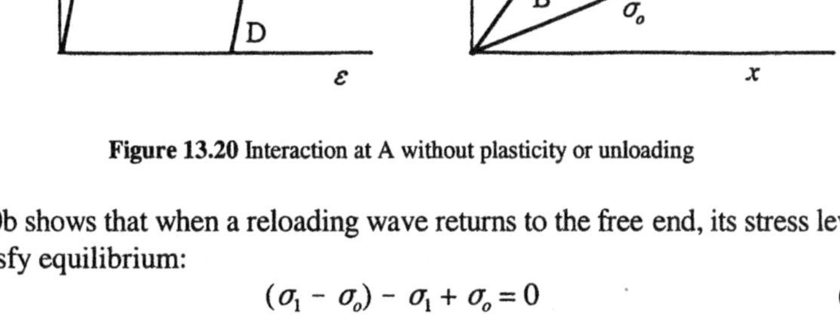

(a) (b)

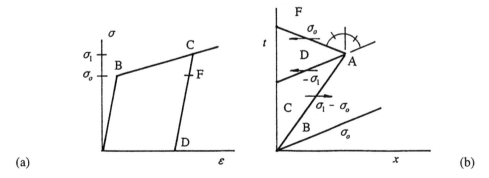

Figure 13.20 Interaction at A without plasticity or unloading

Figure 13.20b shows that when a reloading wave returns to the free end, its stress level must be σ_o to satisfy equilibrium:

$$(\sigma_1 - \sigma_o) - \sigma_1 + \sigma_o = 0 \tag{13.23a}$$

and using eq(13.19b) to ensure velocity compatibility:

$$(\sigma_1 - \sigma_o)\, c_o/c - \sigma_1 - \sigma_o = 0 \tag{13.23b}$$

Solving eqs(13.23a,b)

$$\sigma_1 = \left(\frac{c_o/c + 1}{c_o/c - 1}\right)\sigma_o \tag{13.23c}$$

The particle velocities for the elastic unloading and re-loading waves become $\sigma_o/(\rho c_o)$ and $\sigma_1/(\rho c_o)$ respectively.

Example 13.5 A bar of model material is compressed at its left end to a stress level $3\sigma_o/2$ before being released at time t_1 (see Fig. 13.21a). Given that the gradient of the plastic line is $S = E/81$, determine all the wave fronts when the bar length is: (i) infinite and (ii) chosen to allow the elastic wave under σ_o to reach its fixed right end at time t_1. Show the residual stress and plastic strain distribution for each case.

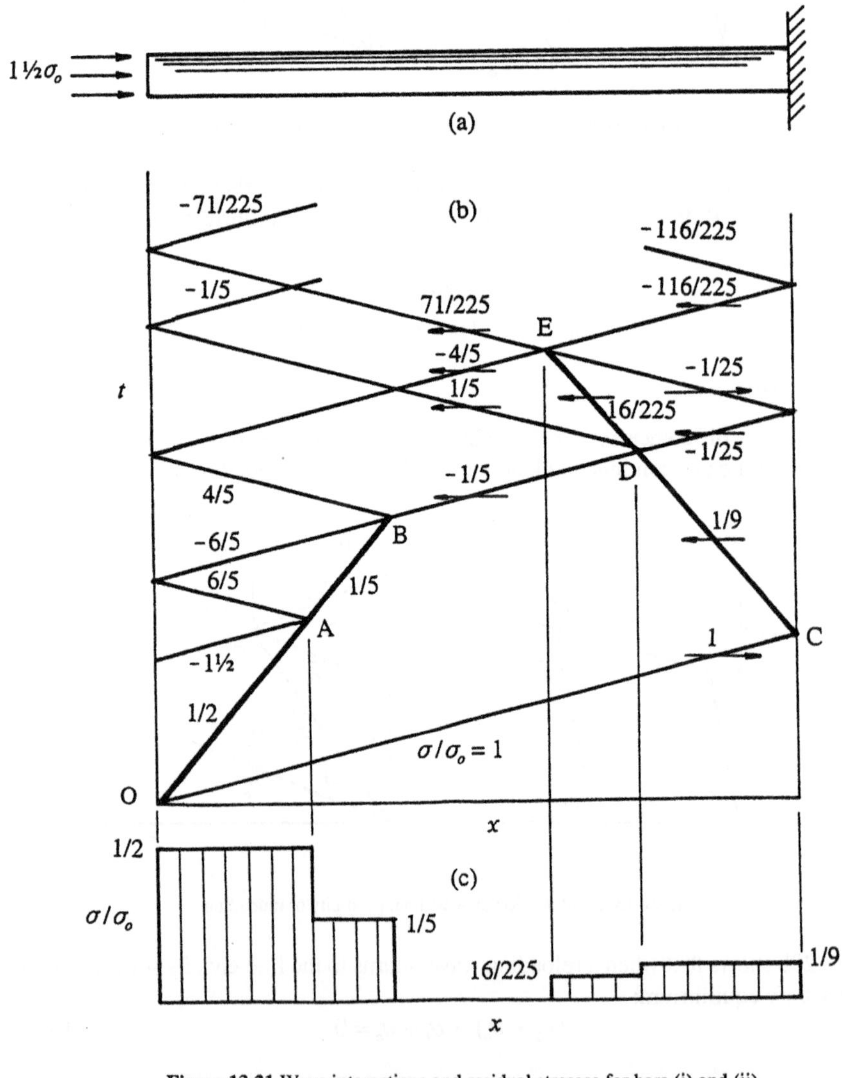

Figure 13.21 Wave interactions and residual stresses for bars (i) and (ii)

Figure 13.21b gives the a combined wave solution for bars (i) and (ii). For bar (i) interactions occur at A and B while the waves transmitted beyond these points do not return (see Figs 13.22a-c). For bar (ii) the wave returns from the fixed end at C resulting in additional interactions at D and E (see Figs 13.23a-c).

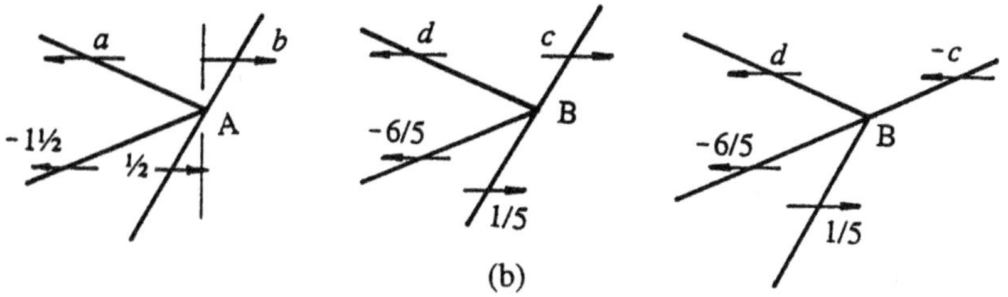

(a) (b) (c)

Figure 13.22 Possible wave interactions at A and B showing plastic stress levels

To determine the wave stresses levels a and b for point A it is necessary to decide which of the three cases above applies. First, assume case I where a weaker plastic wave continues under stress b and re-loads elastically under stress a (see Fig. 13.22a). The equilibrium and compatibility requirements are expressed from eqs(13.20) and (13.21a,b):

$$\tfrac{1}{2} - 1\tfrac{1}{2} + a = b$$
$$(\tfrac{1}{2} \times 9) - 1\tfrac{1}{2} - a = (b \times 9)$$

from which $a = 6/5$ and $b = 1/5$. Similarly, at B, by the same assumption (see Fig. 13.22b):

$$1/5 - 6/5 + d = c$$
$$(1/5 \times 9) - 6/5 - d = (c \times 9)$$

These equations show that the plastic stress increment c is negative, so invalidating case I. Assuming that case II applies to B (see Fig. 13.22c), the force and velocity balance eqs(13.22a,b) yield:

$$1/5 - 6/5 + d = -c$$
$$(1/5 \times 9) - 6/5 - d = -c$$

which gives the valid result: $c = 1/5$ and $d = 4/5$ and this may be applied to the x - t diagram in Fig. 13.21b to reveal that unloading and re-loading waves follow each other along an infinite length without further interactions. Figure 13.21c shows the stress increment levels $\Delta\sigma$ within the plastic wave OAB. Referring to Fig. 13.17a, $\Delta\sigma$ may be converted to plastic strain as follows:

$$\Delta\varepsilon^P = \Delta\sigma/S - \Delta\sigma/E \qquad (13.24)$$

For bar (ii), the plastic wave reflection at the fixed-end point C (see Fig. 13.23a) occurs with zero net velocity. From eq(13.19b):

$$\frac{1}{\rho c_0} - \frac{a}{\rho c_0} \times \left(\frac{c_0}{c}\right) = 0$$

from which $a = 1/9$ given $c_o/c = 9$.

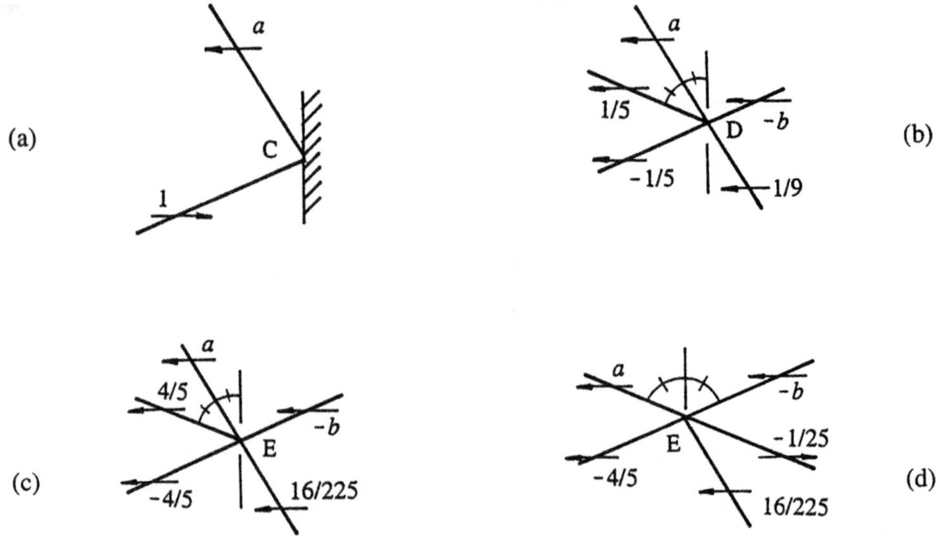

Figure 13.23 Wave interactions at C, D and E

The interaction at point D in Fig. 13.23b is physically tenable since the force and velocity balance equations

$$1/5 - 1/5 + a = 1/9 - b$$

$$1/5 + 1/5 + (a \times 9) = b + (1/9 \times 9)$$

show that $a = 16/225$ and $b = 1/25$ are correctly positive. However, if a similar interaction is assumed for point E (see Fig. 13.23c):

$$-4/5 + 4/5 + a = 16/225 - b$$

$$4/5 + 4/5 + (a \times 9) = (16/225 \times 9) + b$$

it follows that a is unacceptably negative, since it is impossible to have a tensile plastic increment beyond a compressive yield stress! Hence it is assumed that at E the plastic increment is overcome as in Fig. 13.23d. A force and velocity balance then gives:

$$-4/5 + a = -b - 1/25 + 16/225$$

$$4/5 + a = b - 1/25 + (16/225 \times 9)$$

from which $a = 71/225$ and $b = 116/225$ are physically acceptable. All waves that follow remain elastic. Equation (13.24) again provides the conversion from the stress levels within plastic wave CDE to residual plastic strain (see Fig. 13.21c).

Example 13.6 A free bar of length l is suddenly subjected to a compressive stress $\sigma_1 = 4\sigma_o$ at its left end for a time $T = l/c_o$ and thereafter removed. Taking a bi-linear stress-strain curve, with $S = E/9$, draw the $x - t$ diagram for a period $t = 0$ to $t = 5T/2$ s. Indicate the stress levels and particle velocities within each region of an x-t diagram.

The x - t diagram (Fig. 13.24) may be constructed from assuming that a plastic wave continues beyond the first interaction point A since $\sigma_1 - \sigma_o = 3$ is large. Also intersecting at this point is the tensile wave following its reflection from the right free end as shown.

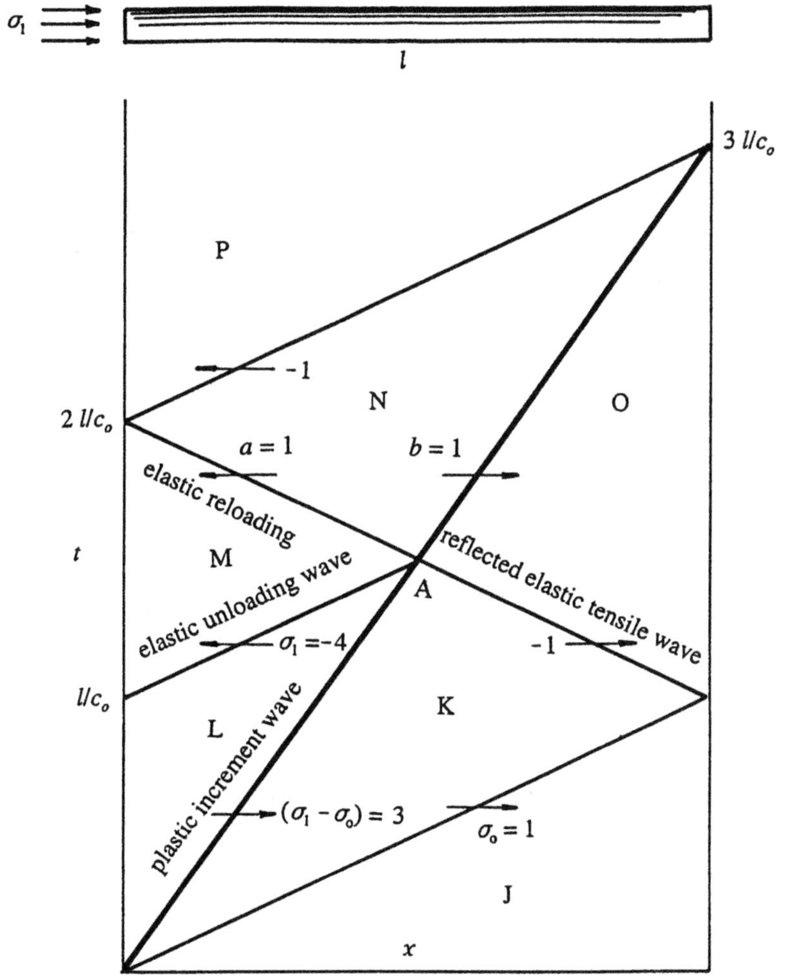

Figure 13.24 The x - t diagram for a free-end bar

Thus, with $c_o/c = \sqrt{(E/S)} = 3$, and with the stress signs and velocity directions given, the force and velocity balance equations for point A become

$$a + 3 - 4 = b - 1$$
$$(3 \times 3) - 4 - a = 3b + 1$$

from which $a = b = 1$ are acceptable. The re-loading compressive wave a is reflected in tension at the right free end and meets the transmitted compressive plastic wave b at the right end at a time $3l/c_o = 3T$. Thereafter, only the the elastic wave reflects. Within the regions J, K, L etc, for this diagram the net stress and absolute velocities are given in Table 13.4.

Table 13.4 Normalised wave stresses and velocities for bar in Fig. 13.24

Region	J	K	L	M	N	O	P
σ / σ_o	0	1	4	0	1	0	0
$v \times \rho c_o$	0	1	10	6	5	8	4

13.6 Plastic Stress Levels

In practice it is necessary to know how to determine the impact stresses. When a plastic impact occurs (i.e. $\sigma_1 > \sigma_o$) between a bar with known speed v and a rigid target, we may assume reflections within the bar shown in Figs 13.25a,b.

(a)

(b)

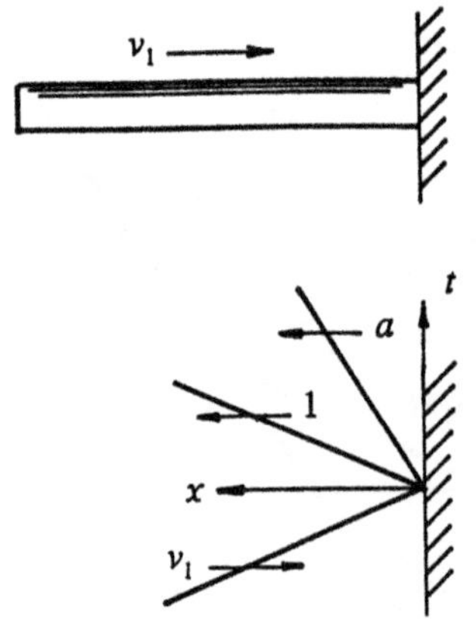

Figure 13.25 Reflections from an impact between bar and rigid target

The impact stress is split into two compressive wave reflections: an elastic wave under σ_o and a plastic increment wave under $a = \sigma_1 - \sigma_o$. Within Fig. 13.25b, the velocities balance:

$$v - \frac{\sigma_o}{\rho c_o} - \frac{(\sigma_1 - \sigma_o)}{\rho c_o}\left(\frac{c_o}{c}\right) = 0 \tag{13.25}$$

Hence, from knowing: v_1, σ_o, ρ, E and S, it follows that σ and a can be calculated. Alternatively, a plastic impact can occur between bars of different sizes. Let the stationary bar 1 be impacted by a larger bar 2 with an initial velocitiy v. The plastic stress levels upon impact become σ_1 and σ_2 respectively, as shown in Fig. 13.26a. Both elastic and plastic waves reflect into each bar from the interface under the stress levels indicated in Fig. 13.26b. The force equilibrium condition and a common particle velocity at impact ensure:

$$A_1 \sigma_1 = A_2 \sigma_2 \tag{13.26a}$$

$$v - \frac{\sigma_o}{\rho c_o} - \frac{(\sigma_2 - \sigma_o)}{\rho c_o}\left(\frac{c_o}{c}\right) = \frac{\sigma_o}{\rho c_o} + \frac{(\sigma_1 - \sigma_o)}{\rho c_o}\left(\frac{c_o}{c}\right) \tag{13.26b}$$

(a)

(b)

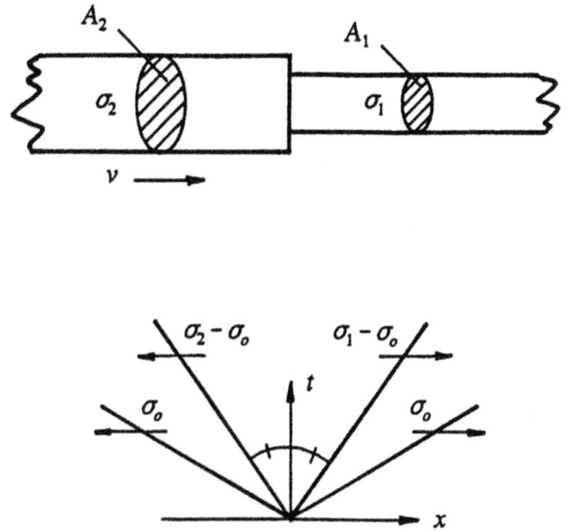

Figure 13.26 Plastic impact between bars of different areas

Equations (13.26a,b) allow for the calculation of σ_1 and σ_2, as the following example shows.

Example 13.7 A striker of length l, moving with a velocity 20 m/s, strikes a stationary rod half its diameter and twice its length (see Fig. 13.27a). Both striker and rod are of steel with a modulus $E = 207$ GPa and density $\rho = 7900$ kg/m^3. Using a a bi-linear stress-strain curve in which $\sigma_o = 465$ MPa and $S = E/16$, draw the x - t diagram from the moment of impact to a later time $t = 7l/c_o$. Plot the distribution of residual stress occurring within this interval.

Since $A_1/A_2 = (d_1/d_2)^2 = ¼$, it follows from eq(13.26a) that $\sigma_2 = \sigma_1/4$. Firstly, assume that both bars become plastic. Substituting into eq(13.26a,b) leads to

$$\frac{\sigma_1}{\sigma_o} = \frac{1}{5}\left(6 + \frac{v\sqrt{E\rho}}{\sigma_o}\right)$$

Using the material constants gives: $\sigma_1/\sigma_o = 1.55$, from which $\sigma_2/\sigma_o = 0.387$. These show that bar 2 remains elastic while bar 1 is plastic. Since this contravenes the impact assumption it is necessary to re-calculate stress ratios under these conditions. Equation (13.26b) is modified to become

$$v - \frac{\sigma_2}{\rho c_o} = \frac{\sigma_o}{\rho c_o} + \frac{(\sigma_1 - \sigma_o)}{\rho c_o}\left(\frac{c_o}{c}\right)$$

Re-arranging with eq(13.26a) as before:

$$\frac{\sigma_1}{\sigma_o} = \frac{4}{17}\left(3 + \frac{v\sqrt{E\rho}}{\sigma_o}\right)$$

for which $\sigma_1/\sigma_o = 1.115$ and $\sigma_2/\sigma_o = 0.278$. Figure 13.27b shows the x - t diagram based upon this elastic-plastic impact. The full solution provides the wave stress components at each intersection between reflections from the ends and from the contacting interface. A sample of the interactions involved appears within Figs 13.28a-d. Only the physically possible solutions for each point are given, i.e. no alternative interactions are possible.

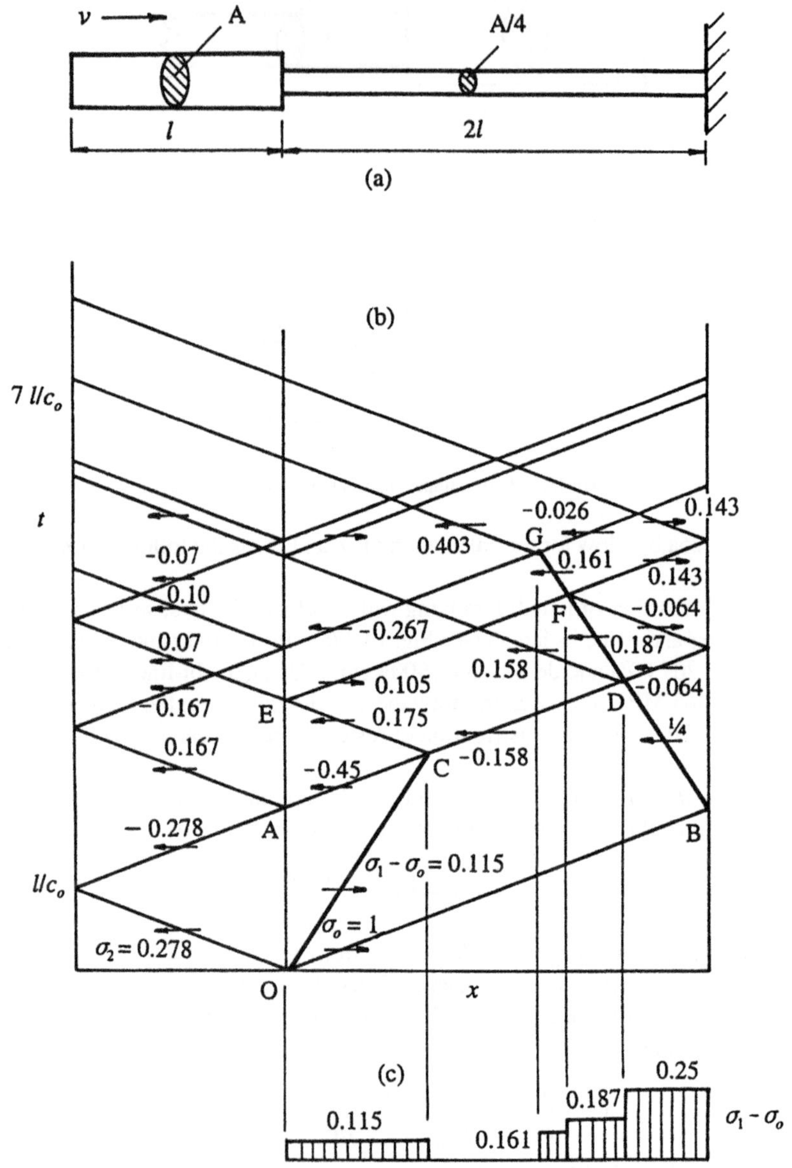

Figure 13.27 The x - t diagram for an elastic-plastic impact between bars

At A: (see Fig. 13.28a). The reflected tensile wave is split elastically at the interface. Applying eqs(13.22a,b):

$$-(0.278 + b)A = -aA/4$$
$$-0.278 + b = -a$$

from which $a = 0.45$ and $b = -0.167$ (i.e. it becomes tensile).

At B: We use the condition that $v_B = 0$ in Fig. 13.28b, to give

$$\frac{1}{\rho c_o} - \frac{a}{\rho c_o}\left(\frac{c_o}{c}\right) = 0$$

from which $a = c/c_o = 1/4$.

At C: Assume that the elastic wave is overcome, as in Fig. 13.28c:

$$b + 0.115 - 0.450 = -a$$
$$-(0.115 \times 4) + 0.450 + b = a$$

from which $a = 0.158$, $b = 0.175$.

At D: Let Fig. 13.28d be the wave interaction. This reveals

$$-0.158 + 0.158 + a = 0.25 - b$$
$$0.158 + 0.158 + 4a = b + (4 \times 1/4)$$

from which $a = 0.187$ and $b = 0.064$.

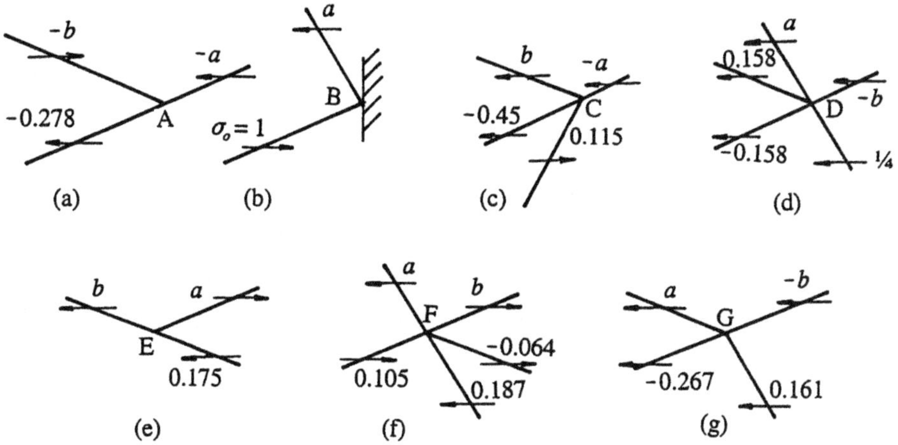

Figure 13.28 Wave stresses at interaction points A-G within Fig. 13.27

At E: Assume a reflected wave b with the same sign (see Fig. 13.28e):

$$bA = \tfrac{1}{4} A\,(a + 0.175)$$
$$b = 0.175 - a$$

from which $a = 0.105$ and $b = 0.07$.

At F: Assume the wave interaction shown in Fig. 13.28f:

$$a + 0.105 = b + 0.187 - 0.064$$
$$0.105 - 4a = b + 0.064 - (4 \times 0.187)$$

from which $a = 0.161$ and $b = 0.143$.

At G: The interaction given in Fig. 13.28f is assumed:

$$a - 0.267 = 0.161 - b$$
$$a + 0.267 = (4 \times 0.161) + b$$

from which $a = 0.403$ and $b = 0.026$. This shows that no plastic wave is transmitted beyond G and only elastic waves will follow after a time $7l/c_o$. The projection, Fig. 13.27c, shows the residual plastic stress distribution, derived from the plastic limbs OC and BDFG. The incremental proportions of the yield stress σ_o shown are readily converted to strain using eq(13.24) with the given modulus.

13.7 Concluding Remarks

We have seen that the approximate solution to the wave equation lends itself to a useful graphical interpretation of stress waves within impacting model materials. Stress, strain and velocity, within travelling and intersecting waves, are provided by this method. Stress levels in excess of the elastic limit are contained most conveniently within a single plastic modulus. More accurately, over a wide plastic stress range, the approach permits a more accurate multi-linear representation of the flow curve. The worked examples and exercises show the nature of the solution furnished by this approach when dealing with the longitudinal impact of a bar with both a rigid solid and another bar of different sectional area and material.

References

1. Kolsky H. *Stress Waves in Solids*, 1953, Clarendon, Oxford.
2. Goldsmith W. *Impact*, 1960, Edward Arnold.
3. Batchelor J. and Davies R. *Surveys in Mechanics*, 1956, Cambridge, pp. 64-138.
4. Abramson J. H., Plass and Ripperger E. A. *Advances in Appl Mech*, 1958, **5**, Academic Press, New York.
5. Cottrell A. H. *Chartered Mech Eng*, Nov 1957, I. Mech. E., London
6. Rinehart J. S and Pearson J. *Behaviour of Metals under Impulsive Loads*, 1965, Dover Pubs, New York.
7. Lee E. H. and Symonds P. S. *Plasticity, Proc 2nd Symposium Naval Structural Mechs*, 1960, Brown University, Pergamon.
8. Redwood M. *Mechanical Waveguides*, 1960, Pergamon Press, London.
9. Kolsky H. and Douch J. *Jl Mech Phys Sols*, 1962 **10**, 195.
10. Johnson W. *Impact Strength of Materials*, 1972, Arnold.
11. Taylor G. I. *Proc Roy Soc*. 1948, **A194**, 289.
12. Bell J. F. *The Physics of Large Deformation of Crystalline Solids*, 1968, Springer-Verlag, New York.

Exercises

13.1 A steel bar of length 510 mm and sectional area 645 mm^2 travels with a velocity 3.65 m/s to collide axially with a similar stationary bar of length 127.5 mm. Determine the time at which the impacting surfaces separate and find the final translational velocity of each bar. Determine the events subsequent to impact if the bar material is known to fracture under a tensile impact speed of 1.83 m/s.

13.2 An explosive charge is detonated at the free end of a uniform concrete beam. This causes 5 scabs of lengths 240 mm, 165mm, 190 mm, 330 mm and 620 mm, to be thrown off at the other end. Assuming a steep-fronted pulse, construct a curve showing the variation of stress in the applied pulse against distance travelled within the concrete. What is the possible error in the stress ordinates of the curve? Take the ultimate strength of concrete as 2.1 MPa.

13.3 The compressive stress pulse in Fig. 13.29a is applied at time $t = 0$ to the left-hand end of the concrete and steel composite bar, shown in Fig. 13.29b.

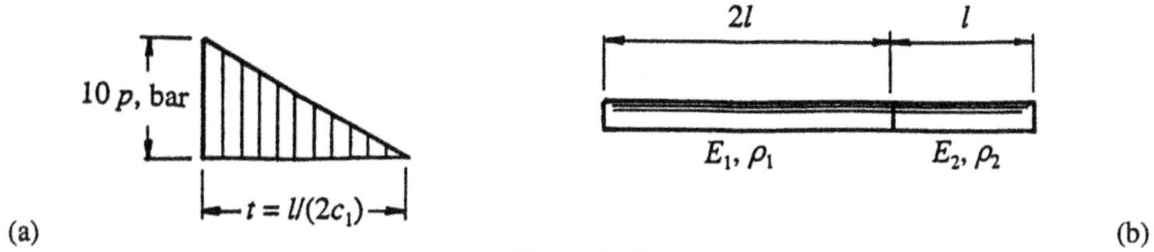

Figure 13.29

Find the sequence of events during time $0 < t < (2l/c_1 + 2l/c_2)$ where subscripts 1 and 2 refer to steel and concrete respectively, c_1 and c_2 are the propagation velocities within these materials. Illustrate the answer with sketches of the transmitted pulses at specific times within this interval. Take the maximum allowable tensile stress in the concrete as $-p$ MPa.

13.4 A tension impact specimen of length l is held by two loading bars, as shown in Fig. 13.30. The end X is pulled suddenly and maintained at a constant velocity v_o. The resulting, square-fronted, wave of tensile stress amplitude σ_o croses section PP at $t = 0$. Derive stress-time curves for the sections PP, QQ and RR from $t = 0$ to $t = 4l/c_o$. Assume both bar materials are similar for which the yield stress is not exceeded. Take the ends to be sufficiently remote for no reflections to occur during the time interval stated.

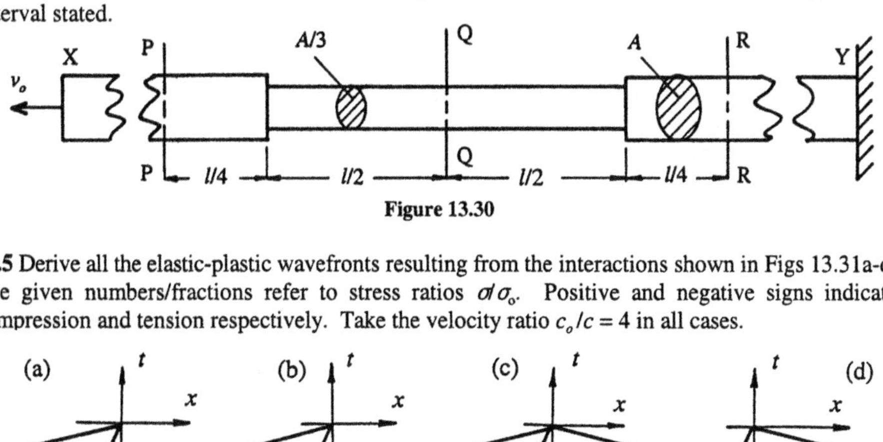

Figure 13.30

13.5 Derive all the elastic-plastic wavefronts resulting from the interactions shown in Figs 13.31a-d. The given numbers/fractions refer to stress ratios σ/σ_o. Positive and negative signs indicate compression and tension respectively. Take the velocity ratio $c_o/c = 4$ in all cases.

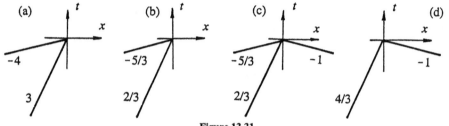

Figure 13.31

13.6 A thin rod of uniform cross-section and length l is rigidly clamped at one end. A tensile stress $25\sigma_o/12$ is suddenly applied at the other end for an interval l/c_o and then removed. Given $S = E/16$, plot on an x - t diagram the resulting stress increments for time $t = 0$ to $t = (5l)/(2c_o)$. The yield stress of the rod is σ_o.

13.7 A thin bar of area A and length $3L$ is fixed to a rigid support at one end. The other end is free and flanged as shown in Fig. 13.32. A tube of similar material, with section area $2A$ and length L, slides freely along the bar to strike the flange with an impact velocity 160 m/s. Draw the x - t diagram from $t = 0$ to $t = 6l/c_o$ given the following: $E = 207$ GPa, $\sigma_o = 275$ MPa, $S = E/16$ and $\rho = 7890$ kg/m^3.

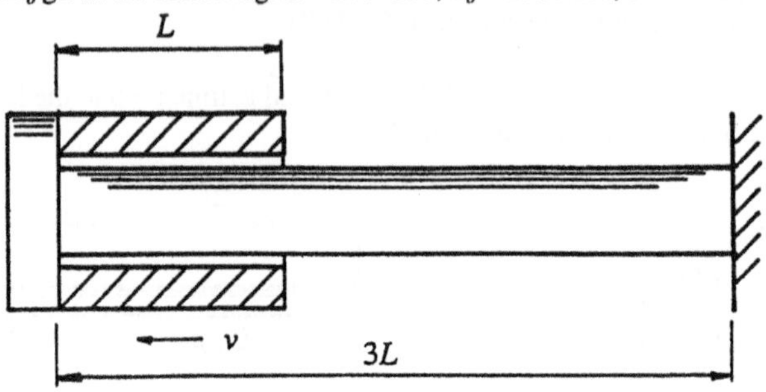

Figure 13.32

13.8 The large diameter end of a light alloy flagpole (see Fig. 13.33) is dropped 2.75m on to a solid rigid base plate. Construct the x - t diagram from the time of impact $t = 0$ to $t = 7l/c_o$. Show that the residual plastic strain distribution following impact is then completed. Take $g = 9.81$ m/s^2 with the following bar properties: $E = 124$ GPa, $S = E/16$, $\sigma_o = 198$ MPa, and $\rho = 3320$ kg/m^3.

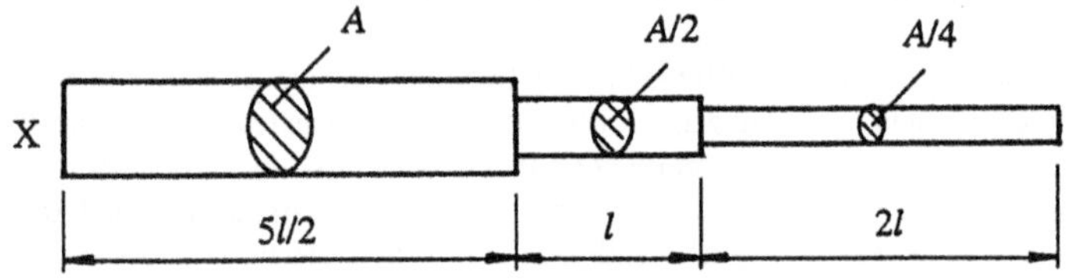

Figure 13.33

13.9 A uniform bar, 255 mm long, travelling at 76.5 m/s, strikes a rigid perpendicular surface. Plot the x - t diagram for the period in which plasticity occurs and sketch the residual plastic strain distribution. Assume a bi-linear stress-strain curve for the bar in which $\sigma_o = 138$ MPa, $E = 6.9$ GPa, $S = E/64$ and $\rho = 8304$ kg/m^3.

13.10 A uniform bar 255 mm long is projected axially at constant speed to strike a rigid perpendicular target. If the initial impact stress within the striking end of the bar is initially 69 MPa, find the striking velocity and plot the x - t diagram, showing the magnitude of each wave front for the duration in which plasticity occurs. Sketch the residual plastic strain distribution after this period. Assume the following: $E = 13.8$ GPa, $\sigma_o = 55$ MPa, $S = E/100$ and $\rho = 7200$ kg/m^3. [Answer: $v_o = 19.56$ m/s]

13.11 A uniform bar of length l and area A is rigidly clamped at one end. A second bar of the same material with length $l/3$ and area $2A$ impacts the clamped bar at its free-end with velocity of 14.75 m/s. Plot the x - t diagram showing the stress magnitudes in both bars from time $t = 0$ to $t = 5l/c_o$. Discuss whether any further plastic wave fronts are generated after this time interval for a material with the following properties: $E = 207$ GPa, $\sigma_o = 345$ MPa, $S = E/16$ and $\rho = 7890$ kg/m^3.

13.12 A stepped circular bar consists of two lengths L and $3L$ with respective areas $2A$ and A. The smaller end strikes a rigid perpendicular target at 12.7 m/s. Plot the location-time diagram showing the magnitudes and direction of the wavefronts for as long as plasticity continues to occur. Plot the residual plastic strain distribution. Assume the following: $E = 207$ GPa, $\sigma_o = 345$ MPa, $S = E/16$ and $\rho = 7890$ kg/m^3.

CHAPTER 14

PRODUCTION PROCESSES

14.1 Introduction

In our earlier work on slip lines and limit loading, analyses were made of the extrusion and rolling processes. We shall now look at these again among a number of large scale bulk forming processes where temperature, friction and strain rate are controlling parameters. These include forging, extrusion, rolling and machining. Alternative analyses of each of these shaping methods adopted here employ work conservation and slab equilibrium principles. Within this some previous results from SLF and bounding theory will be employed. The introductory treatment given will serve to extend the range of methods available for the analyses of forming and shaping of metals within the plastic range.

14.2 Hot Forging

In the case of hot forging, a billet of material is compressed between flat, parallel dies successively so that thickness is reduced by the required amount. Uniform bars and rings may be produced by this method, either by rapid, repeated hammering or by a slower hydraulic pressure. More complex shapes, e.g. connecting rods and propellor blades, may be produced by forcing material into die cavities. Working material into such shapes is inhibited by its loss in temperature and an increase in surface friction, both of which increase the compressive force required.

14.2.1 Geometrical Constraints

Figure 14.1 shows the process of deforming a volume of material beneath dies that reduce the thickness from a cogging process. Depending upon the final dimensions required, there will usually be some degree of spreading and bulging in the width. That is, $w_1 > w_o$, but for each cogging step the material preserves its initial volume:

$$w_o b\, t_o = w_1 b_1 t_1 \tag{14.1a}$$

in which b_1 is the formed length of material beneath the die breadth b. Equation (14.1a) is written as

$$\ln (w_1 / w_o) + \ln (b_1 / b) + \ln (t_1 / t_o) = 0 \tag{14.1b}$$

A *coefficient of spread* s is defined as

$$s = \frac{\ln (w_1 / w_o)}{\ln (t_o / t_1)} \tag{14.2a}$$

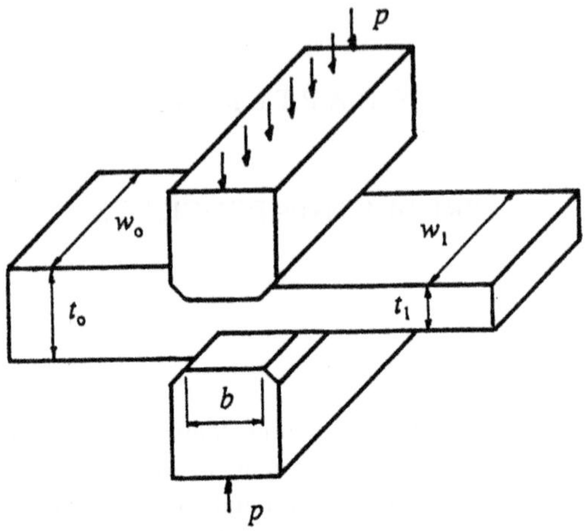

Figure 14.1 Forging between flat dies

A *coefficient of elongation* follows from eqs(14.1b) and (14.2a) as

$$1 - s = \frac{\ln(b_1/b)}{\ln(t_o/t_1)} \tag{14.2b}$$

The two coefficients in eqs(14.2a,b) give width and breadth ratios:

$$\frac{w_1}{w_o} = \left(\frac{t_o}{t_1}\right)^s \quad \text{and} \quad \frac{b_1}{b} = \left(\frac{t_o}{t_1}\right)^{1-s} \tag{14.3a,b}$$

where s is regarded as an empirical constant that depends upon the bite ratio b/w_o. If $b = w_o$, eqs(14.2a,b) give $s = 0.5$. Where w_o is large, giving $b/w_o \approx 0$, then $s = 0$ to give no spread under plane strain. Conversely, if the w_o dimension is small then $s = 1$. A practical formula, that embodies these three cases, defines s as

$$s = \frac{(b/w)_o}{1 + (b/w_o)} \tag{14.4}$$

Equations (14.1) - (14.4) apply to a forging pass in which the squeeze ratio t_o/t_1 is constant. The following recommendations [1] are made to avoid a central region of unworked material, excessive distortion and overlapping at the die edges:

$$\frac{t_o}{t_1} \le 1.3, \quad \frac{b}{t_o} \ge \frac{1}{3} \quad \text{and} \quad \frac{t_o}{w_o} \le \frac{3}{2}$$

14.2.2 Die Pressure and Loading

As a first approximation it is assumed that the material beneath the die (see Fig. 14.1) deforms in plane strain. Thus, when the width w remains unaltered, the die pressure will be given by the SLF solutions discussed in Section 6.7. These solutions are combined within Fig. 6.50 to give a normalised die pressure appropriate to the b/t ratio in lubricated dies. However, in the presence of friction, SLF theory underestimates die pressure. The accuracy of the latter is improved when it is assumed that the shear yield stress k of the material has been reached at the interface, as shown in Fig. 14.2.

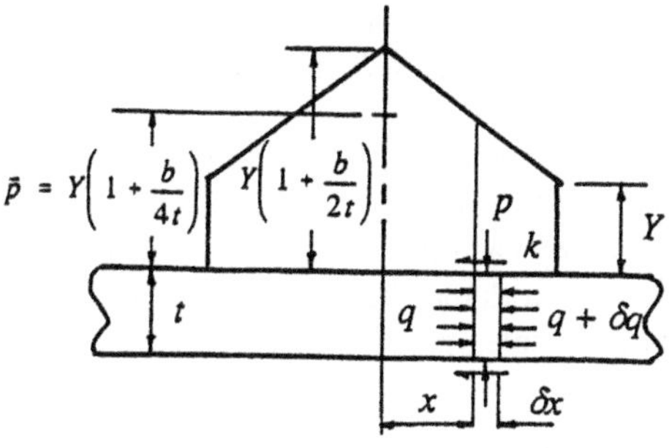

Figure 14.2 Forging of strip showing a material element and the friction hill

Plane strain conditions are assumed so that horizontal equilibrium of a strip with a unit width ($w = 1$) reveals

$$(2k \times \delta x \times 1) + (t \times \delta q \times 1) = 0 \tag{14.5}$$

Within the strip, the two compressive principal stresses p and q conform to the Tresca yield criterion:

$$p - q = Y = 2k \tag{14.6}$$

Combining eqs(14.5) and (14.6) leads to the pressure gradient for the die half:

$$\mathrm{d}p/\mathrm{d}x = \mathrm{d}q/\mathrm{d}x = -\ 2k\ /\ t \tag{14.7a}$$

Integrating gives the die pressure

$$p = -\ Yx\ /\ t + C \tag{14.7b}$$

in which the constant $C = Y + Yb/2t$ is found from the condition that $p = Y$ for $x = b/2$ at the die edge. Hence, the die pressure is distributed as:

$$p = Y\left(1 + \frac{b}{2t} - \frac{x}{t}\right) \tag{14.8a}$$

Equation (14.8a) defines a linear pressure distribution over one half of the die face, as shown in Fig. 14.2. The so-called *friction hill* estimates the forces required to forge both square and circular prisms [2]. The three-dimensional pressure required is distributed within a cone and a pyramid respectively over the die face. To describe a simple cogging operation, with $b/t > 1$, the mean pressure is adequate. Equation (14.8a) gives this as

$$\bar{p} = Y(1 + b/4t) \tag{14.8b}$$

The following empirical expressions [1] provide a closer estimate for each range of die breadth to thickness ratio:

$$p = Y\,(\,0.797 + 0.203t/b\,)\ \text{ for }\ b/t < 1 \tag{14.9a}$$

$$p = Y\ (0.750 + 0.250b/t\,)\ \text{ for }\ b/t > 1 \tag{14.9b}$$

Example 14.1 A block, 60 mm square and 300 mm long, is to be hot forged into a thinner strip, 20 mm thick, 60 mm wide by 900 mm long, by repeated cogging steps between a die whose breadth b is to minimise the plastic work required. Given that the b/t ratios involved with the reduction in thickness from 60 mm to 20 mm involve die pressures expressed by both eqs(14.9a,b), find the required value of b. Neglect work hardening and spread.

Since $60 < b < 20$ mm, eq(14.9a) applies initially with $t_o = 60$ mm for a reduction to $t = b$ (i.e. $b/t \le 1$). Thereafter, eq(14.9b) will apply for $t_o = b$ for final reductions to $t_1 = 20$ mm (i.e. $b/t \ge 1$). The work done by each die is the product of the force F applied and the displacement y it produces. This work may be referred to a current strip thickness t with an appropriate change to the limits of integration. Hence, the work done by the die pair per step becomes:

$$W = 2\int_0^{(t_o - t)/2} F\,\mathrm{d}y = \int_t^{t_o} F\,\mathrm{d}t$$

Substituting from eqs(14.9a,b), with $F = pwb$ and multiplying by the number of cogging steps l_o/b (l_o is the initial length), gives the total work done:

$$\sum W = Ywl_o\left[\int_b^{t_o}\left(0.797 + 0.203\frac{t}{b}\right)\mathrm{d}t + \int_{t_1}^{b}\left(0.750 + 0.250\frac{b}{t}\right)\mathrm{d}t\right]$$

$$= Ywl_o\left[0.797(t_o - b) + \frac{0.203}{2b}(t_o^2 - b^2) + 0.75(b - t_1) + 0.25b\ln\left(\frac{b}{t_1}\right)\right]$$

The die breadth which minimises the total work follows from:

$$\frac{\mathrm{d}\sum W}{\mathrm{d}b} = 0 = -0.797 - \frac{0.203t_o^2}{2b^2} - \frac{0.203}{2} + 0.75 + 0.25\left[1 + \ln\left(\frac{b}{t_1}\right)\right]$$

$$0 = 0.1015\left(1 - \frac{t_o^2}{b^2}\right) + 0.25\ln\left(\frac{b}{t_1}\right) = 0$$

Substituting $t_o = 60$ mm and $t_1 = 20$ mm, provides the required die width: $b = 37.56$ mm.

14.3 Cold Forging

Small parts may be cold forged to improve their surface finish and hardness. The process is usually conducted at rapid rates under a drop hammer using dies to produce the complicated shapes required, e.g. in crankshaft and coin production. Forging may be employed for non-ferrous alloys and carbon steels when the interface frictional stress is less than the shear yield stress k in the contact zone. Assuming Coulomb friction $\tau = \mu p \le k$ implies that the material beneath the dies slides under a constant frictional coefficient μ. The strip equilibrium equation (14.7a) becomes

$$\frac{\mathrm{d}p}{\mathrm{d}x} = -\frac{2\mu p}{t} \tag{14.10a}$$

Integrating eq(14.10a) leads to

$$\ln p = -2\mu x / t + C \tag{14.10b}$$

Along the edges of the die ($x = \pm b/2$) friction is absent and $p = 2k$ is reached. This gives $C = \mu b/t + \ln 2k$ and eq(14.10b) becomes

$$p = 2ke^{\frac{\mu b}{t}\left(1 - \frac{2x}{b}\right)} \tag{14.10c}$$

Equation (14.10c) describes the friction hill distribution shown in Fig. 14.3a.

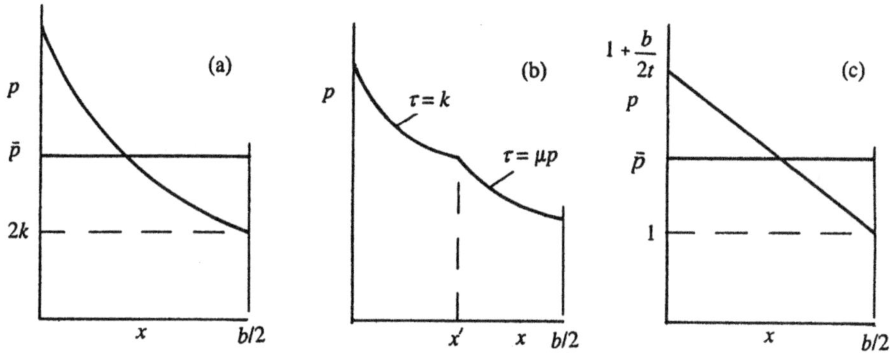

Figure 14.3 Friction hills for cold forging

From this, the mean pressure $\bar{p}$ in Fig. 14.3a may be found:

$$\bar{p} \times \frac{b}{2} = \int_0^{b/2} p\,\mathrm{d}x = 2k\int_0^{b/2} e^{\frac{\mu b}{t}\left(1 - \frac{2x}{b}\right)}\mathrm{d}x$$

$$\frac{\bar{p}}{2k} = \frac{t}{\mu b}\left(e^{\mu b/t} - 1\right) \tag{14.11}$$

The peak pressure for the hill in Fig. 14.3a is restricted to where $\mu p = k$, and so, with a widening of the die, the pressure in the central region becomes $p = k/\mu$. If μ remains constant in this region the friction hill is modified with a flat plateau. Figure 14.3b shows a more likely pressure distribution under a varying μ. Eq(14.10b) gives the corresponding position $x = x'$ at which edge sliding under $\tau = \mu p$ becomes central sticking under $\tau = k$:

$$x' = \frac{b}{2}\left[1 - \frac{t}{\mu b}\ln\left(\frac{1}{2\mu}\right)\right] \tag{14.12}$$

If x' is to lie within the die's half-width ($0 < x' < b/2$), it follows that $\ln(1/2\mu) < \mu b/t$, from eq(14.12). When $x' < b/2$, the pressure within the central zone ($0 \le x \le x'$) is found from integrating eq(14.7a)

$$p = C - \frac{2kx}{t}$$

where C follows from setting p to eq(14.10b) with $x = x'$ from eq(14.12). This leads to

$$p = \frac{2k}{t}\left(\frac{b}{2} - x\right) + \frac{k}{\mu}\left(1 - \ln\frac{1}{2\mu}\right) \qquad (14.13)$$

At the friction limit $\mu = ½$ and eq(14.12) gives $x' = b/2$. Correspondingly, eq(14.13) shows that the pressure then becomes linearly distributed over the half-width (see Fig. 14.3c) as

$$p = \frac{2k}{t}\left(t + \frac{b}{2} - x\right)$$

for which the mean value is

$$\frac{\bar{p}}{2k} = 1 + \frac{b}{4t}$$

In the case of Fig. 14.3b, the mean pressure is

$$\frac{\bar{p}}{2k} = \frac{t}{\mu b}\left\{\frac{1}{4\mu}\left[(1 + \alpha)^2 + 1\right] - 1\right\}$$

where $\alpha = \mu b/t - \ln[1/(2\mu)]$. The present theory is regarded as acceptable where $b/t < 3$ and $\mu < 0.3$. Slip-line field solutions (see Section 6.7) will provide more reliable estimates of die pressure in the case of $b/t > 3$, particularly where μ is high.

14.4 Extrusion

We have discussed SLF solutions to various extrusion geometries and friction conditions earlier. Making an assumption of plane strain restricts the application to rectangular dies and billets. Consequently, the axi-symmetric extrusion of bars and tubes remains to be analysed. These products may be produced by direct and indirect methods. In conventional hot extrusion (see Fig. 6.24) the ram must exert a force of several hundred tonnes to overcome friction at the container walls. This method is favoured for the production of various bar sections in non-ferrous alloys where glass has been used as a lubricant. With even higher temperatures required to extrude steels, the molten glass solidifies upon the surface of the extrudate. A less convenient, inverted extrusion process (Fig. 14.4a), reduces the forces required, since the billet remains stationary in the container under a moving die.

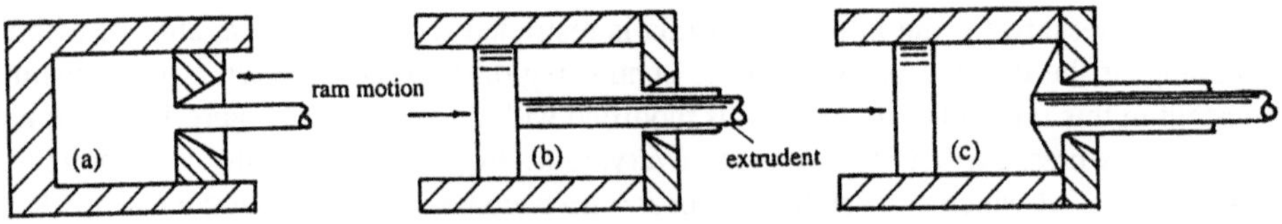

Figure 14.4 Extrusion processes

Direct extrusion with a ram and mandrel is used for tube production by either method shown in Figs 14.4b,c. In the preferred design, the mandrel remains fixed in position by webbing to the port hole die (see Fig. 14.4c). This provides for better support of the extruded tube and thereby impoves the concentricity between tube diameters.

Taking an annular strip between mandel and container wall in Fig. 14.5 allows a similar analysis for each process.

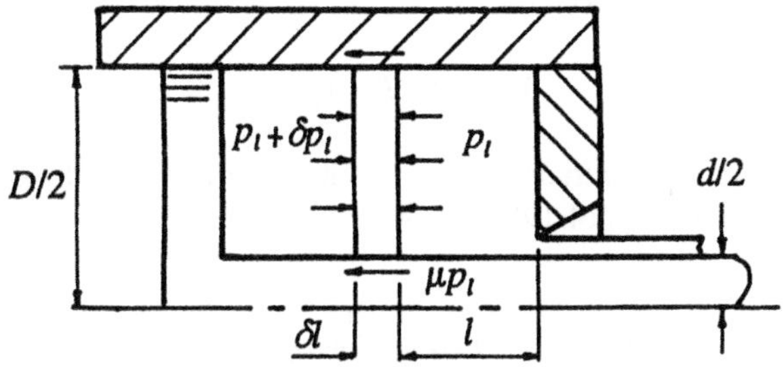

Figure 14.5 Analysis of extruded tube half-section

At the distance l from the die face, the equilibrium of this strip element is expressed as

$$\frac{\pi}{4}(D^2 - d^2) \times \delta p_l = \pi(D + d)\,\delta l \times \mu\, p_l \tag{14.14}$$

Separating the variables p and l within eq(14.14) allows an integration to

$$\frac{p_l}{p} = \exp\left(\frac{4\mu l}{D - d}\right) \tag{14.15}$$

where p is taken as a mean extrusion pressure, revealed by SLF analyses:

$$p/2k = a + b \ln R \tag{14.16}$$

where $R = H/h$ is the extrusion ratio and a and b are constants, which have been found [3] for this process as $a = 0.47$ and $b = 1.2$. Combining eqs(14.15) and (14.16) gives

$$\frac{p_l}{Y} = (0.47 + 1.2 \ln R)\exp\left(\frac{4\mu l}{D - d}\right) \tag{14.17}$$

When the mean extrusion pressure is found for two billets of lengths l_1 and l_2, eq(14.17) may be re-arranged to provide the friction coefficient

$$\mu = \frac{(D - d)}{4(l_1 - l_2)} \ln\left(\frac{p_{l_1}}{p_{l_2}}\right) \tag{14.18}$$

A numerical value $\mu \approx 0.03$ was found from eq(14.18) for a range of extruded lengths [3]. The work done for an incremental displacement is

$$\delta W = p_l \times A \times \delta l$$

Substituting p_l from eq(14.17) and integrating for the total work done:

$$W = AY_m(0.47 + 1.2 \ln R)\int_0^l \exp\left(\frac{4\mu l}{D - d}\right) \mathrm{d}l$$

$$= AY_m(0.47 + 1.2 \ln R)\frac{(D - d)}{4\mu}\exp\left(\frac{4\mu l}{D - d} - 1\right) \tag{14.19}$$

In eq(14.19), a mean yield stress Y_m is introduced to allow for the influence of a temperature rise from the work required to overcome friction. This is

$$W = mC\,(T - T_o) \tag{14.20}$$

where m is the mass of the unextruded billet, C is its specific heat and T_o is its initial, pre-heated temperature. Equating (14.19) and (14.20) gives the temperature rise $\Delta T = T - T_o$:

$$\Delta T = \frac{AY_m}{\rho A l C}(0.47 + 1.2 \ln R)\frac{4\mu}{(D - d)}\exp\left(\frac{4\mu l}{D - d} - 1\right) \tag{14.21}$$

where ρ is the material density and l is the unextruded length. Equation (14.21) allows the maximum extrusion ratio $R \approx D/d$ to be found. To determine Y_m, the mean strain rate at the extrusion temperature must be estimated. For example in a conventional extrusion, assume that a ring of dead metal exists in the corner of the vessel. The deformation then becomes confined to the shaded conical volume in Fig. 14.6a.

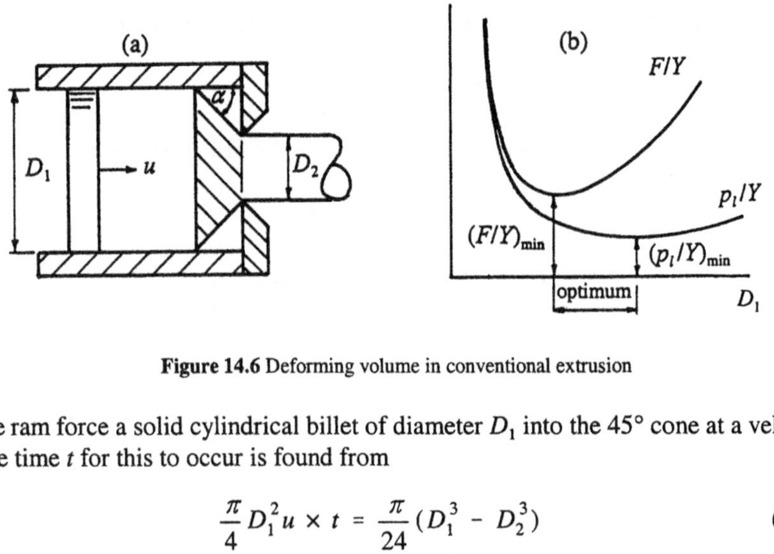

Figure 14.6 Deforming volume in conventional extrusion

Let the ram force a solid cylindrical billet of diameter D_1 into the 45° cone at a velocity of u. The time t for this to occur is found from

$$\frac{\pi}{4}D_1^2 u \times t = \frac{\pi}{24}(D_1^3 - D_2^3) \tag{14.22a}$$

Taking the mean strain in the cone as $\varepsilon_m = \ln R$, where $R = D_1/D_2$, the mean strain rate in time t may be approximated from eq(14.22a):

$$\dot{\varepsilon}_m = \frac{\varepsilon_m}{t} = \frac{6D_1^2 u}{D_1^3 - D_2^3}\ln R \approx \frac{6u}{D_1}\ln R \tag{14.22b}$$

Equation (14.17) may be used to minimise the extrusion pressure and force with respect to the container diameter D_1 [3]. For a solid circular extrusion the minimum force and pressure are not coincident (see Fig.14.6b). The curves show that it would be advisable to select a value of D_1 lying between these minima. When the above theory is applied to an extrusion within conical dies some modifications are necessary. Firstly, for an inverted extrusion process the coefficients a and b in eq(14.16) depend upon the semi-cone angle

$$a = a' - a''\cot\alpha \quad \text{and} \quad b = b' + b''\cot\alpha \tag{14.23a,b}$$

where a', a'', b' and b'' are constants. Secondly, for direct extrusion through conical dies the mean pressure is further raised to overcome additional wall friction:

$$\frac{p}{Y} = \frac{(a + b \ln R)}{(1 - 4\mu/D_1)}$$

where a and b are given by eqs(14.23a,b).

Example 14.2 Show that for a simple homogenous, cylindrical extrusion the force acting upon the ram is given by: $F = YA_r \ln (A_r/A_e)$ where A_r is the area of the ram, A_e the area of the extruded bar and Y is yield stress for rigid-plastic material. In a two-stage, extrusion process a 20 mm billet is first produced from a 10% reduction to its original area. Thereafter, the billet is extruded to 18 mm. Given that the true stress-natural strain law for the annealed material is expressed by a Hollomon law: $\sigma = 980\varepsilon^{0.19}$, determine the force necessary for extrusion accounting for work hardening. How is the force expression altered in the case of a Ludwick description of the flow curve?

Let the ram force $F = YA_r$ move through an incremental distance δl to give the work increment

$$\delta W = YA_r \delta l = \frac{YV}{l}\delta l \tag{i}$$

where V is the volume displaced in a length l. With a homogenous deformation of rigid plastic material, occurring between an initial length l_r and a final length l_e, the work/unit volume becomes

$$\frac{W}{V} = Y\int_{l_r}^{l_e} \frac{dl}{l} = Y \ln\left(\frac{l_e}{l_r}\right) = Y \ln\left(\frac{A_r}{A_e}\right) \tag{ii}$$

where $A_r l_r = A_e l_e$ provides the reduction in area ratio. On the ram side, the specific work is

$$\frac{W}{V} = \frac{F l_r}{A_r l_r} = \frac{F}{A_r} \tag{iii}$$

Equating (ii) and (iii) gives the ram force:

$$F = YA_r \ln\left(\frac{A_r}{A_e}\right) \tag{iv}$$

In the first stage, the area reduction is

$$\frac{A_o - A_1}{A_o} = 0.1 \quad \Rightarrow \quad \frac{A_o}{A_1} = 1.11$$

and from this, the true strain is

$$\varepsilon = \ln\left(\frac{l_1}{l_o}\right) = \ln\left(\frac{A_o}{A_1}\right) = \ln(1.11) = 0.105$$

From the Hollomon law, the corresponding flow stress is $\sigma = 638.6$ MPa. In the second stage the strain increment is similarly found:

$$\Delta\varepsilon = \ln\left(\frac{l_e}{l_r}\right) = \ln\left(\frac{A_r}{A_e}\right) = \ln\left(\frac{20}{18}\right)^2 = 0.211$$

The total strain after stage 2 is $\varepsilon = 0.105 + 0.211 = 0.316$, for which the stress is 787.3 MPa. The average stress for stage 2 is $\sigma_m = 712.97$ MPa and from this the extrusion force in eq(iv) may be modified to account for work hardening:

$$F = \sigma_m A_r \ln\left(\frac{A_r}{A_e}\right) = 712.97 \times \frac{\pi}{4}(20)^2 \ln\left(\frac{20}{18}\right)^2 = 47.2\,\text{kN}$$

Using Ludwick's hardening law (9.43a), the extrusion force follows from eq(iii) $F = A_r w$, in which the specific work $w = W/V$ in eq(i) becomes:

$$w = \int \sigma \, d\varepsilon = \int_{\varepsilon_r = 0}^{\varepsilon_e} (Y + A\varepsilon^n)\, d\varepsilon$$

$$= Y\varepsilon_e + \frac{A\varepsilon_e^{n+1}}{n+1} = Y \ln\left(\frac{A_r}{A_e}\right) + \frac{A}{n+1}\left[\ln\left(\frac{A_r}{A_e}\right)\right]^{n+1}$$

14.5 Hot Rolling

This continuous cogging process may be performed hot or cold between rolling mills arranged as 2-high, 4-high, clustered, planetary and pendulum (see Figs 14.7a-e).

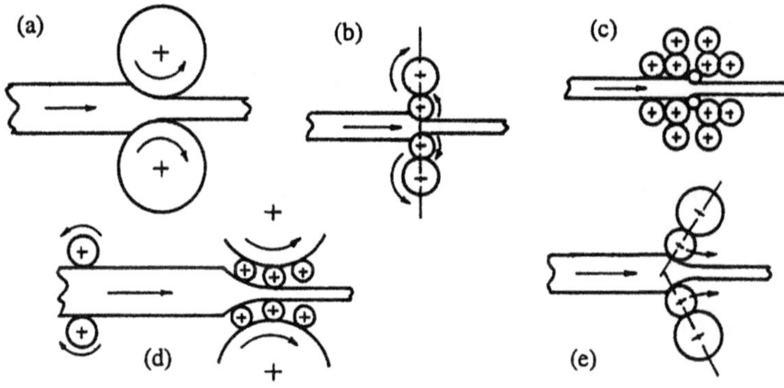

Figure 14.7 Rolling processes

The aim of all rolling is to reduce sheet thickness in a continuous process while maintaining uniform thickness and width throughout [4]. The roll force and torque will dictate the reduction, which may be increased with more sophisticated mills, involving a number of smaller rollers in contact with the sheet.

14.5.1 Wide Sheet

The theory of hot rolling assumes plane strain in which no spread occurs, i.e. transverse cross-sections remain plane and no roll deformation occurs. Hence the theory is similar to that for cold rolling strip (p. 230). In Fig. 14.8a, a 2-high mill transmits the roll radii R to the sheet within an arc of contact. At the entry and exit planes the thicknesses are h_1 and h_2 and axial stresses on these planes are zero when front and back tensions are absent.

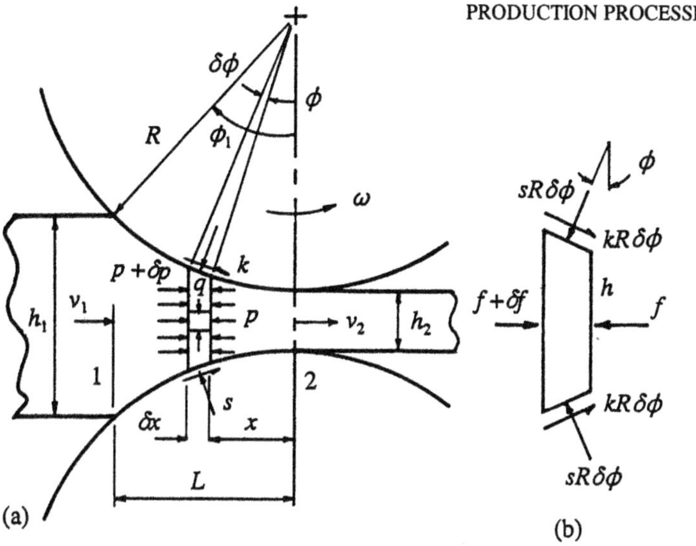

Figure 14.8 Geometry and forces within simple hot rolling

The volume rate remains constant, so for sections at entry, exit and at one between these:

$$b_1 h_1 v_1 = bhv = b_2 h_2 v_2$$

where v_1, v_2 and v are the respective velocities at these positions. Since $b_1 \approx b \approx b_2$, it follows that $v_2/v_1 = h_1/h_2 > 1$. The tangential roll velocity $v_R = \omega R$ lies between v_1 and v_2, i.e. $v_R > v_1$ at entry and $v_R < v_2$ at exit. Where v_R equals the strip speed there is no relative motion and at this *neutral point*, the frictional force between contacting surfaces changes direction. In a hot rolling process the shear yield stress k is attained by the friction present. The direction of k at the intermediate position shown applies to where the roll speed is greater than the strip. This is coupled with a normal pressure s along the arc interface and a distribution in the axial compression from p to $p + \delta p$. The forces acting upon the elemental strip, distance x from the vertical symmetry line, are shown in Fig. 14.8b. The balance between horizontal forces from pressure ($f = ph$) and those from the roll is expressed for a unit of width as:

$$(p + \delta p)(h + \delta h) + 2kR\delta\phi\cos\phi = ph + 2sR\delta\phi\sin\phi \qquad (14.24a)$$

Expanding (14.24a) and neglecting the product ($\delta p \times \delta h$) leads to

$$\frac{\mathrm{d}}{\mathrm{d}\phi}(ph) = 2R(s\sin\phi \pm k\cos\phi) \qquad (14.24b)$$

in which $-\ k\cos\phi$ applies to the horizontal friction force component at the entry side of the neutral point and $+\ k\cos\phi$ applies to the exit side of this point. Strictly, eq(14.24b) requires a numerical solution to p but a good approximate closed solution follows from assuming a rolling reduction between flat, parallel dies. This gives $\phi = 0$, $h_1 = h_2$ and $\delta x = R\delta\phi$ and eq(14.24b) becomes

$$h_2\,\mathrm{d}p\,/\,\mathrm{d}x = \pm\,2k \qquad (14.25a)$$

Both p and s are compressive and assuming $s \approx q$ for the strip element shown, the plane strain yield criterion (14.6) is simply

$$s - p = 2k \qquad (14.25b)$$

Combining eqs(14.25a,b) gives

$$h_2 \frac{ds}{dx} = \pm\, 2k \tag{14.25c}$$

Integrating eq(14.25c) for the exit side:

$$s = s_2 + \frac{2k}{h_2} x \tag{14.26a}$$

where s_2 applies to the exit plane ($x = 0$) where $p = 0$. Hence eq(14.25b) gives $s_2 = 2k$ and eq(14.26a) becomes

$$s = 2k\left(C + \frac{x}{h_2} \right) \tag{14.26b}$$

From experiment, C is taken as $\pi/4 \approx 0.8$ and not unity [5]. Equation (14.26b) defines the *friction hill* for hot rolling shown in Fig. 14.9.

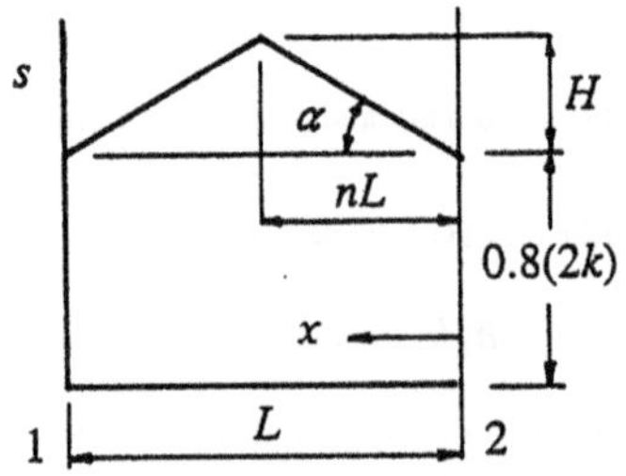

Figure 14.9 Friction hill for hot rolling

Equation (14.26b) gives the gradient, $\tan \alpha = H/(nL)$, of the hill as

$$\frac{ds}{dx} = \frac{2k}{h_2} \tag{14.26c}$$

in which the neutral point is assumed to bisect the projected arc length L. The area beneath the hill gives the roll force/unit width

$$P = \int s\,dx = 2 \times 2k \int_0^{L/2} \left(C + \frac{x}{h_2} \right) dx = 2kL\left(C + \frac{L}{4h_2} \right) \tag{14.27a}$$

which is comparable with eqs(14.9a,b) for a cogging operation. Each roll torque/unit width may be approximated from eq(14.27a) as

$$T = P \times L/2 \tag{14.27b}$$

Now, from the geometry in Fig. 14.8

$$L \approx R\phi_1 \quad \text{and} \quad x \approx R\phi \tag{14.28a}$$

$$h_1 - h_2 = 2R\,(1 - \cos\phi_1) \approx R\phi_1^{\,2} \tag{14.28b}$$

$$h - h_2 = 2R\,(1 - \cos\phi) \approx R\phi^2 \tag{14.28c}$$

and setting the ' draft ' $\delta = h_1 - h_2$, eqs(14.28a,b) give the geometry of the contact arc:

$$\phi_1 = \sqrt{\delta / R} \quad \text{and} \quad L = \sqrt{R\,\delta} \tag{14.28d,e}$$

Thus, knowing R, δ, h_2 and k, the roll force and torque follow from eqs(14.27a,b). Sims [6] modified the yield criterion (14.25b) to correspond with $C \approx 0.8$:

$$p = s - \frac{\pi}{4}(2k) \tag{14.29}$$

and set $\sin\phi \approx \phi$ and $\cos\phi \approx 1$ in eq(14.24b) for rolling mills with a small arc of contact. Substituting from eqs(14.28c) and (14.29), the equilibrium eq(14.24b) becomes

$$\frac{\mathrm{d}}{\mathrm{d}\phi}\left(\frac{s}{2k} - \frac{\pi}{4}\right) = \frac{\pi R\phi}{2(h_2 + R\phi^2)} \pm \frac{R}{h_2 + R\phi^2} \tag{14.30}$$

Equation (14.30) may be integrated, using the appropriate limits for ϕ between exit and entry (where $s = (\pi/4)(2k)$ at each position), to give respective closed solutions

$$\frac{s^+}{2k} = \frac{\pi}{4}\ln\left(\frac{h}{h_2}\right) + \frac{\pi}{4} + \sqrt{\frac{R}{h_2}}\tan^{-1}\left(\sqrt{\frac{R}{h_2}}\,\phi\right) \quad \text{(exit)} \tag{14.31a}$$

$$\frac{s^-}{2k} = \frac{\pi}{4}\ln\left(\frac{h}{h_1}\right) + \frac{\pi}{4} + \sqrt{\frac{R}{h_2}}\tan^{-1}\left(\sqrt{\frac{R}{h_2}}\,\phi_1\right) - \sqrt{\frac{R}{h_2}}\tan^{-1}\left(\sqrt{\frac{R}{h_2}}\,\phi\right) \quad \text{(entry)} \tag{14.31b}$$

The neutral point $\phi = \phi_n$ is found from equating (14.31a,b) for $h = h_n = h_2 + R\phi_n^2$:

$$\frac{\pi}{4}\ln\left(\frac{h_2}{h_1}\right) = 2\sqrt{\frac{R}{h_2}}\tan^{-1}\sqrt{\frac{R}{h_2}}\,\phi_n - \sqrt{\frac{R}{h_2}}\tan^{-1}\sqrt{\frac{h_1 - h_2}{h_2}} \tag{14.32}$$

The roll force, acting normal to a unit width of roll, is found from eq(14.32) as follows:

$$P = R\int_0^{\phi_1} s\,\mathrm{d}\phi = R\int_0^{\phi_n} s\,\mathrm{d}\phi + R\int_{\phi_n}^{\phi_1} s\,\mathrm{d}\phi$$

$$= 2kR\left[\frac{\pi}{2}\sqrt{\frac{h_2}{R}}\tan^{-1}\sqrt{\frac{h_1 - h_2}{h_2}} - \frac{\pi\phi_1}{4} - \ln\left(\frac{h_n}{h_2}\right) + \frac{1}{2}\ln\left(\frac{h_1}{h_2}\right)\right] \tag{14.33}$$

where $\surd[(h_1 - h_2)/h_2] = \surd(R/h_2)\phi_1$. Simms [6] interpretated eq(14.33) graphically by writing $P = 2kLQ_P$ in which Q_P depends upon R/h_2 and h_1/h_2. The total roll torque/unit width may also be estimated from eq(14.27b) simply as $T = PL$ or, more accurately, from k as:

$$T = 2\int_{\phi_n}^{\phi_1/2} kR^2\,\mathrm{d}\phi = 2kR^2\left(\phi_1/2 - \phi_n\right) \tag{14.34a}$$

The limits of integration apply to the range of ϕ over which the direction of k is not opposed within the adjacent region. Since ϕ_n appears in eq(14.32) and $\phi_1 = \surd[(h_1 - h_2)/R]$, eq(14.34a) may be re-written as

$$T = 2kR^2Q_T \tag{14.34b}$$

The dependence of Q_T upon R/h_2 and h_1/h_2 has appeared in graphical form [6]. The shear yield stress k in eqs(14.33) and (14.34b) must apply to the mean strain rate for the process.

We may take $2k$ to be the mean value of the plane strain yield stress for rolling. For this a stress-strain curve is required at the mean strain rate for rolling (see p. 453). The curve allows a mean stress $2\bar{k}$ to be read or, where work-hardening is evident (e.g. see Fig. 14.10), as the ordinate of a rectangle with an equal enclosed area:

$$2\bar{k}\,\varepsilon_m = \int_0^{\varepsilon_m} 2k\,\mathrm{d}\varepsilon \tag{14.34c}$$

where $\varepsilon_m = \ln(h_1/h_2)$ is the exit strain simulated, say, from a plane strain compression test (see p. 287).

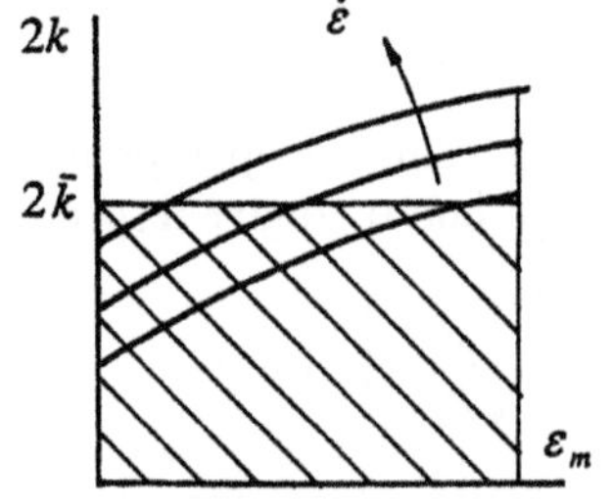

Figure 14.10 Mean stress and strain in plane strain

Example 14.3 Estimate the roll force and torque required to roll a 300 mm wide, annealed sheet from 4 mm to 3.5 mm thick between two rolls each of radius 175 mm. How is the force altered when the original area had been attained by a previous reduction of 10%? The plane strain, true stress versus natural strain curve is given by a Swift law (9.44b) in which $\sigma_o = 2k = 200$ MPa, $\varepsilon_o = 0.05$ and $n = 0.25$.

The compressive strain involved in a reduction from 4 to 3.5 mm is

$$\varepsilon = \int_{h_2}^{h_1} \frac{dh}{h} = \ln\left(\frac{h_1}{h_2}\right) = \ln\left(\frac{4}{3.5}\right) = 0.134$$

Applying eq(14.34c), gives a mean flow stress $2\bar{k}$ for the strain range $0 \le \varepsilon \le 0.134$ (compression implied) as

$$2\bar{k} \times 0.134 = 200 \int_0^{0.134} (1 + 20\varepsilon)^{0.25}\,\mathrm{d}\varepsilon \approx 32.78$$

from which $2\bar{k} = 244.6$ MPa. Equation (14.28e) gives the arc length as

$$L = \sqrt{R(h_1 - h_2)} = \sqrt{175(4 - 3.5)} = 9.35\text{ mm}$$

from which eqs(14.27a,b) supply each roll force and the torque as:

$$P = 2\bar{k}Lw\left(\frac{\pi}{4} + \frac{L}{4h_2}\right) = 244.6 \times 9.35 \times 300 \times 10^{-3}\left(0.7854 + \frac{9.35}{4 \times 3.5}\right) = 997.1\text{ kN}$$

$$T = PL/2 = 997.1 \times (9.35 \times 10^{-3})/2 = 4.66\text{ kNm}$$

In the case of a 10% pre-reduction in area ($A_o/A_1 = l_1/l_o = 1.11$), the lower limit of integration in eq(14.34c) is: $\ln(1.11) = 0.105$ and the upper limit is: $0.105 + 0.134 = 0.239$.

14.5.2 Narrow Strip

Rolling narrow strip modifies the friction hill (Fig. 14.9) between the planes of entry and exit. The hill top slopes at β to sheet edges but extends toward the centre of the rolls retaining its flat plateau, to give the *friction roof* in Fig. 14 11.

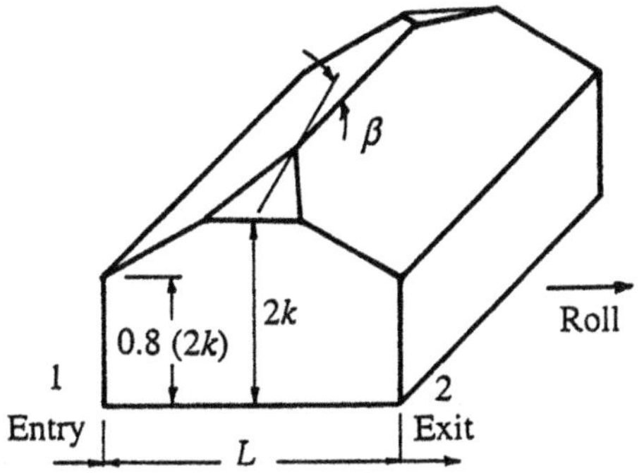

Figure 14.11 Friction roof for rolling narrow strip

The total roll force is found from the volume beneath the roof [7]:

$$P_T = 2k\bar{b}L\left[C + \frac{g}{2} + \frac{\bar{h}}{3\bar{b}g}(g - 0.2)^3\right] \qquad (14.35)$$

where $g \approx L/2h_2$ when the neutral point is central. The mean thickness and breadth, $\bar{h}$ and $\bar{b}$ respectively, are based upon a parabola describing the shape of the contact arc and the spread between entry and exit. These give:

$$\tan\beta = \frac{2k}{\bar{h}}, \quad \bar{h} = \frac{1}{3}(2h_2 + h_1) \quad \text{and} \quad \bar{b} = \frac{1}{3}(2b_2 + b_1)$$

As with hot forging, a *spread factor* S and a *coefficient of spread* s (see eq(14.4)) can be defined for strip rolling:

$$S = \frac{b_2}{b_1} \quad \text{and} \quad s = \frac{L/b_1}{1 + L/b_1} \approx \frac{1}{1 + b_1/\sqrt{R\delta}}$$

In a two-roll mill, eqs(14.27b) and (14.28e) approximate the total roll torque as:

$$T_T = 2\left(\frac{P_T L}{2}\right) = P_T\sqrt{R\delta}$$

14.5.3 Strain Rate in Hot Rolling

A consideration of the strain rate in rolling becomes important when estimating the appropriate shear yield stress k in eqs(14.27a) and (14.35). For a wide strip, the through-thickness strain increment is $\delta\varepsilon = \delta h/h$, giving its rate as

$$\dot{\varepsilon} = \frac{1}{h}\frac{dh}{dt} = \frac{1}{h}\left(\frac{dh}{d\phi} \times \frac{d\phi}{dx} \times \frac{dx}{dt}\right) \qquad (14.36)$$

where, from the geometry in Fig. 14.8,

$$x = R\sin\phi \quad \Rightarrow \quad \frac{\mathrm{d}x}{\mathrm{d}\phi} = R\cos\phi \tag{14.37a}$$

$$h = h_2 + 2R(1 - \cos\phi) \quad \Rightarrow \quad \frac{\mathrm{d}h}{\mathrm{d}\phi} = 2R\sin\phi \tag{14.37b}$$

$$\frac{\mathrm{d}x}{\mathrm{d}t} = -\,v = -\frac{V}{bh} \tag{14.37c}$$

where $V = bhv$ is the volume of material rolled per second. Substituting eqs(14.37a-c) into eq(14.36) gives

$$\dot{\varepsilon} = -\frac{2V}{bh^2}\tan\phi \tag{14.38}$$

Equation (14.28c) approximates the thickness at the x-plane to

$$h = h_2 + 2R\,(1 - \cos\phi) \approx h_2 + R\phi^2 \tag{14.39a}$$

from which we may write an approximation for $\tan\phi$:

$$\tan\phi \approx \phi = \sqrt{\frac{h - h_2}{R}} \tag{14.39b}$$

Substituting eq(14.39b) into (14.38) gives

$$\dot{\varepsilon} = -\frac{2V}{bh^2}\sqrt{\frac{h - h_2}{R}} = -\frac{2V}{b}\sqrt{\frac{h_2}{R}}\left[\frac{\sqrt{(h/h_2) - 1}}{h^2}\right] \tag{14.40}$$

Equation (14.40) allows the strain rate distribution within the pass to be determined. This shows a near maximum rate at the entry plane $h = h_1$, falling to zero at the exit plane $h = h_2$. A mean strain rate $\dot{\varepsilon}_m$ may be identified with the ordinate of a rectangle with the same area as that beneath the curve of $\dot{\varepsilon}$ versus h:

$$(h_1 - h_2)\,\dot{\varepsilon}_m = \int_{h_1}^{h_2} \dot{\varepsilon}\,\mathrm{d}h$$

Under plane strain conditions, the flow in the transverse direction is zero. Hence $\dot{\varepsilon}_m$ applies to the mean rate of thinning, this equalling the mean rate of axial extension beneath the rolls.

14.6 Cold Rolling

Dimensional stability, hardness and improved surface finish may be achieved with a final cold rolling of pre-tensioned sheet. This ensures that the yield stress for the sheet material is far greater than that for all previous hot passes. Cold rolling raises the normal pressure s in Fig. 14.8 and elastically flattens the roll itself by altering its radius from R to R'. As with cold forging, the interfacial shear stress is lowered to μs and this cannot exceed the shear yield stress k for the material. Thus, the horizontal force equilibrium condition for a wide sheet modifies eq(14.24a) to

$$\frac{\mathrm{d}}{\mathrm{d}\phi}(ph) = 2R's(\sin\phi \pm \mu\cos\phi)\;{}^{+\ exit}_{-\ entry} \tag{14.41}$$

Bland and Ford's [8] simplifying assumptions, $s \approx q$, $\sin\phi \approx \phi$ and $\cos\phi \approx 1$, allowed the plane strain yield criterion: $q - p = 2k$, to be applied to a plastic hardening material. Here, the plane strain yield stress $2k$ depends upon the plastic strain level, and eq(14.41) becomes

$$\frac{\mathrm{d}}{\mathrm{d}\phi}[(s - 2k)h] = 2R's(\phi \pm \mu)\ {}^{+\ exit}_{-\ entry}$$

$$2kh\frac{\mathrm{d}}{\mathrm{d}\phi}\left(\frac{s}{2k} - 1\right) + 2\left(\frac{s}{2k} - 1\right)\frac{\mathrm{d}}{\mathrm{d}\phi}(kh) = 2R's(\phi \pm \mu)$$

$$2kh\frac{\mathrm{d}}{\mathrm{d}\phi}\left(\frac{s}{2k}\right) + 2\left(\frac{s}{2k} - 1\right)\frac{\mathrm{d}}{\mathrm{d}\phi}(kh) = 2R's(\phi \pm \mu) \qquad (14.42)$$

With the exception of cold rolling an annealed material [8], the second term on the left-hand side of eq(14.42) is negligible, i.e. the product kh is almost a constant and the bracket term is small. Thus, with the substitution from eq(14.39a), the integration follows as

$$2k(h_2 + R'\phi^2)\frac{\mathrm{d}}{\mathrm{d}\phi}\left(\frac{s}{2k}\right) = 2R's(\phi \pm \mu)$$

$$\int\frac{\mathrm{d}(s/2k)}{(s/2k)} = \int\frac{2R'\phi\,\mathrm{d}\phi}{h_2 + R'\phi^2} \pm \int\frac{2R'\mu\,\mathrm{d}\phi}{h_2 + R'\phi^2}$$

$$\ln\left(\frac{s}{2k}\right) = \ln\left(h_2 + R'\phi^2\right) \pm 2\mu\sqrt{\frac{R'}{h_2}}\tan^{-1}\sqrt{\frac{R'}{h_2}}\,\phi + \ln C$$

$$\ln\left(\frac{s}{2khC}\right) = \pm\mu Q$$

$$s = 2khCe^{\pm\mu Q} \qquad (14.43)$$

in which $Q = 2\sqrt{(R'/h_2)}\tan^{-1}[\sqrt{(R'/h_2)}\phi]$. Constant C is found from the boundary conditions for a pre-tensioned strip. Let these be t_1 ($= -p_1$) and t_2 ($= -p_2$) at entry and exit, where $\phi = \phi_1$ and $\phi_2 = 0$ respectively. Applying the plane strain yield criterion to each position gives, correspondingly, $s_1 = 2k_1 - t_1$ and $s_2 = 2k_2 - t_2$. With these substitutions into eq(14.43), the roll pressures become

$$s^+ = \frac{2kh}{h_2}\left(1 - \frac{t_2}{2k_2}\right)e^{\mu Q} \qquad (14.44a)$$

$$s^- = \frac{2kh}{h_1}\left(1 - \frac{t_1}{2k_1}\right)e^{\mu(Q_1 - Q)} \qquad (14.44b)$$

where Q_1 applies to the entry condition $\phi = \phi_1$. The exponential terms in eqs(14.44a,b) describe friction hills with both front and back tensions, t_1 and t_2 respectively, in Fig. 14.12. In the absence of t_1 and t_2, the friction hill (shaded) is superimposed upon the plane strain yield stress, the latter increasing from $2k_1$ to $2k_2$ in the contact zone.

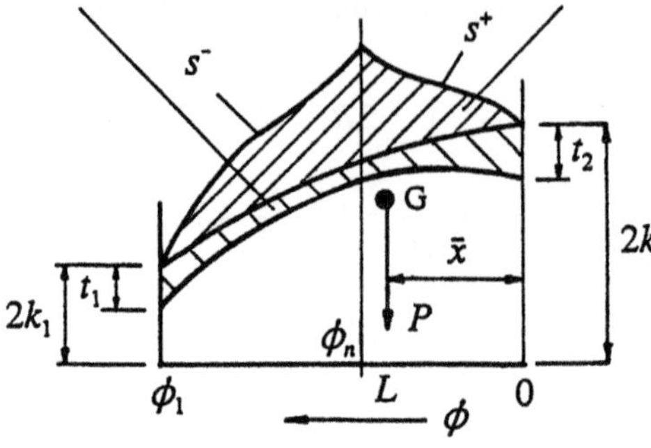

Figure 14.12 Roll pressure showing the influence of strip tensions

The yield stress variation may be replaced with a mean value, found from eq(14.34c). When present, pre-tensions reduce both the roll presure and the roll force for a given reduction. When $t_1 > t_2$ these tensions shift the neutral point ϕ_n toward the exit plane and, conversely, toward the entry plane when $t_2 > t_1$. The position ϕ_n follows from equating (14.44a,b):

$$\phi_n = \sqrt{\frac{h_2}{R'}}\tan\left(\frac{Q_n}{2}\sqrt{\frac{h_2}{R'}}\right)$$

where

$$Q_n = \frac{Q_1}{2} - \frac{1}{2\mu}\ln\left\{\frac{h_1[1 - (t_2/2k_2)]}{h_2[1 - (t_1/2k_1)]}\right\}$$

The roll force/unit width of strip is given by the enclosed area in Fig. 14.12:

$$P = R'\left(\int_0^{\phi_n} s^+ \mathrm{d}\phi + \int_{\phi_n}^{\phi_1} s^- \mathrm{d}\phi\right) \tag{14.45}$$

where s^+ and s^- are given by eqs(14.44a,b). Each roll torque/unit width may be found [9] from the sum of the torques due to s, t_1 and t_2:

$$T = \mu R R' \int_0^{\phi_1} s\,\mathrm{d}\phi + \frac{R}{2}(t_1 h_1 - t_2 h_2) \tag{14.46}$$

and this may be written as $T = P\bar{x}$. Taking the centroid $\bar{x}$ for the pressure distribution as approximately one half the contact length, i.e. $\bar{x} \approx L/2$ in Fig. 14.12, gives a reasonable estimate of the torque upon a single roll. The deformed roll radius in eqs(14.45) and (14.46) may be estimated by replacing the friction hill with an elliptical distribution of pressure along the arc of contact [10]:

$$R' = R\left(1 + \frac{KP}{\delta}\right) \tag{14.47}$$

where, typically, $K = 10.8 \times 10^{-6}$ m^2/MN for hardened steel rolls. The error involved in estimating R' from eq(14.47) is likely to be less than the error arising from neglecting elastic strain when cold rolling thin, hard strip [9].

14.7 Wire and Strip Drawing

14.7.1 Frictionless Drawing

The simplest analysis of cold drawing a wire through frictionless dies assumes an homogenous deformation of rigid, perfectly-plastic, incompressible material. For example, Fig. 14.13a shows the reduction in cross-sectional area from A_o to A_1 for drawing wire in conical dies. This method can be used to achieve successive reductions from a 10-12 mm stock diameter, usually at the maximum reduction possible (about 60%) in a single pass. Multiple passes are employed to obtain greater reductions and, with an intermediate anneal, wire as fine as 0.02 mm diameter may be cold drawn.

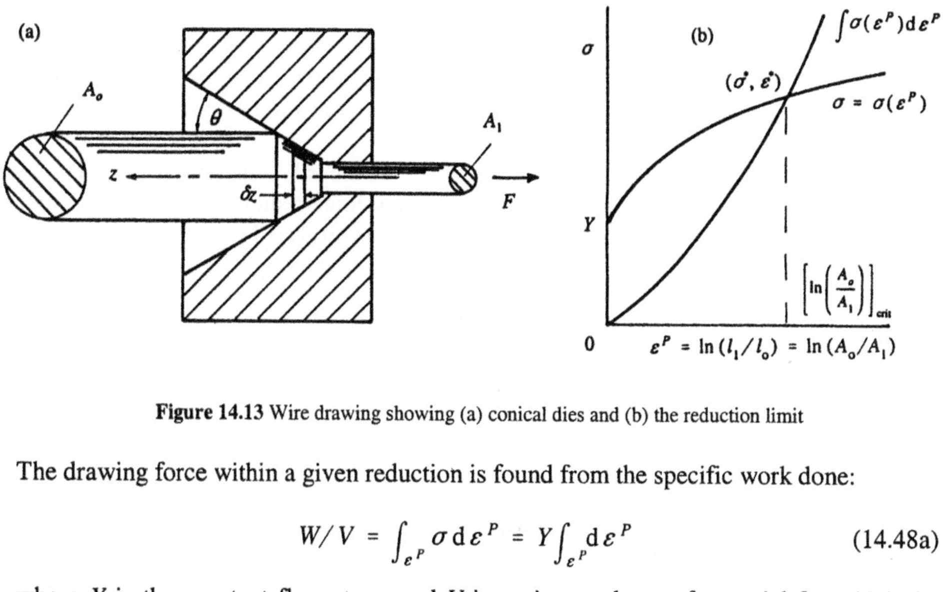

Figure 14.13 Wire drawing showing (a) conical dies and (b) the reduction limit

The drawing force within a given reduction is found from the specific work done:

$$W/V = \int_{\varepsilon^P} \sigma \, d\varepsilon^P = Y \int_{\varepsilon^P} d\varepsilon^P \tag{14.48a}$$

where Y is the constant flow stress and V is a given volume of material for which the corresponding lengths before and after reduction are l_o and l_1. As $V = A_o l_o = A_1 l_1$, eq(14.48a) gives

$$W/V = Y \int_{l_o}^{l_1} dl/l = Y \ln(l_1/l_o) = Y \ln(A_o/A_1) \tag{14.48b}$$

Also, the work done in drawing a length l_1 under a force F is $W = F l_1$. Hence

$$W/V = (F l_1)/(A_1 l_1) = F/A_1 \tag{14.49}$$

Equating (14.48b) and (14.49) gives the wire drawing force:

$$F = A_1 Y \ln(A_o/A_1) \tag{14.50a}$$

The drawing stress follows from eq(14.50a) as

$$\sigma_w = F/A_1 = Y \ln(A_o/A_1) = Y \ln(l_1/l_o) \tag{14.50b}$$

Now, σ_w should be less than Y if the wire is not to break. Setting $\sigma_w = Y$ in eq(14.50b) gives $A_o/A_1 = 2.718$, from which the maximum percentage reduction in area is 63.2%. For a hardening material $\sigma = \sigma(\varepsilon^P)$ in eq(14.48a), giving the drawing stress as $\sigma_w = \int \sigma(\varepsilon^P) d\varepsilon^P$. Since the latter defines the area beneath the flow curve, it permits the geometrical interpretation of the failure condition shown in Fig. 14.13b. The point of intersection gives the maximum possible reduction. For example, using Swift's law (9.44b), with intersection co-ordinates (σ^*, ε^*), gives:

$$\sigma_o \left(1 + \frac{\varepsilon^*}{\varepsilon_o}\right)^n = \sigma_o \int_0^{\varepsilon^*} \left(1 + \frac{\varepsilon}{\varepsilon_o}\right)^n d\varepsilon = \frac{\sigma_o \varepsilon_o}{(n+1)} \left[\left(1 + \frac{\varepsilon^*}{\varepsilon_o}\right)^{n+1} - 1\right] \quad (14.51a)$$

Setting $\sigma^*/\sigma_o = (1 + \varepsilon^*/\varepsilon_o)^n$ in eq(14.51a) provides an equation in the normalised stress σ^*/σ_o:

$$\left(\frac{\sigma^*}{\sigma_o}\right)^{\frac{n+1}{n}} - \frac{(n+1)}{\varepsilon_o}\left(\frac{\sigma^*}{\sigma_o}\right) - 1 = 0 \quad (14.51b)$$

which depends upon the hardening exponent n and, more strongly, upon the strain history term ε_o. Alternatively, we may use an average yield stress for the process within eq(14.50a) to account more simply for a history of hardening, as the following example shows.

Example 14.4 The true stress-strain relationship for an annealed material is given by the Swift law: $\sigma = 75(1 + 30\varepsilon)^{0.65}$. Determine the draw force required to reduce a 3 mm diameter wire to 2.75 mm, given that the 3 mm wire was produced originally from a 20% area reduction. What is the greatest reduction from 3 mm in a single pass?

Let the sequence of processes be 0-1-2 where 0 is for the annealed material, 1 applies to 3 mm wire with a previous 20% area reduction on entry to the die and 2 applies to the wire at exit, following a further area reduction of $100(3^2 - 2.75^2)/3^2 = 15.9\%$. The natural strains associated with the 20% and 15.9% reductions are:

$$\varepsilon_{0\text{-}1} = \ln(l_1/l_0) = \ln(A_0/A_1) = \ln(1/0.80) = 0.223$$

$$\varepsilon_{1\text{-}2} = \ln(l_2/l_1) = \ln(A_1/A_2) = \ln(1/0.84) = 0.174$$

Thus, the natural strains at entry to and exit from the die are 0.223 and 0.397. The Swift law gives the corresponding stresses as 282.4 MPa and 395.5 MPa. Using a mean yield stress $Y_m = 339$ MPa for the draw process, eq(14.50a) is written as

$$F = A_2 Y_m \ln(A_1/A_2) = 5.94 \times 339 \times \ln(7.069/5.94) = 350.4 \text{ N}$$

The greatest reduction, based upon the flow stress in the drawn material, is found from

$$350.4 = 395.5 \ln(A_1/A_2)$$

giving $A_1/A_2 = 2.425$ and $d_2 = 1.926$ mm.

14.7.2 Friction and Die Angle

There have been many refinements to the simple wire drawing analysis, notable among these are the accounts of friction, geometry and redundant work, reviewed in [11]. A theoretical account of the first two takes a disc of perfectly-plastic material within the contact zone (see Fig. 14.13a) to reveal the normal pressure p and friction stress μp around the conical interface. These are equilibrated by a change to the axial stress $\delta\sigma_z$ over thickness δz. Let the position z increase with D, as shown in Fig. 14.14a.

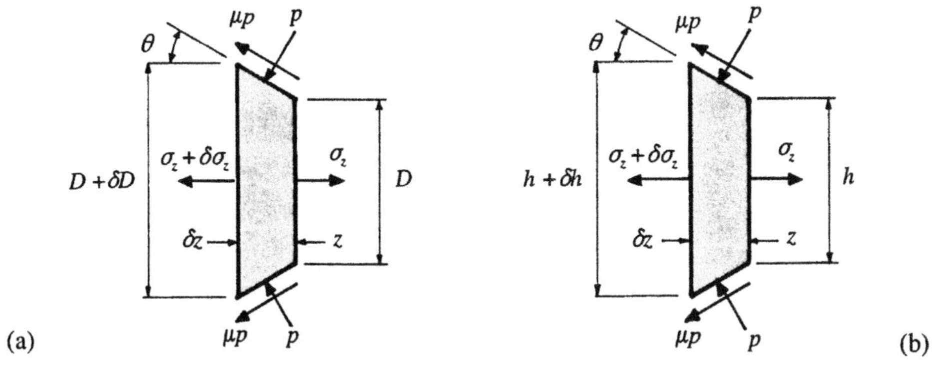

Figure 14.14 Stress state in contact region for (a) wire and (b) strip drawing

Horizontal forces will remain balanced when:

$$(\sigma_z + \delta\sigma_z)\frac{\pi}{4}(D + \delta D)^2 - \sigma_z\frac{\pi D^2}{4} + \left(\frac{p\,\pi D\,\delta z}{\cos\theta}\right)\sin\theta + \left(\frac{\mu p\,\pi D\,\delta z}{\cos\theta}\right)\cos\theta = 0 \tag{14.52a}$$

Setting $\delta z = \delta D/(2\tan\theta)$ in eq(14.52a) and ignoring small products upon expansion:

$$D\,\delta\sigma_z + 2\sigma_z\delta D + 2p(1 + \beta)\,\delta D = 0 \tag{14.52b}$$

where $\beta = \mu/\tan\theta$. The radial and circumferential strain increments within the cone are the same since $\delta\varepsilon_\theta^{\,P} = (\pi\delta D)/(\pi D) = \delta\varepsilon_r^{\,P}$. Substituting this condition into the Levy-Mises eqs(6.1a-c) show that $\sigma_\theta = \sigma_r$ for a cylindrical reduction. As a consequence both the Tresca and von Mises yield criteria reduce to $\sigma_z - \sigma_r = Y$ when the frictional stress μp is ignored. Moreover, with a small semi-cone angle θ, we can set $\sigma_r \approx -p$ to give

$$p = Y - \sigma_z \tag{14.53}$$

Combining eqs(14.52b) and (14.53) gives

$$\frac{2\,\delta D}{D} = \frac{\delta\sigma_z}{\beta\sigma_z - (1 + \beta)Y} \tag{14.54a}$$

Integrating eq(14.54a) between exit position 1 and entry position 0 leads to

$$2\int_{D_1}^{D_0}\frac{\mathrm{d}D}{D} = \int_{\sigma_{z1}}^{\sigma_{z0}}\frac{\mathrm{d}\sigma_z}{\beta\sigma_z - (1 + \beta)Y}$$

$$\left(\frac{D_1}{D_0}\right)^{2\beta} = \frac{\beta\sigma_{z1} - (1 + \beta)Y}{\beta\sigma_{z0} - (1 + \beta)Y} \tag{14.54b}$$

Re-arranging eq(14.54b) gives a wire draw stress $\sigma_w = \sigma_{z1}$ in the presence of back tension σ_{z0}:

$$\sigma_w = \frac{Y(1 + \beta)}{\beta}\left[1 - \left(\frac{D_1}{D_0}\right)^{2\beta}\right] + \sigma_{z0}\left(\frac{D_1}{D_0}\right)^{2\beta} \tag{14.54c}$$

In the absence of σ_{z0} the drawing stress and force required to give a reduction in area ratio

$$r = (A_o - A_1)/A_o = 1 - (D_1/D_o)^2 \tag{14.55a}$$

are, respectively:

$$\sigma_w = Y(1 + 1/\beta)\left[1 - (1 - r)^\beta\right], \quad F = A_1\,\sigma_w \tag{14.55b,c}$$

The greatest reduction ratio $r = r_{max}$ applies when $\sigma_w = Y$ in eq(14.55b):

$$r_{max} = 1 - 1/(1 + \beta)^{1/\beta} \tag{14.56}$$

Equation (14.53) shows that $\sigma_z = Y - p$ and $\delta\sigma_z = -\delta p$. Substituting into eq(14.54a), the pressure distribution follows with an integration from the exit position 1:

$$2\int_{D_1}^{D} \frac{\mathrm{d}D}{D} = \int_{p_1}^{p} \frac{\mathrm{d}p}{Y + \beta p}$$

from which

$$p = (p_1 + Y/\beta)(D/D_1)^{2\beta} - Y/\beta \tag{14.57}$$

where $p_1 = Y - \sigma_1$ and $\sigma_1 = \sigma_w$ is given by the draw stress eq(14.55b).

In the case of a strip drawing analysis (see Fig. 14.14b) the strip thickness h replaces the wire diameter D. The horizontal force equilibrium equation becomes

$$(\sigma_z + \delta\sigma_z)(h + \delta h)\,w - \sigma_z h\,w + \left(\frac{2pw\,\delta z}{\cos\theta}\right)\sin\theta + \left(\frac{2\mu pw\,\delta z}{\cos\theta}\right)\cos\theta = 0 \tag{14.58a}$$

where w is the strip width. With $\delta z = \delta h/(2\tan\theta)$, eq(14.58a) simplifies to

$$h\,\delta\sigma_z + \sigma_z\,\delta h + p(1 + \beta)\,\delta h = 0 \tag{14.58b}$$

where $\beta = \mu/\tan\theta$. As the width strain is zero, eq(6.1c) gives the width stress as $(\sigma_z - p)/2$. Both Tresca and von Mises give a plane strain yield criterion $\sigma_z + p = Y$. Hence, the integration of eq(14.58b) leads to a draw stress without back pull and a maximum reduction identical to eqs(14.55b) and (14.56). With a reduction in thickness from h_o to h_1, the reduction ratio is $r = 1 - h_1/h_o$.

An alternative, semi-empirical account [12] of work lost to friction and shear employs the mean radial pressure upon the conical interface. This is determined when friction and back pull are absent. Thus, in Fig. 14.15a, a horizontal force balance reveals

$$p_m = \frac{A_1\,\sigma_w^*}{A_o - A_1} \tag{14.59a}$$

It is assumed that the same p_m exists when friction of low μ is present. As the draw stress is altered to σ_w (see Fig. 14.15b), the new equilibrium equation becomes:

$$\mu p_m\left(\frac{A_o - A_1}{\sin\theta}\right)\cos\theta + p_m\left(\frac{A_o - A_1}{\sin\theta}\right)\sin\theta = \sigma_w A_1 \tag{14.59b}$$

in which the bracket term is the sloping surface area. Combining eqs(14.59a,b) leads to

$$\sigma_w = \sigma_w^*\,(1 + \mu\cot\theta) \tag{14.60a}$$

When σ_w^* is defined by eq(14.50b) and a redundant shear work factor $\phi > 1$ is introduced, the drawing stress eq(14.60a) is modified to

$$\sigma_w = (1 + \mu \cot\theta)\, \phi\, Y_m \ln\left(A_o / A_1\right) \tag{14.60b}$$

where, for a hardening material, Y_m is the mean flow stress in the given reduction. As ϕ increases with increasing die angle θ, the term $(1 + \mu \cot\theta)$ in eq(14.60b) decreases when μ is constant. This agrees with experimental data [12] showing that an optimum die angle minimises σ_w.

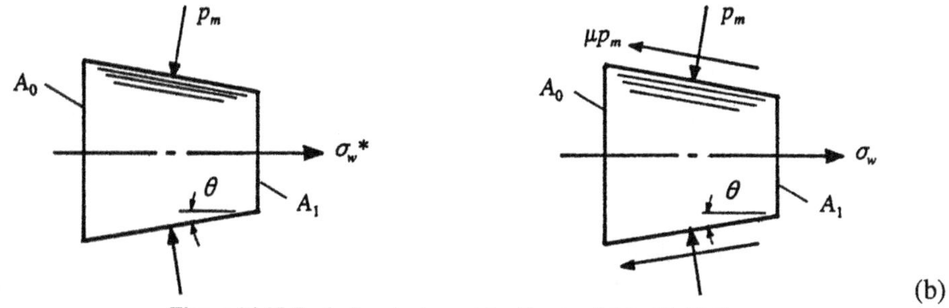

Figure 14.15 Conical contact area (a) without and (b) with friction

14.8 Orthogonal Machining

In a simplified analysis of machining, the cutting edge of the tool face is taken to lie perpendicular to the tool axis and this axis lies at right angles to the workpiece surface. After setting the *depth of cut b*, the rake face of the hardened tool removes workpiece material under a constant tool velocity, i.e. the *feed v*. The workpiece remains stationary during planing, but it rotates at a constant rotational speed when turning. The *rake angle* α is crucial to the type of chip produced and the *clearance angle* δ avoids excessive rubbing. Depending upon the material and tool geometry, the chip formation may be discontinuous or continuous, both of with can form a *built-up edge* upon the tool (see Figs 14.16a-c). In brittle workpiece materials, a periodic rupture occurs ahead of the tool to produce the discontinuous segments in Fig. 14.16a. In ductile materials the continuous chip occurs by shear whilst the building of an edge between chemically similar materials in contact is due to friction welding (see Figs 14.16b,c).

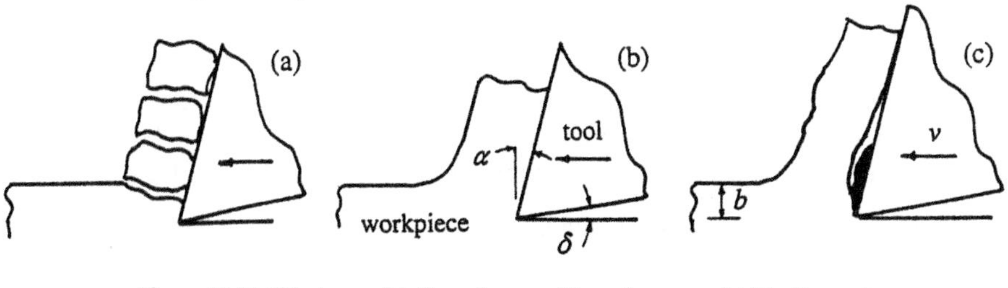

Figure 14.16 Chip types: (a) discontinuous, (b) continuous and (c) built-up edge

Under a very rapid shear strain rate ($\approx 10^4$ s^{-1}) the edge breaks away beyond a critical size. Its fragmented particles embed in the workpiece so imparing surface finish and geometry. The scheme of orthogonal machining along a *shear plane* AB is shown in Fig. 14.17.

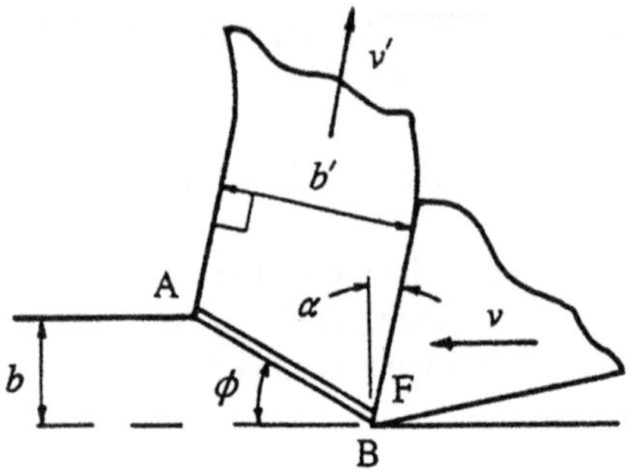

Figure 14.17 Orthogonal cutting geometry

The orientation ϕ of the shear plane to the cutting direction is found from applying a constant volume condition to material removal:

$$l\,w\,b = l'w'b' \tag{14.61a}$$

Plane strain involves no side flow and so $w = w'$. This allows the chip length and thickness ratios to be equated:

$$r_t = l'/\,l = b\,/\,b' \tag{14.61b}$$

The length of the shear plane AB is found from the geometry shown in Fig. 14.17:

$$AB = \frac{b}{\sin\phi} = \frac{b'}{\cos(\phi - \alpha)} \tag{14.62}$$

Combining eqs(14.61b) and (14.62) gives

$$\tan\phi = \frac{r_t \cos\alpha}{1 - r_t \sin\alpha} \tag{14.63a}$$

The most accurate way to find r_t is from chip weight measurement. From eqs(14.61a,b)

$$r_t = \frac{\rho l'bw}{\rho lbw} = \frac{\rho bw}{\text{chip weight/unit length}} \tag{14.63b}$$

where ρ is the workpiece density, b defines the depth of cut and w is the tool width.

14.8.1 Stress, Strain and Work

During machining an equilibrium condition is maintained between the tool force and the workpiece reaction R. Following Merchant [13], these forces are taken to be co-linear so that the resolved components of R form the three force systems shown in Figs 14.18.

(a) parallel and perpendicular to the rake face:

$$F = R\sin\beta \tag{14.64a}$$
$$N = R\cos\beta \tag{14.64b}$$

(b) parallel and perpendicular to the shear plane:

$$F_S = R\cos(\phi + \beta - \alpha) \tag{14.65a}$$
$$F_N = R\sin(\phi + \beta - \alpha) \tag{14.65b}$$

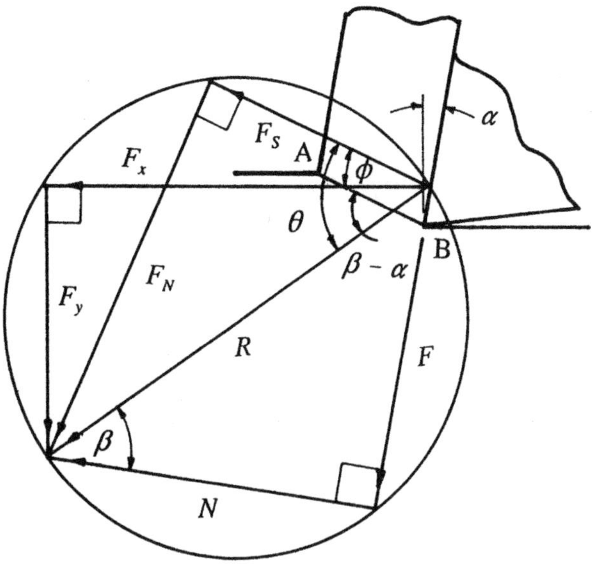

Figure 14.18 Resolved reaction forces showing Merchant's circle

(c) parallel and perpendicular to the tool motion:

$$F_x = R\cos(\beta - \alpha) \tag{14.66a}$$
$$F_y = R\sin(\beta - \alpha) \tag{14.66b}$$

Forces in eqs(14.66a,b) are measurable with a dynanometer and may be used to determine R and the other force components in eqs(14.64a,b) and (14.65a,b). Since R is common to each expression it is identified with the diameter of a *Merchant's circle* that superimposes the three force systems (a), (b) and (c), as shown in Fig. 14.18. The conversion of F_S and F_N to shear plane stress components τ_s and σ_s respectively, requires the division of these forces by the surface area of the shear plane, i.e. $wb/\sin\phi$. This gives

$$\begin{aligned}\tau_s &= \frac{F_S}{wb/\sin\phi} = \frac{R}{wb}\cos(\phi + \beta - \alpha)\sin\phi \\ &= \frac{F_x}{wb}\sec(\beta - \alpha)\cos(\phi + \beta - \alpha)\sin\phi\end{aligned} \tag{14.67a}$$

$$\begin{aligned}\sigma_s &= \frac{F_N}{wb/\sin\phi} = \frac{R}{wb}\sin(\phi + \beta - \alpha)\sin\phi \\ &= \frac{F_x}{wb}\sec(\beta - \alpha)\sin(\phi + \beta - \alpha)\sin\phi\end{aligned} \tag{14.67b}$$

in which R was eliminated from eqs(14.66a,b). At the tool tip, the shear zone has a finite thickness, i.e. length FG, within the enlarged scale of Fig. 14.19. Let a particle displace from F to F′ with shear distortion as it crosses this plane. This defines the shear strain γ as

$$\gamma = \frac{\mathrm{BF'}}{\mathrm{FG}} = \frac{\mathrm{BG}}{\mathrm{FG}} + \frac{\mathrm{GF'}}{\mathrm{FG}} = \tan(\phi - \alpha) + \cot\phi \tag{14.68}$$

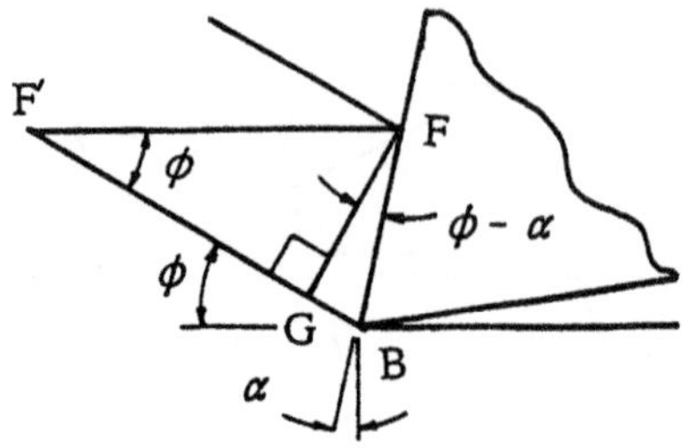

Figure 14.19 Shear zone displacement

The product of eqs(14.67a) and (14.68) provides the work done/unit volume of sheared material. With a cutting tool velocity v (m/s), the rate of shear work becomes

$$\dot{W}_s = \tau_s \gamma \times wbv = \frac{F_S}{(wb/\sin\phi)} \gamma \times wbv = F_S \gamma v \sin\phi \qquad (14.69a)$$

In addition, work is done to overcome friction along the rake face. In force system (a) above, frictional work is the product of the force F and the chip velocity v':

$$\dot{W}_f = F \times v' = F \times r_t v \qquad (14.69b)$$

where r_t in eq(14.61b) also expresses the ratio of cut to uncut lengths in a given time, i.e. the velocity ratio $r_t = b/b' = v'/v$ (Fig. 14.17). Adding eqs(14.69a,b), gives the total work rate:

$$\dot{W} = \dot{W}_s + \dot{W}_f = (F_S \gamma \sin\phi + F r_t)\, v \qquad (14.70)$$

14.8.2 Orientation of the Shear Plane

(a) *Maximum Shear Stress*

The previous analyses assume that the shear angle ϕ is known. Equations (14.63a,b) allow ϕ to be found experimentally. Alternatively, a purely theoretical estimate for ϕ may be found from assuming that τ_s is the maximum shear stress within the shear band. From eq(14.67a), the condition that τ_s is a maximum is given by

$$\frac{d\tau_s}{d\phi} = \cos(\phi + \beta - \alpha)\cos\phi - \sin(\phi + \beta - \alpha)\sin\phi = 0$$

$$\tan\phi = \cot(\phi + \beta - \alpha) = \tan\left[\frac{\pi}{2} - (\phi + \beta - \alpha)\right]$$

$$\therefore \quad \phi = \pi/4 - (\beta - \alpha)/2 \qquad (14.71a)$$

So, for a given β, the effect of increasing α and ϕ will lower both γ and $\dot{W}_s$. Conversely, for a given α and increasing β, ϕ is lowered while both γ and $\dot{W}_s$ are raised. Equation (14.71a) implies: (i) the shear plane is not material dependent and (ii) a friction coefficient μ applies to the force system (a) above, as follows:

$$\mu = \frac{F}{N} = \tan\beta \;\Rightarrow\; \beta = \tan^{-1}\mu = \tan^{-1}\frac{F}{N} \qquad (14.71b)$$

Equation (14.71a) describes a line with a gradient − ½ and an intercept of $\pi/4$ in axes of ϕ versus $\beta - \alpha$ (see Fig. 14.20).

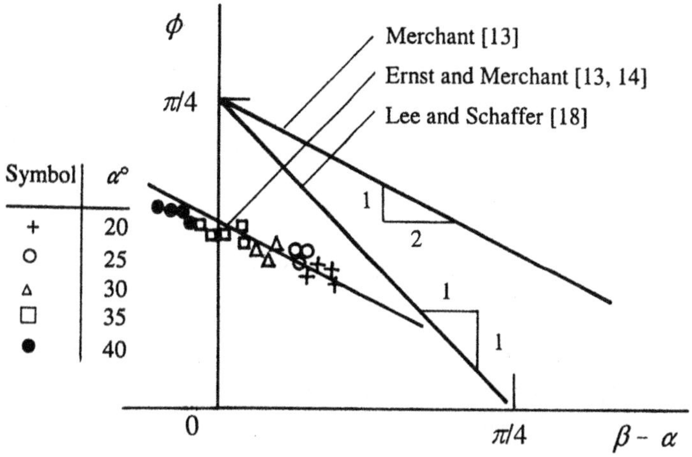

Figure 14.20 Theoretical predictions to shear angle ϕ. Data from Eggleston et al [15]

Because experiment has not confirmed the prediction of ϕ from eq(14.71a), the theory has been modified in various ways. Merchant [13) and Ernst [14] postulated that τ_s and σ_s within eqs(14.67a,b) were linearly dependent:

$$\tau_s = \tau_o + K\sigma_s \tag{14.72a}$$

where K is a constant and $\tau_s = \tau_o$ for $\sigma_s = 0$ within eqs(14.67a,b). Substituting this condition into eq(14.72a) and differentiating for a maximum shear stress as before, leads to a modified straight-line relationship:

$$\phi = \frac{1}{2}\cot^{-1}K - \frac{1}{2}(\beta - \alpha) \tag{14.72b}$$

from which the intercept $\frac{1}{2}\cot^{-1}K$ can match any observed value different from $\pi/4$. Data for machining brass with a range of rake angles from 20° to 40° [15] is given as an example in Fig. 14.20. These conform quite well to eq(14.72b) when $K = 37°$. However, experiment [16] has shown that the shear yield stress is not as sensitive to a superimposed hydrostatic stress as implied by eqs(14.72a,b). Pugh [17] preferred to express observed data empirically through a linear equation

$$\phi + s(\beta - \alpha) = c$$

where s and c were found to be dependent upon the workpiece material, e.g. for an aluminium alloy $s = 1.5$ and $c = 25°$ were found.

(b) *Minimum Work*

An alternative approach to locating ϕ minimises the total work done in machining. This requires that the two work rate components in eq(14.70) appear in terms of ϕ. Thus, forces F and F_S are written in terms of F_x from eqs(14.64a) and (14.66a) then substituted with r_t and γ from eqs(14.63a) and (14.68). This leads to a normalised work expression:

$$w = \frac{\dot{W}}{F_x v} = [\sec(\beta - \alpha)\cos(\phi + \beta - \alpha)\sin\phi][\cot\phi + \tan(\phi - \alpha)] + \sin\beta\sec(\beta - \alpha)\frac{\sin\phi}{\cos(\phi - \alpha)} \tag{14.73a}$$

In this purely geometric approach the two terms on the right-hand side of eq(14.73a) sum to unity for any combination of α, β and γ. Hence $w = 1$ and a minimum $\dot{W}$ ($= F_x v$) corresponds to minimising F_x with respect to ϕ. From eqs(14.65a) and (14.66a):

$$F_x = \frac{F_S\cos(\beta - \alpha)}{\cos(\phi + \beta - \alpha)} = \frac{\tau_s w b\cos(\beta - \alpha)}{\sin\phi\cos(\phi + \beta - \alpha)} \tag{14.73b}$$

Taking τ_s to be the constant shear yield strength of the material, the differentiation of eq(14.73b), i.e. $\mathrm{d}F_x/\mathrm{d}\phi = 0$, again leads to eq(14.71a). In practice, there are non-uniform shear and normal stress distributions upon the flank face, due to a combination of sliding and sticking. The minimum work condition then yields a different ϕ value (see Example 14.5).

14.8.3 Slip-Line Field

Making the usual assumptions of a rigid plastic workpiece, the existence of a shear plane and a stress-free chip, Lee and Schaffer [18] produced two possible slip line fields for orthogonal machining, given in Figs 14.21a,b.

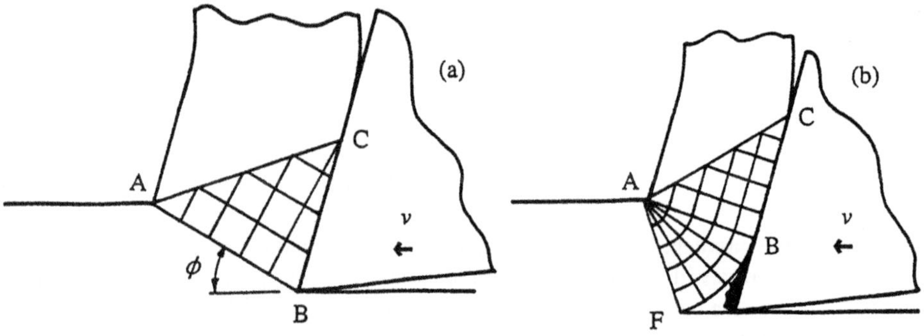

Figure 14.21 Slip-line fields for machining

In the the simpler of these (Fig. 14.21a), the stress free line AC limits the field so that the slip plane AB is one of a number of orthogonal slip lines within the deformation zone that all meet AC at 45°. The zone between AB and AC moves as a rigid block, providing the shear plane orientation*:

$$\phi = \frac{\pi}{4} - (\beta - \alpha) \tag{14.74a}$$

Equation (14.74a) describes the line in Fig. 14.20 with a similar intercept to Merchant but with a gradient of − 1, this often being in better agreement with experimental data. The shear and normal stress on AB are each k, this corresponding to force components, from eqs(14.65a,b) and (14.66a,b), as follows:

* Revealed from a Mohr's circle construction for the stress state at point B (see Fig. 14.28b).

$$F_S = F_N = \frac{kbw}{\sin\phi} \tag{14.74b}$$

$$F_{\substack{x(+)\\y(-)}} = kbw(\cot\phi \pm 1) \tag{14.74c}$$

where k is the shear yield stress. Equations (14.74b,c) become invalid when $\beta - \alpha = \pi/4$ (since cot 0 = ∞), a condition associated with the formation of a built-up edge. In the corresponding slip line field solution (see Fig. 14.21b) two shear planes AB and AF contain a continuous deformation zone ABF ahead of the rigid motion of block ABC. While this field matches conditions for an edge to form, strictly it is not an appropriate field for materials that violate perfect plastic behaviour under very rapid rates of machining.

14.8.4 Work Hardening

Palmer and Oxley [19] added further integral tems $\int(\partial k/\partial\beta)\mathrm{d}\alpha$ and $\int(\partial k/\partial\alpha)\mathrm{d}\beta$ respectively to Hencky's equations (6.8a,b), to allow for the influence of hardening along an orthogonal pair of α, β slip lines. These *modified Hencky equations* confirm a fan of slip lines within the shear zone and a consequent chip curvature. To simplify the analysis they extended the single shear zone either side of AB using parallel lines CD and EF, as shown in Fig. 14.22a.

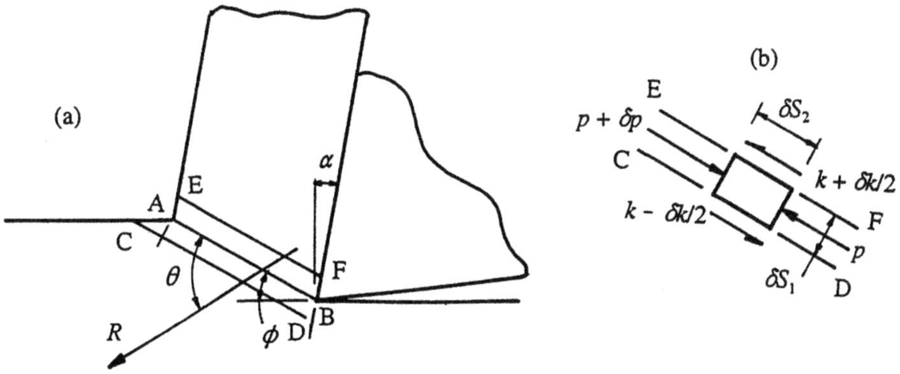

Figure 14.22 Simple approximation to a shear zone

Across an element $\delta S_1 \times \delta S_2$ within this zone (Fig. 14.22b), the shear and hydrostatic stress increase by amounts δk and δp. The following equilibrium equation applies along AB:

$$(p + \delta p)\delta S_1 + (k - \delta k/2)\delta S_2 = p\,\delta S_1 + (k + \delta k/2)\delta S_2$$

$$\delta p = \frac{\delta k}{\delta S_1}\delta S_2 \quad \Rightarrow \quad p = \int \frac{\partial k}{\partial S_1}\mathrm{d}S_2 + C_{S_2} \tag{14.75a,b}$$

This hydrostatic pressure p also lies normal to AB. C_{S_2} is an integration constant pertaining to a known pressure along S_2. Taking the partial derivative to be constant between parallel lines CD and EF, eq(14.75b) shows, for an integration with respect to S_2, that p increases in proportion to the length of AB. Since the area of shear plane AB is $bw/\sin\phi$, the average normal and shear forces follow from stresses p and k:

$$F_N = \frac{(p_A + p_B)bw}{2\sin\phi} \tag{14.76a}$$

$$F_S = \frac{kbw}{\sin\phi} \tag{14.76b}$$

The inclination θ of the resultant force R to the shear plane is found from eqs(14.76a,b):

$$\tan\theta = \frac{F_N}{F_S} = \frac{p_A + p_B}{2k} \tag{14.77a}$$

where θ and ϕ are related to α and β in Fig. 14.18:

$$\theta = \phi + \beta - \alpha \tag{14.77b}$$

Further relationships involving θ, α and ϕ are required (e.g. eq(14.76b)) to enable each of these orientations to be separated.

The Lee and Schaffer solution [18] corresponds to setting $p_A = p_B$ with $\theta = 45°$. This implies that AB lies in the direction of maximum shear so that R aligns with the maximum principal stress direction. Otherwise, the pressure at B is known from applying the modified Hencky equations to the rotation $\pi/4 - \phi$ at point C. Knowing that one principal stress is zero at C, the shear stress must meet the free surface at 45° (see Fig. 14.23). This gives

$$p_A = k\left[1 + 2\left(\frac{\pi}{4} - \phi\right)\right] \tag{14.78a}$$

Applying eq(14.75a), the pressure at B becomes

$$p_B = p_A - \left(\frac{\partial k}{\partial S_1}\right)\frac{b}{\sin\phi} \tag{14.78b}$$

and, since $\partial k/\partial S_1$ is a constant between parallel lines CD and EF, the variation in p between A and B becomes linear. Figure 14.23 shows that two hydrostatic pressure distributions along AB are possible when the pressure ordinates at A and B are defined by eqs(14.78a,b).

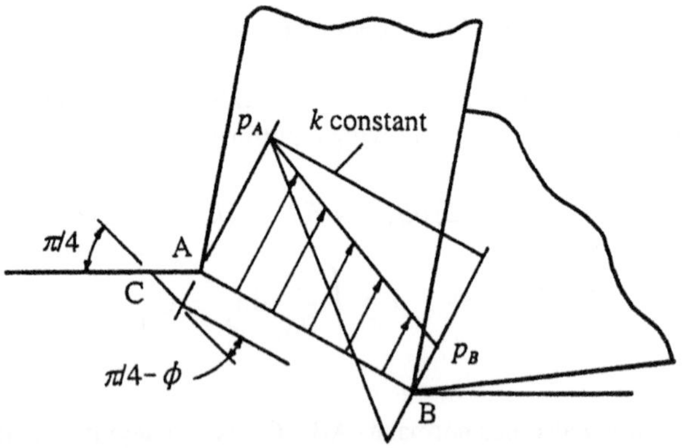

Figure 14.23 Hydrostatic pressure distribution along AB

At B, the pressure may become tensile if the rate of hardening $\partial k/\partial S_1$ across the strip is sufficiently large. Here, a discontinuous chip is likely when θ and ϕ are low within eq(14.77b), given that $\beta - \alpha$ is reasonably constant. The amount of hardening will depend upon the material, the shear rate and the temperature arising from interface friction.

14.8.5 Shear Strain and Strain Rate

The shear strain across a narrow deformation band for machining has been previously defined in eq(14.68). Figure 14.24 shows a hodograph for orthogonal machining. The orientations α and ϕ appear in the physical plane (Fig. 14.24a) for which v is the cutting speed and v' is the chip velocity.

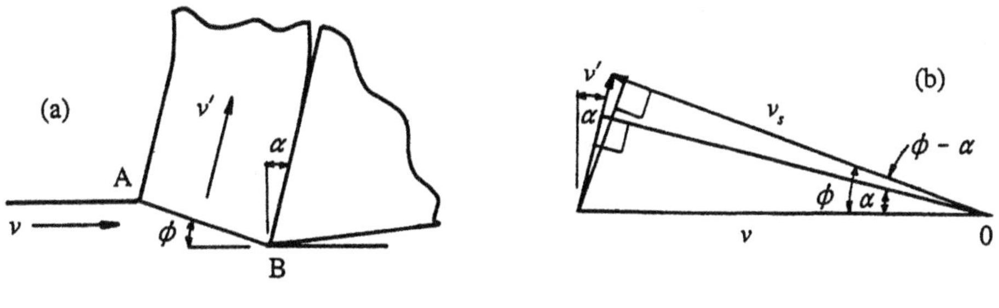

Figure 14.24 Hodograph for orthogonal machining

In the hodograph (Fig. 14.24b), v_s is the velocity along the shear plane AB. This velocity is found by ensuring continuity of normal and tangential velocities across the rake face:

$$v \cos\alpha = v_s \cos(\phi - \alpha) \tag{14.79a}$$

$$v \sin\phi = v' \cos(\phi - \alpha) \tag{14.79b}$$

Equations (14.79a,b) supply the chip velocity and shear strain. Figure 14.24b provides the latter in terms of the velocities within eq(14.79a) as

$$\gamma = \frac{v_s}{v \sin\phi} = \frac{\cos\alpha}{\sin\phi \cos(\phi - \alpha)} \tag{14.80a}$$

The reader should confirm, from the previous derivation of shear strain for machining, that eqs(14.80a) and (14.68) are indeed identical. The mean shear strain rate along AB is defined as

$$\dot{\gamma} = \frac{v_s}{\delta S_1} = \frac{v}{\delta S_1} \times \frac{\cos\alpha}{\cos(\phi - \alpha)} \tag{14.80b}$$

where δS_1 is the perpendicular distance between EF and CD (see Fig. 14.22a).

Example 14.4 The conditions for cutting a work hardening material are: rake angle $\alpha = 33°$, cutting speed $v = 3.75$ m/s, depth of cut $b = 0.1$ mm and width of cut $w = 4.3$ mm. Assume an experimental value for the shear angle $\phi = 25.5°$ for a rectangular shear zone of aspect ratio 10:1. Calculate the shear strain, the shear strain rate, the friction angle β and the cutting forces F_x and F_y. The initial shear yield stress k and the linear hardening gradient $m = \mathrm{d}k/\mathrm{d}\gamma$ depend upon the mean shear strain rate in the manner of Fig. 14.25.

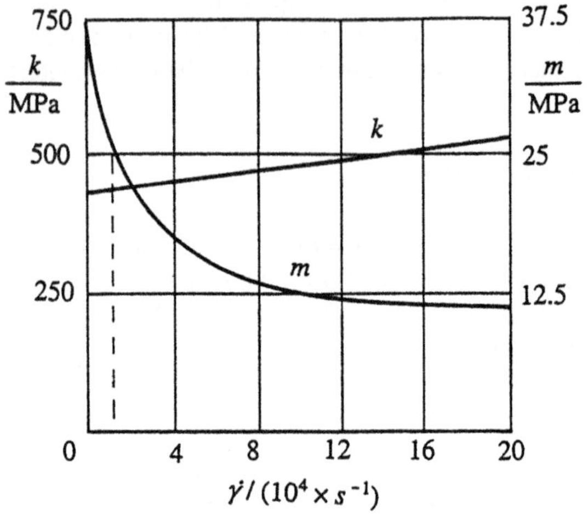

Figure 14.25 Flow rate dependent properties k and m

Firstly, the shear zone thickness is estimated from Fig. 14.22a as

$$\frac{b}{\delta S_1 \sin\phi} = 10, \;\Rightarrow\; \delta S_1 = \frac{b}{10\sin\phi} = \frac{0.2}{10\sin 25.5^\circ} = 0.023 \text{ mm}$$

Equations (14.80a,b) supply the shear strain and shear strain rate:

$$\gamma = \frac{\cos\alpha}{\sin\phi\cos(\phi - \alpha)} = \frac{\cos 33^\circ}{\sin 25.5^\circ \cos(25.5^\circ - 33^\circ)} = 1.965$$

$$\dot{\gamma} = \frac{v\cos\alpha}{\delta S_1 \cos(\phi - \alpha)} = \frac{(3.75 \times 10^3)\cos 33^\circ}{0.023\cos(25.5^\circ - 33^\circ)} = 138 \times 10^3\, s^{-1}$$

From Fig. 14.25, the initial yield stress on plane CD (Fig. 14.22b) and the work hardening gradient are

$$k - \frac{\delta k}{2} = 435.4 \text{ MPa} \quad \text{and} \quad m = 25.2 \text{ MPa}$$

Since $\delta k = m\gamma$, the shear stress k upon plane AB is found from

$$k - \frac{m\gamma}{2} = 435.4 \;\Rightarrow\; k = 435.4 + \frac{(25.2 \times 1.965)}{2} = 460.2 \text{ MPa}$$

The normal pressures upon the slip plane at end points A and B follow from eqs(14.78a,b):

$$\frac{p_A}{k} = 1 + 2\left(\frac{\pi}{4} - \phi\right) = 1 + 2\left(\frac{\pi}{4} - \frac{(25.5 \times \pi)}{180}\right) = 1.681$$

$$\frac{p_B}{k} = \frac{p_A}{k} - \frac{\delta k}{k} \times \frac{b}{\delta S_1 \sin\phi} = 1.681 - \frac{10 \times 49.6}{460.2} = 0.603$$

Then, from eqs(14.77a,b),

$$\tan\theta = \frac{p_A + p_B}{2k} = \frac{1.681 + 0.603}{2} = 1.142 \quad \Rightarrow \quad \theta = 48.79°$$

$$\beta - \alpha = \theta - \phi = 48.79 - 25.50 = 23.29°$$

from which the friction angle $\beta = 56.29°$ and the shear angle $\phi = 25.5°$ is confirmed. Finally, the cutting forces F_x and F_y follow from eliminating R within Merchant's two force systems (14.65a) and (14.66a,b). Setting $F_S = kbw/\sin\phi$, these give

$$F_x = \frac{kwb\cos(\beta - \alpha)}{\sin\phi\cos(\phi + \beta - \alpha)} = \frac{460.2 \times 4.3 \times 0.1\cos(56.29° - 33°)}{\sin 25.5°\cos(25.5° + 56.29° - 33°)} = 640.8 \text{ N}$$

$$F_y = \frac{kwb\sin(\beta - \alpha)}{\sin\phi\cos(\phi + \beta - \alpha)} = \frac{460.2 \times 4.3 \times 0.1\sin(56.29° - 33°)}{\sin 25.5°\cos(25.5° + 56.29° - 33°)} = 275.9 \text{ N}$$

Note, from this example, that if b is reduced so too is ΔS_1. Consequently, $\dot{\gamma}$ and k are raised and with them p_A and p_B are increased. This implies that a *size-effect* exists in machining.

14.8.6 Friction

In an account of friction upon the rake face, eq(14.71b) shows that we may either employ the friction coefficient $\mu = F/N$, i.e. the ratio between shear and normal forces, or the friction angle $\beta = \tan^{-1}(F/N)$. It is assumed that μ, or β, is independent of (i) the contacting area between the rake face and the chip and (ii) the chip speed v'. Both forces F and N are the resultant of the respective shear and normal stress distributions (τ, σ) along the rake face.

(a) *Empirical Approach*

The normal stress σ reaches a maximum at B and falls to zero at the free surface. Between these Zorev [20] assumed a parabolic σ distribution. The shear force was assumed to be spread over two zones: (i) sticking, where $\tau = \tau_B$ = constant, within the nose region and (ii) sliding, where μ is constant but τ decreases to zero at the free surface (see Fig. 14.26).

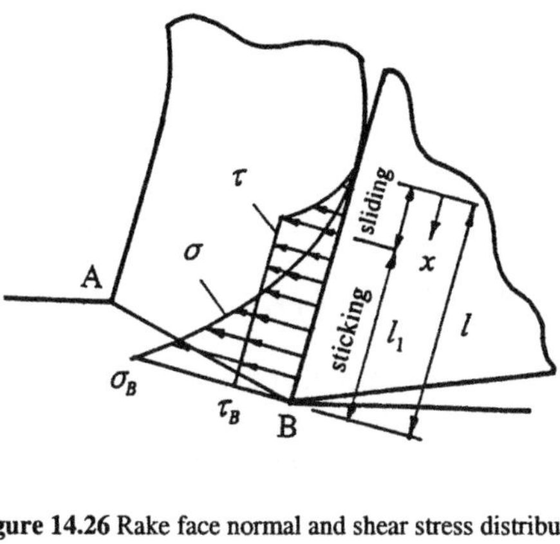

Figure 14.26 Rake face normal and shear stress distributions

With the x-co-ordinate as shown, the following σ relationships hold for $l \le x \le 0$:

$$\sigma = p x^{q}, \qquad \sigma_B = p l^{q}, \qquad \frac{\sigma}{\sigma_B} = \left(\frac{x}{l}\right)^{q} \tag{14.81a,b,c}$$

where p and q are constants. For $0 \le x \le (l - l_1)$, the sliding shear stress becomes

$$\tau = \mu \sigma = \mu \sigma_B \left(\frac{x}{l}\right)^{q} \tag{14.82a}$$

For $(l - l_1) \le x \le l$, the constant, sticking shear stress at B is given from eq(14.82a):

$$\tau = \tau_B = \mu \sigma_B \left(\frac{l - l_1}{l}\right)^{q} \tag{14.82b}$$

where μ in eqs(14.82a,b) is the constant friction coefficient for the sliding zone. Integrating eqs(14.81c) and (14.82a,b) with respect to flank area $w\delta x$, gives N and F:

$$N = w \int_0^l \sigma \, dx = \frac{\sigma_B l w}{1 + q} \tag{14.83a}$$

$$F = w \int_0^l \tau \, dx = \tau_B w \left[l_1 + \frac{l - l_1}{1 + q} \right] \tag{14.83b}$$

Setting $\tau_B = k$, where k is the shear yield strength, and dividing eqs(14.83a,b) leads to the simplified friction angle:

$$\beta = \tan^{-1} \left[\frac{k}{\sigma_B} \left(1 + \frac{q l_1}{l} \right) \right] \tag{14.84a}$$

With no sliding set $l_1 = l$ in eq(14.84a) and assume k remains constant in sticking to give:

$$\beta = \tan^{-1} \left[\frac{k(1 + q)}{\sigma_B} \right] = \tan^{-1} \left(\frac{k}{\sigma_m} \right) \tag{14.84b}$$

where, from eq(14.83a), the mean normal stress is $\sigma_m = N/(wl)$. Under these conditions eq(14.84b) shows that the friction angle depends solely upon the mean normal stress.

Example 14.5 Using Zorev's analysis of the flank face stresses, determine the orientation of the shear plane, based upon a minimum work condition in a non-hardening material, when: $\alpha = 5°$, $l_1/l = 1/2$, $q = 2$ for $l/b = 1, 2, 3 \ldots. 6$.

Let subscript s refer to the shear plane and f to the flank face. The total work rate is simply:

$$\dot{W}_t = F_s v_s + F_f v_f \tag{i}$$

To find the forces F_s and F_f, it is assumed that the shear yield stress k remains constant all along the shear plane and along a portion of the flank face defined by l_1 in Fig. 14.26. The two velocity components v_s and $v_f = v'$ are found from the hodograph in Fig. 14.24b. Thus, the two work terms in eq(i) appear as

$$F_s v_s = \frac{kwb}{\sin\phi} \times \frac{v\cos\alpha}{\cos(\phi - \alpha)} \tag{ii}$$

$$F_f v_f = \left[kwl_1 + \frac{kw}{(1+q)}(l - l_1) \right] \frac{v\sin\phi}{\cos(\phi - \alpha)} \tag{iii}$$

Substituting eqs(ii) and (iii) into eq(i) leads to the normalised work rate expression

$$\frac{\dot{W}_t}{kwbv} = \frac{\cos\alpha}{\sin\phi\cos(\phi - \alpha)} + \frac{l}{b}\left[\frac{l_1}{b} + \frac{(1 - l_1/l)}{(1+q)} \right] \frac{\sin\phi}{\cos(\phi - \alpha)} \tag{iv}$$

Setting $\alpha = 5°$ and $q = 2$ with $l_1/l = 1/2$ for $l/b = 1, 2, 3 \ldots 6$, we may plot eq(iv) as a function of ϕ, as shown in Fig. 14.27.

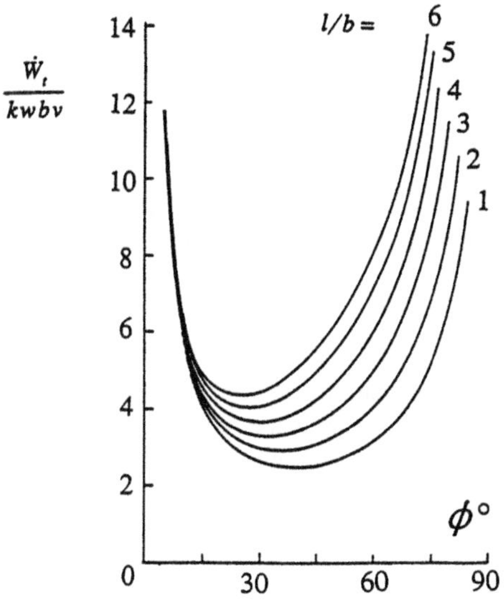

Figure 14.27 Minimum work condition

With an increasing orientation ϕ of the shear plane the work rate decreases within the first term but increases for the flank face in the second term. The result is that their sum, i.e. the normalised total work rate, attains a minimum value under these cutting conditions. As the chip length to thickness ratio increases from 1 to 6 the shear angle ϕ, corresponding to the minimum work condition, diminishes successively from 40° to 25° as shown. This analysis is in qualitative agreement with the many factors known to influence the work rate under practical cutting conditions [21].

(b) *Analytical Approach*

Following [19], the stress states (σ, τ) for the rake face and the shear plane are related by the equations of stress transformation at the tool tip. Figure 14.28a shows the rake face stress state at point B as σ_B and τ_B. Correspondingly, upon the shear plane AB, the normal pressure is p_B and the shear stress is k.

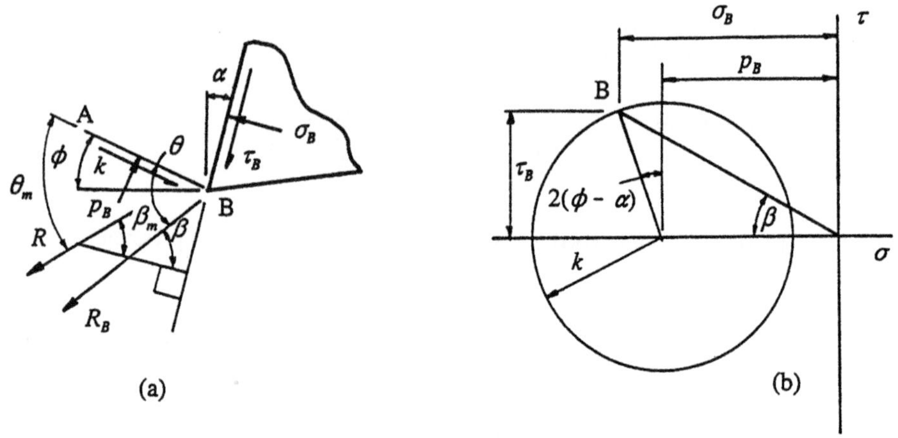

Figure 14.28 Mohr's circle for point B at tool tip

For orientations α and ϕ of the two planes shown in Fig. 14.28a, the Mohr's circle is given in Fig. 14.28b. The geometry of this circle reveals

$$[\,p_B + k \sin 2(\phi - \alpha)\,]\tan \beta_B = k \cos 2(\phi - \alpha) \tag{14.85a}$$

Combining eqs(14.77a) and (14.78a) and eliminating p_B from eq(14.85a) gives the orientation θ of the resultant force R_B at B relative to the shear plane AB (see Fig. 14.28a):

$$\theta = \tan^{-1}\left[\frac{1}{2} + \frac{\pi}{4} - \phi + \frac{\cos 2(\phi - \alpha)}{2\tan\beta} - \frac{\sin 2(\phi - \alpha)}{2}\right] \tag{14.85b}$$

where β applies to point B only. Equation (14.85b) needs a modification to include the mean values θ_m and β_m across the whole shear plane AB. We take the distributions in σ and τ along the rake face to be similar to those shown in Fig. 14.26. These provide the average normal and shear forces:

$$N = \sigma_B l w/2 \tag{14.86a}$$

$$F = \tau_B l_1 w + \frac{\tau_B w}{2}\left(l - l_1\right) = \frac{\tau_B w}{2}\left(l + l_1\right) \tag{14.86b}$$

Dividing eqs(14.86a,b) reveals a mean friction angle for the flank face:

$$\tan\beta_m = \frac{F}{N} = \frac{\tau_B}{\sigma_B}\left(1 + \frac{l_1}{l}\right) = \tan\beta\left(1 + \frac{l_1}{l}\right) \tag{14.87a}$$

Substituting $\tan\beta$ from eq(14.87a) into eq(14.85b) gives the orientation θ_m of the resultant force $R = \sqrt{(N^2 + F^2)}$ as:

$$\theta_m = \tan^{-1}\left[\frac{1}{2} + \frac{\pi}{4} - \phi + \frac{(1 + l_1/l)\cos 2(\phi - \alpha)}{2\tan\beta_m} - \frac{\sin 2(\phi - \alpha)}{2}\right] \tag{14.87b}$$

where F and N are measured forces. Taken with eq(14.77b), the shear angle ϕ may be calculated from eq(14.87b) for a given rake angle α and an assumed ratio $l_1/l < 1$.

14.9 Concluding Remarks

Various analyses of metal shaping processes are available. The simplest work conservation formulae, as used for extrusion and wire drawing, apply only to homogenous deformation. They ignore contributions to work done from friction and redundant shearing. Consequently, the applied forces required to extrude and draw are underestimated. The strip equilibrium method assumes a constant friction coefficient in non-hardening material. This theory allows corrections based upon mean rates and stresses to be made when a process work hardens a material. The convenient closed-form solutions afforded by these analyses allow rapid estimates of the forces involved compared, say, to the slip line field method. The latter is exact in the case of plane strain plasticity involving a plastic rigid material. The library of SLF solutions (see Chapter 6) applies mainly to forming by extrusion and cogging (repeated indentation). Orthogonal machining is predominantly a plastic shearing process but is complicated by rapidly varying strain rates and temperature gradients. Simplifying assumptions allow for estimates of the cutting forces, the shear angle, work hardening and friction. In practice these measures are used to offset adverse cutting conditions that lead to unacceptable rates of wear and poor surface finish.

References

1. Wistreich J. G. and Shutt A. *Jl Iron and Steel Inst*, 1959, **6**, 132.
2. Bishop J. F. W. *Jl Mech Phys and Solids*, 1958, **6**, 132-144.
3. Hirst S. and Ursell D. H. Proceedings: *Technology of Engineering Manufacture*, 1958, p. 58, I. Mech. E, London.
4. Larke E. C. *The Rolling of Strip, Sheet and Plate*, 1957, Chapman and Hall, London.
5. Orowan E. *Proc. I. Mech. E*, 1943, **150**, 140.
6. Sims R. B. *Proc. I. Mech. E*, 1954, **168**, 191.
7. Orowan E. and Pascoe K. J. *Iron and Steel Inst Rpt*, 1946, **34**, 124.
8. Bland D. R. and Ford H. *Proc. I. Mech. E*, 1948, **159**, 144.
9. Bland D. R. and Ford H. *Jl Iron and Steel Inst*, 1952, **171**, 245.
10. Hitchcock J. *Roll Neck Bearings*, 1935, ASME Res Pubs.
11. Avitzur B. *Metal Forming: Processes and Analysis*, 1968, McGraw-Hill.
12. Wistreich J. G. *Proc. I. Mech. E*, 1955, **169**, 659.
13. Merchant M. E. *Jl Appl. Physics*, 1945, **16**, 267.
14. Ernst H. *Annals New York Acad. Sci*, 1951, **53**, 936.
15. Eggleston D.M., Herzog R. P. and Thomsen E. G. *Trans ASME* (*B*), 1959, **81**, 263.
16. Crossland B. *Proc. I. Mech. E*, 1954, **168**, 935.
17. Pugh H. Ll. D. *Jl Mech. Eng. Sci*, 1964, **6**, 4.
18. Lee E. H. and Schaffer B. W. *Jl Appl. Mech*, 1951, **18**, 405.
19. Palmer W. B. and Oxley P. L. B. *Proc. I. Mech. E*, 1959, **173**, 623.
20. Zorev N. N. Proc: *Int. Conf. Prod. Eng. Res*, 1963, p. 43, Pittsburg.
21. Rowe G. W. and Spick P. T. *Trans ASME*, 1967, **89B**, 530.

Exercises

14.1 A piece of lead 25 mm × 25 mm × 150 mm, with a yield stress 7 MPa, is forged between flat dies to a final aproximate size: 6 mm × 104 mm × 150 mm. Assume that a friction coefficient $\mu = 0.25$ prevails with sliding between the lead and the full width of the dies in the contact region. Determine the pressure distribution that the dies exert upon the lead and the total forging force required.

14.2 Confirm that the mean pressure for the friction hill shown in Fig. 14.3b is

$$\frac{\bar{p}}{2k} = \frac{t}{\mu b}\left\{\frac{1}{4\mu}\left[(1 + \alpha)^2 + 1\right] - 1\right\}$$

where $\alpha = \mu b/t - \ln(1/2\mu)$.

14.3 Compare the work required to hot forge a 0.75 m long billet, with 150 mm square cross-section, into a strip 50 mm thick, 150 mm wide and 2.25 m long, using dies of breadth (a) 50 mm, (b) 100 mm and (c) 150 mm. Hence show that the minimum work is done when the die breadth lies between (a) and (b). Neglect work hardening and spread and assume that small reductions are made so that the existence of the transitional shape at the edges of the die may be neglected. Take the load L at any thickness to be given by $L = YAC$ in which Y is the yield stress. The contact area is $A = wb$, where b is the die width and w the billet width. For this cogging process take constant $C = 0.797 + 0.203t/b$ for $b/t < 1$ and $C = 0.75 + 0.25b/t$ for $b/t > 1$.

14.4 In forging the ring shown in Fig. 14.29 a plastic hinge is formed opposite the forging platens. If the shear yield stress of the ring is constant at k, its mean radius is R and the other dimensions are as shown, prove that the additional pressure on the platens required to form the plastic hinge is approximately $p' = kt/4R$, given that the plastic hinge moment is $kwt^2/2$. If the final dimensions of the ring are to be $t = 125$ mm, $R = 1.075$ m and $w = 175$ mm for $k = 40$ MPa and a maximum available force of 1.5 MN, determine the maximum internal diameter of the ring at which forging can begin. Assume plane strain conditions and take the total pressure upon the platens to be $p = 2k[C + t/(8R)]$ where $C = 0.797 + 0.203t/b$.

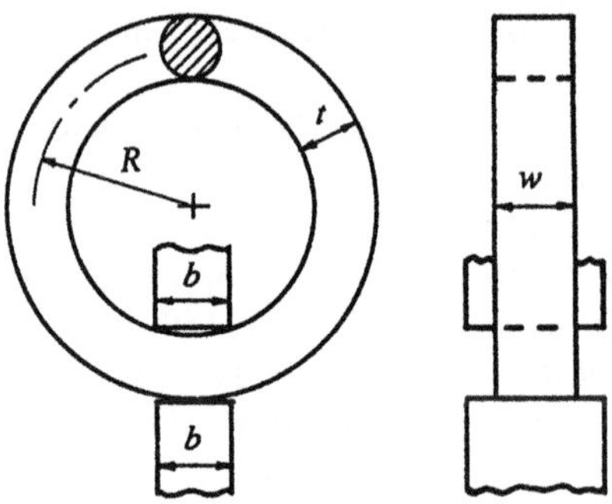

Figure 14.29

14.5 A cylindrical component of original height $h_o = 50$ mm is forged between flat platens to a height $h_1 = 40$ mm. The lower platen is rigid whilst the upper one descends with a velocity $v_o = 3$ m/s. Show that the true mean strain rate for the process is given by the following expression and from this calculate its magnitude.

$$\dot{\varepsilon} = -\frac{v_o \ln(h_o/h_1)}{2(h_o - h_1)}$$

14.6 Determine the pressure on a ram required to reduce a billet from 40 mm diameter to 37 mm diameter in a homogenous extrusion process . Assume that the 40 mm diameter billet is the result of a previous process where the sectional area was reduced by 15% for material in an annealed state, obeying a Hollomon flow law: $\sigma = 670\varepsilon^{0.5}$.

14.7 What drawing stress is required to reduce a rod from 100 mm to 70 mm in a conical die with 30° included angle? Assume that the friction coefficient is 0.05 and that the material is rigid-perfectly plastic with a tensile yield stress of 160 MPa. What is the maximum reduction that can be achieved under these conditions? [Answer: 127 MPa]

14.8 Show that the maximum possible change in section area for wire drawing a rigid, perfectly plastic material is 63%. If 3 mm diameter wire is to be drawn at maximum reduction in each of three sequencial processes, what is the smallest diameter possible after the third process? Examine whether an optimum die angle exists from a consideration of friction and contact length assuming an homogenous deformation within the tapered die.

14.9 In a homogenous wire drawing operation upon hardening material, the drawing force F may be estimated as:

$$F = Y_m A_1 \ln\left(\frac{A_0}{A_1}\right)$$

where A_0 is the original cross-sectional area, A_1 is area of the drawn wire and Y_m is the average yield stress for material hardened by the process. Using an average yield stress between entry and exit, determine the drawing force required to reduce a wire from 3 mm to 2.75 mm, given that a 20% reduction in area had previously been applied in reaching the 3 mm wire size. These reductions should be referred to the annealed state for which the Hollomon flow law is: $\sigma = 622\varepsilon^{0.32}$.

14.10 What drawing stress is required to draw a 500 mm wide, annealed steel strip from 2.5 to 2.4 mm thick in dies with 15° included angle and a friction coefficient 0.1? What is the influence of a previous 20% reduction in area arising from rolling the annealed material given that it conforms to a Ludwik law: $\sigma = 200 + 100\varepsilon^{0.8}$?

14.11 What drawing stress is required to draw a 150 mm wide × 6 mm thick strip with a 20% reduction in area? This strip had been reduced in thickness by 30% in a previous process from its annealed condition. What drawing force would be required for a further 20% reduction in area? The die angle is 15° and the friction coefficient is 0.05. The true stress-strain behaviour of annealed strip material shows natural (logarithmic) strains of 0.356, 0.580 and 0.803 for stress levels 350, 430 and 490 MPa respectively. [Answer: 73 kN, 69 kN]

14.12 Orowan [5] proposed a solution to the problem of hot rolling narrow flat strip with a width/thickness ratio at which the decrease in roll pressure at the edges is not negligible. Assuming that the neutral point lies midway between exit and entry, show that the total roll force is given by:

$$P = kbL\left[0.8 + \frac{L}{4h_o} - \frac{2hh_o}{3bL}\left(\frac{l}{2h_o} - 0.2\right)^3\right]$$

where k is the plane strain yield stress, b is the mean strip width, h is the mean thickness of strip within the roll gap, h_o is the final thickness of the strip and L is the projected length over the arc of contact.

14.13 What do you understand by the term friction hill in forging and rolling operations? Outline the various frictional conditions which can occur in hot and cold working and sketch the associated friction hills. Describe the geometrical factors which affect the friction hill in such operations, particularly spread. Define the term plane strain and explain the phrase 'plane sections remain plane'. Prove, when plane strain forging a strip of thickness t, with the breadth b of the flat platens less than

$$b = \frac{t}{\mu}\ln\left(\frac{1}{2\mu}\right),$$

the frictional shear stress between the strip and platens will never attain the shear yield stress of the strip material. The coefficient of friction in the contact zone is μ.

14.14 Consider simple cold rolling of a thin strip when both rolls and strip remain elastic. The distribution of roll pressure is approximately parabolic, being given by

$$s = \frac{E_s(a^2 - x^2)}{(1 - \nu_s^2)R'h}$$

where a is the half-length of the constant arc, E_s and ν_s are the strip's elastic constants, R' is the deformed roll radius and x is the distance from the axis of symmetry. The value of R' may be derived from Hitchcock's analysis [10], namely

$$\frac{1}{R'} = \frac{1}{R} - \frac{4(1 - \nu_r^2)P}{E_r \pi a^2}$$

where R is the undeformed roll radius, P is the total roll force, E_r and ν_r are the elastic constants for the roll material. It may be assumed that at the centre of the strip, on the axis of symmetry, a longitudinal compressive stress p is induced due to the combined effect of frictional stresses acting on the surface of the strip and the mean of the back and front strip tensions $\bar{t}$ from:

$$p = \frac{\mu P}{h} - \bar{t}$$

where μ is the Coloumb friction coefficient. Assumung a yield criterion: $s - p = mY$ where Y is the tensile yield stress and m is a factor lying between 1 and $2/\sqrt{3}$, show that the minimum thickness which could be rolled plastically is approximately:

$$h_{\min} = (mY - \bar{t})\left[\frac{14.22}{E_s}\mu^2 R(1 - \nu_s^2) + \frac{9.05}{E_r}\mu R(1 - \nu_r^2)\right]$$

14.15 Metal cutting is basically a process of plastic deformation. Discuss an orthogonal cutting operation on a practical work material emphasising the role of the important geometrical, physical and metallurgical factors involved and their inter-relationships. What are the parameters of particular importance to the conduct and control of machining?

14.16 Explain the unconstrained geometry of plastic deformation in orthogonal machining. Discuss the limitations of the friction angle expression: $\tan\beta = F/N$ when applied to work/tool interface friction in machining. Given that the shear strain γ depends upon the shear angle ϕ and the rake angle α in the form: $\gamma = \cot\phi + \tan(\phi - \alpha)$, show that there will be a greater tendency for the chip to break away when the rake angle is negative rather than positive.

14.17 Discuss briefly the formulation and implementation of two shear plane models of the orthogonal metal cutting process indicating their limitations. Explain how a model which admits work hardening of the workpiece provides a more realistic explanation of machining behaviour in practice.

14.18 The conditions when orthogonal machining a work hardening material are: rake angle $\alpha = 10°$, cutting speed $v = 0.5$ m/s and depth of cut $b = 0.2$ mm. Assume an experimental value $\phi = 25°$ for the shear angle with a rectangular shear zone of aspect ratio 10:1. Calculate the shear strain, the shear strain rate, and the inclinations of the resultant force R to the flank face and to the shear plane. Use the flow properties for the material given in Fig. 14.25.

14.19 A work hardening material is machined with rake angle 8°, cutting speed 1.8 m/s, depth of cut 0.25 mm, width of cut 6.5 mm and chip thickness 0.4 mm. Assume a rectangular shear zone of aspect ratio 10:1. Calculate the shear strain γ, the shear strain rate $\dot{\gamma}$, the friction angle β and the two cutting forces F_x and F_y. The dependence of the initial shear yield stress k and the work hardening gradient $m = dk/d\gamma$ upon the shear strain rate are as given in Fig. 14.25.

CHAPTER 15

APPLICATIONS OF FINITE ELEMENTS

15.1 Introduction

Up to now the analyses given of plasticity theory together with its applications have resulted, largely, in closed solutions across a wide range of topics. The reader will no doubt be aware of the availability of commercial finite element (FE) packages that provide numerical solutions to many of these topics involving bulk plasticity and sheet metal forming. Consequently it is pertinent to end this book with an overview and a few applications of the FE technique. Firstly, we shall describe the essence of the FE method as it applied to elasticity. This is intended to aid an understanding of how the linear elastic theory is extended incrementally to the elastic-plastic regime. Specific details can be found in many of the more dedicated FE texts [1-4]. The theory upon which FE depends: the yield criterion, constitutive relations, the flow and hardening rules, can all be found in our earlier chapters on classical plasticity. We shall show how these are transcribed into their matrix forms for efficient programming. Various FE codes are applied here to sheet metal forming problems. The particular package employed is described and its outputs are selected for comparison with available experimental data. In the first application, an example is given of hydrostatic bulge forming within circular and elliptical dies. This particular test provides FE users with a validation check in which the predicted plot between pressure and displacement is overlaid with experimental data. The agreement found between the two plots inspires a trust in an FE simulation of how the deformation evolves in forming more complex shapes. In the second application, the Erichsen test is simulated in order to provide strain distributions under a ball indentation - where an analytical solution does not exist. Again, the predictions are overlaid with experimental data from this test. Finally, it is shown how FE is used as a tool for preserving the integrity of a pressed sheet metal product at the die try-out stage of manufacture. Among the more sophisticated outputs that FE can provide here are strain contours and signatures, these showing the proximity of the strain state at any position in the pressing to the sheet material's forming limit.

15.2 Elastic Stiffness Matrix

Elastic FE applies the theory of elasticity within the sub-division of a body into smaller elements inter-connected throughout at nodal points. This technique of discretisation enables an assembly of elements to describe the initial unstrained shape of the body. Thus, the behaviour of the entire body under stress may be computed from the known elastic behaviour of its elements. When FE does this for a frame lying within the continuum, it must provide for equilibrium and compatibility of its internal stress and strain distributions and match external boundary conditions, usually of known force and displacement.

In the *stiffness or displacement method* of finite elements, the displacements at the nodal points are the unknowns. These displacements are solved numerically from their relation to the nodal forces. This relation is established by assembling a stiffness matrix to connect the ' vectors ' of nodal point forces and displacements for each element. An overall stiffness matrix **K** is then assembled, where for a given mesh $\mathbf{f} = \mathbf{K}\,\boldsymbol{\delta}$, in which **f** and $\boldsymbol{\delta}$ represent forces and displacements at its many nodes as column matrices. The displacements are found numerically from a matrix inversion $\boldsymbol{\delta} = \mathbf{K}^{-1}\mathbf{f}$. Strains follows from a strain-displacement relation and the stress from its stress-strain (constitutive) relation. While such relations are well established for elasticity, the procedure must assume a function expressing the displacements of a node upon its co-ordinates. Thus, from applying the displacement function across all elements, the stiffness method provides numerical solutions to the stress and strain distributions within a loaded body. In the *force or flexibility method* of finite elements, the displacements are found from an assumed stress distribution. Though less popular, this method has proven a better choice for statically indeterminate framed structures [1], but it will not be expounded here.

In essence, the stiffness FE method aims to compile and invert the matrix, connecting force to displacement, for an assembly of elements in a computationally efficient manner. There are many different finite elements, both plane and with axial symmetry, including triangular, rectangular, bar, shell and volume types. These elements appear within many well documented application of FE including beam and plate bending, torsion of prismatic bars, plates and bars under plane stress and strain, pressure vessels and rotating discs.

In static FE codes, the common approach is to derive the element stiffness matrix $\mathbf{K}^e$, connecting its nodal point force and displacement vectors, $\mathbf{f}^e$ and $\boldsymbol{\delta}^e$, respectively:

$$\mathbf{f}^e = \mathbf{K}^e\,\boldsymbol{\delta}^e \tag{15.1}$$

The method for doing this will be shown briefly in selected examples that follow: simple tension, torsion and bending, before considering an element used for plane stress and strain in more detail. Bold, upper-case Roman letters **A, B, C, K**, denote matrices and bold, lower case Roman and Greek symbols $\mathbf{f}, \boldsymbol{\alpha}, \boldsymbol{\varepsilon}, \boldsymbol{\delta}, \boldsymbol{\sigma}$, denote column matrices or column vectors. The word 'vector' is reserved for quantities that take on a physical meaning, as with force **f** and displacement $\boldsymbol{\delta}$. We begin with a summary of the matrices $\mathbf{K}^e$ in eq(15.1) for simple bar elements in tension, torsion and bending.

15.2.1 Tension

Consider, for simple tension of a bar element of length L, with uniform cross-sectional area A, connecting nodal points 1 and 2, as shown in Fig. 15.1.

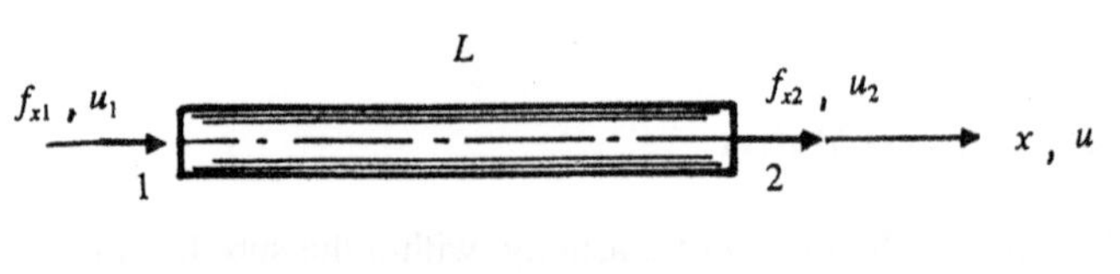

Figure 15.1 Tension bar element

The axis of the bar is aligned with the x-direction. The nodal forces f_{x1} and f_{x2} are taken to act at nodes 1 and node 2 respectively in the positive x-direction. These forces produce nodal displacements u_1 and u_2, aligned with positive x. Here, the stiffness matrix $\mathbf{K}^e$ is obvious from the uni-axial relationship between force and displacement: $f = (AE/L)u$. When written in the matrix form of eq(15.1) this becomes

$$\begin{bmatrix} f_{x_1} \\ f_{x_2} \end{bmatrix} = \frac{AE}{L}\begin{bmatrix} 1 & -1 \\ -1 & 1 \end{bmatrix}\begin{bmatrix} u_1 \\ u_2 \end{bmatrix} \tag{15.2}$$

where $\mathbf{f}^e = [\, f_{x1} \;\; f_{x2}\,]^T$ (in which $f_{x2} = -f_{x2}$) and $\boldsymbol{\delta}^e = [\, u_1 \;\; u_2\,]^T$ are force and displacement vectors. Equation (15.2) reveals a symmetrical, 2 × 2 element stiffness matrix $\mathbf{K}^e$, as would be expected when each of the two nodes has a single degree of freedom. The element's stresses and strains follow applying the relationships $\sigma = E\varepsilon = Eu/L$ and $\varepsilon = \partial u/\partial x = \sigma/E$, given a displacement function $u = u(x)$.

15.2.2 Torsion

In Fig. 15.2 the uniform, solid circular shaft element of length L and radius r is subjected to nodal torques t_1 and t_2, as shown, which twist nodes 1 and 2 by the amounts θ_1 and θ_2.

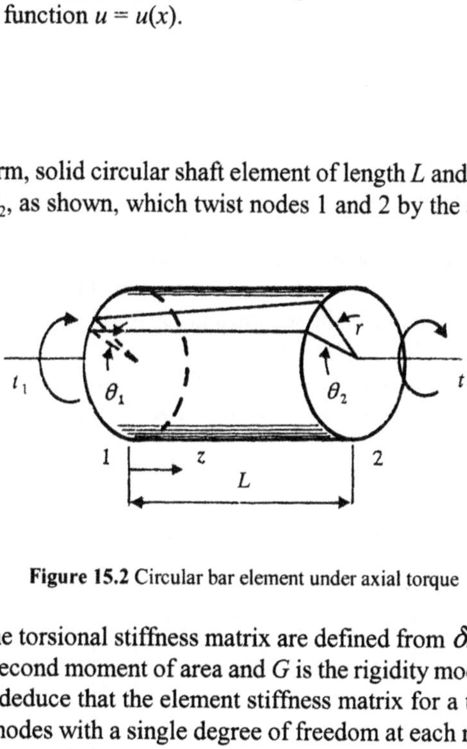

Figure 15.2 Circular bar element under axial torque

The components of the torsional stiffness matrix are defined from $\delta t/\delta\theta = JG/L$, in which $J = \pi r^4/2$ is the polar second moment of area and G is the rigidity modulus. As with a tension bar element, we may deduce that the element stiffness matrix for a torsion bar element is 2 × 2; there being two nodes with a single degree of freedom at each node. Thus, $\mathbf{K}^e$ follows directly from the stiffness relation as

$$\begin{bmatrix} t_1 \\ t_2 \end{bmatrix} = \frac{JG}{L}\begin{bmatrix} 1 & -1 \\ -1 & 1 \end{bmatrix}\begin{bmatrix} \theta_1 \\ \theta_2 \end{bmatrix} \tag{15.3a}$$

Comparing eqs(15.3a) with in eq(15.1), we identify $\mathbf{f}^e = [\, t_1 \;\; t_2\,]^T$ and $\boldsymbol{\delta}^e = [\theta_1 \;\; \theta_2\,]^T$ as the torque and twist vectors respectively. Once the nodal twists have been found from a function $\theta = \theta(z)$, the element's nodal shear strains $\boldsymbol{\gamma}^e = [\,\gamma_1 \;\; \gamma_2\,]^T$ and stresses $\boldsymbol{\tau}^e = [\,\tau_1 \;\; \tau_2\,]^T$ follow proportionately from the elastic relation $\gamma = \tau/G = r\theta/L$ as

$$\boldsymbol{\gamma}^e = (r/L)\,\boldsymbol{\delta}^e, \quad \boldsymbol{\tau}^e = (Gr/L)\,\boldsymbol{\delta}^e \tag{15.3b,c}$$

15.2.3 Beam Bending

The element shown in Fig. 15.3a is used to determine the displacements, strains and stresses for a beam under bending. Shear force q must accompany bending moment m, each of which are taken to differ between nodes 1 and 2. Such actions produce both a vertical deflection v and a rotation θ, in the positive directions shown.

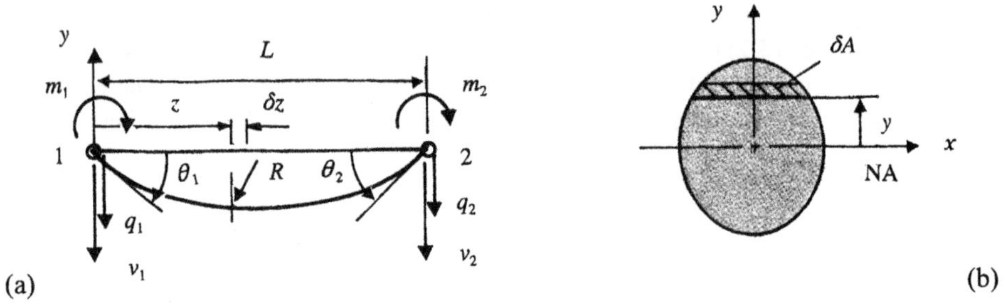

Figure 15.3 Beam element, showing arbitrary cross-section

The respective nodal force and displacement vectors now become: $\mathbf{f} = [\, q_1 \;\; m_1 \;\; q_2 \;\; m_2 \,]^T$ and $\boldsymbol{\delta} = [\, v_1 \;\; \delta_1 \;\; v_2 \;\; \delta_2 \,]^T$. The beam cross-section, which may be of any shape, is uniform along the length. Co-ordinate axes, x and y in the section and z in the length, pass through the centroid of the section (see Fig. 15.3b). Under bending, the longitudinal axis z becomes the neutral axis; this deflects to a radius of curvature R but suffers no stress or strain. With two degrees of freedom existing at each node the stiffness matrix $\mathbf{K}^e$, connecting $\mathbf{f}$ to $\boldsymbol{\delta}$, has a dimension 4×4 with the following symmetry [5]:

$$\begin{bmatrix} q_1 \\ m_1 \\ q_2 \\ m_2 \end{bmatrix} = \frac{EI}{L^3} \begin{bmatrix} 12 & 6L & -12 & 6L \\ 6L & 4L^2 & -6L & 2L^2 \\ -12 & -6L & 12 & -6L \\ 6L & 2L^2 & -6L & 4L^2 \end{bmatrix} \begin{bmatrix} v_1 \\ \theta_1 \\ v_2 \\ \theta_2 \end{bmatrix} \qquad (15.4)$$

Once the nodal displacements v and rotations $\theta = \mathrm{d}v/\mathrm{d}z$ have been found from a displacement function $v = v(z)$, the element's strain $\boldsymbol{\varepsilon}^e = [\, \varepsilon_1 \;\; \varepsilon_2 \,]^T$ and stress $\boldsymbol{\sigma}^e = [\, \sigma_1 \;\; \sigma_2 \,]^T$ follow from bending theory: $\varepsilon = y/R = y(\mathrm{d}^2v/\mathrm{d}z^2)$ and $\sigma = E\varepsilon$.

15.3 Energy Methods

The foregoing review has shown that the FE stiffness method is resolved into finding the stiffness matrix $\mathbf{K}^e$ for a given element. Clearly, bar and beam elements allow their stiffness matrices, eqs(15.2) - (15.4), to be expressed quite simply. However, a derivation of each stiffness component K_{ij}^e of $\mathbf{K}^e$ is required for most other elements. The basis for this is to employ a convenient energy method; either the principle of virtual work (PVW) or the principal of stationary potential energy (SPE) may be used [5]. These principles show that K_{ij}^e can be expressed in terms of the product of matrices in the nodal point co-ordinates and the elastic constants for the element material.

15.3.1 Principle of Virtual Work

The stiffness method of FE imposes virtual displacements at the nodes of a deformable element where real forces are applied. Conversely, virtual forces and real displacements are used with the flexibility method of FE. For the former method, the system of real forces, f_k ($k = 1, 2, 3 \ldots$) is in equilibrium with its internal stress state σ_{ij}. When the real forces experience virtual, in-line displacements δ_k^v, the corresponding virtual strains are ε_{ij}^v. Superscript v denotes that displacements and strains are virtual. The principal of virtual work states that no net work can be done by imposing a displacement upon an equilibrium force system. When displacement due to self-weight is neglected, we can express this principle in two useful, equivalent ways:

$$f_k\,\delta_k^v - \int \sigma_{ij}\,\varepsilon_{ij}^v\,\mathrm{dV} = 0 \quad \Rightarrow \quad (\mathbf{f}^e)^T\,\boldsymbol{\delta}^{ev} - \int \boldsymbol{\sigma}^T\,\boldsymbol{\varepsilon}^v\,\mathrm{d}V = 0 \qquad (15.5a,b)$$

$$\delta_k^v\,f_k - \int \varepsilon_{ij}^v\,\sigma_{ij}\,\mathrm{dV} = 0 \quad \Rightarrow \quad (\boldsymbol{\delta}^{ev})^T\,\mathbf{f}^e - \int (\boldsymbol{\varepsilon}^v)^T\,\boldsymbol{\sigma}\,\mathrm{d}V = 0 \qquad (15.6a,b)$$

In the matrix notation, eq(15.5b) expresses $\mathbf{f}$, $\boldsymbol{\delta}$, $\boldsymbol{\sigma}$ and $\boldsymbol{\varepsilon}$ as column matrices. The reversal to the order of their multiplication within eq(15.6b) governs which matrix to transpose. The resulting scalar products remain the same: $\mathbf{f}^T\,\boldsymbol{\delta} = \boldsymbol{\delta}^T\,\mathbf{f}$ and $\boldsymbol{\varepsilon}^T\boldsymbol{\sigma} = \boldsymbol{\sigma}^T\,\boldsymbol{\varepsilon}$. The additional superscript (e) is used when applying the principle specifically to an element's nodal forces and displacements. Thus, components of the nodal force vector $\mathbf{f}^e$ and the internal stress matrix $\boldsymbol{\sigma}$ are real and in equilibrium. Components of the nodal displacements vector $\boldsymbol{\delta}^{ev}$ and the internal strain matrix $\boldsymbol{\varepsilon}^v$ are compatible and virtual, i.e. independent of the real force-stress system. Hence we can integrate the strain independently of stress over a volume V, within which the stress and strain can vary. Actually, eq(15.6b) is more convenient for deriving $\mathbf{K}^e$ in the FE analysis that follows. Here, $(\boldsymbol{\delta}^{ev})^T$ is the transpose of the element's virtual, nodal displacement vector $\boldsymbol{\delta}^{ev}$. The generic word ' displacement ' refers to deflection, rotation and twist at the nodal points. Moreover, the generic ' force ' vector $\mathbf{f}^e$ will contain the real nodal forces, moments and torques. Should forces be distributed, the virtual work principle allows $\mathbf{f}^e$ to be compiled from an equivalent system of concentrated nodal forces [6].

15.3.2 Stationary Potential Energy

With both stress and strain being real, a stationary value of the total potential energy P will apply to an equilibrium condition. Now P is the sum of the internal strain energy stored U within a deformable element and the negative work of its external forces $-V$. Hence, a stationary P will lie in the condition:

$$\mathrm{d}\left[\int\int \boldsymbol{\sigma}^T\,\mathrm{d}\boldsymbol{\varepsilon}\,\mathrm{d}V - (\mathbf{f}^e)^T\,\boldsymbol{\delta}^e\right] = 0 \qquad (15.6a)$$

With proportionality between stress and strain, the first integration over strain, reduces eq(15.6a) to

$$\mathrm{d}\left[\tfrac{1}{2}\int \boldsymbol{\sigma}^T\,\boldsymbol{\varepsilon}\,\mathrm{d}V - (\mathbf{f}^e)^T\,\boldsymbol{\delta}^e\right] = 0 \qquad (15.6b)$$

Equation (15.6b) is applied as a partial differentiation with respect to the element's displacement vector $\boldsymbol{\delta}^e$. Using the relationship $\boldsymbol{\sigma}^T\,\boldsymbol{\varepsilon} = \boldsymbol{\varepsilon}^T\,\boldsymbol{\sigma}$, the stationary PE gives

$$\partial P/\partial \boldsymbol{\delta}^e = \tfrac{1}{2}\int_V [\,\boldsymbol{\sigma}^T(\partial\boldsymbol{\varepsilon}/\partial\boldsymbol{\delta}^e) + \boldsymbol{\varepsilon}^T(\partial\boldsymbol{\sigma}/\partial\boldsymbol{\delta}^e)]\,dV - (\mathbf{f}^e)^T = 0 \qquad (15.6c)$$

15.4 Plane Triangular Element

Let us illustrate the application of each energy method (PVW and SPE) to find the stiffness matrix for a plane triangular element in Fig. 15.4a. This element applies to bodies in plane stress and plane strain, i.e. those loaded in the plane x-y, where, respectively, the z-dimension can be both small (through a plate's thickness) and large (along a bar's axis).

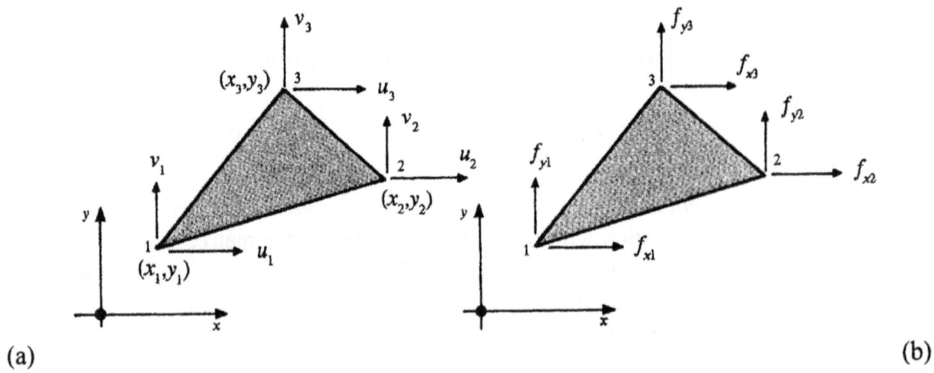

Figure 15.4 Forces and displacements at the nodes of a plane triangular element

15.4.1 Nodal Displacements

Firstly, the displacements (u, v) for any point (x, y) in the element are assumed to obey the linear displacement functions:

$$u(x, y) = \alpha_1 + \alpha_2 x + \alpha_3 y \qquad (15.7a)$$

$$v(x, y) = \alpha_4 + \alpha_5 x + \alpha_6 y \qquad (15.7b)$$

where coefficients α_1, α_2 ... α_6 are to be found. In matrix form, eqs(15.7a,b) appear as a displacement vector $\boldsymbol{\delta} = [u \;\; v]^T$:

$$\boldsymbol{\delta} = \begin{bmatrix} u \\ v \end{bmatrix} = \begin{bmatrix} 1 & x & y & 0 & 0 & 0 \\ 0 & 0 & 0 & 1 & x & y \end{bmatrix} \begin{bmatrix} \alpha_1 \\ \alpha_2 \\ \alpha_3 \\ \alpha_4 \\ \alpha_5 \\ \alpha_6 \end{bmatrix} \qquad \boldsymbol{\delta} = \mathbf{A}\,\boldsymbol{\alpha} \qquad (15.7c,d)$$

where $\alpha = [\alpha_1\ \alpha_2\ \alpha_3\ \alpha_4\ \alpha_5\ \alpha_6]^T$ is a column matrix of six components, matching the number of degrees of freedom and $\mathbf{A}$ is a 6 × 2 matrix in co-ordinates (x, y). The displacements for nodes 1, 2 and 3 may be found from substituting the co-ordinates (x_1, y_1), (x_2, y_2) and (x_3, y_3) into eq(15.7a,b). This gives the three nodal point displacement vectors: $\delta_1 = [u_1\ v_1]^T$, $\delta_2 = [u_2\ v_2]^T$ and $\delta_3 = [u_3\ v_3]^T$, which re-appear in the element's nodal point displacement vector δ^e:

$$\delta^e = \begin{bmatrix} \delta_1 \\ \delta_2 \\ \delta_3 \end{bmatrix} = \begin{bmatrix} u_1 \\ v_1 \\ u_2 \\ v_2 \\ u_3 \\ v_3 \end{bmatrix} = \begin{bmatrix} 1 & x_1 & y_1 & 0 & 0 & 0 \\ 0 & 0 & 0 & 1 & x_1 & y_1 \\ 1 & x_2 & y_2 & 0 & 0 & 0 \\ 0 & 0 & 0 & 1 & x_2 & y_2 \\ 1 & x_3 & y_3 & 0 & 0 & 0 \\ 0 & 0 & 0 & 1 & x_3 & v_3 \end{bmatrix} \begin{bmatrix} \alpha_1 \\ \alpha_2 \\ \alpha_3 \\ \alpha_4 \\ \alpha_5 \\ \alpha_6 \end{bmatrix} \qquad \delta^e = \mathbf{A}^e \alpha \qquad (15.8a,b)$$

where $\delta^e = [\delta_1\ \delta_2\ \delta_3]^T$ and $\mathbf{A}^e$ is a 6 × 6 matrix of nodal co-ordinates. It now follows that the coefficients $\alpha_1, \alpha_2 \ldots \alpha_6$, are found from inverting matrix $\mathbf{A}^e$ in eq(15.8b):

$$\alpha = (\mathbf{A}^e)^{-1}\,\delta^e \qquad (15.9a)$$

where, the inversion of $\mathbf{A}^e$ in eq(15.8a) gives

$$(\mathbf{A}^e)^{-1} = \frac{1}{2\Delta}\begin{bmatrix} x_2y_3 - x_3y_2 & 0 & x_3y_1 - x_1y_3 & 0 & x_1y_2 - x_2y_1 & 0 \\ y_2 - y_3 & 0 & y_3 - y_1 & 0 & y_1 - y_2 & 0 \\ x_3 - x_2 & 0 & x_1 - x_3 & 0 & x_2 - x_1 & 0 \\ 0 & x_2y_3 - x_3y_2 & 0 & x_3y_1 - x_1y_3 & 0 & x_1y_2 - x_2y_1 \\ 0 & y_2 - y_3 & 0 & y_3 - y_1 & 0 & y_1 - y_2 \\ 0 & x_3 - x_2 & 0 & x_1 - x_3 & 0 & x_2 - x_1 \end{bmatrix} \qquad (15.9b)$$

A square matrix relation $\mathbf{A}^e(\mathbf{A}^e)^{-1} = \mathbf{I}$ applies to eq(15.9b), where $\mathbf{I}$ is a unit matrix and Δ gives the area of the element as $\Delta = ½\,[(x_1 - x_2)(y_2 - y_3) - (x_2 - x_3)(y_1 - y_2)]$. Combining eqs(15.7b) and (15.9a) expresses the element's general displacement vector $\delta = [u\ \ v]^T$ in terms of its three nodal displacement vectors $\delta^e = [\delta_1\ \delta_2\ \delta_3]^T$:

$$\delta = \mathbf{A}\,(\mathbf{A}^e)^{-1}\,\delta^e = \mathbf{N}\,\delta^e \qquad (15.10)$$

where the matrix product $\mathbf{A}\,(\mathbf{A}^e)^{-1}$ provides the element's *shape function* (see Exercise 15.1).

15.4.2 Strain-Displacement Matrix

Taking the displacement derivatives from eqs(15.6a,b) provides compatible, infinitesimal strains in the x, y plane:

$$\varepsilon_x = \partial u/\partial x = \alpha_2,\ \varepsilon_y = \partial v/\partial y = \alpha_6 \text{ and } \gamma_{xy} = \partial u/\partial y + \partial v/\partial x = \alpha_3 + \alpha_5 \qquad (15.11a\text{-}c)$$

Writing eqs(15.11a-c) in matrix form:

$$\begin{bmatrix} \varepsilon_x \\ \varepsilon_y \\ \gamma_{xy} \end{bmatrix} = \begin{bmatrix} 0 & 1 & 0 & 0 & 0 & 0 \\ 0 & 0 & 0 & 0 & 0 & 1 \\ 0 & 0 & 1 & 0 & 1 & 0 \end{bmatrix} \begin{bmatrix} \alpha_1 \\ \alpha_2 \\ \alpha_3 \\ \alpha_4 \\ \alpha_5 \\ \alpha_6 \end{bmatrix} \qquad \boldsymbol{\varepsilon} = \mathbf{C}\boldsymbol{\alpha} \qquad (15.11d,e)$$

The symbolic notation used in eq(15.11e), shows that the 6 × 3 matrix **C** connects the column matrices $\boldsymbol{\varepsilon} = [\, \varepsilon_x \;\; \varepsilon_y \;\; \gamma_{xy} \,]^{\mathrm{T}}$ and $\boldsymbol{\alpha} = [\alpha_1 \;\; \alpha_2 \;\; \alpha_3 \;\; \alpha_4 \;\; \alpha_5 \;\; \alpha_6 \,]^{\mathrm{T}}$. Substituting eq(15.9a) into eq(15.11e) gives strains $\boldsymbol{\varepsilon} = [\varepsilon_x \;\; \varepsilon_y \;\; \gamma_{xy} \,]^{\mathrm{T}}$ at any point (x, y) from the nodal point displacements $\boldsymbol{\delta}^e$:

$$\boldsymbol{\varepsilon} = \mathbf{C}\,(\mathbf{A}^e)^{-1}\boldsymbol{\delta}^e = \mathbf{B}\,\boldsymbol{\delta}^e \qquad (15.12a)$$

where the matrix multiplication $\mathbf{B} = \mathbf{C}\,(\mathbf{A}^e)^{-1}$ employs eqs(15.9b) and (15.11d):

$$\mathbf{B} = \frac{1}{2\Delta}\begin{bmatrix} y_2 - y_3 & 0 & y_3 - y_1 & 0 & y_1 - y_2 & 0 \\ 0 & x_3 - x_2 & 0 & x_1 - x_3 & 0 & x_2 - x_1 \\ x_3 - x_2 & y_2 - y_3 & x_1 - x_3 & y_3 - y_1 & x_2 - x_1 & y_1 - y_2 \end{bmatrix} \qquad (15.12b)$$

We see that **B** in eq(15.12b) appears only in terms of the nodal point co-ordinates, thereby revealing a ' constant strain ' triangle. This arises from the linear displacement functions assumed in eqs(15.6a,b), which leads to matrix **B** (and **C**) being independent of x and y.

15.4.3 Constitutive Relations

Plane stress-strain relations may be expressed in symbolic notation as, $\boldsymbol{\varepsilon} = \mathbf{P}\,\boldsymbol{\sigma}$ where **P** is a (3 × 3) elasticity matrix, connecting column matrices of stress $\boldsymbol{\sigma} = [\, \sigma_x \;\; \sigma_y \;\; \tau_{xy}]^{\mathrm{T}}$ and strain $\boldsymbol{\varepsilon} = [\varepsilon_x \;\; \varepsilon_y \;\; \gamma_{xy}\,]^{\mathrm{T}}$. Alternatively, since $\boldsymbol{\sigma} = \mathbf{P}^{-1}\,\boldsymbol{\varepsilon}$, we may write from eq(15.12a):

$$\boldsymbol{\sigma} = \mathbf{D}\,\boldsymbol{\varepsilon} = \mathbf{D}\,\mathbf{B}\,\boldsymbol{\delta}^e \qquad (15.13a)$$

where $\mathbf{D} = \mathbf{P}^{-1}$ is a 3 × 3 inverse elasticity matrix, whose plane stress components are:

$$\mathbf{D} = \frac{E}{(1 - \nu^2)}\begin{bmatrix} 1 & \nu & 0 \\ \nu & 1 & 0 \\ 0 & 0 & \tfrac{1}{2}(1 - \nu) \end{bmatrix} \qquad (15.13b)$$

and whose plane strain components are:

$$\mathbf{D} = \frac{E(1-\nu)}{(1+\nu)(1-2\nu)}\begin{bmatrix} 1 & \nu/(1-\nu) & 0 \\ \nu/(1-\nu) & 1 & 0 \\ 0 & 0 & \tfrac{1}{2}(1-2\nu)/(1-\nu) \end{bmatrix} \quad (15.13c)$$

As with the strain components, the three stress components σ_x, σ_y, τ_{xy} are also constant throughout a triangular element for given nodal displacement vector $\boldsymbol{\delta}^e$. When plotting a triangular element's stress and strain distributions their magnitudes would normally be assigned to the triangle's centroid or to the mid-position of one side.

15.4.4 Element Stiffness Matrix

Finally, the element stiffness matrix $\mathbf{K}^e$ may be derived. This is a square matrix of dimension 6×6, connecting the nodal point force and displacement vectors:

$$\mathbf{f}^e = \mathbf{K}^e\,\boldsymbol{\delta}^e \quad (15.14a)$$

The element's force vector $\mathbf{f}^e = [\,f_{x1}\;\; f_{y1}\;\; f_{x2}\;\; f_{y2}\;\; f_{x3}\;\; f_{y3}\,]^T$ may be shortened to

$$\mathbf{f}^e = [\,\mathbf{f}_1\;\; \mathbf{f}_2\;\; \mathbf{f}_3\,]^T \quad (15.14b)$$

in which the nodal forces are: $\mathbf{f}_1 = [\,f_{x1}\; f_{y1}\,]^T$, $\mathbf{f}_2 = [\,f_{x2}\; f_{y2}\,]^T$ and $\mathbf{f}_3 = [\,f_{x3}\; f_{y3}\,]^T$ (see Fig. 15.4b). Correspondingly, the element's displacement vector $\boldsymbol{\delta}^e$ is shortened:

$$\boldsymbol{\delta}^e = [\,\boldsymbol{\delta}_1\;\; \boldsymbol{\delta}_2\;\; \boldsymbol{\delta}_3\,]^T \quad (15.14c)$$

in which the nodal displacements are: $\boldsymbol{\delta}_1 = [u_1\;\; v_1]^T$, $\boldsymbol{\delta}_2 = [u_2\;\; v_2]^T$ and $\boldsymbol{\delta}_3 = [u_3\;\; v_3]^T$ (see Fig. 15.4a). Now $\boldsymbol{\delta}^e$ follows from inverting $\mathbf{K}^e$ in eq(15.14a):

$$\boldsymbol{\delta}^e = (\mathbf{K}^e)^{-1}\,\mathbf{f}^e \quad (15.14d)$$

The components K_{ij} of $\mathbf{K}^e$ in eq(15.14a) follow from nodal point co-ordinates (x_i, y_i, for i = 1, 2, 3) and the elastic constants E and ν, i.e. these appear within the components of matrices **D** and **B**, given in eqs(15.12b) and (15.13b,c). To find the relation between $\mathbf{K}^e$, **D** and **B** the principal of virtual work or stationary potential energy is used.

(a) *Virtual Work*

On substituting eqs(15.12a) and (15.13a) into eq(15.5b), the PVW gives

$$(\boldsymbol{\delta}^{ev})^T\,\mathbf{f}^e = \int (\mathbf{B}\,\boldsymbol{\delta}^{ev})^T\,(\mathbf{D}\,\mathbf{B}\,\boldsymbol{\delta}^e)\,\mathrm{d}V = (\boldsymbol{\delta}^{ev})^T\,(\,\mathbf{B}^T\,\mathbf{D}\,\mathbf{B}\,\boldsymbol{\delta}^e)\int \mathrm{d}V \quad (15.15a)$$

where neither **B** nor **D** depend upon x and y. Note that $(\mathbf{B}\boldsymbol{\delta}^{ev})^T = (\boldsymbol{\delta}^{ev})^T\,\mathbf{B}^T$, this allowing $(\boldsymbol{\delta}^{ev})^T$ to be cancelled:

$$\mathbf{f}^e = \mathbf{B}^T\,\mathbf{D}\,\mathbf{B}\,\boldsymbol{\delta}^e\,V \quad (15.15b)$$

where $\int \mathrm{d}V = V = At$, where A and t are, respectively, the area and thickness of the triangle. It follows from eqs(15.14a) and (15.15b) that this element's stiffness matrix becomes

$$\mathbf{K}^e = \mathbf{B}^T \mathbf{D}\,\mathbf{B}\,V \tag{15.15c}$$

where $\mathbf{K}^e$ is a (6 × 6) symmetrical matrix.

(b) *Stationary Potential Energy*

Alternatively, eq(15.15c) may be derived from substituting eqs(15.12a) and (15.13a) into eq(15.6c):

$$\frac{1}{2}\int_V \left[(\mathbf{D}\,\boldsymbol{\varepsilon})^T \frac{\partial}{\partial \boldsymbol{\delta}^e}(\mathbf{B}\,\boldsymbol{\delta}^e) + \boldsymbol{\varepsilon}^T \frac{\partial}{\partial \boldsymbol{\delta}^e}(\mathbf{D}\,\mathbf{B}\,\boldsymbol{\delta}^e)\right] \mathrm{d}V - (\mathbf{f}^e)^T = 0$$

$$\frac{1}{2}\int_V [(\mathbf{D}\,\boldsymbol{\varepsilon})^T \mathbf{B} + \boldsymbol{\varepsilon}^T \mathbf{D}\,\mathbf{B}]\, \mathrm{d}V - (\mathbf{f}^e)^T = 0$$

Using the relation $\mathbf{D} = \mathbf{D}^T$:

$$(\mathbf{f}^e)^T = \int_V \boldsymbol{\varepsilon}^T \mathbf{D}\,\mathbf{B}\,\mathrm{dV} = \int_V (\mathbf{B}\,\boldsymbol{\delta}^e)^T \mathbf{D}\,\mathbf{B}\,\mathrm{d}V$$

this can be written as

$$(\mathbf{f}^e)^T = (\boldsymbol{\delta}^e)^T (\mathbf{B}^T \mathbf{D}\,\mathbf{B})\,V$$

Taking the transpose of both sides:

$$\mathbf{f}^e = [(\boldsymbol{\delta}^e)^T (\mathbf{B}^T \mathbf{D}\,\mathbf{B})]^T V = (\mathbf{B}^T \mathbf{D}\,\mathbf{B})^T \boldsymbol{\delta}^e V = (\mathbf{D}\,\mathbf{B})^T \mathbf{B}\,\boldsymbol{\delta}^e V = (\mathbf{B}^T \mathbf{D}^T \mathbf{B}\,V)\,\boldsymbol{\delta}^e$$

when again, eq(15.15c) follows.

Either of these stiffness matrix derivations allow expressions for the components K_{ij} of $\mathbf{K}^e$ to be found from the matrix multiplication: $\mathbf{B}^T\mathbf{D}\,\mathbf{B}$. For example, the first of the diagonal and off-diagonal matrix components become

$$K_{11}^e = [t/(4\Delta)][D_{11}(y_2 - y_3)^2 + D_{33}(x_3 - x_2)^2]$$

$$K_{12}^e = K_{21}^e = [t/(4\Delta)][D_{12}(x_3 - x_2)(y_2 - y_3) + D_{33}(x_3 - x_2)(y_2 - y_3)]$$

where D_{11}, D_{33} and $D_{12} = D_{21}$ depend upon the plane problem, as defined in eqs(15.13b,c).

15.4.5 Overall Stiffness Matrix

Finally, the overall stiffness matrix **K** is assembled from the individual element stiffness matrices $\mathbf{K}^e$. The dimension of **K** may be established in advance. To illustrate the assembly, consider the four element, plane stress cantilever, shown in Fig. 15.5. With 6 nodes, each with 2 degrees of freedom, **K** becomes a 12 × 12 matrix.

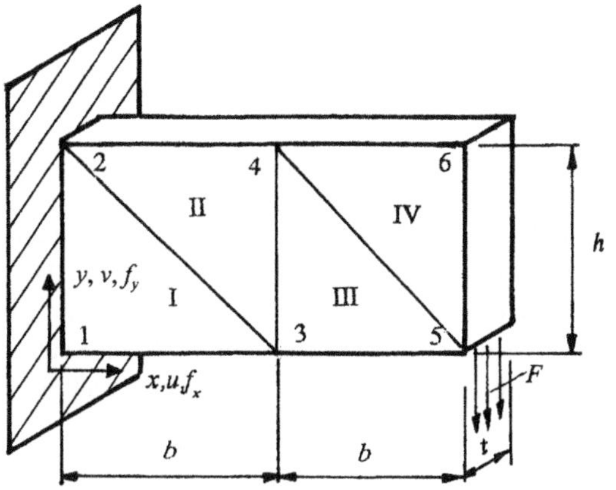

Figure 15.5 Plane stress cantilever composed of four triangular elements

The dimension of **K** can be reduced to (6 × 6) using sub-matrices $\mathbf{M}_{ij}^{e}$ with the shortened force and displacement notation adopted within eqs(15.14b,c). Writing the element number e = I, II, III and IV, the sub-matrices express **K** more compactly as:

$$\begin{bmatrix} \mathbf{f}_1 \\ \mathbf{f}_2 \\ \mathbf{f}_3 \\ \mathbf{f}_4 \\ \mathbf{f}_5 \\ \mathbf{f}_6 \end{bmatrix} = \begin{bmatrix} \mathbf{M}_{11}^{I} & \mathbf{M}_{12}^{I} & \mathbf{M}_{13}^{I} & \cdot & \cdot & \cdot \\ \mathbf{M}_{21}^{I} & \mathbf{M}_{22}^{I}+\mathbf{M}_{22}^{II} & \mathbf{M}_{23}^{I}+\mathbf{M}_{23}^{II} & \mathbf{M}_{24}^{II} & \cdot & \cdot \\ \mathbf{M}_{31}^{I} & \mathbf{M}_{32}^{I}+\mathbf{M}_{32}^{II} & \mathbf{M}_{33}^{I}+\mathbf{M}_{33}^{II}+\mathbf{M}_{33}^{III} & \mathbf{M}_{34}^{II}+\mathbf{M}_{34}^{III} & \mathbf{M}_{35}^{III} & \cdot \\ \cdot & \mathbf{M}_{42}^{II} & \mathbf{M}_{43}^{II}+\mathbf{M}_{43}^{III} & \mathbf{M}_{44}^{II}+\mathbf{M}_{44}^{III}+\mathbf{M}_{44}^{IV} & \mathbf{M}_{45}^{III}+\mathbf{M}_{45}^{IV} & \mathbf{M}_{46}^{IV} \\ \cdot & \cdot & \mathbf{M}_{53}^{III} & \mathbf{M}_{54}^{III}+\mathbf{M}_{54}^{IV} & \mathbf{M}_{55}^{III}+\mathbf{M}_{55}^{IV} & \mathbf{M}_{56}^{IV} \\ \cdot & \cdot & \cdot & \mathbf{M}_{64}^{IV} & \mathbf{M}_{65}^{IV} & \mathbf{M}_{66}^{IV} \end{bmatrix} \begin{bmatrix} \boldsymbol{\delta}_1 \\ \boldsymbol{\delta}_2 \\ \boldsymbol{\delta}_3 \\ \boldsymbol{\delta}_4 \\ \boldsymbol{\delta}_5 \\ \boldsymbol{\delta}_6 \end{bmatrix} \qquad (15.16)$$

Here individual stiffness components appear within the sub-matrix addition. For example, the diagonal element $\mathbf{M}_{33}^{I} + \mathbf{M}_{33}^{II} + \mathbf{M}_{33}^{III}$ in eq(15.16) represents

$$\mathbf{M}_{33}^{I} + \mathbf{M}_{33}^{II} + \mathbf{M}_{33}^{III} = \begin{bmatrix} K_{33}^{I} & K_{34}^{I} \\ K_{43}^{I} & K_{44}^{I} \end{bmatrix} + \begin{bmatrix} K_{33}^{II} & K_{34}^{II} \\ K_{43}^{II} & K_{44}^{II} \end{bmatrix} + \begin{bmatrix} K_{11}^{III} & K_{12}^{III} \\ K_{21}^{III} & K_{22}^{III} \end{bmatrix}$$

The final step is to ensure that nodes 1 and 2 provide zero displacement at supports, i.e. at the cantilever fixing points $u_1 = v_1 = u_2 = v_2 = 0$. Numerically, this is achieved by multiplying the appropriate stiffness components by a large number, such that upon matrix inversion the deflection is eliminated at the required position.

Following this assembly of elements, the resulting stress, strain and displacement relations are applied in the same sense as the element's node numbering. The nodal displacement vector can then be found from the following inversion process:

$$\mathbf{f} = \mathbf{K}\boldsymbol{\delta} \;\Rightarrow\; \boldsymbol{\delta} = \mathbf{K}^{-1}\mathbf{f} \qquad (15.17a,b)$$

Thus, knowing $\boldsymbol{\delta}^{e}$, the three, constant strain and stress components are referred to the centroid of each element. They follow from the matrix multiplications:

$$\boldsymbol{\varepsilon}^e = \mathbf{B}^e\,\boldsymbol{\delta}^e \tag{15.18a}$$

$$\boldsymbol{\sigma}^e = \mathbf{D}\,\mathbf{B}^e\,\boldsymbol{\delta}^e = \mathbf{H}^e\,\boldsymbol{\delta}^e \tag{15.18b}$$

In eqs (15.18a,b) $\mathbf{B}^e = \mathbf{B}$ (from eq 15.12b) as it depends only upon the nodal point co-ordinates and $\mathbf{H}^e = \mathbf{D}\,\mathbf{B}^e$ (from eq 15.13a) depends upon the elastic constants, according to the plane condition. The assembly of an overall stiffness matrix for several hundred elements requires a computer. With efficient node-numbering, we see from the cantilever example that the symmetrical matrix **K** is formed with its non-zero elements lying within a narrow diagonal band. Both the nodal ordering and the partitioning of **K** into sub-matrices, as indicated, serve to the reduce the computer memory required [7]. Such features would be incorporated within modern automated meshing practice.

15.5 Elastic-Plastic Stiffness Matrix

15.5.1 Vector Representation

The matrix form of the Prandtl-Reuss eq(4.3b) appears as:

$$\mathrm{d}\mathbf{E} = \mathrm{d}\lambda\,\mathbf{T}' + \frac{(1-2\nu)}{3E}\,\mathbf{I}\,\mathrm{tr}\,(\mathrm{d}\mathbf{T}) + \frac{1}{2G}\,\mathrm{d}\mathbf{T}' \tag{15.19}$$

where **E** and $\mathbf{T}'$ are, respectively, (3 × 3) matrices of total strain and deviatoric stress and **I** is a unit matrix. Alternatively, column matrices (vectors) of stress and strain are often more convenient to use within the application of finite elements to incremental plasticity. Correspondingly, in a vector representation of eq(15.19), column matrices of the deviatoric stress and plastic strain increment tensor components are used:

$$\boldsymbol{\sigma}' = [\sigma_{11}'\ \ \sigma_{22}'\ \ \sigma_{33}'\ \ \sqrt{2}\,\sigma_{12}'\ \ \sqrt{2}\,\sigma_{23}'\ \ \sqrt{2}\sigma_{13}'\,]^{\mathrm{T}} \tag{15.20a}$$

$$\mathrm{d}\boldsymbol{\varepsilon}^P = [\mathrm{d}\varepsilon_{11}{}^P\ \ \mathrm{d}\varepsilon_{22}{}^P\ \ \mathrm{d}\varepsilon_{33}{}^P\ \ \sqrt{2}\mathrm{d}\varepsilon_{12}{}^P\ \ \sqrt{2}\mathrm{d}\varepsilon_{23}{}^P\ \ \sqrt{2}\mathrm{d}\varepsilon_{13}{}^P]^{\mathrm{T}} \tag{15.20b}$$

where $\mathrm{d}\varepsilon_{12}{}^P$, $\mathrm{d}\varepsilon_{23}{}^P$ and $\mathrm{d}\varepsilon_{13}{}^P$ tensor shear strains. If engineering shear strain components are preferred, then re-writing eq(15.20b) with: $\mathrm{d}\varepsilon_{12}{}^P = \mathrm{d}\gamma_{12}{}^P/2$, $\mathrm{d}\varepsilon_{13}{}^P = \mathrm{d}\gamma_{13}{}^P/2$ etc, gives:

$$\mathrm{d}\boldsymbol{\varepsilon}^P = [\mathrm{d}\varepsilon_{11}{}^P\ \ \mathrm{d}\varepsilon_{22}{}^P\ \ \mathrm{d}\varepsilon_{33}{}^P\ \ \mathrm{d}\gamma_{12}{}^P/\sqrt{2}\ \ \mathrm{d}\gamma_{23}{}^P/\sqrt{2}\ \ \mathrm{d}\gamma_{13}{}^P/\sqrt{2}\,]^{\mathrm{T}} \tag{15.20c}$$

Others [8] make use of column matrices of engineering stress and strain components, as in eq(15.13), but these do not always convert correctly into the quantities we have met earlier in our tensor notation of plasticity. For example, it follows from eqs(15.20a,b) that the corresponding expressions, given in Sections 9.2 and 9.9, for equivalent stress, equivalent plastic strain increment and the plastic work increment, appear correctly as

$$\bar{\sigma}^2 = \frac{3}{2}\sigma_{ij}'\sigma_{ij}' = \frac{3}{2}\boldsymbol{\sigma}'^{\mathrm{T}}\boldsymbol{\sigma}' \tag{15.21a}$$

$$\left(\mathrm{d}\bar{\varepsilon}^P\right)^2 = \frac{2}{3}\mathrm{d}\varepsilon_{ij}^P\,\mathrm{d}\varepsilon_{ij}^P = \frac{2}{3}\left(\mathrm{d}\boldsymbol{\varepsilon}^P\right)^{\mathrm{T}}\mathrm{d}\boldsymbol{\varepsilon}^P \tag{15.21b}$$

$$dW^P = \sigma_{ij}\, d\varepsilon_{ij}^P = \boldsymbol{\sigma}^T d\boldsymbol{\varepsilon}^P = (d\boldsymbol{\varepsilon}^P)^T \boldsymbol{\sigma}$$

$$dW^P = \sigma_{ij}'\, d\varepsilon_{ij}^P = \boldsymbol{\sigma}'^T d\boldsymbol{\varepsilon}^P = (d\boldsymbol{\varepsilon}^P)^T \boldsymbol{\sigma}' \qquad (15.21c,d)$$

In particular, writing eq(15.21a) in full, the matrix multiplication gives the correct scalar quantity for equivalent stress:

$$\bar{\sigma}^2 = \frac{3}{2}\begin{bmatrix}\sigma_{11}' & \sigma_{22}' & \sigma_{33}' & \sqrt{2}\sigma_{12}' & \sqrt{2}\sigma_{23}' & \sqrt{2}\sigma_{13}'\end{bmatrix}\begin{bmatrix}\sigma_{11}' \\ \sigma_{22}' \\ \sigma_{33}' \\ \sqrt{2}\sigma_{12}' \\ \sqrt{2}\sigma_{23}' \\ \sqrt{2}\sigma_{13}'\end{bmatrix} \qquad (15.22a)$$

$$\bar{\sigma}^2 = \frac{3}{2}\left\{(\sigma_{11}')^2 + (\sigma_{22}')^2 + (\sigma_{33}')^2 + 2[(\sigma_{12}')^2 + (\sigma_{13}')^2 + (\sigma_{23}')^2]\right\} \qquad (15.22b)$$

15.5.2 Deviatoric Stress

The conversion from absolute stress to deviatoric stress in this notation will appear as:

$$\begin{bmatrix}\sigma_{11}' \\ \sigma_{22}' \\ \sigma_{33}' \\ \sqrt{2}\sigma_{12}' \\ \sqrt{2}\sigma_{23}' \\ \sqrt{2}\sigma_{13}'\end{bmatrix} = \begin{bmatrix}1&0&0&0&0&0\\0&1&0&0&0&0\\0&0&1&0&0&0\\0&0&0&1&0&0\\0&0&0&0&1&0\\0&0&0&0&0&1\end{bmatrix}\begin{bmatrix}\sigma_{11} \\ \sigma_{22} \\ \sigma_{33} \\ \sqrt{2}\sigma_{12} \\ \sqrt{2}\sigma_{23} \\ \sqrt{2}\sigma_{13}\end{bmatrix} - \frac{1}{3}\begin{bmatrix}1\\1\\1\\0\\0\\0\end{bmatrix}\begin{bmatrix}1 & 1 & 1 & 0 & 0 & 0\end{bmatrix}\begin{bmatrix}\sigma_{11} \\ \sigma_{22} \\ \sigma_{33} \\ \sqrt{2}\sigma_{12} \\ \sqrt{2}\sigma_{23} \\ \sqrt{2}\sigma_{13}\end{bmatrix} \qquad (15.23a)$$

Matrix multiplication in eq(15.23a) gives: $\sigma_{11}' = \sigma_{11} - ⅓(\sigma_{11} + \sigma_{22} + \sigma_{33})$ etc and $\sigma_{12}' = \sigma_{12}$ etc, consistent with our usual tensor notation $\sigma_{ij}' = \sigma_{ij} - ⅓\delta_{ij}\,\sigma_{kk}$. It is seen that eq(15.23a) may be re-written as:

$$\boldsymbol{\sigma}' = \mathbf{I}\,\boldsymbol{\sigma} - ⅓\,\mathbf{a}\,\mathbf{a}^T\boldsymbol{\sigma} = [\mathbf{I} - ⅓\,\mathbf{a}\,\mathbf{a}^T]\,\boldsymbol{\sigma} = \boldsymbol{\mu}\,\boldsymbol{\sigma} \qquad (15.23b)$$

where $\mathbf{a} = [1\ 1\ 1\ 0\ 0\ 0\,]^T$. Also, the matrix $\boldsymbol{\mu}$ follows from eqs(15.23a,b) as:

$$\boldsymbol{\mu} = \mathbf{I} - ⅓\,\mathbf{a}\,\mathbf{a}^T \qquad (15.24a)$$

$$\boldsymbol{\mu} = \begin{bmatrix} 2/3 & -1/3 & -1/3 & 0 & 0 & 0 \\ -1/3 & 2/3 & -1/3 & 0 & 0 & 0 \\ -1/3 & -1/3 & 2/3 & 0 & 0 & 0 \\ 0 & 0 & 0 & 1 & 0 & 0 \\ 0 & 0 & 0 & 0 & 1 & 0 \\ 0 & 0 & 0 & 0 & 0 & 1 \end{bmatrix} \tag{15.24b}$$

Since $\boldsymbol{\mu} = \boldsymbol{\mu}^{\mathrm{T}}$, the following relationships hold:

$$\boldsymbol{\mu\mu} = \boldsymbol{\mu\mu}^{\mathrm{T}} = \boldsymbol{\mu}^{\mathrm{T}}\boldsymbol{\mu} = \boldsymbol{\mu}^{\mathrm{T}}\boldsymbol{\mu}^{\mathrm{T}} \tag{15.25a}$$

The product $\boldsymbol{\mu\mu}$ also appears from eq(15.24a) as:

$$\begin{aligned} \boldsymbol{\mu\mu} &= [\mathbf{I} - \tfrac{1}{3}\,\mathbf{a}\,\mathbf{a}^{\mathrm{T}}]\,[\mathbf{I} - \tfrac{1}{3}\,\mathbf{a}\,\mathbf{a}^{\mathrm{T}}] \\ &= \mathbf{I}\,[\mathbf{I} - \tfrac{1}{3}\,\mathbf{a}\,\mathbf{a}^{\mathrm{T}}] - \tfrac{1}{3}\,\mathbf{a}\,\mathbf{a}^{\mathrm{T}}[\mathbf{I} - \tfrac{1}{3}\,\mathbf{a}\,\mathbf{a}^{\mathrm{T}}] \\ &= [\mathbf{I} - \tfrac{1}{3}\,\mathbf{a}\,\mathbf{a}^{\mathrm{T}}] - \tfrac{1}{3}\,\mathbf{a}\,\mathbf{a}^{\mathrm{T}} + \tfrac{1}{3}\,\mathbf{a}\,(\tfrac{1}{3}\,\mathbf{a}^{\mathrm{T}}\mathbf{a})\,\mathbf{a}^{\mathrm{T}} \\ &= \mathbf{I} - \tfrac{1}{3}\,\mathbf{a}\,\mathbf{a}^{\mathrm{T}} = \boldsymbol{\mu} \end{aligned} \tag{15.25b}$$

which employs the fact that $\mathbf{a}^{\mathrm{T}}\mathbf{a} = 3$.

15.5.3 Constitutive Relations

In converting the Prandtl-Reuss theory to this vector notation, use is made of the fact that the sum of the final two terms in eq(15.19) make up the elastic increment of strain. More commonly, this sum expresses the incremental, elastic constitutive relations (see p. 327):

$$\begin{aligned} \mathrm{d}\varepsilon_{11}^{\,e} &= (1/E)[\mathrm{d}\sigma_{11} + \nu(\mathrm{d}\sigma_{22} + \mathrm{d}\sigma_{33})] \\ \mathrm{d}\varepsilon_{22}^{\,e} &= (1/E)[\mathrm{d}\sigma_{22} + \nu(\mathrm{d}\sigma_{11} + \mathrm{d}\sigma_{33})] \\ \mathrm{d}\varepsilon_{33}^{\,e} &= (1/E)[\mathrm{d}\sigma_{33} + \nu(\mathrm{d}\sigma_{11} + \mathrm{d}\sigma_{22})] \end{aligned}$$

where E, G and ν are the elastic constants. The latter appear within an elastic flexibility matrix, linking the column matrices of strain and stress increments, as follows:

$$\begin{bmatrix} \mathrm{d}\varepsilon_{11}^{e} \\ \mathrm{d}\varepsilon_{22}^{e} \\ \mathrm{d}\varepsilon_{33}^{e} \\ \sqrt{2}\,\mathrm{d}\varepsilon_{12}^{e} \\ \sqrt{2}\,\mathrm{d}\varepsilon_{23}^{e} \\ \sqrt{2}\,\mathrm{d}\varepsilon_{13}^{e} \end{bmatrix} = \begin{bmatrix} 1/E & -\nu & -\nu & 0 & 0 & 0 \\ -\nu & 1/E & -\nu & 0 & 0 & 0 \\ -\nu & -\nu & 1/E & 0 & 0 & 0 \\ 0 & 0 & 0 & 1/(2G) & 0 & 0 \\ 0 & 0 & 0 & 0 & 1/(2G) & 0 \\ 0 & 0 & 0 & 0 & 0 & 1/(2G) \end{bmatrix} \begin{bmatrix} \mathrm{d}\sigma_{11} \\ \mathrm{d}\sigma_{22} \\ \mathrm{d}\sigma_{33} \\ \sqrt{2}\,\mathrm{d}\sigma_{12} \\ \sqrt{2}\,\mathrm{d}\sigma_{23} \\ \sqrt{2}\,\mathrm{d}\sigma_{13} \end{bmatrix} \tag{15.26a}$$

which is simply written as

$$\mathbf{d\varepsilon}^e = \mathbf{P}^e \mathbf{d\sigma} \tag{15.26b}$$

The incremental plastic component of strain is written from eqs(10.14) and (10.18) as

$$d\varepsilon_{ij}^P = \frac{3\bar{\sigma}\, d\bar{\sigma}\, \sigma_{ij}'}{2E^P \bar{\sigma}^2} \tag{15.27a}$$

where $E^P = d\bar{\sigma}/d\bar{\varepsilon}^P$ has now replaced H' as the plastic modulus, i.e. the gradient to the flow curve, where the latter has had the elastic components of strain removed. If we employ a Ludwik or a Swift law to describe the plastic flow curve from a tension test, then E^P becomes the derivative of eq(9.43a) or (9.44a). Alternatively, using total strain, the gradient E^t of the tangent to a uni-axial stress-strain curve may be written as

$$E^t = \frac{d\bar{\sigma}}{d\varepsilon^t} = \frac{d\bar{\sigma}}{d\bar{\varepsilon}^P + d\varepsilon^e} \tag{15.27b}$$

Substituting $d\bar{\varepsilon}^P = d\bar{\sigma}/E^P$ and $d\varepsilon^e = d\bar{\sigma}/E$ into eq(15.27b) provides a relationship between the elastic, plastic and the tangent moduli:

$$E^P = \frac{E^t}{1 - E^t/E} \tag{15.27c}$$

The product $\bar{\sigma}d\bar{\sigma}$ in eq(15.27a) is found from eq(15.21) as

$$\bar{\sigma}\, d\bar{\sigma} = \frac{3}{2}\sigma_{ij}'\, d\sigma_{ij}' = \frac{3}{2}\boldsymbol{\sigma}'^{\mathrm{T}} \mathbf{d\sigma}' \tag{15.28a}$$

The matrix form of eq(15.27a) follows from eqs(15.20), (15.21a) and (15.28a):

$$\mathbf{d\varepsilon}^P = \frac{3\boldsymbol{\sigma}'^{\mathrm{T}} \mathbf{d\sigma}' \boldsymbol{\sigma}'}{2E^P(\boldsymbol{\sigma}'^{\mathrm{T}} \boldsymbol{\sigma}')} \tag{15.28b}$$

Since $\boldsymbol{\sigma}'^{\mathrm{T}}\mathbf{d\sigma}'$ is scalar, the product term in the numerator of eq(15.28b) may be written as

$$(\boldsymbol{\sigma}'^{\mathrm{T}} \mathbf{d\sigma}')\boldsymbol{\sigma}' = \boldsymbol{\sigma}'(\boldsymbol{\sigma}'^{\mathrm{T}} \mathbf{d\sigma}') = (\boldsymbol{\sigma}'\boldsymbol{\sigma}'^{\mathrm{T}})\mathbf{d\sigma}'$$

Substituting from eq(15.23b) gives the absolute stress form:

$$(\mu\sigma)\,(\mu\sigma)^{\mathrm{T}}\mathrm{d}\,(\mu\sigma) = \mu\sigma\sigma^{\mathrm{T}}(\mu^{\mathrm{T}}\mu)\mathrm{d}\sigma = \mu\sigma\sigma^{\mathrm{T}}\mu\mathrm{d}\sigma = (\mu\sigma)(\mu\sigma)^{\mathrm{T}}\mathrm{d}\sigma = \sigma'\sigma'^{\mathrm{T}}\mathrm{d}\sigma$$

from which eq(15.27c) is written as

$$\mathbf{d\varepsilon}^P = \frac{3\boldsymbol{\sigma}'\boldsymbol{\sigma}'^{\mathrm{T}} \mathbf{d\sigma}}{2E^P\boldsymbol{\sigma}'^{\mathrm{T}}\boldsymbol{\sigma}'} = \mathbf{P}^P \mathbf{d\sigma} \tag{15.28c}$$

Equation (15.28c) may be combined with eq(15.26b) to give the total incremental strain from Prandtl-Reuss theory, in a simplified matrix form:

$$\mathbf{d}\boldsymbol{\varepsilon}^{t} = \left(\mathbf{P}^{e} + \mathbf{P}^{P}\right)\mathbf{d}\boldsymbol{\sigma} \tag{15.28d}$$

The symmetrical plasticity matrix $\mathbf{P}^P$ in eq(15.28c) is formed from a matrix product $\boldsymbol{\sigma}'\boldsymbol{\sigma}'^{\mathrm{T}}$:

$$\boldsymbol{\sigma}'\boldsymbol{\sigma}'^{\mathrm{T}} = \begin{bmatrix} \sigma_{11}'^{2} & \sigma_{11}'\sigma_{22}' & \sigma_{11}'\sigma_{33}' & \sqrt{2}\sigma_{11}'\sigma_{12}' & \sqrt{2}\sigma_{11}'\sigma_{13}' & \sqrt{2}\sigma_{11}'\sigma_{23}' \\ \sigma_{22}'\sigma_{11}' & \sigma_{22}'^{2} & \sigma_{22}'\sigma_{33}' & \sqrt{2}\sigma_{22}'\sigma_{12}' & \sqrt{2}\sigma_{22}'\sigma_{13}' & \sqrt{2}\sigma_{22}'\sigma_{23}' \\ \sigma_{33}'\sigma_{11}' & \sigma_{33}'\sigma_{22}' & \sigma_{33}'^{2} & \sqrt{2}\sigma_{33}'\sigma_{12}' & \sqrt{2}\sigma_{33}'\sigma_{13}' & \sqrt{2}\sigma_{33}'\sigma_{23}' \\ \sqrt{2}\sigma_{12}'\sigma_{11}' & \sqrt{2}\sigma_{12}'\sigma_{22}' & \sqrt{2}\sigma_{12}'\sigma_{33}' & 2\sigma_{12}'^{2} & 2\sigma_{12}'\sigma_{13}' & 2\sigma_{12}'\sigma_{23}' \\ \sqrt{2}\sigma_{13}'\sigma_{11}' & \sqrt{2}\sigma_{13}'\sigma_{22}' & \sqrt{2}\sigma_{13}'\sigma_{33}' & 2\sigma_{13}'\sigma_{12}' & 2\sigma_{13}'^{2} & 2\sigma_{13}'\sigma_{23}' \\ \sqrt{2}\sigma_{23}'\sigma_{11}' & \sqrt{2}\sigma_{23}'\sigma_{22}' & \sqrt{2}\sigma_{23}'\sigma_{33}' & 2\sigma_{23}'\sigma_{12}' & 2\sigma_{23}'\sigma_{13}' & 2\sigma_{23}'^{2} \end{bmatrix} \tag{15.29a}$$

Alternatively, an absolute stress form for $\mathbf{P}^P$ follows from eq(15.28c) as

$$P^{P} = \frac{3\boldsymbol{\mu}\boldsymbol{\sigma}(\boldsymbol{\mu}\boldsymbol{\sigma})^{\mathrm{T}}}{2E^{P}(\boldsymbol{\mu}\boldsymbol{\sigma})^{\mathrm{T}}(\boldsymbol{\mu}\boldsymbol{\sigma})} = \frac{3(\boldsymbol{\mu}\boldsymbol{\sigma}\boldsymbol{\sigma}^{\mathrm{T}}\boldsymbol{\mu})}{2E^{P}\boldsymbol{\sigma}^{\mathrm{T}}\boldsymbol{\mu}\boldsymbol{\sigma}} \tag{15.29b}$$

Matrix multiplication within eq(15.28c) leads to the plastic strain increments, typically:

$$d\varepsilon_{11}^{P} = \frac{\begin{gathered}[(2\sigma_{11} - \sigma_{22} - \sigma_{33})d\sigma_{11} + (2\sigma_{22} - \sigma_{11} - \sigma_{33})d\sigma_{22} + (2\sigma_{33} - \sigma_{11} - \sigma_{22})d\sigma_{33} + \\ 6(\sigma_{12}d\sigma_{12} + \sigma_{13}d\sigma_{13} + \sigma_{23}d\sigma_{23})]\left[\sigma_{11} - \tfrac{1}{2}(\sigma_{22} + \sigma_{33})\right]\end{gathered}}{E^{P}[(\sigma_{11} - \sigma_{22})^{2} + (\sigma_{22} - \sigma_{33})^{2} + (\sigma_{11} - \sigma_{33})^{2} + 6(\sigma_{12}^{2} + \sigma_{13}^{2} + \sigma_{23}^{2})]}$$

$$\sqrt{2}\,d\varepsilon_{12}^{P} = \frac{\begin{gathered}[(2\sigma_{11} - \sigma_{22} - \sigma_{33})d\sigma_{11} + (2\sigma_{22} - \sigma_{11} - \sigma_{33})d\sigma_{22} + (2\sigma_{33} - \sigma_{11} - \sigma_{22})d\sigma_{33} + \\ 6(\sigma_{12}d\sigma_{12} + \sigma_{13}d\sigma_{13} + \sigma_{23}d\sigma_{23})](3/\sqrt{2})\sigma_{12}\end{gathered}}{E^{P}[(\sigma_{11} - \sigma_{22})^{2} + (\sigma_{22} - \sigma_{33})^{2} + (\sigma_{11} - \sigma_{33})^{2} + 6(\sigma_{12}^{2} + \sigma_{13}^{2} + \sigma_{23}^{2})]}$$

to which are added the incremental elastic strain components from matrix $\mathbf{P}^e$, in eq(15.26a). Setting $E^P = d\bar{\sigma}/d\bar{\varepsilon}^P$ and substituting from eqs(15.22b) and (15.28a), the incremental plastic strains, according to the Levy-Mises theory, are recovered (see Section 10.3.1):

$$d\varepsilon_{11}^{P} = \frac{d\bar{\sigma}}{E^{P}\bar{\sigma}}\left[\sigma_{11} - \frac{1}{2}\left(\sigma_{22} + \sigma_{33}\right)\right] \quad \text{etc}$$

$$d\varepsilon_{12}^{P} = \frac{d\bar{\sigma}}{E^{P}\bar{\sigma}}\left(\frac{3}{2}\sigma_{12}\right) \quad \text{etc}$$

which confirm the particular combination of eqs(10.18a) and (10.19a).

15.5.4 Virtual Work

If eq(15.28d) is now inverted to give $d\sigma = \mathbf{D}^{ep}\, d\varepsilon^t$, where $\mathbf{D}^{ep} = (\mathbf{P}^e + \mathbf{P}^P)^{-1}$, then the PVW may be applied in its incremental form to provide a tangential stiffness matrix for elastic-plastic deformation. Omiting superscript e (for element) the principle becomes

$$(\Delta\boldsymbol{\delta}^{\,v})^{\mathrm{T}}\,\Delta\mathbf{f} - \int_V (\Delta\boldsymbol{\varepsilon}^{\,v})^{\mathrm{T}}\,\Delta\boldsymbol{\sigma}\,\mathrm{d}V = 0 \qquad (15.30a)$$

where

$$\Delta\boldsymbol{\sigma} = (\mathbf{P}^e + \mathbf{P}^P)^{-1}\Delta\boldsymbol{\varepsilon}^t = \mathbf{D}^{eP}\Delta\boldsymbol{\varepsilon}^t \qquad (15.30b)$$

giving

$$\Delta\mathbf{f} = \mathbf{K}\,\Delta\boldsymbol{\delta} \qquad (15.30c)$$

in which **K** is a variable stiffness, corresponding to the variable E^P. Combining eqs(15.30a,b) with eq(15.12a):

$$(\Delta\boldsymbol{\delta}^{\,v})^{\mathrm{T}}\,\Delta\mathbf{f} - \int_V (\mathbf{B}\,\Delta\boldsymbol{\delta}^{\,v})^{\mathrm{T}}\,(\mathbf{D}^{eP}\,\mathbf{B}\,\Delta\boldsymbol{\delta})\,\mathrm{d}V = 0$$

Cancelling $(\Delta\boldsymbol{\delta}^{\,v})^{\mathrm{T}}$ and removing $\Delta\boldsymbol{\delta}$ from the integrand:

$$\Delta\mathbf{f} = \left\{\int_V \mathbf{B}^{\mathrm{T}}\,\mathbf{D}^{eP}\,\mathbf{B}\,\mathrm{d}V\right\}\Delta\boldsymbol{\delta} \qquad (15.31a)$$

Comparing eq(15.31a) with eq(15.30c) provides the tangent stiffness matrix:

$$\mathbf{K} = \int_V \mathbf{B}^{\mathrm{T}}\,\mathbf{D}^{eP}\,\mathbf{B}\,\mathrm{d}V \qquad (15.31b)$$

In general, for most plane and axi-symmetric elements, the co-ordinates x, y appear within matrix **B** and so the matrix product in eq(15.31b) must be integrated over the volume. The fact that the elements K_{ij} must be updated continually to form a tangent stiffness matrix for each increment of plastic strain, is the essential difference to the constant K_{ij} components for the linear-elastic matrix, as previously described.

15.5.5 FE Codes

The analysis above shows that the elastic-plastic and elastic FE program structures are essentially the same, but solving $\Delta\boldsymbol{\delta} = \mathbf{K}^{-1}\,\Delta\mathbf{f}$, with each non-linear strain increment, is equivalent to a complete linear analysis, repeatedly applied. Consequently, the cost of running an elastic-plastic FE analysis will be greater, given that the stiffness matrix needs to be inverted and convergence criteria applied to each new increment of strain. Such repetitions, administered within static codes (as described above), tend to be slow, though alternative, faster FE simulations of sheet metal forming are now available commercially. For example, the matrix inversion is avoided within an incremental, explicit dynamic code. In offsetting the need to invert **K**, the more recent dynamic codes reduce storage and avoid the often slow, convergent-dependent solutions found with implicit static codes [9]. Alternatively, incremental procedures would be avoided altogether when using an approximate, one-step FE method, based upon Hencky's total strain, constitutive relations (see p. 95). The user needs to choose between these codes when using brick or shell elements to simulate the required shape. Mesh refinement in regions of high strain is now automated and the option to choose between alternative yield criteria for orthotropic sheet metals obviates the need to write specific user subroutines. Though developed for plasticity under dynamic impacts, the dynamic codes are also efficient for analyses of non-linear quasi-static processes, including sheet metal forming, extrusion and rolling. The explicit code requires an increase to the real speeds of each process followed by a check that the outputs have remained unaffected, i.e. they remain within the steady state solution pertaining to a range of speeds. In particular, the accuracy of spring-back predictions in sheet metal is especially important to monitor, where it may be necessary to revert to the static code [10]. A rigid punch/die combination is usually assumed, and with a geometry that is often axi-symmetric, the meshing need only be applied to one quadrant of an initially flat sheet. This imposes certain boundary conditions that must be met, i.e. the constraining of points that do not displace during the forming operation. Where the sheet makes contact with its punch/die, a friction coefficient is required to attest the interfacial shear arising from forming forces. In the region of metal-to-metal contact, the punch/die geometry is also required. The latter are normally taken to be rigid as the sheet deforms in an elastic-plastic manner between them. Plastic incompressibility limits the element choice. Solid elements will require fewer nodes to enforce volume constancy; say at each corner node when simulating forging with brick elements. Shell elements, which define the full thickness, require fewer integration points. Inputs include the sheet's elastic constants and its flow curve. The latter may be expressed in co-ordinates of true stress and natural strain to allow for a piecewise linear approximation, or an exact fit to this curve. Codes which admit anisotropy require the r and n values, the strength coefficient and the pre-strain, as in Swift's law, eq(9.44a,b).

There are essentially three incremental FE formulations for non-linear plasticity, depending upon the severity of deformation [4]: (i) small displacements and small strains, (ii) large displacements with moderate strains and (iii) large displacements with finite strains. Within (i), the elastic and plastic strains are of similar magnitudes and analysis used is that described above. In (ii) and (iii), more generally, the appropriate stress and strain definitions replace engineering measures. In (ii) the true stress and natural strain are adequate to provide for the deformation found, typically, in the plane of sheets formed with large displacements. Here, the natural strain is continually converted to engineering strain when applying the FLD. The particular finite measures chosen for combination (iii) are the second Piola-Kirchoff stress and the Lagrangian (Green's) strain, which were defined in Chapter 1. Other elements such as the yield criterion and the hardening rule remain the same. Outputs from FE are usually displayed as contours of equivalent stress and strain and the displacements found provide an image of the distortion. In sheet metal forming, the safeness of forming zones is assessed within coloured contour maps and by the superimposition of strain states and strain signatures upon the FLD, some of which are now described.

15.6 FE Simulations

In the three applications of FE that follow we shall consider stretching of thin sheet provided by hydraulic bulging, ball indentation and the pressing of a panel into the shape of a flanged box. Three different, available FE codes are applied to provide the simulations: the first two are incremental, implicit codes, as described above, and the third is a rapid, one-step solution used for the die try-out stage of new product designs. The elements employed are respectively: shell, 2-layered brick and single-layered rectangles with 8-nodes. In every simulation of sheet metal forming a criterion of fracture is essential to limit the strains in the sheet to safe working levels. Invariably, the forming limit diagram (FLD) is employed for this purpose (see Fig. 12.25). The simulated in-plane strains are placed at their appropriate positions within the FLD's principal, engineering strain axes. This locates the proximity of the forming strains to each branch of the FLD. The branches are usually taken as the lower lines in a band of scatter to provide the required margin of safety all for working strain levels. Points for strain combinations lying beneath the FLD are safe while points found to lie on or above the FLD are unsafe. The FE simulation thus locates the most likely site of failure, providing the critical load and displacement correspondingly. The many different strain combinations that arise during a forming operation produce a spread in points across the two quadrants of strain that continually changes. Consequently, the monitoring of strains in this way is integrated within the simulation, given that the FLD for the sheet material is available. Normally, a FLD is found from laboratory tests to ensure representative strain paths.

15.6.1 Bulging of Sheet Metal

The bulge test has the advantage of eliminating contact friction; where oil under pressure is applied to the underside of a clamped and sealed disc, forcing it to bulge into a top die (see Fig. 9.18). Consequently, bulging in a circular die may be used as a convenient check upon FE simulations of stretch forming processes. Given that an analytical bulge theory is available for both isotropic and anisotropic sheet (see Sections 9.7 and 12.6), the theoretical relation between the pressure and pole displacement provides the required check upon an FE prediction. The theoretical plot (see Fig. 15.6a) between these two parameters shows a rising curve that reaches a pressure maximum at the inception of a diffuse strain, pole instability. That such a behaviour is confirmed by experiment and reproduced quite well by FE indicates that other outputs from the simulation can be believed.

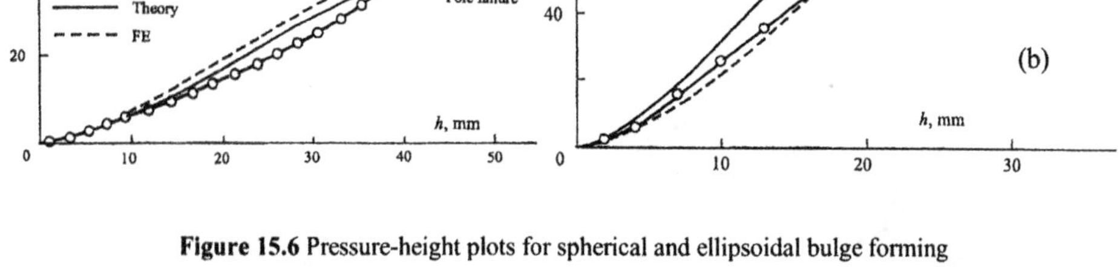

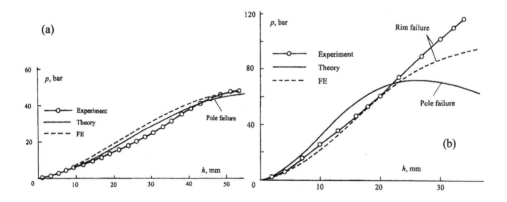

Figure 15.6 Pressure-height plots for spherical and ellipsoidal bulge forming

Actually, an experimentally measured pressure-height plot provides a better check upon an FE prediction because the theory assumes that bulge sections are circular arcs and that pole deformation is equi-biaxial. In practice, some drawing of rim material occurs over the locking bead [11] with the sharp bending of the sheet at this position. Moreover, for a severely rolled sheet, the pole may become oblated [12, 13]. In this FE simulation of the bulge test, shell elements were used within an implicit, static code. The simulation also included bulging with an elliptical die in which rim failures are known to occur [14]. Not surprisingly here, the pressure-height plot is reproduced better by FE than the pole failure theory (see Fig. 15.6b).

The full field strain predictions (see Figs. 15.7a and 15.8a) reveal the most highly strained regions and their proximity to the forming limit. They show that, while a pole failure is expected for a circular bulge, a rim failure will occur in a narrow ellipsoidal bulge with an axis ratio of 0.42. The accompanying FLD's (see Figs 15.7b and 15.8b) reveal that these failures correspond to the equi-biaxial and plane strain positions respectively. The latter correspond to a point lying at the base of the V in the FLD. A constraining of the minor strain is the plane strain condition, known to be responsible for many service failures.

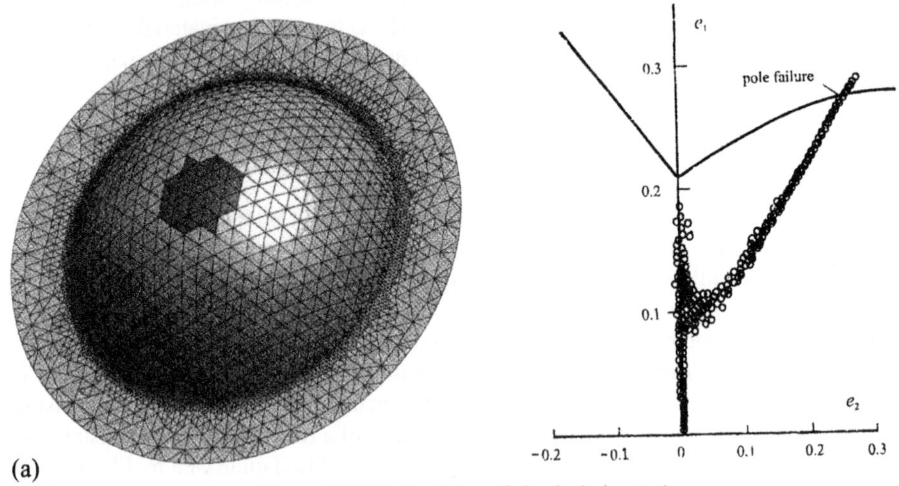

(a) (b)

Figure 15.7 FE simulation of circular bulge strains

Such predictions have been confirmed by experiment [14] in the case of a 1.22 mm aluminium alloy (6016). Strains for circular bulging lie within the first quadrant of the FLD at positions dependent upon the site in the bulge wall. The pole of a circular bulge approximates to a state of high, equi-biaxial strain. Here we see, from Fig. 15.7b, that a pole failure is predicted when, of all the strains at different locations from rim to pole, the pole's equi-biaxial strain path is the first to meet the FLD. On the other hand, for the narrow ellipsoidal bulge, shown in Fig. 15.8, it is the plane strain path at the indicated position on the rim that first reaches the FLD; the simulation thus locating this position as the failure site.

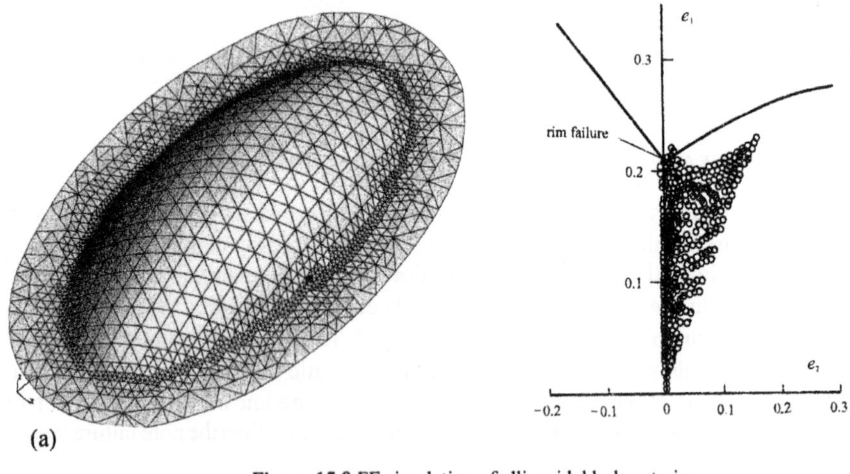

(a) (b)

Figure 15.8 FE simulation of ellipsoidal bulge strains

The strain predicted in the bulge wall can offer an alternative basis for the confirmation of FE output, provided strains can be measured accurately (see Figs 15.9a,b). These strains are calculated from after-test optical measurements of the distortions to a small circular grid pattern etched to the sheet surface. Typically, the hoop ε_θ and meridional natural strain ε_r distributions along principal axes, from rim to pole of a bulge, provide one basis for the comparison. Modern methods employ full-field, digital imaging of the deforming grid to assess the accuracy of strain maps provided by an FE simulation [11].

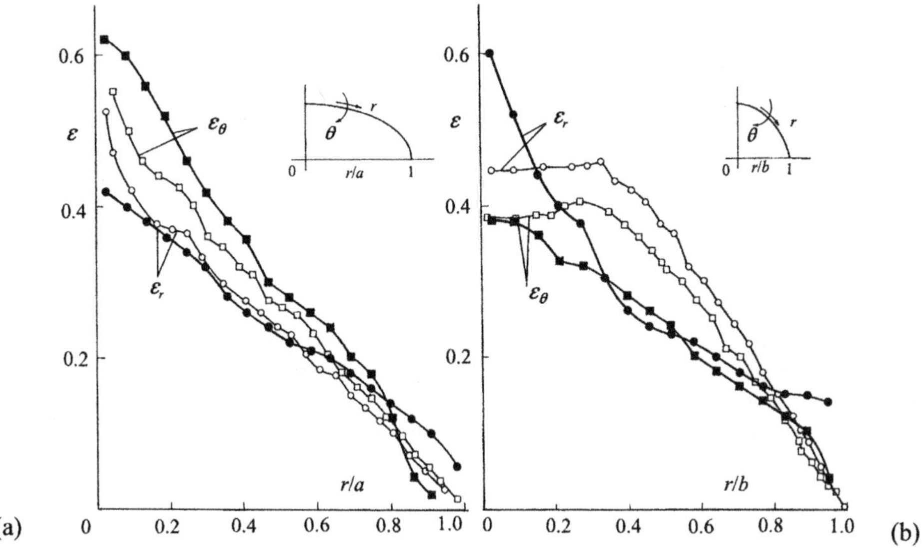

Figure 15.9 FE predictions (open symbols) to principal, ellipsoidal bulge strains along (a) major, (b) minor axes

15.6.2 Indentation Test

In the Erichsen test (see Fig. 15.10) no die is used as this allows the material in contact with the punch to stretch and thin under a tensile, biaxial stress state. The Erichsen number is simply the maximum height of the indentation at the initiation of failure. The comparative ductility measure provided by this number is particularly useful when assessing the suitability of different sheet metals for stretch forming, a fact universally recognised with their now being many international standards for conducting the test [15 - 19].

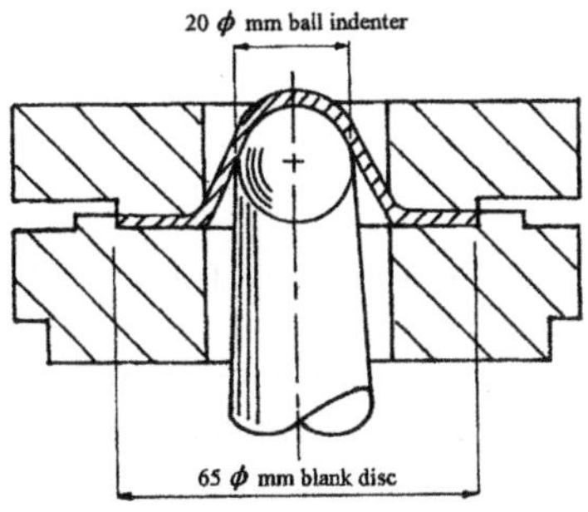

Figure 15.10 Erichsen indentation test geometry

In the simulation of an Erichsen test, as the punch velocity does not increase beyond 10 mm/s, the slower, static code is adequate. This test provides its measure of formability from test conditions (applied here) in which a 20 mm ball indents the centre of a 65 mm dia disc, clamped at 15.3 bar around its periphery. With an axial symmetry present, only one quarter of the initially flat disc need be meshed. An implicit Lagrangian code is used to provide for in-plane natural strains (< 80 %) from large punch displacements. This code employs two layers of 8-noded, solid elements to build the sheet thickness [20]. The in-plane, principal strain distributions, as predicted for the indented disc, will apply to any meridional line drawn from pole to rim if the sheet is initially isotropic. However, with the rolled steel sheet tested here (1.24 mm thick, En 3B steel), the principal strains can vary with direction to the roll; this being a consequence of the different *r* values within the plane of the sheet. The FE predictions allow for plastic anisotropy by adopting Hill's 1948 yield criterion (see eq 11.2). This criterion requires three *r* values (0.95, 1.1 and 1.28) with their respective flow curves for the 0°, 45° and 90° directions. The flow curves for the rolled steel sheet were expressed with Swift's law (9.44b) in which σ_o, n and ε_o connect true stress to natural strain (average values used were: 245 MPa, 0.17 and 0.0057 respectively). Elastic anisotropy is negligible by comparison, this allowing averaged elastic moduli E and ν (200 GPa and 0.27 respectively) to be taken from tensile tests in the three directions. The Coloumb friction coefficient for either lubricated or dry contact between sheet and punch is also required. Here, respective values of 0.05 and 0.2 were used in which the former value applies to lubrication with a thin film of polyethylene. The natural, radial and hoop strains were calculated from after-test radial displacements Δr between concentric circles, whose diameters had changed by Δd:

$$\varepsilon_r = \ln(1 + \Delta r/r), \quad \varepsilon_\theta = \ln(1 + \Delta d/d)$$

Figure 15.11a shows these principal strains at position *s* from the pole along the roll meridion for a lubricated test. The comparisons show that measured strain distributions are predicted reasonably well by FE. Inevitably, errors arise in the measurements of indentation distortion to a grid of initially concentric circles applied to the small flat discs.

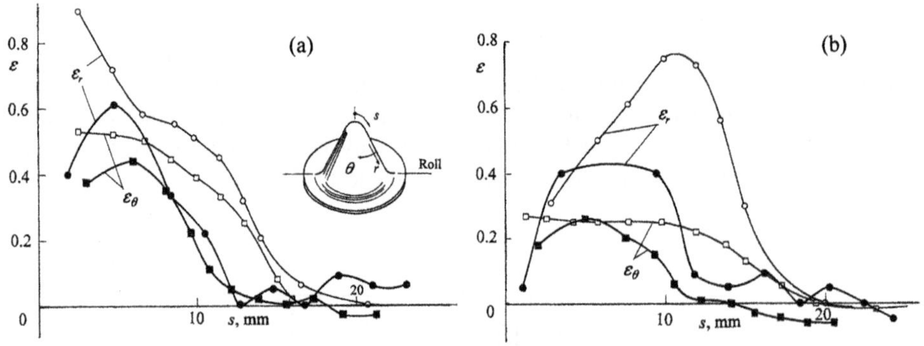

Figure 15.11 FE predictions (open symbols) to Erichsen principal strains (a) lubricated, (b) dry

Within the lubricated contact zone ($s < 10$ mm) strains are approximately equi-biaxial, reaching their maxima at the pole. Greater differences occur in these strains for unlubricated contact (see Fig. 15.11b). In the non-contact region ($s > 10$ mm), the smaller strains differ, falling toward zero at the clamp end. Unsupported material, lying beyond the point of tangency with the ball to the edge of the rim clamp, is formed into a conical shape. In this region, friction is absent and a near plane strain condition prevails in which the clamp constrains the circumferential strain. In the annular region of flat material lying under the rim clamp ($s > 18$ mm), the contact friction under the normal clamping pressure opposes the sheet sliding motion. Some drawing-in is evident where ε_θ has become compressive. While these three distinct regions may be analysed separately [21], it is essential that they should marry properly in transition. Here, an FE simulation of the Erichsen test can assist, given that the simulation has been validated for certain test parameters, of which the most obvious is the force versus displacement relation.

15.6.3 Pressed Panel

In simulating the deformation within a pressed panel, say for a car body, the FE analysis often begins with the final shape required. The 8-noded rectangular mesh is applied to the panel design and an edge tension introduced to avoid wrinkling [22]. The approximate, one-step FE solution simply flattens the panel to ascertain its strains and the sheet size required. Of course, as the one-step method does not track strain history, it is prone to error for positions in the panel that underwent non-radial strain paths. There has been much debate upon the influence of non-radial paths upon the FLD, including experimental work that shows a translation in the position of this diagram [23]. The strains from simpler radial paths may be superimposed upon a fixed FLD to examine their spread and safety. Ideally, all the material should be strained without violating the FLD. If the true FLD is not available use may be made of a construction, in which geometrically similar FLDs are assumed. Each plane strain intercept is derived from the known influences of the strain hardening exponent and thickness of the material [24, 25], as is shown in Fig. 15.12.

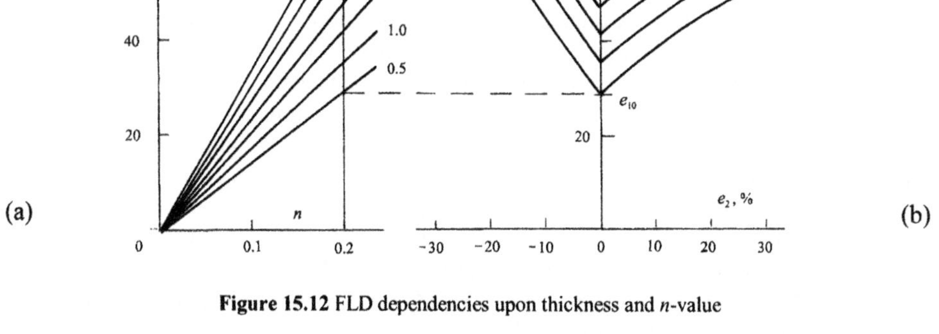

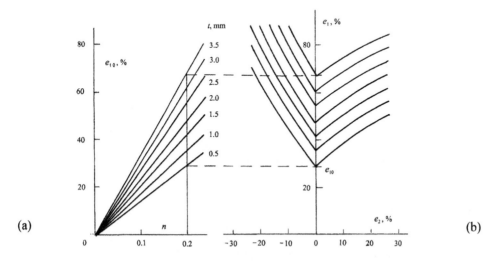

Figure 15.12 FLD dependencies upon thickness and n-value

The rapid, one-step simulation assists with die try-out, the selection of a material grade, blank sizing and assessing the feasibility of manufacturing new panel shapes. To achieve a properly worked sheet, a change to the panel design, tooling and material choice can be made at the try-out stage. To facilitate a re-design, the FE outputs consist of contour plots showing forming zones, safety zones, equivalent strains and principal strain signatures.

Figure 15.13 gives an example of a forming zone display in which all in-plane strain combinations for a box-shaped panel design lie safely beneath the lower FLD band.

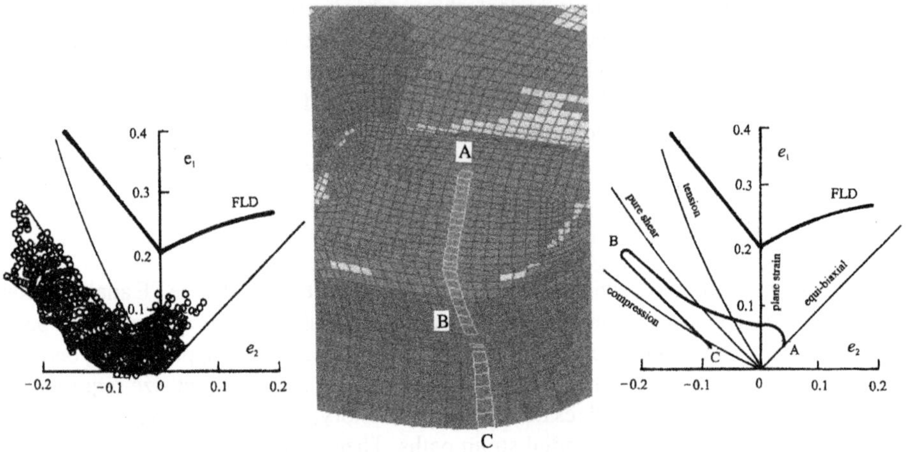

Figure 15.13 Predictions to principal strains in a pressed panel

Also shown in Fig. 5.13 is a contour line ABC running from the flat top, down the vertical wall and along the flanged rim. Shown inset is this contour's strain signature, formed from connecting the principal, engineering strain combinations from point to point along it. This reveals that the full gamut of strain paths arise when forming a corner: from biaxial tension at the top, though plane strain and pure shear in the wall, to a simple compression at the rim. The signature may be verified with an economical circle grid analysis, i.e. by applying a grid only to selected regions known to suffer high strain levels.

15.7 Concluding Remarks

It has been shown how FE is applied incrementally to the regime of elastic-plastic deformation. The examples cited have applied various static FE codes to quasi-static sheet metal forming processes. Commercial FE codes adopt both implicit and explicit formulations, depending upon the nature of the plasticity. For gradual rates of loading the implicit (static) method fulfills equilibrium requirements in meeting its convergence criteria with each numerical time step. The latter implies a new strain increment for a quasi-static process and this increment may be made large to reduce the computing time, say, where hardening is relatively linear. On the other hand, implicit codes involve a far greater computation time to attain large strain with non-linear hardening, where accuracy demands smaller increments. In this case the use of an explicit code obviates the need for matrix inversion at each time step, through its use of the central difference solution to the equation of motion. Where the straining process is rapid, the time step in a simulation becomes physically meaningful. Here the explicit codes can match dynamic plastic processes including slower metal forming processes. If the time step is to be increased to reduce computation time it should not exceed a stability limit. In the absence of an instability the forming process time can be reduced, say from simulating an increase in punch speed, provided this is known not to alter the material response.

References

1. Zeinkiewicz O. C. *The Finite Element Method*, McGraw-Hill, 1977.
2. Owen D. R. J. and Hinton E. *Finite Elements in Plasticity*, Pineridge Press, 1980.
3. Belytschko T., Liu W. K. and Moran B. *Nonlinear Finite Elements for Continua and Structures*, Wiley, 2000.
4. Bathe K-J. *Finite Element Procedures*, Prentice-Hall, 1996
5. Rees, D. W. A. *Mechanics of Solids and Structures*, I.C. Press, 2000.
6. Spencer W. J. *Fundamental Structural Analysis*, Macmillan, 1988.
7. Fenner R. T. *Finite Element Methods for Engineers*, Macmillan, 1975.
8. Rowe G. W., Sturgess C. E. N., Hartley P. and Pillinger I. *Finite-Element Plasticity and Metalforming Analysis*, Cambridge University Press, 1991.
9. Wenner M. L. Proceedings: *Numisheet 2005*, CP778A, pp 3-7 (eds Smith L. M et al).
10. Wang C.T. Proceedings: *Numisheet 2002*, pp 13-24 (eds Yang D. Y et al).
11. Hu C. Q. and Rees D. W. A. Proceedings: *SAE 2006*, 06M-311.
12. Storakers B. *Int Jl Mech Sci*., 1996, **8**, 619.
13. Shang H. M. and Hsu T. C. *Trans ASME, Jl Eng for Industry*, 1979, **101**, 341.
14. Rees D. W. A. *Jl Strain Analysis*, 2000, **35**(2), 109.
15. BS 3855: 1965, 1983, *Modified Erichsen Cupping Test for Sheet and Strip Metal.*
16. DIN 50 101 / 50 102.
17. ISO 8490, R 149 - 60.
18. ASTM 643 - 84.
19. Euronorm 14 - 58/67.
20. Rees D. W. A. and Solen T. Proceeding *9th ESAFORM Conference*, Glasgow, 2006.
21. Kaftanoglu B. and Alexander J. M. *Jl Inst Metals*, 1961-2, **90**, 457.
22. Rees D. W. A. and Power R. *Jl Mats Proc Tech*, 1998, **77**, 134-144.
23. Laukonis J. A. and Gosh K. *Met Trans*, 1978, **A9**, 1849.
24. Hiam J. and Lee A. *Sheet Metal Industries*, 1978, **55**, 631.
25. Rees D. W. A. Proceeding *4th ESAFORM Conference*, Liège, 2001, Vol 2, 293.

Exercises

15.1 Using the principle of superposition between the two stiffness matrices (15.2) and (15.3a), find the stiffness matrix for a circular bar element subjected to combined axial tension and torsion.

15.2 Using eq(15.4), construct the stiffness matrix for two adjacent beam elements when each of their three nodes has two degrees of freedom.

15.3 The load applied to a single beam element is distributed uniformly from node 1 to node 2 at q/unit area of the top surface of length l and width w. Examine ways in which to construct the element's equivalent nodal force vector. Note, that the exact method, based upon the shape factor, gives the equivalent nodal force vector (i.e. including both shear forces and bending moments) as:
$\mathbf{f}^e = (qlw/12)[6 \quad l \quad 6 \quad -l]^T$.

15.4 Assemble the overall stiffness matrix for the 4-element cantilever beam shown in Fig. 15.5, for $F = 10$ kN. Hence show that the vertical deflection beneath the load is 0.127 mm. Take dimensions $b = 120$ mm, $h = 80$ mm and $t = 10$ mm respectively, with elastic moduli $E = 200$ GPa and $\nu = 0.3$.

15.5 Show that, for a plane triangular element, the displacements u and v may be expressed as

$$u = N_i\, u_i = N_1\, u_1 + N_2\, u_2 + N_3\, u_3 \quad \text{and} \quad v = N_i\, v_i = N_1\, v_1 + N_2\, v_2 + N_3\, u_3$$

in which $N_i = a_i + b_i x + c_i y$ ($i = 1, 2, 3$) represent the shape functions:

$$N_1 = a_1 + b_1 x + c_1 y,\ N_2 = a_2 + b_2 x + c_2 y \text{ and } N_3 = a_3 + b_3 x + c_3 y$$

Using the nodal displacements $u = u_1$, $v = v_1$ for $x = x_1$ and $y = y_1$ etc, show that coefficients a_i, b_i and c_i appear in terms of nodal point co-ordinates and the element's area $\Delta = ½(x_1 - x_2)(y_3 - y_3) - ½(x_2 - x_3)(y_1 - y_2)$, as:

$$a_1 = (x_2 y_3 - x_3 y_2)/(2\Delta),\ b_1 = (y_2 - y_3)/(2\Delta),\ c_1 = (x_3 - x_2)/(2\Delta)$$
$$a_2 = (x_3 y_1 - x_1 y_3)/(2\Delta),\ b_2 = (y_3 - y_1)/(2\Delta),\ c_2 = (x_1 - x_3)/(2\Delta)$$
$$a_3 = (x_1 y_2 - x_2 y_1)/(2\Delta),\ b_3 = (y_1 - y_2)/(2\Delta),\ c_3 = (x_2 - x_1)/(2\Delta)$$

15.6 When a uniformly distributed normal pressure is applied to the side 2-3 of the triangular element in Fig. 15, the principle of virtual displacements gives an equivalent nodal force vector as:

$$\mathbf{f}^e = t\left(\mathbf{A}^{e-1}\right)^{\mathrm{T}} \int \mathbf{A}^{\mathrm{T}} \mathbf{p}\, \mathrm{d}l$$

where $\mathbf{p} = [\, p_x \quad p_y \,]^{\mathrm{T}}$ and $\mathbf{f}^e$, $\mathbf{A}^e$ and $\mathbf{A}$ are given in eqs(15.14b), (15.8a) and (15.7c) respectively. Show, from the matrix multiplication and integration, that the equivalent nodal force components satisfy equilibrium: $f_{x1} + f_{x2} + f_{x3} = p_x l_{23} t$ and $f_{y1} + f_{y2} + f_{y3} = p_y l_{23} t$. Take $\mathrm{d}l = l_{23}\, \mathrm{d}x/(x_2 - x_3)$, in which the side length $l_{23} = \sqrt{[(x_2 - x_3)^2 + (y_3 - y_2)^2]}$

15.7 Examine other ways of expressing the elastic-plastic matrix $\mathbf{D}^{ep}$ in $\Delta\sigma = \mathbf{D}^{eP}\Delta\varepsilon$ from eqs(15.28) and (15.29). Hint, see references [2, 8].

15.8 Show that if real, nodal body forces exist (vector **b**) in addition to real, nodal external forces then the principal of virtual elastic displacements modifies eqs(15.5) and (15.6) into the following forms:

$$f_k\, \delta_k^{\mathrm{v}} + b_k\, \delta_k^{\mathrm{v}} - \int_V \sigma_{ij}\, \varepsilon_{ij}^{\mathrm{v}}\, \mathrm{dV} = 0 \quad \Rightarrow \quad (\mathbf{f}^e)^{\mathrm{T}}\, \boldsymbol{\delta}^{ev} + (\mathbf{b})^{\mathrm{T}}\, \boldsymbol{\delta}^{ev} - \int_V (\boldsymbol{\sigma})^{\mathrm{T}} \boldsymbol{\varepsilon}^{\mathrm{v}}\, \mathrm{d}V = 0$$

$$\delta_k^{\mathrm{v}} f_k + \delta_k^{\mathrm{v}} b_k - \int_V \varepsilon_{ij}^{\mathrm{v}}\, \sigma_{ij}\, \mathrm{dV} = 0 \quad \Rightarrow \quad (\boldsymbol{\delta}^{ev})^{\mathrm{T}}\, \mathbf{f}^e + (\boldsymbol{\delta}^{ev})^{\mathrm{T}}\, \mathbf{b} - \int_V (\boldsymbol{\varepsilon}^{\mathrm{v}})^{\mathrm{T}}\, \boldsymbol{\sigma}\, \mathrm{d}V = 0$$

15.9 Show that the principal of virtual displacements can be modified as follows in the case of incremental plasticity with body forces distributed as a self weight/unit volume.

$$(\Delta\boldsymbol{\delta}^{ev})^{\mathrm{T}}\, \mathbf{f}^e + \int_V [\, (\Delta\boldsymbol{\delta}^{\mathrm{v}})^{\mathrm{T}}\, \mathbf{b} - (\Delta\boldsymbol{\varepsilon}^{\mathrm{v}})^{\mathrm{T}}\, \boldsymbol{\sigma}\,]\, \mathrm{d}V = 0$$

15.10 At the die try-out stage, the strain signature, taken across the corner ABCD of a pressed steel box, is found to lie in the danger region of the FLD (see Fig. 15.14). List the most likely modifications that could be made to make the pressing safe. Given that the original sheet thickness is 1.5 mm and the *n*-value is 0.2, employ the construction in Fig. 15.12 to recommend possible material changes.

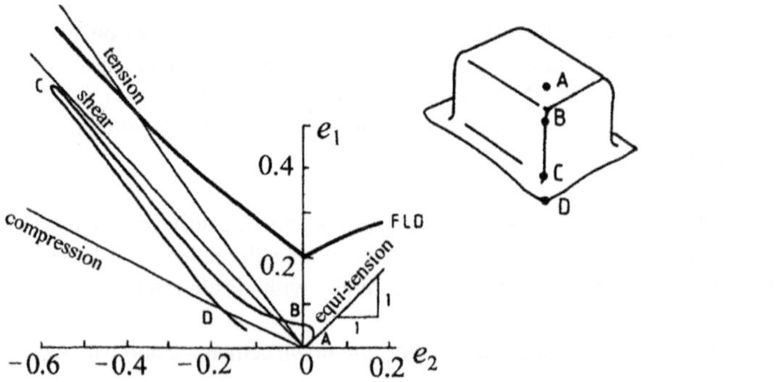

Figure 15.14

INDEX